Quality of Service Architectures for Wireless Networks:
Performance Metrics and Management

Sasan Adibi
Research In Motion (RIM), Canada

Raj Jain
Washington University in St. Louis, USA

Shyam Parekh
Bell Labs, Alcatel-Lucent, USA

Mostafa Tofighbakhsh
AT&T Labs, USA

Information Science REFERENCE

INFORMATION SCIENCE REFERENCE

Hershey · New York

Director of Editorial Content:	Kristin Klinger
Director of Book Publications:	Julia Mosemann
Acquisitions Editor:	Dave DeRicco
Development Editor:	Christine Bufton
Publishing Assistant:	Kurt Smith
Typesetter:	Carole Coulson
Quality control:	Jamie Snavely
Cover Design:	Lisa Tosheff
Printed at:	Yurchak Printing Inc.

Published in the United States of America by
Information Science Reference (an imprint of IGI Global)
701 E. Chocolate Avenue
Hershey PA 17033
Tel: 717-533-8845
Fax: 717-533-8661
E-mail: cust@igi-global.com
Web site: http://www.igi-global.com/reference

Library of Congress Cataloging-in-Publication Data

Quality of service architectures for wireless networks : performance metrics and management / Sasan Adibi ... [et al.], editors.
p. cm.
Includes bibliographical references and index.
Summary: "This book further explores various issues and proposed solutions for the provision of Quality of Service (QoS) on the wireless networks"-- Provided by publisher.
ISBN 978-1-61520-680-3 (hardcover) -- ISBN 978-1-61520-681-0 (ebook) 1.
Wireless LANs--Quality control. 2. Network performance (Telecommunication) 3.
Wireless Internet. I. Adibi, Sasan, 1970-
TK5105.78.Q36 2010
004.6'5--dc22
2009040024

British Cataloguing in Publication Data
A Cataloguing in Publication record for this book is available from the British Library.

List of Reviewers

Abdel Karim Al-Tamimi, *Washington University in Saint Louis, USA*
Cagatay Buyukkoc, *AT&T Labs, USA*
Mustafa Ergen, *WiChorus, USA*
Nada Golmie, *National Institute of Standards and Technology, USA*
Ehsan Haghani, *New Jersey Institute of Technology, USA*
Libin Jiang, *University of California, Berkeley, USA*
Jiwoong Lee, *University of California, Berkeley, USA*
Jeonghoon Mo, *Yonsei University, Korea*
Subhas Chandra Mondal,*Wipro Technologies, India*
Nikhil Shetty, *University of California, Berkeley, USA*
Biplab Sikdar, *Rensselaer Polytechnic Institute, USA*
Chakchai So-In, *Washington University in St. Louis, USA*

Table of Contents

Section 2
Resource Management

Section 3
Mobility

Detailed Table of Contents

Emergence of all IP based wired and wireless networks for mobile services calls for new innovations and architectural approach. Coexistence of legacy and emerging networks such as different generations of networks based on 3GPP and 3GPP2 specifications, Wi-Fi and WiMAX, have posed new challenges to guarantee acceptable quality of experiences to the users. Different user environments such as fixed, nomadic, and vehicular have brought about new Quality of Service (QoS) practices and have introduced policies to best optimize the network resources and enhance user experiences.

Mobile systems and particularly UMTS are growing fast. These systems convey data based services in addition to customary voice services. Quality of service is a function of data rate, delay and signal to noise plus interference ratio in these systems. In this Chapter first the authors pay attention to UMTS and its QoS architecture, then to service categorization due to QoS. Afterwards they review some QoS parameters. Then they study Layer 2 QoS parameters and general concepts about Transport channels. Then the authors review TCP effects on the throughput in the air interface. The authors introduce HS-DPA in the next section. Finally they pay attention to data traffic models and their effects on the system capacity and Erlang capacity and delay in the system.

WiMAX technology, based on the IEEE 802.16 standard, is a promising broadband wireless technology for the upcoming 4G network. WiMAX has excellent QoS mechanisms to enable differentiated Quality of service of various applications. QoS in broadband wireless access network such as WiMAX is a difficult and complicated task, as it adds unpredictable radio link, user and traffic demand. WiMAX supports end-to-end QoS provisioning to allow various applications and services. This chapter aims to provide a detailed overview of the QoS in WiMAX, the current and the future. Various air-interface and network mechanisms that enable the end-to-end QoS provisioning are then discussed. Finally, the novel mechanisms to improve the QoS provisioning in the next generation WiMAX system are also discussed.

WiMAX is the most promising technology of recent years; it can be the technology that resolves some problems related to the spread of wireless service. Thinking of the concept of service, the most important related issue is the QoS (Quality of Service). Behind WiMAX, there is the IEEE 802.16 protocol (IEEE 802.16, 2004), which provides some basic mechanisms to guarantee QoS. This chapter aims to explore these mechanisms, but it also attempts to highlight the absence of some elements in the protocol or those components in it that can be improved. The protocol can be optimized and in the last part of chapter the authors show how to improve it using a set of algorithms collected by literature. Finally, it is explained how instruments, not designed to be applied to the world of wireless, such as games theory or fuzzy logic, can be used to deal with wireless issues.

Although simulation studies show performance increases when load sharing wireless integrated networks, these studies assume a limited, defined, configuration. Simulation examples of load sharing consider only performance of specific scenarios, and do not estimate capacity or other benefits for a generic network. This study discusses other potential benefits of a load shared network, such as flexibility, survivability, modularity, service focus, quality of service, and autoreconfigurability. The authors evaluate these other benefits by developing mathematical models and measurements to quantify a set of potential benefits of load sharing. In addition, the authors consider capacity considerations against a best-case model. Varied overflow algorithms are then simulated assuming standard HSPA+ and WLAN data rates. The results are compared to the estimated and best-case performance metrics.

Section 2
Resource Management

In this chapter, the authors present a delay-sensitive MAC adaptation scheme combined with an admission control mechanism. The proposed solution is based on thorough analysis of the tradeoff existing between high network utilization and achieving bounded QoS metrics in operated 802.11-based networks. First, the authors derive an accurate delay estimation model to adjust the contention window size in real-time basis by considering key net-work factors, MAC queue dynamics, and application-level QoS requirements. Second, the authors use the abovementioned delay-based CW size adaptation scheme to derive a fully distributed admission control model that provides protection for existing flows in terms of QoS guarantees.

This chapter briefly reviews fundamental protocol engineering aspects and presents resource allocation approaches for wireless relay networks. Important cooperative diversity protocols and their typical applications in different wireless network environments are first described. Then, performance analysis and QoS provisioning issues for wireless networks using cooperative diversity are discussed. Finally, resource allocation in wireless relay networks through power allocation for both single and multiuser scenarios are presented. For the multi-user case, the authors consider relay power allocation under different fairness criteria with or without user minimum rate requirements. When users have minimum rate requirements, the authors develop a joint power allocation and admission control algorithm with low-complexity to circumvent the high complexity of the underlying problem. Numerical results are then presented, which illustrate interesting throughput and fairness tradeoff and demonstrate the efficiency of the proposed power control and admission control algorithms.

Call admission control in mobile cellular networks has become a high priority in network design research due to the rapid growth of popularity of wireless networks. Dozens of various call admission policies have been proposed for mobile cellular networks. This chapter proposes a classification of user based

IEEE 802.16 WiMAX (Worldwide Interoperability for Microwave Access) is a major standard technology for Wireless Metropolitan Area Networks (Wireless MAN). Quality-of-service (QoS) scheduling class and mobility management are two main issues for supporting seamless high-speed data and media-stream communications. Previous works on WiMAX handoff however have mainly addressed a particular scenario or a single QoS class. This chapter first presents an overview of the QoS scheduling classes supported by the IEEE 802.16 standard, followed by a survey of major related works proposed to enhance 802.16e handoffs. Next, it will present a new context-sensitive handoff scheme that supports the five 802.16 QoS scheduling classes, and is energy-aware – it may switch to energy-saving mode during handoff. It will then illustrate performance evaluation, which will show that, compared to three existing methods, the proposed scheme successfully supports the five QoS classes in both layers 2 and 3 handoff, decreases end-to-end handoff delay, delay jitter, and service disruption time; it also increases throughput and energy efficiency. Finally, key implementation and cost issues are discussed. The authors believe that this chapter is a significant contribution for providing high-quality, seamless data and media streaming over 802.16 as well as LTE (Long-Term Evolution) cellular networks, and would be a valuable part of QoS architectures in the wireless networking domain.

Vehicular communication networks (VCNs) have emerged as a key technology for next-generation wireless networking. DSRC/WAVE as a leading technology for VCN provides a platform for Intelligent Transportation System (ITS) services, as well as multimedia and data services. Some of these services such as active safety and multimedia services have special requirements for QoS provision. However, when providing QoS, the VCN characteristics are the cause for several new issues and, especially when vehicles travel at high speeds of up to 200 km/h. These issues are addressed in the context of roadside networks and vehicular ad hoc (unplanned) networks (VANETs), including vehicle-to-vehicle (V2V) and vehicle-to-roadside (V2R) communications. As one result, plenty of solutions for provisioning QoS in VCNs have been classified in regards to VANETs and roadside networks, as well as a focus on layer-2 and layer-3. Following those results, several QoS solutions, including medium access and routing protocols, are presented and discussed. Additionally, open research issues are discussed, with an objective to spark new research interests in the presented field.

Section 4
Multimedia

limitations of conventional QoS measuring when evaluating quality from the service and user point of view. In terms of media, compressed video usually constitutes a very interdependent structure degrading in a non-graceful manner when exposed to Binary Erasure Channels (BEC), like the Internet or wireless networks. Accordingly, not only the type of encoder and its major encoding parameters (e.g. transmission rate, image definition or frame rate) contribute to the quality of a received video, but also QoS parameters are usually a cause for different types of decoding artifacts. As a result of this, several worldwide standard entities have been evaluating new metrics for the subjective assessment of video transmission over IP networks. In this chapter the authors are especially interested in explaining some of the best practices available to monitor, evaluate and assure good levels of QoE in packet oriented networks for rich media applications like high quality video streaming. For such applications, service requirements are relatively loose or difficult to quantify and therefore specific techniques have to be clearly understood and evaluated. By the mid of the chapter the reader should have understood why even networks with excellent QoS parameters might have QoE issues, as QoE is a systemic approach that does not relate solely to QoS but to the ensemble of components composing the communication system.

This chapter describes several advanced techniques for estimating the video distortion deriving from multiple video packet losses. It provides different usage scenarios, where the Peak to Signal Noise Ratio (PSNR) video metric may be used for improving the end user quality. The key idea of the presented applications is to effectively use the distortion information associated to each video packet. This allows one to perform optimal decisions in the selection of the more suitable packets to transmit. During the encoding process, the encoder estimates first the loss impact (for instance the amount of error propagation) of each packet. Afterwards, it generates side information as a "hint" for making video content aware transmission decisions. In this way, it is possible to define new scheduling schemes that give more priority to the packets with higher loss impact, and to assign fewer resources to the packets with lower loss impact. To this end, the usage of hint tracks, introduced in the MPEG-4 systems part, provides a syntactic means for storing scheduling information about media packets that significantly simplifies the operations of a streaming server. Moreover, the prioritization scheme may be used to minimize the overall error propagation under the delay constraint imposed by the video presentation deadline.

In this chapter, the authors describe the intrinsic needs to effectively integrate interactive vocal conversations over heterogeneous networks including packet- and circuit- based networks. The requirement to harmonize transport networks is discussed and a foreseen architecture multi -operators and -services is presented. Moreover, envisaged remedies to the ever increasing network complexity are also summarized. Subjective and objective methodologies to evaluate voice quality under listening and conversational conditions are thoroughly described. In addition, software- and emulation- based frameworks developed in order to evaluate and improve voice quality are rigorously described. This chapter stresses parametric model-based assessment algorithms due to their ability to be useful for on-line network management. In particular, the authors describe parametric assessment algorithms over last-hop wireless Telecom networks and packet-based networks. The last part of this chapter describes several management applications which consider users' preferences and providers' needs.

The 3GPP IMS defines a network architecture that allows rapid provisioning of rich multimedia services. While standardization of the IMS core architecture is largely complete, there are several areas that are still to be addressed before effective deployment can be realized. In particular a QoS framework is required that efficiently manages scarce network resources, ensures reliability and differentiates IMS services from web-based services. This chapter reviews the most promising candidate resource management frameworks, performs architectural alignment and defines a set of generic terms and elements to provide a convenient point of departure for future research. This harmonization of standardized architectures is critical to avoid interoperability concerns that could cripple deployment. Further challenges are discussed, in particular the vertical and horizontal co-ordination of resources, and current research works that address these challenges are presented.

Section 5
Ad-Hoc/Mesh

A Mobile Ad hoc NETwork (MANET) consists of a collection of mobile nodes. They communicate in a multi-hop way without a formal infrastructure. Owing to the uniqueness such as easy deployment and self-organizing ability, MANET has shown great potential in several civil and military applications. As MANETs are gaining popularity day-by-day, new developments in the area of real time and multimedia applications are increasing as well. Such applications require Quality of Service (QoS) evolving with respect to bandwidth, end-to-end delay, jitter, energy etc., Consequently, it becomes necessary for MANETs to have an efficient routing and a QoS mechanism to support new applications. QoS provisioning for MANET can be achieved over different layers, starting from the physical layer up to the application

The wireless mesh network (WMN) is an economical solution to enable ubiquitous broadband services due to the advantages of robustness, low infrastructure costs, and enhancing coverage by low power. The wireless mesh network also has a great potential for realizing green communications since it can save energy and resources during network operation and deployment. With short-range communications, the transmission power in the wireless mesh networks is lower than that in the single-hop networks. Nevertheless, wireless mesh network should face scalability issue since throughput enhancement, coverage extension, and QoS guarantee are usually contradictory goals. Specifically, the multi-hop communications can indeed extend the coverage area to lower the infrastructure cost. However, with too many hops to extend coverage, the repeatedly relayed traffic will exhaust the radio resource and degrade the quality of service (QoS). Furthermore, as the number of users increases, throughput and QoS (delay) degrade sharply due to the increasing contention collisions. In this chapter, from a network architecture perspective the authors investigate how to overcome the scalability issue in WMNs, so that the tradeoff between coverage and throughput can be improved and the goal of QoS provisioning can be achieved. The authors discuss main QoS-related research directions in WMNs. Then, the authors introduce two available scalable mesh network architectures that can relieve the scalability issue and support QoS in WMNs for the wide-coverage and dense-urban coverage. The authors also investigate the optimal tradeoff among throughput, coverage, and delay for the proposed WMNs by an optimization approach to design the optimal system parameters.

T. Nyandeni, Council for Scientific and Industrial Research (CSIR), Defence, Peace, Safety and Security (DPSS), South Africa
C. Kyara, Council for Scientific and Industrial Research (CSIR), MERAKA, South Africa
P. Mudali, University of Zululand, South Africa
S. Nxumalo, University of Zululand, South Africa
N. Ntlatlapa, Council for Scientific and Industrial Research (CSIR), MERAKA, South Africa
M. Adigun, University of Zululand, South Africa

Routing is an essential mechanism for proper functioning of large networks and routing protocols make use of routing metrics to determine optimal paths. The design of routing metrics is critical for achieving high throughput and the authors begin this chapter by proposing the design principles for routing metrics. These design principles are for ensuring the proper functioning of the network and achieving high throughput. The authors continue by giving a detail analysis of the existing routing metrics. They also look at the pitfalls of the existing routing metrics. The authors conclude the chapter by outlining the future research directions.

Section 6
Future

Cognitive Radio (CR) is a next-generation wireless communication technology that improves the utilization of the overall radio spectrum through dynamic adaptation to local spectrum availability. In CR networks, unlicensed or Secondary Users (SUs) may operate in underutilized spectrum owned by licensed or Primary Users (PUs) conditional upon the PU encountering acceptably low interference levels. A Cognitive Wireless Ad Hoc Network (CWAN) is a multi-hop self-organized and dynamic network that applies CR technology for ad-hoc mode wireless networks that allow devices within range of each other to discover and communicate in a peer-to-peer fashion without necessarily involving infrastructure such as base stations or access points. Research into Quality of Service (QoS) in CWAN is still in its infancy. To date, there is only a perfunctory attempt to improve the data-link and network layers of the Open Systems Interconnection (OSI) reference model for CR hosts, and so this is the focus of this chapter. The authors present a discussion on the architecture, open issues and design approaches related to QoS provisioning in CWAN. Their discussion aims to establish a foundation for further research in several unexplored, yet promising areas in CWAN.

Next Generation Mobile Networks (NGMNs) constitute the evolution of mobile network architectures towards a common IP based network. One of the main research topics in wireless networks architectures is QoS control and provisioning. Different approaches to this issue have been described. The introduction of the NGMNs is a major trend in telecommunications, but the heterogeneity of wireless accesses increases the challenges and complicates the design of QoS control and provisioning. This chapter provides an overview of the standard architectures for QoS control in Wireless networks (e.g. UMTS, WiFi, WiMAX, CDMA2000), as well as, the issues on this all-IP environment. It provides the state-of-the-art and the latest trends for converging networks to a common architecture. It also describes the challenges that appear in the design and deployment of QoS architectures for heterogeneous accesses and the available solutions. The Evolved Core from 3GPP is analyzed and described as a suitable and promising solution addressing these challenges.

Preface

This book provides design considerations and guidelines for implementing Quality of Service (QoS) within emerging 4G networks. QoS best practices are recommended by the contributing authors, and new innovative concepts, solutions, and research results are presented in depth.

The editors originally came together about four years ago as the core team of the Application Architecture Task Group of the WiMAX Forum (which later became the Application Working Group) to facilitate various applications over the mobile broadband access networks. We pursued development of best practices and guidelines to encourage the industry towards unified solutions for better interoperability and performance. Given that there was not a good reference that looked at the performance requirements of the existing and emerging voice, video and data applications in the context of the architectural constraints of the mobile broadband networks, we decided to pull together the present book to fill that void.

A recurring subtext in this book is that the wired and wireless networks have a key difference in how the QoS required for different applications can be supported over them. Although a number of intelligent solutions have been developed to manage QoS over the wired networks, because of the commoditization of the underlying resources, we find that more often than not the service providers resort to "throwing bandwidth" at the QoS issues for resolving them. Wireless networks cannot afford such a luxury. These networks not only have tight limits on how much bandwidth they can offer due to the spectrum scarcity, they need to manage interference and congestion dynamics in presence of mobility. This accompanied with the explosion of new applications over the mobile broadband networks (e.g., plethora of new Blackberry and iPhone applications) has made it critical that efficient QoS management solutions are implemented to ensure widespread success of the mobile broadband networks. The ongoing debate on net neutrality necessitates that the QoS management solutions continue to provide open access while supporting and encouraging adoption of new QoS intensive services.

Emergence of all IP based wired and wireless networks for mobile services calls for new innovations and architectural approach. Coexistence of legacy and emerging networks such as different generations of networks based on 3GPP and 3GPP2 specifications, Wi-Fi and WiMAX, have posed new challenges to guarantee acceptable quality of experiences to the users. Different user environments such as fixed, nomadic, and vehicular have brought about new Quality of Service (QoS) practices and have introduced policies to best optimize the network resources and enhance user experiences.

Additional challenges come from emergence of complementary technologies such as ad hoc and cellular networks. The demand for heterogeneous access increases the difficulty in providing consistent end-to-end QoS control mechanisms. The authors believe new and innovative QoS mechanisms must include convergence of multi-radio and multi access solutions with the state-of-the-art QoS control capabilities. The focus also needs to be on standardization of common practices to unify and provide consistent experience when users move from one network to another. Seamless roaming, seamless handoff, and selective session persistence may be the subject of discussion over the next few years. New

QoS architectures for heterogeneous access will need to make certain assumptions with respect to end devices capabilities. New industry standards may be required to accommodate source as well as network initiated requests, including the ones for QoS renegotiations. Solutions may include location, behavior and resource aware admission control, policy-based management and cross-layer optimization.

The Internet is transforming from a network with the fixed best-effort packet delivery architecture to the mobile services architecture. The recent trend shows wide deployment of networked business applications with specific QoS requirements. In current mobile Internet, traffic flows are typically supported on the Best Effort basis while relying on upper layer protocols like TCP for resource sharing. This approach does not account for the diverse QoS requirements for different applications, time varying availability of radio resources and differentiation among the users. Many proposals, including the ones presented in this book, are being evaluated by the industry. For example, dynamic QoS support and intelligent controls including adaptive traffic prioritizations are proposed to be injected into the networks, applications and end devices to enable increased Quality of Experience (QoE) and lower usage of the radio resources. Application adaptation roll-out is expected from the developers of the emerging mobile intelligent applications, while network adaptation is expected through the mechanisms provided by the service providers.

The contributed chapters are categorized in following broad areas: (1) Broadband Networks, (2) Resource Management, (3) Mobility, (4) Multimedia, (5) Ad Hoc and Mesh Networks, and (6) Future. The Broadband Networks area considers the QoS architectures of representative networks. Next, the Resource Management and Mobility areas consider management of the scarce radio resources as well as handover controls in mobile scenarios for satisfying the QoS requirements. The Multimedia area considers various applications, including most demanding real-time voice and video applications that drive the QoS management expected from the new generation of mobile networks. Finally, Ad Hoc and Mesh Networks as well as Future areas focus on the promising evolution of the wireless technologies and include discussion on the QoS issues in the networks based on such technologies.

Sasan Adibi, Research In Motion
Raj Jain, Washington University in St. Louis
Shyam Parekh, Bell Labs, Alcatel-Lcuent
Mostafa Tofighbakhsh, AT&T Labs

Acknowledgment

The editors of this book would like to thank the technical and professional individuals who helped us in the organization, review, and editing of this book.

First of all we would like to thank the following reviewers: Abdel Karim Al-Tamimi (Washington University in St. Louis), Cagatay Buyukkoc (AT&T Labs), Mustafa Ergen (WiChorus), Nada Golmie (National Institute of Standards and Technology), Ehsan Haghani (New Jersey Institute of Technology), Libin Jiang (University of California, Berkeley), Jiwoong Lee (University of California, Berkeley), Jeonghoon Mo (Yonsei University), Subhas Chandra Mondal (Wipro Technologies), Nikhil Shetty (University of California, Berkeley), Biplab Sikdar (Rensselaer Polytechnic Institute) and Chakchai So-In (Washington University in St. Louis). We appreciate their help and support very much.

We are also very thankful to the following IGI Global staff: Christine Bufton, Erika Carter, David DeRicco, Jan Travers, Jennifer Weston and Neely Zanussi. Their positive attitude and patience are greatly appreciated.

We would like to sincerely thank our respective management at Research In Motion, Washington University in St. Louis, Alcatel-Lucent and AT&T Labs for their encouragement and support.

Lastly, and most importantly, we are indebted to our families. Their invaluable and relentless support, encouragement, and love are without doubt the most important reasons behind all our achievements.

Sasan Adibi, Research In Motion
Raj Jain, Washington University in St. Louis
Shyam Parekh, Bell Labs, Alcatel-Lucent
Mostafa Tofighbakhsh, AT&T Labs

Chapter 1
Introduction

Sasan Adibi
Research In Motion (RIM), Canada

Raj Jain
Washington University in St. Louis, USA

Shyam Parekh
Bell Labs, Alcatel-Lucent, USA

Mostafa Tofighbakhsh
AT&T Bell Labs, USA

OVERVIEW

Emergence of all IP based wired and wireless networks for mobile services, calls for new innovations and architectural approaches. Coexistence of legacy and emerging networks such as different generations of networks based on 3GPP and 3GPP2 specifications, Wi-Fi and WiMAX, have posed new challenges to guarantee acceptable Quality of Experience (QoE) to the users. Different user environments such as fixed, nomadic, and vehicular have brought about new Quality of Service (QoS) practices and have introduced policies to best optimize the network resources and enhance user experience.

Additional challenges come from emergence of complementary technologies such as ad hoc and cellular networks. The demand for heterogeneous access increases the difficulty in providing consistent end-to-end QoS control mechanisms. We believe new and innovative QoS mechanisms must include convergence of multi-radio and multi access solutions with the state-of-the-art QoS control capabilities. The focus also needs to be on standardization of common practices to unify and provide consistent experience when users move from one network to another. Seamless roaming, seamless handoff, and selective session persistence may be the subject of discussion over the next few years. New QoS architectures for heterogeneous access will need to make certain assumptions with respect to end devices capabilities. New industry standards may be required to accommodate source as well as network initiated requests, including the ones for QoS renegotiations. Solutions may include

DOI: 10.4018/978-1-61520-680-3.ch001

location, behavior and resource aware admission control, policy-based management and cross-layer optimization.

The Internet is transforming from a network with the fixed best-effort packet delivery architecture to the mobile services architecture supporting differentiated QoS. The recent trend shows wide deployment of networked business applications with specific QoS requirements. In current mobile Internet, traffic flows are typically supported on the best effort basis while relying on upper layer protocols like TCP for resource sharing. This approach does not account for the diverse QoS requirements for different applications, time varying availability of radio resources and differentiation among the users. Many proposals, including the ones presented in this book, are being evaluated by the industry. For example, dynamic QoS support and intelligent controls including adaptive traffic prioritizations are proposed to be injected into the networks, applications and end devices to enable increased QoE and lower usage of the radio resources. Application adaptation roll-out is expected from the developers of the emerging mobile intelligent applications, while network adaptation is expected through the mechanisms provided by the service providers.

Standardization Bodies

International standardization bodies are responsible to develop new standards and maintain existing ones. The following standardization bodies are just examples of that operate within the various areas of communications, including Quality of Service (QoS) for the current and next generation networks.

Institute of Electrical and Electronics Engineers (IEEE) – IEEE is an international and professional organization that hosts many high caliber research and development activities in various fields of electrical engineering, including IEEE 802.11 standards representing Wireless Local Area Networks (WLAN) or Wi-Fi standards and IEEE 802.16 standards representing Wireless and Wired Wide Area Networks or WiMAX (Worldwide Interoperability for Microwave Access) standards. Both Wi-Fi Alliance and WiMAX Forum are global non-profit industry associated organizations promoting the advancements for Wi-Fi and WiMAX technologies through various certifications programs, certifying products that pass minimum conformance and performance tests.

Internet Engineering Task Force (IETF) – IETF is responsible for the development of Internet Standards through Request for Comments (RFCs). IETF and IEEE collaborate on different levels and once a standard is proposed through IEEE publications, further higher layer protocols related advancements may be carried out through various IETF RFCs.

The 3rd Generation Partnership Project (3GPP) Roadmap – 3GPP is a collaboration among various telecommunications association groups promoting a globally applicable third generation (3G) mobile systems. 3GPP's specifications are within the scope of the International Telecommunication Union (ITU)'s International Mobile Telecommunications (ITU-2000) project, which are based on Global System for Mobile Communications (GSM) specifications evolutions, including Universal Mobile Telecommunications System (UMTS), High Speed Packet Access (HSPA), Long Term Evolution (LTE), and LTE-Advanced (LTE-A). Another variation of 3GPP also exists: 3GPP2, which should not be confused with 3GPP. 3GPP2 specifies standards for another 3G technology based on Code Division Multiple Access (CDMA or IS-95), also known CDMA2000.

Book Organization

The contributed chapters are categorized in the following six broad areas: (1) Broadband Wireless Networks, (2) Resource Management, (3) Mobility, (4) Multimedia, (5) Ad Hoc and Mesh

Networks, and (6) Future Wireless Networks. The Broadband Networks area considers the QoS architectures of representative networks. Next, the Resource Management and Mobility areas consider management of the scarce radio resources as well as handover controls in mobile scenarios for satisfying the QoS requirements. The Multimedia area considers various applications, including most demanding real-time voice and video applications that drive the QoS management expected from the new generation of mobile networks. Finally, Ad Hoc and Mesh Networks as well as Future areas focus on the promising evolution of the wireless technologies and include discussion on the QoS issues in the networks based on such technologies.

1. QOS IN BROADBAND WIRELESS NETWORKS

Tetherless communication from anywhere, at anytime, and for any application is now in the era of high expectations. With its feasibility no longer in question, current focus has been on issues such as extensions of underlying protocols for enhanced user experience and improved spectral efficiency for economic competitiveness. In order to appreciate how far we have traveled in achieving the promise of wireless communication, in this opening section of the book, we have included case studies of QoS architectures of some of the most adopted wireless technologies today. UMTS and its variations are the example of a most widely deployed commercial wireless technology worldwide. WiMAX represents a culmination of the aggregate learning through the long journey of wireless technology evolution. Hence, we have included in this section articles on QoS architectures of UMTS and WiMAX. In the context of load distribution, Wireless LANs (WLANs) are also discussed briefly.

J. Chimeh describes the QoS architecture in UMTS networks in the chapter entitled "Quality of Service in UMTS Mobile Systems." In recent years, UMTS has enjoyed rapid growth with worldwide deployments. UMTS is a dominant 3G technology with a detailed QoS support. This chapter provides an overview of UMTS with special focus on its QoS architecture. In particular, service categories and the corresponding QoS parameters are discussed at length. Readers may find coverage of the interesting topics such as: TCP performance, traffic models and system capacity. The chapter also introduces the HSDPA technology.

Benefits of wireless load distribution are explored in the chapter entitled "Quantifying Operator Benefits of Wireless Load Distribution," by S. Lincke and J. Brandner. It is argued that the methods and approaches used in practice to establish the benefits of wireless load sharing are unsatisfactory due to the lack of richness in the scenarios considered. In addition to the performance benefits, advantages related to capacity enhancements, flexibility, survivability, modularity and reconfigurability are emphasized. These additional benefits are evaluated through analytical methods, measurements and simulations of HSPA+ and WLAN.

WiMAX has been heralded as a 4G broadband wireless technology that would truly extend the internet to the wireless domain. WiMAX networks are based on the IEEE 802.16 standard that provides elaborate support for different levels of QoS required by different applications. The chapter entitled "QoS Architecture of WiMAX" by R. Vannithamby and M. Venkatachalam provides a detailed overview of current QoS support in WiMAX and its likely evolution. The chapter also discusses the mechanisms required for enabling QoS and provides ideas for improvements.

F. Rango, A. Malfitano and S. Marano discuss strengths and weaknesses of the QoS mechanisms of WiMAX in the chapter entitled "Cross-Layer QoS Architecture: The WiMAX Point of View." After discussing the QoS mechanisms in detail, the chapter focuses on how best to improve WiMAX.

To this end, a number of optimized algorithms are identified. Novel approaches for improving WiMAX performance are discussed next.

2. RESOURCE MANAGEMENT FOR QOS

Resource management is the key to QoS provisioning. Resource management consists of deciding whether to accept the request for a new flow and then to manage flow servicing so that the QoS guarantees are met. These two components of the resource allocation are called "admission control" and "scheduling." This book covers both of these topics in good detail. We cover a variety of wireless networks including IEEE 802.11 LANs, IEEE 802.16 metropolitan area networks, Cellular networks, and Satellite networks. The recent developments in multi-antenna systems have also been addressed.

IEEE 802.11 networks use a CSMA/CA access method. The probabilistic nature of the contention mechanism makes it difficult to make QoS guarantees and to make admission control decisions. The QoS depends upon several time varying factors including the number of active flows and the active traffic volume for each traffic class, etc. One potential solution to this problem is to use a virtual MAC that helps estimate the delay in real-time. The virtual MAC does not really transmit any packets; it works in parallel to the real MAC and helps in admission control by estimating whether the channel will be able to sustain the traffic of the new connection. Such a scheme has been discussed in Chapter 6.

WiMAX is one of the newer wireless standards that stresses multiple class of service and defines several QoS parameters. Since the standard does not specify scheduling and admission control algorithms to achieve the QoS, a number of algorithms have been proposed. A survey of such algorithms is presented in Chapter 9. It is concluded that majority of the proposed algorithms do not cover all the classes specified in the standard or attempt to meet all the QoS parameters allowed by the standard. In particular, the resource allocation for ertPS class has received little attention.

As mentioned earlier, admission control is an important aspect of the QoS provisioning. Its purpose is to limit the load on the networks so that the guaranteed QoS can be met. For this, the cellular network users need to specify their QoS requirements and traffic characteristics to the base stations. In the past, almost all of the QoS requiring traffic was voice and therefore the QoS requirements and traffic characteristics were implicitly specified. With the introduction of data and video in these networks, this is no longer true and explicit declaration of traffic characteristics and QoS is required. A classification of such efforts in the cellular networks is presented in Chapter 8. The proposed classification is expected to help develop new call admission control policies by comparing existing approaches.

For multimedia broadcasting, satellite communication offers special advantages over terrestrial wireless in that they can economically cover a wide geographic area of the earth surface. However, unidirectional nature of this communication and the long delays cause problems that have been subject of research for several decades. Therefore, combining terrestrial networks with satellite networks offers a viable alternative. Packet scheduling in such hybrid satellite-terrestrial networks (HSTN) is critical for achieving QoS and is a topic of a comprehensive survey in Chapter 10. A number of scheduling techniques have been compared in terms of QoS provisioning and fairness.

Much of the recent research in wireless is related to multi-antenna or multiple-input multiple output (MIMO) systems. MIMO systems using multiple antennas at the mobile devices are difficult due to their size and power requirements. Therefore, recently virtual MIMO systems, in which multiple devices and intermediate relay stations cooperate to provide multiple parallel transmissions of the same signal. In these cooperative diversity systems, the sender transmits the signal to the destination as well as multiple relay stations. Then these relay stations retransmit a modified form of the signal to the destination if necessary. Resource allocation in such adoptive cooperative systems is the topic of a survey in Chapter 7. A smart radio resource allocation algorithm should guarantee good overall network throughput while providing a fair access to the relay nodes by various mobile devices.

3. QOS ISSUES IN MOBILITY

QoS can only be achieved by orchestrating protocols and mechanisms in a synchronized fashion. Signaling, QoS mechanisms, call admission control algorithms, bandwidth granting algorithms, handoff and adaptive modulation algorithms can be addressed and resolved by using a wide range of solutions and architectures. Building an efficient end-to-end solution is a balancing act to best utilize the scarce infrastructure resources including radio resources and power. Furthermore, IP-based networks such as the Internet in its original form do not provide any QoS or mobility support, therefore as it stands, the existing Internet cannot be used to deploy IP based mobile networks. The flexibility as well as other benefits of deploying IP based mobile networks, has lead to numerous research activities in developing QoS and mobility mechanisms for the Internet. On the other hand, there are strong incentives for mobile wireless networks to move towards IP technology. The most prevailing of them is to capitalize on the success of Internet applications but also to provide a common forwarding and management plane where convergence of the different wireless networks can be built.

Bhuvaneswari Chellappan, Teng-Sheng Moh, and Melody Moh illustrate two major areas of research works in the IEEE 802.16 networks QoS and fast mobility support. They describe the basic handoff scheme of 802.16e, and define 3 levels of QoS for handling management traffic: 1. *QoS Classification*, 2. *QoS Scheduling*, 3. *Service flows*. Furthermore they describe the survey of handoff mechanisms including layer-2 handoff schemes, aimed to reduce the delay by avoiding unnecessary scanning of neighboring base stations (BSs). Layer-3 schemes surveys are mostly based on Fast Handovers for Mobile IPv6 (FMIPv6) protocol. Under this scheme, the handoff procedure of FMIPv6 has been reused to suit the 802.16 link layer technology. It uses the primitives proposed by IEEE 802.21 for performing a handoff. In this scheme, there are two modes of handoff, called predictive mode and reactive mode. Various handoff QoS modes defined where each mode supports one or more QoS scheduling services are: *Mode 1 - Conversational Mode (for UGS), Mode 2 - Streaming Mode (for rtPS and ErtPS), Mode 3 - Interactive Mode (for nrtPS), Mode 4 - Background Mode (for nrtPS and BE)* and *Mode 5 - Standby Mode (No traffic)*. Later they design a fast uplink service for Mode 1 (Conversational mode). A fast downlink service is adopted for Modes 1 and 2 (Conversational and Streaming modes), and two low-power handoff operations are designed for Modes 4 and 5 (Background and Standby modes) while a basic Layer 3 Handoff (L3HO) framework is adopted, with extension for concurrent Layer-2 Handoff (L2HO) support, in both predictive and reactive modes.

Hamid Aghvami, Dev Pragad, and Vasilis Friderikos propose mobility solutions such as Mobile IP to support the movement of IP enabled mobile devices. Micro mobility solutions such as Proxy Mobile IPv6 and Hierarchical Mobile IPv6 were developed to provide seamless handover support to ongoing sessions. On the other hand, QoS mechanisms such as IntServ, DiffServ were developed to ensure that a stable level of QoS is maintained during a session and QoS routing is designed to ensure the path with best available QoS resource is selected for a given session. The mobility and QoS mechanisms were developed in isolation to address respective requirements. However, independent functioning of mobility and QoS mechanisms might not lead to the optimal performance. They address the provisioning of QoS in micro mobility enabled wireless access networks. Micro mobility management ensures that during handover the disruption caused to the live sessions are kept to a minimum.

Though many micro mobility and QoS mechanisms have been proposed to solve their respective aspects of network operation, they often have interactions with each other and can lead towards network performance degradation. This chapter focuses specifically on the issues of interaction between micro mobility and QoS mechanisms. Special focus is given to the relatively unexplored area of Mobility Agents' impact on the wireless access network. Mobility agents play a central role in providing micro mobility support. Micro mobility management enhances seamless communication link within access networks. A profusion of micro mobility protocols have been introduced to deal with frequent binding updates. Furthermore, the Hierarchical Mobile IPv6 (HMIPv6) introduces a Mobile IPv6 (MIPv6) node called the Mobility Anchor Point (MAP) which can be located at any level in a hierarchical topology including Access Routers (AR). This primary function of the MAP is to reduce the signaling outside the local subnet or access network and thereby reduce the large delays which occur in normal Mobile IP handovers. Proxy Mobile IPv6 (PMIPv6) intends to provide network based mobility support for mobile nodes (MNs) without the need for direct participation of MNs. PMIPv6 is based on MIPv6 and uses many of the signaling of MIPv6 as well as Home Agent (HA) functionalities. This chapter explored a variety of mobility and QoS interactions and covered the research activities over the recent past in this area. Following which, the impact MA based micro mobility solutions can have on the routing (QoS and traditional routing) of a network was explored.

Robil Daher and Djamshid Tavangarian present a detailed investigation of the current state-of-the-art of QoS-mechanisms, protocols and models for standards for vehicular communication networks (VCNs). They explore real-time applications and their QoS requirement for vehicular environments. Then, the main issues and challenges for adopting QoS in VCNs are addressed. Their work classifies the solutions in accordance with roadside networks and vehicular ad hoc networks, and as layer-2 and layer-3. Consequently, they present several QoS solutions, including medium access and routing protocols, to reflect the state of the art in this field. They show when providing QoS, the VCN characteristics are the cause for several new issues and especially when vehicles travel at high speeds of up to 200 km/h (125 miles/h). These issues are addressed in the context of roadside networks and vehicular ad-hoc networks (VANETs), including vehicle-to-vehicle (V2V) and vehicle-to-roadside (V2R) communications. At the end they present several QoS solutions, including medium access and routing protocols along with a set of open research issues with an objective to spark new research interests in the presented field.

Vehicular communication networks (VCNs) have emerged as a key technology for next-generation wireless networks. Both Dedicated Short Range Communication (DSRC) and Wireless Access in Vehicular Environments and (WAVE) are leading technologies for VCNs and provide a platform for Intelligent Transportation System (ITS) services, as well as multimedia and data services. Some of these services, such as active safety and multimedia services, have special requirements for QoS provision. Following those results, several QoS solutions, including medium access and routing protocols, are presented and discussed in Chapter 14. Additionally, open research issues are discussed, with an objective to spark new research interests in the presented field. Issues and challenges for providing QoS are addressed by various modes: ad hoc mode point-to-point (P2P) for vehicle-to-vehicle (V2V) communications and cell-based mode point-to-multipoint (P2MP) for vehicle-to-roadside (V2R) communications. The QoS requirements of all modes are shown with different communication characteristics

Ramón M. Rodríguez-Dagnino and Hideaki Takagi address dynamic mobility management for wireless cellular networks. The handover process is a complex function of many factors including size of wireless cells, user's mobility path, and call patterns. Early works in this direction have assumed exponential distributions for both the Call Holding Time, or equivalently Inter-Call Time, and the Cell Residence Time. Besides its importance in dimensioning wireless networks, counting the number of cell crossing boundaries is also important for location of mobile users in a specific location area. The main goal in the location algorithms is to minimize the signaling cost resulting from the users updates in the database serving the area. Even if the user is not active in a conversation, it is necessary to keep track of it by updating the database. They attempt to find an optimal cost to reduce signaling traffic and database loads. Further, they show that dynamic schemes are preferable where the mobile terminal takes the decision of when to update. Some regular or periodic events are used in these dynamic schemes.

Richard Good, David Waiting, and Neco Ventura address the issues rich multimedia QoS provisioning for IMS. Effective deployment can be best realized when QoS framework can efficiently manage scarce network resources. This and the ability to differentiate IMS services from web-based services are fundamental arguments in this chapter. The chapter reviews resource management frameworks and architectural alignment in an effort to harmonize standardized architectures for increase interoperability. The IP Multimedia Subsystem (IMS) promises to revolutionize inter-personal communication and enable convergence of wired and wireless services. Multimedia enriched services can be delivered over multi radio access technologies. IMS is seen as complementary infrastructure for carriers' in-house applications and services. Despite all progress, there are several hurdles to overcome before circuit-switched technologies can be moth-balled once and for all. The advent of the intelligent network improved the operators' ability to provide enhanced services both in voice and data communications while rapidly expanding requirements of customers wishing to make use of rich multimedia Internet applications. The popularity of Internet-based VoIP applications has shown that the expected quality of legacy must at least be matched. The chapter describes the interfaces standardized as part of Long-Term Evolution (LTE) and Evolved Packet Core (EPC) in 3GPP. The Release 7 and beyond policy and charging control (PCC) architecture calls for Application Function (AF), a Policy and Charging Rules Function (PCRF) and a Policy and Charging Enforcement Function. Each of these functions is described in details.

Mihai Ivanovici and Răzvan Beuran address issues and challenges related to correlating quality of experience and quality of service for network applications. User perceived experience as compared to network quality of experience is fundamentally tightly coupled. This chapter points out, there is a significant difference between what a network application experiences as quality at network level, and what the user perceives as quality at application level. From the network point of view, applications require certain delay, bandwidth and packet loss bounds to be met – ideally zero delay and zero loss. However, users should not be directly concerned with network conditions, and furthermore they are usually neither able to measure, nor capable to predict them. Users only expect good application performance, i.e., a fast and reliable file transfer, high quality for voice or video transmission, etc., depending on the application being used. This is true in both wired as well as wireless networks.

In order to understand network application behaviors, as well as the interaction between the application and the network, one must perform a delicate task – the one of correlating the Quality of Service (QoS), i.e., the degradation induced at network level (as a measure of what the application experiences), with the Quality of Experience (QoE), i.e., the degradation perceived by the user at application level (as a measure of the user-perceived quality). This is done by simultaneously measuring the QoS degradation and the application QoE on an end-to-end basis. These measures must be then correlated by taking into account their temporal relationship. Assessing the correlation between QoE and QoS makes it possible to predict application performance given a known QoS degradation level, or to determine the QoS bounds that are required in order to attain a desired QoE level.

4. QOS FOR MULTIMEDIA

F. Babich, M. D'Orlando, and F. Vatta consider the issues faced in video streaming over wireless networks in the chapter entitled "Video Distortion Estimation and Content-Aware QoS Strategies for Video Streaming over Wireless Networks". After discussing the characteristics and QoS requirements of multimedia applications, the chapter presents the benefits of content-aware strategies including packet scheduling and other QoS techniques. In this context, the utility of the hint tracks adopted by MPEG-4 is discussed. The chapter further reviews the state-of-the-art techniques for estimating video distortion and presents various mechanisms for improving the end user perceived quality. An evaluation testbed for streaming video is also presented.

The chapter entitled "Quality of Experience versus QoS in Video Transmission" by A. Marquet, I. Monteiro, N. Martins and M. Nunes presents the importance of Quality of Experience (QoE) metrics for video and differentiates them from the QoS metrics commonly used by an IP network for quantifying the treatment received by the packets as they traverse the network. It is argued that the diversity among traffic distribution platforms, video traffic characteristics and video display devices poses new challenges in assessing the user perceived QoE. The chapter reviews the best practices available today to evaluate and assure robust QoE, both subjective and objective, for multimedia applications like streaming video. The ongoing efforts by various standards organizations for defining the QoE metrics are also presented.

S. Jelassi, H. Youssef, and G. Pujolle consider perceptual quality of voice conversations over wireless networks in their chapter entitled "Perceptual Quality Assessment of Packet-Based Vocal Conversations over Wireless Networks: Methodologies and Applications." After reviewing the QoS provisioning approaches that impact the perceived quality, the chapter presents the methodologies for evaluating these. They include both subjective and objective methodologies based on both simulated and emulated evaluation approaches. The chapter also considers the parametric model-based techniques for predicting the perceived quality. Furthermore, examples for managing networks as well as end-devices based on voice quality measurements are also presented.

5. QOS IN AD-HOC AND MESH NETWORKS

Mobile ad-hoc networks (MANETs) are networks comprised of randomly positioned wireless nodes, which are able to move freely in a wireless domain. Different categories for MANETs are based on their topologies and functions. From the functional point of view, MANET routing protocols are categorized as: *Power-aware, QoS-aware, Security-Aware,* and *Multicast* routing protocols. From the topological point of view, the main three MANET routing protocols categories are: *Flat, Hierarchical* and *Geographic Position Assisted* routing protocols.

The first topological routing protocol is the flat routing protocol. There is only one layer (tier) in this type of routing protocol. Therefore, all nodes are processed without any specific order or groupings. Flat routing protocols are further subcategorized into: *Reactive, Proactive* and *Hybrid* routing protocols.

Reactive (On-Demand) Routing Protocols: In a reactive routing protocol, nodes do not have any information about the current availability of other nodes except for the nodes which were involved in previous communications. Once the source node is ready to transmit data to an unknown destination, the route to this new destination has to be learned and saved in the routing table. Routes to new destinations are determined as they are required. This is performed through the route discovery mechanism, by transmitting Route REQuest (RREQ) and Route REPly (RREP) messages. The transmitting node constructs an RREQ packet and transmits to all its immediate neighboring nodes. If the destination node is among the neighboring nodes, the RREP packet is transmitted back to the source; otherwise, all the neighboring nodes also transmit the RREQ packet to their own neighbors, excluding the incoming RREQ interface. This process is continued until the RREQ reaches the destination. Then the destination node will include its information and reply back to the source node. The RREP will include all the intermediate nodes information between the destination and source nodes, from which the source node will store the route to that specific destination.

Various reactive routing protocols differ in several aspects including the route selection and forwarding functionalities. These variations may contribute to the QoS measures provided by the ad-hoc protocol. The main advantage in reactive routing protocol schemes is the fact that they generate relatively less amount of traffic overhead as compared to proactive protocols, since they do not require constant updates. The main disadvantage of reactive routing protocols is the additional delay in new route calculations. An example of reactive routing protocols is Ad-Hoc On-demand Distance Vector routing (AODV) and Dynamic Source Routing (DSR).

Proactive (table-driven) routing protocols: In proactive routing protocols, all nodes are constantly involved in information updates which are stored in routing tables. Therefore, all nodes are supposed to know the locations of other nodes in the wireless domain. When a source node requires transmitting to a destination within the wireless domain, the route is already known and fetched from the routing table. The major advantage of proactive routing protocols, compared to reactive routing protocols, is that they require virtually no time to find a route to any destination within the wireless domain. The disadvantage is the higher bandwidth requirement and extra overhead because of the constant route updates. This disadvantage of proactive routing protocols makes reactive routing protocols better choices in most of the MANET applications. An example of proactive routing protocols is Optimized Link State Routing (OLSR) and Destination Sequenced Distance Vector routing (DSDV).

Hybrid (proactive/reactive) routing protocols: These routing protocols take advantage of both reactive and proactive functions to sort out routing issues. Hybrid routing protocol example includes Core Extraction Distributed Ad-hoc Routing (CEDAR).

Hierarchical routing protocols: In hierarchical routing protocols, as opposed to flat routing protocols, nodes are often grouped in various tiers distinguished by hierarchy of functions. In the simplest form, a two-tier hierarchical routing protocol could consist of two groups of nodes, the main group and the subgroup, where each group has a specific function that needs to be done before the next group starts continuing the process of the previous group. A feature of hierarchical routing protocols is that the local traffic (i.e., updates, inquiries) related to each group is kept locally and is usually not transmitted to the other groups. This way, a separation of traffics can be achieved to reduce the amount of overhead and required bandwidth. The main disadvantages of hierarchical routing protocols include: suboptimal routes with the vulnerability of single point of failure and bottlenecks. Also as the number of tiers increases, the dynamic node management issue becomes a challenge. An example of this type of routing protocol is Hierarchical State Routing (HSR) protocol.

Geographic position assisted routing protocols: In this type of routing protocols, nodes are usually equipped with some sort of node location identification mechanism, similar to that of a Global Positioning System (GPS), which makes the location of all nodes readily available. The advantage of this type of routing protocol is similar to proactive routing protocol, which is the fact that nodes' location discovery requires virtually no time, resulting in a reduction of control overhead. The disadvantage, however, is the cost associated to the location determination system deployment. An example of this type of routing protocol is Location Aided Routing (LAR) protocol.

Functional routing protocols, as opposed to topological routing protocols, are mostly concerned with the functions associated to the nodes, which are subcategorized as follows.

Power-aware routing protocols: In this type of routing protocol, the most important factor is the limitations of the nodal power consumption. Every effort is made to ensure power consumption is kept relatively low, which is a major challenge in wireless nodes running on scared battery power. An example of this type of routing protocol is Power-Aware Source Routing (PSR).

QoS-aware routing protocols: A wide range of network parameters including delay, bandwidth, and packet drop ratio are considered Quality of Service (QoS) related parameters. Any MANET protocol offering optimizations in regards to these parameters is considered a QoS-aware routing protocol. An example of this type of routing protocol is QoS-aware source-initiated ad-hoc routing (QuaSAR).

Security-aware routing protocols: In this type of routing protocols, providing security is the main objective. Private information communications, anonymous routing, node and data authentication are examples of the services offered by these types of routing protocols. An example of this type of routing protocol is Secure Efficient Distance Vector routing for mobile wireless ad-hoc networks (SEAD).

Multicast routing protocols: Multicasting involves the transmission of packets from one source to many destinations, reducing the costs for communications involving multiple recipients. An example of a multicast routing protocol includes Multicast Ad-Hoc On-Demand Distance Vector (MAODV).

Multipath and load-balancing routing protocols: In multipath routing protocols, multiple paths between a source node and a destination node are established. Load-balancing across multiple paths is a mechanism by which the traffic load from a source to a destination is widely distributed among various paths to try to avoid congestions and provide optimum routing.

Mesh networking: Mesh networking is a collection of wireless (usually fixed) nodes, which are included in a communication scenario to ensure any two nodes can communicate and transmit data. This may cause major challenges since stationary wireless nodes are not moveable nodes and reconfiguration around broken or blocked paths may often be required. MANET nodes may be utilized in mesh networking if mobility is restricted.

There are several chapters discussing MANET and mesh networking scenarios in this book, which are introduced in the following subsection.

QoS and Energy-Aware Routing for Wireless Sensor Networks by Shanghong Peng Simon X. Yang and Stefano Gregori discusses details of QoS- and energy-aware wireless sensor networks. It considers the challenges about various network resources including energy, memory capacity, computation capability, and maximum data rate, while maintaining QoS supports. This chapter features a novel bio-inspired flavor of QoS- and energy-aware routing algorithm based on an ant colony optimization idea to meets QoS requirements in an energy-aware fashion, balancing the node energy utilization to maximize the network lifetime. Various network-related parameters are evaluated through extensive simulation results.

Quality of Service (QoS) Routing in Mobile Ad hoc Networks by R. Asokan and A. M. Natarajan concentrates on the problem of QoS provisioning at the network layer. QoS routing aims at finding a feasible path, which satisfies QoS constraints on bandwidth, end-to-end delay, jitter, energy, etc. This chapter provides a detailed survey of major contributions to the QoS routing in MANETs. A few proposals on the QoS routing using optimization techniques and inter-layer approaches have also been addressed. It concludes with a discussion on the future directions and challenges in QoS routing support in MANETs.

Queuing Delay Analysis of Multi-Radio Multi-Channel Wireless Mesh Networks by Chengzhi Li and Wei Zhao contains a study featuring multi-radio and multi-channel wireless mesh networks with IEEE 802.11e based ingress access points for local clients and point-to-point wireless links over non-overlapping channels for wireless mesh network backbones. A set of algorithms is provided to analyze the performance of such wireless mesh networks with wideband fading channels in various office building and open space environments and commonly-used Regulated and Markov On-Off traffic sources. The goal is to establish a theoretical framework to predict the probabilistic end-to-end delay bounds for real-time applications over such wireless mesh networks.

Scalable Wireless Mesh Network Architectures with QoS Provisioning by Jane-Hwa Huang Li-Chun Wang and Chung-Ju Chang presents a study concerning Wireless Mesh Network (WMN) from a network architecture perspective to investigate ways of overcoming the scalability issue in WMNs. The purpose is to improve the tradeoff between coverage and throughput to achieve QoS provisioning objectives. In this chapter, main QoS-related research directions in WMNs are discussed, following an introduction to two available scalable mesh network architectures that can relieve the scalability issue and support QoS in WMNs for the wide-coverage and dense-urban coverage. Then an optimal tradeoff among throughput, coverage, and delay for the proposed WMNs by an optimization approach to design the optimal system parameters is discussed.

Towards Designing High-Throughput Routing Metrics for Wireless Mesh Networks by T. Nyandeni, C. Kyara, P. Mudal, S. Nxumalo, N. Ntlatlapa, and M. Adigun deals with the analysis of existing routing metrics to determine optimal paths. This chapter proposes the design principles for routing metrics, which are critical for achieving high throughput. These design principles ensure proper functioning of the network and enable high throughput. Discussions on the pitfalls of the existing routing metrics are also included and the chapter concludes with an outline of the future research directions.

6. QOS IN FUTURE WIRELESS NETWORKS

In the discussion of future directions, the term "Next Generation Networking (NGN)" is coined to describe a few key architectural evolutions for telecommunication core and access technologies being developed in the near future. The general idea behind NGN is to provide an evolutional framework consisting of network transport entities offering advancements to various multimedia communications (voice, data, video, etc) on All-IP communication infrastructures.

The NGN approaches for mobile networks include LTE and LTE-A (LTE-Advanced) enhancements. LTE is an access technology, based on the UMTS evolution covered under the 3GPP Release 8. LTE is considered a 3G technology with nominal uplink peak rates of at least 50 Mbps and downlink rates of 100 Mbps. However it does not meet the requirements for 4G (aka IMT Advanced), where data rates of up to 1

Gbps is expected. LTE-A, an enhancement of LTE designed to meet this requirement, is considered one of the 4G candidates.

Other advancements expected in the NGN are further progression of ad-hoc and peer-to-peer mobile computing implementations including cognitive radio technologies, which is a wireless communication paradigm for a wireless node, where transmission and reception parameters are changed to accommodate the environment interferences and to communicate efficiently.

A. Diez Albaladejo and F. Gouveia M. Corici provide a comprehensive overview of the QoS control architecture being adopted by the new generation of wireless networks in the chapter entitled "Evolution of QoS Control in Next Generation Mobile Networks." The chapter first introduces the QoS control architecture of UMTS, CDMA2000, LTE, Wi-Fi and WiMAX networks. It then describes the up-to-date trends for converging to a common architecture. The challenges faced in specifying the convergent QoS control architecture for integration of diverse access networks are presented. Furthermore, the architectural support needed for end-to-end QoS assurance is also discussed.

Quality of Service (QoS) Provisioning in Cognitive Wireless Ad Hoc Networks: Architecture, Open Issues and Design Approaches by Kok-Lim Alvin Yau, Peter Komisarczuk and Paul D. Teal discusses Cognitive Radio (CR) in a next-generation wireless communication technological context that aims to improve the utilization of the overall radio spectrum through dynamic adaptation to local spectrum availability. A Cognitive Wireless Ad Hoc Network (CWAN) is a multi-hop self-organized and dynamic network that applies CR technology for ad-hoc mode wireless networks that allow devices within range of each other to discover and communicate in a peer-to-peer fashion without necessarily involving infrastructure such as base stations or access points. The aim of this chapter is to present a discussion on the architecture, open issues and design approaches related to QoS provisioning in CWAN and to establish a foundation for further research in several unexplored, yet promising areas in CWAN.

Section 1
Broadband

Chapter 2
Quality of Service in UMTS Mobile Systems

Jahangir Dadkhah Chimeh
Iran Telecommunication Research Center, Iran

1.1 INTRODUCTION

Mobile systems and particularly UMTS are growing fast. These systems convey data based services in addition to customary voice services. Quality of service is a function of data rate, delay and signal to noise plus interference ratio in these systems. In this Chapter first the author pays attention to UMTS and its QoS architecture, then to service categorization due to QoS. Afterwards he reviews some QoS parameters. Then he studies Layer 2 QoS parameters and general concepts about Transport channels. Then he review TCP effects on the throughput in the air interface. he introduces HSDPA in the next section. Finally he pays attention to data traffic models and their effects on the system capacity and Erlang capacity and delay in the system.

1.2 UMTS Architecture

Figure 1 shows the layered UMTS architecture and protocols as outlined in 3GPP TS 23 107 (2007). The figure shows the UMTS architecture in terms of its entities User Equipment (UE), UTRAN and Core Network. The respective reference points Uu (Radio Interface) and Iu (CN-UTRAN interface) are shown. The protocols over Uu and Iu interfaces are divided into two structures: User plane protocols and Control plane protocols.

This figure illustrates furthermore the high-level functional blocks into the Access Stratum (AS) and the Non-Access Stratum (NAS).

The Access Stratum offers services through the following Service Access Points (SAP) to the Non-Access Stratum:

- General Control (GC) SAPs
- Notification (Nt) SAPs
- Dedicated Control (DC) SAPs

The SAPs are marked with circles in Figure 1.

The NAS protocols enable the transfer of information between the UE and CN. The information can be either user or control information carrying all the signaling required to set-up or tear down the service connection as well as to perform other functionalities specific to a mobile network (e.g. mobility management). This information is almost

DOI: 10.4018/978-1-61520-680-3.ch002

Figure 1. Layered UMTS architecture (Prez-Romero et al., 2005; TS 25 401, 2007)

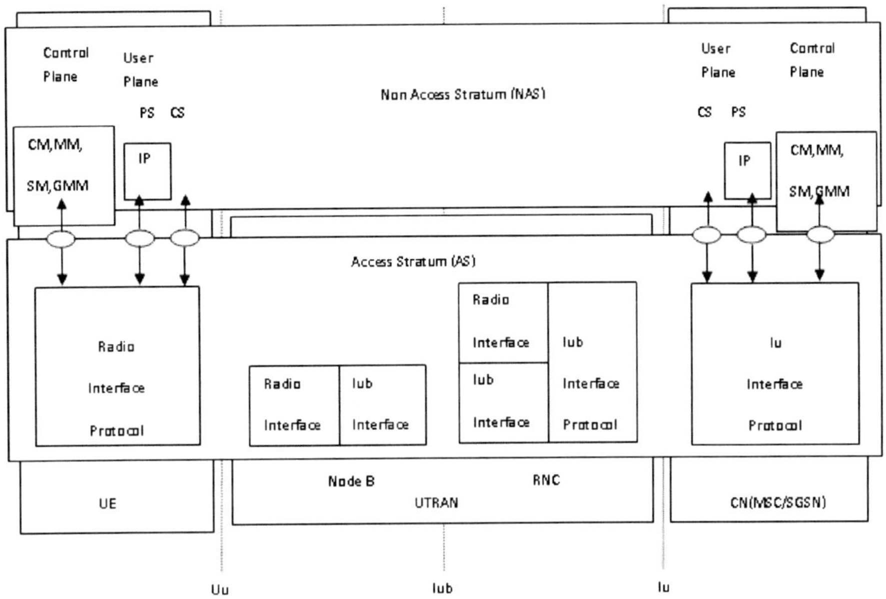

independent of the underlying layers of the protocol architecture and of the elements of the access network that are traversed in the path between the UE and the Core Network.

Two examples of NAS functions in the control plane are the Connection Management (CM) and Session Management (SM) functions which are responsible for the establishment and release of the connections or sessions for an UE respectively. Other examples are Mobility Management (MM) and GPRS Mobility Management (GMM) functions which are responsible for mobility functions at the network layer (e.g. subscriber location area updating, routing area updating, paging, etc.). In the user plane, the main NAS function at the network layer for packet switched services is the IP protocol while for the circuit services, information comes directly from the source without the need for a network function.

NAS functions rely on the AS functions to exchange information between the UE and the CN, as shown in Figure 1. The AS consists of a group of functions that are specific to the access network being used (3GPP TS 23 107, 2007). This means that even if the NAS functions are the same for a UMTS or a GSM/GPRS access network, the AS protocols that allow the transfer of these messages through the different nodes may be different. In the UMTS architecture, the AS includes three different protocol stacks, namely the radio interface protocol Uu, the Iub interface protocol and the Iu interface protocol which may be specific for data or circuit switched connections. The radio interface protocol stack establishes communication between the UE and the UMTS access network (UTRAN). Note that the protocols at the upper layers terminate in the UE and RNC, while the lower layers terminate in the UE and Node B. Iub interface protocols involve the communication of the lower layers of the RNC and the Node B. Iu interface protocols allow communication between the RNC and the CN. Iu is divided into two protocols Iu-CS and Iu-PS. Iu-CS is responsible for the communication between RNC and MSC and the Iu-PS is responsible for communication between RNC and SGSN.

The AS provides the NAS with a service of information transfer between the UE and the CN

Figure 2. Radio access bearer concept

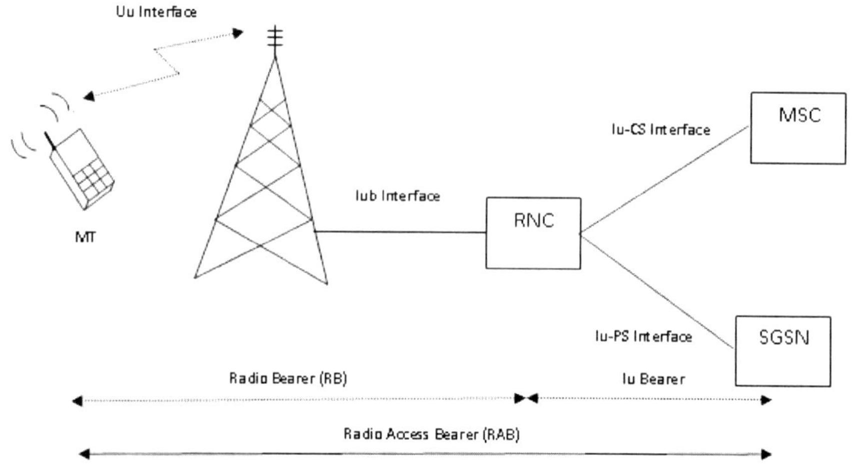

in what is named a Radio Access Bearer (RAB). A RAB consists of two parts, the Radio Bearer, corresponding to the Radio Access Network between the UE and the RNC, and the Iu Bearer, defined between the RNC and the MSC or SGSN (see Figure 2).

1.3 Qos Architecture of UMTS

We consider network services from a Terminal Equipment (TE) to another TE or end-to-end. An End-to-End Service may have a certain Quality of Service (QoS) which is provided for the user of a network service. It is the user who decides whether or not he is satisfied with the provided QoS. To realize a certain network QoS a Bearer Service with defined characteristics and functionalities is to be set up from the source to the destination of a service end point. A bearer service includes all aspects to enable the provision of a contracted QoS. These aspects are among the control signaling, user plane transport and QoS management functionality. A UMTS bearer service layered architecture is depicted in Figure 3. Each bearer service on a specific layer offers it's individual services using services provided by the layers below.

1.3.1 Bearer Services

Here we briefly explain each bearer service (BS). The End-to-End Service on the application level uses the bearer services of the underlying network(s) and it may be conveyed over several networks (not only UMTS). As it is shown in Figure 3 End-to-End-Service used by the TE will be realized using a TE/MT Local Bearer Service, a UMTS Bearer Service, and an External Bearer Service. It is the UMTS Bearer Services (UMTS BS) that provide the UMTS QoS. The UMTS Bearer Service is constituted from two parts, the Radio Access Bearer Service and the Core Network Bearer Service. The Radio Access Bearer Service provides confidential transport of signaling and user data between MT and CN Edge Node with the QoS adequate to the negotiated UMTS Bearer Service or with the default QoS for signaling. The Core Network Bearer Service of the UMTS core network connects the UMTS CN Edge Node with the CN Gateway to the external network. The role of this service is to efficiently control and utilize the backbone network in order to provide the contracted UMTS bearer service. The UMTS packet core network shall support different backbone bearer services for variety of QoS. The Core Network Bearer Service uses a generic Back-

Figure 3. UMTS QoS architecture (3GPP TS 23 107, 2007)

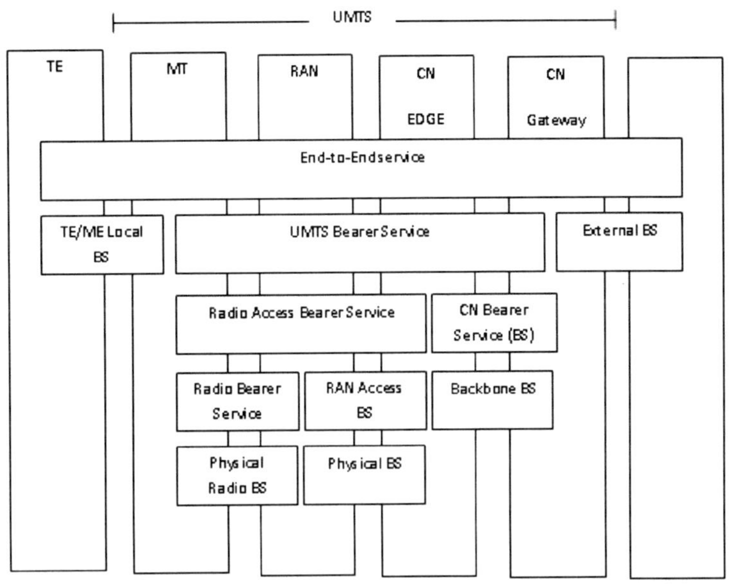

bone Network Service. The Backbone Network Service covers the layer1/Layer2 functionality and is selected according to operator's choice in order to fulfill the QoS requirements of the Core Network Bearer Service. The Backbone Network Service is not specific to UMTS but may reuse an existing standard.

As we saw the Radio Access Bearer Service is realized by a Radio Bearer Service and a RAN Access -Bearer Service. The Radio Bearer Service covers all the aspects of the radio interface transport. This bearer service is provided by the UTRAN FDD/TDD or the GERAN. The RAN Access Bearer Service together with the Physical Bearer Service provides the transport between RAN and CN. RAN Access bearer services for packet traffic shall provide different bearer services for variety of QoS.

The Radio Bearers used for transferring signaling messages are called Signaling Radio Bearers (SRBs). The SRBs are defined as:

- SRB1 is used to carry RRC signaling performed in support of Access Stratum specific needs (RLC operates in unacknowledged mode)
- SRB2 is used to carry RRC signaling performed in support of Access Stratum specific needs (RLC operates in acknowledged mode)
- SRB3 is used to carry RRC signaling performed in support of Non-Access Stratum specific needs (RLC operates in acknowledged mode)
- SRB4 is used to carry RRC signaling performed in support of Non-Access Stratum specific needs (RLC operates in acknowledged mode)

1.3.2 QoS Management Functions in the Network

The purpose of this section is to give an overview of functionality needed to establish, modify and maintain a UMTS Bearer Service with a specific QoS. The QoS management functions of all UMTS entities together shall ensure the provision of the negotiated QoS of the service between the access points of the UMTS bearer service. The end-to-end service is provided by translation/mapping with UMTS external services. We describe the QoS management functions in two different control and

Figure 4. QoS management functions for UMTS bearer service in the control plane (3GPP TS 23.107 Version 7.1.0, 2007)

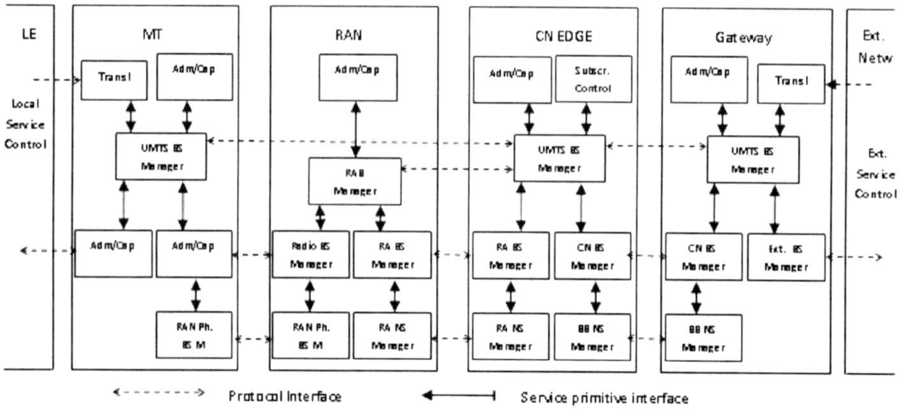

user planes. In the control plane these functions include Service Manager, Translation function, Admission/Capacity control and Subscription Control. Service Management co-ordinates the functions of the control plane for establishing, modifying and maintaining the services it is responsible for and, it provides all user plane QoS management functions with the relevant attributes. Translation functions performs the converting between UMTS bearer service attributes and QoS attributes of the external networks service control protocol (e.g. between IETF TSPEC and UMTS service attributes). The service manager may include a translation function to convert between its service attributes and the attributes of a lower layer service it is using. The Admission/Capacity control maintains information about all available resources of a network entity and about all resources allocated to UMTS bearer services. The function checks also the capability of the network entity to provide the requested service, i.e. whether the specific service is implemented and not blocked for administrative reasons. Subscription Control checks the administrative rights of the UMTS bearer service user to use the requested service with the specified QoS attributes.

In the user plane QoS management functions maintain the signaling and user data traffic within certain limits, defined by specific QoS attributes. These functions include Mapping functions, Classification functions, Resource Manager, Traffic conditioner. Mapping functions provides each data unit with the specific marking required to receive the intended QoS at the transfer by a bearer service. Classification functions assigns data units to the established services of a MT according to the related QoS attributes if the MT has multiple UMTS bearer services established. Resource Manager distributes the available resources between all services sharing the same resource. The resource manager distributes the resources according to the required QoS. Resource management performs scheduling, bandwidth management and power control for the radio bearer. Traffic conditioner provides conformance between the negotiated QoS for a service and the data unit traffic. Traffic conditioning is performed by policing or by traffic shaping.

1.4 Service Categories

According to 3GPP TS 22-105 Version 7.1.0 (2006), telecommunication services are divided into the basic and supplementary services. Basic services are also divided into bearer and teleservices (Figure 6)

Bearer services carry signals between each access nodes. Telecommunication services which include terminal functionalities make end-to-end connections between end users.

Figure 5. QoS Management functions for UMTS bearer service in the user plane (3GPP TS 23.107 Version 7.1.0, 2007)

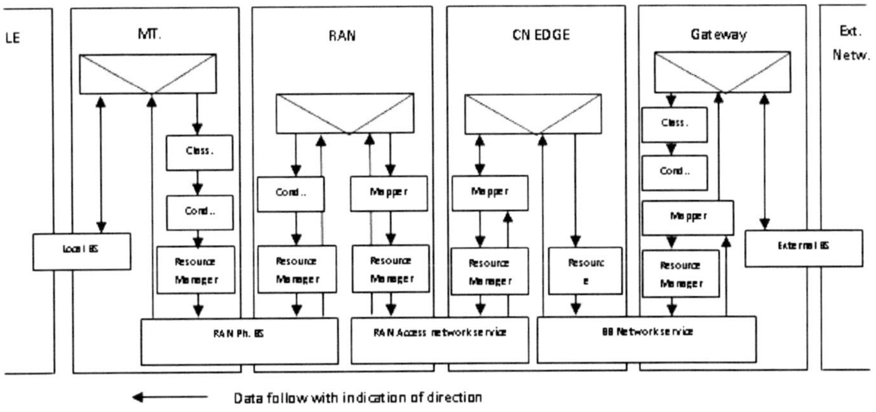

Figure 6. Basic telecommunication services in PLMN (3GPP TS 22 105 Version 7.1.0, 2006)

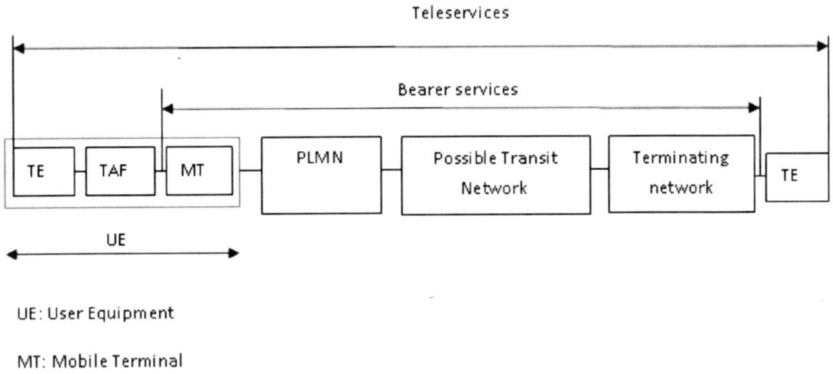

Supplementary services change and supplement basic services but they are not standalone by themselves, it means they are only used with basic services. According to 3GPP TS 23.107 Version 7.1.0 (2007), we have four classes for different kinds of services as:

- Conversational classes
- Streaming classes
- Interactive classes
- Background classes

The main difference between these traffic classes is how much delay sensitive they are: Conversational traffic class is very delay sensitive while background class is the most delay insensitive traffic class. Conversational and Streaming classes are mainly intended to be used to carry real-time traffic flows. Conversational real-time services, like video telephony, are the most delay sensitive applications and those data streams should be carried in Conversational class.

Conversational traffics include bidirectional voice calls, video telephony, video conferencing, remote sensing, telemedicine, interactive games. Streaming traffics include downloading multimedia while displaying. Interactive traffics include Web services, e.g., Web browsing, transactional services, e.g., electronic commerce, chat, data base access, system monitoring and control, etc. Background services include file transfer protocol (FTP), Email, Automatic information distribution.

Besides, regarding to the service sensitivity and non-sensitivity to error we have two services as Error tolerable service such as Conversational (voice and video), Voice messaging, Audio and video streaming, fax; and error non-tolerable services such as Telnet, Interactive games, E-commerce, WWW browsing, Still images, File transfer, and email (3GPP TS 22 105 Version 7.1.0, 2006).

Conversational and streaming services are called guaranteed bit rate, i.e., data rates for these service users should be guaranteed. Circuit switching traffics are in these domain. In contrary, interaction and background service classes are in non-guaranteed bit rate domain in which constant and guaranteed bit rates are not mandatory for them. It means service data rate may vary according to the system characteristics.

1.4.1 UMTS Bearer Service Attributes

UMTS bearer service attributes describe the service provided by the UMTS network to the user of the UMTS bearer service. A set of QoS attributes (QoS profile) specifies this service.

These attributes (parameters) include the Maximum bit rate (kbps), the Guaranteed bit rate (kbps), Delivery order (y/n), the maximum SDU size (octets), SDU format information (bits), SDU error ratio, Residual bit error ratio, Delivery of erroneous SDUs (y/n/-), Transfer delay (ms), Traffic handling priority, Allocation/Retention Priority, Source statistics descriptor ('speech'/'unknown'), Signaling Indication (Yes/No). The defined Radio Access Bearer attributes and their relevancy for each bearer traffic class are summarized in Table 1.

A list of finite attribute values or the allowed value range is defined for UMTS Bearer Services and Radio Access Bearer services. The value list/value range defines the values that are possible to be used for an attribute considering every possible service condition. Further limitations may appear when a service is defined as a combination of different attributes; for example the shortest possible delay may not be possible to use together with the lowest possible SDU error ratio.

Table 2 lists the value range of the UMTS bearer service attributes for four different traffic classes in summary.

Table 1. Radio access bearer attributes defined for each bearer traffic class (3GPP TS 23.107 Version 7.1.0, 2007)

Traffic Class	**Conversational Class**	**Streaming Class**	**Interactive Class**	**Background Class**
Maximum bit rate	×	×	×	×
Delivery order	×	×	×	×
Maximum SDU rate	×	×	×	×
SDU format information	×	×		
SDU error rate	×	×	×	×
Residual bit rate ration	×	×	×	×
Delivery of erroneous SDUs	×	×	×	×
Transfer delay	×	×		
Guaranteed bit rate	×	×		
Traffic handling priority	×		×	
Allocation retention priority	×	×	×	×
Source statistics descriptor	×	×		
Signaling indication			×	

Table 2. Value ranges for UMTS bearer service attributes (3GPP TS 23.107 Version 7.1.0, 2007)

Traffic Class	Conversational Class	Streaming Class	Interactive Class	Background Class
Maximum bit rate	<=256000	<=256000	<=256000	<=256000
Delivery order	Yes/No	Yes/No	Yes/No	Yes/No
Maximum SDU size (octet)	<=1500 or 1502	<=1500 or 1502	<=1500 or 1502	<=1500 or 1502
SDU format information				
Delivery of erroneous SDUs	Yes/No/-	Yes/No/-	Yes/No/-	Yes/No/-
Residual BER	5*10-2, 10-2 5*10-3 10-2, 10-4, 10-5, 10-6	5*10-2, 10-2 5*10-3 10-2, 10-4, 10-5, 10-6	4*10-3, 10-5 6*10-8	4*10-3, 10-5 6*10-8
SDU error rate	10-2, 7*10-3, 10-3, 10-4, 10-5	10-1, 10-2, 7*10-3, 10-3, 10-4, 10-5	10-3, 10-4, 10-5	10-3, 10-4, 10-5
Transfer delay (ms)	100, maximum value	300, maximum value		
Guaranteed bit rate (kbps)	<=256000	<=256000		
Traffic handling priority			1,2,3	
Allocation retention priority	1,2,3	1,2,3	1,2,3	1,2,3
Source statistics descriptor	Speech/unknown	Speech/unknown		
Signaling indication			Yes/No	

1.4.2 QoS Requirements

It shall be possible for one application to specify its QoS requirements to the network by requesting a bearer service with any of the specified traffic type, maximum transfer delay, delay variation, bit error ratios and data rates. Table 3 indicates the range of values that shall be supported. These requirements are valid for both connection and connectionless traffic. It shall be possible for the network to satisfy these requirements without wasting resources on the radio and network interfaces due to granularity limitations in QoS.

Table 3. QoS requirements for various environments (3GPP TS 22 105 Version 7.1.0, 2006).

	Real Time (constant Delay)	Non Real Time (variable Delay)
Operating environment	BER/ Max transfer delay	BER/ Max transfer delay
Satellite (Terminal relative speed to ground up to 1000 km/h for plane)	Max Transfer Delay less than 400 ms BER 10-3 to 10-7 Note 1	Max Transfer Delay 1200 ms or more, Note 2 BER 10-5 to 10-8
Rural outdoor (Terminal relative speed to ground up to 500 km/h for plane), Note 3	Max Transfer Delay 20-300 ms BER 10-3 to 10-7 Note 1	Max Transfer Delay 150 ms or more, Note 2 BER 10-5 to 10-8
Urban/ suburban outdoor (Terminal relative speed to ground up to 120 km/h)	Max Transfer Delay 20-300 ms BER 10-3 to 10-7 Note 1	Max Transfer Delay 150 ms or more, Note 2 BER 10-5 to 10-8
Indoor/ Low range outdoor (Terminal relative speed to ground up to 10 km/h)	Max Transfer Delay 20-300 ms BER 10-3 to 10-7 Note 1	Max Transfer Delay 150 ms or more, Note 2 BER 10-5 to 10-8

Note 1: There is likely to be compromise between BER and delay

Note 2: The Max Transfer Delay should be here regarded as the target value for %95 of the data.

Note 3: The value of 500 km/h as the maximum speed to be supported in the rural outdoor environment was selected in order to provide service on high speed vehicle (e.g. trains). This is not meant to be the typical value for this environment (250 km/his more typical).

1.5 Other QoS Parameters

After R99, releases 4 to 8 were developed. They have higher capabilities and higher data rates. They differ in the data rates, link set up and tear down times and data and voice transmission via IP network. One of the newest air interface, R5 and later uses High Speed Data Packet Access (HSDPA). As an example HSDPA can provide 14.4 Mbps and 5.76 Mbps data rates in R6. Release 8 named Long Term Evolution (LTE) used SC-FDMA in the uplink and OFDMA in the downlink (3GPP TS 36.104 version 8.3.0, n.d.).

In a modern telecommunication network such as UMTS, the aim of the operator is to offer high Quality of Service (QoS) to the users.

Generally, QoS is the collective effect of service performances, which determine the degree of satisfaction of a user of a service. Under the general heading of quality of experience (QoE) one of the more noticeable points faced by the user is the apparent delay in set up or channel allocation times for different connections. The set up and channel allocation delay can be defined as the time interval from the instant the user initiate a connection request until the complete message indicating the channel allocation is received by the calling terminal or by the application server. When establishing a connection, the user due to this delay, may think that the connection has not gone through or the network is not responding which may prompt the user to re-dial, reconnect or even in some cases to abandon the connection attempt. From the service provider's perspective, improving the quality of service is very important giving their users a good perception of the network performance and efficiency.

There are some mechanisms to improve the connection establishment times. The delay in set up or channel allocation times can be attributed to:

- Processing time in the UTRAN
- Processing time in the Core network
- Processing time in UE
- Call setup and alerting phase in the core network
- UTRAN and CN Protocols and associated overhead including protocol conversion
- Signalling delay on the air interface
- NAS procedures

To evaluate the above attributes we may

- Review the CS and PS Call and session Setup and channel allocation procedures in UMTS
- Highlight the improvements where call and session setup process can be improved and consider impacts the relevant specifications
- Highlight the improvements to the existing RRC state transitions
- Identify possible ways to enhance call and session setup performance whilst keeping in mind R99 backwards compatibility
- Put forward change request relevant to specifications
- Focus on the reduction of delay caused by RAN related aspects
- Review performance requirements for e.g. RRC procedures
- Review network RRM strategies

1.6 QoS in Layer Two

Figure 7 depicts radio interface protocol architecture which introduces layers 1 to 3 of UTRAN (3GPP, R6, TS 125.301, 2005). The radio interface is layered into three protocol layers; Physical layer (L1), Data link layer (L2) and Network layer (L3). We describe layer two in this section.

Layer 2 is split into the following sublayers: Medium Access Control (MAC), Radio Link Control (RLC), Packet Data Convergence Protocol (PDCP) and Broadcast/Multicast Control (BMC).

Layer 3 and RLC are vertically divided into Control (C-) and User (U-) planes. PDCP and

Figure 7. Radio interface protocol architecture (3GPP, R6, TS 125.301, 2005).

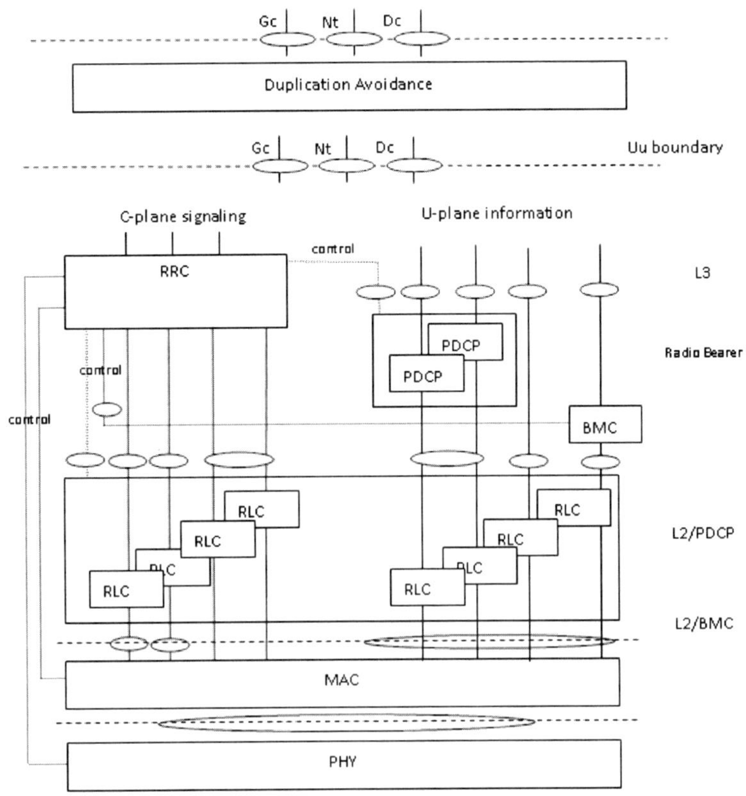

BMC exist in the U-plane only.

In the C-plane, Layer 3 is partitioned into sublayers where the lowest sublayer, denoted as Radio Resource Control (RRC), interfaces with layer 2. The next sublayer provides 'Duplication avoidance' functionality.

Each block in Figure 7 represents an instance of the respective protocol. Service Access Points (SAP) for peer-to-peer communications are marked with circles at the interface between sub-layers. The SAP between MAC and the physical layer provides the transport channels. The SAPs between RLC and MAC sublayer provide the logical channels. The RLC layer provides three types of SAPs, one for each RLC operation mode (UM, AM, and TM). PDCP and BMC are accessed by PDCP and BMC SAPs, respectively. The service provided by layer 2 is referred to as the radio bearer. The C-plane radio bearers, which are provided by RLC to RRC, are denoted as signaling radio bearers. In the C-plane, the interface between 'Duplication avoidance' and higher L3 sublayers (CC, MM) is defined by the General Control (GC), Notification (Nt) and Dedicated Control (DC) SAPs.

Besides, there are connections between RRC and MAC as well as RRC and L1 providing local inter-layer control services. An equivalent control interface exists between RRC and the RLC sub-layer, between RRC and the PDCP sublayer and between RRC and BMC sublayer. These interfaces allow the RRC to control the configuration of the lower layers. For this purpose separate control SAPs are defined between RRC and each lower layer (PDCP, RLC, MAC, and L1).

In summary some functions of RLC include (3GPP, R6, TS 125.301, 2005):

- **Segmentation and reassembly:** This function performs segmentation/reassembly of variable-length upper layer RLC Protocol Data Units (PDUs) into/from smaller RLC PDUs. The RLC PDU size is adjustable to the actual set of transport formats. Note: RLC Service Data Unit (SDU) is higher layer packet received in lower layer, before segmenting into RLC PDUs.

- **Concatenation:** If the contents of an RLC SDU cannot fill by one RLC PDU, the first segment of the next RLC SDU may be put into the RLC PDU in concatenation with the last segment of the previous RLC SDU.
- **Padding:** When concatenation is not applicable and the remaining data to be transmitted does not fill an entire RLC PDU of given size, the remainder of the data field shall be filled with padding bits.
- **Transfer of user data:** This function is used for conveyance of data between users of RLC services. RLC supports acknowledged, unacknowledged and transparent data transfer. QoS setting controls transfer of user data.
- **Error correction:** This function provides error correction by retransmission (e.g. Selective Repeat, Go Back N, or a Stop-and-Wait ARQ) in acknowledged data transfer mode.
- **In-sequence delivery of upper layer PDUs:** This function preserves the order of upper layer PDUs that were submitted for transfer by RLC using the acknowledged data transfer service. If this function is not used, out-of sequence delivery is provided.
- **Duplicate detection:** This function detects duplicated received RLC PDUs and ensures that the resultant upper layer PDU is delivered only once to the upper layer.
- **Flow control:** This function allows an RLC receiver to control the rate at which the peer RLC transmitting entity may send information.
- **Sequence number check:** This function is used in unacknowledged mode and guarantees the integrity of reassembled PDUs and provides a mechanism for the detection of corrupted RLC SDUs through checking sequence number in RLC PDUs when they are reassembled into a RLC SDU. A corrupted RLC SDU will be discarded.
- **Protocol error detection and recovery:** This function detects and recovers from errors in the operation of the RLC protocol.
- **Ciphering:** This function prevents unauthorized acquisition of data. Ciphering is performed in RLC layer for non-transparent RLC mode.
- **SDU discard:** This function allows an RLC transmitter to discharge RLC SDU from the buffer.

The MAC sublayer is made up of several different MAC entities, MAC-d, MAC-c/sh/m, MAC-hs, MAC-es/MAC-e and MAC-m.

The MAC-hs entity provides Hybrid ARQ functionality, and is only used on the HS-DSCH channel.

The MAC-es/MAC-e entities provide Hybrid ARQ functionality, and are only used with E-DCH channel.

The MAC-m entity provides selection combining functionality for multimedia broadcast/multicast traffic channel (MTCH) from different cells. MAC-m is only used for FACH carrying MTCH and multimedia broadcast/multicast schedule channel (MTCH). In summary some functions of MAC include (3GPP, R6, TS 125.301, 2005):

- **Mapping between logical channels and transport channels.**
- **Selection of appropriate Transport Format for each Transport Channel depending on instantaneous source rate**
- **Priority handling between data flows of one UE**

- **Priority handling between UEs by means of dynamic scheduling**
- **Identification of UEs on common transport channels**
- **Hybrid ARQ functionality for High Speed Downlink Shared Channel (HS-DSCH) and Enhanced Dedicated transport Channel (E-DCH) transmission.** The MAC-hs and MAC-e entities are responsible for establishing the HARQ entity in accordance with the higher layer configuration and handling all the tasks required to perform HARQ functionality. This functionality ensures delivery between peer entities by use of the ACK and NACK signaling between the peer entities.
- **In-sequence delivery and assembly /disassembly of higher layer Protocol Data Units (PDUs) on HS-DSCH channel.** The transmitting MAC-hs entity assembles the data block payload for the MAC-hs PDUs from the delivered MAC-d PDUs. The MAC-d PDUs that are assembled in any one MAC-hs PDU are the same priority, and from the same MAC-d flow. The receiving MAC-hs entity is then responsible for the reordering of the received data blocks according to the received TSN, per priority and MAC-d flow, and then disassembling the data block into MAC-d PDUs for in-sequence delivery to the higher layers.
- **In-sequence delivery and assembly / disassembly of higher layer PDUs on E-DCH.** The transmitting MAC-es/MAC-e entity assembles the data block payload for the MAC-e PDUs from the delivered MAC-d PDUs. The receiving MAC-es entity is then responsible for the reordering of the received data blocks according to the received TSN and Node-B tagging information, per re-ordering queue, and then disassembling the data block into MAC-d PDUs for in-sequence delivery to the higher layers.

1.6.1 Error Recovery Mechanisms in Layer Two

As explained before, the data link layer is Layer 2 of the seven-layer OSI model as well as of the five-layer TCP/IP reference model. It responds to service requests from the network layer and issues service requests to the physical layer. The data link layer is split into MAC and LLC sub-layers. The uppermost sub-layer is the Logical Link Control (LLC) one. This sub-layer multiplexes protocols running at top of the data link layer, and optionally provides flow control, acknowledgment, and error recovery. Media Access Control (MAC) is blow LLC. This refers to the sub-layer that determines who is allowed to access the media at any time (usually CSMA/CD). The Media Access Control sub-layer also determines where a frame of data ends and the next frame starts.

Error detection is the ability to detect errors caused by noise or other impairments during transmission from the transmitter to the receiver. Error correction has an additional feature that enables identification and correction of the errors. When a sender transmits a frame, it might be corrupted or lost. The data link layer at destination checks the received frame for error and uses an ARQ mechanism to send back its status to the sender. There are two ways for this mechanism; Stop and Wait ARQ and Continuous ARQ (Elahi, 2000). In the Stop and Wait ARQ mechanism the sender sets a timer to a definite time after sending a frame and waits for receiving an ACK message for that duration. If this message doesn't arrive at the sender or if the timer times out, the frame will be retransmitted. This method can be implemented in Half-Duplex communication. In the Continuous ARQ mechanism the transmitter sends frames continuously and the receiver sends back ACK or NACK messages from a distinct channel (Full Duplex). The sending process continues by a number of frames specified by a window *size* even without receiving an ACK packet from the receiver. The continuous ARQ mechanism is implemented in two forms: Go-Back-N ARQ and Selective Repeat ARQ. In the first form there is a buffer at the sender in which a copy of the transmitted frames resides and will not be deleted before they are not received correctly. When a NACK(n) is received at the source, the frames will be retransmitted one after the other from the frame *n*. In the second form, when some of the received frames are not correct the receiver requests the sender to retransmit only the unsent frames. So, the receiver should be capable of reordering the received frames.

1.6.2 General Concepts About Transport Channels

These channels are subdivided into dedicated transport channels (DCH and E-DCH) and common transport channels (BCH, PCH, RACH, FACH, HS-DSCH). All Transport Channels are defined as unidirectional (i.e. only uplink or downlink). This means that a UE can have simultaneously (depending on the services and the state of the UE) one or several transport channels in the downlink and one or more other transport channels in the uplink.

Some DCHs can be multiplexed and mapped onto one or several Dedicated Physical Channels (DPCH) on the physical layer. A DPCH consists of two parts, Dedicated Physical Control Channel (DPCCH) and Dedicated Physical Data Channel (DPDCH). A DPCCH carries control information which is generated internally on L1. A DPDCH carries the encoded bits of the DCH transport channels (Fig 8). Table 4 illustrates how the bit mapping is done in normal transmission mode in this layer. There are several different slot formats defined with different split of data and control bits. At establishment of a downlink DPCH, one of the permitted slot formats is selected and applied. MAC delivers Transport Block or a Transport Block Set to the physical layer every Transmission Time Interval (TTI). TTI is always a multiple of the minimum interleaving period (e.g. 10ms, the length of one radio frame). A TTI can be 2, 10, 20, 40 or 80 ms in duration.

Figure 8. Frame structure for downlink dedicated physical channels (DPCH) (3GPP TS 25.211, 2007).

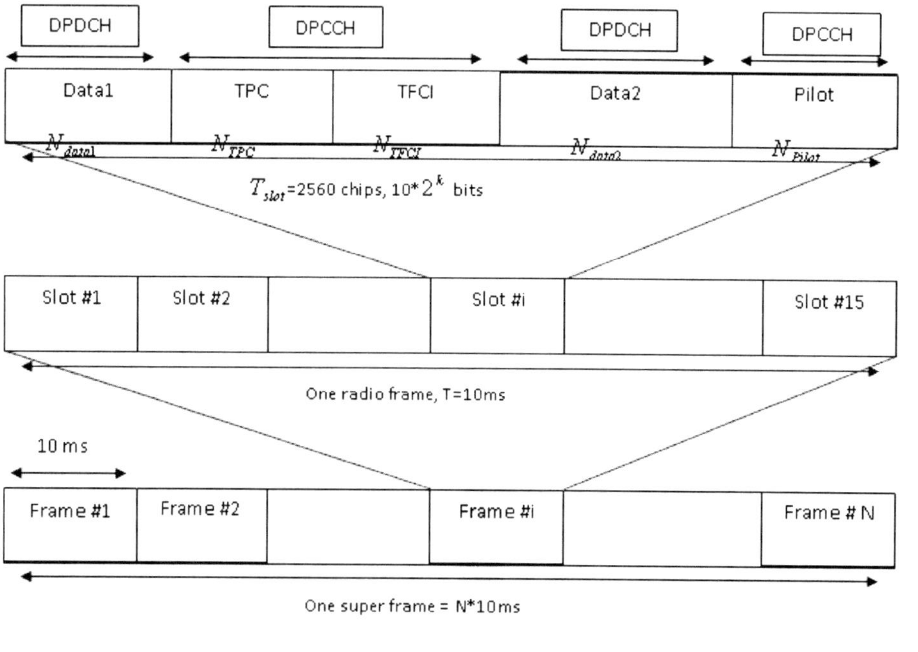

Figure 9. Exchange of MAC PDU between MAC and L1 (3GPP TS 125.302, 2002)

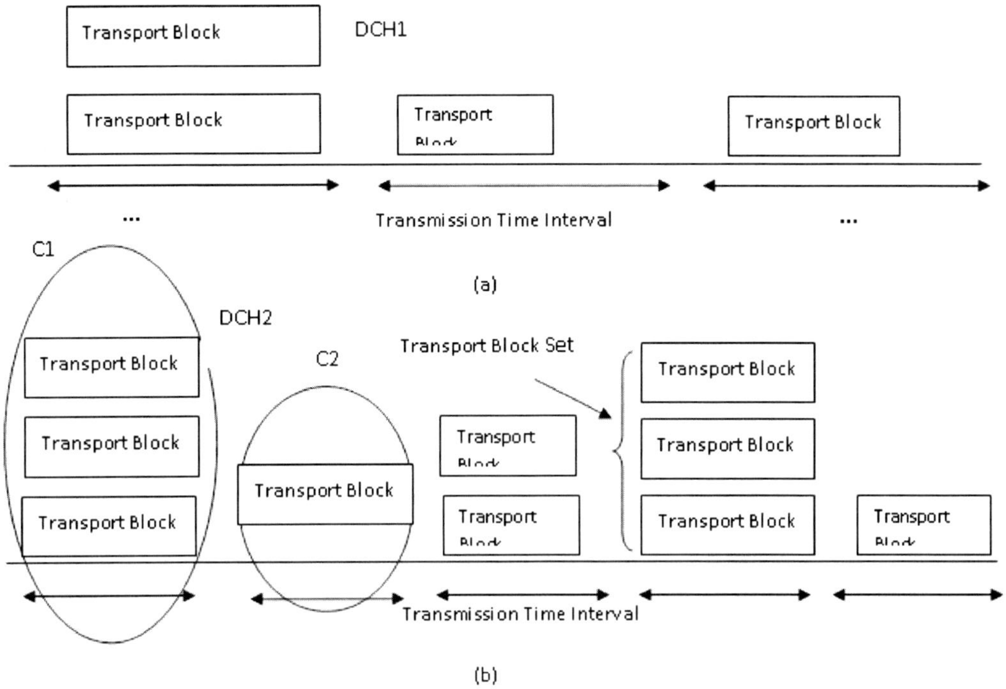

Table 4. Downlink DPCH slot formats in normal transmission mode (3GPP TS 25.211, 2007)

Slot Format # i	Channel Bit Rate (kbps)	Channel Symbol Rate (kbps)	SF	Bits/Slot	DPDCH Bits/ Slot		DPCCH Bits/Slot			Transmitted slots per radio frame N_{Tr}
					N_{Data1}	N_{Data2}	N_{TPC}	N_{TFCI}	N_{Pilot}	
0	15	7.5	512	10	0	4	2	0	4	15
1	15	7.5	512	10	0	2	2	2	4	15
2	30	15	256	20	2	14	2	0	2	15
3	30	15	256	20	2	12	2	2	2	15
4	30	15	256	20	2	12	2	0	4	15
5	30	15	256	20	2	10	2	2	4	15
6	30	15	256	20	2	8	2	0	8	15
7	30	15	256	20	2	6	2	2	8	15
8	60	30	128	40	2	28	2	0	4	15
9	60	30	128	40	6	26	2	2	4	15
10	60	30	128	40	6	24	2	0	8	15
11	60	30	128	40	6	22	2	2	8	15
12	120	60	64	80	12	48	4	8	8	15
13	240	120	32	160	28	112	4	8	8	15
14	480	240	16	320	56	232	8	8	16	15
15	960	480	8	640	120	488	8	8	16	15
16	1920	960	4	1280	240	1000	8	8	16	15

Figure 10. Data flow for non-transparent RLC and transparent MAC [11]

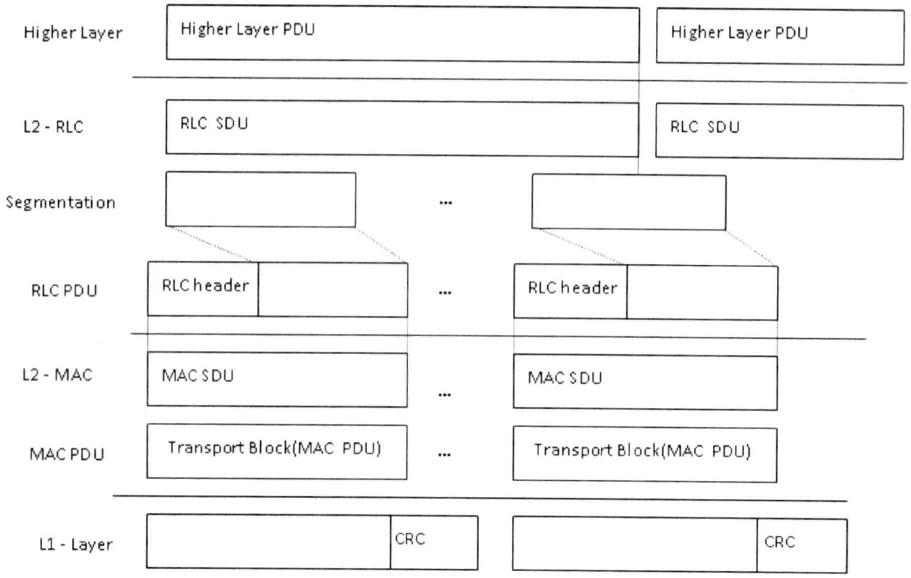

Figure 9 shows an example in which at a certain time instances Transport Blocks are exchanged between MAC and L1 via two parallel transport channels DCH1 and DCH2. Transmission Time Interval, i.e. the time between consecutive deliveries of data between MAC and L1, is also illustrated in that figure. Transport Block (equal to a MAC PDU) is the basic unit exchanged between L1 and MAC, for L1 processing. Transport Block Set is defined as a set of Transport Blocks which are exchanged between L1 and MAC at the same time instance using the same transport channel. Data on each transport channel is organized in Transport Blocks. Depending on the requested QoS variable numbers of transport blocks with variable lengths can be transmitted in each TTI i.e. one or more TBs can be inserted into one TTS. Transport Block Size is defined as the number of bits in a Transport Block Set. In the Figure 9(a) TTI=20ms and transport block length varies TTI by TTI. In the Figure 9(b) TTI=10ms and both the length and the number of TBs varies.

1.6.3. Throughput Evaluation

Data flow mechanisms through UMTS Layer 2 are characterized by the applied data transfer modes in RLC (acknowledged, unacknowledged and transparent transmission) in combination with the data transfer type on MAC, i.e. whether or not a MAC header is required. The case where no MAC header is required is referred to as "transparent" MAC transmission. Acknowledged and unacknowledged RLC transmission modes both require a RLC header. In unacknowledged transmission, only one type of unacknowledged data PDU is exchanged between peer RLC entities. In acknowledged transmission, both data PDUs and control PDUs are exchanged between peer RLC entities. This reduces the throughput but helps the link not to be disconnected in acknowledged transmission mode relative to unacknowledged transmission mode.

There are some different combinations of data flows in Layer 2 as: transparent RLC with transparent and non-transparent MAC transmission, non-transparent RLC with transparent and non-transparent MAC transmission [11]. For lack

Figure 11. (a) An end-to-end system model (b) end to end protocol stack of a Web browsing user plane

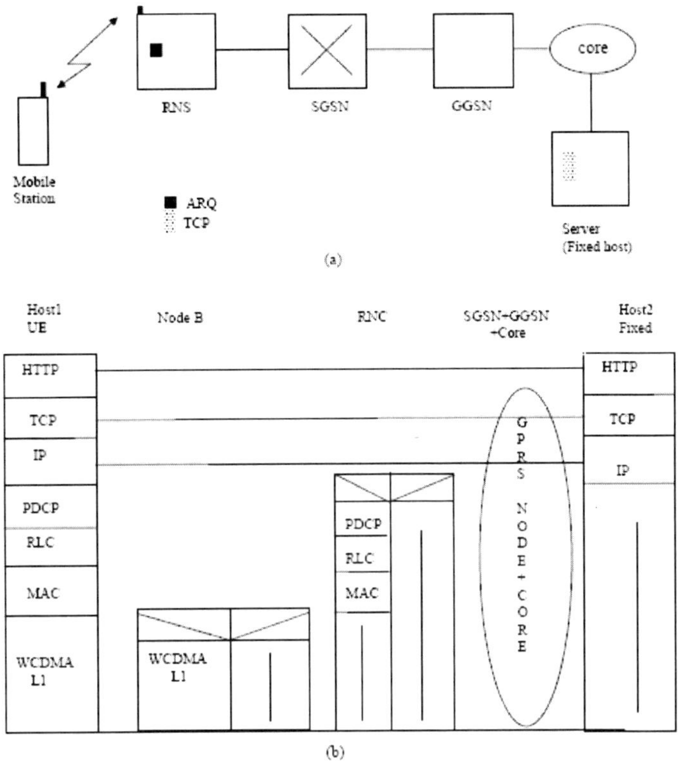

of space we only illustrate non-transparent RLC with transparent MAC transmission in Fig .10. A number of MAC PDUs shown in the figures may comprise a transport block set. Note, however that in all cases a transport block set must not necessarily match with only one RLC SDU. The span of a transport block set can be smaller or larger than an RLC SDU (Figure 10). The received PDUs can be reassembled by simply concatenating all RLC PDUs included in a transport block set as implied by the used transport format (TF).

Now in a scenario we consider a TCP connection between two hosts such that the first link on the end-to-end path from the sender to the receiver is a wireless radio link and the second link is a wired link and connected to a server (fixed host). In Figure 11(a) a Web browsing user has attempted to connect to a server in a public Internet network and intends to download a file from the remote server. Such a scenario is common in mobile communications. The protocol stack on the way is illustrated in the Figure 11(b). We want to evaluate effects of PDU retransmission due to packet error on the wireless link performance. We consider UE, Node B and RNC nodes in the network and assume UTRAN with AM data transfer service (Figure 11).

We assume that RLC is in acknowledged mode and MAC is in transparent mode. Therefore, RLC requires a header but MAC requires no header (Figure 10). In acknowledge transmission mode, both data PDUs and control PDUs are exchanged between peer RLC entities. We assume RLC SDU has been received in RNC from a server and converted to four AMD PDUs. Here, four RLC headers are added to them. Then, they pass through RNC/MAC by putting them into transport blocks transparently (MAC PDUs) and then pass them through the physical layer and arrive at NODE B.

Figure 12. Timing diagram of data transfer in RLC AM with only one error packet and one retransmission time

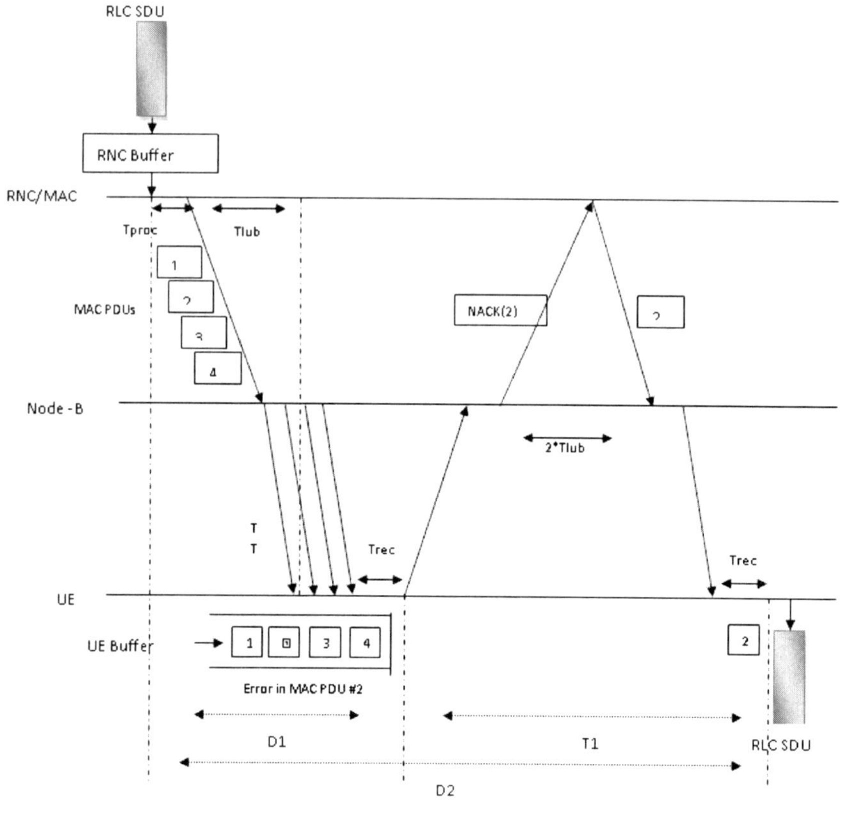

Figure 12 illustrates the timing diagram of the data transfer from the server (RNC) to the MS. In this scenario we assume that only one MAC PDU is in error in UE [12]. Here, we have shown the processing and propagation delay times distinctly. T_{proc} is the SDU processing time needed for segmentation, T_{lub} is the propagation delay time which is independent from TB. T_{rec} is the time after receiving the last PDU in UE and before transmitting the status PDU message [13]. We assumed RLC SDU is segmented to four L2 RLCs and the headers are added to them to constitute RLC PDUs (Figs. 10, 12). Besides, we see that RLC PDUs are the same as MAC PDUs in the transparent MAC mode (Figure 10). We assume a simple Selection Repeat ARQ protocol. In every SDU, the transmitter entity polls the receiver for a status report. According to the Selective Repeat ARQ, the receiver sends a PDU containing the status report indicating that the PDUs are received correctly and the ones to be retransmitted. When the Status PDU is received in the sender, PDUs buffered in the retransmission buffer of the sender entity are deleted or retransmitted according to the status report. Every PDU can be retransmitted at most k_{max} times. When this number is reached, the transmitter entity discards that PDU and all PDUs belonging to the same SDU (Vacirca et al., 2003). Thus, when a packet encounters an error the effective throughput reduces to $d/((d+h)\,/\,R+RTT)$ in which d and h are numbers of data and header bits, respectively, and RTT is the round trip time of a PDU between RNC and UE.

The delay from the time we send a NACK until a correct PDU is received is referred to as the round trip time (RTT). This is equal to the transmission time of a NACK plus 2 times the propagation delay, transmission time of a PDU and the recovery time. Assuming no error occurs, we calculate transmission time D_1 as the sum of the processing time, propagation delay T_{Iub}, PDU transmission time to the receiver and recovery time. If D_1 is the SDU transmission time from RNC to UE, we have

$$D_1 = T_{proc} + T_{Iub} + mTTI + T_{rec} \quad (1)$$

where m is the number of TTIs necessary to convey a RLC SDU and the header. Assuming an error occurs, T_1 equals the sum of twice the propagation delay, recovery time (T_{rec}), a NACK transmission time (Time Slot) and transmission time of a PDU as follows:

$$T_1 = 2T_{Iub} + T_{rec} + TimeSlot + NTTI \quad (2)$$

In the above formula we assumed that we can transfer a NACK by a time slot.

If for a lost PDU we need *k* duplicate transmissions, then the total time of the transmission RLC PDUs and the final correct reception of MAC PDU, D_k, is:

$$D_k = D_1 + (k-1)T_1 \quad (3)$$

Now we can calculate the effective throughput as [15]

$$effective \quad throughput\,(k) = \frac{Correct \quad Transferred \quad Data}{D_k} \quad (4)$$

If we define the efficiency of the protocol as

$$efficiency = \frac{effective \quad throughput}{link \quad bit \quad rate} \quad (5)$$

then we have

$$efficiency = \frac{Correct \quad Transferred \quad Data}{D_k.R} \quad (6)$$

The MAC PDU size may be between 126 and 32766 bytes (3GPP TS 125.302, 2002). Now, we assume RLC SDU + headers =12600 bits and segmentation parameter = 4, then we find that RLC PDU=3150 bits which are transferred by one transport block. We also assume each of the 4 MAC PDUs contains d=3100 data bits and h = 50 header bits. In Layer 1 after CRC attachment, Turbo coding R=1/3, tail bit attachment and rate matching for forward link we found 9500 bits in the physical layer which must be transferred by a super frame (Figure 8).

For the slot format 13 in Table 1, the channel bit rate R and spreading factor SF are 240 kbps and 16, respectively. We have Nd1+Nd2 = 140 bits in each slot or 140*15=2100 bits /10ms. So, the number of frames in a super frame is N = 4.5 (we assume N=5). Besides, the transmission time duration of 9500 bits is 5*10 = 50ms and the number of TTIs necessary to convey a RLC SDU is m = 4*N (4 is the segmentation number). Here we assumed a TB will be transferred in a TTI (see also C2 in the Figure 9).

If we assume a cell radius of 10km, TTI = 10ms, T_{proc} = 10ms and T_{prop} = 0.3ms and also assume only one block is in error in a SDU (Figure 12), the sketch of the throughput of a forward link as a function of the retransmission times (k) is plotted in Figure13.

Figure 13 shows PDUs (TBs) retransmission effects on the throughput when a RLC SDU is segmented to 4, 12, and 16 segments respectively. We see each additional retransmission causes a degradation in the throughput, but for large k the throughput degrades more when segmentation parameter is less. This is because the larger is a segmentation parameter, the smaller is a segment length. Thus when an error occurs, a smaller segment can be transferred faster than a longer one. D_k also indicates the total delay of k retransmis-

Figure 13. Effective throughput of UMTS system versus the number of MAC PDU transmissions

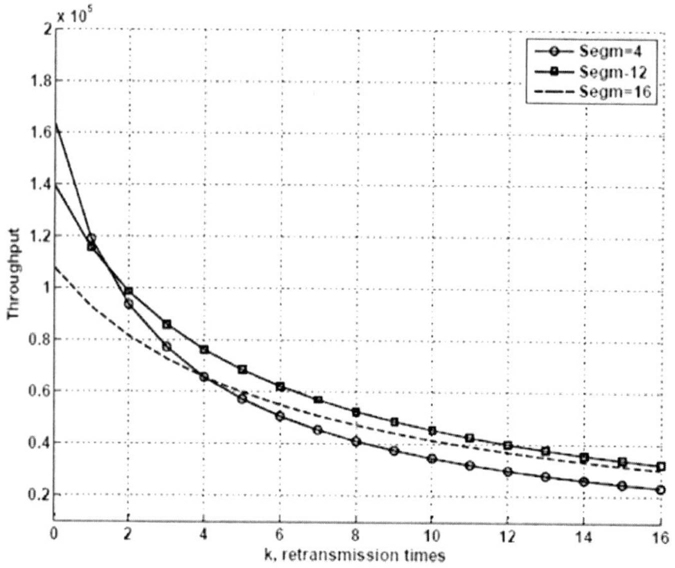

sion times of an erroneous MAC PDU which we encounter it. It evidences that the retransmission results in delay and therefore lower data rate.

We also vary the data block length between 0 to 10^5 bits and find the throughput efficiency of a forward link with i.i.d. errors versus retransmission times and data lengths as shown in Figure 14 (see also C1 in Figure 9). We see when we have more TBs in a TTI, the efficiency approaches to one. Finally we consider the state in which a MAC PDU is in error.

1.7 TCP/Layer 2 Effects on the Air Interface Throughput

TCP is an end-to-end transport protocol in the Internet Protocol (IP) suite which is widely used in popular applications like SMTP, ftp and http. TCP guarantees reliable and in-sequence delivery of packets. TCP performance gets severely affected when used on channels which are typically characterized by high error rates (e.g., wireless channels). Although TCP has been designed, optimized and tuned in wired networks to react to the packet loss due to congestion, in wireless systems service degradation can be due to bit (packet) errors. In UMTS, TCP and ARQ protocols operate against loss and error in wired and wireless sections respectively.

TCP in a wireless network experiences several challenges. One of the issues is how to deal with the spurious timeout caused by the abruptly increased delay, which triggers unnecessary retransmission and congestion control. It is known that the link-layer error recovery scheme, the channel scheduling algorithm, and handover often make the link latency very high. Bandwidth of the wireless link often fluctuates because the wireless channel scheduler assigns a channel for a limited time to a user. Thus, the variance of inter-packet arrival time becomes high, which may result in spurious timeout. The Eifel algorithm has been proposed to detect the spurious timeout and to recover by restoring the connection state saved before the timeout [18, 19].

Although the packet loss rate of the wireless link has been reduced due to link-layer retransmission and Forward Error Correction (FEC), losses still exist because of the poor radio conditions and mobility. Therefore, non-congestion errors could sharply decrease the TCP sending rate. Packet reordering at the TCP layer may be caused by

Figure 14. Throughput efficiency of a forward link with i.i.d. errors versus retransmission times and data length

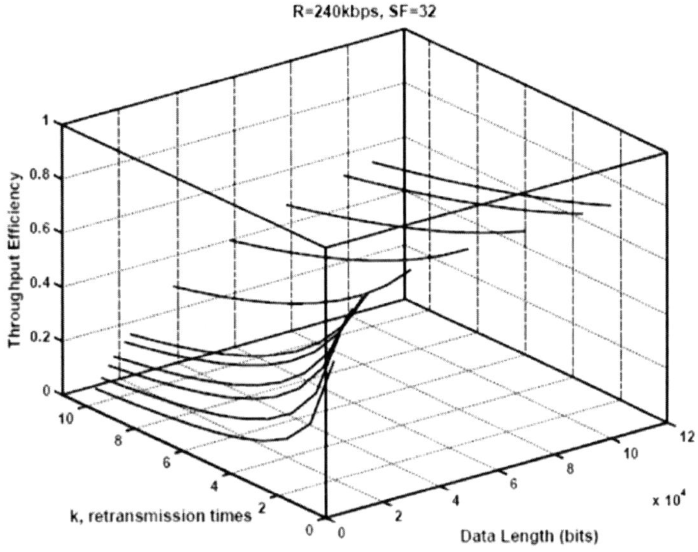

link-layer retransmission, which also results in unnecessary retransmission and congestion. In the wireless network, in general, bandwidth and latency at uplink and at downlink directions are different. Hence, the throughput over downlink may be decreased because of ACK congestion at the uplink (Lee, 2006).

Now we consider a TCP connection between two hosts such that the first link on the end-to-end path from the sender to the receiver is a wireless radio link [20, 21]. Such a scenario is common in mobile communication and is illustrated in Figure 11(a). The protocol stack on the way from mobile host to fixed host is illustrated if Figure 11(b).

We assume there is no packet loss due to congestion on the wireless link but some packets may be corrupted under adverse radio link conditions. In our study, we consider that the bit error patterns on the radio link are independent. On the wired network, packets may only get lost when congestion occurs.

As described in [20] we assume that TCP sends one cumulative ACK_{TCP} for b consecutive TCP segments and is always in congestion avoidance. Besides, Packet loss is detected in one of the two ways, either upon reception of a triple-duplicate ACK_{TCP} (denoted by TD), or upon expiration of a Time-Out (denoted by T0). In case of a TD, window size is decreased by half, while upon expiration of a T0, it is decreased to 1. Moreover, we assume that the loss behavior is bursty, i.e., packet losses are correlated within a back-to-back transmission. Hence, when a packet is lost, all remaining packets in the same round are lost as well [20]. Furthermore, under the assumption that rounds are separated by each TCP round trip time, RTT_{TCP}, loss in one round is independent of loss in other rounds.

We consider TCP Reno version. Let T0 denote the TCP time-out and p denote the loss rate dkue to congestion in the wired portion of the network. For the steady-state of TCP throughput, in a wired context only we have [21]

$$Th(p) = \frac{1}{RTT_{TCP}\sqrt{\frac{2bp}{3}} + T_0 \min(1, 3\sqrt{\frac{3bp}{8}})p(1+32p^2)} \tag{7}$$

and for an end to end protocol consisted of wireline and wireless channel we have

$$[Th(p, PER)]^{-1} = RTT_{TCP}\sqrt{\frac{2b(p+PER-p*PER)}{3}} + T_0 \min\left(1, 3\sqrt{\frac{3b(p+PER-p*PER)}{8}}(p+PER-p*PER) \times \left(1+32\left(p+PER-p*PER\right)^2\right)\right) \quad (8)$$

which is the same as (7) in which p is substituted by

*Global Average Packet Loss Rate= 1-(1-PER)(1-p)=p+PER-p*PER* (9)

In addition we have packet error rate as in equation $PER = 1-\left(1-FER\right)^n$ in which FER is frame error rate in wireless section and n is the number of frames in a packet. Equation (8) is for a complete (wireline and wireless) system without ARQ mechanism.

If we consider a Go Back-N mechanism in layer 2 and also the case of independent and identically distributed bit errors (i.i.d.), we have the throughput as (Wennstrom et al., 2004):

$$[Th\left(p, FER\right)]^{-1} = \left(RTT_{wire} + nbD_{ARQ} + RTT_{wless} + ND_{ARQ}\frac{FER\,(nb-1)}{1-FER}\sqrt{\frac{2bp}{3}}\right) + T_0 \min\left(1, 3\sqrt{\frac{2bp}{8}}\right)p\left(1+32\,p^2\right) \quad (10)$$

in which D_{ARQ} is the constant delay component as a result of ARQ frame processing and RTT_{wless} are round trip times for ARQ (air channel from UE to RNC and vice versa) and wired sections respectively.

Now we consider a mobile channel where a subscriber moves in it. It is surrounded by some obstacles which incident rays strike them. We compute BER as described in [21] which used in MATLAB as a reference for fading channel simulation. There, it is modeled as a linear FIR filter with tap weights given by

$$g_n = \sum_k h_k \sin c(\tau_k/T - n) \quad for \quad -N_1 \le n \le N_2 \quad (11)$$

where

- The summation has one term for each major path.

Figure 15. Effect of air channel on TCP throughput (kbps)

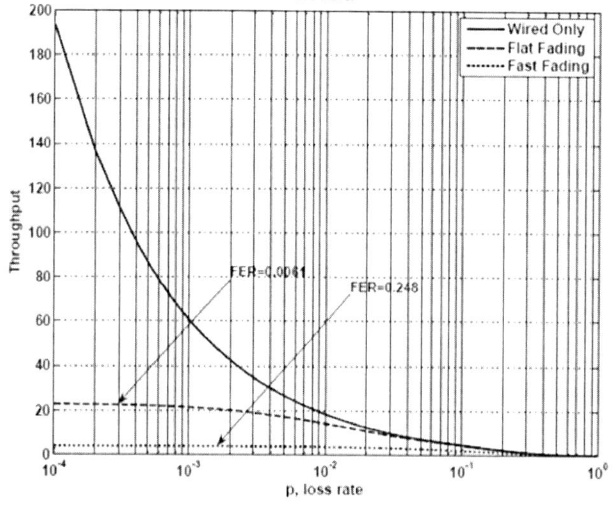

- $\{\tau_k\}$ is the set of path delay.
- T is the input sample period.
- N_1 and N_2 are chosen so that $|g_n|$ is small when n is less than $-N_1$ or greater than N_2.
- $\{h_k\}$ is the set of complex path gain which are not correlated with each other.

Suppose we use RLC in acknowledged mode (AM) and Go-Back-N mechanism of it. For a TCP protocol we assume TCP segment is split into n frames with the parameters $RTT_{wire} = 0.2 \quad b = 10 \quad T_0 = 0.4$ and for GO-back-N ARQ protocol with parameters $RTT_{ARQ} = 0.005 \quad D_{ARQ} = 0.001 \quad n = 10 \quad N = 10$ Besides we assume different values for FER and use MATLAB and find Figs. 15 and 16. In Figure 15 we see the throughput reduction in a wired and in a complete (fixed and wireless) network without ARQ protocol. We assumed two fixed values FER=0.0061 and FER=0.248 in the wireless link.

In Figure 16 we plotted the throughput versus FER in a wireless system with and without ARQ protocol.

1.8 HSDPA

In order to avoid downlink channelization code shortage, a DSCH has been specified for WCDMA Release 99 system, and has been designed for enabling high data rate packet transmission. Further, WCDMA Release 5 introduces HSDPA to realize higher speed data rate together with lower round-trip times. The HSDPA concept can be seen as a continued evolution of the R99 DSCH and a new transport channel targeting packet data transmissions, the high speed DSCH (HS-DSCH) [17].

W-CDMA technology which provides the air interface for UMTS and the 3G system defined by the 3GPP (Third Generation Partnership Project), can in perfect conditions deliver peak data rates of up to 2 Mbps. But in typical network deployment, a cell will have a maximum capacity of around 1 Mbps shared between the cell's users. Peak user data rates are limited to 384 kbps.

Release 5 of the 3GPP W-CDMA specification adds HSDPA in an effort to make the system more efficient for packet data applications by increasing peak data rates and reducing packet latency. Although the theoretical peak data rate

Figure 16. Throughput vs. FER in a wireless system

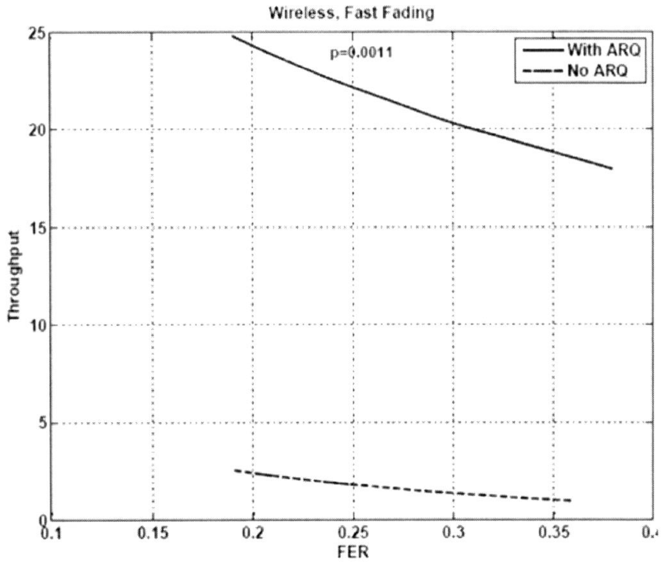

Table 5. Comparison of basic properties of DSCH and HSDSCH

Feature	R99 DSCH	R5 HS-DSCH
Variable spreading factor	Yes (4 - 258)	No(16)
Fast power control	Yes(1500 Hz)	
Fast rate control	No (QPSK, TC=1/3)	Fast link adaptation and adaptive modulation and coding (AMC)
Fast HARQ	No	Yes
HARQ with soft combining	No	CC or IR
TTI	10 or 20ms	2ms
Location of MAC	RNC	Node-B
CRC attachment	Per Transport Channel	Per TTI
Peak data rate	~2Mbps	~ 14 Mbps

for HSDPA is approximately 14 Mbps, the actual rates achieved will be much lower than that. The performance of HSDPA depends largely on the cell size. In macro cell applications, HSDPA may improve on W-CDMA data capacity only by perhaps 30 percent, with sustainable peak data rates for one user of maybe 1 Mbps. But in micro and pico cell deployments where co-channel interference is minimal, HSDPA is capable of delivering much higher performance over basic W-CDMA. The exact improvement is very hard to predict since it depends on actual channel conditions and the real-time capabilities of the BTS – neither of which are standardized. However, some credible estimates for Release 5 suggest a cell capacity of up to 3 Mbps rising to 5 Mbps in Release 6, which includes a more advanced UE receiver and improved BTS packet scheduling. Peak user data rates might reach 3.6 Mbps for short periods of time but are unlikely to be sustainable.

HSDPA technology is backwards-compatible with 3GPP Release 99, so voice and data applications developed for W-CDMA can still run on the upgraded networks, and the same radio channel will support W-CDMA and HSDPA services simultaneously. The result of adding HSDPA to W-CDMA is similar to that of adding E-GPRS to GSM: that is, the improvement in peak data rates and the overall increase in system capacity, particularly in small cells.

1.8.1 Changes in HSDPA

To improve W-CDMA system performance, HSDPA makes a number of changes to the radio interface, which mainly affects the physical and transport layers:

- Shorter radio frame
- New high-speed downlink channels
- Use of 16QAM modulation in addition to QPSK modulation
- Code multiplexing combined with time multiplexing
- A new uplink control channel
- Fast link adaptation using adaptive modulation and coding (AMC)
- Use of hybrid automatic-repeat-request (HARQ)
- Medium access control (MAC) scheduling function moved to Node B

The fundamental characteristics of the HS-DSCH and the DSCH are compared in Table 5.

1.9 Traffic Effects on the System Capacity

A traffic source model can be modeled as an ON/OFF model. Figure 17 shows an example of time-based ON/OFF trajectory of the traffic activity.

Figure 17. An illustration of time-based ON/OFF trajectory of traffic activity.

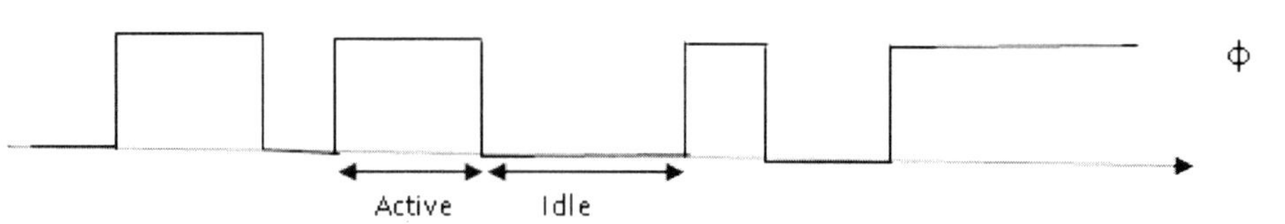

A traffic source usually alternates active and idle periods. Indeed activity factor represents the fraction of the time that the source is generating traffic. In the OFF (Idle) time the source doesn't generate any packet.

The random characteristic of traffic activity is assumed to be represented by the mean of traffic activity, called the traffic activity factor. The activity factor of voice or data traffic, α, is defined as the probability that the state is ON and can be given as

$$\alpha = \frac{E[ON \quad duration]}{E[ON \quad duration] + E[OFF \quad duration]} \tag{12}$$

We now calculate the activity factor α for the traffic types Telnet, WWW, E-mail for a 384Kbps data traffic. In this calculation we use third column of the Table 6.

We show these calculations for the Web browsing traffic as follows:

According to that table we have 5 packet calls per a session, inter arrival time of 120s between packet calls, 25 packets per packet call, the average packet size of 480 bytes and inter arrival time of 0.067s between packets. We first calculate the whole OFF time in a WWW session. The average time of a packet is 480×8/348000=0.01s and the OFF time between two consecutive packets is 0.067-0.01=0.057s and OFF time in a packet call is (25-1) ×0.057=1.368s.

The whole OFF time between packet calls is 5×1.368=6.84s. So the whole OFF time in a session is $4 \times 120 + 6.84 = 486.84s$.

Now we calculate ON time of a WWW session in a packet call as $25\times0.01=0.25s$.So we find that the whole ON time in a WWW session is $5\times0.25=1.25s$.

Finally the activity factor α is calculated from (1) as α=1.25/(1.25+486.84)=0.00256.

We have also calculated this parameter for two other services. The results are listed in Table 7. Besides, these calculations are done for the data rate 64kbps and the results are listed in the Table 8.

1.9.1 Capacity Calculations

We assume there are N user groups in reverse link. One group is for voice service, and the other groups are for various data services, for examples Telnet and web browsing, etc. Users in one group have the same SNR and information data rate requirement. We define the power received by the BS as $S_{v,i}$ for the ith voice user in the voice user group and $S_{d_j,n}$ for the *n*th user in the data user group *j* (j = 1, 2, ..., N-1), and define the information data rates as R_v for the voice user group and R_{d_j} for the data user group *j*. The received *Eb/Not*

Table 6. Traffic models and their characteristics (Tripathi, 2001)

Model Parameter	Telnet	WWW	ftp	Email	Fax
No. of Packet calls per Session	Geometric (mean of 114)	Geometric (mean of 5)	1	Geometric (mean of 2)	Geometric (mean of 3)
Inter Arrival Time or reading Time between Packet Calls (sec)	1	Geometric (mean of 120)	-	Pareto (mean of 90)	Weibull (mean of 30)
No. of Packet per Packet Call	1	Pareto (mean of 25)	Pareto (mean of 62)	Weibull (mean of 15)	Pareto (mean of 15)
Packet Size(bytes)	Geometric (mean of 90)	480	480	480	480
Inter Arrival Time between Packet (sec)	Geometric (mean of 1)	Geometric (mean of 0.067)	Geometric (mean of 0.067)	Geometric (mean of 0.067)	Geometric (mean of 0.27)

Table 7. Activity Factors based on the Table 1 and data rate 384kbps

	ON duration(sec)	OFF duration(sec)	Activity Factor
Telnet	0.217	112.79	0.0019
www	1.25	486.84	0.00256
E-mail	0.3	90.8	0.0033

Table 8. Activity Factors based on the Table 1 and data rate 64kbps

	ON duration(ses)	OFF duration(sec)	Activity Factor
Telnet	1.28	111.73	0.0113
www	7.5	480.84	0.015
E-mail	1.8	90.1	0.0019

for i^{th} data user in the data user group j is (Kim & Koo, 2005):

$$\left(\frac{E_b}{N_{0t}}\right)_{d_j,i} \cong \frac{W}{R_{d_j}} \cdot \frac{S_{d_j,i}}{\sum_{k=1}^{N_v} \alpha_v S_{v,k} + \sum_{j=1,j\neq i}^{N-1} \sum_{n=1}^{N_{d_j}} \alpha_{d_j} S_{d_{j,n}} + I + \eta_k W}$$
$$for \quad i = 1,2,\ldots,N-1$$

According to the perfect power control, we have $S_{v,k} = S_v$, $S_{d,j} = S_{d_j}$ and $S_{d,n} = S_{d_j}$ for all k and n. Then (12) is approximately modified to

$$\left(\frac{E_b}{N_{0t}}\right)_{d_j} \cong \frac{W}{R_{d_j}} \cdot \frac{S_{d_j}}{\alpha_v N_v S_v + \left(N_{d_j} - 1\right)\alpha_{d_j} S_{d_j} + \sum_{j=1,i\neq j}^{N-1} N_{d_j} \alpha_{d_j} S_{d_j} + I + \eta_k W}$$

1.10 Data Traffic Considerations

Data traffic can be conveyed either through circuit switch systems or packet switch systems. Circuit switch systems have usually constant bit rates, while packet switch systems may have variable bit rates. So far there are some Erlang tables that are only pertinent to voice traffic in circuit switch systems.

Now to handle the packet switch traffic we consider a queuing model which contains m servers. The system includes K customers (including the customers in service). Besides, we assume that the population is infinite (M/M/m/K). We can use this model for non-real time services because when all m channels are busy, upon reception of a new call

Figure 18. State transition-rate diagram for m servers, finite storage K and infinite population (M/M/m/K)

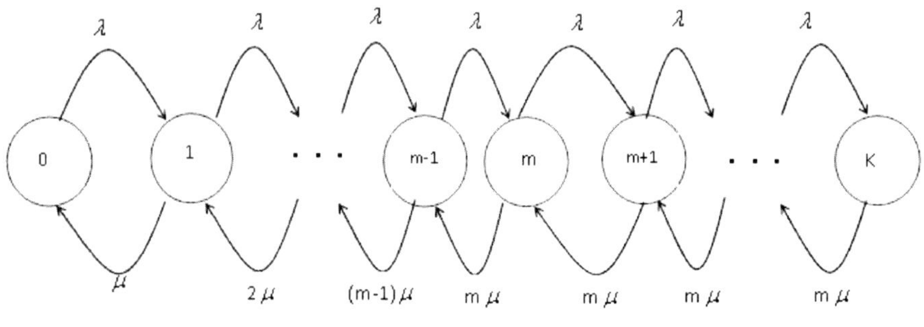

attempt, it will be inserted into the queue before it is lost. The birth and death coefficients in this situation are as follow (see also Figure 18)

$$\lambda_n = \begin{cases} \lambda & n < K-1 \\ 0 & n \geq K \end{cases} \qquad (15)$$

and

$$\mu_n = \begin{cases} n\mu & n \leq m \\ m\mu & m < n \leq K \\ 0 & K < n \end{cases} \qquad (16)$$

in which n is the number of subscribers in the queue whom their attempts have been accepted.

P_n is the probability of being in the state n (or there exist n subscribers in the system). On the other hand indicates the percentage of the time that the system contains n subscribers (Kleinrock, 1975).

$$P_n = C_n P_0 \qquad (17)$$

in which

$$C_n = \begin{cases} \left(\frac{\lambda}{\mu}\right)^2 \frac{1}{n!} & n < m \\ \left(\frac{\lambda}{\mu}\right)^n \frac{1}{m!} \frac{1}{m^{n-m}} & m \leq n \leq K \\ 0 & K < n \end{cases} \qquad (18)$$

and

$$P_0 = \begin{cases} \dfrac{1}{1 + \sum_{q=1}^{n-1} \left(\frac{\lambda}{\mu}\right)^q \frac{1}{q!}} & n < m \\ \dfrac{1}{1 + \sum_{n=1}^{m-1} \left(\frac{\lambda}{\mu}\right)^n \frac{1}{n!} + \frac{1}{m!} \sum_{n=m}^{K} \left(\frac{\lambda}{\mu}\right)^n \frac{1}{m^{n-m}}} & m \leq n \leq K \end{cases} \qquad (19)$$

Now if we assume there is not any queue in the system so that when all servers are busy the call attempts are lost, the above formula will change to Erlang B formula as

$$P_m = \frac{\left(\lambda/\mu\right)^m / m!}{\sum_{k=0}^{m} \left(\lambda/\mu\right)^k / k!} \qquad (20)$$

P_m describes the fraction of time that all servers are busy.

Calls have a (memory-less) exponential duration distribution with λ, the arrival rate of new calls (birth rate) per unit time and h=1/ μ, h, Busy Hour Traffic (BHT), is time duration (in the above unit time) of a call during the busiest period of operation (we have assumed a call terminates with rate μ). We can show (20) by $B = \left(N, \lambda/\mu\right)$ and write

$$B(N, A) = \frac{(A)^N / N!}{\sum_{k=0}^{N} (A)^k / k!} \qquad (21)$$

where B is probability of blocking, N is the number of trunks (channels) and A=λh total amount of the traffic offered in Erlang.

Because of the similarity in the traffic statistical models of the incoming and outgoing voice and data traffic users, we can use (17), (18) and (19) for computing the new Erlang D table for packet switch traffics such as Telnet, www, Email. Now from (17) if we assume n=K we can calculate the blocking probability as

$$P_K = \frac{A^K \frac{1}{m!} \frac{1}{m^{K-m}}}{\sum_{q=0}^{K-1} \frac{A^q}{q!} + \frac{1}{m!} \frac{A^K}{m^{K-m}}} \qquad (22)$$

1.10.1 Delay Calculations

Delay is another important quality of service factor in the mixed traffic systems. According to Kim and Koo (2005), an end-to-end delay must not exceed 100ms and 200ms for voice and video services respectively. For non-real time data traffic if W_q is the average long term waiting time of a subscriber in the queue then from (17) and in accordance with the rule of Littel we can write

$$L_q = \lambda W_q \tag{23}$$

in which L_q is the average number of subscribers in the queue in the long term and is equal to

$$\begin{aligned} L_q &= \sum_{n=0}^{\infty} nP_n \\ &= \frac{P_0}{m!}\left(\frac{\lambda}{\mu}\right)^m \cdot \frac{r}{\left(1-r\right)^2}\left[1 - r^{K-m+1} - \left(1-r\right)\left(K-m+1\right)r^{K-m}\right] \end{aligned} \tag{24}$$

with $r = \lambda / m\mu$.

Thus from (23) and (24) we can find W_q which must be greater than the above thresholds.

REFERENCES

3GPP, R6, TS 125.301 (2005). *Radio interface protocol architecture.*

3GPP TS 125.302. (2002, March). *Services provided by the physical layer, V5.0.0.*

3GPP TS 22 105 Version 7.1.0 (2006, December). *UMTS; Services and Services Capabilities.*

3GPP TS 23 107. (2007, October). *Quality of Service (QoS) concept and architecture.*

3GPP TS 23.107 Version 7.1.0. (2007, June). *UMTS; Multimedia Messaging Service (MMS).*

3GPP TS 25.211. (2007, October). *Physical channels and mapping of transport channels onto physical channels (FDD), V7.3.0.*

3GPP TS 36.104 version 8.3.0 (Release 8). (n.d.). *Base Station (BS) radio transmission and reception.*

3GPP TS 23.110: UTRAN access stratum: services and functions, V6.4.0 (2004-12).

Canton, A. F., & Chahed, T. "End-to-End Reliability in UMTS: TCP over ARQ", IEEE, 2001.

Che-Sheng Chiu. Chen-chiu Lin, "Comparative Downlink Shared Channel Performance Evaluation of WCDMA Release 99 and HSDPA", 2004 IEEE International Conference on Networking, Sensing & Control, Taipei, Taiwan, March 21-23, 2004.

Dadkhah Chimeh, J., Hakkak, M., Bakhshi, H., & Azmi, P. "Throughput Evaluation in UMTS", 2008 Second International Conference on Future Generation and Networking, 2008.

Elahi, A. (2000). *Network Communication Technology*. London: Thomson Learning.

Jeruchim, M. C., Balaban, P., & Chanmugan, K. S. Simulation of Communication Systems: modeling, methodology and techniques, Kluwer Academic/Plenum Publishers, 2000.

Kim, K., & Koo, I. (2005). *CDMA Systems Capacity Engineering*. Norwood, MA: Artech House.

Kleinrock, L. (1975). *Queueing systems Vol. I: theory*. Hoboken, NJ: John Wiley & Sons.

Lee, Y. (2006). Measured TCP Performance in CDMA 1x EV-DO Network. In *PAM conference*.

Li, J., Montuno, D. Y., Wang, J., & Zhao, Y. Q. "Performance Evaluation of the Radio Link Control Protocol in 3G UMTS", *Proc. Of the 4th IASTED International Multi-Conference, Wireless and Optical Communication*, Banf Canada, pp. 524-529, July, 2003.

Prez-Romero, J. Sallent, O., Agusti, R., & Diaz-Guerra, M. A. (2005). *Radio Resource Management Strategies in UMTS*. New York: Wiley.

Tripathi, N. D. (2001). Simulated Base analysis of the radio interface performance of an IS-2000 system for various data services. *IEEE.*

TS 25 401. (2007, October). *UTRAN overall description.*

Vacirca, F., Vendictis, A. D., & Baiocchi, A. (2003). Investigating Interactions between ARQ Mechanisms and TCP over Wireless Links. *IEEE GLOBECOM.*

Vacirca, F., Vendictis, A. D., Todini, A., & Baiocchi, A. "On the Effects of ARQ Mechanisms on TCP Performance in Wireless Environments", *Globecom, IEEE*, pp. 671-671, 2003.

Wennstrom, A., Alferedsson, S., & Brunstorm, A. (2004). TCP over Wireless networks. Karlstad, Sweden: Karlstad University Press.

Chapter 3
QoS Architecture of WiMAX

Rath Vannithamby
Intel Corporation, USA

Muthaiah Venkatachalam
Intel Corporation, USA

ABSTRACT

WiMAX technology, based on the IEEE 802.16 standard, is a promising broadband wireless technology for the upcoming 4G network. WiMAX has excellent QoS mechanisms to enable differentiated Quality of service of various applications. QoS in broadband wireless access network such as WiMAX is a difficult and complicated task, as it adds unpredictable radio link, user and traffic demand. WiMAX supports end-to-end QoS provisioning to allow various applications and services. This chapter aims to provide a detailed overview of the QoS in WiMAX, the current and the future. Various air-interface and network mechanisms that enable the end-to-end QoS provisioning are then discussed. Finally, the novel mechanisms to improve the QoS provisioning in the next generation WiMAX system are also discussed.

1. INTRODUCTION TO WIMAX QOS ARCHITECTURE

Recently, IEEE 802.16 (IEEE 802.16e-2005, 2006) based mobile WiMAX has become a very attractive candidate for 4G wireless systems. With Orthogonal Frequency Division Multiple Access (OFDMA) technology and mobility support, mobile WiMAX promises superior spectral efficiency and capacity, allowing mobile stations (MS) to access voice and various IP services through broadband wireless metropolitan area network. WiMAX technology is broadly based on the radio layers developed in IEEE 802.16 working group. Specifically, WiMAX Release 1.0 (WiMAX Forum, n.d.a). and Release 1.5 (WiMAX Forum, n.d.b). are based on IEEE 802.16e (IEEE 802.16e-2005, 2006; IEEE P802.16Rev2/D4, 2008). The next generation WiMAX Release 2.0 currently under development, will be based on IEEE 802.16m standard. Note that we interchangeably use the terms WiMAX and IEEE 802.16 in this chapter.

WiMAX airlink has a centralized medium access control (MAC) layer. All required bandwidth

DOI: 10.4018/978-1-61520-680-3.ch003

for UL applications have to be scheduled and granted by BS on the air interface. Hence, in order to satisfy the end-to-end quality of service (QoS) constraints of heterogeneous applications in WiMAX networks, the UL scheduling on the air link plays an important role. When a MS needs to transmit to BS in the UL, the bandwidth allocation is obtained via bandwidth request/grant process between MS and BS. Corresponding to the traffic characteristics of different services, five types of scheduling services have been defined for the WiMAX airlink: unsolicited grant service (UGS), real-time polling service (rtPS), non-real-time polling service (nrtPS), extended real-time polling service (ertPS) and best effort (BE) service. Among them, UGS, rtPS and ertPS are mainly used for real-time (RT) traffic and interactive traffic such as VoIP, video and online gaming, while nrtPS and BE are usually utilized for non-real-time traffic such as file transfers, emails, and web browsing.

The WiMAX network has been designed to support the WiMAX airlink QoS. The WiMAX network provides mechanisms for the applications (both operator hosted and external web based applications) to negotiate the required QoS for the application in question. The overall QoS framework is very efficient in supporting various types of traffic such as VoIP, Video streaming, online gaming, file transfers, web browsing etc.

The WiMAX forum has developed the concept of USI (WiMAX Forum, n.d.b), which is an API that can be exposed by the WiMAX operator to the external world, wherein the vast majority of web based applications in the external world such as YouTube video, Skype voice, online gaming etc can use this interface to request the required QoS for their services from the WiMAX network.

Interesting surveys, analysis and simulations studies on QoS support over WiMAX networks were published recently. A survey on the basics of Mobile WiMAX networks is given in Li et al. (2007). A survey of scheduling research on Mobile WiMAX network is provided in Chakchai et al. (2009). Filin et al. (2008) introduces an efficient and fast QoS guaranteed adaptive transmission algorithm for Mobile WiMAX. Talwalkar & Ilyas (2008) focuses on analysis of QoS in WiMAX networks. Neves et al. (2008) provides a simulation study of QoS differentiation support in WiMAX networks.

In this chapter we detail the operation of QoS in the WiMAX network and the usage of USI to setup QoS enabled VoIP calls. We then detail the WiMAX airlink QoS mechanisms and then move on to the latest and the greatest innovations happening in the area of QoS for the next generation WiMAX airlink.

2. END-TO-END WIMAX NETWORK ARCHITECTURE AND THE SUPPORT OF QOS

In this section, we will describe the End to End operation of QoS in a WiMAX network (WiMAX Forum, n.d.a) right from the MS to the base station (BS) to the ASN-GW to the core network (CSN/NSP). We will provide insights on how E2E QoS is provisioned, setup and torn down in a WiMAX network as well as the other associated procedures for QoS in the WiMAX network.

WiMAX defines a QoS framework for the air interface. This consists of the following key elements:

- Connection-oriented service
- Five data delivery services at the air interface, namely, UGS, RT-VR, ERT-VR, NRT-VR and BE (IEEE 802.16e-2005, 2006)
- Provisioned QoS parameters for each subscriber
- A policy requirement for admitting new service flow requests

A WiMAX QoS subscription could be associated with a number of service flows characterized by QoS parameters. This information is presumed to be provisioned in a subscriber management system like an AAA database, or a policy server. 2 different subscription models are possible. Under the static service model, the subscriber station is not allowed to change the parameters of provisioned service flows or create new service flows dynamically. Under the dynamic service model, an MS or BS may create, modify or delete service flows dynamically. In this case, a dynamic service flow request is evaluated against the provisioned information to decide whether the request could be authorized. The following steps detail the service flow creation:

a. Permitted service flows and associated QoS parameters are pre-provisioned for each subscriber via the management plane.
b. A service flow request initiated by the MS or BS (as detailed in section 3) is evaluated against the provisioned information, and the service flow is created if permissible.
c. A service flow thus created transitions to an admitted, and finally to an active state either due to BS action (this is possible under both static and dynamic service models). Transition to the admitted state involves the invocation of admission control in the BS and (soft) resource reservation, and transition to the active state involves actual resource assignment for the service flow. The service flow can directly transit from provisioned state to active state without going through admitted state.
d. A service flow can also transition in the reverse from an active to an admitted to a provisioned state.
e. A dynamically created service flow can also be modified or deleted at a later point in time.

2.1 WiMAX Network Architecture to Support QoS

Figure-1 shows the E2E WiMAX network architectural model to support QoS (WiMAX Forum, n.d.a). In the figure, the key entities are the MS, the ASN, the visited and home and Network service provider (NSP). The NSP is equivalent to the core network and is also known as the CSN. The visited NSP is the same as the home NSP if the MS is not roaming. The ASN is equivalent to the radio access network (RAN). The ASN consists primarily of the BS and the ASN-GW.

The home NSP contains the AAA server, the policy function (PF) and the associated policy databases. Maintained information includes H-NSP's general policy rules as well as application dependant policy rules as well as users QoS profile. The AAA may, in addition, provision the PF's database with user's QoS profile and associated policies. The PF is in charge to evaluate QoS service requests against these policies. The NSP may also contain the application function (AF), which in essence is an operator hosted application – which can trigger the QoS service requests.

The BS in the ASN contains a QoS function called the Service Flow Management (SFM) logical entity. The SFM entity is responsible for the creation, admission, activation, modification and deletion of 802.16 service flows. It consists of an Admission Control (AC) function, and associated local resource information. The AC is used to decide whether a new service flow can be admitted based on existing radio and other local resource usage. WiMAX does not mandate a specific admission control mechanism and several mechanisms are possible based on vendor differentiation.

The ASN-GW in the ASN consists of the QoS function called the Service flow Authorization (SFA). In case the user QoS profile is downloaded from the AAA into the SFA at network entry phase, the SFA is typically responsible for evaluating any service request against user QoS profile. The SFA

Figure 1. E2E QoS Functional elements in a WiMAX network

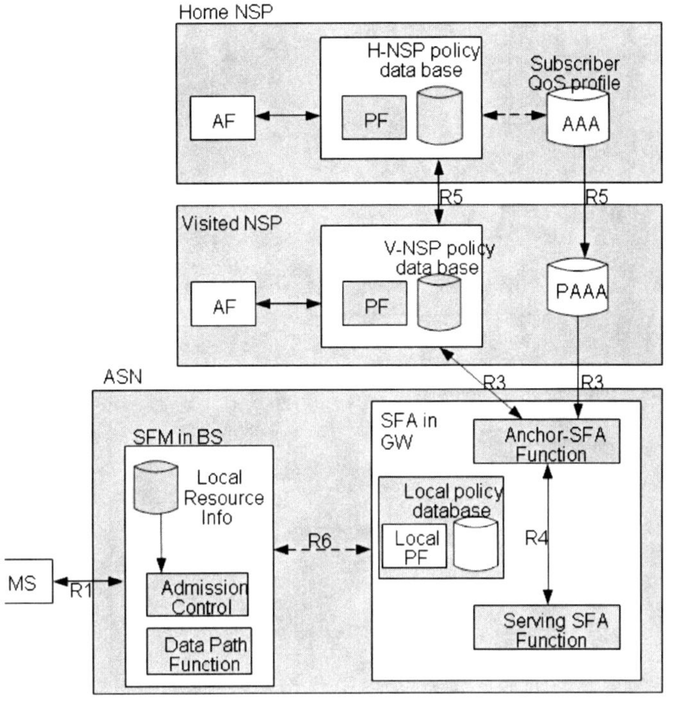

may also perform ASN-level policy enforcement using a local policy database and an associated local policy function (LPF). The LPF can also be used to enforce admission control based on locally available resources.

There is also a network management system (not shown in the figure) that allows administratively provisioning service flows.

Based on WiMAX service provider requirements, the provisioned information may include additional parameters such as user priority, which are used to enforce relative priorities (e.g., gold, silver, and bronze) across users. For example, the user priority may be taken into account in situations where the service flow requests across all users exceed the radio resource capacity and therefore a subset of those has to be rejected.

2.2 WiMAX QoS Setup Procedures

Figure 2 shows the basic procedure for setting up or modifying QoS service flows for a given WiMAX user.

In the above figure service flows will be setup once the user completes his initial network entry into the WiMAX network. The QoS policies for the user are downloaded from the AAA server in the core network to the SFA in the ASN. The SFA then applies these policies to set up the service flows. It sends wimax control messages to the SFM in the BS. It is to be noted that there could be other SFAs in the path that may relay this message to the SFM in the serving BS. The SFM can then apply local admission control policies for the available radio resources and then use the WiMAX airlink signaling with the MS to establish these service flows. The airlink mechanisms for QoS are detailed in the next section. The same procedure can be used to also modify existing service flows of a given user. The service flows can basically be of any of the QoS type that is defined for the WiMAX airlink such as UGS, rtPS, ertPS etc.

Figure 2. Service flow creation/modification for a given user

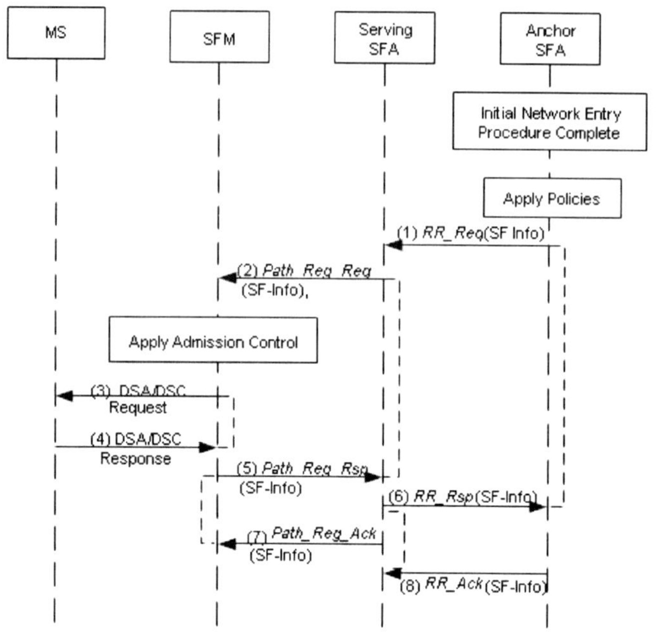

2.3 Universal Services Interface (USI) to Enhance QoS for the Web Based Applications

The WiMAX network provides mechanisms for the applications (both operator hosted and external web based applications) to negotiate the required QoS for the application in question.

These days, supporting QoS for the external web based applications by the wireless operator can be a good thing, provided the operator can get some revenue out of it. In order to do this, WiMAX forum has developed the concept of Universal Services Interface (USI) (WiMAX Forum, n.d.b) that can be exposed by the WiMAX NSP to the external world. The vast majority of web based applications in the external world such as YouTube video, Skype voice, online gaming etc can use this interface to request the required QoS for their services from the WiMAX network.

Figure 3 shows the USI system. The USI system resides in the NSP (aka core network or CSN). The iASP in the figure refers to the Internet Application Service Provider such as video streaming services

Figure 3. USI system

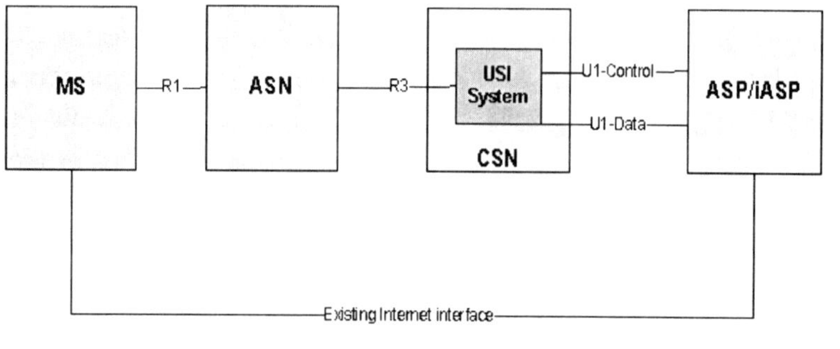

Figure 4. USI QoS session creation

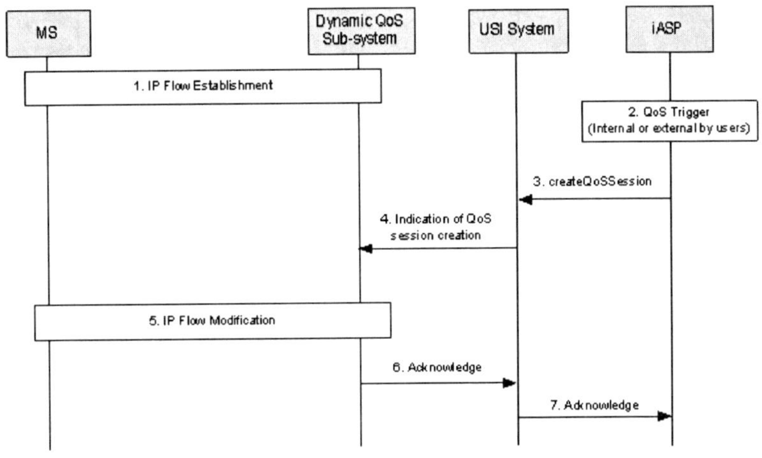

like YouTube etc. A new interface called the U1 is defined between the iASP and the USI. This U1 interface can have 2 parts to it, the data part (U1-data) and the control part (U1-control). The U1 interface is basically a web based interface so that it can be easily used by the huge gamut of web applications today.

Using the U1-conrol interface, the iASP can request the needed QoS for its applications from the WiMAX operator. The WiMAX operator may in turn charge the iASP for granting the requested QoS based on the business model that is employed.

Figure 4 shows the procedure for the iASP to request QoS from the USI system.

As can be in the figure, the iASP triggers the QoS to the USI system using create QoS Session command. The USI system then talks to the QoS subsystem in the WiMAX network to setup the required QoS for the external application. Then the USI system responds back to the iASP with the acknowledge command. The charging and billing commands may also be exchanged between the USI and iASP (not shown) either during or after the QoS transaction, based on the business model employed.

2.4 Setting up QoS Enabled VoIP Calls from the Internet via USI

Typically VoIP has been an application that is deployed by the operator itself. With the changing landscape of VoIP, several VoIP service providers have started providing VoIP for users on the Internet (examples include Skype and GoogleTalk to name a couple). However, these types of VoIP services are not nearly "commercial VoIP" with regards to the WiMAX end user, due to the following reasons:

a. QoS on the WiMAX access link for such VOIP calls is neither negotiated nor guaranteed
b. Emergency calling support is typically not available

USI can be used to address the above 2 shortcomings and set up QoS enabled VoIP calls from the Internet using applications such as Skype and GoogleTalk.

VoIP calls can be established in 2 ways:

a. **MS originated:** Where the MS initiates the outgoing call

b. **MS terminated:** Where the MS receives an incoming call

MS-Originated VoIP Call Establishment

The call establishment is shown in Figure 5. Robust Header Compression (ROHC) (RFC 3095, n.d.). or other header compression mechanisms may be used to optimize the data path for the VoIP call.

Here, the user identification is performed upon the registration of the MS. At some point in time after user identification, the MS signals to the Voice Service provider (aka VSP e.g.: Skype) at the application layer to establish a VoIP call. VSP then authorizes the request with the USI in the NSP. As part of this step, it also requests proper QoS to be set up for the VoIP call. USI then contacts the AAA server for authorization for the VoIP and ROHC request. AAA then triggers the ROHC function in the ASN-GW for establishing the ROHC enabled SF. The ROHC enabled SF is then created. Upon successful authorization, the USI requests the QoS to be setup for the VoIP call from the Dynamic QoS sub-system. Then, the successful setup of the QoS is sent from the Dynamic QoS sub- to the USI and onto VSP. At this point, QoS enabled, ROHC compressed VoIP call is established. Accounting update is then performed.

MS-Terminated VoIP Call Establishment

Figure 6 shows this scenario of VoIP call establishment.

Here, the user identification is performed upon the registration of the MS. At some point in time after user identification, the VSP receives a call for the MS. The VSP then contacts the MS to set up this call. As part of this step, if the MS is in idle (aka power save) state, the MS may be paged to exit the idle state, in a manner that is transparent to the VSP, as detailed in (WiMAX Forum, n.d.a). The VSP then authorizes the request with the USI in the NSP. As part of this step, it also requests proper QoS to be set up for the VoIP call. The USI then contacts the AAA server for

Figure 5. MS-originated VoIP call establishment

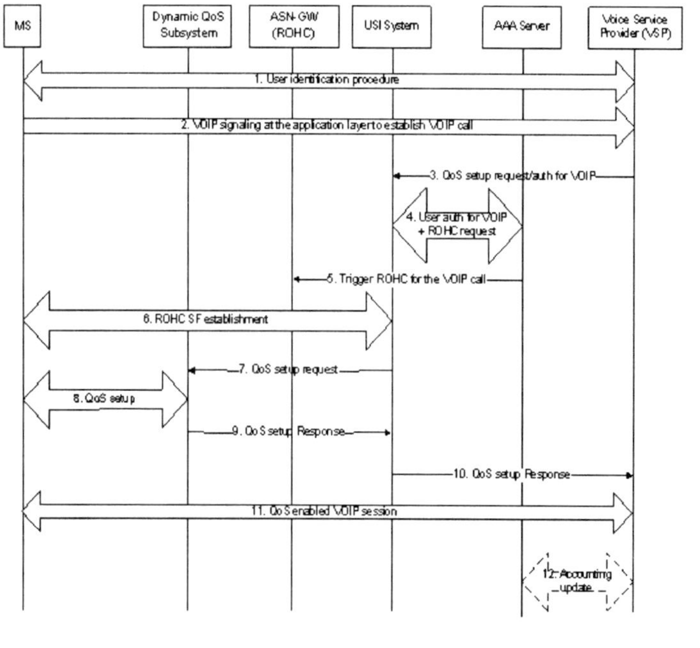

Figure 6. MS-terminated VoIP call establishment

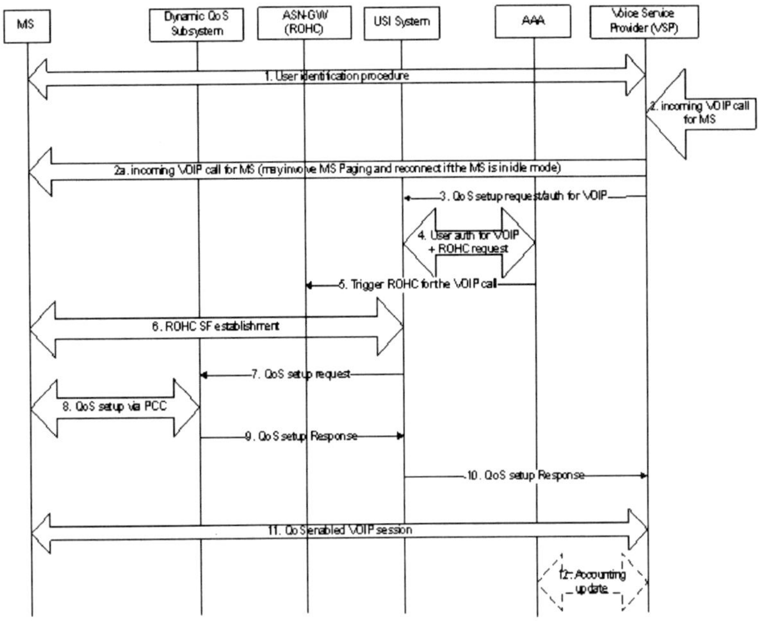

authorization for the VoIP and ROHC request. The AAA, triggers the ROHC function in the ASN-GW for establishing the ROHC enabled SF. Upon successful authorization, the USI requests the QoS to be setup for the VoIP call from the dynamic QoS sub system in the WiMAX network. QoS setup happens via this QoS sub system and the successful setup of this QoS is indicated to the USI and onto VSP. VoIP call is successfully established at this point.

3. IEEE 802.16 AIR INTERFACE MECHANISMS TO SUPPORT QOS

This section gives an overview of the QoS mechanisms incorporated into the IEEE 802.16 air interface standard. The PHY and MAC layer functions are described in detail with regards to their QoS aspects. The relations and interactions of these QoS mechanisms are described to give an understanding of how QoS can be achieved in supporting different applications with various QoS requirements over WiMAX.

This section also describes the practical challenges in enabling current and future applications and how the features are designed in WiMAX to support the required QoS for such applications. As an example, support of VoIP is illustrated.

3.1 Physical Layer Functions for QoS support

This section explains how physical layer functions such as link adaptation, HARQ, channel aware scheduling, channel quality feedback, localized and distributed resource patricians, finer resource allocation granularity, power control and power adaptation, etc. are designed to contribute to support the required QoS.

One of the main enabler of the high data rate transmission possible in the WiMAX systems is the capability of channel quality feedback. This feedback allows the BS to perform channel aware scheduling, opportunistic scheduling that schedules the users when the channel is good. When multiple users need to be scheduled, the BS can pick the user with the best channel at any point in time. Indeed, there are several scheduling strategies such as proportional fair for various optimizations. WiMAX supports resource allocation in both downlink and uplink on a per-frame basis. The data packets are associated to service flows with well defined QoS parameters in the MAC layer so that the scheduler can correctly determine the packet transmission ordering over the air interface. The resource allocation is delivered in MAP (IEEE 802.16e-2005, 2006) messages at the beginning of each frame. Therefore, the resource allocation can be changed frame-by-frame in response to traffic and channel conditions. Additionally, the amount of resource in each allocation can range from one slot to the entire frame. The fast and fine granular resource allocation allows superior QoS for data traffic.

WiMAX supports wideband channel quality feedback as well as narrowband feedback. The narrowband feedback allows the BS to schedule the user transmissions on the best frequency-time resource units. WiMAX supports frequency-diverse sub-channels such as PUSC permutation, where sub-carriers in the sub-channels are pseudo-randomly distributed across the bandwidth, sub-channels are of similar quality. Frequency-diversity scheduling can support a QoS with fine granularity and flexible time-frequency resource scheduling. WiMAX also supports contiguous permutation such as AMC permutation; the sub-channels may experience different attenuation. The frequency-selective scheduling can allocate mobile users to their corresponding strongest sub-channels. The frequency-selective scheduling enhances the QoS guaranteeing capability and system capacity with a moderate increase in channel feedback overhead in the uplink.

WiMAX supports power boosting feature. Basically, it allows the capability of adjusting the transmit power to enhance the data packet detection and decoding probability. The level of boosting can be chosen based on the QoS requirement. To support delay sensitive applications, power boosting is an elegant feature that can manage the delay and packet error requirements of the application.

WIMAX also supports HARQ feature. This feature allows for physical layer retransmissions for error recovery. For delay sensitive applications that have stringent delay bound, it may not be possible to recover the error MACX layer ARQ due to the latency associated with the ARQ protocol, but it is possible for few rounds of HARQ retransmissions within the delay bound with the help of a fast HARQ feedback channel. In addition, HARQ feature allows for the transmitter to optimize transmit power and/or the data transmission rate by aggressively choosing lower transmit power and/or higher data rate with the understanding that quick retransmission is possible via HARQ. For QoS sensitive applications, it is possible to choose conservative power and/or data rates during retransmissions for successful packet transmission within the given delay bound.

The dynamic nature of the resource allocation allows handling jitter. For example, VoIP packets can reach the BS with jitter in the order of tens of ms. It is possible the BS can expedite the transmission of such packets with priority, lower order MCS (to reduce the potential of HARQ retransmissions rounds) boosted transmit power, etc.

3.2 MAC Layer Functions for QoS Support

This section explains the MAC layer functions and procedures that enable QoS support in WiMAX system.

In the Mobile WiMAX MAC layer, QoS is provided via service flows. This is a unidirectional flow of packets that is provided with a particular set of QoS parameters. Before providing a certain type of data service, the BS and MS first establish a unidirectional logical link between the peer MACs called a connection. The MAC then associates packets traversing the MAC interface into a service flow to be delivered over the connection. The QoS parameters associated with the service flow define the transmission ordering and scheduling on the air interface. The connection-oriented QoS therefore, can provide accurate control over the air interface. Since the air interface is usually the bottleneck, the connection-oriented QoS can effectively enable the end-to-end QoS control. The service flow parameters can be dynamically managed through MAC messages to accommodate the dynamic service demand. The service flow based QoS mechanism applies to both DL and UL to provide improved QoS in both directions. Mobile WiMAX supports a wide range of data services and applications with varied QoS requirements.

In the downlink, since the BS scheduler knows the channel condition and the traffic demand, it can schedule the transmission in such a way the QoS requirements for the connection is met as long as the total traffic demand at the BS does not exceed the limit of the air link. However, in the uplink, only the mobile station knows the traffic demand.

WiMAX is packet based system; there is no dedicated channel for the MS to send data. WiMAX uses a bandwidth request mechanism that allocates a small portion of each transmitted frame as a contention slot. With this contention slot, a subscriber station can enter the network by asking the BS to allocate an uplink slot. The BS evaluates the subscriber station's request in the context of the subscriber's service-level agreement and allocates a slot in which the subscriber station can transmit uplink packets.

Next, the service flow classification and establishments are described. We explain the other MAC features such as header compression, silence suppression, specific scheduling strategies, Idle mode and Paging strategies and optimized handover in Section 3.4 that illustrates how these features support VoIP application.

3.2.1 Service Flow Classification and Dynamic Service Establishment

A service flow provides unidirectional transport of packets either to uplink packets that are transmitted by the MS or to downlink packets that are transmitted by the BS. It is characterized by a set of parameters as a Service Flow identifier (SFID), service class name (UGS, rtPS, ertPS, nrtPS, or BE), and QoS parameters (such as Maximum sustained traffic rate, minimum reserved traffic rate, and maximum latency).

There are three kinds of service flow management messages. Dynamic Service Addition (DSA) for the addition of a new service flow, Dynamic Service Change (DSC) for the modification of service flow parameters, and Dynamic Service Delete (DSD) for the deletion of an existing flow service. Service flows are created, changed, or deleted using DSA, DSC, and DSD. The DSA messages create a new service flow. The DSC messages change an existing service flow. The DSD messages delete an existing service flow.

3.2.2 Bandwidth Request and Grants

The ability to quickly transmit the data and control information as it is generated and arrived to the transmit buffer at the MS is a major requirement to support the needed QoS, especially, on uplink. WiMAX has defined bandwidth request mechanism that can make the resources available for data transmissions based on the resources need and the current BS loading.

It is essential for the uplink to feedback accurate and timely information as to the traffic conditions and QoS requirements to make an efficient resource allocation and provide the desired QoS in the uplink. Multiple uplink bandwidth request mechanisms, such as bandwidth request through ranging channel, piggyback request and polling are designed to support UL bandwidth requests. The UL service flow defines the feedback mechanism for each uplink connection to ensure predictable UL scheduler behavior.

WiMAX defines five QoS classes: Unsolicited Grant Service (UGS), real-time Polling Service (rtPS), extended real-time Polling Service (ertPS), non-real-time Polling Service (nrtPS), and Best Effort (BE). The five defined QoS classes and the associated differences in the bandwidth request mechanisms are described below:

- **UGS** supports real-time service flows that have fixed-size data packets on a periodic basis. The BS provides grants in unsolicited manner. The UGS subscribers are prohibited from using contention request opportunities.
- **rtPS** supports real-time service flows that have variable size data packets on a periodic basis. The BS periodically provides unicast request opportunities in order to allow the user to specify the desired bandwidth allocation. The user is prohibited from using contention request opportunities.
- **ertPS** supports real-time service flows. It is built on the efficiency of both UGS and rtPS. The BS provides unicast grants in an unsolicited manner like UGS. Whereas the UGS allocations are fixed in size, the ertPS allocations are dynamic. Then, the MS can request to change the size of grants by sending bandwidth change request.
- **nrtPS** is designed to support non real-time service flows that have variable size data packets on a periodic basis. The MS can use contention request opportunities to send a bandwidth request with contention. The MS can also provide unicast request opportunities.
- **BE** is used for best effort traffic where no throughput or delay guarantees are provided. The MS can use unicast request opportunities as well as contention request opportunities. When the BS or the SS creates a connection, it associates the connection with a service.

3.3 QoS Support for VoIP

This section illustrates how the WiMAX features are contributing to the QoS support for VoIP in the WiMAX system as an example.

3.3.1 Link Adaptation

In a mobile environment, it is possible the channel condition at the MS can change with time. In order to be spectrally efficient, the MCS used for data transmission and reception needs to be adjusted according to the channel variation. Adjustment in MCS requires changes in the allocated resources. The dynamic nature of the WiMAX channel quality feedback, scheduling and the packet transmissions allows for appropriate MCS adaptation based on the channel variation due to VoIP user mobility and environmental changes, otherwise, the user needs to be supported with the lowest order MCS that can consume a lot more radio resources for data transmissions. Without this feature, larger number of VoIP users with QoS guarantees cannot be supported.

3.3.2 HARQ

In addition to link adaptation through channel quality feedback and adaptive modulation and coding, HARQ is enabled in 802.16e using the "Stop and Wait" protocol, to provide fast response to packet errors at the PHY layer. Chase combining HARQ is supported to improve the reliability of a retransmission when a PDU error is detected. A dedicated ACK channel is also provided in the uplink for HARQ ACK/NACK signaling. Uplink ACK/NACKs are piggybacked on DL data. Multi-channel HARQ operation with a small number of channels is enabled to improve efficiency of error recovery with HARQ. Mobile WiMAX also provides signaling to allow asynchronous HARQ operation for robust link adaptation in mobile environments. The one-way delay budget for VoIP on the downlink or the uplink is limited between 50 and 80ms. This includes queuing and retransmission delay. Enabling HARQ retransmissions for error recovery significantly improves the ability of the system to meet the stringent delay budget requirements and outage criteria for VoIP.

3.3.3 Packet Header Compression

The speech payload from the AMR vocoder operating at 12.2Kps is 33 bytes every 20ms in the active state and 7 bytes every 160ms in the inactive state. This payload is typically carried over RTP (Real-time Transport Protocol), UDP (User Datagram Protocol), and IP (Internet Protocol). Protocol headers associated with RTP, UDP and IP constitute 40bytes with IPv4 and 60 bytes with IPv6. Excluding the 6 byte MAC header and 2 byte HARQ CRC, it can be seen that a significant portion of the VoIP packet transmitted over the air interface includes protocol overheads. The fraction of overhead from protocol headers is even greater for VoIP packets carrying speech samples from codecs operating at lower bit rates (7.95 Kbps) such as EVRC or G.729.

To reduce the protocol header overhead, header compression techniques are typically used for VoIP. With Robust Header Compression (ROHC), the protocol headers are compressed to about 3-4 bytes prior to transmission. Mobile WiMAX enables header compression with support for ROHC.

3.3.4 Silence Suppression and Bandwidth Request

In the absence of silence suppression, service requirements for VoIP flows in 802.16e are ideally served by the Unsolicited Grant Service (UGS), which is designed to support flows that generate fixed size data packets on a periodic basis. The fixed grant size and period are negotiated during the initialization process of the voice session.

Service flows such as VoIP with silence suppression generate larger data packets when a voice flow is active, and smaller packets during periods of silence. The Real Time Polling Service (rtPS) is designed to support real-time service flows that generate variable size data packets on a periodic basis. rtPS requires more request overhead than UGS, but supports variable grant sizes.

In conventional rtPS, a Bandwidth Request Header is sent in a unicast request opportunity to allow the SS to specify the size of the desired grant. The desired grant is then allocated in the next UL subframe. Although the polling mechanism of rtPS facilitates variable sized grants, using rtPS to switch between VoIP packet sizes when the SS switches between the talk and silent states introduces access delay. rtPS also results in MAC overhead during a talk spurt since the size of the VoIP packet is too large to be accommodated in the polling opportunity, which only accommodates a Bandwidth Request Header. The delay between the bandwidth request and subsequent bandwidth allocation with rtPS could violate the stringent delay constraints of a VoIP flow. rtPS also incurs a significant overhead from frequent unicast polling that is unnecessary during a talk spurt.

The ertPS scheduling algorithm improves upon the rtPS scheduling algorithm by dynamically decreasing the size of the allocation using a grant management sub-header or increasing the size of the allocation using a bandwidth request header. The size of the required resource is signaled by the MS by changing the Most Significant Bit (MSB) in the transmitted data.

3.3.5 Persistent Scheduling

The basic idea behind individual persistent scheduling is that a user is assigned a set of resources for a period of time and the necessary information for the packet transmission are sent only once at the beginning of the assignment. For the rest of the period of allocation, the MS is assumed to know all the information for data reception on the DL and data transmission on the UL. Note that the allocation period can be infinite. In other words, persistent scheduling is in effect until updated. In the case of dynamic scheduling, a MAP element is required to specify resource allocation information every time a VoIP packet is scheduled. On the other hand, in the case of persistent scheduling, resource allocation information is sent once in a persistent MAP element and not repeated in the subsequent frames. The additional resource that becomes available due to MAP overhead reduction can be used to increase VoIP capacity while the QoS guarantees are maintained to al the VoIP users in the WiMAX system.

3.3.6 Fast Connection Setup via Paging During Idle Mode

WiMAX system uses idle mode operation when there is no activity to save power. In this mode, the BS and MS use a pre-negotiated pattern of alternating available and unavailable interval for any potential for any potential paging for a mobile terminated call. In WiMAX, the paging is designed in such a way that the user can be connected quickly regardless of how much the user roamed around within the WiMAX network to answer the incoming call. Basically, every time the user crosses a paging zone boundary, the MS needs to report to the paging controller via BS so that the MS location can be tracked for the potential paging. Without this feature, VoIP call connection setup time requirement cannot be guaranteed while the allowing the MS to save power during inactive.

3.3.7 Optimized HO Mechanism

During handover from one BS to other BS the VoIP service can be interrupted since there is a time gap to establish a connection at the new BS, and it takes time to move the context information from the old BS to the new BS. WiMAX supports optimized handover mechanism that takes only few tens of ms and the VoIP service interruption is seamless.

4. NEXT GENERATION WIMAX AIR INTERFACE AND NETWORKS

The sections so far have described the QoS mechanisms existing in the current WiMAX airlink and the network. However, advanced QoS mechanisms are currently being developed for the next generation WiMAX airlink in IEEE 802.16m-08/003r7 (2009). This section provides an overview of such mechanisms. An example of such advanced mechanism is the Adaptive Granting and Polling mechanism (aGPS) that dynamically adjusts the polling and granting intervals based on the traffic activity so that the polling overhead can be optimized. Another example is Group Scheduling mechanism for VoIP users. In addition, the next generation system supports MIMO mode, beam forming and interference mitigation efficiently. These features contribute to VoIP capacity and quality.

As described in section 3, WiMAX has a centralized medium access control (MAC) layer. All required bandwidth for UL applications have to be scheduled and granted by BS. Hence, in order to satisfy the end-to-end quality of service (QoS) constraints of heterogeneous applications in WiMAX networks, the UL scheduling plays an important role. When a MS needs to transmit to BS in the UL, the bandwidth allocation is obtained via bandwidth request/grant process between MS and BS. Corresponding to the traffic characteristics of different services, five types of scheduling services have been defined in WiMAX as described in section 3. Among them, UGS, rtPS and ertPS are mainly used for real-time (RT) traffic, while nrtPS and BE are usually utilized for non-real-time traffic. Using aGPS, all the existing QoS classes such as UGS, rtPS, ertPS, nrtPS and BE can be realized. Hence in other words, aGPS becomes more of an "umbrella framework" for QoS in WiMAX.

The concept of aGPS is quite simple and elegant. During the traffic on-period, the granting and polling happens similar to the current ertPS or rtPS. During the traffic off-period detection and handling, two possibilities arise.

- **Implicit:** BS itself adjusts the grant or polling configuration adaptively. The adaptive algorithm can be optimized with different functions for different applications.

- **Explicit:** BS adapts grant or polling configuration with MS's assistance.

Apart from aGPS, other aspects of QoS are also being enhanced in 802.16m. Some examples include the design of an ultra fast contention channel with very minimal latency, so that the end user can request incremental bandwidth as needed for his bursty application.

As described in Section 3, Persistent Scheduling mechanism supports higher VoIP capacity; however, it has issues with VoIP data packet packing efficiency and link adaptation capability. The next generation WiMAX system will support Group Scheduling for VoIP users. Group scheduling is basically a persistent scheduling mechanism not just for one user but for a group of users. Groups can be generated based on the users channel conditions, the codec used, etc. Once the users are allocated in a group the individual user location is not fixed in the OFDMA resource area, however, the relative position in the group is fixed if all the users are active. If the group carries both active and silent user, a bitmap is needed to specify which the active and silent users are. Based on the bitmap and knowing the fact that for the users with the same MCS and same codec the amount of resources that needs for each user in that group is the same. This mechanism allows for complete resource packing and very efficient in resource utilization.

Overall, the next generation of WiMAX airlink promises a lot of innovations that will help the end user experience real time and high interactive services with excellent Quality of Service.

5. SUMMARY

Several features of the WiMAX protocol ensure robust QoS protection for services such as streaming audio and video. As with any other type of network, users have to share the data capacity of a WiMAX network, but WiMAX's QoS features allow service providers to manage the traffic based on each subscriber's service agreements on a link-by-link basis. Service providers can charge a premium for guaranteed audio/video QoS, beyond the average data rate of a subscriber's link. The next generation WiMAX system will incorporate additional features to support excellent QoS for variety of services including VoIP, Video and Gaming.

REFERENCES

Chakchai, S.-In., Jain, R., & Tamimi, A. K. (2009). Scheduling in IEEE 802.16e mobile WiMAX networks: key issues and a survey. *IEEE Journal on Selected Areas in Communications*, *27*(2), 156–171. doi:10.1109/JSAC.2009.090207

Filin, S. A., Moiseev, S. N., & Kondakov, M. S. (2008). Fast and Efficient QoS-Guaranteed Adaptive Transmission Algorithm in the Mobile WiMAX System. *IEEE Transactions on Vehicular Technology*, *57*(6), 3477–3487. doi:10.1109/TVT.2008.919930

IEEE802.16e-2005. (2006, February 28). IEEE Standard for Local and Metropolitan Area Networks Part 16: Air Interface for Fixed and Mobile Broadband Wireless Access Systems - Amendment 2: Physical and Medium Access Control Layers for Combined Fixed and Mobile Operation in Licensed Bands and Corrigendum 1.

IEEE P802.16Rev2/D4. (2008, May). *Draft Standard for Local and Metropolitan Area Networks Part 16: Air Interface for Fixed and Mobile Broadband Wireless Access Systems.*

IEEE 802.16m-08/003r7. (2009). *The IEEE 802.16m System Description Document.*

Li, B., Qin, B. Y., Low, C. P., & Gwee, C. L. (2007, December). A Survey on Mobile WiMAX (Wireless Broadband Access). *IEEE Communications Magazine*, *45*(12), 70–75. doi:10.1109/MCOM.2007.4395368

Neves, P., Fontes, F., Monteiro, J., Sargento, S., & Bohnert, T. M. (2008). Quality of service differentiation support in WiMAX networks. *International Conference on Telecommunications (ICT) 2008.*

RFC 3095. (n.d.). *ROHC Framework and four profiles: RTP, UDP, ESP, and uncompressed.*

Talwalkar, R. A., & Ilyas, M. (2008). Analysis of Quality of Service (QoS) in WiMAX networks. In *16th IEEE International Conference on Networks (ICON).*

WiMAX Forum. (2006, August). *Mobile WiMAX - Part I: A Technical Overview and Performance Evaluation* [White Paper]. Retrieved from http://www.wimaxforum.org/news/downloads/Mobile_WiMAX_Part1_Overview_and_Performance.pdf

WiMAX Forum. (n.d.a). *Network Working Group Document, Release 1.0.*

WiMAX Forum. (n.d.b). *Universal services interface (USI): An Architecture for Internet+ Service Model.* Network Working Group Document, Release 1.5 (draft).

KEY TERMS AND DEFINITIONS

AAA: Authentication, Authorization, and Accounting

AF: Application Function

AMC: Adaptive Modulation and Coding

ARQ: Automatic Repeat request

ASN: Access Service Network

ANS-GW: Access Service Network Gateway

BS: Base Station

CC: Chase Combining (also Convolutional Code)

CDF: Cumulative Distribution Function

CID: Connection IDentifier

CQICH: Channel Quality Indicator CHannel

CRC: Cyclic Redundancy Check

CSN: Connectivity Service Network

DL: Downlink

ertPS: extended real time polling service

GMH: Generic MAC Header

HARQ: Hybrid Automatic Repeat reQuest

HSPA: High Speed Packet Access

LPF: Local Policy Function

MAC: Medium Access Control

MCS: Modulation and Coding Scheme

MIMO: Multiple Input Multiple Output (Antenna)

MPDU: MAC Packet Data Unit

MS: Mobile Station

NSP: Network Service Provider

nrtPS: Non-Real-Time Packet Service

OFDMA: Orthogonal Frequency Division Multiple Access

PDU: Packet Data Unit

PER: Packet Error Rate

PF: Policy Function

PHY: PHYsical layer

PUSC: Partially Used Sub-Channelization

QoS: Quality of Service

RAN: Radio Access Network

rtPS: Real-Time Polling Service

SFA: Service Flow Agreement

UGS: Unsolicited Grant Service

UL: Uplink

USI: Universal Service Interface

VoIP: Voice over Internet Protocol

VSP: VoIP Service Provider

WiMAX: Worldwide Interoperability for Microwave Access

Chapter 4
Cross-Layer Architecture: The WiMAX Point of View

Floriano De Rango
University of Calabria, Italy

Andrea Malfitano
University of Calabria, Italy

Salvatore Marano
University of Calabria, Italy

ABSTRACT

WiMAX is the most promising technology of recent years; it can be the technology that resolves some problems related to the spread of wireless service. When thinking of the concept of service, the most important related issue is the QoS (Quality of Service). Behind WiMAX, there is the IEEE 802.16 protocol (IEEE 802.16, 2004), which provides some basic mechanisms to guarantee QoS. This chapter aims to explore these mechanisms, but it also attempts to highlight the absence of some elements in the protocol or those components in it that can be improved. The protocol can be optimized and in the last part of chapter we show how to improve it using a set of algorithms collected by literature. Finally, it is explained how instruments not designed to be applied to the world of wireless, such as games theory or fuzzy logic, can be used to deal with wireless issues.

INTRODUCTION

This chapter deals with a particular aspect of 802.16 protocol, i.e. the QoS point of view. An overview of the mechanisms related to the QoS is given in this chapter, differentiating the mechanisms on the basis of the specific operating mode of 802.16 protocol. The 802.16 can operate in two mode: PMP (Point-to-Multipoint) mode and the mesh mode, and for each of them, the concepts related to QoS will be introduced and commented, and in doing this, a cross-layer approach will be used. Various gaps in protocol have been left in a voluntary way, this gives greater flexibility to the protocol, since the implementers have the opportunity to create algorithms that are optimized according to their objectives and their application scenarios. In order to better understand how the protocol can be improved and enriched, what is present in the literature will be observed.

Finally, at the end of the chapter some specific cases of integration of WiMAX networks with other

DOI: 10.4018/978-1-61520-680-3.ch004

technologies will be discussed, and also it will be seen how to solve problems of traditional wireless networks using specific theories.

The purpose of this chapter is to make the reader aware of how the protocol guarantees the QoS and what basic concepts the protocol provides for this. To ensure that these concepts do not remain only as theoretical concepts, in discussions of the open issues of the protocol, practical solutions proposed in the literature will be included.

BACKGROUND

The main topic of this chapter is the QoS. QoS is very important, indeed essential, to any type of network taken into account. The quality is a concept closely related to the type of services provided to users. In fact, once certain restrictions for a quality service are set, it means meeting customer expectations and hence their satisfaction. QoS can be defined in different ways depending on the point of view and the level of abstraction that is considered. The user point of view is higher and is more abstract, everything is done in a transparent manner. The network point of view, or more correctly, the protocol point of view is undoubtedly the most complex. Each protocol uses its own set of mechanisms to ensure QoS at different layers of the protocol stack. In this chapter the QoS term is approached to WiMAX technology. This choice is linked to the fact, that the WiMAX technology is suitable for a wide variety of scenarios and can solve a great number of problems. In the literature there are several works that describe different scenarios for the applicability of this technology:

- (Diamond, Hossain & Niyato, 2007)a telemedicine application scenario
- (De Rango, Malfitano & Marano, 2006) an application of 802.16 protocol to an HAP (High Altitude Platform) scenario
- (Gheorghisor & Leung, 2008; Matolak, Sen, & Wan, 2007) an airport scenario
- (Hempel, Sharif, Wang, Mahasukhon & Zhou, 2008) a railroad application
- A military application instead is described in Ganz & Wongthavarawat (2003)

Other papers, (Andrews, Chen, Ghosh & Wolter, 2005) instead analyze the potentialities and the future of this promising technology.

WiMAX is the acronym for Worldwide Interoperability for Microwave Access, and behind this label there is a non-profit-making society whose purpose is to accelerate the introduction of wireless devices with 802.16 technology. The IEEE 802.16 protocol is the name that identifies the standard proposed by IEEE. This protocol has a considerable number of opportunities for improvements that are related to specific processes, such as the call admission control process, the bandwidth allocation and others. In support of this, a list of interesting works in the literature can be found (Hossain & Niyato, 2006; Chou, Lin & Liu, 2008) related to the scheduling problem, related to call admission control (Agrawal, Li & Wang, 2005; Chang, Chen & Chou, 2007) and inherent handoff issue (Kwon, Park & Suh, 2006).

Issues, Controversies, Problems

Even if the QoS, in the considered protocol, is something that is well defined by a series of constraints, it is affected by the goodness of solutions designed to enrich the protocol and to improve and optimize certain aspects. The completion of protocol can be made by adding call admission control algorithms, bandwidth granting algorithms, handoff and adaptive modulation algorithms, these can be addressed and resolved by using a wide range of solutions and architectures.

To enrich the protocol is possible to create integrated solutions which aim to achieve optimized solutions. In works of Hossain & Niyato (2007) and of Geetha & Jayaparvathy (2007), for example, the authors consider an integrated solution for both call admission control and bandwidth problems.

Each of the various issues under consideration, may also be dealt with in a single protocol stack layer, or it is possible to consider a completely different approach which is called cross-layer. This type of approach can be safely applied to each of the issues taken into consideration and considers the solution from the perspective of two or more layers of the protocol stack. Whether or not one chooses a resolution method which considers the "interaction" of the various issues (for example, one can consider the interaction between the call admission control and bandwidth allocation process), the most interesting resolution mode, of course, is the cross-layer. In the literature there are many cross-layer solutions to the introduced problems. Examples related to the centralized scheduling problem are papers of Passas, Salkintzis & Xergias (2008) and of Chen, Tseng, Wang & Wu (2009), or in paper of Chen & Hsieh (2007) there is a cross-layer handoff solution to the problem.

Solutions and Recommendations

As for the solutions to the issues introduced, the best way to create a solution, from the authors' point of view, is without doubt cross-layer architecture. There follows a detailed explanation of how this mode works to build solutions. To resolve a problem, a cross-layer solution can be designed, but in addition, a number of processes can be identified that can interact with each other in performing their duties. Once the problem to be solved is identified, the stack protocol layers that are involved must be individuated, along with the issues to be considered for each layer, identifying what are the objectives to be achieved for each layer and what are the parameters to optimize.

The optimization made in the individual protocol layer is essential to building an efficient cross-layer algorithm. An example may be to create an algorithm of cross-layer handoff to work at the MAC and network layer. If the MAC component of the algorithm is not optimized and it introduces high delays, surely the cross-layer solution will not be characterized by low delays and thus it is not a good solution.

WIMAX: QOS IN A MIXED MOBILE-FIXED NETWORK

The IEEE 802.16 protocol defines guidelines to provide wireless broadband services in a wide area. There are different releases of previous protocols that define the physical (PHY), medium access control (MAC) layer protocol and also each management aspect; the first layer defines five air interfaces and the second one allows itself to be interfaced with the IP (Internet Protocol) or ATM (Asynchronous Transfer Mode) upper layer protocol. In the last IEEE 802.16 version, precisely in 802.16e (IEEE 802.16, 2005; IEEE 802.16, 2007), user mobility is also contemplated and this mobile capability makes it more interesting and useful for practical applications. This protocol allows wireless multimedia services to be provided to a wide area; the "wide" term brings many advantages: both economic and practical. Even more complex heterogeneous scenarios are possible and realistic, in that WiMAX protocol can operate in synergy with other protocols such as Wi-Fi or UMTS. These scenarios have obviously a great series of advantages but also they present a long list of problems to be addressed, it is necessary to meet the need of users, ensuring user satisfaction.

The protocol, to guarantee QoS describes various mechanisms dependent on network topology, PMP (Point-to-Multipoint) or mesh mode, and it lays the foundations for their extension and coordination at different protocol stack layers.

QoS Mechanisms Offered by IEEE 802.16

IEEE 802.16 defines a single-level MAC (Medium Access Control) with various modifications and

Figure 1. IEEE 802.16 protocol stack

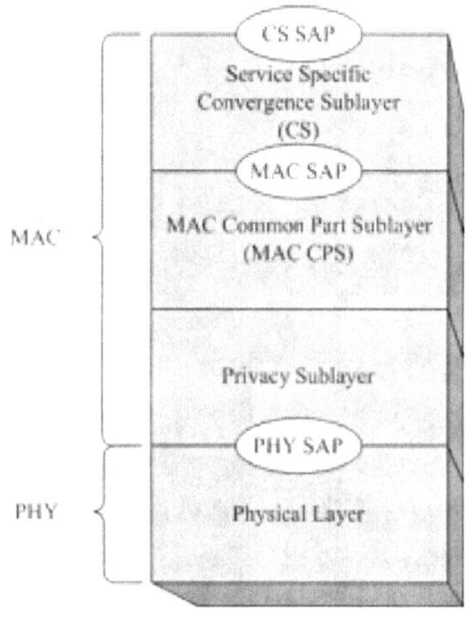

improvements published in various steps, which adds various physical layer specifics, covering both licensed and license-exempt bands. The IEEE 802.16 protocol was specified through a stack architecture, visible in Figure 1.

The various sublayers can interact with each other and access the services of the lower layers through the SAP (Service Access Point), so for example, the Convergence Sublayer provides a set of services to higher layers through the CS-SAP, and in turn it enjoys the services of the Common Part Sublayer through the SAP called MAC-SAP. The whole in order to allow communications between equal entities, typical of a protocol defined by a stack architecture.

The protocol offers QoS mechanisms at both MAC and PHY protocol layer. In the following paragraphs, the different mechanisms of QoS offered by the protocol and how they can cooperate with each other in order to achieve the common goal of quality of service are explained.

MAC Layer Point of View

In this section, the MAC point of view will be illustrated and we are going to introduce what are the mechanisms offered by this layer in order to guarantee well-defined levels of QoS.

The protocol supports two different modes, the Point-to-Multipoint mode and the optional mesh mode. To correctly distinguish the two modes the different entities that come into play in a WiMAX network are defined:

- **BS:** Base station
- **SS:** Subscriber station
- **MSS:** Mobile subscriber station

The SSs and the MSSs are the users stations, the latter are equipped with mobility capabilities, while the BS is the base station and has a central role for different reasons in both operational modes.

In the case of PMP mode, the only connecting links existing between the various entities, are the links of BS with the various user stations, fixed or mobile. No direct link is possible between the various user stations. The BS is a central entity for the bandwidth allocation and the user stations registration. In the case of the mesh mode there is also the possibility of creating links between the various SSs. In practice, a user station which does not fall within the range of a BS, can reach it by

Figure 2. Representations of PMP (a) and mesh (b) model

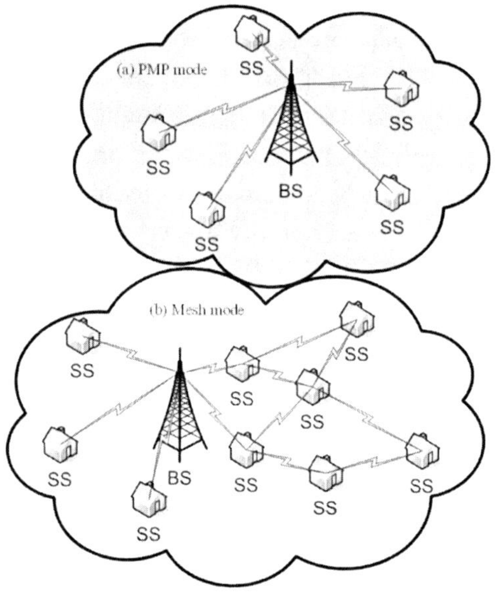

exploiting the presence of any link with nearby user stations. This is not applicable to MSSs stations that continue to be bound to the BS. In figure 2 it is possible to see the two different modes. Figure 2.a represents the PMP mode, instead Figure 2.b shows the more complex mesh mode.

In mesh mode the BS loses the central role but retains a certain importance because it is the only station to have access to the "rest of the world", taking the role of gateway to the Internet. The distinction between the two operational modes is necessary because in both cases the QoS, is managed and assured in a different way and using different MAC mechanisms. Before proceeding with the discussion, briefly the various protocol layers and the structure of a MAC PDU are introduced, which will make it easier to understand what is stated later in the chapter.

The MAC layer, as visible in Figure 1, is divided into 3 different sublayers, the upper layer is the Convergence Sublayer (CS). The main task of the Convergence Sublayer is to ensure to different types of higher protocol layers the ability to communicate with the lower stack layers.

The central sublevel is the Common Part Sublayer (CPS). It performs typical tasks of the medium access control layer, thus providing algorithms to ensure efficient coordination between the various entities that require transmission bandwidth allocation. The last sublevel that includes the MAC is the Privacy sublayer that gives a strong protection from theft of service to service providers. Moreover, it protects the data flow from unauthorized access by strengthening the encryption of the flows passing through the network.

The MAC PDU is shown in Figure 3, and consists of a fixed length header equal to 6 bytes, a payload that can contain one or more SDU (Service Data Unit) or SDU fragments or even can be absent, and finally, optionally, the CRC (Cyclic Redundancy Check) field can appear. The represented header is characterized by fixed length and contains several fields:

- **HT:** Header type, which is used to distinguish between a generic header and bandwidth request header used in PMP mode;
- **EC:** Encryption Control, which is used to indicate if the payload is encrypted;
- **Type:** It is used to indicate if the payload contains one or more subheaders;
- **Rsv:** Not used;
- **CI:** It indicates if the payload end with a CRC portion;
- **EKS:** It indicates the payload encryption key;
- **LEN:** The length of the PDU, including header and CRC;
- **CID:** It is the connection identifier, it in mesh mode contains link and network identifier;
- **HCS:** Header check sequence, it is used to detect header errors.

The payload of a MAC PDU, can carry both data and management messages. The format of the management message is constituted of two parts:

- "Management message type": type of message conveyed;
- "Management message payload": actual message.

PMP Mode QoS Introduction

The PMP mode of 802.16 protocol is strongly connection oriented and each connection is identified by a 16-bit CID. In downlink, the BS is the only station that is able to transmit in a broadcast way without coordination with the other stations, and each user station retains only what is directly to itself. The various user stations should instead share the uplink channel.

BS can allocate bandwidth to SSs, periodically, in order to send bandwidth requests. This mechanism is called polling and it can be of two types:

Figure 3. MAC PDU and generic MAC header

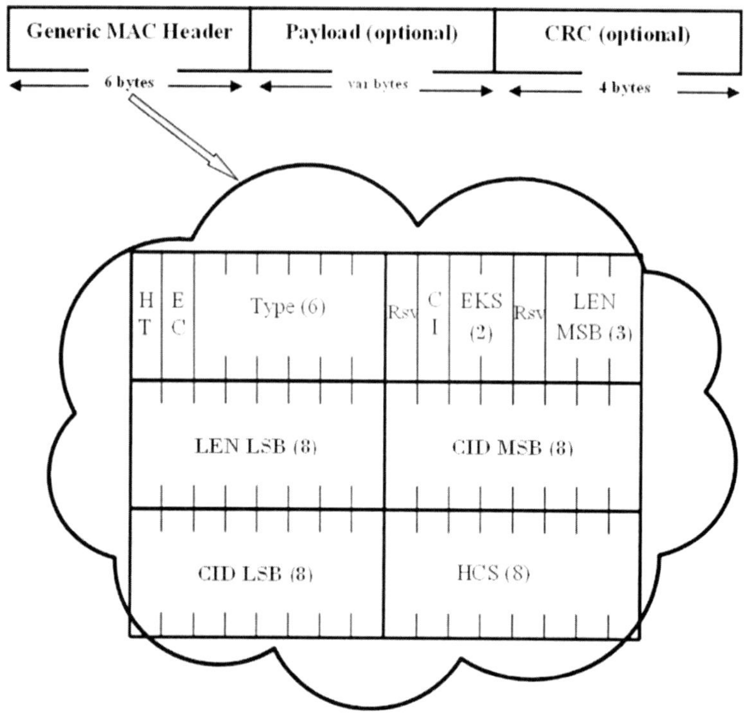

- Broadcast polling
- Unicast polling (including the Poll Me bit: PM)

Using broadcast polling a collision may happen; in which case the contention resolution method is the use of the exponential backoff. Once the various stations are sent the bandwidth requests to the BS, it can allocate the bandwidth in two ways:

- **Grant-per-connection (GPC):** The BS allocates bandwidth to the single connection
- **Grant-per-SS (GPSS):** The BS includes all the bandwidth requests, made by the same SS for all its connections, and gives to the SS a single aggregate grant, thus the user station can divide the granted bandwidth among the various connections.

Very interesting in IEEE protocol is the polling-based MAC layer that is more deterministic than the contention-based MAC used by 802.11. What makes IEEE 802.16 a strong protocol, in this regard, are well-defined concepts, such as the connection, the scheduling data service and service flow. The connection is the basic mechanism, the foundations that allows the existence of various concepts that form the architecture of the QoS protocol. The QoS parameters are linked to the service flow, but a service flow cannot exist unless associated with a connection. A single SS may provide services to an entire building, as a result, each SS can embrace all types of different user traffic, with the same characteristics, within a single connection. So everything revolves around the concept of connection and service flow. The connections can operate in a dynamic way, they can be created, their parameters can be changed and finally, a connection can be deleted.

The mapping of the SDU over the corresponding connection, contributes to the QoS classification, because in this way, the non-delay tolerant SDU will never be mapped on a connection that carries the best effort traffic and SDUs of delay-tolerant application will not be mapped on a connection that can handle traffic with stringent delay constraints. The QoS diversification is visible also in management messages traffic, in fact, between SS and BS three different management connections will be instantiated with different QoS levels:

- Basic management connection: used to exchange short urgent messages
- Primary management connection: carrying longer messages and delay tolerant
- Secondary management connection: used to carry standards-based delay tolerant messages

Each connection is associated with a single scheduling data service (UGS, rtPS, nrtPS and BE) and each data service is associated with a set of QoS parameters that quantify aspects of its behavior. Moreover, each scheduling data service is associated with specific bandwidth request mechanisms that allow it to respect qualitative constraints imposed by the specific application.

The framework will be completed and will appear in all its beauty with the service flow concept description. It represents the points of contact with the structure of the real and practical applications constraints. The scheduling data services realize a qualitative classification of traffic classes, instead the service flow, will 'dirty its hands' with the real constraints of user applications.

The QoS in IEEE 802.16 protocol is closely linked to the service flow concept: a service flow is a bi-directional flow of packets that provides a particular QoS. Each service flow is characterized by specific qualitative constraints. A service flow is enabled between an SS and a BS and the necessary characteristics are assigned to it for the particular type of transmission required by the SS; once activated, one and only one connection will be associated with it. In this way, all communications will take place between SS and BS, with certain restrictions, can be sent in a single connection within a single service flow. Service flows of various kinds can be created:

- Provisioned, are the provided service flow that are not bandwidth reserved to flow. These service flows are activated in a deferred way
- Admitted: service flows that are not activated, but with reserved bandwidth
- Activated: they are active

When a service flow is admitted it is characterized by a given CID. Only an activated service flow may forward packets. A service flow is characterized by the following attributes:

- Service flow ID
- Connection ID
- A QoS parameter set

and the QoS parameters set defining the QoS for the particular services are:

- **MSR:** Maximum sustained rate
- **MRR:** Minimum reserved rate
- Maximum-latency
- Maximum jitter
- Priority

The MRR acts as the "guarantee", while the MSR serves to limit a connection. In 802.16 all service flows have a 32-bit service flow identifier (SFID). Since multiple service flows may need to share a common set of QoS parameters, the protocol developers have introduced the concept of service classes or service class names (SCN). A service class is an optional object that may be implemented at the BS.

Mesh Mode QoS Introduction

In mesh mode all those mechanisms set in the PMP mode to guarantee QoS cease to exist. In the mesh mode there is the ability to create and manage direct links between the SSs stations. Each entity is generically named "node" and new concepts are introduced:

- **Neighbor:** A node with a direct link with the considered node
- **Neighborhood:** Is the set of all neighbors
- **Extended neighborhood:** In addition to neighboring nodes, it contains all the neighbors of the neighborhood.The BS loses the central role that characterizes the PMP mode, and in fact, the basic principle that governs the mesh network is the following:

no one node can transmit on its own initiative, including the BS node, without coordinating its transmission within its extended neighborhood.

In a network that operates in mesh mode, there are two different ways to allocate bandwidth according to a kind of distributed or centralized scheduling. The distributed scheduling, in turn, can be either coordinated or uncoordinated. In the distributed coordinated scheduling, all stations must coordinate their transmissions in their extended neighborhood. This type of scheduling uses all or a portion of the scheduling control subframe, to send its regular schedule and to propose changes of the said in a PMP mode, i.e. the messages used in this phase are sent in a broadcast way.

All stations in a network use the same channel to transmit the schedule information. This information will be issued in format requests-grants. The distributed coordinated scheduling ensures that all the transmissions will take place without having to rely on the base station. The uncoordinated scheduling can ensure communications with fast setup on the basis of individual links. Both modes of distributed scheduling, use a three-way-handshake protocol. Figure 4 shows the three-way-handshake process and the three messages:

- Request
- Grant
- Ack

which are exchanged between two generic nodes to request and obtain bandwidth. It is also possible to note the frame division into two different subframes. This is explained in more detail in the next section.

The second mode of bandwidth allocation is based on centralized scheduling. In this case, the BS determines the flow assignments on the basis of requests received by SSs. The BS works as in the PMP mode, the only difference is that in this case not all the SSs can rely on a direct

Figure 4. Three-way-handshake process

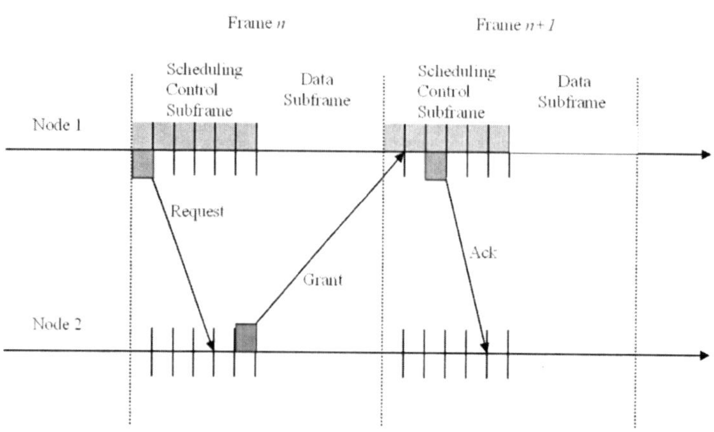

connection with the BS, hence the requests-grants message must be issued within the system in the broadcast mode.

The scheduling mechanisms described above, use a series of messages that are exchanged within the node extended neighborhood. These messages can be grouped into two sets:

- Scheduling control messages (MSH-DSCH, MSH-CSCH, MSH-CSCF)
- Network control messages (MSH-NCFG)

The network control messages can be sent in the network control subframe and therefore cannot be present in every frame, because this protocol alternates frames containing the network control subframe and scheduling control subframe. The dispatch of each control messages is made in a collision free mode and this is granted by the presence of two fields in the message, these fields allow the calculation of the next transmission time of each neighbor node:

- *xmt holdoff exponent*
- *next xmt mx*

Each node, at the instant in which it sends a message, will calculate its next transmission instant and expresses it in a range using the two previous mentioned terms. In practice, the node does not tell to the neighbors the next transmission instant, but sends an interval time in which the next transmission take place, this interval is defined by the following constraints:

$$\text{next xmt time} > 2^{xmt\ holdoff\ exponent} * \text{next xmt mx} \quad (1)$$

$$\text{next xmt time} <= 2^{xmt\ holdoff\ exponent} * (\text{next xmt mx} + 1). \quad (2)$$

Between a transmission and the next one, a node must wait in silence for an interval time equal to:

$$xmt\ holdoff\ time = 2^{(xmt\ holdoff\ exponent + 4)}. \quad (3)$$

When a node sends a network control or a scheduling control message, in addition to sending information about itself, it will also send information about its neighborhood, so each node, collecting the information received from all the neighbors will be able to reconstruct information about the 2-hop neighborhood. Within the extended neighborhood and in a certain slot, only one node can transmit.

Packet by Packet QoS Application

In mesh mode the QoS must be guaranteed, packet-by-packet, in the link context. It must be the node, within the constraints of the distributed bandwidth allocation algorithm, to ensure compliance with the quality constraints of the individual application.

To satisfy QoS constraints, the protocol defines specific fields within the PDU header. The generic header of a MAC PDU contains a 16-bit CID field. In the PMP mode this field contains the identifier of the BS - SS connection, rather in the mesh mode, the CID field is split into two parts, the first portion of 8-bits is the logical network identifier, the second portion of the same size contains the link identifier. This is true in the case of the MAC management broadcast message. If the MAC PDU contains a data payload, the first 8-bits portion of the CID is redistributed over four fields used to implement the QoS policies. The fields are:

- **Type:** Indicates whether the PDU is a management message or an IP datagram
- **Reliability:** Indicates the number of admitted retransmissions for the MAC PDU in question
- **Priority/class:** It indicates the priorities associated with the membership class of the message

Figure 5. Holdoff and mesh contention period

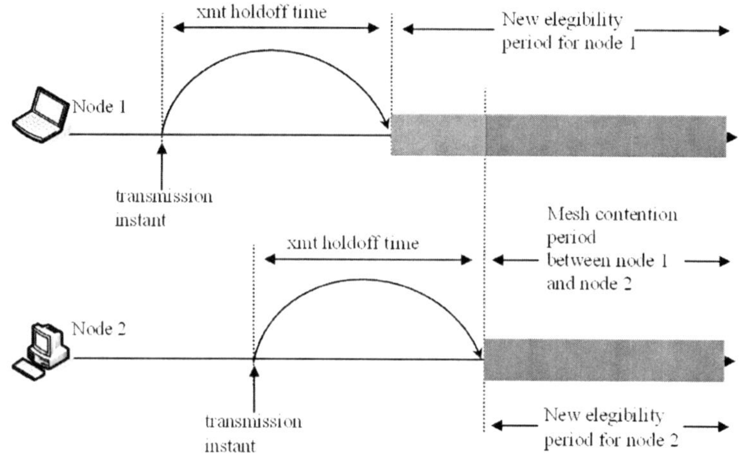

- **Drop precedence:** A message with a high drop precedence value has a high probability of being eliminated in case of network congestion

The presence of these fields provides the protocol with the capabilities of creating service classes in which to map the various user applications, defining a priority and providing nodes with the capability of dropping a packet belonging to a particular class, according to its weight.

The implementer, to provide QoS management, has other mechanisms available under the protocol. These factors are closely linked to the nature and structure of the frame designed in the protocol. A frame in mesh mode consists of a control subframe and a data subframe type. The control subframe can be of two types. The first is used to create and maintain the cohesion of the structure and it is called (see figure 5) Network Control subframe. The second type is used to coordinate the scheduling, centralized and/or distributed within the network, and is defined Schedule Control subframe. A frame can contain a network control subframe or a scheduling control subframe, in an alternate way. The second type of subframe is more frequent than the first one. The periodicity of the subframe type, the number of transmission opportunities to put a certain type of messages and other network parameters are derived from Network Descriptor contained in the network configuration messages; it can be transmitted within the Network Control subframe. Except for the first transmission opportunities, which can be used to send only for new entry nodes messages. The frames that contain the Schedule Control subframe are authorized to carry only messages related to scheduling information. The alternation between the two types of subframe is not static and the protocol is able to define how many frames containing the scheduling control subframe occur between two frames containing the network control subframe. Of course, the idea is to find a value that represents a good compromise of alternation between the two types of subframe. A small number of scheduling control subframes make the bandwidth allocation a slow process, therefore, it should collide with the efforts to obtain certain levels of QoS. On the other hand a great number of scheduling control subframes, could make the responses to requests for network reconfiguration excessively slow, because this would decrease the transmission opportunity for configuration messages and for new entry node request messages.

Another factor which affects the QoS is the behavior of the *xmt holdoff exponent* parameter. This parameter determines the ineligibility time of a node, that is, determines the time of silence between a scheduling information message transmission and the next. High values of this parameter makes a node too slow to make bandwidth requests, consequently, the node that will make any grants is slow in responding. Optimization is very important to calculate the range of *xmt holdoff time* value. Looking at the equation proposed in the protocol that allows the calculation of that interval (3), the presence of the "4" as a fixed part of exponent, can be noted. This fixed part can lead to continued growth of silence, which is the time interval between two successive transmissions. Figure 5 shows the representation of the holdoff interval of two nodes. In particular, the new eligibility period for each node and the mesh contention period can be seen, in which the two nodes must reach a deal to decide who can transmit. The deal is reached by the use of mesh election procedure explained in detail in the protocol (IEEE 802.16, 2004). An algorithm that can calculate the exponent value in a dynamic and adaptive way would be an interesting solution.

PHY Layer Point of View

The physical layer is the lowest layer found in the protocol stack. In particular, the protocol defines a single 802.16 MAC layer but different air interfaces. Any system implementing this layer, must respect the constraints set in terms of transmission techniques, supported modulation and many other specific characteristics. The protocol provides for the possibility of using both single carrier modulation techniques and multi-carrier modulation techniques such as OFDM (Orthogonal Frequency Division Multiplexing). The presence of such different air interfaces make the transmission robust and adaptable to the type of scenario in which the network devices are operating. Consider the single carrier modulation, it is perfect for an environment where there is no high impact of multipath fading, and therefore an environment characterized by a non-frequency selective transmission channel is considered, while the OFDM modulation is the best solution for frequency selective transmission channel. Figure 6 shows the interfaces provided by protocol.

The supported modulations are BPSK, QPSK and from 16 to 256 QAM with the possibility of obtaining different data rates to vary the encryption type. 802.16 technologies support both the TDD and FDD mode, allowing greater flexibility in deploying the network. In the TDD mode, downlink

Figure 6. Air interfaces supported by PHY layer

Designation	Applicability	Duplexing Alternative
WirelessMAN-SC	10-66 GHz	TDD FDD
WirelessMAN-SCa	Below 11 GHz Licensed bands	TDD FDD
WirelessMAN-OFDM	Below 11 GHz Licensed bands	TDD FDD
WirelessMAN-OFDMA	Below 11 GHz Licensed bands	TDD FDD
WirelessMAN-HUMAN	Below 11 GHz License-exempt bands	TDD

and uplink operating in the same frequency band at different times, alternating transmission of the downlink and uplink frame. As stated above, the TDD is used for services that have an asymmetric traffic into the two different links. In FDD mode downlink and uplinks signals are transmitted simultaneously on two different frequency channels, and this results in an inefficient usage of resources, where the traffic is asymmetric, because the downlink and uplink spectra are unused for a long time. Therefore, while the TDD is more appropriate in the case of asymmetric traffic or in scenarios where there is no pair of channels, the FDD on the other hand, is more appropriate in the case of symmetric traffic (VoIP).

AMC: Adaptive Modulation and Coding

All the 802.16 technologies use AMC (Adaptive Modulation and Coding). This feature allows one to improve performance, and optimize the throughput and the range of coverage. The AMC provides a dynamic range of modulation and code rate for each user, depending on the condition of the radio link. When the received signal is low, the system automatically selects a more robust but less efficient modulation in terms of capacity, in order to keep the probability of error equal to the target level. When the signal level received is high, then high modulation are chosen without increasing the probability of error. If the base station is unable to establish a stable connection to a remote user using the modulation scheme of the highest level, 256 QAM, the modulation level is reduced to 16 QAM or QPSK with reduction of supply of throughput, but with increased efficiency over the distance.

The so-called Adaptive Modulation and Coding (AMC) technique, has been proposed in order to choose the most effective scheme based on the state of the channel. The 802.16 standard can achieve its high data rate and efficiency by using multiple orthogonal carrier signals (OFDM) instead of a single carrier approach.

IEEE 802.16 Mobility QoS Introduction

The last version of the IEEE 802.16 protocol was devised at the end of 2005 (see IEEE 802.16, 2005; IEEE 802.16, 2007). This version identified by 802.16e, is designed for the use of WiMAX in scenarios where users have mobility. The main novelties introducer with this version are the following:

- The considered frequencies are 2.3 GHz, 2.5 GHz, 3.3 GHz, 3.5 GHz and 5.8 GHz
- Scalability of the channels on the basis of the availability of bandwidth
- Support for adaptive antennas, the radio beam is realized not by mechanical but by electronic mode
- Handover management
- Roaming management

The introduction of mobility can include the possibility of allowing users the opportunity to enjoy the network services traveling by bus, by car or train, within certain constraints of speed. Unfortunately, the presence of mobility brings also negative aspects. Mobility introduces difficulties for the management of QoS thanks to the presence of a phenomenon known as handoff. The term "handoff" denotes the process of switching from one BS to another during an ongoing call or a data session when the user is moving. In traditional cellular networks, handoff of a terminal from one base station to another is a critical function to support mobile devices. Since handoff is handled primarily at protocol layers 3 and 4, it is not directly supported by the IEEE 802 standards, which specify only layers 1 and 2. The handoff procedure can be both soft or hard. In the first case the connection to the "old" BS is interrupted only after establishing a connection with the "new"

BS, in the latter case, the connection to the "old" BS is interrupted before the user has established a connection with the "new "BS.

QoS and Handover

With the 802.16e version of the protocol, there is the attempt to allow a terminal, on a vehicle in motion, to stay connected (transferring data) up to a speed of 120 km/h; this limit is dictated by the characteristics of the handover protocol. The choice between soft and hard handoff can be related to the QoS: this occurs because the soft handoff reduces latency, it is more appropriate for real time services such as VoIP, while the hard handoff is more suitable for real time services such as data services.

The handover process, as defined in protocol IEEE 802.16e, may be used in a number of situations. Some example are the following:

- When the MSS moves and (owing to signal fading, interference levels, etc.) needs to change the BS to which it is connected in order to provide a higher signal quality
- When the MSS can be serviced with higher QoS by another BS

So the handoff process and the exchange of BS can also be launched to select a BS capable of providing a service with a higher QoS level. If a cross-layer architecture (Chen & Hsieh, 2007) is taken into account, considering levels two and three of the protocol stack, two different types of handoff can be identified: one that occurs in the data link layer and the second is happening in the network layer.

On the MAC layer of the 802.16 protocol, the presence of an IP level can be considered, therefore, where the process of handoff starts at level 2, if the mobile terminal remains within the same IP subnet, it has only to restore contact with the new BS at level 2, of course without changing their IP configuration. Within the 802.16 protocol, the handoff was conceived more as a hard handoff, because the mobile terminal has cut the bridges with the current station before registering itself with the new BS, and the data flow can be resumed only after completing the phase of registration with the new BS, once defined the BS objective. If the new BS resides in a different IP subnet, then the mobile terminal should re-establish two connections: one at level 2 and the second at level 3, and this occurs in order to get a new IP configuration (new IP address, default router, etc.). In this case at layer 3 the handoff must be completed by the Mobile IPv6 (MIPv6). Bearing in mind our goal of quality of service, it can be said that the MIPv6 does not solve the problems of latency handoff, since MIPv6 acts as a location path management protocol rather than a handoff protocol. The situation improves with the Fast MIPv6 (FMIPv6) acting proactively, because it tries to anticipate the steps listed, and thus before the cut of the current connection has already occurred. Exploring the protocols and moving from the 802.11 to 802.16 protocol there is an improvement; this novelty can be highlighted noting that in the 802.16 protocol, the handoff process is designed with the QoS as a reference point. In fact, the choice of the base station can be effected using the concept that the mobile terminal can choose the new BS on the basis of QoS level provided by the BS.

Figure 7 shows the detailed handoff procedure. In conclusion, it can certainly be said that a good solution to the handoff problem is a cross-layer architecture; while the protocol layers from the first to fourth may come into the issue. The cross-layer nature is due to the fact that the mobile terminal, subject to this process, cannot ignore the context in which it operates and that is the architecture and protocol with which it interacts.

WHAT'S MISSING?

All the mechanisms offered by the IEEE 802.16 protocol to guarantee QoS have been described.

Figure 7. Handoff procedure

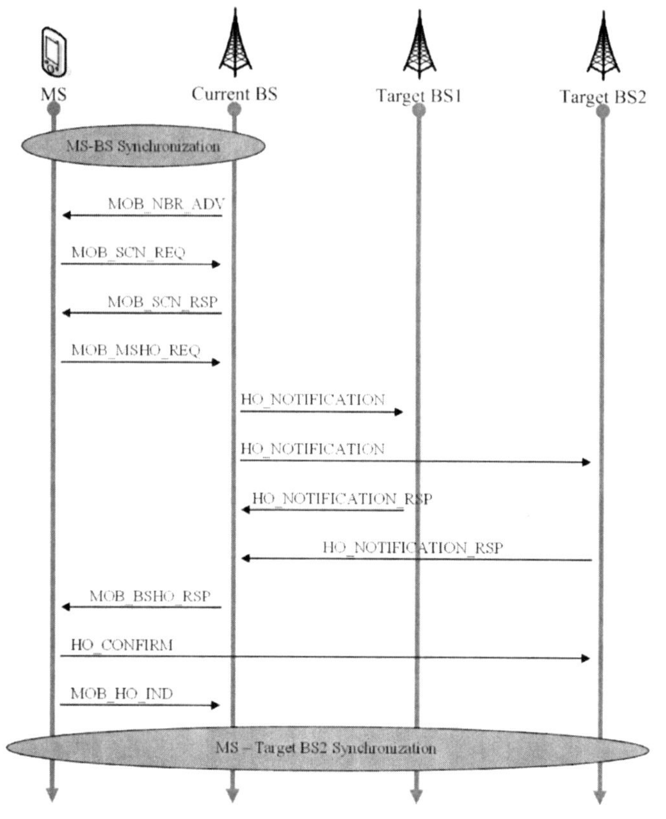

The PHY and the MAC layer characteristics in both operation modes, PMP and mesh mode have been explored. But, as can be seen, no one mechanism or algorithm defined in detail was introduced. This is not owing to an oversight or superficiality in treating the issues considered. In fact, the IEEE 802.16 protocol, as well as many other protocols, is simply limited to providing the mechanisms and guidelines for those that are the different gears that create the protocol. Consequently, for both the MAC and PHY layer, the designers of the protocol leave ample freedom to act within the limits imposed by the guidelines, in order to extend the protocol with algorithms that can optimize it from every point of view.

MAC Layer Missing Elements

Once the guidelines of the protocol are defined, and considering the QoS point of view, the two major missing elements of the 802.16 protocol, are represented by Scheduling and Call Admission Control algorithms.

For example, the protocol does not specify what is to be the BS behavior when a new request, for a new connection, arrives at BS? Should the BS accept or reject the new request? And if it accepts the new connection, how much bandwidth can it make available to the new connection? To give an answer to these questions, there is need to implement a series of algorithms that characterize the BS behavior and the generic mesh node behavior if a network is considered operating in the mesh mode. In the following a set of "solutions" are introduced as proposed in the literature.

Scheduling Algorithm

The absence of a scheduling algorithm is a voluntary omission. The adopted strategy is therefore to leave the door open to experimentation, in order to find solutions that optimize the QoS and management of available bandwidth. The issue of scheduling is none other than the decision regarding the bandwidth size to be allocated to the various entities that come into play in a wireless network. In other words, a scheduling algorithm decides, at any given moment, who to and how long it can transmit, and hence decides the amount of bandwidth to be allocated.

Among the mechanisms used for the task of providing QoS, without a doubt the scheduling plays a very important role. A scheduling algorithm can be structured in various ways and considering various QoS metrics. Later in this section solutions to this problem, as proposed in the literature, are considered.

PMP Mode

The scheduling algorithm used by BS, which operate in a PMP mode, have to guarantee the QoS constraints and fairness among the connections. In PMP mode all the scheduling algorithms proposals have the task of creating an efficient mechanism that allows each kind of scheduling data service, the capability to respect the QoS constraints imposed on them. In the following some examples of scheduling algorithms are illustrated.

In a work (Hossain & Niyato, 2006) the authors present an analytical discussion of the issue of bandwidth allocation. The proposed idea includes the concepts of priority, related to the scheduling data service, and the concept of threshold, linked to the instantaneous size of the code. In fact, the authors present a "queue-aware" solution. There are two schemes at the basis of allocation, one identified as Complete Partitioning (CP) and the second as Complete Sharing (CS). CP considers a static allocation of the band, giving higher priority to UGS. In the second case, instead, a dynamic allocation of bandwidth is made, giving high priority to the UGS scheduling data services. But in this case there is the following behavior: if the bandwidth required by a UGS connection is less than a certain threshold, then the remaining available bandwidth is allocated to other types of traffic. The authors analyze the performance of the proposed mechanism, which has just been introduced in an easy way, and demonstrate that when the tail of services is stable, the algorithm can maintain the average delay at a low and constant level maximizing the utilization levels of the band.

The paper of Chou, Lin & Liu (2008) appear most interesting. Since, in this work, the authors implement a set of scheduling algorithms collected in the literature in a WiMAX simulator, realized with the NS2 tool (Network Simulator). This is a work in which the proposal is an evaluation of scheduling performances; these scheduling proposals take into account different policies, such as EDF (Earliest Deadline First), WRR(Weighed Round Robin), WFQ (Weighted Fair queuing), RIO (RED In/Out, i.e. Random Early Detection with In and Out) and others. In this work the reader can find clear explanations about all these schemes. The work does not propose a new algorithm, but it is interesting to see what is the behavior of different solutions proposed in the literature, once compared on the same scenario. The authors conclude the paper saying that if there is need for best throughput performances then the best solution is the RIO scheme, but there are cases in which this scheme obtains negative results caused by the presence of a network bottleneck.

Another kind of scheduling scheme present in the literature is described in work of Tian & Yuan (2007), it considers a cross-layer architecture to decide the amount of bandwidth provided to scheduling data services. The goal is to guarantee the QoS constraints by taking into account the channel condition.

Mesh Mode

For networks operating in mesh mode, the scheduling algorithms, to ensure certain level of QoS, do

Figure 8. Bandwidth requests collection and grants sending in centralized scheduling

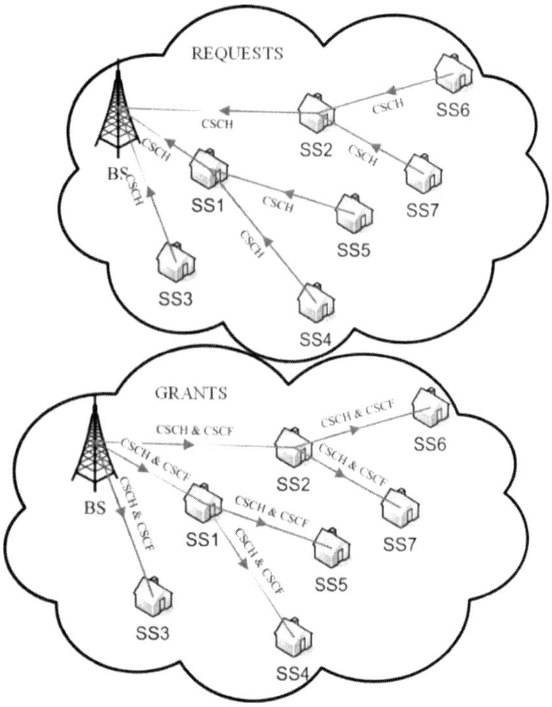

not have the possibilities to use the mechanisms available to the PMP mode, and thus, they can only use some fields present in the MAC PDU header, and precisely in the first 8-bit portion of the CID field. These fields must be used to make a kind of classification among the various streams that travel on the network, even if there is not a flow concept in the mesh mode, the classification can be made only packet-by-packet.

Distributed Algorithm

In distributed mesh mode everything has to happen in a distributed manner, this may seem a disadvantage, but it also has its positive side. Two neighboring nodes can make a quick setup of a connection, avoiding the hop by hop delays in requests/grants mechanism of the centralized mode. The distributed algorithm proposals in the literature, are not numerous, and there are also some possible improvements, since in mesh mode, a number of parameters come into play that require some attention in their setting values.

In particular, a proposal that can be considered interesting, is the work of Cao, Ma, Wang & Zhang (2007). In this paper the authors propose a stochastic model for a distributed scheduler. This model is used to evaluate the scheduler performance. The scheduler is tested under various configurations and the results of the analytical model are compared with those obtained from a simulator built using the NS2 tool. The model is really interesting because it accurately maps the simulation results.

Also another work (Cao, Ma, Wang, Zhang & Zhu, 2005) presents an analytical modeling of a distributed scheduler, and the results are verified by simulations. Even in such case resulting in highly accurate analytical modeling. The authors also have another important result: a distributed scheduling, optimized to achieve certain levels of quality of service, it is able to change, dynamically, the *xmt holdoff exponent* parameter of individual node.

Centralized Algorithm

In mesh mode with centralized scheduling, there is an entity that oversees the bandwidth allocation: the mesh BS. All requests must be channeled to the BS and in a subsequent phase, the BS will distribute the various grants.

Passas, Salkintzis & Xergias (2008) describe how the 802.16 protocol can distribute multimedia traffic in a mesh topology, ensuring well-defined quality of service constraints. At the basis of the centralized scheduling there is the concept of a covering tree. The covering tree is a logical structure that consists of a subset of network links. In the links of this logical structure, are collected the requests and distributed the grants of bandwidth. Each centralized scheduling algorithm refers to this structure. The covering tree and the messages exchanged in a typical centralized mesh algorithm can be seen in figure 8.

The basic concept of work of Passas, Salkintzis & Xergias (2008) is the creation of an Enhanced Frame Registry Tree Scheduler (E-FRTS). The proposed scheduling, prepares the time frame in advance. In this way it tries to avoid the short time available between two successive frames. The basic concept of the scheduler is the introduction of a flexible data structure that maps the decisions of the scheduler. The structure in question is the E-FRTS, and it is a data structure tree, which collects all the information needed to build the next frame. In this way, each data packet can be scheduled in advance and before its deadline. Please note that it can be defined as a cross-layer algorithm, because it takes into account also the variability of PHY parameters. The algorithm introduced in Passas, Salkintzis & Xergias (2008) is really interesting owing to the number of concepts that are involved and for the results achieved. A detailed study of previous work is suggested, as an example of a well-formulated centralized scheduling algorithm under mesh mode.

Also work of Chen, Tseng, Wang & Wu (2009) is interesting in several aspects. The authors propose a centralized cross-layer scheduler, where different considerations at different protocol layers are introduced. The authors elaborate considerations at the network layer inherently to the construction of the routing tree, they look at the shared resource in the MAC layer and the channel reuse in the PHY layer.

Another work (Dastis, Hollick, Mogre, Schwingenschlogl & Steinmetz, 2006) however, even if it does not propose a scheduling algorithm, elaborates a test of the features provided by the protocol to implement a centralized scheduling. It is useful to understanding how to set the protocol configuration parameters in an optimal way.

Call Admission Control Algorithm

When the BS receives the request from the SS to create a new connection, has to decide whether to admit and then to activate the new connection. Obviously, the BS has to decide how much bandwidth to be allocated to the new connection for the lifetime of the service. The previous decision of the BS can be divided into two steps:

- The first is the admission decision, i.e. whether the BS decides to accept the new connection or not
- The second is inherent to the bandwidth to grant to the SS for the admitted connection

Both the decisions are inherent to the bandwidth utilization in the network and also the QoS concepts are involved. In fact, the creation of a new connection can modify the allowed bandwidth to the existing connections; thus, all the QoS constraints must be reviewed. Therefore there is a "risk" in this choice, because admitting a new connection, the possibility of worsening the provided QoS to the old connections must be accepted. The first of the previously listed process decisions is called call admission control, and this decision influences the network band utilization for a long time, i.e. it is a long-term decision. The second, instead, is a short-term decision.

PMP Mode

In the PMP mode, the only entity that has to decide about call admission control is the BS. In the literature there is a great number of proposed solutions, a small part of these have been chosen to show to the reader the methods used to resolve this problem.

This issue, in work of Agrawal, Li & Wang (2005) is addressed with the proposal of a simple but efficient algorithm. The authors consider the classification of the scheduling data services provided by the protocol. The various service classes are organized by the authors using a priority, thus, the services can be listed following the priority order: UGS, rtPS, nrtPS and BE. Each SS has an amount of fixed bandwidth "B", which is allowed to it by the BS. A portion of this it is reserved to UGS connections, this portion is called "U". The UGS connection can require an amount of bandwidth equal to b_{UGS}. When BS receives a request for a new connection, to keep a decision, the BS uses a degradation model and it follows these conditions:

- If the new request is a UGS connection: if the sum of the bandwidth allocated to the existing connections and the new b_{UGS} request is equal or less than "B", then the new UGS connection is admitted, otherwise it is refused;
- If the new request is an rtPS connection: if the sum of the bandwidth allocated to the existing connections and the new b_{rtPS} request is less or equal than the "B" – "U" amount of bandwidth, then the new connection is admitted. Otherwise all the amount of bandwidth, allowed to existing nrtPS connections, are decremented by size "d". This new bandwidth is now available for the new connection and if it is sufficient to meet the new request, then the connection is accepted. Otherwise the decrementing steps can continue until a threshold is reached, this threshold is related to QoS constraints, and the admission decision is finally taken based on the availability of bandwidth obtained in this process;
- If the new request is an nrtPS connection the same steps for the rtPS connection is followed;
- All the BE connections, instead, are always admitted but, they can transmit only when the other connections are in silence.

This algorithm is an example of how the issue of call admission control can be solved. It obtains good results about call blocking probability and bandwidth utilization.

Another work (Chang, Chen & Chou, 2007) proposes a particular call admission control for polling service. The authors decide to optimize the admission control for polling service because this type of service is related to a delay in polling. The polling is deterministic but also characterized by the delay. The authors propose an adaptive polling scheme with cost-based call admission control. The goal is to optimize the bandwidth utilization, and thus three different basics concepts are present:

- Hierarchical polling
- Cost based
- Call admission control

The hierarchical concept allows polling services characterized by a priority. The cost-based function utilizes the ideas of residual bandwidth and an optimized cost-based function that are introduced in other works (Chang & Liang, 2004; Chang, Hsiao & Hwang, 2007).

Another way to resolve this problem is to elaborate a mathematical analysis (Hossain & Niyato, 2007). They make a mathematical elaboration to obtain the optimal solution to the call admission control problem, and in particular they apply an optimization theory, used in the operative research issue.

Mesh Mode

The call admission control algorithms, related to the mesh mode, are rare in the literature. A work that presents an analysis in this field is Lee, Narlikar, Pal, Wilfong & Zang (2006). The authors, as a first step propose the construction of a coverage tree in a centralized manner and then propose two quality constraints for the admission of a new call. The bonds are defined in relation to the transmission rate and delay. In the mesh mode, especially in the distributed scheduling, the presence of control algorithms for the acceptance of a call, is not as widespread as in PMP.

The presence of an algorithm for call admission control is closely related to scheduling algorithm.

In fact there is in the literature a work (Li, Lin, Liu, Tao, & Zeng, 2005) that considers an integrated call admission control and bandwidth allocation algorithm. The basic idea is very simple: there is the presence of a threshold. When a new bandwidth request arrives to a node, the call is not refused if the actual bandwidth utilization is less than threshold value.

PHY Layer Lacks

The physical layer of the 802.16 protocol has gaps or aspects that could certainly be improved or optimized. The guidelines of the protocol, offer the possibility of using adaptive modulation, but of course, these techniques can be enhanced by creating algorithms that take into account a number of interesting points.

QoS-Based AMC Algorithm

The protocol in question provides a variety of modulation techniques that can be used when the channel condition varies, in fact it provides robust techniques, such as the QPSK, and less robust technique, such as the 256 QAM.

The protocol in PMP mode allows a burst transmission, where in each burst it is possible to change the modulation and other physical parameters. This is possible also in the mesh mode where notices of changes are transmitted in the MSH-NCFG messages. So one can consider the possibility of creating cross-layer algorithms which take into account the MAC level constraints and the behavior of the transmission channel. One way is to consider an algorithm, characterized by a component, which not only is able to make "measurements" of the channel state, but also is able to model the channel behavior. Some examples of the channel model are present in the literature (De Rango, Malfitano & Marano, 2006; De Rango, Malfitano & Marano, 2007; De Rango, Malfitano & Marano, 2008).

An interesting algorithm can be considered that consists of two components. The first one is able to make an analysis of the channel, and by modeling it, this component does not perform a simple measures of the state of the channel, but it forecasts the channel behavior. These forecasts, which can be expressed in terms of probability of losing a packet, can be used as input to the second component of the algorithm. This second component could take as input the previous estimation of packet loss and the quality constraints imposed by the MAC layer. Regarding this point with the input data, the algorithm may be able to make decisions about improving the quality of transmission. This approach to the problem is taken into consideration in the work of De Rango, Malfitano & Marano (2009).

Cross-Layer QoS Architecture

The need for optimization and protocol performances improvement conduct to increasing interest on cross-layer solutions. Now we can see how born this idea. Consider a mesh node, to make data transmission, must be present a protocol entity to manage the transmission techniques, this task is developed by physical layer. The node, before to transmit, has to consider the interference due to the presence of neighboring nodes and to estimate the interference entity, the node should know the number of neighboring nodes. Thus become essential the presence of an entity for the medium access management. The next step is to find the destination, in fact if destination does not belong to neighborhood, the source has to start a process to individuate a route toward destination node. This task can be related to network layer. We can continue the description until each protocol layer is defined and introduced, but now it is very interesting to note how the introduction of a cross-layer architecture appears so natural. If we want to optimize the route choice, we can consider for example an interference-aware routing algorithm in which network layer and physical layer collaborate to individuate the best choice. Another simple cross-layer scheme can be constituted by MAC and PHY cooperation, in fact the presence of neighboring nodes can introduce interference, the MAC can manage the communication with these nodes and can communicate to PHY an adjustment to improve SNR (Signal-to-Noise ratio) value. Considering the inherent characteristics of wireless communication and networking, the traditional layered network architecture can be considered inadequate to achieve the full potential of the networks. Cross-layer design approaches have been proposed to improve and optimize the network performance by breaking the layer boundaries and explicitly passing information from one layer to others. Cross-layer design refers to a paradigm that exploits inter-relations between network layers to improve the efficiency and quality.

The term cross-layer therefore does not refer only to a specific layers set but may be associated with any level of the protocol stack. For example, situations can be considered in which (Lin, May & Yang, 2007) the cooperation take place between MAC and IP level, providing, for example QoS mapping features between two or more layers, even in work realized by Chen, Guo & Jiao (2005) there is the proposal of a cross-layer architecture in order to provide a mapping of InterServ and Diffserv services. Or a cooperation of functionalities offered by the physical and MAC layer can be considered.

To continue an overview of published works, in a paper Kaloxylos, Passas, Salkintzis & Triantafyllopoulou (2007) propose and study a cross-layer mechanism that can improve real-time QoS provisioning over IEEE 802.16 metropolitan area networks. This mechanism utilizes information provided by the physical and MAC layers and using a heuristic algorithm it derives new operational parameters for the physical and application layers, which can improve the performance of real-time applications. Also another work (Kaloxylos, Passas & Triantafyllopoulou, 2007) presents a MAC-PHY cross-layer mechanism to provide multimedia services of high quality.

Other works are present in the literature and may be cited here as examples of well-defined cross-layer mechanisms in order to achieve clearly defined objectives for quality of services offered to the users. The whole taking into account certain characteristics of the scenarios under consideration, such as mobility issues and consequently handover, such as in paper of Kuo & Yao (2006), or issues of mitigation effects and deterioration of the signal.

FUTURE QOS CHALLENGES

WiMAX technology is a very promising technology and it is characterized by a series of advantages. Certainly, it was conceived with the prospect of becoming the technology that could eliminate the digital divide problem.

Nevertheless, the solutions that are attracting increasing interest, are the integrated architecture, in which two or more technologies can be integrated and can cooperate in order to guarantee high quality of services over large areas and to a large number of users.

End-to-End QoS in Heterogeneous Architecture

The chances to create integrated architectures are different; cooperation such as WiMAX – Wi-Fi, WiMAX - UWB, or WiMAX – 3G or other kind of cooperation can be considered. Each of these integrated architectures is proposed to be applied in specific scenarios and to achieve well-defined objectives.

Each protocol is characterized by its own mechanisms to ensure QoS in a network segment. But what happens when a data stream of a user must go through more than one segment of the integrated network? Once a protocol of a specific segment, admits a new call, and once the call has been moved to another segment, how can the QoS levels guaranteed at the instant of the admission call be maintained? The problem is to guarantee an end-to-end QoS.

The problem of guaranteeing the QoS is also related to handoff procedure. If it is important to resolve the handoff problem in a simple network architecture, it is more important to resolve the same problem in integrated networks. In integrated networks, the handoff process, can be classified into two different types:

- **Vertical,** if it takes place between two different protocols
- **Horizontal,** if it takes place between two base stations of the same protocol

The vertical handoff one does not take place triggered by received signal strength from base station, but may be due to a variety of balancing and/or optimization traffic choices.

WiMAX and Wi-Fi

The coexistence and interoperability between two technologies, such as WiMAX and Wi-Fi, certainly brings a number of problems to solve. On the one hand there is the frame-based WiMAX, on the other there is the contention-based Wi-Fi, which until the advent of IEEE 802.11e version of the protocol, had serious difficulties in guaranteeing QoS.

The QoS for delay-sensitive applications, in integrated architecture, can be guaranteed and does not become a problem only if there is an efficient mapping between the QoS mechanisms of the two protocols. The standard 802.11a/b/g provides any type of service in the same way, giving bandwidth to it in a best effort mode. There has been a change with the advent of the 802.11e version, which introduces a central coordinator (HC: Hybrid Coordinator) and the possibility of diversifying the traffic by a Traffic Specification (TSPEC). TSPEC describes the traffic characteristics and its requirements in terms of QoS. TSPEC provides the HC with a mechanism to implement a call admission control. There are two ways to characterize the QoS:

- Prioritized QoS: the MAC PDU is associated with a priority
- Parameterized QoS: the QoS is set with the help of some parameters

The steps followed in the 802.11e protocol to guarantee the QoS, are the following:

- Identification of a TSPEC on the basis of required traffic type
- Setup of the traffic stream
- The MAC PDU management on the basis of the QoS setup in negotiating phase

In the literature, there are some proposals that allow the two protocols to interoperate, ensuring a certain quality of service.

In work of Gakhar, Gravey & Leroy (2005) the following scenario is considered: there is a Radio Gateway (RG), which has the role of an SS and it is able to establish connections to a WiMAX BS. It also acts as QAP (QoS Access Point) for the Wi-Fi network. The authors propose a specific mechanism to carry out a kind of mapping between the services offered by the two protocols. In the paper, the creation of a set of traffic classes, C_1 .. C_4, is proposed:

- CBR with real time traffic (audio / video), C_1
- VBR with real time traffic (video on demand), C_2
- VBR with precious data (exchange of files, delay tolerant), C_3
- Unspecified traffic, C_4

In each of these classes a set of well-defined parameters constraints of the 802.16 and 802.11

protocols are matched; so the C_i classes, have the role of interface between the flow classes of the two protocols. A certain flow of the 802.11 protocol, corresponds to a C_i flow, and the C_i flow will corresponds to a 802.16 protocol scheduling data service, with defined constraints.

A different way to solve the problem is that described in another paper (Berlemann, Hiertz, Hoymann & Mangold, 2006). In this work the authors introduce a central BSHC entity. BSHC was conceived from the BS of WiMAX and HC of Wi-Fi. In this solution, BSHC is a hybrid node with both network interfaces. The interoperability is based on the integration of the 802.11 transmission sequences in the structure of the 802.16 MAC frame.

WiMax and 3G

Another important integrated scenario could be a WiMAX-3G scenario. In 3G cellular networks the base stations are connected with the base station controllers via point-to-point links. These links (T1/E1) are not suitable for the development of existing wireless networks, because they are symmetrical links. Traffic flows such as the Internet traffic are asymmetric and mapped on a symmetrical link, would lead to an architecture characterized by an inefficient bandwidth exploitation.

Using a network based on the 802.16 protocol would allow the creation of a symmetrical link between the base station and the 3G RNC (Radio Network Controller). The 802.16 protocol supports the TDD mode and consequently, the possibility of carrying asymmetric traffic efficiently. This type of integration, which does not allow the terminal the capability to switch from a WiMAX connection to 3G connection or vice versa, is rather an integration to optimize the well-tested 3G architecture. The 802.16 protocol is able to perform this role of backhaul through point-to-point link in a good way, and also it is able to guarantee good performances. An example of this scenario can be found in work of Bu, Chan & Rainjee (2005).

One more fascinating scenario than previous ones is the scenario where the two technologies can operate providing the user the ability to perform handoff between the two technologies according to the occurrence of certain events. These events, for example, may be caused by user requests to change QoS constraints, or this event could be a decision taken by the operator in order to balance the carried traffic and/or to optimize the bandwidth usage. A similar scenario is taken into account in work (Nakhjiri, 2007), where the authors propose a new key exchange protocol that can ensure a particular degree of security. Obviously, in such scenarios the QoS, especially, must be one of the main objectives. In fact, considering a scenario where a user can switch from one 3G connection to a 802.16 connection, the first thing that comes to mind is the need for a vertical handoff process that is optimized to reduce the delays. In his paper Kim (2006) proposes a scenario similar to the previous one, enriched with other new proposals to ensure the reduction of latency typical of vertical handoff.

New Way to Resolve WiMAX QoS Problem

In the various sections of this chapter, an overview of the IEEE 802.16 protocol and of the mechanisms that the protocol offers to bring quality services that meet well-defined constraints has been given. The various scheduling, call admission control and handoff algorithms reviewed, are based on traditional approaches to solving problems. Recently a new trend has emerged as part of research. This trend is to use techniques, theories or instruments that were not conceived or developed to solve typical wireless network problems.

The following sections provide a look at two of these fascinating theories that are applied to the issues addressed in this chapter, in order to obtain good results also in an elegant way. One of these is the games theory, it was conceived for applications in economic or social studies. The second theory under consideration is fuzzy logic, which was developed mainly for use with artificial intelligence and personal computer development.

Games Theory on End-to-End WiMAX Scenario

The game theory was developed primarily for use in economic issues, but it has had great success in various disciplines. In the world of telecommunications it is used to model and solve problems concerning the management of radio resources.

The Games Theory is the mathematical science that analyzes conflict situations and it researches cooperative and competitive solutions. The game theory investigates individual decisions in situations where there are interactions between different actors; thus, it studies and describes the methods to solve the games, that is, to calculate the outcomes of the interactions represented by the games. Various situations can be involved in the definition of game, thus, it is necessary to group the games into subclasses; each subclass includes similar situations from the point of view of the rules of the game.

A first distinction that can be made between the games is as follows:

- Cooperative Games: a game is cooperative if there is the possibility, for certain subsets of players called eligible coalitions, to enter into binding agreements, which may give some benefit to individual players.
- Non-cooperative games: a game is non-cooperative, when there are no feasible binding agreements between the players.

A game, in general, is defined only by creating a set of rules. The rules of the game have to specify at least the following aspects of the game itself:

- The players
- When it is the turn of player, i.e., when a player can make their own actions
- What the actions are which each player can choose when it is his turn to play
- Information available to each player
- What the possible outcomes of the game are
- What is the utility that each player obtains (it is defined payoff)

Important concepts of game theory are the strategy and the solution to the game. A strategy is a complete plan of action of the player. In order to calculate the solution, it is essential that each player acts in a rational way. There are different ways to get a solution for a game, one of them is to calculate the solution called the Nash equilibrium. Hence it is an equilibrium point, no one player wants to move himself from this point, because no one benefit can be obtained going away from this point.

In the literature there are several works that propose solutions obtained with the application of game theory to the radio resource allocation problem. The following examples are useful to understood how to apply this theory to wireless problems.

Hossain & Niyato (2007) presents a model of game theory for the bandwidth grant and call admission control, inherently to two types of service defined by the IEEE 802.16 standards: rtPS and nrtPS. The objective is to find the equilibrium point between rtPS and nrtPS connections, which are responsible for providing bandwidth to a new connection so that the QoS requirements of existing connections and new connections are met. The authors, represent the payoff, as the user utility calculated as a function of delay and throughput perceived by connections. Among the available strategies of both types of connections, the Nash equilibrium is determined, and the decision on the admission control is carried out according to perceived QoS performance and to the balance of Nash. The players are all the rtPS and nrtPS connections, while the strategy for each of the players is the bandwidth offered to the new connection.

The payoff for the rtPS/nrtPS connections is the total utility of the existing rtPS/nrtPS connections plus the utility gained by the new connection; the payoff for the BS is the sum of all the connections payoff.

In this scenario a conflict arises because the existing rtPS and nrtPS connections want to maximize their QoS performance, while the BS wants to maximize its overall utility. The Nash equilibrium indicates the amount of bandwidth that the BS must take away from all existing rtPS and nrtPS connections to provide bandwidth to the new connection. The solutions derived using game theory are interesting and are compared, by authors, with results obtained by static and adaptive allocation and call admission control schemes. The framework for the bandwidth grants, based on games theory, may provide a slighter delay for the rtPS connections.

Other interesting papers present in the literature and that deal with the application of game theory to QoS issues in WiMAX are the work of Hossain and Niyato (2007) and the other one of Geetha & Jayaparvathy (2007). The first, was elaborated by the same authors of the previous described work, but in their last work, the authors further develop the concepts introduced in their first work, however, they take the QoS services objectives into account. Geetha & Jayaparvathy (2007) use game theory concepts to create a traffic classification based on the available bandwidth. They consider multimedia traffic, evaluate an analytical model and make a comparison between analytical and simulations results.

Fuzzy Logic, What Idea to Guarantee QoS

Fuzzy logic is a logical extension of Boolean logic in which one can assign a truth degree value between 0 and 1 to each proposition. It is strongly linked to the theory of fuzzy sets. When speaking about truth degree or belonging value, we mean that a property may be true or false as in classical logic, but fuzzy logic introduces the possibility that the truth degree can assume intermediate values. Fuzzy logic modifies the notion of binary logic, according to which, the predicates can take only two states: true and false.

An example of application of fuzzy logic to the wireless issues is work of Nie, Wen & Zeng (2007). This paper is interesting both for the use of fuzzy logic and for the specific scenario considered. The scenario under consideration is constituted of integrated WiMAX and Wi-Fi protocols and the objective of the authors is the creation of a vertical and horizontal handoff algorithm. In this case, fuzzy logic will be introduced, but also a re-examination of the handoff issue related to the quality of service. The vertical handoff is not triggered by the received signal strength, but it must be based on other metrics. The authors propose a bandwidth scheme based on the adaptive fuzzy logic algorithm. The scenario under consideration is the following: there is a number of WiMAX cells in which there are Wi-Fi cells. The proposed scheme works in this way: when the mobile user is in a WMAN cell, it tries to verify through threshold mechanisms whether it is possible to migrate to the WLAN cell, otherwise it checks the option to switch to another WMAN cell; if the mobile unit is in a WLAN cell it checks first whether it is possible to switch to another WLAN cell, otherwise it verifies the possibility of transiting to a WMAN cell. The fuzzy logic is used to implement an additional module; the authors, using classical logic, were not able to describe the different speed and traffic levels.

FUTURE RESEARCH DIRECTIONS

As for the scheduling algorithms and call admission control for the PMP mode, the future trend is to create new more efficient algorithms, which can have a vertical characterization. The term vertical is obviously intended as a multi-protocol approach to the issues introduced. This can include the opportunity to ensure the QoS and optimized packets transmission, using cross-layer mechanisms integrating channel models (which take into account the different effects of attenuation of the signal, capable to foresee the behavior of the channel) with MAC layer algorithms.

Moreover, researchers seek to solve the problems of traditional wireless networks using theories not directly belonging to the world of wireless. A typical example, in the literature is the applications of the game theory, conceived in sociological and economic areas, or applications of fuzzy logic, or genetic theories to problems of scheduling, call admission control and routing.

Another interesting field is the mobility. The last amendment inherent to mobility in WiMAX (802.16e), does not allow MSS mobile units the ability to operate in the mesh mode. The mechanisms for updating the neighboring nodes, as they exist, are unsuitable for this eventuality, as a result, future studies are to examine in what way and how to improve the mechanism as now offered by the protocol. The same issue of handoff, as part of mobility, provides excellent starting points for research, both in terms of the physical layer, which needs to change in order to become the fastest possible and in terms of other protocol layers.

Consider, in addition the development of dynamic scheduling algorithms that are capable "of working together" with the handoff procedures; this is a useful perspective in network architectures in which the handoff does not happen only because of the user mobility, but also to optimize the bandwidth allocation, diverting terminal traffic toward other networks integrated with WiMAX.

CONCLUSION

The chapter has been realized in a descriptive way, where every argument is integrated and motivated by a number of references from the literature. The inclusion of these references is intended to make the reading of the chapter even more interesting. A number of issues, all related to the quality of service, have been analyzed. For every problem, in fact, the relation with the QoS was emphasized.

Finally, in the chapter, some interesting scenarios have been taken into consideration, which are characterized by integration of different technologies. This section, together with the section which describes applications of particular theories to the problems of the world of wireless, makes the chapter also attractive and interesting from an educational perspective. The two cases represent two examples of integration between different disciplines.

To conclude the discussion, it can certainly be said that all the mechanisms and algorithms dealt with, represent the delicate gears of the complex machine that is the 802.16 protocol. Without the perfect development and cooperation of all the gears, ensuring well-defined levels of QoS becomes a difficult challenge.

REFERENCES

Agrawal, D. P., Li, W., & Wang, H. (2005). *Dynamic admission control and QoS for 802.16 Wireless MAN*. Paper presented at the Wireless Telecommunications Symposium 2005.

Andrews, J. G., Chen, R., Ghosh, A., & Wolter, D. R. (2005). Broadband wireless access with WiMax/802.16: current performance benchmarks and future potential. *IEEE Communications Magazine*, *43*(2), 129136.

Berlemann, L., Hiertz, G. R., Hoymann, C., & Mangold, S. (2006, May). *Coexistence and interworking of IEEE 802.16 and IEEE 802.11(e)*. Paper presented at the Vehicular Technology Conference 2006 (VTC 2006), Melbourne, Australia.

Bu, T., Chan, M. C., & Rainjee, R. (2005, March). *Designing wireless radio access networks for third generation cellular networks*. Paper presented at the 24th Annual Joint Conference of the IEEE Computer and Communications Societies.

Cao, M., Ma, W., Wang, X., & Zhang, Q. (2007). Analysis of IEEE 802.16 mesh mode scheduler performance. *IEEE Transactions on Wireless Communications, 6*(4), 1455–1464. doi:10.1109/TWC.2007.348342

Cao, M., Ma, W., Wang, X., Zhang, Q., & Zhu, W. (2005, May). *Modelling and performance analysis of the distributed scheduler in IEEE 802.16 mesh mode*. Paper presented at the MobiHoc '05, Urbana-Champaign, Illinois, USA.

Chang, B. J., Chen, Y. L., & Chou, C. M. (2007). *Adaptive hierarchical polling and cost-based call admission control in IEEE 802.16 WiMAX networks*. Paper presented at the WCNC 2007 Conference.

Chang, B. J., Hsiao, W. C., & Hwang, R. H. (2007, February). *Multiple classes of QoS guarantee in distributed multicast routing*. Paper presented at the 6th International Conference on Advanced Communication Technology.

Chang, B. J., & Liang, Y. H. (2004, October). *Analysis of OVSF code tree for code assignment in WCDMA cellular communications*. Paper presented at the IEEE/ACM MASCOTS 2004.

Chen, I., Guo, Q., & Jiao, W. (2005, December). *An integrated QoS control architecture for IEEE 802.16 broadband wireless access systems*. Paper presented at the Global Telecommunications Conference 2005 (GLOBECOM '05).

Chen, L. W., Tseng, Y. C., Wang, D. W., Wang, Y. C., & Wu, J. J. (2009). Exploiting Spectral Reuse in Routing, Resource Allocation, and Scheduling for IEEE 802.16 Mesh Networks. *IEEE Transactions on Vehicular Technology, 58*(1), 301–313. doi:10.1109/TVT.2008.923685

Chen, Y. W., & Hsieh, F. Y. (2007, March). *A cross layer design for handoff in 802.16e network with IPv6 mobility.* Paper presented at the WCNC 2007, Kowloon, Hong Kong.

Chou, C. L., Lin, J. C., & Liu, C. H. (2008, March). *Performance evaluation for scheduling algorithms in WiMAX network*. Paper presented at the 22nd International Conference on Advanced Information Networking and Applications.

Dastis, V., Hollick, M., Mogre, P. S., Schwingenschlogl, C., & Steinmetz, R. (2006, May). *Performance analysis of the real-time capabilities of coordinated centralized scheduling in 802.16 mesh mode*. Paper presented at the Vehicular Technology Conference 2006 (VTC 2006-Spring).

De Rango, F., Malfitano, A., & Marano, S. (2006, November). *PER evaluation for IEEE 802.16 - SC and 802.16e protocol in HAP architecture with user mobility under different modulation schemes*. Paper presented at the Global Telecommunications Conference 2006 (GLOBECOM '06).

De Rango, F., Malfitano, A., & Marano, S. (2007, December). *Parametric Markov chain model in HAP architecture with IEEE 802.16 protocol.* Paper presented at the Int. Symposium on Wireless Personal Multimedia Communications, Jaipur, India.

De Rango, F., Malfitano, A., & Marano, S. (2008, November). *Performance comparison of Markov chain based models in a IEEE 802.16e scenario*. Paper presented at the MILCOM 2008, San Diego.

De Rango, F., Malfitano, A., & Marano, S. (2009). *Instant weighed probability model to guarantee QoS in IEEE 802.16e scenario*. Paper presented at the WCNC 2009, Budapest, Hungary.

Diamond, J., Hossain, E., & Niyato, D. (2007). IEEE 802.16/WiMAX-based broadband wireless access and its application for telemedicine/e-health services. *IEEE Wireless Communications, 14*(1), 72–83. doi:10.1109/MWC.2007.314553

Gakhar, K., Gravey, A., & Leroy, A. (2005, October). *IROISE: A new QoS architecture for IEEE 802.16 and IEEE 802.11e interworking*. Paper presented at the 2nd International Conference on Broadband Networks (BroadNets 2005), Boston, MA.

Ganz, A., & Wongthavarawat, K. (2003, October). *IEEE 802.16 based last mile broadband wireless military networks with quality of service support.* Paper presented at the Military Communications Conference, MILCOM 2003.

Geetha, S., & Jayaparvathy, R. (2007, November). *Resource allocation and game theoretic scheduling in IEEE 802.16 fixed broadband wireless access systems*. Paper presented at the 4th International Conference Innovations in Information Technology, Dubai.

Gheorghisor, I., & Leung, K. H. (2008, May). *Broadband wireless networks for airport surface communications.* Paper presented at the Integrated Communications, Navigation and Surveillance Conference (ICNS 2008), Bethesda, MD.

Hempel, M., Sharif, H., Wang, W., Mahasukhon, P., & Zhou, T. (2008, May). *Throughput vs. distance tradeoffs and deployment considerations for a multi-hop IEEE 802.16e railroad test bed.* Paper presented at the Vehicular Technology Conference (VTC Spring 2008), Singapore.

Hossain, E., & Niyato, D. (2006). Queue-aware uplink bandwidth allocation and rate control for polling service in IEEE 802.16 broadband wireless networks. *IEEE Transactions on Mobile Computing*, *5*(6), 668–679. doi:10.1109/TMC.2006.85

Hossain, E., & Niyato, D. (2007). QoS-aware bandwidth allocation and admission control in IEEE 802.16 broadband wireless access network: A non-cooperative game theoretic approach. *Computer Networks: The International Journal of Computer and Telecommunications Networking*, *51*(11), 3305–3321.

Hossain, E., & Niyato, D. (2007). Radio resource management games in wireless networks: an approach to bandwidth allocation and admission control for polling service in IEEE 802.16. *IEEE Wireless Communications*, *14*(1), 27–35. doi:10.1109/MWC.2007.314544

Hossain, E., & Niyato, D. (2007). Radio resource management games in wireless networks: an approach to bandwidth allocation and admission control for polling service in IEEE 802.16. *IEEE Wireless Communications Journal*, *14*(1), 27–35. doi:10.1109/MWC.2007.314548

IEEE802.16-2004. (2004). IEEE Standard for Local and metropolitan area networks Part 16: Air Interface for Fixed Broadband Wireless Access Systems.

IEEE802.16e-2005. (2005). IEEE Standard for Local and metropolitan area networks Part 16: Air Interface for Fixed and Mobile Broadband Wireless Access Systems Amendment for Physical and Medium Access Control Layers for Combined Fixed and Mobile Operation in Licensed Bands.

IEEE 802.16 Amendment 4: mobility enhancements. (2007). *Draft amendment to IEEE standard for local and metropolitan area network.* IEEE 802.16e-03/07.

Kaloxylos, A., Passas, N., Salkintzis, K., & Triantafyllopoulou, D. K. (2007). A heuristic cross-layer mechanism for real-time traffic over IEEE 802.16 networks. *International Journal of Network Management*, *17*(5), 1–14.

Kaloxylos, A., Passas, N., & Triantafyllopoulou, D. K. (2007, April). *A cross-layer optimization mechanism for multimedia traffic over IEEE 802.16 networks.* Paper presented at the European Wireless 2007.

Kim, P. S., & Kim, Y. J. (2006). New authentication mechanism for vertical handovers between IEEE 802.16e and 3G wireless networks. *International Journal of Computer Science and Network Security*, *6*(9B), 138–143.

Kuo, G. S., & Yao, H. J. (2006 October). *An integrated QoS-aware mobility architecture for seamless handover in IEEE 802.16e Mobile BWA Networks*. Paper presented at the Military Communications Conference 2006 (MILCOM 2006).

Kwon, D. H., Park, J., & Suh, Y. J. (2006, September). *An integrated handover scheme for fast mobile IPv6 over IEEE 802.16e systems.* Paper presented at the Vehicular Technology Conference 2006 (VTC-2006 Fall), Montreal, Canada.

Lee, S., Narlikar, G., Pal, M., Wilfong, G., & Zhang, L. (2006, April). *Admission control for multihop wireless backhaul networks with QoS support.* Paper presented at the IEEE WCNC 2006, Las Vegas NV.

Li, Q., Lin, Z., Liu, F., Tao, J., & Zeng, Z. (2005, July) *Achieving QoS for IEEE 802.16 in Mesh Mode.* Paper presented at the 8th Internal Conference on Computer Science and Informatics, Salt Lake City, Utah.

Lin, Y. H., Mai, Y. T., & Yang, C. C. (2007, February). *Cross-layer QoS framework in the IEEE 802.16 network.* Paper presented at the 9th International Conference on Advanced Communication Technology.

Matolak, D. W., Sen, I., & Wang, B. (2007, April). *Performance of IEEE 802.16 OFDMA Standard Systems in Airport Surface Area Channels.* Paper presented at the Integrated Communications, Navigation and Surveillance Conference.

Nakhjiri, M. (2007, September). *Use of EAP-AKA, IETF HOKEY and AAA mechanisms to provide access and handover security and 3G-802.16M interworking.* Paper presented at the 18th Annual IEEE International Symposium on Personal, Indoor and Mobile Radio Communications (PIMRC'07).

Nie, J., Wen, J., & Zeng, L. (2007, November). *A bandwidth based Adaptive fuzzy logic handoff in IEEE 802.16 and IEEE 802.11 hybrid networks.* Paper presented at the International Conference on Convergence Information Technology.

Passas, N., Salkintzis, A. K., & Xergias, S. (2008). Centralized resource allocation for multimedia traffic in IEEE 802.16 mesh networks. *Proceedings of the IEEE, 96*(1), 54–63. doi:10.1109/JPROC.2007.909929

Tian, C., & Yuan, D. (2007, September). *A novel cross-layer scheduling algorithm for IEEE 802.16 WMAN.* Paper presented at the I International Workshop on Cross Layer Design (WCLD '07).

ADDITIONAL READING

Agharebparast, F., Alnuweiri, H. M., Fallah, Y. P., Leung, V. C. M., & Minhas, M. R. (2008). Analytical modeling of contention-based bandwidth request mechanism in IEEE 802.16 wireless networks. *IEEE Transactions on Vehicular Technology, 57*(5), 3094–3107. doi:10.1109/TVT.2007.914474

Bae, J., Peterson, R., Berryl, R., Honig, M. L., & Visotsky, E. (2008). *On the uplink capacity of an 802.16j system.* Paper presented at the Wireless Communications and Networking Conference, WCNC 2008.

Barka, E., Chamas, H., & Shuaib, K. (2008, November). *Impact of IPSec on the Performance of the IEEE 802.16 Wireless Networks.* Paper presented at the New Technologies, Mobility and Security (NTMS '08).

Boavida, F., Curado, M., Fontes, F., Leao, G., Neves, P., Palma, D., et al. (2008, May). *The cost of using IEEE 802.16d dynamic channel configuration.* Paper presented at the IEEE International Conference on Communications, ICC '08.

Chen, K. C., & De Marca, J. R. B. (2008). *Mobile WiMAX.* Hoboken, NJ: John Wiley & Sons Inc.

Choi, S., & Choi, Y. (2007, May). *Service charge and energy-aware vertical handoff in integrated IEEE 802.16e/802.11 networks.* Paper presented at the 26th IEEE International Conference on Computer Communications (INFOCOM 2007).

Das, D., Kalle, R. K., & Lele, A. (2007, October). *On the performance of triple play over 802.16e based networks for rural environments.* Paper presented at the Asia-Pacific Conference on Communications (APCC 2007).

Das, D., Panigrahy, D., Raj, M., Ritesh Kumar, K., & Vatsa, O. J. (2007, January). *Adaptive Power Saving Algorithm for Mobile Subscriber Station in 802.16e*. Paper presented at the 2nd International Conference on Communication Systems Software and Middleware (COMSWARE 2007).

Djukic, P., & Valaee, S. (2007, June). *Towards Guaranteed QoS in Mesh Networks: Emulating WiMAX Mesh over WiFi Hardware*. Paper presented at the 27th International Conference on Distributed Computing Systems Workshops (ICDCSW '07).

Eastwood, L., Gupta, V., Migaldi, S., & Xie, Q. (2008). Mobility using IEEE 802.21 in a heterogeneous IEEE 802.16/802.11-based, IMT-advanced (4G) network. *IEEE Wireless Communications*, *15*(2), 26–34. doi:10.1109/MWC.2008.4492975

Fang, Y., & Zhou, Y. (2006, October). *Security of IEEE 802.16 in mesh mode*. Paper presented at the Military Communications Conference, MILCOM 2006.

Hollick, M., Mogre, P. S., Schott, C., & Steinmetz, R. (2007, June). *Slow and Steady: modelling and performance analysis of the network entry process in IEEE 802.16*. Paper presented at the Fifteenth IEEE International Workshop on Quality of Service.

Hongcheng, Z., Junkai, Z., Suili, F., & We, Y. (2008, December). *MAC performance evaluation of IEEE 802.16j*. paper presented at the International Symposium on Information Science and Engineering (ISISE '08).

Jing, X., Mau, S. C., Matyas, R., & Raychaudhuri, D. (2005, December). *Reactive cognitive radio algorithms for co-existence between IEEE 802.11b and 802.16a networks*. Paper presented at the Global Telecommunications Conference (GLOBECOM '05).

Kim, E., Kim, K. S., & Kim, W. K. (2008, May). *Location Aided Location Update in IEEE 802.16e Wireless MANs*. Paper presented at the Vehicular Technology Conference (VTC Spring 2008).

Leung, K. K., Mukherjee, S., & Rittenhouse, G. E. (2005, August). *Protocol and control mechanisms to save terminal energy in IEEE 802.16 networks*. Paper presented at the Pacific Rim Conference on Communications, Computers and signal Processing (PACRIM 2005).

Matolak, D. W., Sen, I., & Wang, B. (2007, September). *Performance evaluation of 802.16e in vehicle to vehicle channels*. Paper presented at the 66th Vehicular Technology Conference (VTC-2007 Fall).

Osborne, M. J. (2003). *An introduction to game theory*. UK: Oxford University Press.

Patil, B. (2008). *WiMAX: End to end network architecture*. Hoboken, NJ: John Wiley & Sons Inc.

Robertazzi, T. G. (2000). *Computer Networks & Systems: Queuing theory and performance evaluation*. Berlin: Springer Verlag.

Shida, L., & Zisu, L. (2008, December). *An effective admission control for IEEE 802.16 WMN*. Paper presented at the Second International Symposium on Intelligent Information Technology Application.

Silvanandam, S. N., Sumathi, S., & Deepa, S. N. (2006). *Introduction to fuzzy logic using Matlab*. Berlin: Springer-Verlag.

So, J. W. (2008). Performance analysis of VoIP services in the IEEE 802.16e OFDMA system with inband signaling. *IEEE Transactions on Vehicular Technology*, *57*(3), 1876–1886. doi:10.1109/TVT.2007.909261

Teyao, C. D. (2007, January). *On the analysis of using 802.16e WiMAX for point-to-point wireless backhaul*. Paper presented at the IEEE Radio and Wireless Symposium, 2007.

Yan, J., Ryan, M., & Power, J. (1995). *Using fuzzy logic: Towards intelligent systems*. Upper Saddle River, NJ: Prentice-Hall.

Chapter 5
Quantifying Operator Benefits of Wireless Load Distribution

S. J. Lincke
University of Wisconsin-Parkside, USA

J. Brandner
University of Wisconsin-Parkside, USA

ABSTRACT

Although simulation studies show performance increases when load sharing wireless integrated networks, these studies assume a limited, defined configuration. Simulation examples of load sharing consider only performance of specific scenarios, and do not estimate capacity or other benefits for a generic network. This study discusses other potential benefits of a load shared network, such as flexibility, survivability, modularity, service focus, quality of service, and auto-reconfigurability. We evaluate these other benefits by developing mathematical models and measurements to quantify a set of potential benefits of load sharing. In addition, we consider capacity considerations against a best-case model. Varied overflow algorithms are then simulated assuming standard HSPA+ and WLAN data rates. The results are compared to the estimated and best-case performance metrics.

1. INTRODUCTION

As a number of wireless networks, such as the cellular networks (e.g., GSM, GPRS, EDGE, UMTS, HSDPA), wireless local area networks (IEEE 802.11a/b/g), and wireless broadband networks (WiMAX) all become deployed, integrating these networks in order to overflow traffic between them makes sense. A number of papers have shown the benefits of load sharing traffic between various wireless networks from a capacity performance perspective. However, other benefits also exist for load sharing. This paper investigates a broad set of potential benefits, as well as methods to quantify or measure these benefits.

Mobile terminals (MTs) benefit from 'Always Best Connected' service, but vertical handovers across diverse network can also be used by operators to load share traffic between networks. A number of papers have focused on the performance improvements of load sharing or load balancing traffic between diverse Radio Access Networks

DOI: 10.4018/978-1-61520-680-3.ch005

(RANs). While improved performance is an interesting benefit, other benefits should also be considered, including modularity, survivability, flexibility, auto-reconfigurability, and quality of service (QoS). In some cases, load sharing will not be advantageous, when considering the issue of service focus. This paper considers these other benefits as well.

Network operators who wish to implement integrated networks need to conduct a feasibility or cost/benefit analysis of how such network integration would perform. Since complex networks are diverse in function, nature, and technology, for ease of use these metrics should be relatively quick to calculate and independent of technology. Our proposed model for performance offers simple but approximate metrics that are easier to work with than complex simulations or analytic models, and help to lead to a generalized understanding of this complex problem. The model is analytic and independent of any particular technology, but depends instead on statistical averages, which are provided by commercial or research literature. Our simulations also are not physical, but depend on reported statistical averages, and are implemented on a discrete event simulator. The particular statistical averages used in this paper involve an HSPA+ cell overflowing to a WLAN.

In literature it is accepted that the MS selects its preferred network. However, in current implementation, the cellular user prioritizes their preferred Public Land Mobile Network (PLMN). This proposal assumes that the user selects the service provider, and the service/network provider has final control over the radio network to serve the user: be it cellular, WLAN, WiMAX, etc. This control occurs via vertical handover (i.e., between RANs). Secondly, we assume the Common Radio Resource Management (CRRM) implementation is distributed, instead of centralized, preventing bottleneck and single source failure problems, and reducing network communications to support the algorithm.

Since actual results depend on the overflow algorithms selected, we include a variety of overflow algorithms. We propose a best-case performance model, then evaluate proposed algorithms against this potential performance. In other papers, we have proposed a Substitution technique to achieve high levels of load sharing. In this paper we show how Substitution can achieve best-case performance, while being relatively easy to implement.

In section 2 we provide background information. In section 3 we define performance metrics and their models. Section 4 provides two packet-oriented simulations comparing overflow algorithms against an optimum capacity model. Section 5 considers industry trends and our simulation results to evaluate the remaining metrics discussed in the paper. Section 6 concludes the paper.

2. BACKGROUND

CRRM studies have generally focused on capacity, and generally measure differences in blocking, packet drop rates, and throughput (Lampropoulos et al., 2006; Lincke, 2005; Perez-Romero et al., 2006; Song et al., 2007). We propose how CRRM can be applied to various business scenarios in (Lincke, 2007), but have not previously quantified these diverse benefits with metrics.

Quantification or qualification of CRRM networks has focused on how to characterize the CRRM network. It is likely that an operator may simultaneously support Global System for Mobile Communication/Enhanced Data for Global Evolution (GSM/EDGE), High Speed Packet Access Evolution (HSPA+), and Long Term Evolution (LTE), and could potentially support other WLAN protocols. Because RANs are so diverse, combining them into an integrated network is complex to describe and simulate. Both Serrador et al. (2006) and Gozalvez et al. (2007) show that a full characterization of a simulation requires many levels of scenarios and details as propagation, traffic, equipment, network, and planning, etc. Chen and Chan (2006) characterize geographic traffic coverage and traffic allocation algorithms. Whether the network is loosely- or tightly-coupled and the goal of load balancing or load sharing will affect the implementation (Song et al., 2007). Radio Access Technologies have become increasingly complex, and the integration of these networks increases complexity even further.

We manage this complexity by working with statistical averages for radio access network capacities. Table 1: "Peak Traffic for Various Radio Access Technologies" reflects capacities for commercial wireless radio access networks, from a variety of sources. In some cases, Table 1 provides results from published simulations. When conflicting values were found from different sources, conservative values were taken. The UMTS figures are taken from an actual "Operator X" in Hong Kong (Tan et al, 2007). The authors found varying total capacities for different cell sites for this operator from 500-2400 kbps downlink, across the country. While Table 1 does reflect accurate results for some defined scenarios, real capacities often vary based on MT location and speed, data rates, radio equipment sophistication, and other interference-related criteria. Because of this complexity, simulating or implementing a solution provides results only for the specific conditions of the test.

To control CRRM complexity, we standardize capacity by defining some basic statistics used in the model, including:

- **Average peak capacity (c):** The average composite throughput that a RAN's cell achieves at full capacity when servicing its most efficient service.
- **Average peak capacity for service X (p_x):** The average composite throughput for the cell at capacity, when carrying only Service X.
- **Data rate for service X:** Average data rate in bps that a user experiences for a particular service.
- **Effective data rate for service X (f_x):** Percent of p_x that a single session utilizes, or the effective data rate in bps of one session relative to c.

In Table 1 we quantify how some services are more efficient carried on some RANs versus other RANs, via the f_x statistic. These statistics will be used in our simulations and metrics below.

3. Model of Load Sharing Benefits

Load distribution may offer a number of benefits. We define these benefits, before discussing their metrics below:

- **Capacity/availability:** Engineering traffic across networks increases efficiency of scale, thereby maximizing frequency and BS equipment investments.
- **Flexibility:** Flexible networks support a wide variance in traffic loads for different services that arise due to time of day, day of week, special events/emergencies, and over time.
- **Survivability:** When partial network equipment failure occurs, recovering lost sessions towards remaining equipment can increase survivability, reputation, and income.
- **Modularity:** The variety of radio access technologies can be viewed as radio resources that can be mixed and matched to support capacity, instead of the view that new technologies must replace old technologies.
- **Service focus:** Operators should prioritize and support those services that serve the business model.
- **Quality of service (QoS):** Sharing network loads can lead to an optimum balance of quality of service metrics, such as blocking, call dropping, packet delay, bandwidth degradation, and bit error rate.
- **Auto-reconfigurability:** In order to take advantage of flexibility, integrated networks must automatically and dynamically adjust to traffic patterns.

Table 1. Peak traffic for various radio access technologies

	Speech	Video	Data
GSM/EDGE (3GPP TS 05.02 2001)	8 per transceiver (full rate) 16 per transceiver (half rate)		14.4, 28.8, 32, 43.2, 59.2 kbps per timeslot 32 kbps x 4 timeslots = 128 kbps Peak: 59.2kbps x 4 timeslots = 236.8 kbps
UMTS (actual) (Tan et al, 2007)	54 users x 12 kbps = 648 kbps f_s=2% => 23.3 kbps	18 x 64 kbps = 1152 kbps f_v=5.6% = 70 kbps	4 x 315 kbps = 1260 (downlink) 14 x 57 kbps = 800 (uplink) Range: 501-1500 kbps
HSUPA (Rel. 6) (actual, simulated) (GSM Assoc. 2007; Holma et al, 2007)	82 users (VoIP 12.2 kbps) f_s=1.2% or 44 kbps		Downlink: 3.6 Mbps
HSPA+ (Rel. 7) (simulated) (GSM Assoc. 2007; Holma et al, 2007; 3G Americas, 2006)	120 users (VoIP 12.2 kbps) f_s=0.83% or 60 kbps		Downlink: 7.2 Mbps Microcell: 8-9.2 Mbps (Macrocell, Round Robin Scheduling, 20 active users:) Avg. User: 200 kbps, User Range: 50-700 kbps
LTE (simulated) (Sanchez et al, 2007)			16 users x 3.7 Mbps = 60 Mbps
IEEE 802.11B – DCF (actual) (Chan & Liew, 2007)	12 VoIP users f_s=8% or 458 kbps		11 Mbps (theoretical): 5.5 Mbps each direction
IEEE 802.11 A/G – DCF (actual) (Chan & Liew, 2007)	56-60 VoIP users f_s=1.7% or 466 kbps		54 Mbps (theoretical): 27 Mbps each direction

To model these benefits, we must first describe our assumptions.

Capacity

Capacity quantification provides an estimate of the potential capacity increase that load sharing provides. While it could be measured as a number of QoS metrics specific to services, here we define it as a percentage of additional capacity available to a RAN via overflow to other RANs' cells. The Maximum Capacity Increase Percentage (MCI%) after load distribution is constrained by:

$$MCI\%_1 = \frac{c_1 + c_2 + \ldots + c_n}{c_1} - 1.0 \quad (1)$$

where c_n is the average peak capacity of a cell from RAN n, before load distribution.

Equation (1) assumes that cells 2..n carry no traffic of their own, or that cell one's traffic fully has priority over other cells' traffic. A more realistic 'Potential Capacity Increase Percentage' (PCI%) considers that cell 1's traffic can only overflow into unused or spare radio resources of other cells.

$$PCI\%_1 = \frac{c_1 + s_2 + \ldots + s_n}{c_1} - 1.0 \quad (2)$$

Above, s_n is the average spare capacity on cell n. This depends on the percent utilization (u_n) of radio resources on cell n, compared to the average peak:

$$s_n = (1 - u_n)\, c_n \quad (3)$$

Because capacity can vary based on physical factors, an average capacity in bits per second can be assumed. However, results are more usefully applied by instead considering the c_n and s_n as the average number of sessions the network usually supports.

Equation (2) shows that the greater the value of s_n are relative to c_1, the greater the potential for capacity gains in the integrated network. Thus, overflowing from larger to smaller capacity networks would offer very little benefit, while overflowing from smaller to same sized or larger cells could offer considerable benefit.

In this case of capacity, we will go beyond defining the model to also consider the model's optimized value. The potential increase in capacity is constrained by the ability of traffic to be overflowed or handed over between different network types. The rate of overflow between two RANs depends upon the flexibility rate (F), which is the percentage at which MTs are compatible with the origination (o) and destination (d) RAN types. $F_{o\text{->}d}$ is the intersection of compatibility percentages, including S=Service-compatible, T=Technology-compatible, I=In coverage, M=Member, and P=Speed-compatible:

$$F_{o\text{->}d} = S * T * I * M * P \tag{4}$$

The overflow of cell 1 to cell 2 depends upon the transference of the excess traffic (not carried by cell 1):

$$offered_1 = c_1 + e_1 \tag{5}$$

where e_1 is the excess or potential overflow traffic. We define the Potential Carried (PC) as the most possible traffic carried by a RAN given its ability to overflow to other RANs. PC is a best-case metric, and in its simplest interpretation, PC equals the offered rate, since the best performance a RAN can achieve is obviously to service all traffic offered.

Taking into account the flexibility rate would restrict the ability for sessions to leave the home RAN. If an overflow handover can occur based only on the compatibility of the last arriving session, PC is limited further by the flexibility rate:

$$PC_1(F_{1\text{->}2}, \text{Last Arrival}) = c_1 + F_{1\text{->}2}(e_1) \tag{6}$$

This equation shows that load sharing using this algorithm can achieve at best a linear increase in performance depending on the flexibility rate, between c_1 and $offered_1$.

Better performance can be achieved when flexible sessions are requested to overflow for arriving inflexible sessions. The best performance can be achieved using a Substitution algorithm, which considers all offered sessions for overflow:

$$PC_1(F_{1\text{->}2}, \text{Substitution}) = c_1 + F_{1\text{->}2}(offered_1) \tag{7}$$

with the constraint that PC never exceeds eq. (5), nor the system capacity $c_1 + s_2$. With full load sharing it is assumed that overflow can occur bi- or multidirectionally: the model should work in both directions since both cells often have spare capacity some percentage of the time.

The Potential Carried is a useful statistic to estimate the amount of traffic eligible for overflow (or the offered traffic to the overflow network). For circuit-oriented services, the ErlangB calculation for blocking also requires the number of resources or channels available to the traffic, which we will call Expected Capacity. The Expected Capacity (EC) of cell 1 with full load sharing to the 'spare' resources on cell 2 through cell n is calculated as:

$$EC_1 = c_1 + s_2 + \ldots + s_n \tag{8}$$

From this point of view, taking into account the flexibility rate would restrict access to the overflow RAN. Employing a Last Arrival or Substitution type overflow algorithm would limit the increase of the EC of cell 1, similar to what happened to the PC.

$$EC_1(F_{1\text{->}2}, \text{Last Arrival}) = c_1 + F_{1\text{->}2}(s_2) \tag{9}$$

$$EC_1(F_{1\text{->}2}, \text{Substitution}) = c_1 + F_{1\text{->}2}(c_1 + s_2) \tag{10}$$

with the constraint that EC never exceeds eq. (8).

These equations are estimates and don't take into consideration the impact that interference, data session size, or priority can have. For example, related studies (e.g, Perez-Romero et al., 2006) have shown that by directing particular traffic to particular RANs, interference can be reduced and the 'average' bandwidth of a RAN can increase to approach the technology's theoretical maximum bandwidth. With directed retry it is possible for the average peak capacity of neighbor cells to decrease due to additional interference. In addition, small bandwidth sessions can crowd out higher bandwidth sessions (Lincke, 2005). The priority of services can also modify the performance of the model, by enabling a high-priority service to expand beyond spare capacity.

Capacity quantification has shown that the more capacity that is available on the overflow destination network, the greater the increase in expected capacity. However, Capacity quantification assumes only one point in time, and traffic demands vary by time and day.

Network Flexibility

Network Flexibility describes the traffic overflow relationship that RANs have with each other over time: complimentary or conflicting. It quantifies the change in demand over time for the services of the different RANs. For example, in a cellular environment including Global System for Mobile Communication (GSM) and Universal Mobile Telecommunications System (UMTS), GSM speech may be busy during the busy hour, while UMTS data may be most busy during the afternoon and evening hours.

A correlation coefficient (cc) (Donnelly, 2004) can measure the traffic distributions of the various RANs, to determine if they coincide or compliment each other. The cc could be used to measure the traffic demand for RAN x (D_x) for each of n=24 hours in the day:

$$cc = \frac{n\sum D_1D_2 - (\sum D_1)(\sum D_2)}{\sqrt{[n\sum D_1^2 - (\sum D_1)^2][n\sum D_2^2 - (\sum D_2)^2]}} \quad (11)$$

The cc ranges between +1 and -1, where higher positive numbers represent a higher correlation and less flexibility, and negative numbers represent higher flexibility.

To measure the network flexibility, three factors are important: 1) the correlation in demand for services (cc); 2) the potential capacity for carrying traffic (PCI); and 3) the demand resulting in potential overflow. Demand is important because load sharing is only effective when there is overflowing traffic: measuring the cc is irrelevant when there is little traffic. Thus, it is more effective to measure the correlation only during busy hours, which can be defined to be any hour with a blocking or drop rate exceeding a threshold, for any RAN. The potential demand can be quantified as the number of busy hours for which load sharing is useful (#). Network Flexibility (NF) coefficient can be noted as:

$$NF = (\#, cc) \quad (12)$$

Survivability

Tipper et al. (1999) define survivability of a network as the capability remaining in a network following a network failure. If a failure occurs unexpectedly, sessions may be unavoidably lost. However, if these dropped sessions can be manually resumed successfully, then the shared network is highly survivable. We measure survivability in terms of the percentage of active sessions that would lose and not regain service during an emergency situation.

Survivability requires that sessions on failed equipment have the technological flexibility to move to an alternate, overflow network ($F_{e->o}$), and that spare capacity be available on that overflow network (s_o). A Potential Survivability (PS) metric is defined as the ratio of carrying capacity over the emergency traffic demand. In the case of a full cell failure, the maximum probability of survival is equivalent to the flexibility rate (13). Below, t_e is the traffic demand (during emergency):

$$PS_{cell1}(F_{e->o}, \text{Full Failure}) = \frac{s_o}{t_e * F_{e->o}} \quad (13)$$

In the case of a partial cell failure, such as loss of a GSM transceiver, the maximum survival rate can be achieved by moving flexible sessions to alternate networks and supporting inflexible sessions on surviving transceivers (14). The potential survivability is in this case augmented by the remaining capacity on the cell in emergency. Below, c_e' is the reduced capacity at the emergency cell:

$$PS_{cell1}(F_{e->o}, \text{Partial Failure}) =$$

$$\frac{c_e'}{t_e[(1-F_{e->o})+(1-j_e)(F_{e->o})]} + \frac{c_e'}{t_e[(1-F_{e->o})+(1-j_e)(F_{e->o})]}\frac{s_o}{j_e * t_e * F_{e->o}} + \frac{s_o}{j_e * t_e * F_{e->o}} \quad (14)$$

The first fraction is most important because the cell in emergency must support all inflexible traffic and as much flexible traffic as possible; while the overflow cells need only carry flexible traffic. Survivability Percentage actually measures the survived traffic (t_e') over total emergency traffic for all cells:

$$SP_{RegionN}(\text{FailureY}) =$$

$$\frac{SurvivedTraffic}{TotalTraffic} = \frac{\sum t_e'}{\sum t_e} \quad (15)$$

Resulting capacity issues can be estimated using previous measures. The Network Flexibility coefficient indicates the overall capacity of a network to survive capacity crunches, regardless of network failure timing. The Capacity quantification can indicate the approximate level of carried traffic the survived network would support.

Modularity

Modularity ensures that when additional equipment is needed, the least-cost equipment can be selected. For example, legacy equipment is free and often already installed. To minimize cost of new technology equipment, operators purchase only sufficient equipment to support the demand for the new service to be sold. A 'Building Block' approach recognizes that cellular technologies vary in their capacities that are often measured in bps/Hz/sector or in cost/megabyte (3G Americas, 2006). The cost per megabyte/day is reducing steadily, from several dollars, to at best $1 in 2006, to projected forecasts of $0.10 for future technologies. Because of the differences in cost of coverage between macro, micro, and picocell technologies, a revised Modularity (M) metric considers bps/km^2/cost (where $ represents cost):

$$M(RAT_A) = (\text{bps}, \text{km}^2, \$) \quad (16)$$

For example, network providers can combine RATs in different ways to achieve loading goals, either as 1) a large 2.5G cell with 802.11 WLAN hot spots, or 2) multiple smaller 3G cells, or 3) a medium sized 3G with larger 2.5G cell. Modularity must be considered in tandem with the applications the RAN supports.

With geographic maps, regions can be evaluated for their traffic and technology needs using M. Due to diverse demands for services, data rates, and technology/multimode capabilities, regions can be categorized as to their technological flexibility and quantified indicating approximate required bandwidth per service. A Modularity Demand for Technology (MDT) or Services (MDS) measures the demand for each technology or service, including considering flexible (multimode) traffic:

$$\text{MDT(GSM, UMTS, 802.11B, Flexible)} = (bps_{GSM}, bps_{UMTS}, bps_{WLAN}, bps_{Flex}) \quad (17)$$

$$\text{MDS(speech, 64 kbps, 128 kbps)} = (bps_{speech}, bps_{64}, bps_{128}) \quad (18)$$

The cost of configuring the region requires evaluating each optional configuration:

$$\text{RegionalCost}_{Region1}(\text{OptionX}) = \sum \$Equipment + \$Installation + \$... \quad (19)$$

Chen and Chan (2006) use a weighted BS-MS graph to show how mobiles can map to cells. Optimal techniques to assign equipment to regions are beyond the scope of this paper.

Service Focus

Service Focus puts a price on each service according to its derived benefit and carrying costs. Given limited radio capacity, some services are given priority over others. Telecommunications operators, store/hotel owners, and companies operate their wireless networks for different business ends. For example, a university or coffee shop may find it cost effective to offer free web access, but not free VoIP calls, to customers on a WLAN. The customer too may find it cost effective to place a specific speech call if the call is free, but not otherwise. When traffic is load shared, the issue of where and whether to carry certain services arises. As Table 1 indicates, some sessions are carried most efficiently on specific RANs.

Every session carried offers a derived benefit and cost. The derived benefit is the revenue or business value that carrying the session produces, while the cost is the price of carrying the session. By calculating a margin for each class of service on each RAN, it becomes obvious which session types should be carried where. If the derived benefit is the same regardless of where the session is carried, the carrying cost is minimized by selecting the lowest priced RAN. However, the derived benefit can vary depending on where the session is carried: e.g., a higher bandwidth may attain a higher price. Overflowing sessions may change the margin and not be cost-effective. The carrying cost is calculated by dividing the cost of operating the cell by the effective percentage utilization of that cell by that service.

$$\text{CarryingCost}_{ServiceX} = \frac{CostOfOperatingCell}{t * f} \quad (20)$$

$$\text{Margin}_{RANn,RegionM}(\text{Service}_X, \text{QoS}_Y) = \text{Derived-Benefit} - \text{CarryingCost} \quad (21)$$

In (21), the QoS could reflect the provided bandwidth related to one or more technologies. Table 1 can then be extended to support a cost and derived benefit column.

Quality of Service

Every commercial telecommunications operator is concerned with QoS measures: blocking, packet delay, packet drop rate, bandwidth degradation, bit error rate, and provided data rate. These statistics must be tracked per RAN, per service, and possibly per tier (bronze, silver, gold) in order to determine the effect of load distribution. Load sharing usually results in improved performance, but can result in worse performance for high-bandwidth services (Lincke, 2005). Also, fairness must be considered in overflow. For example, it would not be fair if a dual mode phone is overflowed to a poor-quality network in order to accommodate an arriving single mode session into a high-quality network. Thus, load sharing (as with any telecommunications network management) must be carefully managed to ensure that the results are beneficial and fairly applied.

A QoS tiering system can measure an organization's QoS standard. Table 2 demonstrates example QoS tiers, defined with threshold QoS levels for each tier. The average QoS per hour can be used to determine the QoS tier for each cell. Although some RANs are higher bandwidth, the response time that they provide to a single user for the services they carry may still be comparable to a lower bandwidth network.

The QoS tier is used in combination with the maximum provided user data rate, since cellular networks may have equivalent blocking rates, but provide different user data rates, and still achieve an identical QoS tier.

Auto-Reconfigurability

Auto-reconfigurability measures the ability of the integrated network to automatically and dynamically load share to provide high-quality service, regardless of the carrying RAN. Auto-reconfigurability is advantageous to minimize traffic engineering planning costs and rapidly alleviate spikes due to external emergency conditions. The Auto-Reconfigurability Metric (ARM) measures the difference in QoS Tiers between the cells with the highest and lowest tier values (MaxT and MinT), for a given time period. Since higher numbers represent greater auto-reconfigurability, ARM is biased by the number of tiers:

$$\text{ARM} = \text{NrTiers} - (\text{MaxT} - \text{MinT}) \tag{22}$$

4. CAPACITY OPTIMIZATION SIMULATION

In this section we perform two simulations that evaluate how various overflow algorithms perform compared to the best Potential Carried, defined by (6) and (7). In order to determine the effectiveness of the overflow algorithms, we force considerable overflow by overloading one network and underloading the other. Realistically, this would occur in cases of modularity due to cost effectiveness, or if an emergency condition resulted in abnormal loads (e.g., a road accident results in traffic backups and high speech traffic levels).

Two of the algorithms use a Substitution policy, which allows any flexible session to overflow to another network to accommodate an arriving inflexible session. Three algorithms use a Last-Arrival (LA) policy which can only overflow the last arriving session if it is flexible (dual mode), when the network to which it's offered is already at capacity. Each overflow policy is combined with a return policy to generate five algorithms:

- **LA late return (LT):** In this LA algorithm, overflowed sessions remain on the overflow RAN until the overflow RAN overflows these sessions back.
- **LA early return (RE):** LA overflowed sessions return as soon as possible to their home network if/when resources become available.
- **LA no return (NO):** LA overflowed sessions never return to their home network.
- **Substitute late return (LA):** In this Substitution algorithm, overflow sessions

Table 2. QoS tier table

	Conversational	Packet Delay	Packet Drop Rate
Tier0	Block%<=1%	< 100 ms	PDR<=2%
Tier1	Block%<=2%	< 250 ms	PDR<=4%
Tier2	Block%<=5%	< 500 ms	PDR<=8%
Tier3	Block%>5%	> 500 ms	PDR>8%

only return if/when the other network overflows.

- **Substitute early return (SB):** Sessions overflow using Substitution, and return as soon as possible to their home network, when resources become available.

HSPA+ -> WLAN Model

For Modularity purposes, a network operator is considering expanding a mature HSPA+ microcell with an IEEE 802.11G WLAN network in a business hotspot. The Service Focus is such that the management of the hotspot desires higher bandwidth, predominantly WLAN access, and high Survivability.

This test simulates two cells handling queued packet data of the 'Interactive' class. The WLAN assumes a single channel operating at 27 Mbps in the downlink direction. The HSPA+ microcell network supports 16 simultaneous users at 500 kbps for a combined downlink throughput of 8 Mbps (from Table 1). The maximum queue length is set to an equivalent of 8 users or 4 megabits of data or a half second delay. Packets will only overflow to another network if the home network is transmitting at capacity and its queue is full. No real-time sessions are supported in this test (but are instead assumed to be carried by the GSM cellular network).

The HSPA+ network is offered 14 Mbps, causing it to overflow to the proposed WLAN. Since no WLAN currently exists, its offered traffic should be zero, but a potential new market for WLAN data is estimated at 2 Mbps. Due to the large disparity between the two networks' capacities, the MCI% of the HSPA+ is 338% and the PCI% is likewise high at 313%. The main issue is whether the flexibility rate (i.e., dual mode and in-coverage of WLAN) is sufficient to load share traffic between the networks.

Our analytic model is based on the Potential Carried equations, (6) and (7), reflecting expected best potential performance. All analytic and simulation models assume that each resource approximately represents twice the Mbps that the network supports in the downlink direction. Thus, we assume one session uses 0.5 Mbps theoretical capacity, with exponential inter-arrival and service times.

Since there is a difference in the number of channels and data rates of the HSPA+ and WLAN networks, we performed a theoretical model before our actual one. In the theoretical model (test 1), we assumed that the WLAN supports 54 simultaneous channels at the same rate of the HSPA+ network: 500 kbps, offering the total 27 Mbps. In the actual model (test 2), we assume the WLAN has one channel operating at 27 Mbps.

To ensure accuracy of simulation results, we also implemented a Markov Chain state model for the two tests, as shown in Figures 1 and 2. These models assume the WLAN carries no traffic of its own, but is only used for HSPA+ overflow (justified since the WLAN home traffic is negligible.) Both figures assume that the states (0,0,0) to (16,0,0) are actively transmitting states, while states (16,0,0) to (16,8,0) reflect added queued calls. States at (17,8,x) and above reflect overflow calls. Thus, the three state variables reflect the total number of active sessions, the number of queued sessions, and the number of overflow sessions, respectively. Flexibility is set to 100%, indicating full load shared performance. The Markov Chain models reflect an Early Return algorithm. Transitions model arrival @(Arr) and departure @(Dpt) rates. The Graphic Markov Chain Modeler tool is freely accessible at www.uwp.edu/staff/lincke/Markov, (Lincke, 2009).

HSPA+ -> WLAN Results and Analysis

Figure 3 shows the simulation results for all overflow algorithms and the theoretical model. PC_{Sub} is the top line, reflecting best-case performance for different Flexibility rates. The Substitution Late Return algorithm comes closest to achieving that

Figure 1. Markov State Diagram for HSPA-WLAN overflow, 54 channels

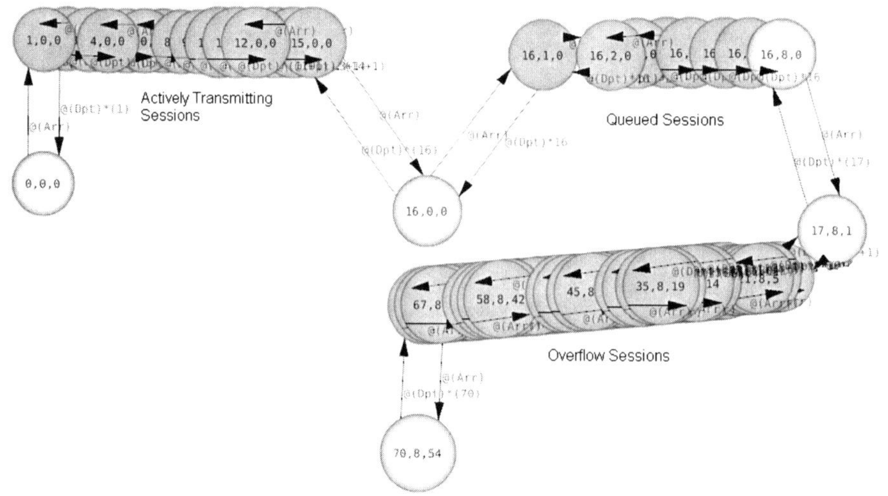

Figure 2. Markov State Diagram for HSPA-WLAN overflow, 1 channel

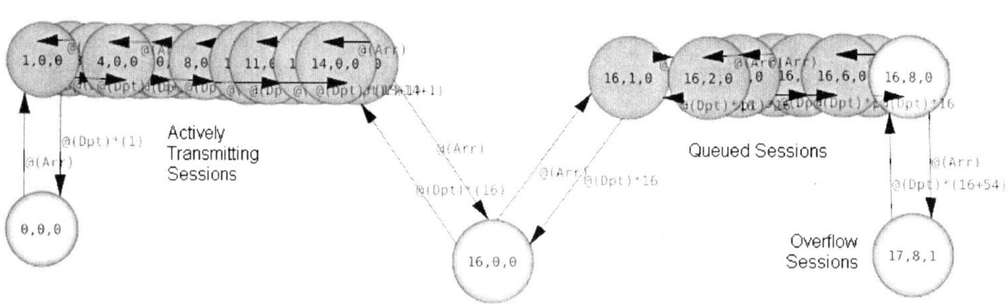

line. The PC_{LA} shows a linear increase in capacity, coinciding with the Last Arrival No Return and Late Return line.

The Late Return algorithm performs very well, but apparently not as well as the best-case (although very nearly so). Since high flexibility rates do match the best-case carried traffic statistic of full load sharing, the difference must be due to an insufficient availability of flexible sessions at the lower flexibility rates, before full load sharing. This may occur because flexible sessions can overflow and be served first, whereas inflexible sessions remain queued for processing. This would result in the rate of flexible sessions queued or served at any point in time being lower than the flexibility rate of the incoming traffic. This is also indicated by the PC_{LA} line coinciding with the Last Arrival No Return and Late Return lines, since the Last Arrival algorithm overflow only depends on the status of the last arrival, and cannot be impacted by queuing. Figure1 not only demonstrates the effectiveness of each algorithm, but also demonstrates that the PC equations accurately predict general performance, at least when channel sizes are consistent.

In our second test, we consider that a WLAN has one channel operating at 27 Mbps, much faster than any of the HSPA+ channels. Since there is only one channel, operating 54 times as fast, Figure 2 shows results for PC_{Sub} assuming 1 and 54 serving channels. The algorithms do not perform to the best-case PC_{Sub} for 54 channels, but do perform much better than PC_{Sub} for one. This implies that our first test configuration per-

Figure 3. Total Carried HSPA+ versus WLAN (56 channels)

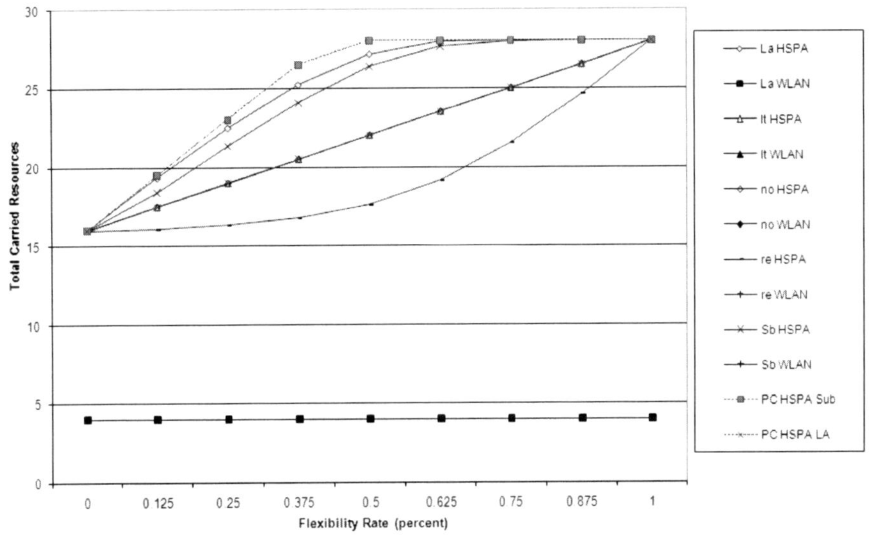

forms better than the second configuration. This is surprising, since it makes sense that one fast communications line provides an average faster response time than multiple slower lines of the same combined speed.

The Markov Chain model confirms simulation results for both tests. Using the Figure 2 one-channel model, overflow traffic is carried 14.6% of the time, resulting in an effective carrying capacity of 7.9 overflow calls for a total of 23.9 total connections. This result is only slightly above the 23.8 result provided by simulation in Figure 4. Likewise, the Figure 1 Test 1 model results in better performance at 28 total connections carried.

Since there is no error in our simulation model, we must accept that the 54-channel model performs better than the 1-channel cell operating 54 times faster. The potential problem is likely with the overflow algorithm, and not the PC_{Sub} best-case equation. One problem with our algorithm is that overflow can only occur if an overflow channel is available for transmission. Thus, when the one-channel WLAN is currently transmitting, overflow arrivals are being rejected. The algorithm can be optimized to achieve better performance by adding queuing on the overflow cell.

It is quite clear that our current algorithms do not approach PC_{Sub} for a one-channel WLAN overflow configuration (although they do for a multi-channel overflow cell). Our future work will involve defining further enhancements to our algorithms, to achieve near-best-case performance, at least to the performance level of the 54-channel model.

5. CONSIDERATION AND INTERPRETATION OF METRICS

While the previous section focused on how the algorithms match up against capacity potential, this section considers the effects of the other benefits and metrics of load sharing. We first consider how published material from the real world impacts the load shared integrated network, before considering our example scenario.

Real World Considerations

The flexibility rate impacts the survivability and capacity. In 2008, there were 160 million W-CDMA subscriptions, with GSM customers "approaching" 3 billion. Thus, approximately

Figure 4. Total Carried HSPA+ versus WLAN (one channel)

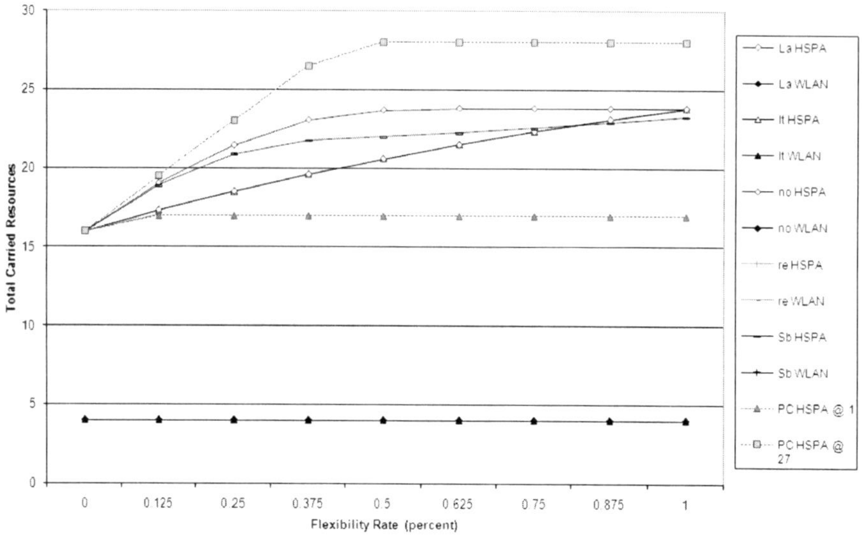

5-6% of phones supported W-CDMA, although this may vary by area. However, if 5-6% of phones are generating sessions with twenty times as much bandwidth as speech calls, then the carrying needs of the data network approach that of the speech network. GSM Arena reports that the most popular HSPA handsets in the summer of 2008 support GSM, EDGE, GPRS, and HSDPA, with some also supporting IEEE 802.11B/G. Thus, the flexibility rate of handsets that can overflow from GSM to a UMTS network is low, while overflow in the other direction is very high. Transfer from an HSPA network to a WLAN is feasible with some HSPA handsets and many laptops (with HSPA network cards), but transfer from WLAN to HSPA is likely very low. With current low rates of HSPA users, GSM and WLAN networks seem feasible to increase survivability.

Network flexibility considers how mismatches in time for the demand for diverse services and networks increase capacity. While it is true that rush hours and lunch hours are likely to be high for speech, data is likely to be high during the work and evening hours. Thus, GSM and HSPA networks are likely to be very compatible (i.e., with low correlation). WLAN networks seem less flexible, since their small cell footprints can only be used when customers are in range. WLANs in business areas are likely to be used only during work hours, while those in residential areas will be used mainly in the evenings and weekends. Thus, for short-term capacity planning, the WLAN should be seen mainly as a source of overflow traffic for network flexibility purposes. As a long term solution, the WLAN serves as a modular 'expansion unit' for heavily-loaded cellular networks during peak hours.

Modularity considers how to most cost-effectively allocate available bandwidth (in the long term). In sales brochures, the cost of a network is measured in $ per gigabyte (GB) and reportedly decreases as new technologies support higher bandwidths. In 2006, one vendor announced reaching 400 GB per day, achieving $1 per GB (3G Americas, 2006). Base stations support multiple protocols including GSM, UMTS/HSPA, and in the future, the Long Term Evolution (LTE) protocol. The System Architecture Enhancement (SAE) integrates WLANs into the cellular network, via an SAE anchor to support mobility. Thus, it is planned that diverse technologies (including WiMAX) can be integrated into a single integrated network, and implemented according to modularity or service focus benefits.

Metrics for the Example Scenario

In addition to modeling the Capacity, we may consider other metrics for our scenarios. We will consider two flexibility rates: 12.5% and 50% (for a new and mature implementation.) Our WLAN configuration may be one Modularity technique in achieving extra capacity (although we do not consider here other configurations.) From a Service Focus perspective, we can consider that the WLAN is being marketed to a shopping mall or doctors' medical building. The mall or office management may or may not desire to share its public access by manual selection. The network provider may offer a reduced fee if cellular overflow traffic is carried by the semi-private network within an agreed-upon limit.

Survivability. As part of the semi-private network agreement, it is feasible that in case of an emergency, either network would attempt to carry the other network's traffic. For survivability, the WLAN and cellular equipment are distinct – if the cellular base station fails to operate, the WLAN still provides ongoing service to an important customer – and vice versa. Assuming an HSPA+ cell failure and a 12.5% or 50% flexibility rate, the HSPA+ would overflow emergency traffic levels of 1.75 or 7 Mbps traffic, respectively, to the WLAN. The Potential Survivability (PS) (13) becomes 23/1.75 = 13.14 and 23/7 = 3.29, which indicates in both cases that HSPA+ flexible traffic has well over 100% chance of being carried on the lightly-loaded overflow network. The percent of surviving traffic (15) would be equal to the Flexibility rate: 12.5% and 50%.

If the WLAN fails (assuming one WLAN), the survivability rate depends on the Flexibility rate, or dual mode capability, of the office traffic. Since the spare capacity of the cellular network is 0%, the PS is also 0%. Thus, the network operator must plan for service degradation to create spare capacity for the emergency WLAN traffic. The combined traffic totals 16 Mbps on the HSPA+ network that supports approximately 8 Mbps. Each user will get a maximum throughput of half the regular rate, or approximately 250 kbps.

QoS quantification and auto-reconfigurability statistics for our tests can be derived from our tests. The HSPA+ packet drop rate (PDR) at the preferred rate of 500 kbps per user without WLAN load sharing is 43%. This unacceptable rate is well beyond Table 2's standard for even the worst tier, Tier 3. In the hypothetical test where the overflow WLAN has 54 channels, the PDR equals 0% at a 62.5% flexibility rate. At 12.5% flexibility, the PDR for Substitution Late Return is 31% (Tier 3), while at 50% flexibility the PDR drops to 3.12%, or Tier 1. For the single-channel WLAN test, the PDR only achieves 15%, which is still in the Tier 3 category. Therefore, the Auto-Reconfigurability Metric (22) reflects a change of two tiers with load sharing, but only with a 50% flexibility rate, using the Substitution Late Return algorithm, in the 54-channel WLAN model.

6. CONCLUSION

Metrics were introduced to quantify seven different benefits of load sharing. A best-case capacity model was provided, which was compared against a number of overflow algorithms. Many of the seven categories of metrics, as well as the best-case capacity model, were evaluated against two scenarios involving two radio access networks which overflow sessions. In one scenario, the Substitution Late Return algorithm was shown to achieve near-best-case performance, and the performance difference compared to the theoretically potential expectations was explained. In the second scenario, best-case results were not achieved, but algorithm enhancements were proposed to improve performance similar to the first scenario. Our results also show that load sharing a HSPA+ network with a WLAN is not effective until moderate flexibility rates can be achieved. We also reviewed pertinent information from selected recent network literature, related to the integrated network implementation.

Our quantification metrics have the strength of being relatively simple to estimate performance for a range of situations. However their simplicity is also a weakness in that the models don't predict the outcome of every situation exactly. Factors such as the overflow and queuing algorithm employed, the actual network hardware implementation, and the effects of the other metrics quantified in this paper, can alter the actual statistics for a given system. However, the capacity and other metrics herein defined offer a good benchmark for the estimation of a variety of benefits of implementing an integrated network.

7. REFERENCES

3GAmericas. (2006). *Mobile Broadband: The Evolution of UMTS/HSPA, 3GPP Release 7 and Beyond.* Retrieved June 30, 2008, from http://www.3gamericas.org

Arena, G. S. M. (2009). Retrieved June 30, 2008 from http://www.gsmarena.com

Association, G. S. M. (2007, February). *HSPA mobile broadband today*. Retrieved June 30, 2008, from http://www.gsmworld.com/hspa

Chan, A., & Liew, S. C. (2007). VOIP Capacity over Multiple IEEE 802.11 WLANs. In *IEEE International Conf. on Communications (ICC)* (pp. 3251-3258). Washington, DC: IEEE.

Chen, B. B., & Chan, M. C. (2006). Resource Management in Heterogeneous Wireless Networks with Overlapping Coverage. In *IEEE Int. Conf. on Communications Systems Software and Middleware (Comsware)* (pp. 1-10). Washington, DC: IEEE.

Donnelly, R. A., Jr. (2004). *The Complete Idiot's Guide to Statistics*. New York: Alpha Books.

Giles, T., Markendahl, J., Zander, J., & Zetterberg, P. (2004). Cost Drivers and Deployment Scenarios for Future Broadband Wireless Networks – Key research problems and directions for research. In *IEEE 59th Vehicular Technology Conf.* (Vol. 4, pp. 2042-2046). Washington, DC: IEEE.

Gozalvez, J., Martin-Sacristan, D., Lucas-Estan, M., Monserrat, J. F., & Gonzalez-Delicado, J. J. Gozalvez, & D., Marhunda, M. (2007). SPHERE –A Simulation Platform for Heterogeneous Wireless Systems. In *IEEE TridentCom* (pp.1-10). Washington, DC: IEEE.

3GPP. (2001). *Digital cellular telecommunications system (Phase 2+): Multiplexing and multiple access on the radio path (3GPP TS 05.02 version 8.9.0 Release 1999)*. Retrieved from http://www.3gpp.org

Holma, H., Toskala, A., KRanta-aho, K., & Pirskanen, K. (2007). High Speed Packet Access Evolution in 3GPP Release 7. *IEEE Communications, 45*(12), 29-35.

Lampropoulos, G., Kaloxylos, A., Passas, N., & Merakos, L. (2006.) A Seamless Service Continuity Scheme for Enhanced Network Performance in UMTS/WLAN Networks. In *Proc. 17th IEEE Int. Symp. On Personal, Indoor and Mobile Radio Comm.* (pp.1-5). Washington, DC: IEEE.

Lincke, S. J. (2005). Wireless Load Sharing with Heterogeneous Services and Adaptive Placement. In *Proc. IEEE Wireless Telecommunications Symp.* (pp. 79-84). Washington, DC: IEEE.

Lincke, S. J. (2007). An Investigation of Seamless Mobility from the Operational Perspective. In *Proc. World Wireless Congress* (pp. 42-47).

Perez-Romero, J., Sallent, O., Agusti, R., Garcia, N., et al. (2006). Network Controlled Cell-Breathing for Capacity Improvement in Heterogeneous CDMA/TDMA Scenarios. In *IEEE Wireless Communications and Networking Conf.* (Vol. 1, pp. 36-41). Washington, DC: IEEE.

Sanchez, J. J., Morales-Jimenez, D., Gomez, G., & Enbrambasaguas, J. T. (2007). Physical Layer Performance of Long Term Evolution Cellular Technology. In *16th IST Mobile & Wireless Communications Summit* (pp. 1-5). Washington, DC: IEEE.

Serrador, A., Galvano, G., Ferreira, L., & Correia, L. M. (2006). Parameters for the Definition of Scenarios for CRRM Performance Evaluation. In *IEEE MELECON.* (pp. 575-578). Washington, DC: IEEE.

Song, W., Zhuang, W., & Cheng, Y. (2007, January-February). Load Balancing for Cellular/WLAN Integrated Networks. *IEEE Network*, 27–33. doi:10.1109/MNET.2007.314535

Tan, W. L., Lam, F., & Lau, W. C. (2007). An Empirical Study on 3G Network Capacity and Performance. In *INFOCOM 2007. 26th IEEE International Conference on Computer Communications* (pp. 1514-1522). Washington, DC: IEEE.

Taylor-Fuller, D., & Lincke, S. J. (in press). The Development of a Graphic Multidimensional Markov Chain Modeler to Diagram Large State Spaces. In *Proc. 41st Annual Simulation Symposium*, Association of Computing Machinery (ACM).

Tipper, D., Ramaswamy, S., & Dahlberg, T. (1999). PCS Network Survivability. In *1999 IEEE Wireless Communication & Networking Conf.* (Vol. 2, pp. 1028). Washington, DC: IEEE

Section 2
Resource Management

Chapter 6
Delay-Based Admission Control to Sustain QoS in a Managed IEEE 802.11 Wireless LANs

A. Ksentini
University of Rennes 1, France

A. Nafaa
University College Dublin, Ireland

1. ABSTRACT

In this chapter, we present a delay-sensitive MAC adaptation scheme combined with an admission control mechanism. The proposed solution is based on thorough analysis of the trade-off existing between high network utilization and achieving bounded QoS metrics in operated 802.11-based networks. First, we derive an accurate delay estimation model to adjust the contention window size in real-time basis by considering key net-work factors, MAC queue dynamics, and application-level QoS requirements. Second, we use the abovementioned delay-based CW size adaptation scheme to derive a fully distributed admission control model that provides protection for existing flows in terms of QoS guarantees.

2. INTRODUCTION

During the last decade, multimedia services such as VoIP and Video have gained an increased success in the 802.11-based wireless network as this latter continuously adds capacity to support more and more bandwidth-hungry services. This has, in turn, opened new business opportunities for Network Operators (NOs) that are now offering new multimedia services over IEEE 802.11-based wireless networks (IEEE 802.11, 1999). The deployment of this kind of application is facilitated by the promise of both new 802.11's physical layers that provides high data rate (100 Mbps), and the new IEEE 802.11 QoS-based standard (IEEE 802.11e, 2005). In fact, the IEEE 802.11e standard is designed to support different sensitive multimedia applications (such as: Voice over IP, Video streaming), besides the classical best effort traffics. Still, the current version of the IEEE 802.11e standard doesn't provide firm QoS guarantees with efficient Admission Control (AC) protocol, the way traditional wired networks do. In fact, it is difficult to maintain QoS for admitted multimedia flows in 802.11-based networks without using an AC protocol. This poses tremendous viability problems on any carrier-grade multimedia services

DOI: 10.4018/978-1-61520-680-3.ch006

Figure 1. DCF access mechanism

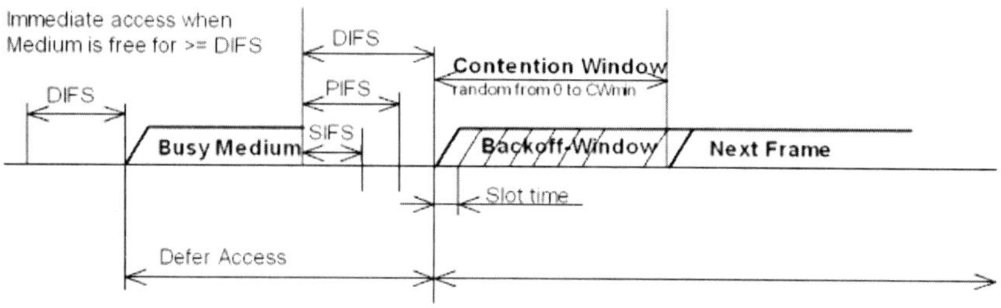

provisioning over 802.11-based networks.

This chapter is organised as follows: section 3 reviews existing works on QoS support and admission control in IEEE 802.11. In section 4, we introduce our Delay-based Admission Control. Simulation results covering the performance of the proposed AC are given in section 5. Finally, we conclude in section 6.

3. QUALITY OF SERVICE SUPPORT AND ADMISSION CONTROL IN IEEE 802.11

In this section, we provide background material on the 802.11 MAC and QoS enhancements. Related works on AC algorithms in 802.11 networks are also reviewed in this section.

3.1. IEEE 802.11 Basic Access Mechanism: DCF

The IEEE 802.11 MAC defines two transmission modes for data packets: the Distributed Coordination Function (DCF) based on Carrier Sense with Multiple Access (CSMA/CA) and the contention-free Point Coordination Function (PCF), where the Access Point (AP) controls all transmissions based on a polling mechanism. The popularity of IEEE 802.11 wireless LAN (WLAN) is mainly due to DCF, whereas the PCF is barely implemented in today's products due to its complexity and inefficiency in common network deployment setup, despites its limited QoS support. PCF may cause unpredictable beacon delays and unbounded transmission latencies (Mangold, 2002). On the other hand, DCF is the basic mechanism for IEEE 802.11 that employs a CSMA/CA algorithm (see Figure 1) and allow for a fully distributed wireless medium sharing. Before sending a packet, a wireless station first senses the medium for a duration equivalent to Distributed Inter-Frame Space (DIFS). If the medium is idle for that duration, the wireless station starts sending immediately. Otherwise, if the wireless station senses the medium as busy, the wireless station backs off for a certain number of time slots (see eq. 1).

$$Backoff = Random\ (0,\ CW\text{-}1)\ *\ SlotTime \qquad (1)$$

Collisions can only occur in the case where two terminals start transmitting on the same slot. For each unsuccessful transmission the Contention Window (CW) is exponentially increased as follows:

$$CW_{new} = \left(CW_{\min} \times 2^{i}\right) \qquad (2)$$

where i is the number of unsuccessful transmission attempts usually referred to as the backoff stage.

Note that, after each successful transmission the CW is initialised with the CWmin.

In order to guarantee undisturbed transmission even in presence of hidden wireless stations, an RTS/CTS (Request to Send/Clear to Send) mechanism is used. When this sender/receiver synchronization mechanism is enforced, the contention

Figure. 2. EDCA traffic classification and mapping

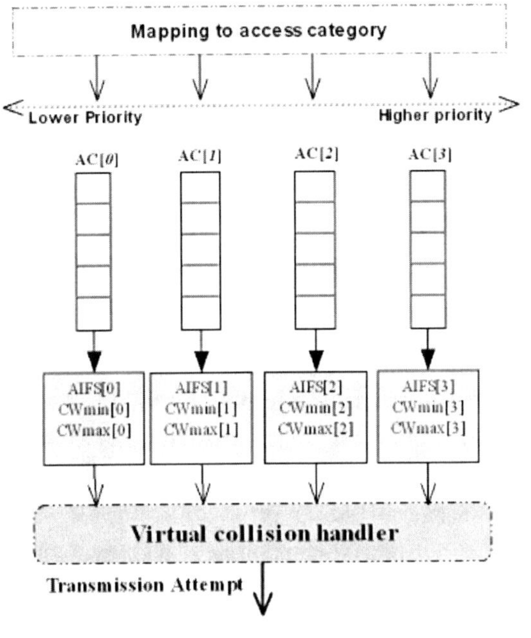

winner does not transmit the data immediately. Instead, it sends an RTS frame in to the receiver that replies with a CTS frame. This ensures that all terminals in the range of either the sender or the receiver are not only aware that a transmission will take place, but they are also aware of the duration of the transmission and medium business. In this case, terminals remain silent during the entire transmission, while only the sender is allowed to transmit frames. While the two extra messages present additional overhead, the mechanism is particularly useful in case of large data frames. In fact, only RTS and/or CTS packets can cause collisions, data frames are aware from collisions and no retransmission (data) is required. Thus, in case of large data frames, the channel utilization is considerably increased.

3.2. QoS Support in IEEE 802.11 Networks: 802.11e EDCA

The need for better access mechanisms supporting service differentiation has led task group E of IEEE 802.11 to propose an extension of the current IEEE 802.11 standard. The 802.11e standard introduces the Hybrid Coordination Function (HCF) that uses concurrently a contention-based mechanism and a pooling-based mechanism, EDCA and HCF Controlled Channel Access (HCCA), respectively. Like DCF, EDCA is most likely to be the dominant QoS-capable channel access mechanism in WLANs because it features a distributed and easy to deploy mechanism. Various 802.11 chips manufacturers are committed to provide 802.11e-campliance in their future product releases. In the following, we rather focus on EDCA; for more details on HCCA please refer to (IEEE 802.11e, 2005).

The QoS support in EDCA is realized with the introduction of Traffic Categories (TCs) concept to distinguish between different traffic classes, giving them different medium access priorities. Each TC has its own transmission queue and its own set of channel access parameters (see Figure. 2).

The service differentiation between TCs is enforced by setting different CWmin, CWmax, Arbitrary Inter-frame Space (AIFS) (see Figure 3) and the optional Transmission Opportunity duration's limit (TXOPlimit). A high-priority TC would typically use smaller AIFS, CWmin or CWmax, which gives it a higher probability to seize the medium more frequently and carry a higher offered load. TC3 and TC2 are generally reserved for real-time applications (such as voice or video transmission), while other TCs (TC1, TC0) are dedicated to best effort and background traffics with no QoS requirements.

3.3. Admission Control Mechanisms for IEEE 802.11 Networks

It is usually crucial to restrict the volume of traffic in WLANs in order to maintain service quality of current serving traffic. If there are no restrictions to limit the volume of traffic being introduced to the service set, performance degradation will result due to higher backoff time and collision rate.

Figure 3. EDCA access mechanism

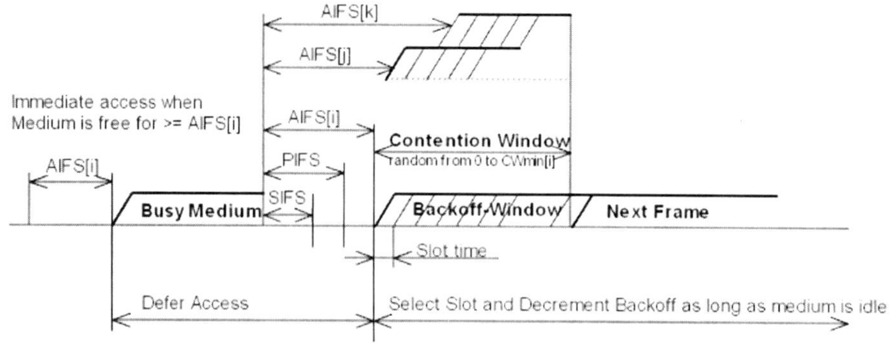

An effective resource allocation in IEEE 802.11 is difficult to achieve due to the intrinsic nature of the CSMA/CA scheme. Unlike traditional wired networks (or point-coordinated wireless networks) where bandwidth provisioning can only be managed by using bandwidth-availability information, flows' admission control in distributed 802.11 networks asks for additional parameters and more advanced models. For the same overall offered load, the network may exhibit widely different performances (i.e., availability levels) depending on the number of competing flows and their respective bit rates. For instance, the network contention level (collision) involved by ten 100Kbps-rate active flows would be different than the one involved by two 500Kbps-rate active flows – although the overall traffic volume is the same for both cases. The difficulty, with distributed 802.11 networks, lies in estimating the achievable QoS performance in the WLAN; this estimation depends on several time-varying factors including the number of active stations, the offered traffic volume for each TC, etc.

Many recent works on 802.11 network dimensioning (Bianchi, 2000; Pong, 2003; Ziouva & Antonakopolous; 2002) have rather focused on analysis of throughput and delay in saturated conditions. Besides considering a single traffic class, these works derived models assuming balanced traffic distribution between active wireless stations. If these analyses are to be used for admission control, flows admission in the network would be achieved in terms of number of active stations rather than in terms of single flows.

Distributed Bandwidth Allocation/Sharing/Extension (DBASE) protocol (Sheu & Sheu, 2001) addresses the problem of resource control in DCF-based mode by splitting the contention period into two sub periods, a period for contention between real-time stations and another one for contention between non-real-time stations. This protocol allows voice-station to a-priory reserve bandwidth using specific messages along with an updated network reservation table maintained at each station to coordinate between competing stations. The differentiation between these two contention periods is based on different AIFSs values for real-time and non-real-time traffic. Besides leading to substantial traffic overhead during reservation process, DBASE is unable to effectively separate between different traffic classes when the network gets fairly loaded since both traffic classes still use the exponential backoff algorithm; non-real-time traffic can draw small backoff interval values and end-up frequently access the network, wasting valuable bandwidth.

Based on local network measurements, authors in (Zhai, 2004) propose to control the arrival rate at each station to achieve a given objective, such as maximum throughput, maximum delay, jitter or loss rate in the network. The developed analytical model is able to assess the capability of the 802.11 to support major QoS metrics. The model is further extended in (Chen, 2005) to control the admission of network flows based on a new metric (channel busyness ratio) as a good indicator of the network state; channel busyness ratio is used to derive rate control algorithm (CARC- Call Admission and Rate Control). Besides not being applicable to 802.11e-like protocol where several traffic classes (having several requirements) may simultaneously operate in the network and even coexist at a single station, CARC try to find the optimal network utilization (maximize the throughput), while barely considering delays fluctuations.

A common drawback of the above introduced techniques (Sheu & Sheu, 2001;Zhai, 2004; Veres, 2001; Ziouva & Antonakopolous; 2002) resides in the fact that it is not possible to support different traffic classes at a given station since the different stations use a common and unique admission criterion. This severely limits the flexibility for a realistic deployment of multimedia streams with different requirements. That is, the network stations should be either voice station or best effort station. Designing advanced admission control mechanisms is clearly very important to operate future value-added services in WLANs where the network operator is able to fragment its quality of service offer into different service classes that can be simultaneously supported on any active station.

Distributed Admission Control (DAC) and Two-level Protection and Guarantee Mechanisms (Xiao & Li, 2004; Xiao, 2004) are combined to address the abovementioned issues. DAC is a measurement-based admission control mechanism that was considered by 802.11e working group. In this algorithm, the resource budget for each TC is periodically announced by the AP in the beacon frame, so that each station may decide whether to accept or not new flows. A new stream to be admitted tries first to access to network - it thereafter rejects itself after a certain period if its requirements are not met by the network; the stream is then locally accepted if it reaches its targeted throughput. With this algorithm, the residual network resources are fairly distributed among the competing streams (resp., streams seeking admission in the network) at different stations in the sense that different TC's (in different stations) compete to accommodate their new entering streams. This fact may lead to non-optimal resources exploitation because there may be situations where there are enough resources to admit one additional stream, but due to the algorithm fairness and absence of coordination several streams can compete for admission And none single stream get accepted, leaving the available bandwidth unused.

Another shortcoming of DAC algorithm resides in the lack of protection to existing flows when the network load is too heavy. If the network resources are not sufficient to admit the new stream, as this latter entering stream try to access the medium and reach its requirements, the performance degradation will affect all active TC's streams (as much as it does for other streams belonging to the same TC of the new entering stream). This is due to the fact that entering streams are aggregated with other active streams in the same TC queue. The abovementioned phenomenon is usually referred to as "spill over" effect in WLAN – when traffic is overloaded in a TC, performance in other TCs will also be affected. Still, the major problem with DAC-based approaches consists in the fact that the overall network bandwidth is statically allocated among different TCs, so each TC receives a fixed share of bandwidth that cannot be exceeded. This may severely affect the flexibility of the admission control mechanism since it is very difficult to beforehand forecast the per-TC traffic volume in realistic multimedia-dedicated WLANs. Therefore, streams from a given TC may be rejected while some bandwidth stay unfilled in other TCs, which means bandwidth wasting or additional revenue loss for network operator. Another side effect is that the admission decision depends only on local measurements collected at the admitting station level. However, the stream admission may have different impacts at different stations (resp. flows) depending on the load of each active station. The stream admission may actually cause QoS violation at certain stations while not effecting at all other active stations in the network; this is particularly prevalent for high-bit-rate stations, which usually cannot carry the load in sufficiently timely manner as the load (resp. medium access delay) increases.

It is readily realized that it is essential for an admission control (AC) model to be able to a-priory estimate the achievable application-level QoS metrics. This way, the admission decision does not affect the existing flows. Authors in (Pong, 2003) propose to estimate the achievable throughput under saturation conditions to ultimately control the flow admission in EDCA-based 802.11e networks. Besides the limitations inherent to saturations assumption, this scheme delivers only throughput guarantees without considering multimedia delays' requirements. Furthermore, due to the static per-TC parameters used in 802.11e, it is not possible to accommodate an important number of multimedia flows (Nafaa, 2005). More specifically, high-priority flows use a too narrow backoff range, which provoke a high intra-class contention. In (Ksentini, 2007), we presented a new MAC protocol featuring an AC. The proposed MAC protocol uses the TXOPlimit to reserve the network channel. However, the drawback of this mechanism is the degradation of the network performance when the offered load increases. Thus, an AC is proposed to protect QoS of admitted flows. Like (Pong, 2003) the AC estimates the network utilization, and acceptation/rejection decision is done according to the remaining resources. Nevertheless, this AC ensures only throughput guarantees without considering multimedia delays' requirements.

Virtual MAC and Virtual Source Algorithms (Veres, 2001; Barry, 2001) propose a fully distributed VMAC (Virtual MAC) algorithm that operates in parallel to the real MAC in the mobile host, although the VMAC does not handle real packets; rather, it handles "virtual packets." Each station runs a VMAC instance that monitors the capability of the wireless channel and passively estimate whether the channel can support new service demands (e.g., delay and loss). Unlike the case of real packets, VMAC doesn't transmit anything but estimates the probability of collision. When a collision is "detected", the VMAC enters a backoff procedure, just as a real MAC would do. The virtual source (VS) algorithm consists of a virtual application; an interface queue, and the VMAC. The virtual application generates virtual packets like a real application. Packets are time-stamped and placed in a virtual buffer. After a virtual packet has been processed in the VMAC, the total delay is calculated.

VMAC's main criterion to make an admission control decision is based only on delay and collision estimates. It does not rely on any achievable throughout assessment, which is also useful to multimedia applications. The achievable QoS is estimated only at the admitting station, although flow admission may unevenly affect the different backlogged flows, provoking delay violation at certain flows while other flows in the network still experience acceptable delays. As mentioned earlier, the outcome of stream admission should be beforehand assessed at all active stations. In fact, flows belonging to the same TC use roughly the same CWs' ranges, and thus they more or less experience the same packet-service times (i.e., the time needed to successfully transmit the frame located at the front of the queue). Hence, depending on the volume of their offered load, different flows may suffer from widely different en-queuing delays. In other words, admission of a new flow means a slightly increased packet-service time with different outcomes on different active flows. The impact of a stream admission should be therefore assessed at all active stations.

Dynamic Multiple-Threshold Reservation (Chen, 2005) propose an algorithm that is capable of granting differential priorities to different traffic classes in wireless multimedia network with cellular infrastructure. DMTBR generalizes the concept of relative priority and hence give the network operator more flexibility to adjust admission control policy by taking into account the offered load. However, as highlighted earlier, cellular networks are point-coordinated and present widely different characteristics compared to contention-based distributed WLAN. In cellular architecture, the network offered load is not correlated with delays since the resources are centrally managed by the BS and each flow receives fixed transmission slots when admitted in the network.

Obviously, the candidate AC mechanism should be distributed and able to manage different TCs at each single station, while providing high flexibility in respect to the relative (per-class) network load configurations (i.e., AC mechanism that enables all possible per-class load distributions as long as the QoS metrics are not violated). Admission decision may be made based on estimates of the achievable QoS at different active stations rather than only at the admitting station.

4. DELAY-BASED ADMISSION CONTROL

In this section, we begin by studying a measurement-based CW adaptation scheme. The objective of this scheme is to guarantee the same QoS metrics (e.g. loss rate, mean delay, mean jitter) for all flows belonging to the same TC. That is, we aim at maintaining a sustained application-level perceived QoS. In this respect, we set a predefined QoS metric (MAC-level transmission delay) threshold for each supported TC. Based on distributed measurements our protocol is able to guarantee multimedia streams requirements (MaxDelay, MaxLoss, and ensured bit-rate) in different network configurations. A key point to enforce predictable QoS performances resides in the ability of our scheme to accurately modelling the achievable QoS metrics performances. After that, we generalize our achievable QoS assessment model to derive a distributed admission control protocol.

4.1. Delay Sensitive MAC Adaptation Model

Conventional IEEE 802.11 backoff schemes have many shortcomings that make it difficult to provide deterministic guarantees. The exponential CW increasing is more likely to produce probabilistic service assurances and high oscillations in delays (throughput) since the CW is reinitialized to its minimum value (CWmin) with each successful transmission. In order to limit the effect of high inter-TC contention, different AIFS[i]s may be assigned to different traffic classes TC[i]; this would delay transmissions of low-priority flows only when their respective transmission attempts coincide with high priority flow transmission. At this point, managing the contending flows through appropriate CW scheme is a key component to effectively maintain acceptable QoS level for multimedia flows.

4.1.1. Delay-Based CW Adjustment

At MAC layer, packets are serviced with a variable latency that depends on the current CW size, the mean frame size (E[P]), and the mean number of transmission attempts before effectively gaining access to the medium. Besides, the network load (i.e., transmission volume from other nodes) may strongly affect the end-to-end communication latency as a substantial amount of time slots is occupied, which ends-up provoking frequent backoff freezing. Actually, each new packet selects a random backoff interval (E[CW]) that is more or less rapidly decremented depending on the number of time slots where the medium was observed as busy. The packet transmission deferring period depends on the selected backoff interval as much as it does depend on the degree of network load.

We define PST (Packet Service Time) as the time needed to successfully transmit a packet; this delay is defined as the time interval elapsed between the time when a packet arrives at the front of the queue and the time when it is received by the receiver. The delay considers only channel access delay, transmission delay, and associated overhead (i.e., queuing delay is not included).

Let B(T)=B/I be the number (B) of busy time slots over the number (I) of idle slots observed during the last T time slots (T=B+I). The total deferring time for a packet can be approximated by E(CW)•(1+B/I); this delay takes into account both the backoff interval and the freezing period. Compared to the technique that achieves direct measurement of the freezing period at each flow [5], our technique is based on continuous monitoring of the overall network load, which could be better exploited to predict network load trends. Measuring the freezing period for each transmitted packet may exhibit high oscillations, not to mention the involved complexity. Using the overall network occupancy (B/I) to estimate the access delay leads to inherent measurements coordination between different active flows as these latter observe the same network activities at any point of time.

We define E[P] as the mean number of time slots occupied by a single packet transmission including PHY/MAC overhead, SIFSs, and ACK when considering the DCF basic mode. It is worth mentioning that within DCF basic method (without RTS/CTS handshaking), each failing transmission (due to frame collision or bit alteration) occupies roughly the same number of slots as a successful transmission (Bianchi, 1996). In the following we assume a DCF MAC protocol operating without RTS/CTS handshaking. In order to better assess the accuracy of our model with simulations, we assume that packet loss provoked by wireless link interferences (BER) is negligible.

The overall packet service time (PST) may be quantitatively estimated as follows

$$PST = \left[E\left(CW\right)\cdot\left(1+B\left(T\right)+E\left(P\right)\right)\right]\cdot E\left[TransAtt\right] \qquad (3)$$

Here, E[TransAtt] is the mean number of transmission attempts needed to successfully access the medium; this parameter depends on the PER (Packet Error Rate) and the automatic retransmission (ARQ) scheme being used at MAC layer. Generally speaking, a packet is kept in the transmitter queue until either a timer times out (i.e., after 7 failed transmission attempts), or the packet is successfully received and acknowledged by the receiver. Since the backoff process have a geometric distribution with probability of success p, the mean number of transmission attempts E[TransAtt] would be 1/p. At this point, the probability of transmission success, p, can be approximated as the fraction of the number of transmitted frames over the number of transmission attempts. Thus, the mean number of transmission attempts E[TransAtt] can be estimated as

$$E\left[TransAtt\right] = \frac{1}{1-\frac{Collisions}{TransmissAttempts}} = \frac{TransAttempts}{SucceedTransmissions} \qquad (4)$$

Note that E[TransAtt] may return different values depending on the flow's traffic class and its associated AIFS. Obviously, inter-TC collisions are most of the time avoided since flows with the highest priority seizes the medium while other flows enter in differing state.

As B(T) is calculated based on the overall network load, it is inherently coordinated between stations. Each station averages the measurements over the period T required to sense "*CWmax*" idle time slots. By choosing the frequency of measuring B/I in this way, we are ensured that all backlogged flows (regardless of priority) would have attempted to access the medium at least once within this period. Thus, B/I measurements is more accurate by considering all active flows, and also more stable as they are averaged over a long-enough period. Throughout this chapter the value of T is set to 1024 "*idle*" slots. For the same reasons, E[TransAtt] values are also averaged over the period needed to sense 1024 idle time slots.

Figure 4. MAC layer queue for TC i

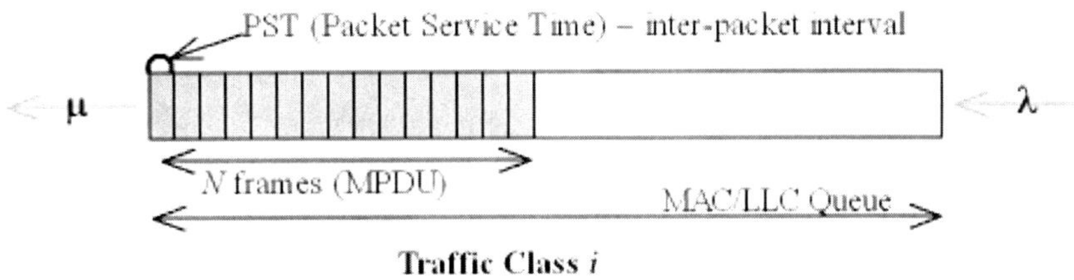

As apparent from Figure 4, each station in the network may have different traffic classes with different requirements in terms of QoS metrics performances. Several MAC queues are indeed implemented within a single station. Each queue supports one TC, behaving similar to a single DCF entity within the 802.11 standard. In this context, the last packet in the queue (packet #N) should not exceed the maximum delay tolerated by the traffic class (TC) to which it belongs. By considering that both the arrivals (λ) and the service (μ) are exponential, the PST will be therefore constrained by

$$PST \le \frac{MaxDelay}{N} \tag{5}$$

The formula above generalize our PST estimation model to estimate the enqueuing time by taking into account the number of packet (N) currently in the MAC queue (the N packets ahead of the last packet entering the queue). From the formulas (3) and (5), and given the queue length (N), the appropriate maximum CW size (CWmax) that would satisfy the delays constraints associated with each service class (regardless its bit-rate) is obtained as follows

$$CW_{\max} = 2.E[CW] \le \frac{2.(MaxDelay - N.E[TranAtt].E[P])}{N.(1 + B(T)).E[TransAtt]} \tag{6}$$

It is commonly accepted (Ziouva & Antonakopoulos, 2002) that WLAN capacity (i.e., channel utilization) decreases with an increasing number (M) of active flows. This is caused by high contention level in which case the medium is often occupied by collisions. In this situation, the mean number of attempts to successfully transmit a frame would grow resulting in additional delays at active flows. The contention CW should be continuously adapted, thereby reacting to changing network conditions while meeting QoS constraints. Actually, when M increases, the CW size is increased to absorb the increasing number of contending flows, and hence minimizing the collision probability for these flows. On the other hand, when M becomes small, the CW size is decreased, which reduces the spacing between successive frame transmissions; large values of CW size may indeed strongly limit the throughput of fewer backlogged flows. As a matter of fact, the current CWsize in use should be always larger than certain variable threshold (CWmin) to avoid network performances collapse. From (Bianchi, 1996), the minimum CW size that maximizes network performances with M contending flows is given by

$$CW_{\min} \ge \left\lceil M \cdot \sqrt{2Tc} \right\rceil, \tag{7}$$

Here, Tc is the average time (in time slots) of channel unavailability upon a collision. Tc is dependent on the physical layer, and is equal to PHYhdr + E[F] + DIFS when RTS/CTS mode is disabled. E[oldCW] is the current mean backoff value. O(T) is the number of slots where the medium was observed as busy out of the previous T slots (B). Like all other network measured parameters (i.e., E[TransAtt] and B(T)), O(T) is weighted in respect to past measures using EWMA

(Exponential Weighted Mean Average).

Although not accurate (i.e., much incertitude still exists due to different flows' priorities and bit rates), the estimate of the number (M) of active flows is quite pertinent since it still precisely reflects the overall trends of the network contention level, which allow readjusting the CW to optimize the network performance. In fact, constraining the contention window by CWmin helps to keep a low collision rate, and hence an acceptable mean transmission attempts (i.e., E[TransAtt] lower than 1.5, which means 3 transmission attempts for 2 successful transmissions). The new CW to be maintained by each TC is given by:

$$newCW_{size} = \frac{CW_{\min} + CW_{\max}}{2}$$

$$\text{with } CW_{max} \geq CW_{min} \qquad (8)$$

If CWmax is smaller than CWmin, we assign CWmin to CWmax. In this case newCWsize is simply re-initialized with CWmin value. This situation does not guarantee MaxDelay; instead, it keeps network collisions within an acceptable level. Using the above introduced CWsize adjustment model, a given flow would use the interval [0, newCWsize] to randomly draw a backoff interval. Note that the parameter CWmin is not necessarily coordinated between flows since its value is, in part, based on current CW size that is maintained by the flow. Accordingly, flows calculate different CWmin values depending on their class of service (MaxDelay constraints) and their offered load as well.

4.1.2. Model Validation

To validate the proposed delay model, we draw a set of simulations. The aim is to evaluate the accuracy of CW adaptation in maintaining bounded MAC queuing delays regardless changes in network load. Throughout our experiment, the relative (per-class) network load is deliberately changed

Figure 5. HP flows' instantaneous delays

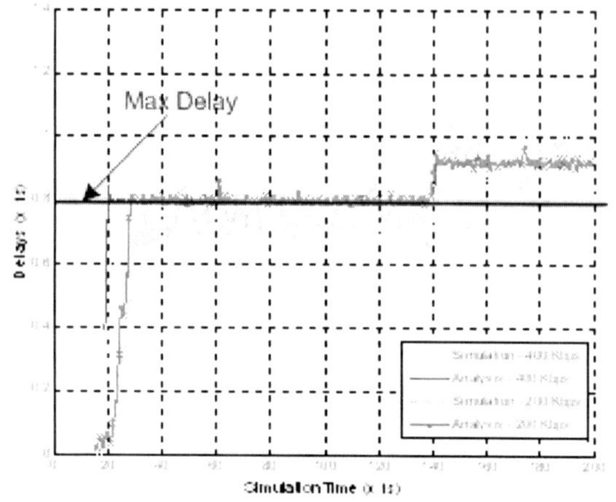

Figure 6. MP flows' instantaneous delay

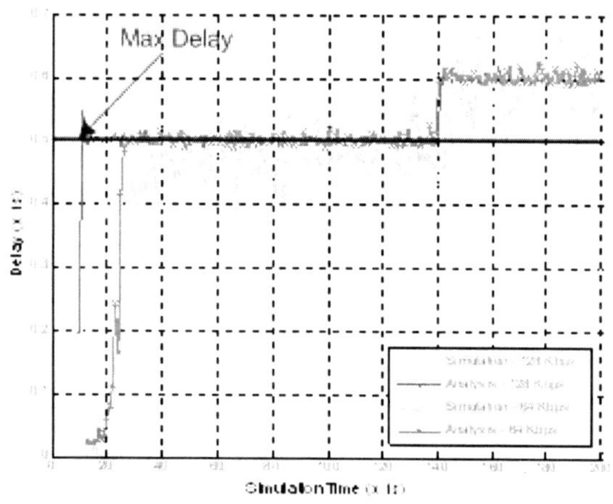

to evaluate the performance of our protocol to suit different network configurations. In order to assess the accuracy of our analysis in terms of estimated achievable delay, Figure 5 and Figure 6 compare the model-predicted enqueuing delays (N*[PST]) with the delays effectively experienced during simulations. All network configurations, flows characteristics, and simulation scenarios used in the simulations are thoroughly discussed in the performance evaluation section (Section 4.1).

Figure 5 illustrates the instantaneous delays experienced by two HP (high priority) flows having different bit rates (128 Kbps and 64 Kbps). Figure 6 gives the instantaneous delays experienced by two MP (medium priority) flows with bit rates of 200 kbps and 400 Kbps, respectively. We give, in each figure, the model-based delays estimated by the four involved flows. The given delays are each time averaged over 1 second. The maximum delay bound to not violate (MaxDelay) is fixed to 0.5 second for HP flows while it is set to 0.8 second for MP flows. Note that the model-estimated delays are calculated at each TC using different MAC parameters as explained in the previous sub-section.

Globally, when the network is sufficiently relaxed, there are no violations of delay thresholds, except some brief spikes that are rather due to (i) short-term fluctuations in collision rate measurements and (ii) disparity between the successively drawn backoff intervals.

As clearly apparent from Figures 5 and 6, our protocol ensures roughly the same delays to TC's flows regardless their respective bitrates. The negligible disparity between delays experienced by different TC's flows is mainly due to slightly different short-term network-measurements (e.g., E[TransAtt] and B(T), O(T)). In fact, for a longer measurement averaging period or resolution (T), the dif-ferent stations would be more coordinated, though, with a seriously reduced responsiveness in face of network load variations.

As apparent from the above figures, the service-level fairness among TC's flows is achieved even when the MaxDelay threshold is violated (between t=140s and t=200s). Although there is sufficient bandwidth available in the network to carry additional offered load, flows experience higher latencies due to higher enqueuing delays induced by an increasing in packet service time (frequent network occupation cause an increasing in the mean number of transmission attempts). At this point, the model-calculated CWmax that would accommodate the delay constraints is actually too low (i.e., CWmax lower than CWmin), which cause the flows to use CWmin as the maximum contention window size (see formula (8)). Since CWmin calculation is mostly based on the currently used CW size, its value is roughly proportional to previous CWmax values. As a consequence, the fairness between the achieved delays is maintained since different flows belonging to the same TC use different CWmin values. Consider that flows with a higher offered load usually maintain lower CWmin in order for them to carry the load during high contention situations.

As above discussed, when the network can no longer guarantee the delay exigencies (between t=140s and t=200s), the CW values of different flows, with different priority, tend to use quite stable values (i.e., CWmin). This explains Figure 7 results between t=140s and t=200s, where all stations are using CWmin as the final CW (new-CWsize) to be used in the backoff process for each packet transmission. This fact causes more important delay violations at HP flows (see Figure 2 and Figure 3) since it is more difficult to ensure delays below 0.5 s under stressed conditions. For more results, and particularly a comparison of the proposed delay-based CW adjustment with EDCA and AEDCF (Romdhani, 2003) can be found in (Nafaa & Ksentini, 2008).

An important observation that came out from the above results is that there is a critical trade-off between the achieved network throughput and delay guarantees for certain flows. Obviously, it is not possible to fully fill the network capacity while still satisfying strict delay requirements. From a practical point of view, increasing flow's throughput, beyond a certain extent, means increasing the enqueuing delays, and thus probably violating delay constraints. The instantaneous transmission delay at a given flow F is a multi-faced problem that depends on the network configuration, i.e., depends on many factors such as the bit-rate of F, the maximum tolerated delay by F, the overall network load, the network contention level (which itself depends on the overall offered load distribution over the different active flows), and finally the delay constraints on the other active flows. Different network configurations (different combination of the abovementioned parameters)

Figure 7. Variation of the Contention Window size (newCWsize)

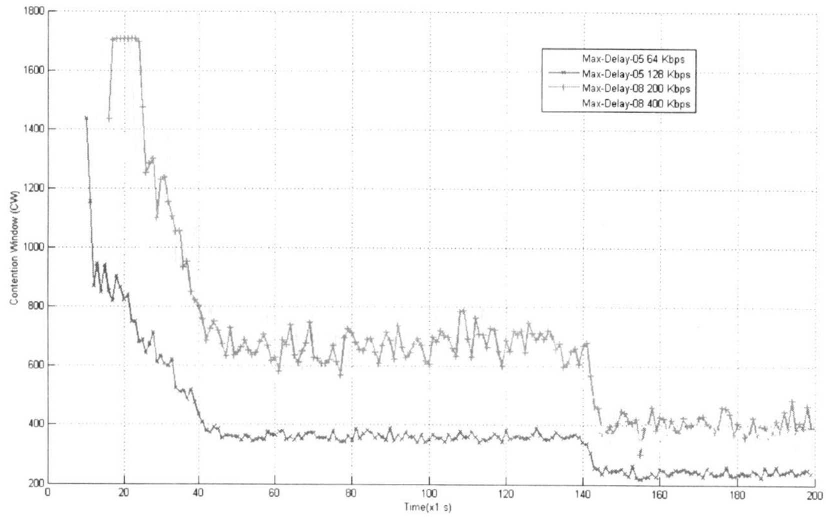

may result in the same overall achieved throughput, though with different achieved delays. From network operator point of view, this situation poses a major problem.

In fact, it is essential to each time find out the optimal network's operation point by maximizing the number of QoS-enabled services in the network, regardless the network configuration. This requires a distributed model able to a-priory (before admitting new services) predict network performances in terms of achievable QoS metrics. The admission control mechanism should allow for various per-class traffic load distributions to allow network operators to optimize their underlying resources and increase their revenues. The difficulty of implementing this approach in 802.11 lies in estimating the consequences, at different active network flows, provoked by streams' admission.

4.2. Multimedia Services Admission and Protection

Since delay estimation is based on inter-packet interval assessment, the achievable throughput together with potential degradations (mean loss rate) may be predictable as well. Using the packet arrival rate (λ), which is a-priory known for a given traffic class (TC), it is possible to capture the queue dynamics based on instantaneous network activities; the packet arrival rate may be for example provided by pre-established Service Level Agreements (SLA). The objective is to predict the impact of new stream's acceptation on the overall network performances. In other words, we assess the consequences resulting from increasing the arrival rate of a given TC/station (i.e., stream admission) before actually admitting any new entering service.

As illustrated in Figure 4, we consider a MAC/LLC queue with a buffer size k. Service is exponential with parameter μ and inter-arrival times are exponential with parameter λ. A loss occurs whenever an arriving packet finds the queue full. The queue occupation rate is thus

$$\rho = \frac{\lambda}{\mu} = \lambda \cdot E\left[PST\right] \tag{9}$$

The queue model is assumed to be a single-server queue with finite waiting room (M/M/1/K). Certainly, the Poisson assumption for the arrivals of packets is not the most realistic, but considering the exponential case reveals essential features of the system and is a fairly appropriate assumption for an aggregate of different streams (TC). The mean loss rate (Lr) of an M/M/1/K queue is given by

$$Lr_i = \frac{(1-\rho)\rho^k}{1-\rho^{k+1}} \tag{10}$$

Since the maximum tolerated loss rate (MaxDrop=Lr) is a-priory known for each TC i, we can numerically fix ρ since the MAC queue size (K) is as well known. In fact, the network operator may propose different levels of QoS guarantees, where each level is characterized by maximum QoS metrics performances bounds (MaxDelay and MaxLoss). For instance, assuming a queue length of k=30 packets and with a maximum tolerated loss rate of MaxDrop=1%, the queue occupation rate ρ should be lower than 0.935. In the same manner, ρ = 0.97 for a maximum tolerated loss rate of MaxDrop=2%. In this chapter, we aim to categorize the traffic into service classes where each service class has a maximum delay and a maximum loss rate to not violate.

Based on the delay analysis (i.e., PST) and the mean tolerated loss rate, we can now determine the appropriate μ (i.e., 1/E[PST]) that satisfies the relation (11). Thus, we analytically figure out the appropriate CW that provides a mean inter-packet transmission interval (E[PST]) necessary to maintain a queue occupation rate at the desired level (ρ). By combining formula (3) and formula (9), we obtain the appropriate contention window size that satisfies the loss requirements associated to a given TC

$$NewCW_{size} = 2 \cdot E\left[CW\right] = 2 \cdot \frac{\frac{\rho}{\lambda} - E\left[P\right] \cdot E\left[TransAtt\right]}{\left(1 + B(T)\right) \cdot E\left[TransAtt\right]} \tag{11}$$

with $CW_{min} \leq newCWsize$ and $newCW_{min} \leq CW_{max}$

While the contention window (CWsize) given by formula (8) ensures an acceptable delay with regards to TC's requirements, formula (11) allows to avoid TC's queue overflow by each time checking if the current PST (i.e., NewCWsize) is able to absorb the packet arrival rate (λ). More precisely, the new CW size ensures that the TC's flow in which the entering stream will be aggregated will not violate its maximum tolerated loss rate. The new calculated CW size (NewCWsize) should be also larger than CWmin. This means that the network is able to accommodate the new-stream's offered load while still meeting delays guarantees (NewCWsize<CWmax) and keeping an acceptable contention level (NewCWsize>CWmin) to avoid network performances collapse.

Combined to the delay-driven CW adjustment introduced in formula (8), the above formula may be used to accept new streams in the network. This consists in assessing if a new stream may be serviced while not interfering with already active flows. As highlighted already, an over-admission will unavoidably affect all currently serviced flows as the medium is shared and an increasing in the contention level affects all flows regardless their bitrates or priorities. As revealed in Figure 7, on the other hand, different active flows may simultaneously maintain widely different CW sizes due to different values of CWmin and CWmax. The maintained CW contention window depends, actually, as much on the flow's offered load as it does on the flow's traffic class. In certain circumstances, an over-admission may cause certain flow to violate its CWmin limit, while other flows still use CW sizes larger than their calculated CWmin; flows with high bitrates are generally the first flows to reach their CWmin limits. At this point, it is readily realized that the impact of new stream admission should be estimated at all stations.

At new stream admission, each flow in the network recalculates the values of CWmin, CWmax, and NewCWsize according to formulas (7), (8), and (11). The new values of these parameters should take into account changes in network availability entailed by admitting a new stream. Accordingly, certain determinant measurement-based parameters such as B(T), O(T), and E[TransAtt] should be reconsidered. While

E[TransAtt] fluctuations are limited by using an appropriate CWmin, both B(T) and O(T) exhibit significant changes that should be considered to accurately re-estimating the new achievable QoS performances.

Again, it is worth mentioning, that λ is actually the arrival rate of streams' aggregate belonging to the same TC. At new stream admission, the overall arrival rate at the TC's queue would increase as follows $\overline{\lambda} = \lambda + \Delta\lambda$, where $\Delta\lambda$ is the packet arrival rate of the new entering stream. In this case, the network load should be updated to reflect the additional load induced by the new stream.

$$\overline{B}(T) = \frac{\overline{B}}{\overline{I}} = \frac{B+\beta}{I-\beta} \text{ with}$$

$$\beta = \overline{\lambda} \cdot T \cdot (20 \cdot 10^{-6}) \cdot L \qquad (12)$$

Here, L is the mean number of time slots occupied by a MAC packet of a given flow, including the overhead involved by acknowledgement. O(T) should be as well updated with the new flow arrival as follows

$$\overline{O}(T) = \frac{\overline{B}}{T} = \frac{B+\beta}{T} \qquad (13)$$

Given the abovementioned parameters, all active stations calculate the new values of CW[i]min, CW[i]max, and NewCW[i]size for each TC i. If the new values satisfy all QoS constraints (CW[i]min <NewCW[i]size<CW[i]max) associated to each TC i, then the station concludes that the entering stream will not affect its already serviced streams. If all stations will not be affected by the entering stream, the AC algorithm may then proceed with stream admission. Otherwise, it means that the stream admission may severely degrade the quality of currently servicing flows, which should lead to rejection of the entering stream.

3.2.2. Admission Control Coordination

The first issue to tackle when designing a distributed AC mechanism is the coordination between competing nodes. In fact, besides necessitating a unified admission model for all stations, we further require to harmonize the estimation of achievable QoS at different station in order to achieve a coordinated admission control decision. Particularly, multiple new real-time streams may be simultaneously admitted by individual nodes if not coordinated, causing "over-admission". To mitigate this problem while keeping the distributed feature of our protocol, we divide the time into admission cycles (epochs) where only one single stream may be accepted in an admission cycle. The network is assumed to operate on "slotted" synchronization epochs, where each epoch is actually equal to a beacon period. This way, the admission cycle is long enough to allow network measurements (E[TransAtt]), at different stations, to converge towards accurate values reflecting the real network conditions before admitting new stream in the next synchronization epoch.

To completely avoid the over-admission problem, we adopt a coordinator-aided admission control scheme. In other words, all admission decisions are made by a coordinating node (CN), which can record the current number of admitted real-time flows and their occupied channel bandwidth in the network; clearly, this will prevent over-admission situations. The coordinator node is also in charge of other responsibilities related to service level agreement (SLA). These additional CN's responsibilities are further discussed in the next sections.

It is important to note that a coordinator is available whether the wireless LAN is working in the infrastructure mode or in the ad hoc mode. If the network is working in the infrastructure mode, the access point is inherently the coordinator. Otherwise, a mobile node can be elected to act as the coordinator in the network using one of many algorithms in the literature (see (Garcia-Molina,

1982), and references therein). A natural solution would be to appoint the node in charge of sending the MAC-level beacon as the CN. As in 802.11 Ad Hoc mode, in case of failure a distributed backoff-based mechanism would design a new node to periodically send the beacon. Further details on the election process are beyond the scope of this chapter.

Each time a station S have a new stream to admit, it should beforehand evaluate locally its impact using new values of B(T) and O(T) as given by formulas (12) and (13). Using formula (11), the station S should as well assess the risk of having overflow by calculating NewCWsize, where λ is replaced by $\lambda + \Delta\lambda$; in Figure 8, λ_i (i=1 to 3) stands for the rate ($\Delta\lambda$) of a new entering stream.

If the new entering stream doesn't affect the locally active TCs' flows, the station S announces the stream's bit-rate (λ) and nominal MSDU size (in terms of time slots) to the CN which, in turn, recalculate the new values of network occupancy parameters (B(T) and O(T)) to be broadcasted. Then, all active stations evaluate the impact of new stream admission (i.e., with new B(T) and O(T) changes) on their TC's flows and eventually deny the admission if the QoS of one of its TCs may degrade. Note that each TC[i]'s flow in the network calculate NewCWsize using its own packet-arrival rate (λ) and maximum queue-occupation-ratio ρ_i corresponding to its traffic class.

Figure 8 illustrates a scenario where in the first beacon period the coordinator receives 3 new streams announcements. The coordinator calculates and broadcasts parameters associated to the first stream (S_1). The admission is then aborted by station n the admission of S_1 interferes with its QoS constraints. In the second beacon period, the coordinator broadcasts S_2 parameters and finish by accepting the stream as no active station have denied the acceptation within the current beacon period. Typically, here S_2 should have a lower packet rate than S_1.

For scalability reasons, AC messages handshake are kept to a minimum by broadcasting CN

Figure 8. Admission control message exchange

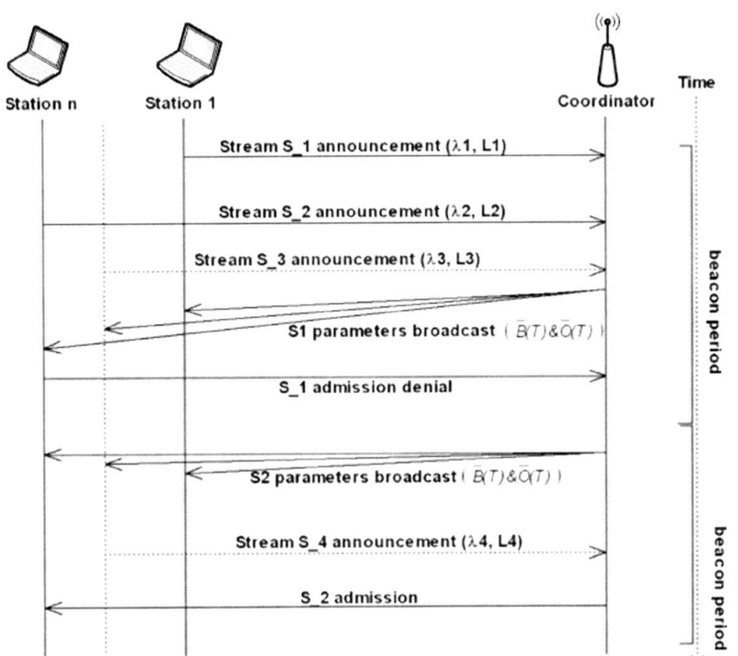

messages (i.e., parameters broadcast and admission messages). Furthermore, response messages (i.e., admission denial message) are sent by an active station only if one of their QoS thresholds, associated to TCs' flows, would be violated with the new stream admission. A single denial message suffices to abort the whole stream admission process, so other stations don't need any more to send denial messages, i.e., all stations overhear AC messages.

To increase the reliability of CN's broadcasted messages, we use efficient basic data rate (1Mbps) usually employed to transmit the beacon, RTS/CTS, and ACK messages. On the other hand, during AC process, all directed messages exchanged between the coordinator node and other stations are fully persistent in the sense that they are retransmitted until successful reception.

Upon a first admission in a given beacon period, the other flows seeking admission in network should differ the announcement to the next beacon period and additional network measurements are carried out before final admission. This allows all stations to take into account the changes in network availability before accepting new streams (i.e., allows the different competing stations to have a coherent perception of the network availability by carrying out measurements during a long-enough period such as a beacon period).

5. PERFORMANCE EVALUATION

In order to evaluate the advantages of the proposed protocol we have constructed a simulation using ns-2 (Network Simulator). We compare our distributed AC protocol scheme using the last IEEE 802.11e standard. Our admission control protocol was implemented atop the last NS2 implementation of IEEE 802.11e that uses a more realistic MAC implementation where the 802.11 nodes are more synchronized thanks to a considerably improved backoff freezing process. We further improved this implementation with a more accurate MAC Timer for better synchronization between flows in respect to network load measurement (i.e., B(T) and O(T) measurements). In this section, we highlight various aspects entailed by deploying effective admission control mechanisms in WLAN, with a special focus on the appropriate brokering strategies[1] to be adopted by network operators.

5.1. Simulation Model

For the simulations, we have created a network consisting of sixteen wireless terminals (WT[i], i=1,..,16). A single coordinator node (CN) is arbitrary chosen among the 16 nodes; the CN is actually the node that periodically send the beacon frame in 802.11 Ad Hoc mode.

Each WT may generate up to two different TC flows at the same time, representing two uniquely prioritized traffic classes: high priority (HP) with a MaxDelay of 500 ms and medium priority (MP) with a MaxDelay of 800ms. In our simulation, we choose to generate only one flow per station, so as to make worse the contention for seizing the medium. In fact, if the backoff counters of two or more TCs collocated in one station elapse at the same time, a scheduler inside the station treat the event as a "virtual" collision without causing network's time-slots waste. In this case, the medium is seized by the TC with the highest priority among the colliding TCs, while other colliding TCs defer their transmissions as if the collision occurred in the real medium

Constant bit rate (CBR) sources are used for all traffics; the properties of these flows are specified in Table 1. CBR sources put more stringent exigencies (e.g., packet rate and en-queuing delay) on network than VBR sources. In fact, multiple CBR sources would require that the network sustains the overall offered load (summation of CBR sources bit rates) throughout the simulation period, which may provoke MAC queues overflows after a fairly long run. In contrast, with multiple VBR sources, the peaks of bit rates are unlikely to occur at the

Table 1. Traffic characteristics

Traffic features	Packet size (Bytes)	Generation Interval (second)	Bit rate (bps)
Max_delay_0.5 (HP)	160	0.02	64000
Max_delay_0.5 (HP)	160	0.01	128000
Max_delay_0.5 (MP)	500	0.02	200000
Max_delay_0.5 (MP)	500	0.01	400000

same time, which allow the network to absorb the brief offered load bursts exhibited, by different traffic sources, at different time scales. In practice, the service level agreement (SLA) between the service costumer and the service provider specifies the service characteristics in terms of fixed mean data rate (and eventually peak data rate) with associated QoS metrics performances bounds. It is extremely difficult, in practice, to precisely characterize the burstiness of a VBR stream.

We performed several simulations runs in order to evaluate the performance of our admission control scheme in respect to different QoS metrics (loss and delay). We also give the evolution of the CW size at each TC flow type as the network configuration changes over the time. Each run consists of 200 seconds of simulated network lifetime with a fixed scenario in terms of per-TC traffic load variation and the order of single flows backlogging. From time t=0s to t=10s, the channel is empty. As from t=10s, new flows of each class are started at three second intervals, and begin competing for the channel. By t=37s each class has five active flows; two 64-Kbps-HP flows, three 128-Kbps-HP flows, three 200-Kbps-MP flows, and two 400-Kbps MP flows. From t=37s to t=140s, the network remains in this state in order for us to asses to what extent our protocol can sustain the quality of service. At t= 140s, four new flows are started at one second intervals as follows: 128-Kbps-HP, 400-Kbps-MP, 64-Kbps-HP, and finally 400-Kbps-MP. At this point, the network is exhibiting a high contention level, which means an increased mean number of unsuccessful transmissions attempts. From t=140s to t=200s, the simulation is completed with sixteen (16) flows backlogged in (16) sixteen different stations.

5.2. Experimental Results

In this section, we especially assess to what extent our admission control scheme is able to protect already active flows. Another important aspect highlighted in this section is the ability of our scheme to keep on admitting new entering flows based on a careful evaluation of their impact on all already active flows.

In this section, we compare the performance of our scheme (BD-bonded delay scheme) when using the admission control mechanism (AC) and without using AC. We refer to those two operation modes as with-AC and without-AC, respectively.

The overall network utilization is shown in Figure 9 in terms of the total achieved throughput (goodput) during the simulation. Clearly, when the network is sufficiently relaxed (before t=140s), there is sufficient bandwidth available and both BDS-AC and BDS achieve similar throughputs, carrying the load as it is offered. However, under stressed conditions, BDS gains a significant advantage over BDS-AC. The goodput gain reaches about 20% when the load is around 2.4 Mbps (between t=140s and t=200s). At this point, the admission control mechanism in BDS-AC rejects three entering flows, 128-Kbps-HP flow at t=140s, 400-Kbps-MP flow at t=141 s, and finally 400-Kbps-MP at t=143s. Meanwhile, a 64-Kbps-HP flow was accepted at t=141s. This

Figure 9. Overall network utilization

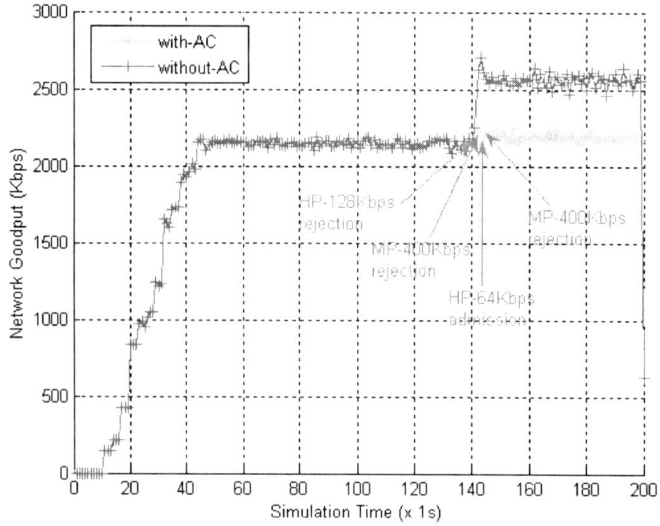

bandwidth gain comes, however, with a serious degradation in the QoS of all active multimedia flows as clearly revealed by delay measurements of single flows.

In Figure 10, Figure 11, Figure 12 and Figure 13, we present the normalized delays for the four flow types. Normalized delay is the value of the instantaneous delay minus the Max Delay allowed for the TC. As can be seen, the four flow types experience high delays as from t=140s when no admission control is applied. The performance degradation starts at t=140s with the over-admission of a 128-Kbps-HP flow. The performance is further degraded with the acceptation of three other flows. Depending on their respective offered load, the different TC flows are differently affected by this increasing in the network contention level.

Although high bit-rate flows maintain quite small contention window sizes compared to other flows, they are still unable to overcome the increasing network offered load and the entailed high PSTs. This decreasing in CW sizes is driven

Figure 10. End-to-end delays for 64-Kbps HP flows

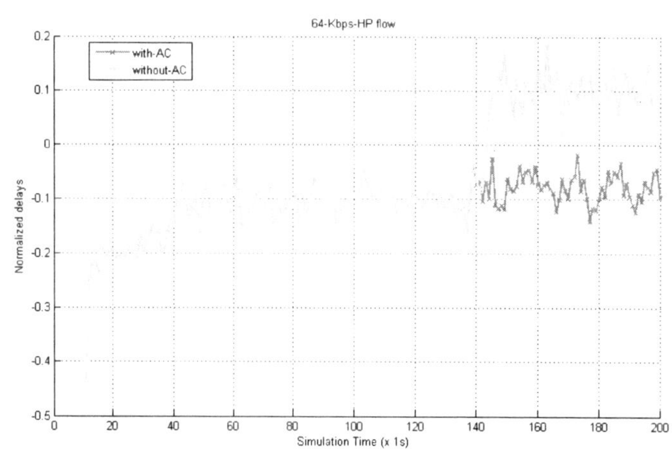

Figure 11. End-to-end delays for 128-Kbps HP flows

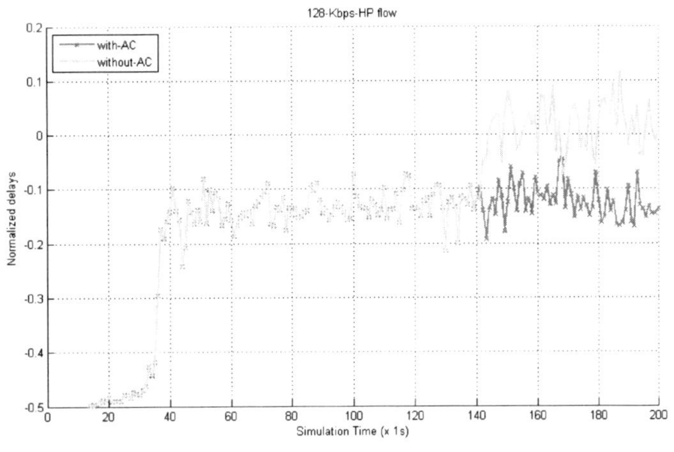

by the delay constraint presented in formula (6) without taking into account the queue overflow risk. As a matter of fact, throughput degradation is mostly caused by excessive packet dropping due to overloaded MAC queues. Here, the advantage of the AC becomes essential by clearly establishing relation between the packet arrival rate and the packet service time, and ultimately assess the achievable QoS before actually admitting any new entering flow. This beforehand flow admission assessment at each active station is done by deriving the ideal CW size (i.e., NewCWsize that ensures that the loss rate constraint will be respected) to be used and comparing it with both (i) CWmin to make sure that the contention is still controlled and (ii) CWmax to make sure that the delay constraints will be respected at each flow active in the network.

It is worth mentioning that the new entering streams were each time rejected by high-bit-rate TC flows (i.e., 400-Kbps MP flows) active in the network. In other words, during the distributed admission control process, stations carrying high-data-rate load rejects the new entering flow. Based on network-based PST measurements, it is much more difficult to maintain an acceptable loss rate if the TC flow is handling a high packet-arrival rate (λ).

Figure 12. End-to-end delays for 200-Kbps MP flows

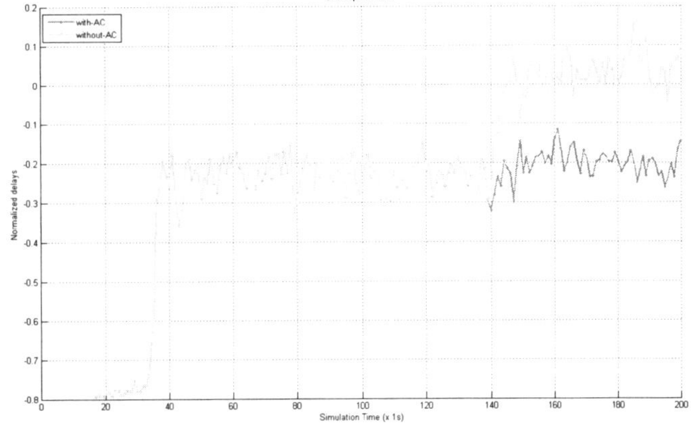

Figure 13. End-to-end delays for 400-Kbps

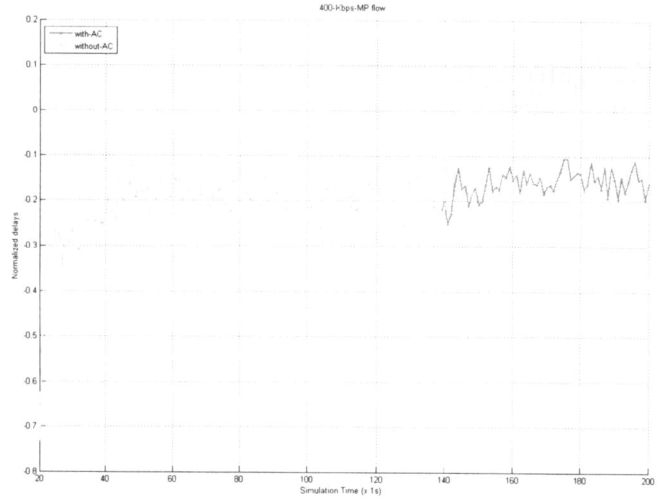

Another important observation to point out is that the results given in the model validation section and those presented in the performance evaluation section are slightly different. For instance, in the above presented results, the achieved delays of different TCs are far below their respective MaxDelay thresholds. This is due to the fact that, with the AC mechanism, the CW size effectively maintained by each flow is generally smaller than the one that would be maintained if AC is not used. In fact, with an additional constraint to avoid MAC queue overflow (i.e., NewCWsize calculation – formula (11)) the actually used CW size is smaller than the one given by formula (8).

For more results concerning the AC's performance reader can refer to (Nafaa & Ksentini, 2008).

6. CONCLUSION

An effective resource allocation in IEEE 802.11 is difficult to achieve due to the intrinsic nature of the CSMA/CA scheme. The difficulty lies in estimating the achievable QoS performance in the WLAN; this estimation depends on several time-varying factors including the number of active flows, the active traffic volume for each AC, etc. Unlike traditional wired networks (or point-coordinated wireless networks) where bandwidth provision can be managed using only bandwidth-availability information, flows' admission control in distributed 802.11 networks asks for additional parameters as well as more advanced models. In this chapter, we begun by introducing a new MAC design featuring a delay-sensitive backoff range adaptation. By monitoring both MAC queue dynamics of each traffic class and the overall network contention level, the MAC adaption scheme reacts based on the degree to which application QoS metrics (delay) are satisfied. We then presented a distributed admission control mechanism that uses the delay-based MAC layer model to accept new flows while protecting the active one. Finally, we validated both delay model and AD through simulation.

REFERENCES

Barry, M., Campbell, A.-T., & Veres, A. (2001). Distributed Control Algorithms for Service Differentiation in Wireless Packet Net-works. In *Proceeding of IEEE INFOCOM 2001*, Anchorage, Alaska, (vol. 1, pp. 582–90).

Bianchi, G. (2000). Performance analysis of the IEEE 802.11 distributed coordination function, *in IEEE Journal of Selected Area on Communication*, vol. 18, pp. 535-547, March 2000.

Bianchi, G., Fratta, L., & Oliveri, M. (1996). Performance evaluation and enhancement of the CSMA/CA MAC protocol for 802.11 wireless LANs, in proceeding of *IEEE PIMRC 1996*, Taipei, Taiwan, Oct. 1996, (pp. 392-396).

Cali, F., Conti, M., & Gregori, E. (2000, December). Dynamic Tuning of the IEEE 802.11 protocol to achieve a theoretical throughput limit. *IEEE/ACM Transaction on Networking*, 8(6), 785-790.

Chen, X., Li, B., & Fang, Y. (2005, March). A dynamic multiple-threshold bandwidth reservation (DMTBR) scheme for QoS provisioning in multimedia wireless networks. *IEEE Transactions on Wireless Communications*, *4*(2). doi:10.1109/TWC.2004.842981

Garcia-Molina, H. (1982). Elections in a distributed computing system. *IEEE Transactions on Computers*, *31*(1). doi:10.1109/TC.1982.1675885

IEEE802.11 Wireless LAN Medium Access Control (MAC) and Physical Layer (PHY) Specifications. (1999). IEEE Standard 802.11, June 1999.

IEEE802.11e Wireless LAN Medium Access Control (MAC) Enhancements for Quality of service (QoS). (2005). Final 802.11e standard, July 2005.

Ksentini, A., Nafaa, A., Guéroui, A., & Naimi, M. (2007, June). ETXOP: A Novel IEEE 802.11 MAC protocol with Admission Control for Sensitive Multimedia Applications. *Elsevier's Performance Evaluation Journal*, *64*(5), 419–443.

Mangold, S., Chois, S., May, P., Klein, O., Hertz, G., & Sibor, L. (2002). IEEE 802.11 wireless LAN for quality of service. In *Proceedings of European Wireless conference*, Florence, Italy, February 2002.

Nafaa, A., & Ksentini, A. (2008). On Sustained QoS Guarantees in Operated IEEE 802.11 Wireless LANs. [TPDS]. *IEEE Transactions on Parallel and Distributed Systems*, (August): 2008.

Nafaa, A., Ksentini, A., Mehoua, A., Ishibashi, A., Iraqi, Y., & Boutaba, R. (2005, July/August). Sliding Contention Window (SCW): Towards Backoff Range-based Service Differentiation over IEEE 802.11 Wireless LAN Networks. *IEEE Network Magazine, Special Issue on Wireless Local Area Networking: QoS provision & Resource Management, 19*(4).

Pong, D., & Moors, T. (2003). Call Admission Control for IEEE 802.11 Contention Access Mechanism. In *Proceeding of IEEE Globecom 2003*, San Francisco, California, USA (pp. 174-8).

Romdhani, L. Ni, Qi., & Turletti, T. (2003). Adaptive EDCF: Enhanced Service Differentiation for IEEE 802.11 802.11 Wireless Ad-Hoc Networks. In *Proceedings of IEEE WCNC 2003*, New Orleans, USA, March 2003.

Sheu, S.-T., & Sheu, T.-F. (2001, October). A bandwidth allocation/sharing/extension protocol for multimedia over IEEE 802.11 ad hoc wireless LANs. *IEEE Journal on Selected Areas in Communications*, *19*, 2065–2080. doi:10.1109/49.957320

Veres, A., Campbell, A-T, Barry, M., & Sun, L (2001). Supporting service differentiation in wireless packet networks using distributed control,(in *IEEE Journal of Selected Area on Communication*, vol. 19, no. 10, October 2001, (pp. 2081-2093).

Xiao, Y., & Li, H. (2004). Evaluation of Distributed Admission Con-trol for the IEEE 802.11e EDCA . *IEEE Communications Magazine*, *42*(9), S20–S24. doi:10.1109/MCOM.2004.1336720

Xiao, Y., Li, H., & Choi, S. (2004). Protection and Guarantee for Voice and Video Traffic in IEEE 802.11e Wireless LANs, proceeding of *IEEE INFOCOM* 2004.

Zhai, H., Chen, X., & Fang, Y. (2004). How Well Can the IEEE 802.11 Wireless LAN Support Quality of Service? *in IEEE Transaction on Wireless Communications*, 2004.

Zhai, H., Chen, X., & Fang, Y. (2006). A Call Admission and Rate Control Scheme for Multimedia Support over IEEE 802.11 Wireless LANs, *in ACM Wireless Networks (Winet)*, July 2006.

Ziouva, E., & Antonakopoulos, T. (2002). CSMA/CA Performance under High Traffic Conditions: Throughput and Delay Analysis . *Elsevier's Computer Communications Journal*, *25*(3), 313–321.

ENDNOTE

1 By brokering strategies, we mean the pricing strategies of network operators in offering QoS-enabled services on the basis of pre-established SLA agreements.

Chapter 7
Resource Allocation and QoS Provisioning for Wireless Relay Networks

Long Bao Le
Massachusetts Institute of Technology, USA

Sergiy A. Vorobyov
University of Alberta, Canada

Khoa T. Phan
University of California, Los Angeles, USA

Tho Le-Ngoc
McGill University, Canada

ABSTRACT

This chapter reviews fundamental protocol engineering aspects and presents resource allocation approaches for wireless relay networks. Important cooperative diversity protocols and their typical applications in different wireless network environments are first described. Then, performance analysis and QoS provisioning issues for wireless networks using cooperative diversity are discussed. Finally, resource allocation in wireless relay networks through power allocation for both single and multi-user scenarios are presented. For the multi-user case, we consider relay power allocation under different fairness criteria with or without user minimum rate requirements. When users have minimum rate requirements, we develop a joint power allocation and addmission control algorithm with low-complexity to circumvent the high complexity of the underlying problem. Numerical results are then presented, which illustrate interesting throughput and fairness tradeoff and demonstrate the efficiency of the proposed power control and addmission control algorithms.

DOI: 10.4018/978-1-61520-680-3.ch007

INTRODUCTION

Emerging broadband wireless applications in most wireless networks require increasingly high throughput and more stringent quality-of-service (QoS) requirements. In this respect, multiple-antenna technologies have been recognized as important solutions for future high-speed wireless networks (Tarokh *et al*, 1998; Tarokh *et al*, 1999; Telatar, 1998). Particularly, employment of multiple antennas at transmitter and/or receiver sides can provide significant multiplexing and/or diversity gains (Zheng & Tse, 2003). The net effects of these gains are the improvements in terms of wireless link robustness (i.e., lower bit error rate (BER)) and network capacity. Unfortunately, the implementation of multiple antennas in most modern mobile devices may be challenging due to their small sizes.

Cooperative diversity has been proposed as an alternative solution where a virtual antenna array is formed by distributed wireless nodes each with one antenna. Cooperative transmission between a source node and a destination node is performed with assistance of a number of relay nodes. In particular, the source and relay nodes collaboratively transmit information to the destination node (Laneman *et al*, 2004; Laneman & Wornell, 2003; Le & Hossain, 2007; Nabar *et al*, 2004; Nosratinia *et al*, 2004; Sendonaris *et al*, 2003a, b; Zhifeng *et al*, 2006). It is intuitive that in order to make cooperative transmission efficient or even possible, the source node has to carefully choose one or several "good" relays and first forward its data to those relays. Then, the source and relays can coordinate their transmissions in such a way that maximum multiplexing/diversity gains can be achieved at the destination node.

Although cooperative diversity is simple in concept, there are many technical issues to be resolved for practical implementation. First, protocol design for cooperative diversity is one of the important research focuses (Azarian *et al*, 2008; Laneman *et al*, 2004; Laneman & Wornell, 2003; Sendonaris *et al*, 2003a, b). Second, it is worth noting that most practical cooperative diversity protocols have two phases: in the first phase, the source node broadcasts its message to assisting relays; in the second phase, the relays collaboratively transmit the received information to the destination. Therefore, cooperative transmission may not be always beneficial or even necessary because direct transmission from the source to the destination node may already be successful. Adaptive cooperative protocols, where nodes cooperate only when necessary and/or they cooperate using incremental transmissions, usually have significantly better performance than "straight-forward" protocols (Azarian *et al*, 2008; Dai & Letaief, 2008; Le *et al*, 2007; Le & Hossain, 2008a; Zhao & Valenti, 2005). In addition, emerging technology such as network coding can be employed to design cooperative protocols (Koetter & Medard, 2003; Li *et al*, 2003; Xiao *et al*, 2007). Finally, other important issues such as relay selection, synchronization among relays' transmissions need to be considered for practical implementation (Beres & Adve, 2008; Bletsas *et al*, 2006; Le & Hossain, 2008b; Lin *et al*, 2006; Ng & Yu, 2007; Tannious & Nosratinia, 2008; Zhao *et al*, 2007).

While most existing works on cooperative diversity in the literature focus on design and performance analysis of cooperative protocols, resource allocation for wireless relay networks receives less attention. However, resource allocation also has significant impacts on system performance (Gunduz & Erkip, 2007; Li *et al*, 2007; Liang *et al*, 2007; Luo *et al*, 2007; Madsen & Zhang, 2005; Yao *et al*, 2005). In fact, assisting relays usually have limited radio resources (e.g., bandwidth and power) and they are shared by several source-destination pairs. Therefore, a smart radio resource allocation for wireless relay networks guarantees both fair access to available relays and good overall network throughput performance. In addition, by using a proper relay selection strategy where each source-destination pair only selects one or a small number of good relays, efficient resource utilization can be achieved with low implementation complexity. Finally, distributed resource allocation algorithms are usually required in wireless relay ad hoc networks because there is no central controller in such applications.

In this chapter, we attempt to provide a brief survey on cooperative protocols, key design issues and resource allocation problems in wireless relay networks. In particular, we describe popular cooperative protocols and their possible extensions and enhancements. We also briefly present typical applications of cooperative communications, namely for multihop cellular and ad hoc networks, and broadcasting applications in ad hoc networks. In addition, we review some existing works on performance analysis and QoS provisioning issues for wireless relay networks. Finally, we introduce a resource allocation framework for single-user and multi-user relay networks including both centralized and decentralized power allocation algorithms.

This chapter is organized as follows. We first describe fundamentals of cooperative diversity techniques and protocols. Then, we overview their typical applications and research issues. We discuss performance analysis and QoS provisioning issues for wireless relay networks. Then, resource allocation problems via power allocation for both single- and multi-user relay networks are presented. Finally, conclusions are stated.

BRIEF OVERVIEW OF COOPERATIVE DIVERSITY

Cooperative diversity protocols allow a number of users to relay signals for one another in such a way that a diversity gain can be achieved. In fact, information theoretic capacity of such a network setting, named a relay channel, has been investigated a few decades ago (Cover & Gamal, 1979). Deep understanding of MIMO systems from both information theoretic and practical system design viewpoints over the past decade has stimulated and attracted significant research efforts in cooperative diversity. In this section, we provide a brief survey on fundamentals of cooperative diversity.

Consider a source node s communicating to a destination node d with the help of m relays, $r_1, r_2, \ldots, r_m$. Let a_{ij} be the channel gain between nodes i and j, and P_i be the transmission power of node i. The signal is corrupted by additive white Gaussian noise. For simplicity, throughout this section, we assume that N is the white Gaussian noise power measured in the signal bandwidth at all nodes. We assume that cooperation among users is performed in phases (i.e., time slots) and users can be synchronized by a common system clock. Figure 1 illustrates a general cooperative diversity protocol where the source broadcasts its message in the first phase and the relays retransmit the message in the second phase. In the following, we describe some popular coopera-

Figure 1. Cooperative protocols

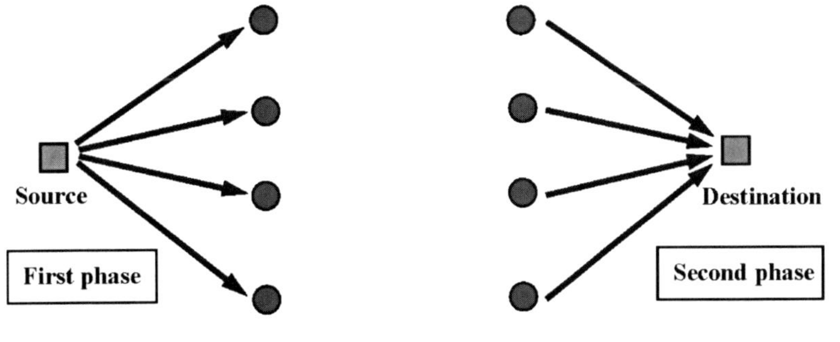

tive diversity protocols and their corresponding performances.

Amplify-and-Forward

In this cooperative protocol, the source broadcasts message x_s in the first phase. The message is received by the destination and relays. Each relay r_i amplifies the received signal in the first phase and transmits to the destination in the second phase. The destination combines the signals received in both phases to decode the message. Specifically, the signal received by relay r_i in the first phase (denoted as y_{r_i}) can be written as

$$y_{r_i} = a_{sr_i} x_s + z_{r_i} \tag{1}$$

where a_{sr_i} is the channel gain for link s - r_i and z_{r_i} denotes Gaussian noise at relay r_i. Suppose each relay normalizes the received signal before transmitting to the destination. Then, the transmitted signal can be written as

$$x_{r_i} = g_{r_i} y_{r_i} \tag{2}$$

where g_{r_i} is the amplifying gain which is given by

$$g_{r_i} = \sqrt{\frac{P_{r_i}}{\left|a_{sr_i}\right|^2 P_s + N}}. \tag{3}$$

Assuming that a maximum-ratio-combiner (MRC) is used at the destination, the source-destination capacity of this protocol is given as (Zhao *et al*, 2007)

$$C_{AF} = \frac{1}{m+1} \log\left(1 + \mathrm{SNR}_s\left|a_{sd}\right|^2 + \sum_{i=1}^{m} \frac{\mathrm{SNR}_s\left|a_{sr_i}\right|^2 \mathrm{SNR}_{r_i}\left|a_{r_i d}\right|^2}{\mathrm{SNR}_s\left|a_{sr_i}\right|^2 + \mathrm{SNR}_{r_i}\left|a_{r_i d}\right|^2 + 1}\right) \tag{4}$$

where $SNR_j = P_j / N$ is the signal-to-noise ratio (SNR) at node $j \in \{s, r_i \mid i = 1, \ldots, m\}$, P_j denotes the power at the souse or relay node, N is the noise power, and a_{sd} is the channel gain of the link *s-d*.

Another important performance measure that is extensively used for investigating the performance of different cooperative diversity protocols is the outage probability. In Rayleigh fading channels, the outage probability of the amplify-and-forward (AF) cooperative protocol can be approximated as (Zhao *et al*, 2007)

$$P_{AF}^{out}\left(SNR, R\right) = \Pr\left[C_{AF} < R\right] \propto \left(\frac{2^{2R} - 1}{SNR}\right)^{m+1}$$

where $SNR = P / N$ and it is assumed that all nodes transmit at power level P. This outage probability shows that AF cooperative protocol achieves diversity order of $m+1$ with m relays.

Decode-and-Forward

For the decode-and-forward (DF) cooperative protocol, relay nodes apply some forms of detection and/or decoding before encoding the information and forwarding it to the destination. Such a cooperative protocol also has two phases (i.e., time slots). In the first phase, the source broadcasts the signal to the relays, which subsequently detect and/or decode it. In the second phase, the relays transmit re-encoded signals to the destination using repetition or space-time codes.

For protocols that require relays to fully decode the received signal in the first phase, the set of relays, which successfully decode the signal at the end of the first phase, is only a subset of all available relays. Let $D(s)$ denote the set of successfully-decoding relays, which will be called a decoding set in the following. For repetition-based coding, the destination receives separate retransmission from each relay $r_i \in D(s)$. Hence, we can write the signal from relay r_i received at the destination d as

$$y_d = a_{r_i d} x_{r_i} + z_d \tag{5}$$

where x_{r_i} denotes the signal transmitted by relay node r_i, $a_{r_i d}$ stands for the channel gain of the link r_i-d, and z_d denotes the Gaussian noise at the destination. If space-time coding is used, the destination will simultaneously receive the superimposed signals from all relays $r_i \in D(s)$ Hence, the received signal at the destination in the second phase can be expressed as

$$y_d = \sum_{r_i \in D(s)} a_{r_i d} x_{r_i} + z_d. \tag{6}$$

It has been shown in (Laneman & Wornell, 2003) that both repetition-based or space-time-coding- based DF protocols achieve full diversity order of $m+1$ in the low rate regime. This diversity gain has been shown to be achievable by a distributed linear dispersion codes (Jing & Hassibi, 2006) and a randomized space-time codes (Sirkeci-Mergen & Scaglione, 2007b). Although both AF and repetition-coding-based DF protocols achieve a full diversity gain, their throughput may degrade because each transmitting relay takes one time slot to transmit to the destination. This limitation can be overcome by enhancing cooperative protocols, namely by using selection/opportunistic or incremental relaying protocol, which will be described subsequently.

Selection/Opportunistic Relaying

Consider m relays available to assist transmission from the source to the destination. Instead of allowing all the relays as in the AF protocol or all the relays in the decoding set as in the DF protocol to transmit in the second phase, selection/opportunistic relay protocols choose one "best" relay to transmit in the second phase (Beres & Adve, 2008; Bletsas *et al*, 2006; Jing & Jafarkhani, 2008; Le & Hossain, 2008b; Lin *et al*, 2006; Ng & Yu, 2007; Tannious & Nosratinia, 2008; Zhao *et al*, 2007). Surprisingly, cooperative protocols based on using smart relay selection strategies usually achieve full diversity order while providing higher throughput than the standard protocols. In fact, the superior throughput performance of selection relaying protocols stems from the fact that they use radio resources (i.e., power and bandwidth) more efficiently than the basic cooperative protocols presented in the previous sections.

Some typical relay selection strategies for both AF and DF based protocols are presented next. Consider an AF protocol with one selected relay, say r_i. From (4), the capacity of the source-destination channel with one relay is

$$C_{SAF} = \frac{1}{2}\log\left(1 + \mathrm{SNR}_s|a_{sd}|^2 + \frac{\mathrm{SNR}_s|a_{sr_i}|^2 \mathrm{SNR}_{r_i}|a_{r_i d}|^2}{\mathrm{SNR}_s|a_{sr_i}|^2 + \mathrm{SNR}_{r_i}|a_{r_i d}|^2 + 1}\right). \tag{7}$$

Therefore, to maximize the capacity, a relay selection strategy would choose a relay that maximizes (Zhao *et al*, 2007)

$$\frac{\mathrm{SNR}_s|a_{sr_i}|^2 \mathrm{SNR}_{r_i}|a_{r_i d}|^2}{\mathrm{SNR}_s|a_{sr_i}|^2 + \mathrm{SNR}_{r_i}|a_{r_i d}|^2 + 1}. \tag{8}$$

For the DF protocol, there is a set of relays which successfully decode the signal in the first phase (i.e., in the decoding set $D(s)$). If relay $r_i \in D(s)$ is chosen for transmission in the second phase, the capacity of the source-destination channel is

$$C_{SDF} = \frac{1}{2}\log\left(1 + \mathrm{SNR}_s|a_{sd}|^2 + \mathrm{SNR}_{r_i}|a_{r_i d}|^2\right). \tag{9}$$

Therefore, to maximize the source-destination capacity, an opportunistic relay selection strategy would choose a relay in the decoding set that maximizes

$$\mathrm{SNR}_s|a_{sd}|^2 + \mathrm{SNR}_{r_i}|a_{r_id}|^2. \tag{10}$$

In (Beres & Adve, 2008; Zhao *et al*, 2007), it has been shown that relay selection strategies in (8) and (10) achieve the full diversity order. Note that these selection metrics require the estimates of $SNR_s|a_{sd}|^2$ and $\mathrm{SNR}_{r_i}|a_{r_id}|^2$. In (Bletsas *et al*, 2006), two simpler relay selection metrics which require only channel gains a_{sr_i} and a_{r_id} have been proposed. Specifically, relay selection strategies that choose a relay such that

$$r_i^* = \arg\max_{r_i}\left\{\min\left\{|a_{sr_i}|^2, |a_{r_id}|^2\right\}\right\} \tag{11}$$

$$r_i^* = \arg\max_{r_i}\left\{\frac{2}{\frac{1}{|a_{sr_i}|^2}+\frac{1}{|a_{r_id}|^2}} = \frac{2|a_{r_id}|^2|a_{sr_i}|^2}{|a_{r_id}|^2+|a_{sr_i}|^2}\right\}. \tag{12}$$

have been developed.

Note that the relay selection criterion in (11) chooses a relay with largest channel gains in both source-relay and relay-destination links. On the other hand, the relay selection rule in (12) maximizes the harmonic mean of channel gains for the source-relay and relay-destination links. It has been shown in (Bletsas *et al*, 2006) that these relay selection criteria provide the optimum diversity-multiplexing tradeoff achieved by the distributed space-time cooperative protocol (Laneman & Wornell, 2003). Other relay selection strategies for orthogonal frequency-division multiple access (OFDMA)-based wireless cellular relay networks and ad hoc networks can be found in (Le & Hossain, 2008b; Ng & Yu, 2007; Tannious & Nosratinia, 2008).

Incremental Relaying

Although selection relaying uses radio resources more efficiently than fixed relaying, both fixed and selection relaying protocols have to always repeat transmission. In fact, direct transmission from the source to the destination may be successful if the corresponding channel condition is not too "bad". Therefore, it can be more efficient if relay transmission is invoked only when direct transmission from the source to the destination in the first phase fails. One simple incremental relaying protocol based on using AF principle which exploits the aforementioned aspect works as follows (Laneman *et al*, 2004). Upon decoding its received signal at the end of the first phase, the destination broadcasts the decoding outcome to the source and relays. If the destination succeeds in decoding the message in the first phase, the source and relays do nothing. Otherwise, all or selected relays amplify their received signals and transmit to the destination. The destination combines all the signals and decodes again.

In fact, incremental relaying protocol can be implemented as an extension of hybrid automatic repeat request (ARQ) protocol (Azarian *et al*, 2008; Dai & Letaief, 2008; Le *et al*, 2007; Le & Hossain, 2008a; Zhao & Valenti, 2005). One possible implementation of ARQ-based relaying can be described as follows (Zhao & Valenti, 2005). Initially, the source node encodes b bits of information into a code-word with length n symbols. The code-word is broken into M blocks, each of which has length n / M. The code can be a simple repetition code, where all blocks are identical, or the blocks can be obtained by puncturing a mother code. The protocol starts by transmitting the first block from the source node. The destination upon decoding the message broadcasts the decoding outcome to all other nodes. If the decoding at the destination is successful, the source proceeds to transmit a new message. Otherwise, either all or one selected relay in the decoding set (i.e., relays that successfully decode the message) re-encode the message and transmit the second block to the destination. The destination combines all the received blocks and attempts to decode again. This procedure continues until the destination is successful in decoding the message or all M blocks are transmitted and the message is discarded.

Incremental relaying has both diversity and throughput advantages because relaying is invoked only when necessary. In (Laneman *et al*, 2004), the authors have shown that incremental relaying using AF principle as presented above achieves the full diversity order. In addition, it can be seen that ARQ-based incremental relaying allows many different code designs, where well-investigated hybrid ARQ protocols can be adapted to the relaying network setting. Also, a combination of incremental relaying, hybrid ARQ and relay selection achieves throughput and energy improvement compared to the standard protocols while still having a full diversity gain.

Other Protocol Enhancements

There are some other possible enhancements of the aforementioned cooperative protocols available in the literature. In particular, network coding can be combined with standard cooperative protocols to improve throughput performance (Koetter & Medard, 2003; Li *et al*, 2003; Xiao *et al*, 2007). The network coding is based on the idea that the users involved in cooperative transmissions can combine their own information with other users's information, e.g., by using linear coding (Li *et al*, 2003), and transmit the combined information in an appropriate manner. This is because through cooperation, users know the messages of their assisted users. This would enhance throughput performance for each user because a single transmission transmits both the user's own message and the message of an assisted user in the combined signal.

Other possible enhancements include combination of adaptive modulation and coding into cooperative protocols (Nechiporenko *et al*, 2009), employing coding in cooperative protocols (Hunter & Nosratinia, 2006), adding power and scheduling considerations for selection of a group of active retransmitting nodes (Ko *et al*, 2009a, b). In (Wei *et al*, 2006), a detection technique for wireless networks, where synchronization of users is impossible, has been proposed. This technique mimics an equalization technique employed in a frequency-selective fading channel.

Finally, relaying transmission concepts can be combined with a medium-access-control (MAC) protocol to improve its throughput performance (Zhu & Cao, 2005). Specifically, through exchanging control information (e.g., RTS/CTS handshake signals), each user can find the optimal transmission strategy between direct transmission and relaying transmission through other relays (i.e., neighboring nodes). By choosing a transmission strategy with higher throughput, the MAC protocol can achieve better overall throughput performance.

Further Discussions

Summarizing the aforementioned cooperative protocols, it is also worth pointing out some important design issues. First, in principle a source-destination pair can be assisted by a large number of relays; however, a small number of "good" relays would be selected for cooperative transmission in most practical applications. Selecting a small number of relays for cooperation would be preferred taking into account both design complexity and overall network performance. Second, cooperative transmission may not be always beneficial especially if the source-destination link is very strong. Therefore, an adaptive cooperative protocols based on using a right amount of cooperation such as incremental relaying protocols would perform better than non-adaptive protocols (e.g., AF and DF protocols).

Due to the distributed nature of cooperative diversity protocols, their employment raises several practical implementation issues. First, synchronization among wireless nodes for implementing the MRC or distributed beamforming may be difficult. In order to resolve this challenge, a receiver detector at the destination node must be able to operate under asynchronous transmissions from a source and relay nodes (Wei *et al*, 2006). In scenarios where space-time coding is employed, the underlying coding strategy should be designed to operate in a decentralized manner (Jing & Hassibi, 2006; Sirkeci-Mergen & Scaglione, 2007b). Second, efficient allocation of radio resources to source and relay nodes in the network should be performed for optimum network performance. We will discuss resource allocation issues in more details in the following sections.

APPLICATIONS AND IMPLEMENTATION OF COOPERATIVE DIVERSITY

Cellular Relay Networks

Cooperative diversity can be employed to enhance throughput and/or improve BER performance of a multi-hop cellular network (Le & Hossain, 2007). In particular, users can take turn to serve as relays for one another. Alternatively, a set of fixed relays can be implemented to assist all the users in each cell. In (Sendonaris *et al*, 2003a, b), a cooperation strategy for a two-user code division multiple access (CDMA) cellular wireless network has been proposed. According to this strategy, each user has two transmission periods where it transmits directly to the base station (BS) in the first period and cooperates with the other user to transmit in the second period. It has been shown that user cooperation indeed increases network throughput and decreases network sensitivity to channel variations.

For multihop cellular networks with fixed relays, transmissions from/to the BS of different users with the help of deployed relays can enhance throughput and BER performances. Since a small number of deployed relays is shared by a large number of users, a relay selection strategy should be employed for cooperative transmissions between users and the BS. In addition, if each fixed relay has several transceivers, which can assist several users simultaneously, power and bandwidth allocation should be performed at these relays to optimize the overall network performance. In general, cooperative transmissions between users and the BS can occur in a multihop fashion (Le & Hossain, 2007). In this case, a joint cluster-based routing and cooperative transmission can be employed as for wireless ad hoc networks. This will be presented in the next subsection.

Cluster-Based Wireless Ad Hoc Networks

In wireless ad hoc networks, a source may want to communicate with a destination that is far away. Hence, a routing protocol is needed to deliver data in a multihop fashion. A traditional routing protocol typically finds a set of wireless links from the source to the destination to establish a multihop route for end-to-end data delivery. Using cooperative diversity, the multihop route can be formed by a set of cooperative and robust abstract "links" instead of simple wireless links (Scaglione *et al*, 2006). In fact, cooperative diversity can be jointly used with a hierarchical routing to enhance end-to-end performance (Hong *et al*, 2002).

For hierarchical routing in wireless ad hoc networks, wireless nodes in the network form clusters each of which is a set of wireless nodes in a neighborhood (Morgenshtern & Bolcskei, 2007). Each cluster has one cluster head. A cluster mimics a cell in wireless cellular network where the cluster head functions similarly to a BS. A hierarchical routing protocol typically finds a set of clusters between the source and the destination. Then, end-to-end routing of information is performed within and between clusters independently. Cooperative diversity can be used for inter-cluster routing as shown in figure 2.

In this cluster-based cooperative routing, a set of wireless nodes between any two neighbor-

Figure 2. Cluster-based cooperative transmission

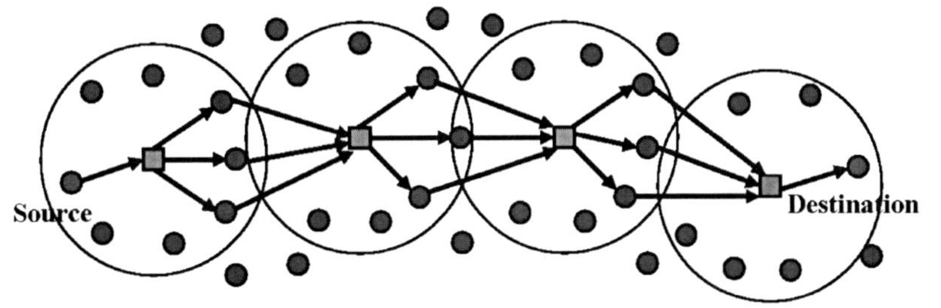

ing clusters is chosen by cluster heads to serve as gateway nodes. The gateway nodes are the relay nodes that assist transmission between two clusters. Therefore, any cooperative protocols presented in the previous sections can be used for inter-cluster transmission. Note that this network architecture can also be used in infrastructure-based wireless mesh networks where mesh routers serving a number of mesh clients can serve as cluster heads. In (Le & Hossain, 2008a), an analytical model has been developed to quantify performance of the aforementioned cluster-based cooperative routing where incremental relaying is employed for inter-cluster transmission.

Cooperative Broadcast in Wireless Ad Hoc Networks

Cooperative diversity can be exploited to enhance broadcast performance in wireless ad hoc networks (Maric & Yates, 2004; Scaglione & Hong, 2003; Sirkeci-Mergen *et al*, 2006; Sirkeci-Mergen & Scaglione, 2007a, b). In broadcast applications, a message is required to be transmitted from a source to all other nodes in the network. By using cooperative diversity, performance improvement in terms of energy consumption or message delivery probability can be achieved by exploiting the fact that each node in the network can collect signals from several simultaneously transmitting nodes. As a special case, cooperative broadcast can be performed in different levels as follows (Sirkeci-Mergen *et al*, 2006). Each node in the network accumulates signals transmitted from other nodes until it achieves high enough SNR to decode the message. After successfully decoding the message, a node broadcasts it into the network. Therefore, by using a smart detection technique, each node can combine signals transmitted from different nodes to enhance the broadcast performance.

PERFORMANCE ANALYSIS AND QOS PROVISIONING FOR WIRELESS RELAY NETWORKS

There is a large body of literature on performance analysis and QoS provisioning for wireless relay networks. In this section, we attempt to review some important research problems and issues along these lines. In fact, there are two important research directions pursued in the literature. The first direction focuses on analyzing performances of cooperative diversity protocols. Performance measures under consideration include ergodic, outage capacity, bit/symbol error rate (B/SER), throughput and packet/frame delay. The second direction concentrates on QoS provisioning, resource allocation and protocol engineering for particular cooperative protocols and applications. In fact, solution approaches for the underlying problems in this direction usually rely on some results in the first direction.

Here, we review some research issues and results for the aforementioned directions. Note that we have discussed ergodic and outage capacity for several important cooperative diversity protocols in the previous sections. In addition, resource allocation for wireless relay networks plays an important role in improving the network performance; therefore, we will treat this topic in more detail in the next section. Regarding B/SER performance analysis of cooperative diversity protocols, there exists many publications which consider different protocols, network settings, e.g., multi-branch, multihop relay networks, (Boyer *et al*, 2004; Ribeiro *et al*, 2005). In general, the BER of a particular cooperative diversity protocol is lower-bounded by the corresponding outage probability.

Exact analysis of B/SER for cooperative diversity protocols (e.g., AF and DF protocols) is usually cumbersome. However, there are some existing works, which consider approximated B/SER analysis for these protocols, e.g., (Ribeiro *et al*, 2005). In particular, the approximated analysis in (Ribeiro *et al*, 2005) gives closed-form expressions of SER for the AF cooperative protocol, which is quite accurate in the high SNR regime. Specifically, consider a scenario where there are m relays helping a source-destination pair. Let γ_{sd}, γ_{sr_i}, $\gamma_{r_i d}$ be the average received SNR for the source-destination, source-relay i, relay i -destination links, respectively. The SER of the AF cooperative protocol can be approximated as (Ribeiro *et al*, 2005)

$$\overline{P}_e \approx C(m,K)\frac{1}{\gamma_{sd}}\prod_{i=1}^{m}\left(\frac{1}{\gamma_{sr_i}}+\frac{1}{\gamma_{r_i d}}\right) \tag{13}$$

where $C(m,K)$ is a constant depending on the number of relays m, modulation scheme, and specular factor K of the Ricean fading channel. The SER in (13) shows that the AF cooperative protocol achieves the full diversity order. Derivation of SER for the DF protocol can be found in (Boyer *et al*, 2004).

Regarding QoS provisioning issues, many wireless applications have delay constraints to guarantee minimum QoS performance besides a common minimum B/SER requirement. In addition, data traffic may be bursty which is usually queued in data buffers upon arriving from the higher layers. Therefore, the total packet delay may consist of queueing and transmission delay components (Cerutti *et al*, 2008). Of course, when data is not buffered, the total packet delay is simply the transmission delay (Narasimhan, 2008). For cooperative diversity protocols that involve several block transmissions for each data packet such as incremental relay protocols, the total packet delay can be controlled by smartly regulating the average number of transmissions. This is similar to controlling the number of transmission attempts in a classical truncated ARQ protocol (Le *et al*, 2007). In general, a cross-layer model should be developed to harmonize and optimize the network performance while meeting delay constraints (Le & Hossain, 2008a).

For emerging applications in multihop wireless networks (e.g., wireless mesh and sensor networks), network/protocol design should be performed to optimize network or QoS performance measures of interest. In (Le & Hossain, 2008b), optimal cross-layer algorithms have been developed to perform joint relay selection, power allocation, and routing to optimize different performance measures including power minimization and rate utility maximization in a general multihop wireless network. In (Khandani *et al*, 2007; Madan *et al*, 2009), centralized and distributed cooperative routing protocols have been proposed to minimize the energy consumption. Finally, relay-selection and power allocation strategies have been proposed to maximize lifetime of a wireless sensor network using the AF cooperative diversity protocol in (Huang *et al*, 2008). These are just few examples where network protocol design and QoS provisioning problems for the corresponding applications are considered. In general, these design problems depend on the specifics of underlying applications which may require very diverse solution approaches to resolve.

RESOURCE ALLOCATION FOR COOPERATIVE WIRELESS NETWORKS

Single-User Resource Allocation

There are quite a few existing works considering resource allocation for single-user cooperative wireless networks (Gunduz & Erkip, 2007; Li *et al*, 2007; Liang *et al*, 2007; Luo *et al*, 2007; Madsen & Zhang, 2005; Yao *et al*, 2005). For the single-user setting, there is only one source communicating to only one destination with the help of one or several relays. Since the capacity of a general relay channel is still an open problem, only some upper and lower capacity bounds are derived in the literature. In (Madsen & Zhang, 2005), lower and upper capacity bounds for different cooperation strategies including time-division relaying and compress-and-forward have been derived. Optimal power allocation schemes that aimed at maximizing these capacity bounds have also been adopted. In (Liang *et al*, 2007), the capacity bounds for parallel relay channels with degraded sub-channels have been derived and optimized through power allocation.

For the practical AF and DF protocols, optimal power allocation methods aiming at maximizing the SNR have been developed in (Li *et al*, 2007; Zhao *et al*, 2007). Using the SNR expression of the AF protocol with m relays (4), the problem of SNR maximization under total and individual relay power constraints can be written as

$$\max_{\{P_{r_i}\}} \quad \mathrm{SNR}_s|a_{sd}|^2+\sum_{i=1}^{m}\frac{\mathrm{SNR}_s|a_{sr_i}|^2\,\mathrm{SNR}_{r_i}|a_{r_id}|^2}{\mathrm{SNR}_s|a_{sr_i}|^2+\mathrm{SNR}_{r_i}|a_{r_id}|^2+1} \tag{14}$$

$$\text{subject to}: \sum_{i=1}^{m} P_{r_i} \le P_T \tag{15}$$

$$0 \le P_{r_i} \le P_i^{\max}. \tag{16}$$

Assuming that white Gaussian noise powers measured in the signal bandwidth at all the relays and the destination are equal to N and the source transmission power P_s are fixed. The optimal relay power in the high SNR regime can be found as (Zhao *et al*, 2007)

$$P_{r_i}=\left[\frac{P_s|a_{sr_i}|^2}{|a_{r_id}|\sqrt{N}}\lambda-\frac{P_s|a_{sr_i}|^2}{|a_{r_id}|^2}\right]_0^{P_i^{\max}} \tag{17}$$

where $[.]_0^{P_i^{max}}$ denotes the projection operation on the interval $[0, P_i^{max}]$, and λ is chosen such that the total relay power is satisfied. Optimal relay power allocation for DF protocol is more involving and depends on the decoding strategy employed at the relays. In (Li *et al*, 2007), optimal relay solutions have been derived for some special cases.

Multi-User Resource Allocation

In this section, we present a resource allocation framework for a multi-user wireless relay network. More details can be found in (Phan *et al*, 2008; Phan *et al*, 2009a, b, c).

System Models

Consider a multi-user relay network in which M source nodes s_i transmit data to their corresponding destination nodes d_i, $i \in \{1,...M\}$. There are also L relay nodes r_j, $j \in \{1,...,L\}$, which are employed to assist transmissions from source to destination nodes. The set of relay nodes assisting the transmission of the source node s_i is denoted by $\mathsf{R}(s_i)$. The set of source nodes using the relay node r_j is denoted by $\mathsf{S}(r_j)$, i.e., $\mathsf{S}(r_j)=\{s_i \mid r_j \in \mathsf{R}(s_i)\}$. Therefore, one particular relay node can forward data for several users[1]. We assume that the AF cooperative scheme is used for re-transmission. Moreover, *orthogonal transmissions* are assumed for simultaneous transmissions among different users by using different channels, e.g., different frequency bands, and time division multiplexing is employed by the AF cooperative scheme for each user. Then, the transmission from a source to a destination node can be described as follows. In the first phase, each source node s_i transmits data to its chosen relays in the set $\mathrm{R}(s_i)$. In the second phase, each relay node amplifies and forwards its received signal to d_i. The corresponding system model is shown in Fig. 3.

The investigated system model is quite general and it covers a large number of applications in different network settings. For example, the model can be applied to cellular wireless networks using relays for uplink with one destination (BS) or downlink with one source (BS) and many destinations. It can also be directly applied to multi-hop wireless networks such as sensor/ad hoc or wireless mesh networks. Moreover, in our model, each source can be assisted by one, several, or all available relays. The presented model, therefore, captures most relay models considered in the literature.

Let P_{s_i} denote the power transmitted by source node s_i; $P_{r_j}^{s_i}$ denotes the power transmitted by relay node $r_j \in \mathrm{R}(s_i)$ for assisting the source node s_i, and $a_{s_i r_j}$ and $a_{r_j d_i}$ denote the channel gains for links s_i - r_j and r_j - d_i, respectively. The channel gains could include the effects of path loss, shadowing and fading. To keep the model in this section general, we assume that the variances of additive circularly symmetric white Gaussian noise (AWGN) at the relay r_j and at the destination node d_i are N_{r_j}, N_{d_i}, respectively. We consider the case when the source-to-relay link is (much) better than the source-to-destination link, which would be an outcome of a typical relay selection strategy employed by each source node. Assuming that MRC is employed at the destination node d_i, the SNR of the combined signal at the destination node d_i can be written as[2]

Figure 3. Multi-user wireless relay network

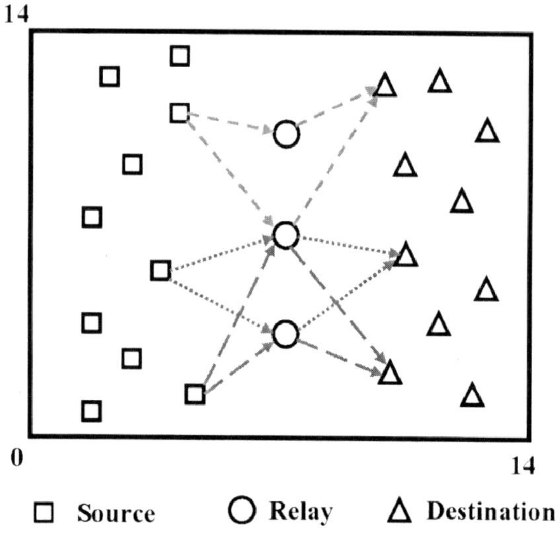

$$\gamma_i = \sum_{r_j \in \mathrm{R}(s_i)} \frac{P_{r_j}^{s_i}}{\alpha_{r_j}^{s_i} P_{r_j}^{s_i} + \beta_{r_j}^{s_i}} \tag{18}$$

where

$$\alpha_{r_j}^{s_i} = \frac{N_{r_j}}{|a_{s_i r_j}|^2 P_{s_i}}, \quad \beta_{r_j}^{s_i} = \frac{N_{d_i} N_{r_j}}{|a_{s_i r_j}|^2 |a_{r_j d_i}|^2 P_{s_i}} + \frac{N_{d_i}}{|a_{r_j d_i}|^2}.$$

It can be verified that the SNR γ_i for user s_i is concave increasing with respect to $P_{r_j}^{s_i}$, $r_j \in \mathrm{R}(s_i)$ Moreover, the rate of user s_i which is defined as $R_i = \log(1 + \gamma_i)$ is also concave increasing.

Formulations of Power Allocation Problem

In general, resource allocation in wireless networks should take into account fairness among users. An attempt to maximize the sum of rates of all the users would generally degrade performance of the worst user(s) significantly. To balance fairness and throughput performance for all the users, we consider two different optimization criteria for power allocation. The first criterion aims at maximizing the minimum rate among all users. In essence, this criteria tries to make rates of all users as equal as possible. For the second criterion, users are given different weights and power allocation is performed to maximize the weighted sum of rates for all users. In the latter case, user(s) in unfavorable conditions could be allocated large weights to prevent severe degradation of their performance. Another possible application for this optimization criterion is to perform QoS differentiation where users of higher service priority can be allocated larger weights. In both optimization criteria, we impose constraints on the total maximum power that each relay can use to assist the corresponding users.

A. Max-Min Rate Fairness Based Power Allocation

We first consider the power allocation problem under max-min rate fairness for the users. Mathematically, it can be formulated as (Phan *et al*, 2009c)

$$\max_{\{P_{r_j}^{s_i} \geq 0\}} \quad \min_{s_i} R_i \tag{19}$$

$$\text{subject to}: \sum_{s_i \in S(r_j)} P_{r_j}^{s_i} \leq P_{r_j}^{\max},\ j = 1, \ldots, L \tag{20}$$

where R_i is the rate of user s_i and $P_{r_j}^{\max}$ is the maximum power of relay r_j. The left-hand side of (20) is the total power that relay r_j allocates to its assisted users which is constrained to be less than its maximum power budget. This constraint is required to avoid overloading the relays in the network.

In general, the power allocation obtained according to the problem (19)-(20) can result in a loss in network throughput because the objective function (19) specifically improves the performance of the worst user(s) that in turn can decrease the overall system throughput. Therefore, this criterion is applicable for networks in which all users are of (almost) equal importance. This is the case, for example, when wireless users pay the same subscription fees, and thus, demand similar level of QoS. It can be seen that the set of linear inequality constraints with positive variables in the optimization problem (19)-(20) is compact and nonempty. Hence, the problem (19)-(20) is always feasible. Moreover, since the objective function $\min_{s_i} R_i$ is an increasing function of allocated powers, the inequality constraints (20) should be met with equality at optimality. Introducing a new variable T, we can equivalently rewrite the optimization problem (19)-(20) in a standard form as

$$\max_{\{P_{r_j}^{s_i} \geq 0,\, T \geq 0\}} \quad T \tag{21}$$

$$\text{subject to:} \quad T - R_i \leq 0,\ i = 1, \ldots, M \tag{22}$$

$$\sum_{s_i \in S(r_j)} P_{r_j}^{s_i} = P_{r_j}^{\max},\ j = 1, \ldots, L. \tag{23}$$

It can be verified that the optimization problem (21)-(23) is convex. Thus, its optimal solution can be obtained using standard convex optimization algorithms (Boyd & Vandenberghe, 2004). In the following, we describe a different formulation for the power allocation problem, which achieves better throughput performance.

B. Weighted-Sum of Rates Fairness Based Power Allocation

As discussed before, max-min rate fairness based power allocation tends to improve performance of the worst user at the cost of overall network throughput degradation. Maximization of the weighted-sum of rates can potentially achieve certain fairness for different users by allocating large weights to users in unfavorable channel conditions while maintaining good network performance in general. Let w_i denote the weight allocated to user s_i. Then, the weighted-sum of rates fairness based power allocation problem can be mathematically posed as (Phan *et al*, 2009c)

$$\max_{\{P_{r_j}^{s_i} \geq 0\}} \quad \sum_{i=1}^{M} w_i R_i \tag{24}$$

$$\text{subject to}: \sum_{s_i \in S(r_j)} P_{r_j}^{s_i} \leq P_{r_j}^{\max}, \; j = 1, \ldots, L. \tag{25}$$

As in the optimization problem (19)-(20), it can be seen that the constraints (25) in the problem (24)-(25) must be met with equality at optimality. Otherwise, the allocated powers can be increased to improve the objective function, and thus, it contradicts with the optimality assumption. In addition, it can be verified that this optimization problem is convex; therefore, its optimal solution can be obtained by any standard convex optimization algorithms.

We would like to note that power allocation schemes based on other fairness criteria can also be considered. For instance, the proportional fairness criterion can be adopted. In terms of system-wide performance metric such as the network throughput, the latter criterion can ensure more fairness than the weighted-sum of rates, while achieving better performance than the max-min fairness (Kelly *et al*, 1998). It can be shown that the objective function to be maximized in the proportional fairness based power allocation scheme is $\prod_{i=1}^{M} R_i$. Consequently, this objective function can be re-formulated as convex function using the log-function.

Distributed Implementation for Power Allocation

To reduce communication overhead and to implement online power allocation for the multi-user relay network, we now develop a distributed algorithm for solving the optimization problem (24)-(25) and show that such a solution converges to the optimal solution. The algorithm is developed based on the dual decomposition approach in convex optimization (Bertsekas, 1999). Applications of this optimization technique for distributed routing, reverse engineering of MAC, and transmission control protocols can be found in (Chiang *et al*, 2007; Kelly *et al*, 1998; Xiao *et al*, 2004).

In dual decomposition method, the original problem is separated into independent subproblems that are coordinated by a higher-level master dual problem. Now, we first write the Lagrangian function by relaxing the total power constraints for the relays as follows

$$L\left(\text{¼}, P_{r_j}^{s_i}\right) = \sum_{i=1}^{M} w_i R_i - \sum_{j=1}^{L} \mu_j \left(\sum_{s_i \in S(r_j)} P_{r_j}^{s_i} - P_{r_j}^{\max} \right) \tag{26}$$

where $\text{¼} = [\mu_1, \mu_2, \ldots, \mu_L]$, $\mu_j \geq 0$, $j = 1, \ldots, L$ are the Lagrange multipliers corresponding to the L linear constraints on the relay powers.

Using the fact that

$$\sum_{j=1}^{L} \mu_j \sum_{s_i \in S(r_j)} P_{r_j}^{s_i} = \sum_{i=1}^{M} \sum_{r_j \in R(s_i)} \mu_j P_{r_j}^{s_i}$$

the Lagrangian in (26) can be rewritten as

$$L\left(\text{¼}, P_{r_j}^{s_i}\right) = \sum_{i=1}^{M} \left[w_i R_i - \sum_{r_j \in R(s_i)} \mu_j P_{r_j}^{s_i} \right] + \sum_{j=1}^{L} \mu_j P_{r_j}^{\max}.$$

The corresponding dual function of the Lagrangian can be written as

$$g(\text{¼}) = \max_{\{P_{r_j}^{s_i} \geq 0\}} \quad L\left(\text{¼}, P_{r_j}^{s_i}\right). \tag{27}$$

Since the original optimization is convex, strong duality holds, and the solution of the underlying optimization problem can be obtained from that of the corresponding dual problem as follows

$$\min_{\{\mu_j \geq 0\}} \quad g(\text{¼}). \tag{28}$$

It can be seen that the dual function in (27) can be found by solving M separate subproblems corresponding to M different users as follows

$$\max_{\{P_{r_j}^{s_i} \geq 0\}} \quad \mathsf{L}_i(¼, P_{r_j}^{s_i}) = w_i R_i - \sum_{r_j \in \mathsf{R}(s_i)} \mu_j P_{r_j}^{s_i} \qquad (29)$$

where $\mathsf{L}_i(¼, P_{r_j}^{s_i})$ corresponds to the ith component of the Lagrangian. Let $\mathsf{L}_i^*(¼)$ be the optimal value of $\mathsf{L}_i(¼, P_{r_j}^{s_i})$ obtained by solving the problem (29), then the dual problem in (28) can be rewritten as

$$\min_{\{\mu_j \geq 0\}} \quad g(¼) = \sum_{i=1}^{M} \mathsf{L}_i^*(¼) + \sum_{j=1}^{L} \mu_j P_{r_j}^{\max}. \qquad (30)$$

A distributed power allocation algorithm can be developed by iteratively and sequentially solving the problems (29) and (30). This algorithm is also known in optimization theory as a primal-dual algorithm. The Lagrange multiplier $\mu_j \geq 0$ represents the pricing coefficient for each unit power at relay j. Therefore, $\mu_j P_{r_j}^{s_i}$ can be seen as the price that user s_i must pay for using $P_{r_j}^{s_i}$ at each relay $r_j \in \mathsf{R}(s_i)$. In particular, the optimization problem (29) can be interpreted as follows. The user s_i tries to maximize its rate minus the total price that it has to pay given the price coefficients at relays. The weight w_i can be seen as a gain coefficient for each unit rate for user s_i.

The details of the distributed power allocation algorithm are as follows. The master dual problem is solved in a distributed fashion at each relay. Specifically, each relay r_j first broadcasts its initial "price" value, i.e., Lagrange multiplier μ_j. These price values are used by the receivers to compute the optimal power levels that the relays should allocate to that particular user. The optimal powers are fed back to the relays, which then updates the next values of the μ_j, $j = 1, \ldots, L$. This procedure is repeated until the so-obtained solution converges to the optimal one.

Note that the dual function $g(¼)$ is differentiable. Therefore, the master dual problem (28) can be solved by using the gradient descent method. The dual decomposition presented in (29) allows each user s_i, for the given μ_j, to find the optimal allocated power $r_j \in \mathsf{R}(s_i)$ as follows:

$$P_{r_j}^{s_i}(¼)\Big|_{opt} = \arg\max \{w_i R_i - \sum_{r_j \in \mathsf{R}(s_i)} \mu_j P_{r_j}^{s_i}\} \qquad (31)$$

which is unique due to the strict concavity.

Due to the fact that the solution of the problem (31) is unique, the dual function $g(¼)$ in the master problem (28) is differentiable, which allows us to use the following iterative gradient method to update the dual variables

$$\mu_j(t+1) = \left[\mu_j(t) - \zeta\left(P_{r_j}^{\max} - \sum_{s_i \in \mathsf{S}(r_j)} P_{r_j}^{s_i}(¼(t))\Big|_{opt}\right)\right]^+ \qquad (32)$$

where t is the iteration index, $[\cdot]^+$ denotes projection onto the feasible set of non-negative numbers, and ζ is the sufficiently small positive step size. The dual variables $¼(t)$ will converge to the dual optimal $¼_{opt}$ as $t \to \infty$, and the primal variable $P_{r_j}^{s_i}(¼(t))\Big|_{opt}$ will also converge to the primal optimal variable $P_{r_j}^{s_i}(¼_{opt})\Big|_{opt}$. Updating $\mu_j(t)$ via (32) can be interpreted as follows. The relay r_j updates its price depending on the requested power levels from its users. The price is increased when the total requested power from users is larger than its maximum limit. Otherwise, the price is decreased. Finally, we summarize the distributed power allocation algorithm as follows.

Distributed Power Allocation Algorithm

Parameters: the receiver of each user estimates/collects its weight coefficient w_i and channel gains of its transmitter-relay and relay-receiver links.

Initialization: set t=0, each relay j initializes $\mu_j(0)$ equal to some nonnegative value and broadcasts this value.

Iterations:

1. The receiver of user s_i solves its problem (31) and then broadcasts the solution $P_{r_j}^{s_i}(¼_{opt})\big|_{opt}$ to its relays.
2. Each relay r_j receives the requested power levels and updates it prices with the gradient iteration (32) using the information received from its assisted users. Then, it broadcasts the new value $\mu_j(t+1)$.
3. Set $t = t + 1$ and go to step 1 until satisfying a predetermined stopping criterion.

The convergence proof of the general primal-dual algorithm can be found in (Bertsekas, 1999). This algorithm only requires message exchange between relays and their assisted receivers. Therefore, it can be easily implemented in a distributed manner with low overhead.

Joint Admission Control and Power Allocation

Here, we consider a scenario in which users have minimum rate requirements. This scenario is important for real-time/multimedia applications, which require certain minimum rates to maintain QoS performance. Because network radio resources may be limited (e.g., limited source and/or relay power), supporting all users with their minimum required rates may not be feasible. Therefore, an admission control mechanism should be employed to determine which users to be admitted into the network. Then, power can be allocated to admitted users in order to ensure that each admitted user achieves the required QoS performance.

Specifically, consider a resource allocation problem that aims at minimizing the total relay power. In addition, each user has a minimum rate requirement. For the above described wireless systems with multiple users and multiple relays, the problem of minimizing the total relay power given a minimum rate constraint for each user can be posed as

$$\min_{\{P_{r_j}^{s_i} \geq 0\}} \quad \sum_{j=1}^{L} \sum_{s_i \in S(r_j)} P_{r_j}^{s_i} \tag{33}$$

$$\text{subject to:} \quad R_i \geq R_i^{\min},\ i = 1,\ldots,M \tag{34}$$

$$\sum_{s_i \in S(r_j)} P_{r_j}^{s_i} \leq P_{r_j}^{\max},\ j = 1,\ldots,L \tag{35}$$

where R_i^{min} denotes the minimum rate requirement for user s_i.

Mathematically, there are instances in which the optimization problem (33)-(35) becomes infeasible. A practical implication of the infeasibility is that it is impossible to serve all M users at their desired QoS requirements. In QoS-supported systems, some users can be dropped or the rate targets can be relaxed as a consequence. We investigate the former scenario and try to maximize the number of users that can be admitted at their minimum rate requirements.

The joint admission control and power allocation problem can be mathematically posed as a two-stage optimization problem (Matskani *et al*, 2007; Matskani *et al*, 2008). All possible sets of admitted users $S_0, S_1, \ldots$ with possibly maximal cardinality (which can be only one or several sets) are found in the first admission control stage, while the optimal set of admitted users S_k is the one among the sets $S_0, S_1, \ldots,$ which requires minimum transmit power in the second power allocation stage. Once the candidate set of admitted users has been determined, the power allocation problem can be shown to be a convex programming problem. However, the admission control problem is combinatorially hard, which introduces high complexity for practical implementation. Therefore, a low-complexity solution approach for the joint admission control and power allocation problem is highly desirable.

A. Reformulation of Admission Control and Power Allocation Problem

The joint admission control and power allocation problem can be equivalently written as a one-stage optimization problem that enables us to develop a low-complexity algorithm to solve the underlying problem. Toward this end, let x_i, $i = 1,\ldots, M$ denote an indicator variable for user s_i where $x_i = 1$ if user i is admitted and $x_i = 0$, otherwise. Given these variables, the underlying problem can be rewritten as (Phan *et al*, 2009b)

$$\max_{\{s_i \in \{0,1\},\, P_{r_j}^{s_i} \geq 0\}} \quad \sum_{i=1}^{M} x_i \tag{36}$$

$$\text{subject to:} \quad R_i \geq R_i^{\min} x_i,\ i = 1,\ldots, M \tag{37}$$

$$\sum_{s_i \in S(r_j)} P_{r_j}^{s_i} \leq P_{r_j}^{\max},\ j = 1,\ldots, L \tag{38}$$

Note that the constraints (37) are automatically satisfied for the users that are not admitted. The indicator variables help to represent the admission control problem in a more compact form. However, the combinatorial nature of the admission control problem still remains due to the binary variables x_i.

Following the conversion steps similar to those used in (Matskani *et al*, 2008), the joint admission control and power allocation problem can be converted to the following one-stage optimization problem

$$\max_{\{x_i \in \{0,1\},\, P_{r_j}^{s_i} \geq 0\}} \quad \varepsilon \sum_{i=1}^{M} x_i - (1-\varepsilon) \sum_{j=1}^{L} \sum_{s_i \in S(r_j)} P_{r_j}^{s_i} \tag{39}$$

$$\text{subject to:} \quad \text{constraints (37),(38)} \tag{40}$$

where ε is a constant chosen to satisfy the following relation

$$\frac{\sum_j P_{r_j}^{\max}}{\sum_j P_{r_j}^{\max} + 1} < \varepsilon < 1. \tag{41}$$

The problem (39)-(40) is a compact mathematical formulation of the joint optimal admission control and power allocation problem. The proof of the equivalence of the one-stage optimization problem and the original two-stage optimization problem can be found in (Matskani *et al*, 2008). Moreover, the one-stage optimization problem is always feasible since in the worst case no users are admitted, i.e., $x_i = 0,\ \forall i = 1,\ldots, M$.

B. Low-Complexity Algorithm

Although the original optimization problem (39)-(40) is NP-hard, its relaxation for which $x_i, i = 1,\ldots, M$ are relaxed to be continuous, can be shown to be a convex programming problem. In the following, we propose a reduced-complexity heuristic algorithm to perform joint admission control and power allocation. The following heuristic algorithm can be used to solve (39)-(40).

Joint Admission Control and Power Allocation Algorithm

1. Set $S := \{s_i \mid i = 1,\ldots, M\}$.
2. Solve convex problem (39)-(40) for the sources in S with x_i being relaxed to be continuous in the interval [0,1]. Denote the resulting power allocation values as $P_{r_j}^{s_i\,*}$, $j = 1,\ldots, M$.

3. For each $s_i \in S$, verify whether $R_i^* \geq R_i^{\min}, \forall s_i \in S$.
   a. If this is the case, then stop and $P_{r_j}^{s_i*}$ are power allocation solutions.
   b. Otherwise, remove the user s_i with largest gap to its target R_i^{min}, i.e., $s_i = \text{argmin}_{s_i \in S} \left\{R_i^* - R_i^{\min} < 0\right\}$ from set S and go to Step 2.

It can be seen that after each iteration, either the set of admitted users and the corresponding power allocation levels are determined or one user is removed from the list of the most likely admitted users. Since there are M initial users, the complexity is bounded above by that of solving M convex optimization problems with different dimensions, where the dimension of the problem depends on the iteration. It is worth mentioning that the proposed reduced complexity algorithm always returns one solution.

Note that the objective function for the considered above joint admission control and power allocation problem was the minimization of the total relay power. However, the principle used to construct the above algorithm can be employed to develop similar algorithms for other objective functions as well (e.g., max-min, weight-sum-rate functions). Due to space constraints, we do not consider these problems here.

Numerical Results for Multi-User Resource Allocation

Consider a wireless relay network as in Fig. 3 with ten users and three relays distributed in a two-dimensional region 14×14 where network sizes are measured with respect to some reference distance. The relays are fixed at coordinates (10,7), (10,10), and (10,12). The source and destination nodes are deployed randomly in the area inside the box area $[(0,0),(7,14)]$ and $[(12,0),(14,14)]$ respectively. In our simulations, each user is assisted by two relays. The noise power is taken to be equal to $N_0 = 10^{-5}$. All users and relays are assumed to have the same minimum rate R^{min} and maximum transmit power $P_{r_j}^{\max}$.

Numerical Results for Power Allocation

We show that by proper weight setting, the weighted-sum of rates maximization based power allocation scheme provides the flexibility required to support users with differentiated services. Particularly, we suppose that users 1 and 2 have higher priority than the others, and set the corresponding weights as $w_1 = w_2 = 5$, $w_3 = w_{10} = 1$ in the optimization problem (24)-(25). Fig. 4 displays the resulting rate of the high-priority users.[3] For reference, we also include in Fig. 4 the corresponding results obtained by equal power allocation (EPA) and by weighted-sum of rates maximization with equal weight coefficients. It can be seen that over the wide range of the relay power limits, the weighted-sum of rates maximization scheme outperforms the EPA. Without much surprise, the performance of the EPA scheme is quite close to that of the weighted-sum of rates maximization with *equal* weight coefficients. On the other hand, the weighted-sum of rates maximization with *unequal* weight coefficients provides noticeable rate enhancement to the high-priority users as compared to the other schemes, especially when the relays have severe power limitation, e.g., a rate gain of about 0.2 b/s/Hz when $P_{r_j}^{\max} = 10$. This figure indicates that the performance difference between different algorithms gets smaller for larger relay power limits. In other words, this reveals an interesting property that when the relays have more (or unlimited) available power, different (relay) power allocation strategies have much less impact on the user rate performance, which is limited by the source transmit power in this case.

Fig. 5 shows the network throughput for the aforementioned power allocation schemes. In the max-min rate fairness based power allocation scheme, there is a significant loss in the network

Figure 4. Rate of high priority users versus $P_{r_j}^{\max}$

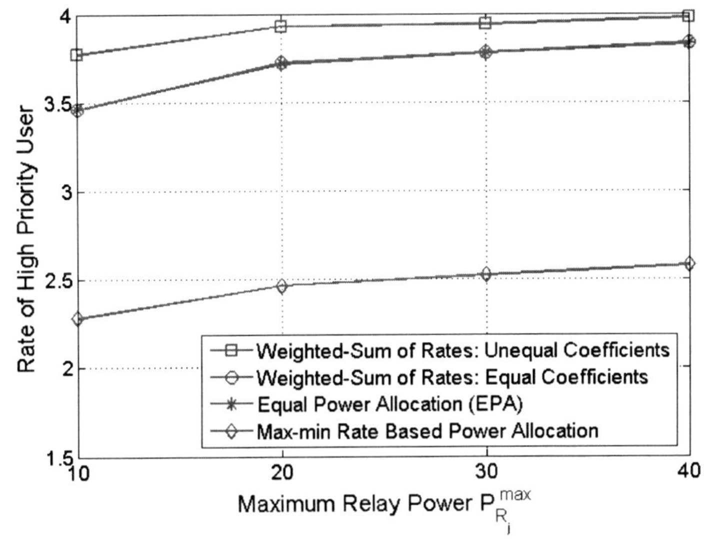

Figure 5. Network throughput versus $P_{r_j}^{\max}$

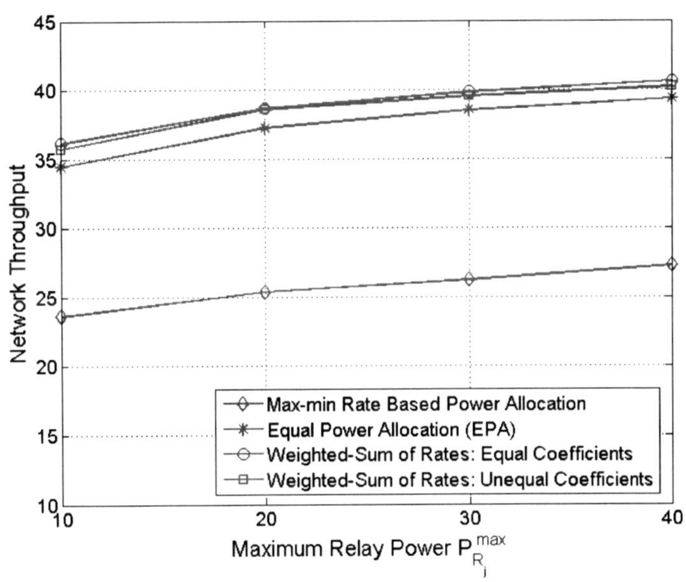

throughput since the objective is to improve the performance of the worst users. This confirms that achieving max-min fairness among users results in a performance loss for the whole system. The weighted-sum of rates fairness based scheme results in maximum throughput. Moreover, the rate gain of the weighted-sum of rates scheme over the EPA scheme is about 1.8 b/s/Hz over the range of the relay power limits. This gain comes at the cost of more complexity in system implementation to optimize the power levels. The weighted-sum of rates based scheme with unequal weights achieves slightly worse performance as compared to its counterpart with equal weights

Figure 6. Evolution of 'price' values and powers allocated at each relay

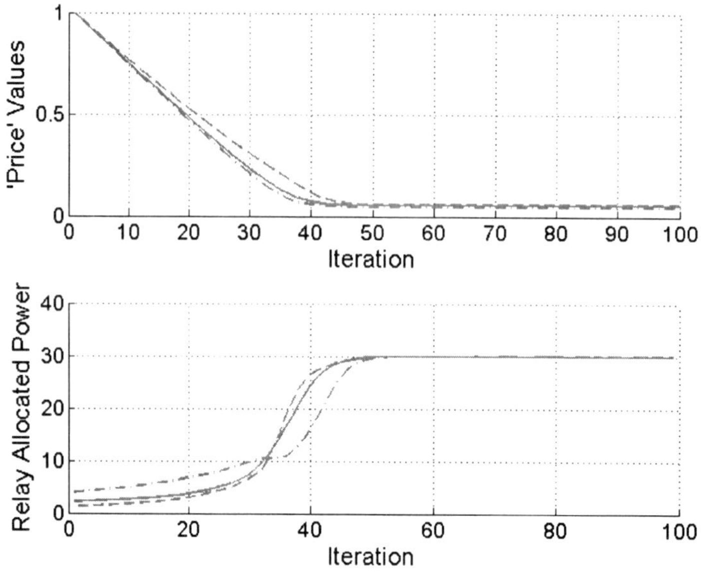

while providing better performance for the high priority users, i.e., users 1 and 2 in Fig. 4.

Figs. 6 and 7 show the evolution of different parameters in the proposed distributed implementation of the power allocation scheme for one particular channel realization. Specifically, Fig. 6 shows the evolution of the price values μ_j, $j = 1, 2, 3$ and the powers at the relays, while Fig. 7 displays the rate for each of the ten users and sum rates of all users. The update parameter ζ was set to 0.001 in this example. With such a choice of the update parameter, we can see that

Figure 7. Evolution of data rate for each user and user sum rate

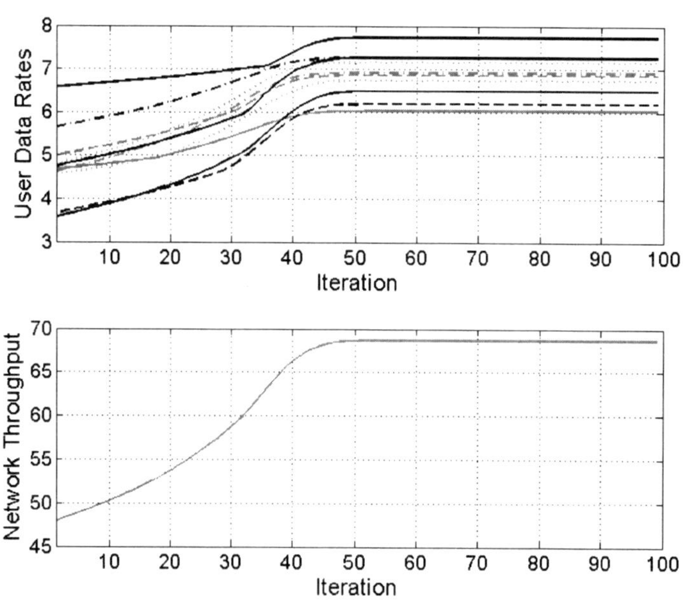

after about 50 updates, the algorithm converges to the optimal solution obtained by solving the proposed optimization problem in the centralized manner.

Numerical Results for Joint Admission Control and Power Allocation

We also investigate the performance of the proposed joint admission control and power allocation algorithm with $P_{s_i} = 1$ and $P_{r_j}^{\max} = 10$. It is assumed that the channel gain is due to the path loss only and the locations of the source and destination nodes are fixed. Different values of $\gamma_i^{min} / R_i^{min}$ have been used. For reference, we also consider the *optimal* admission control and power allocation scheme using exhaustive search over all feasible user subsets. A *feasible* user subset contains the *maximum* possible number of users and is selected as the *optimum* user subset if it requires the *smallest* transmit power. The simulation parameters and the performance results for the *optimal* admission control and power allocation scheme, and the proposed heuristic scheme are recorded in the columns "optimum allocation" and "proposed algorithm" in Table I, respectively.

Note that the running time is measured in seconds. It can be seen that the proposed algorithm determines exactly the optimal number of admitted users in all cases. The transmit power required by our proposed algorithm is just marginally larger than that required by the optimal admission control and power allocation based on exhaustive search. However, the running time for the proposed algorithm is dramatically smaller than that required by the optimal one. This makes the proposed approach attractive for practical implementation. As expected, when γ_i^{min} increases, a smaller number of users is admitted with a fixed amount of power. For example, nine users and four users are admitted with SNR $\gamma_i^{min} = 12$ dB and 14 dB, respectively.

CONCLUSION

In this chapter, we have presented a survey of cooperative diversity and discussed important resource allocation problems in wireless relay networks. Specifically, we have described fundamental cooperative protocols and pointed out

Table 1. Results with $P_{s_i} = 1$, $P_{r_j}^{\max} = 10$ (Running time in seconds)

	Optimum Allocation	**Proposed Algorithm**
SNR/rate	12 dB/4.0746 b/s/Hz	12 dB/4.0746 b/s/Hz
# users served	9	9
Users served	1, 2, 3, 4, 6, 7, 8, 9, 10	1, 2, 3, 4, 6, 7, 8, 9, 10
Transmit power	20.3619	20.4446
Running time	18.72	5.39
SNR/rate	13 dB/4.3891 b/s/Hz	13 dB/4.3891 b/s/Hz
# users served	6	6
Users served	1, 2, 7, 8, 9, 10	1, 2, 7, 8, 9, 10
Transmit power	22.9531	23.0342
Users served	2, 3, 7, 8, 9, 10	-
Transmit power	23.7717	-
Running time	458.07	9.60
SNR/rate	14 dB/4.7070 b/s/Hz	14 dB/4.7070 b/s/Hz
# users served	4	4
Users served	7, 8, 9, 10	7, 8, 9, 10
Transmit power	25.6046	25.6195
Running time	850.28	11.78
SNR/rate	15 dB/5.0278 b/s/Hz	15 dB/5.0278 b/s/Hz
# users served	2	2
Users served	8, 10	8, 10
Transmit power	7.5310	7.5320
Running time	930.11	12.92
SNR/rate	16 dB/5.3509 b/s/Hz	16 dB/5.3509 b/s/Hz
# users served	1	1
Users served	8	8
Transmit power	9.8002	9.8025
Running time	931.11	13.15

some recent enhanced protocols available in the literature. Typical applications of cooperative diversity in multihop cellular networks, cluster-based wireless ad hoc networks and broadcasting in ad hoc networks have been introduced. We have also presented the overview on resource allocation problems for single and multi-user wireless relay networks. For the multi-user case, we have investigated optimal relay power allocation and admission control problems with fairness consideration using centralized and distributed approaches. Simulation results are shown to confirm the theoretical developments.

REFERENCES

Azarian, K., El Gamal, H., & Schniter, P. (2008). On the optimality of the ARQ-DDF protocol. *IEEE Transactions on Information Theory*, *54*(4), 1718–1724. doi:10.1109/TIT.2008.917722

Beres, E., & Adve, R. (2008). Selection cooperation in multi-source cooperative networks. *IEEE Transactions on Wireless Communications*, *7*(1), 118–127. doi:10.1109/TWC.2008.060184

Bertsekas, D. P. (1999). *Nonlinear Programming (2nd Ed.).* Belmont, MA: Athena Scientific.

Bletsas, A., Khisti, A., Reed, D. P., & Lippman, A. (2006). A simple cooperative diversity method based on networh path selection. *IEEE Journal on Selected Areas in Communications*, *24*(3), 659–672. doi:10.1109/JSAC.2005.862417

Boyd, S., & Vandenberghe, L. (2004). *Convex Optimization (1st ed.).* Cambridge, UK: Cambridge University Press.

Boyer, J., Falconer, D. D., & Yanikomeroglu, H. (2004). Multihop diversity in wireless relaying channels. *IEEE Transactions on Communications*, *52*(10), 1820–1830. doi:10.1109/TCOMM.2004.836447

Cerutti, I., Fumagalli, A., & Gupta, P. (2008). Delay models of single-source single-relay cooperative ARQ protocols in slotted radio networks with Poisson frame arrivals. *IEEE/ACM Trans. Networking*, *16*(2), 371–382. doi:10.1109/TNET.2007.900360

Chiang, M., Low, S. H., Calderbank, A. R., & Doyle, J. C. (2007). Layering as optimization decomposition: A mathematical theory of network architectures. *Proceedings of the IEEE*, *95*(1), 255–312. doi:10.1109/JPROC.2006.887322

Cover, T. M., & El Gamal, A. A. (1979). Capacity theorems for the relay channel. *IEEE Transactions on Information Theory*, *25*(5), 572–584. doi:10.1109/TIT.1979.1056084

Dai, L., & Letaief, K. B. (2008). Throughput maximization of ad-hoc wireless networks using adaptive cooperative diversity and truncated ARQ. *IEEE Transactions on Communications*, *56*(11), 1907–1918. doi:10.1109/TCOMM.2008.041164

Gunduz, D., & Erkip, E. (2007). Opportunistic cooperation by dynamic resource allocation. *IEEE Transactions on Wireless Communications*, *6*(4), 1446–1454. doi:10.1109/TWC.2007.348341

Hong, X., Xu, K., & Gerla, M. (2002). Scalable routing protocols for mobile ad hoc networks. *IEEE Network*, *16*(4), 11–21. doi:10.1109/MNET.2002.1020231

Huang, W.-J., Hong, Y.-W. P., & Kuo, C.-C. J. (2008). Lifetime maximization for amplify-and-forward cooperative networks. *IEEE Transactions on Wireless Communications*, *7*(5), 1800–1805. doi:10.1109/TWC.2008.061075

Hunter, T. E., & Nosratinia, A. (2006). Diversity through coded cooperation. *IEEE Transactions on Wireless Communications*, *5*(2), 283–289. doi:10.1109/TWC.2006.1611050

Jing, Y., & Hassibi, B. (2006). Distributed space-time coding in wireless relay networks. *IEEE Transactions on Wireless Communications*, *5*(12), 3524–3536. doi:10.1109/TWC.2006.256975

Jing, Y., & Jafarkhani, H. (2008). Single and multiple relay selection schemes and their diversity orders, In *Proc. IEEE Inter. Conf. on Commun.*, Beijjing, China.

Kelly, F. P., Maulloo, A. K., & Tan, D. K. H. (1998). Rate control in communication networks: shadow prices, proportional fairness and stability. *The Journal of the Operational Research Society*, *49*(3), 237–252.

Khandani, A. E., Abounadi, J., Modiano, E., & Zheng, L. (2007). Cooperative routing in static wireless networks. *IEEE Transactions on Communications*, *55*(11), 2185–2192. doi:10.1109/TCOMM.2007.908538

Ko, Y., Vorobyov, S. A., & Ardakani, M. (2009a). How much multiuser diversity gain is required over large-scale fading? In *Proc. IEEE Inter. Conf. on Commun.*, Dresden, Germany.

Ko, Y., Vorobyov, S. A., & Ardakani, M. (2009b). How much multiuser diversity is required for homogeneous multiuser systems? *IEEE Trans. Wireless Commun.*, revised.

Koetter, R., & Medard, M. (2003). An algebraic approach to network coding. *IEEE/ACM Trans. Netw.*, *11*(5), 782–795.

Laneman, J. N., Tse, D. N. C., & Wornell, G. W. (2004). Cooperative diversity in wireless networks: Efficient protocols and outage behavior. *IEEE Transactions on Information Theory*, *50*(1), 3062–3080. doi:10.1109/TIT.2004.838089

Laneman, J. N., & Wornell, G. W. (2003). Distributed space–time-coded protocols for exploiting cooperative diversity in wireless networks. *IEEE Transactions on Information Theory*, *49*(10), 2415–2425. doi:10.1109/TIT.2003.817829

Le, L. B., & Hossain, E. (2007). Multihop cellular networks: Potential gains, research challenges, and a resource allocation framework. *IEEE Communications Magazine*, *45*(9), 66–73. doi:10.1109/MCOM.2007.4342859

Le, L. B., & Hossain, E. (2008a). An analytical model for ARQ cooperative diversity in multihop wireless networks. *IEEE Transactions on Wireless Communications*, *7*(5), 1786–1791. doi:10.1109/TWC.2008.060798

Le, L. B., & Hossain, E. (2008b). Cross-layer optimization frameworks for multihop wireless networks using cooperative diversity. *IEEE Transactions on Wireless Communications*, *7*(7), 2592–2602. doi:10.1109/TWC.2008.060962

Le, L. B., Hossain, E., & Zorzi, M. (2007). Queueing analysis for GBN and SR ARQ protocols under dynamic radio link adaptation with non-zero feedback delay. *IEEE Transactions on Wireless Communications*, *6*(9), 3418–3428. doi:10.1109/TWC.2007.06020038

Li, S.-Y. R., Yeung, R. W., & Cai, N. (2003). Linear network coding. *IEEE Transactions on Information Theory*, *49*(2), 371–381. doi:10.1109/TIT.2002.807285

Li, Y., Vucetic, B., Zhou, Z., & Dohler, M. (2007). Distributed adaptive power allocation for wireless relay networks. *IEEE Transactions on Wireless Communications*, *6*(3), 948–958. doi:10.1109/TWC.2007.05256

Liang, Y., Veeravalli, V. V., & Poor, H. V. (2007). Resource allocation for wireless fading relay channels: Max-min solution. *IEEE Transactions on Information Theory*, *53*(10), 3432–3453. doi:10.1109/TIT.2007.904996

Lin, Z., Erkip, E., & Stefanov, A. (2006). Cooperative regions and partner choice in coded cooperative systems. *IEEE Transactions on Communications*, *54*(7), 1323–1334. doi:10.1109/TCOMM.2006.877963

Luo, J., Blum, R. S., Cimini, L. J., Greenstein, L. J., & Haimovich, A. M. (2007). Decode-and-Forward cooperative diversity with power allocation in wireless networks. *IEEE Transactions on Wireless Communications*, *6*(3), 793–799. doi:10.1109/TWC.2007.05272

Madan, R., Mehta, N. B., Molisch, A. F., & Zhang, J. (2009). Energy-efficient decentralized cooperative routing in wireless networks. *IEEE Transactions on Automatic Control*, *54*(3), 512–527. doi:10.1109/TAC.2009.2012979

Madsen, A. H., & Zhang, J. (2005). Capacity bounds and power allocation for wireless relay channels. *IEEE Transactions on Information Theory*, *51*(6), 2020–2040. doi:10.1109/TIT.2005.847703

Maric, I., & Yates, R. D. (2004). Cooperative multihop broadcast for wireless networks. *IEEE Journal on Selected Areas in Communications*, *22*(6), 1080–1088. doi:10.1109/JSAC.2004.830912

Matskani, E., Sidiropoulos, N. D., Luo, Z.-Q., & Tassiulas, L. (2007). Joint multiuser downlink beamforming and admission control: A semi-definite relaxation approach. In *Proc. IEEE Inter. Conf. Accous. Speech and Sig. Proc.*, Honolulu, Hawaii, USA.

Matskani, E., Sidiropoulos, N. D., Luo, Z.-Q., & Tassiulas, L. (2008). Convex approximation techniques for joint multiuser downlink beamforming and admission control. *IEEE Transactions on Wireless Communications*, *7*(7), 2682–2693. doi:10.1109/TWC.2008.070104

Morgenshtern, V. I., & Bolcskei, H. (2007). Crystallization in large wireless networks. *IEEE Transactions on Information Theory*, *53*(10), 3319–3349. doi:10.1109/TIT.2007.904789

Nabar, R. U., Bolcskei, H., & Kneubuhler, F. W. (2004). Fading relay channels: Performance limits and space–time signal design. *IEEE Journal on Selected Areas in Communications*, *22*(6), 1099–1109. doi:10.1109/JSAC.2004.830922

Narasimhan, R. (2008). Delay-limited throughput of cooperative multiple access channels with hybrid-ARQ. In *Proc. IEEE Sym. Info. Theory*, Toronto, Canada.

Nechiporenko, T., Phan, K. T., Tellambura, C., & Nguyen, H. H. (2009). (to appear). Capacity of Rayleigh fading cooperative systems under adaptive transmission. *IEEE Transactions on Wireless Communications*.

Ng, T. C.-Y., & Yu, W. (2007). Joint optimization of relay strategies and resource allocations in a cooperative cellular network. *IEEE Journal on Selected Areas in Communications*, *25*(2), 328–339. doi:10.1109/JSAC.2007.070209

Nosratinia, A., Hunter, T. E., & Hedayat, A. (2004). Cooperative communication in wireless networks. *IEEE Communications Magazine*, *42*(10), 68–73. doi:10.1109/MCOM.2004.1341264

Phan, T. K., Le, L. B., Vorobyov, S. A., & Le-Ngoc, T. (2009a). Centralized and distributed power allocation in multi-user wireless relay networks. In *Proc. IEEE Inter. Conf. Commun.*, Dresden, Germany.

Phan, T. K., Le, L. B., Vorobyov, S. A., & Le-Ngoc, T. (2009c). (to appear). Power allocation and admission control in multi-user relay networks via convex programming: Centralized and distributed schemes. *EURASIP J. Wireless Commun. and Networking*.

Phan, T. K., Le-Ngoc, T., Vorobyov, S. A., & Tellambura, C. (2008). Power allocation in wireless relay networks: A geometric programming based approach. In *Proc. IEEE Global Telecommun. Conf.*, New Orleans, LA, USA.

Phan, T. K., Le-Ngoc, T., Vorobyov, S. A., & Tellambura, C. (2009b). (to appear). Power allocation in wireless miltiuser relay networks. *IEEE Transactions on Wireless Communications*.

Ribeiro, A., Cai, X., & Giannakis, G. B. (2005). Symbol error probabilities for general cooperative links. *IEEE Transactions on Wireless Communications*, *4*(3), 1264–1273. doi:10.1109/TWC.2005.846989

Scaglione, A., Goeckel, D., & Laneman, J. N. (2006). Cooperative communications in mobile ad-hoc networks: Rethinking the link abstraction. *IEEE Signal Processing Magazine*, *23*(5), 18–29. doi:10.1109/MSP.2006.1708409

Scaglione, A., & Hong, Y. W. (2003). Opportunistic large arrays: Cooperative transmission in wireless multihop ad hoc networks to reach far distances. *IEEE Transactions on Signal Processing*, *51*(8), 2082–2092. doi:10.1109/TSP.2003.814519

Sendonaris, A., Erkip, E., & Aazhang, B. (2003a). User cooperation diversity-Part I: System description. *IEEE Transactions on Communications*, *51*(11), 1927–1938. doi:10.1109/TCOMM.2003.818096

Sendonaris, A., Erkip, E., & Aazhang, B. (2003b). User cooperation diversity-Part II: Implementation aspects and performance analysis. *IEEE Transactions on Communications*, *51*(11), 1939–1948. doi:10.1109/TCOMM.2003.819238

Sirkeci-Mergen, B., & Scaglione, A. (2007a). On the power efficiency of cooperative broadcast in dense wireless networks. *IEEE Journal on Selected Areas in Communications*, *25*(2), 497–507. doi:10.1109/JSAC.2007.070223

Sirkeci-Mergen, B., & Scaglione, A. (2007b). Randomized space-time coding for distributed cooperative communication. *IEEE Transactions on Signal Processing*, *55*(10), 5003–5017. doi:10.1109/TSP.2007.896061

Sirkeci-Mergen, B., Scaglione, A., & Mergen, G. (2006). Asymptotic analysis of multistage cooperative broadcast in wireless networks. *IEEE Transactions on Information Theory*, *52*(11), 5166–5166. doi:10.1109/TIT.2006.883627

Tannious, R., & Nosratinia, A. (2008). Spectrally-efficient relay selection with limited feedback. *IEEE Journal on Selected Areas in Communications*, *26*(8), 1419–1428. doi:10.1109/JSAC.2008.081008

Tarokh, V., Jafarkhani, H., & Calderbank, A. R. (1999). Space-time block coding from orthogonal designs. *IEEE Transactions on Information Theory*, *45*(5), 1456–1467. doi:10.1109/18.771146

Tarokh, V., Seshadri, N., & Calderbank, A. R. (1998). Space-time codes for high data rate wireless communication: Performance criterion and code construction. *IEEE Transactions on Information Theory*, *44*(2), 744–765. doi:10.1109/18.661517

Telatar, I. E. (1998). Capacity of multi-antenna Gaussian Channels. *European Trans. Telecommunications*, *10*(6), 585–595. doi:10.1002/ett.4460100604

Wei, S., Goeckel, D. L., & Valenti, M. C. (2006). Asynchronous cooperative diversity. *IEEE Transactions on Wireless Communications*, *5*(6), 1547–1557. doi:10.1109/TWC.2006.1638675

Xiao, L., Fuja, T. E., Kliewer, J., & Costello, D. J. Jr. (2007). A network coding approach to cooperative diversity. *IEEE Transactions on Information Theory*, *53*(10), 3714–3722. doi:10.1109/TIT.2007.904990

Xiao, L., Johansson, M., & Boyd, S. (2004). Simultaneous routing and resource allocation via dual decomposition. *IEEE Transactions on Communications*, *52*(7), 1136–1144. doi:10.1109/TCOMM.2004.831346

Yao, Y., Cai, X., & Giannakis, G. B. (2005). On energy efficiency and optimum resource allocation of relay transmissions in the low-power regime. *IEEE Transactions on Wireless Communications*, *4*(6), 2917–2927. doi:10.1109/TWC.2005.858294

Zhao, B., & Valenti, M. C. (2005). Practical relay networks: A generalization of hybrid-ARQ. *IEEE Journal on Selected Areas in Communications*, *23*(1), 7–18. doi:10.1109/JSAC.2004.837352

Zhao, Y., Adve, R., & Lim, T. J. (2007). Improving amplify-and-forward relay networks: Optimal power allocation versus selection. *IEEE Transactions on Wireless Communications*, *6*(8), 3114–3123.

Zheng, L., & Tse, D. (2003). Diversity and multiplexing: A fundamental tradeoff in multiple antenna channels. *IEEE Transactions on Information Theory*, *49*(5), 1073–1096. doi:10.1109/TIT.2003.810646

Zhifeng, P. L., Lin, Z., Erkip, E., & Panwar, S. (2006). Cooperative wireless communications: A cross-layer design. *IEEE Communications Magazine*, *13*(4), 84–92. doi:10.1109/MWC.2006.1678169

Zhu, H., & Cao, G. (2005). rDCF: A relay-enabled medium access control protocol for wireless ad hoc networks. In *Proc. IEEE INFOCOM*, Miami, FL, USA.

ENDNOTES

1. The term user refers to a source-destination pair in this context.
2. We consider the case where source-relay links are much better than the source-destination link. This would be an outcome of a typical relay selection strategy employed by each source node.
3. We observe that users 1 and 2 have indistinguishable performance, so only one curve for each scheme is plotted.

Chapter 8
User Based Call Admission Control Algorithms for Cellular Mobile Systems

Hamid Beigy
Sharif University of Technology, Iran

M. R. Meybodi
Amirkabir University of Technology, Iran

ABSTRACT

Call admission control in mobile cellular networks has become a high priority in network design research due to the rapid growth of popularity of wireless networks. Dozens of various call admission policies have been proposed for mobile cellular networks. This chapter proposes a classification of user based call admission policies in mobile cellular networks. The proposed classification not only provides a coherent framework for comparative studies of existing approaches, but also helps future researches and developments of new call admission policies.

1. INTRODUCTION

The frequency spectrum allocated to the mobile communication networks is very limited. This means that the frequency channels have to be reused as much as possible in order to support the many thousands of simultaneous calls that may arise in any typical mobile communication network (Katzela & Naghshineh, 1996). Thus, the efficient management and sharing of channels among numerous users become an important issue. In cellular networks the geographical area covered by the network is divided into smaller regions called cells. Each cell is serviced by a base station, located at its center. The base station is used to service the users located at that cell. A number of base stations are again linked to a central server called mobile switching center, which also acts as a gateway of the mobile communication network to the existing wire-line networks such as PSTN, or internet. A base station communicates with users (mobile stations) through wireless links and with mobile switching centers through dedicated links. The model of such a network referred to as *cellular network* is shown in figure 1 (Das & Sen & Jayaram, 1998).

We assume that the network uses a fixed channel assignment algorithm, which means that each base station has a fixed number of channels (capacity).

DOI: 10.4018/978-1-61520-680-3.ch008

Figure 1. System model of cellular networks

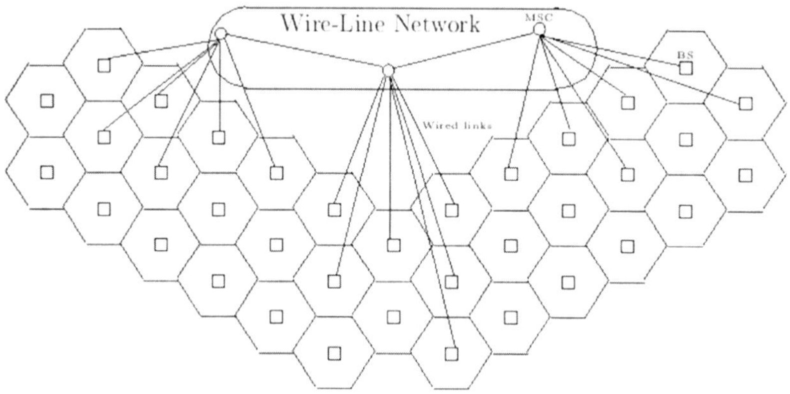

This capacity is interpreted in terms of bandwidth and is independent of used multiple access technology such as FDMA, TDMA, or CDMA. In order for a mobile user to be able to communicate with other user(s), a connection usually must be established between the users. The establishment and maintenance of a connection in cellular networks is the responsibility of the base stations. In order to establish a connection, a mobile user must first specify its traffic characteristics and quality of service (QoS) requirements. This traffic specification may be either implicit or explicit depending on the type of services provided by the network. For example, in a cellular phone network, the traffic characteristics and QoS requirements of voice connections are known a priori to the base station, and therefore, they are usually specified implicitly in a connection request. The next generation wireless networks are expected to eventually carry multi-media traffic such as voice, mixed voice and data, image transmission, email and etc. The traffic characteristics and the QoS requirements of connections for these services may not be known a priori to the base station. In these networks, mobile users must specify explicitly the traffic characteristics and the QoS requirements as a part of the connection request. Then, the base station determines whether it can meet the requested QoS requirements and, if possible, establish a connection.

When a call is originated and attempted in a cell, one channel allocated to the base station is used for the communication between the mobile station and the base station as long as channel is available. When all channels in a cell are in use while a call is attempted, then it will be blocked and cleared from the system. When a call gets a channel, it will keep the channel until its completion, or until it moves out of the cell, in which case the used channel will be released. When the mobile station moves into a new cell while its call is ongoing, a new channel needs to be acquired in the new call for further communication. This process is called *handoff* and must be transparent to the mobile user. During the handoff, if there is no channel is available in the new call for the ongoing call, it is forced to terminate before its completion.

When a user moves from one cell to another, the base station in the new cell must be responsible for all the previously established connections. A significant responsibility involves allocating sufficient resources in the cell for maintaining the QoS requirements of the established connections. If sufficient resources are not allocated to the handoff calls, the QoS requirements may not be met, which in turn may result in forced termination of the connection. Since the forced termination of established connections is usually more objectionable than rejection of a new con-

Figure 2. The call admission control algorithm

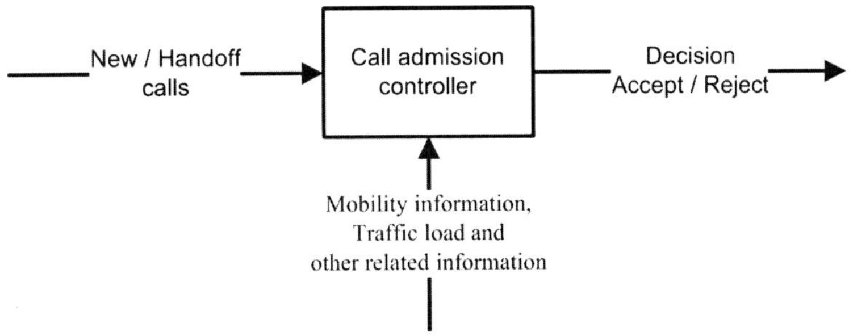

nection request, it widely believed that a cellular network must give a higher priority to the handoff connection requests as compared to new connections requests. Handoff problems are expected to become more and more important since the size of cells in emerging cellular networks tends to be smaller, which implies that handoff would occur more frequently, to attain a higher capacity.

In order to satisfy the QoS requirements, call admission control algorithms are needed, which determine whether a call should be either accepted or rejected at the base station and assign the required channel(s) to the accepted call. This results in a distributed call admission control strategy, which can be applied to every base station. Whenever a new call arrives, the call admission policy takes the call as input and based upon the current traffic conditions of network, decides whether or not to accept the user, as illustrated in figure 2. Call admission control in mobile cellular networks became a high priority in network design and research due to the rapid growth of popularity of wireless networks. A large number of call admission policies have been proposed for mobile cellular networks. However, despite years of research efforts, the call admission problem remains a critical issue and a high priority, especially given the perspectives of continually growing speed and size of future wireless networks. It is often difficult to characterize and compare various features among different policies. A good and detailed classification helps the researches and engineers to understand the similarities and differences among various schemes and decide which techniques are best suited for particular use (Beigy & Meybodi, 2003d; Ghaderi & Boutaba, 2006; Cruz-Perez & Ortigoza-Guerrero, 2007). These classifications not only provides a coherent framework for comparative studies of existing approaches, but also helps in future researches and developments of new call admission policies. This chapter is based on the classification given in (Beigy & Meybodi, 2003d).

The rest of this chapter is organized as follows: Section 2 describes the call admission problem and presents the proposed classification. Section 3 gives the non-prioritized call admission policies and the prioritized call admission policies are given in section 4. Optimal policies are given in section 5 and section 6 concludes the chapter.

2. CALL ADMISSION CONTROL

The challenges in the wireless networks are to guarantee the QoS requirements while taking into account the limited number of channels and interference between them. The study of the different schemes to accept calls in communication networks is known as the call admission control problem. Call admission control for high-speed wire-line networks have been intensively studied in the last few years. There are two major differences between wireless and wire-line networks due to the link characteristics and user mobility. The transmission links for the broadband wire-line networks are characterized by high transmission rates and very low error rates. In contrast, wireless links have a much smaller transmission rates and a much high error rates. The second major difference between the two networks is the user mobility. In wire-line networks, the user-network interface remains fixed throughout the duration of a connection whereas the user-network interface in a wireless environment may change throughout the connection. Due to the user mobility, call admission control becomes much more complicated in the wireless networks than wire-line networks. An accepted call that has not completed in the current cell may have to be handed off to another cell. During the handoff, the call may not be gain a channel in the new cell to continue its service due to the forced call termination. Thus, the new calls and handoff calls to be treated differently in terms of resource allocation. Since users tend to be much more sensitive to forced call termination (call dropping) than to the call blocking, handoff calls are normally assigned higher priority over the new calls.

Call admission control is one method to manage radio resources in order to adapt to the traffic variations. Call admission control denotes the process to make a decision for new admission according to the amount of the available resources versus users QoS requirements, and the effect upon the QoS of the existing calls imposed by new calls. Call admission control plays a very important role in cellular networks because it directly controls the number of users in the network and must be designed to guarantee the QoS requirements. The usual network performance indicators are the blocking probability of new calls, the dropping probability of handoff calls, the computation and communication overheads, and the total carried load. Good call admission control policies have to balance the dropping probability of handoff calls and the blocking probability of new call in order to provide the desired QoS requirements.

There has been much research into call admission control policies for cellular networks. A good call admission control algorithm must have the following features in order of importance.

- Maximize channel utilization in a fair manner to all calls
- Minimize the dropping probability of connected calls
- Minimize the reduction of the QoS for the connected calls
- Minimize the blocking probability of new calls

Call admission control policies can be divided into a number of different categories depending on the comparison basis. For example, when call admission control policies are compared based on decision policies, they can be divided into user (number)-based CAC (NCAC) and interference-based CAC (ICAC) policies (Ishikawa & Umeda, 1997). NCAC policies accept/reject calls based on the number of users in the cell. Using ICAC, a base station, by monitoring the interference on a call-by-call basis, determines whether or not a new call is acceptable. The new call is blocked if the observed interference level exceeds a CAC interference threshold. Each base station should measure the total power of received signals in the spreading bandwidth before dispreading them. ICAC therefore requires overheads for base station hardware and complicates its architecture, while NCAC can be implemented by means of base station software.

Before we start presenting NCAC schemes, we give a general framework for call admission control, which will be used throughout this chapter, is developed. We consider network cells with N classes of calls $W = \{w_1, \dots, w_N\}$, and C full duplex channels. Class $w_i\ (for\ 1 \le i \le N)$ consists of a stream of statistically identical calls with Poisson arrival at rate λ_i and independent identical exponentially distributed call holding times with the same mean $1/\mu$. Assume that all classes need only one channel for each call. Let c denotes the number of busy channels in the cell. The state space S of a cell is given by $S = \{c \mid c \le C\}$. We define the admission policy $u : S \times W \to \{0,1\}$, where *u(x,w)* specifies the probability of acceptance of calls of class *w* when the cell is in state *x*. At any time *t*, the decision to accept or reject calls of class *w* depends only on the current state of the cell or its neighboring cells. From the point of the call admission controller, the process can be modeled as a Markov process, where the transition rates between the states $x, y \in S$ for a call of class *w*, are given by

$$q(x,y) = \begin{cases} \sum_{w=1}^{N} u(x,w)\lambda_w & if\ y = x+1 \\ x\mu & if\ y = x-1 \\ 0 & otherwise \end{cases} \quad (1)$$

Function *u(x,w)* may be deterministic or stochastic (probabilistic), static or dynamic. Based on function *u(x,w)*, the call admission control policies can be divided into non-prioritized, prioritized, and optimal policies, as shown in figure 3. In non-prioritized policies (Hong & Rappaport, 1986), all calls are accepted when the requested channels are free, while in prioritized policies, one group of calls have a higher priority than other groups, for example, the handoff calls have the higher priority than new calls. In prioritized policies, when the requested channels are not available, the call may be queued or rejected. Optimal policies accept/reject calls to maximize throughput of the network.

3. NON-PRIORITIZED CALL ADMISSION CONTROL POLICIES

In these call admission control policies (Hong & Rappaport, 1986), no single class is treated differently than any other classes. This is the simplest scheme and involving checking to guarantee that the requested bandwidth is available for the calls. If the bandwidth requirements can be met, then the call is accepted and the bandwidth is allocated; otherwise the call is blocked. This policy always accepts calls as long as doing so leads to a state

Figure 3. Classification of user based call admission control algorithms

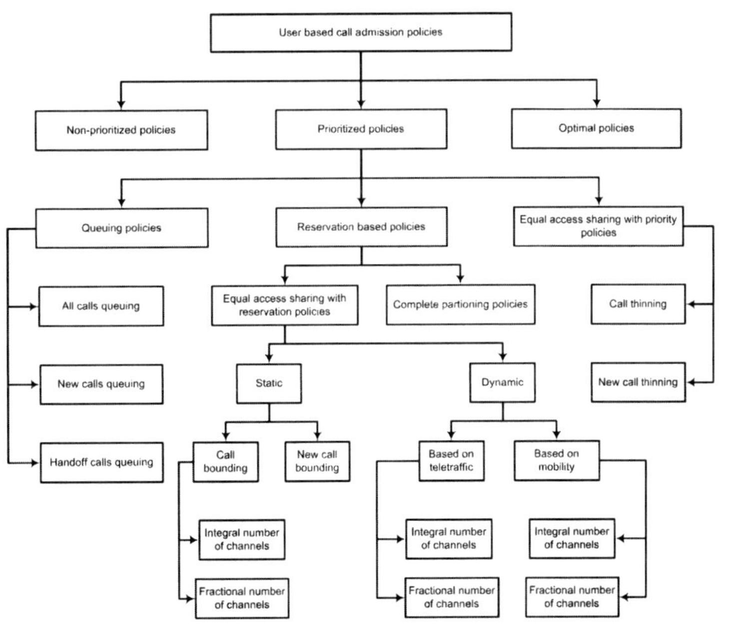

Figure 4. State transition diagram for non-prioritized scheme

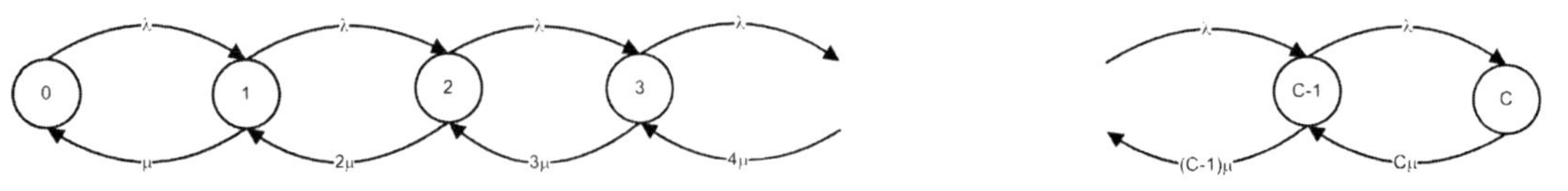

in the state space S, that is,

$$u(x,w) = \begin{cases} 1 & if\ x+1 \in S \\ 0 & otherwise \end{cases} \quad (2)$$

In order to study the performance of this scheme, we consider a homogenous cellular network where all cells have the same number of channels, C, and experience the same arrival rates for all classes of calls. Without loss of generality, we consider two classes of calls: new and handoff calls. We assume that the arrival of new and handoff calls are Poisson distributed with rates λ_n and λ_h, respectively and the call holding time of calls are exponentially distributed with the mean $1/\mu$. Note that the same service rate for both types of calls implies that the base station of a cell does not need to discriminate between new and handoff calls, once they are connected. These assumptions have been found reasonable as long as the number of mobile users in a cell is much greater than the number of channels allocated to that cell. Define the state of a cell at time t by the total number of occupied channels, $c(t)$. Thus, the channel occupancy can be modeled by a continuous time Markov chain with states $0,1,\ldots,C$. Figure 4 shows the state transition diagram of a system with C channels and non-prioritized call admission scheme.

Define the steady state probability $P_n = \lim_{t\to\infty}$ Prob $[c(t)=n]$ as the probability of n channels being occupied. Given this, it is straight forward to derive probability P_n (for $n=0,1,\ldots, C$). The steady state probability P_n that n channels are busy is given by the following expression.

$$P_n = \left(\frac{\rho^n}{n!}\right) P_0, \quad (3)$$

where

$$P_0 = \left[\sum_{k=0}^{C} \left(\frac{\rho^k}{k!}\right)\right]^{-1}, \quad (4)$$

and $\rho = (\lambda_n + \lambda_h)/\mu$. Two commonly used performance measures for cellular networks are: *dropping probability of handoff calls* (B_h) and *blocking probability of new calls* (B_n). The dropping probability of handoff calls represents the probability that a handoff call being dropped during handovers. This probability is defined as the ratio between the number of calls dropped by the system and the total number of admitted calls. The blocking probability of new calls represents the probability that a new call being denied access to the network. This probability is defined as the percentage of calls that are denied access to the network. Given the state probabilities, we can drive the blocking probability of new calls and the dropping probability of handoff calls.

$$B_n = B_h = \frac{\rho^C}{C!} P_0 \quad (5)$$

4. PRIORITIZED CALL ADMISSION CONTROL POLICIES

In prioritized call admission control policies, a priority is assigned to each class of calls. These priorities are implemented through function *u(x,w)*. For example, from the point of view of a mobile user, dropping of an ongoing call is less desirable than blocking of a new call. Therefore, to reduce the chances of unsuccessful handoff calls, the system assigns a higher priority to the handoff calls. Thus the function *u(x,.)* has a higher value for handoff calls than the new calls. The prioritized call admission policies can be divided into three groups: *equal access sharing with priority, reservation based and queuing based policies*, and *queuing priority policies*, which are described in the rest of this section.

4.1 Equal Access Sharing with Priority Policies (EASWP)

In these call admission control policies, all classes of calls have access to all channels but some classes have a higher priority than others. This priority is implemented through the use of function *u(x,w)* *p(x,w)*, where *p(x,w)* is the probability of accepting calls of class *w* when the cell is in state *x*. The reported EASWP policies can be classified as call thinning and new call thinning schemes, which are briefly described below.

4.1.1 Call Thinning Schemes

In these schemes, the state of system, *x*, is the number of busy channels in the cell. Call thinning schemes, in turn can be divided into two subclasses: static and dynamic schemes: In what follows, we explain these schemes for two classes of calls. In static call thinning schemes, *u(x, w)* is determined based on a priori information and remain fixed during the operation of the network. Ho & Lea (1999) proposed a static call thinning scheme and linear programming was used to determine the optimal values of *p(x, w)*. In this scheme, we have

$$u(x,w) = \begin{cases} p(x,w) & if\ x < C \\ 0 & if\ x = C \end{cases} \tag{6}$$

A restricted version of this scheme, which is called *fractional guard channel* (FGC) scheme, was proposed by Ramjee & Towsley & Nagarajan (1997). In this scheme, the handoff calls have higher priority over the new calls. This scheme accepts new calls with certain probability that depends on the channel occupancy of the cell and accepts the handoff calls when the cell has free channels. In this scheme, we have

$$u(x,w) = \begin{cases} 1 & if\ x < C\ and\ w = handoff\ calls \\ p(x) & if\ x < C\ and\ w = new\ calls \\ 0 & if\ x = C \end{cases} \tag{7}$$

Since *p(x)* only appears when new calls arrives, *p(x)*'s are called new call admission probabilities. The idea behind this scheme is to smoothly throttle the new call stream as the network traffic is building up. Thus, when the network is approaching the congestion, the accepted new calls become thinner. Due to the flexible choice of new call admission probabilities, this scheme can be made very general. Figure 5 shows the state transition diagram of a homogeneous network with *C* channels and FGC scheme.

Define the steady state probability $P_n = \lim_{t \to \infty}$ Prob $[c(t)=n]$ as the probability of *n* channels being occupied. The steady state probability P_n that *n* channels are busy is given by the following expression (Ramjee & Towsley & Nagarajan, 1997).

Figure 5. State transition diagram for FGC scheme

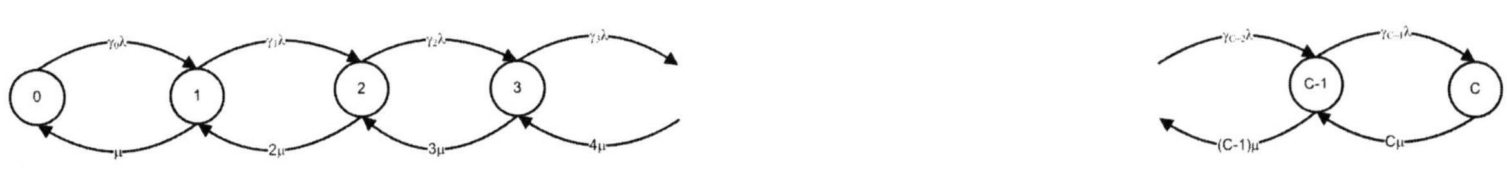

Figure 6. State transition diagram for UFC scheme

$$P_n = \left(\frac{\rho^n}{n!} \prod_{k=0}^{n} \gamma_k \right) P_0, \tag{8}$$

where

$$P_0 = \left[\sum_{k=0}^{C} \left(\frac{\rho^k}{k!} \prod_{k=0}^{n} \gamma_k \right) \right]^{-1}, \tag{9}$$

$\gamma_k = [\alpha + (1-\alpha)p(k)]$, $\alpha = \dfrac{\lambda_h}{\lambda_n + \lambda_h}$, and $\rho = (\lambda_n + \lambda_h) / \mu$. Given these state probabilities, we can drive the blocking probability of new calls and the dropping probability of handoff calls.

$$B_n(C,p) = P_0 (1-\alpha) \sum_{m=0}^{C} \frac{\rho^m}{m!} \prod_{k=0}^{m-1} \gamma_k$$

$$B_h(C,p) = P_0 \frac{\rho^C}{C!} \prod_{k=0}^{C-1} \gamma_k \tag{11}$$

The most disadvantage of this scheme is that no algorithm is given to find p(x)s. In order to find p(x), a restricted version of this scheme called uniform fractional guard channel scheme (UFC) is introduced by Beigy & Meybodi (2004a). In this scheme, the new call admission probabilities are independent of channel occupancy. Thus, in this scheme, we have

$$u(x,w) = \begin{cases} 1 & x < C \; and \; w = handoff \; calls \\ p & x < C \; and \; w = new \; calls \\ 0 & x = C \end{cases} \tag{11}$$

Figure 6 shows the state transition diagram of a homogeneous network with C channels and UFC scheme.

The steady state probability P_n that n channels are busy is given by the following expression:

$$P_n = \left(\frac{(\rho\gamma)^n}{n!} \right) P_0, \tag{12}$$

where

$$P_0 = \left[\sum_{k=0}^{C} \left(\frac{(\rho\gamma)^k}{k!} \right) \right]^{-1}, \tag{13}$$

$\gamma = [\alpha + (1-\alpha)p]$, $\alpha = \dfrac{\lambda_h}{\lambda_n + \lambda_h}$, and $\rho = (\lambda_n + \lambda_h) / \mu$. Given these state probabilities, we can drive the blocking probability of new calls and the dropping probability of handoff calls.

$$B_n(C,p) = 1 - \alpha \left[1 - \frac{(\rho\gamma)^C}{C!} P_0 \right]$$

$$B_h(C,p) = P_0 \frac{(\rho\gamma)^C}{C!} \tag{14}$$

$B_n(C,p)$ and $B_h(C,p)$ have interesting properties, which enable us to design an algorithm for finding the optimal value of parameter p. It was shown that $B_n(C,p)$ and $B_h(C,p)$ are monotonically decreasing and increasing function of p, respectively (Beigy & Meybodi, 2004a). The algorithm 1 is given for finding the optimal value of p and can be described as follows. At first, the algorithm considers the case when all channels are shared between handoff and new calls. If the complete sharing does not satisfy the level of QoS, then the algorithm considers the case when all channels are exclusively used for handoff calls. If the exclusive use of channels for handoff calls does not satisfy the level of QoS, then the number of allocated channels to the cell is not sufficient and the algorithm terminates; otherwise the algorithm searches for the optimal value of p. The search method used in this algorithm is binary search.

Algorithm 1: The algorithm for finding the optimal value of p

```
Algorithm FindUFCParameter
set upper ← 1; lower  ←  0
if(Bh (C,1)≤ Ph)thenreturn 1
end if
if(Bh (C,0) ≥ Ph) then
return 0
end if
while ((upper -lower) < 0.0001)
doset p ← ⌊(upper + lower) /2⌋
if(Bh (C,1)> Ph)thenset upper ←
p
else
set lower ← p
end if
end while
return  p
end Algorithm
```

In dynamic call thinning schemes, $u(x,w)$ is adapted based on information gathered during the operation of the network. Some dynamic call thinning algorithms are reported in (Ayyagari & Empremides, 1999; Wu, & Wong & Li, 2002). A dynamic call thinning scheme for multi-media cellular network is presented Ayyagari & Empremides (1999). In this scheme, calls are classified on the basis of channel requirement and a propriety level is associated with each class of calls. This scheme collects calls in a time period and then accepts calls with the higher priorities. In (Wu, & Wong & Li, 2002), a call admission scheme called stable dynamic call admission control scheme is suggested. The aim of this scheme is to maximize the channel utilization (minimize the new call blocking probability) subject to a hard constraint on the dropping probability of handoff calls. In this scheme, status information is exchanged periodically among neighboring cells, and even next neighboring cells if necessary. The exchanged information includes the channel occupancies and the new call arrival rates. Each cell updates its acceptance ratio (the maximum fraction of new calls to be accepted in the cell) in the next control period at the beginning of that period. The control action is obtained by solving system of equations which specifies the average dropping probability of handoff calls must be equal to the QoS of the system. Beigy & Meybodi (2004b) proposed a learning automaton based algorithm to adjust the value of p, in which a learning automaton is associated to each cell. In this algorithm as shown in Algorithm 2, when a handoff call arrives, it is accepted as long as there is a free channel. If there is no free channel, the handoff call is blocked. When a new call arrives to a particular cell, the learning automaton associated to that cell chooses one of its actions. If action ACCEPT is selected by the automaton and the cell has a free channel, then action ACCEPT is rewarded. If there is no free channel to be allocated to the arrived new call, the call is blocked and the action ACCEPT is penalized. When the automaton selects action REJECT, the algorithm computes an estimation of the dropping probability of handoff calls ($\hat{B}_h$) and uses it to decide whether or not accept new calls. If the current estimate of dropping probability of handoff calls is less than the given threshold p_h and there is a free channel, then the new call is accepted and action REJECT is penalized; otherwise, the new call is rejected and action REJECT is rewarded.

Algorithm 2: The learning automata based algorithm for finding the optimal value of p

```
Algorithm AdaptiveUFC-I
if (NEW CALL) thenif (action of
learning automaton is ACCEPT)
thenif (c(t) < C) then
accept call and reward action
ACCEPT
 else
reject call and penalize action
ACCEPT
end if
else
if (c(t) < C and B̂_h < p_h) then
accept call
 else
reject call
end if
Compute B̂_h if (new call is ac-
cepted and B̂_h < p_h) then
penalize action REJECT
else
reward action REJECT
end if
end if
end if
end Algorithm
```

The simulation results reported in (Beigy and Meybodi, 2004b) shows that this algorithm cannot maintain the specific level of QoS for the dropping probability of handoff calls. This problem may be due to the existence of delay in the cellular network, because the selected action of learning automaton is immediately rewarded / penalized. Since the effect of the estimated new call admission probability is specified after a time period, then the reward/ punishment of learning automaton must be given in the end of that period. In order to overcome this problem, another algorithm is given in (Beigy & Meybodi, 2004b), in which the action probability vector of learning automaton is adjusted upon the arrival the next new call. This algorithm, as shown in Algorithm 3, uses a learning automaton to accept/reject new calls and a pre-specified level of dropping probability of handoff calls is used to penalize/reward the action selected by the learning automaton. This algorithm can be described as follows. When a handoff call arrives, it is accepted as long as there is a free channel. If there is no free channel, the handoff call is dropped. When a new call arrives to a particular cell, the learning automaton associated to that cell chooses one of its actions. If action ACCEPT is selected by automaton and the cell has at least one free channel, the incoming call is accepted and the selected action is rewarded. If there is no free channel to be allocated to the arrived new call, the call is blocked and action ACCEPT is penalized. When the automaton selects action REJECT, then the new call is rejected and the base station computes an estimation of the dropping probability of handoff calls ($\hat{B}_h$) and uses it to reward or punish action REJECT. If the current estimate of dropping probability of handoff calls is less than the given threshold p_h, then action REJECT is penalized; otherwise, action REJECT is rewarded. Beigy & Meybodi (2003a) showed that this algorithm finds the optimal value of the of UFC's Parameter.

Algorithm 3: The learning automata based algorithm for finding the optimal value of p

```
Algorithm AdaptiveUFC-II
if (NEW CALL) thenif (action of
learning automaton is ACCEPT)
thenif (c(t) < C) then
accept call and reward action
ACCEPT
 else
reject call and penalize action
ACCEPT
end if
else
reject call & compute B̂_h if (B̂_h
< p_h) then
penalize action REJECT
else
```

```
reward action REJECT
end if
end if
end if
end Algorithm
```

Figure 7 shows the performance of this under different handoff traffic when the other parameters of the cell are fixed. Note that the level of QoS is maintained by this algorithm for various handoff traffic conditions. Figure 8 shows the B_n and B_h for this algorithm for two typical different handoff traffic loads, which shows that that the admission probability converges to its optimal value.

4.1.2 New Call Thinning Schemes

Below we explain the one new call thinning scheme (Fang & Zhang, 2002). In this scheme, the state of system, x, is the number of ongoing new calls in the cell. For the sake of simplicity assume that we have two classes of calls: new calls and handoff calls. This scheme, which limits the new calls in the system, gives a higher priority to the handoff calls over the new calls. This scheme accepts new calls with certain probability that depends on the number of ongoing new calls in the cell and accepts the handoff calls when the cell has free channels. In this scheme, we have

$$u(x,w) = \begin{cases} 1 & if\ x < C\ and\ w = handoff\ calls \\ p(x) & if\ x < C\ and\ w = new\ calls \\ 0 & if\ x = C \end{cases} \tag{15}$$

4.2 Reservation Based Call Admission Control Policies

In reservation based call admission control policies, some of the channels allocated to the cell are reserved for the higher priority calls. In these policies, we have u(x,w)=0 for some x and w. In these call admission control policies, all classes of calls are accepted equally within a specified bandwidth of the maximum channel capacity that depends on the given class. Once the available channel capacity has been used, only calls that are of a high priority will be accepted to use the remaining (reserved) channels (bandwidth). This has the effect of prioritizing a traffic class above the other traffic classes. In the reservation based policies, classes of calls can be grouped and fix a threshold for each group. When restricted to simple form, these policies dedicate a certain number of channels for each group and the remaining channels are shared among all groups. To define a simple form for these policies, we form W groups, $G_1, \ldots, G_W$, such that each w belongs to only one group G_W. The reservation based policies can be stated as:

Figure 7. Performance of the adaptive UFC algorithm for different handoff traffic

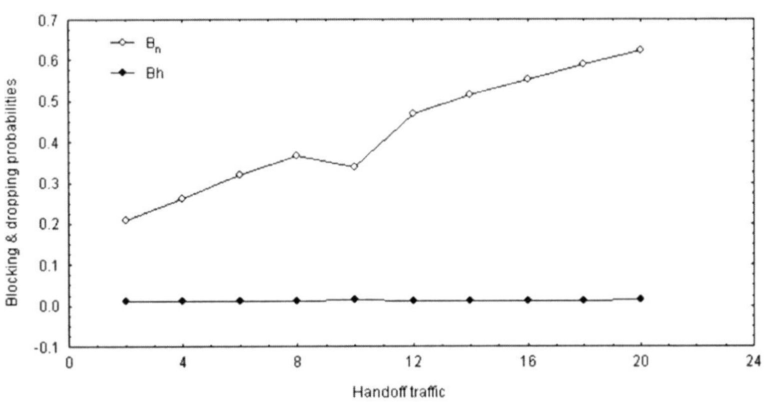

Figure 8. Convergence of the proposed algorithm for different handoff traffic

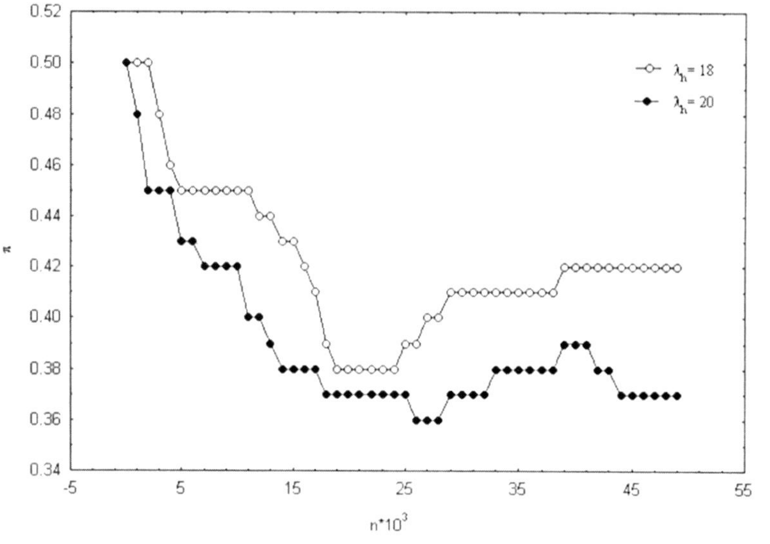

$$u(x,w) = I\{x \le T_w\} \wedge I\{x+1 \in S_{G_W}\} \quad (16)$$

where T_W is the maximum channel capacity for calls of class w in group G_W and S_{G_W} is the set of channels associated to group G_W. These policies can be divided into two main groups: *equal access sharing with priority* (EASWR) and *complete partitioning schemes*, which are explained in the following subsections.

4.2.1 Equal Access Sharing with Reservation (EASWR)

In these call admission control policies, we have $S = S_{G_1} = S_{G_2} = ... = S_{G_W}$. Thus, all classes of calls can use any channel and calls are accepted with equal probabilities within a specified bandwidth of the maximum channel capacity. Once the available channel capacity has been used, only calls that are of high priority will be accepted to use the remaining (reserved) channels. This has the effect of prioritizing one class above the other classes, that is,

$$u(x,w) = I\{x \le T_w\} \wedge I\{x+1 \in S\}, \quad (17)$$

where T_W is the maximum channel capacity for calls of class *w*. Based on the manner used for determination of the values of $T_w s$, the EASWR policies can be divided in two main groups: *static* and *dynamic EASWR* schemes. In static EASWR schemes, values of $T_w s$ are determined based on the a priori information about the network and remain unchanged during the operation of the network while in dynamic EASWR schemes, $T_w s$ are adapted during the operation of the network. The static EASWR schemes can be divided into two main groups: *call bounding* and *new call bounding* schemes. In the call bounding schemes, the call admission is based on the number of ongoing calls (number of busy channels) in the cell while in the new call bounding schemes; the call admission is based on the number of ongoing new calls in the cell. In dynamic EASWR schemes, function *u(x,w)* is adapted according to the some available information. In dynamic EASWR schemes, the number of channels is allocated and reserved dynamically using *traffic analysis* and *prediction of mobile terminal movement*.

Static call bounding schemes: In the call bounding schemes, admission of a new call is based on the number of ongoing calls in the cell, independent of type of calls. In other words, the state x of a cell is defined as the number of busy channels in the cell. Based on the values of $T_w s$, the call bounding schemes can be divided into two schemes: *reserving integral number of channels* and *reserving fractional number of channels*. In the reserving integral number of channels, all $T_w s$ are integer values while in reserving fractional number of channels, at least one of T_w's are fractional numbers. In the reserving integral number of channel schemes, range of function $u(x,w)$ is the set of $\{0,1\}$. When only two groups G_1 and G_2 (one for new calls and the other for handoff calls) are considered, this scheme is referred to as *guard channel policy*, or *cutoff priority policy* in which a fixed number of channels is reserved in each cell exclusively for handoff calls (Hong & Rappaport, 1986). Under such policy, new calls and handoff calls are treated equally on a first-come first-served basis for channel allocation until a predetermined channel utilization threshold is reached. Let T be this threshold. At this point, new calls are simply blocked and only handoff call requests are accepted. In other words, a new call is accepted if $c < C - T$, where $T \geq 0$ is the number of channels reserved specifically for handoff (guard channels), that is,

$$u(x,w) = \begin{cases} 1 & if\ x < C\ and\ w = handoff\ calls \\ 1 & if\ x < T\ and\ w = new\ calls \\ 0 & if\ x = C \end{cases} \quad (18)$$

Figure 9 shows the state transition diagram of a homogeneous network with C channels and guard channel scheme. The system is modeled by a typical M/M/C/C queuing model.

The steady state probability P_n that n channels are busy is given by the following expression.

$$P_n = \begin{cases} \dfrac{\rho^n}{n!} P_0 & if \quad n \leq T \\ \alpha^{-T} \dfrac{(\rho\alpha)^n}{n!} P_0 & if \quad T < n \leq C \end{cases} \quad (19)$$

where

$$P_0 = \left[\sum_{k=0}^{T} \left(\frac{\rho^k}{k!} \right) + \alpha^{-T} \sum_{k=T+1}^{C} \frac{(\rho\alpha)^k}{k!} \right]^{-1}, \quad (20)$$

$\alpha = \dfrac{\lambda_h}{\lambda_n + \lambda_h}$ **and** $\rho = (\lambda_n + \lambda_h) / \mu$. Given these state probabilities, we can drive the blocking probability of new calls and the dropping probability of handoff calls.

$$B_n(C,T) = P_0 \alpha^{-T} \sum_{m=T+1}^{C} \frac{(\rho\alpha)^m}{m!}$$

$$B_h(C,T) = P_0 \alpha^{-T} \frac{(\rho\alpha)^C}{C!} \quad (21)$$

It has been shown that B_n (B_h) is a monotonically decreasing (increasing) function of T and there is an optimal threshold T^* in which the blocking probability of new calls is minimized subject to the hard constraint on the dropping probability of handoff calls. Algorithm 4 can be

Figure 9. State transition diagram for guard channel scheme

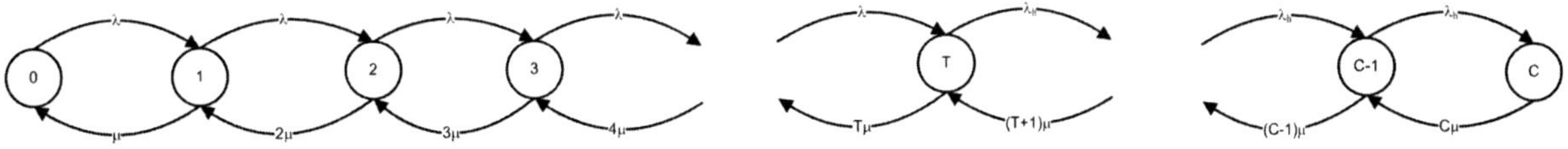

used to find the optimal value of threshold T^* (Ramjee & Towsley & Nagarajan, 1997; Haring & Marie & Puigjaner & Trivedi, 2001).

Algorithm 4: The algorithm for finding the T^*

```
Algorithm FindGCParameter
set upper ← 1; lower  ←  0
if(B_h (C,C)≤ P_h)thenreturn C
end if
if(B_h (C,0) ≥ P_h) then
return 0
end if
while ((upper -lower) < 0.0001)
doset p ← ⌊(upper + lower) /2⌋
if(B_h (C, ⌊p⌋)> P_h)thenset upper
← p
else
set lower ← p
end if
end while
return ⌊p⌋
end Algorithm
```

Chang & Kim (2001) proposed an algorithm to find the optimal number of guard channels in a general multi-cell networks, which minimizes the weighted average of dropping probability of handoff calls in a cluster while satisfying the pre-specified QoS for new calls and co-channel interference constraints. Approximate analyis of guard channel scheme supporting two classes of calls (new and handoff calls) with different average channel holding times were done by Fang & Zhang (200) and Yavuz & Leung (2006). Chen & Lee (2001) considered two traffic classes of voice and transactions and proposed a static guard channel scheme to maintain the upper bound of dropping probability of handoff transaction calls. In this approach, $(C-T)$ guard channels are reserved for handoff transaction calls, but new calls and handoff voice calls have the same priority. Thus, this scheme fails to maintain the upper bound for dropping probability of handoff voice calls. In order to maintain the upper bound for dropping/blocking probability for different classes of calls, call admission schemes with multi-thresholds are introduced.

In (Yin & Li & Zhang & Lin, 2000), dual-threshold reservation (DTR) scheme is given for integrated voice/data wireless networks. In DTR scheme, three classes of calls, data calls (both new and handoff calls), new voice calls and handoff voice calls in increasing order of level of QoS are considered. The basic idea behind the DTR scheme is to use two thresholds, one for reserving channels for handoff voice calls, while the other is used to block data calls into the network in order to preserve the blocking performance of voice calls in terms of the dropping probability of handoff calls and the blocking probability of new calls, that is,

$$u(x,w)=\begin{cases}1 & if\ x<C\ and\ w=handoff\ voice\ calls\\ 1 & if\ x<T_2\ and\ w=new\ voice\ calls\\ 1 & if\ x<T_1\ and\ w=new\ data\ calls\ or\ handoff\ data\ calls\\ 0 & if\ x=C\end{cases} \quad (22)$$

DTR assumes that the bandwidth requirement of voice and data are the same. The equations for blocking probabilities of DTR are derived using a two-dimensional Markov chain and the effect of different values for number of guard channels on dropping and blocking probabilities are studied, but no algorithm for finding the optimal number of guard channels is given. Beigy & Meybodi (2003b) and Beigy & Meybodi (2003c) proposed two algorithms to find the optimal values of T_1 and T_2 for a single cell and multi-cells system, respectively when the average channel holding times for new and handoff calls are the same. Tzeng & Lu (Tzeng & Lu, 2008) designed a call admission control scheme that uses two thresholds; one threshold is used to determine whether or not to accept a new call arrival into a cell, and the other threshold is used to limit the total

Figure 10. State transition diagram for multi-threshold guard channel scheme

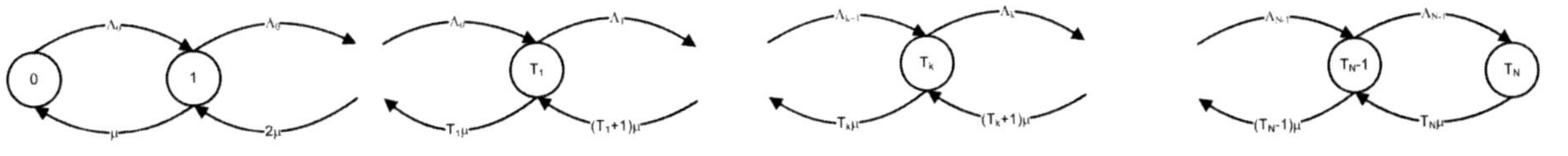

number of calls in a cell. The objective of this scheme aims to satisfy the total completion time requirement of mobile users while maximizes channel utilization.

Beigy & Meybodi (2005a) generalized the idea of two-threshold guard channel scheme to multi-classes and *multi-threshold guard channel scheme* for N classes of calls was introduced. In this scheme, a homogenous cellular network was considered where all cells have the same number channels C and experience the same call arrival rates for all types of calls. In each cell, the arrival of calls of class k ($k=1,...,N$) is Poisson distributed with arrival rate λ_k and the channel holding time of calls of class k is exponentially distributed with the same mean $1/\mu$. Thus, the total call arrival rate is $\Lambda_0 = \lambda_1 + \lambda_2 + ... + \lambda_N$. Assume that the calls of class k has a certain level of QoS such that its blocking probability must be less than q_k. Without loss of generality, it is assumed that $q_1 \geq q_2 \geq ... \geq q_N$. This implies that calls for class k require fewer resources than calls of class $k+1$, i.e. calls for class $k+1$ have a higher priority than calls of class k. To provide the specific level of QoS for calls, the allocated channels of each cell are partitioned into N subsets. In order to partition the channel sets, (N-1) thresholds, $T_1, T_2, ..., T_{N-1}$ ($0 \leq T_1 \leq T_2 \leq ... \leq T_{N-1} \leq C$ are used. For the sake of simplicity, two additional fixed thresholds $T_0 = -1$ and $T_N = C$. The procedure for accepting calls in multi-threshold guard channel scheme is given in equation (23) can be described as follows. A call from class k is accepted when the number of busy channels is smaller than T_k; otherwise the call is blocked.

$$u(x,w) = \begin{cases} 1 & if\ x < T_w \\ 0 & otherwise \end{cases} \tag{23}$$

Let $c(t)$ denote the number of occupied channels in the given cell and $\Lambda_k = \sum_{j=k+1}^{N} \lambda_j$, $\alpha_k = \frac{\Lambda_k}{\Lambda_0}$, and $\rho = \frac{\Lambda_0}{\mu}$. In the multi-threshold guard channel scheme, $c(t)$ is a continuous-time Markov chain (birth-death process) with states $0, 1, ..., C$. Figure 10 shows the state transition diagram of a system with C channels and multi-threshold guard channel scheme. The system is modeled by a typical M/M/C/C queuing model.

The steady state probability P_n that n (for $T_k \leq n \leq T_{k+1}$) channels are busy is given by the following expression:

$$P_n = P_0 \frac{\left(\rho\alpha_k\right)^n}{n!} \prod_{j=1}^{k} \left(\frac{\alpha_{j-1}}{\alpha_j}\right)^{T_j}, \tag{24}$$

where

$$P_0 = \left[\sum_{k=0}^{N-1} \prod_{j=1}^{k} \left(\frac{\alpha_{j-1}}{\alpha_j}\right)^{T_j} \sum_{n=T_k+1}^{T_{k+1}} \frac{\left(\rho\alpha_k\right)^n}{n!}\right]^{-1}. \tag{25}$$

Given these state probabilities, the blocking probability of calls of class k is calculated using the following equation.

$$B_k(T_1, ..., T_N) = P_0 \sum_{n=T_k+1}^{T_N} P_n. \tag{26}$$

Properties of $B_k(T_1, ..., T_N)$ have been studied in (Beigy & Meybodi, 2005a). It was shown that $B_k(T_1,\dots,T_N)$ is a monotonically increasing function of T_k and a monotonically decreasing function of T_j ($j \neq k$). Algorithm 5 can be used for finding the optimal values of thresholds $T_1,\dots,T_{N-1}$ for the following problem: given C channels allocated to a cell, the objective is to find the optimal values of $T_1,\dots,T_{N-1}$ in a such a way that it minimizes $B_1(T_1,\dots,T_N)$ subject to the constraints $B_k(T_1,\dots,T_N) \leq q_k$ (for k=2,...,N).

Algorithm 5: The algorithm for finding the optimal values of $T_1,\dots,T_{N-1}$

```
Algorithm FindMTGCParameters
set T0← -1; T1← T2← T3← ....← TN← C;
if BN(T1,T2,...,TN) ≤ qN then return (T1,T2,...,TN)
end if
for k ← N down to 2 do while Tk-1 > 0 and Bk(T1,T2,...,TN) > qk do if
not MinBlockCheck (T1,T2,...,TN, k+1) then for m ← 1 to k-1 do set
Tm ← Tm-1
end for
end if
end while
end for
if there is at least one class
that QoS is not satisfied then
return 'the number of assigned
channel to this cell is small'
end if
return (T1,T2,...,TN)
end Algorithm
function MinBlockCheck (T1,...,TN, k)
if k = N and Tk-1 < Tk and
Bk(T1,T2,...,Tk-1+1,...,TN) ≤ qk-
then set Tk-1← Tk-1+1
return true else if k < N
and not MinBlockCheck (
T1,...,TN, k+1) and Tk-1 < Tk and
Bk(T1,T2,...,Tk-1+1,...,TN) ≤ qk-
then set Tk-1← Tk-1+1
return true
end if
return false
end function
```

Beigy & Meybodi (2005a) also considered the problem of finding a call admission control scheme that minimizes the number of required channels while preserving the QoS level for all priority levels (all classes of calls) and Algorithm 6 is given to find such optimal number of channels and guard channels.

Algorithm 6: The algorithm for finding the optimal values of $T_1,\dots,T_{N-1}$

```
Algorithm FindMTGCMinChannels
set T0← -1; T1← T2← T3← ....← TN← 0;
while at least one constraint is
not satisfied do
MinChannelCheck (T1,...,TN, 1)
end if
for k ← N down to 1 do while Tk-1
> 0 and Tk-1 < Tk and all con-
straints when set Tk is set to
Tk ← Tk+1 are satisfied do set Tk←
Tk+1
end while
end for
return (T1,T2,...,TN)
end Algorithm
function MinChannelCheck (T1,...,TN, k)
if k = N and BN(T1,T2,...,TN) >
qk then set TN← TN+1
return true else if k < N
and not MinChannelCheck (
T1,...,TN, k+1) and Tk < Tk+1 and
Bk(T1,T2,...,Tk+1,...,TN) > qk-
then set Tk← Tk+1
return true
end if
```

```
return false
end function
```

In reserving integral number of channels, a number of channels are exclusively reserved for highest priority calls which results in less channels available to lowest priority calls and hence the total carried traffic suffers. In these schemes, if only the blocking probability of highest priority calls is considered, these schemes give very good performance, but the blocking probability of lowest priority calls is degraded to a great extent. This effect can be degraded by reserving fractional number of channels.

In schemes that reserve fractional number of channels, the call admission controller has more control on both the dropping probability of handoff calls and the blocking probability of new calls. When only two groups G_1 and G_2 (one for new calls and the other for handoff calls) are considered this policy is referred to as *limited fractional guard channel scheme* (LFG) in which a fractional number of channels is reserved in each cell exclusively for handoff calls (Ramjee & Towsley & Nagarajan, 1997). The LFG scheme uses an additional parameter p and operates the same as the guard channel policy except when T channels are occupied in the cell, in which case new calls are accepted with probability p, that is,

$$u(x,w) = \begin{cases} 1 & if\ x < C\ and\ w = handoff\ calls \\ 1 & if\ x < T\ \ and\ w = new\ calls \\ p & if\ x = T\ \ and\ w = new\ \ calls \\ 0 & if\ x = C \end{cases} \tag{27}$$

Figure 11 shows the state transition diagram of a homogeneous network with C channels and LFG scheme.

The steady state probability P_n that n channels are busy is given by the following expression:

$$P_n = \begin{cases} \dfrac{\rho^n}{n!} P_0 & if \quad n \le T \\ \gamma \alpha^{-(T+1)} \dfrac{(\rho\alpha)^n}{n!} P_0 & if \quad T < n \le C \end{cases} \tag{28}$$

where

$$P_0 = \left[\sum_{k=0}^{T} \left(\frac{\rho^k}{k!} \right) + \gamma \alpha^{-(T+1)} \sum_{k=T+1}^{C} \frac{(\rho\alpha)^k}{k!} \right]^{-1}, \tag{29}$$

$\gamma = \left[\alpha + (1-\alpha)p \right]$, $\alpha = \lambda_h / (\lambda_n + \lambda_h)$ and, and $\rho = (\lambda_n + \lambda_h) / \mu$. Given these state probabilities, we can drive the blocking probability of new calls and the dropping probability of handoff calls.

$$B_n(C,T,p) = \left(1-p\right)\frac{\rho^T}{T!} + \gamma\alpha^{-(T+1)} \sum_{m=T+1}^{C} \frac{(\rho\alpha)^m}{m!}$$

$$B_h(C,T,p) = P_0 \gamma \alpha^{-(T+1)} \frac{(\rho\alpha)^C}{C!} \tag{30}$$

It has been shown that B_n (B_h) is a monotonically increasing (decreasing) function of $T + p$ and therefore there is an optimal pair (T^*, p^*), which minimizes the blocking probability of new calls subject to the hard constraint on the dropping probability of handoff calls. The following algorithm (Algorithm 7) can be used to obtain the optimal pair (T^*, p^*) (Ramjee & Towsley & Nagarajan, 1997).

Algorithm 7: The algorithm for finding the optimal pair *(T^*, p^*)*,

```
Algorithm FindLFGParameter
set upper ← 1; lower ← 0
if(B_h (C,C,0)≤ P_h)thenreturn
(C,0)
end if
```

```
if(B_h (C,0,0) ≥ P_h) then
return (0,0)
end if
while ((upper -lower) < 0.0001)
doset p ← ⌊(upper + lower) /2⌋
if(B_h (C, ⌊p⌋, p-⌊p⌋)> P_h)thenset
upper ← p
else
set lower ← p
end if
end while
return (⌊p⌋, p-⌊p⌋)
end Algorithm
```

Vazquez-Avila & Cruz-Perez & Ortigoza-Guerrero (2006) compared uniform fractional channel scheme, limited fractional channel, scheme, guard channel scheme from different performance criteria.

New call bounding schemes: In new call bounding schemes, new calls are accepted if the number of channels used by new calls is less than a threshold (bound for new call) provided that the cell has enough channels for allocating to the incoming new calls. In other words, the state, x, of a cell is defined as the number of ongoing new calls in the cell. Fang & Zhang (2002) proposed a new call bounding scheme in which $T_w s$ are integers. In this scheme, when a new call arrives, if the number of new calls in a cell exceeds a threshold then the new call is blocked; otherwise it will be accepted and the handoff call is rejected only when all channels in the cell are occupied. The idea behind this scheme is that we would rather accept fewer new calls than dropping the ongoing calls in the future, because customers are more sensitive to the call dropping than the call blocking. In (Chung & Chiu, 2002), a new call bounding scheme is given for integrated voice/data wireless networks. In this scheme, it is assumed that the number of ongoing data calls always is constant. This scheme accepts the incoming voice request if the number of voice connections is less than the voice threshold T_1. Since the number of data connections is fixed, there is no call admission control for data connections. In (Chung & Chiu, 2002), no algorithm is given to determine the optimal value of T_1.

Fang (2003) proposed a call admission scheme, which is a generalization of fractional guard channel and new call bounding schemes for multiple classes of calls. This scheme accepts calls with a certain probability, which is determined by the number of busy channels belonging to the priority level of the arriving call in the cell. Fang (2003) analyzed the blocking probabilities of calls when all classes of calls have the same average channel holding time. Wang & Fang & Pan (2008) are studied two variants of the call admission scheme given in (Fang, 2003) for the case that different classes of calls have arbitrary channel requirements and different average channel holding times and their blocking performance analysis are carried out using multi-dimensional Markov process. The first variant uses the information about the total amount of busy channels (bandwidth units) and the second variant utilizes the number of users belonging to the same priority level.

Dynamic EASWR schemes based on teletraffic analysis: In these schemes, function $u(x,w)$ is adapted based on the estimated traffic. Since all ongoing calls in the neighboring cells are potential handoff calls to the test cell, these schemes estimate the handoff arrival rate as a function of the number of ongoing calls in the neighboring cells. In these schemes, the number of reserved channels can be an integral number or a fractional number.

The *linear weighting scheme* is given in (Acampora & Naghshineh, 1994a; Acampora & Naghshineh, 1994b) uses the mean number of ongoing calls in the neighboring cells, I, within a maximum cell distance d from the test cell in determining of the call admission. Let S_d denotes the set of cells in a maximum cell distance d from the test cell and c_i denotes the number of ongo-

Figure 11. State transition diagram for limited fractional guard channel scheme

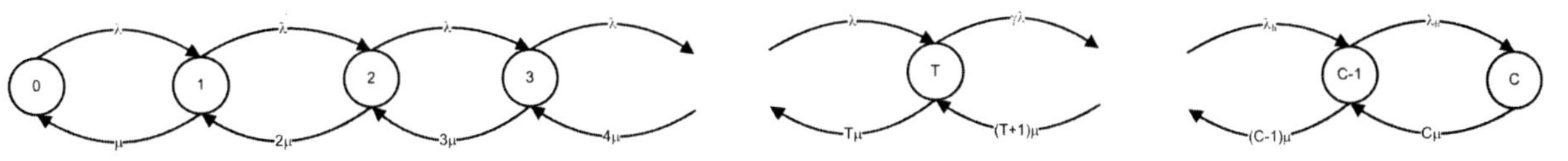

ing calls in the neighboring cell *i*. In this scheme, the state of the system at each time instant is defined as

$$x = \left[\frac{1}{|S_d|} \sum_{i \in S_d} c_i \right] \quad (31)$$

In linear weighting scheme, the new calls are only accepted to the originating cell if

$$u(x,w) = \begin{cases} 1 & \text{if } x < C \text{ and } w = \text{handoff calls} \\ 1 & \text{if } x < T_1 \text{and } w = \text{new calls} \\ 0 & \text{if } x = C \end{cases} \quad (32)$$

Note that the guard channel scheme is a special case of this algorithm where $S_d = i$. Peha & Sutivong (2001) proposed a call admission scheme called *weighted sum scheme,* which uses the weighted sum of the number of ongoing calls in the test cell and in the neighboring cells in determining the admission. Let c_i be the mean number of ongoing calls in the neighboring cells with distance I and p_i be the weight of these cells such that $\sum_{i=1}^{\infty} p_i = 1$ and $p \geq 0 (for\ i \geq 0)$. The state of system in weighted sum scheme at each time instant is defined as $x = \left[\sum_{i=0}^{\infty} p_i c_i \right]$. In this scheme, the new calls are only accepted to the originating cell if

$$u(x,w) = \begin{cases} 1 & \text{if } x < C \text{ and } w = \text{handoff calls} \\ 1 & \text{if } x < T_1 \text{ and } w = \text{new calls} \\ 0 & \text{if } x = C \end{cases} \quad (33)$$

The optimal value of weights p_i can be determined experimentally. The *distributed call admission scheme, proposed by*Naghshineh & Schwartz (1996), does not need the exchange of status information upon the arrival of calls (new and handoff calls). Rather, it only requires the exchange of such information periodically. The admission control algorithm calculates the maximum number of calls that can be accepted in the test cell without violating the QoS of the existing calls in that cell as well as calls in its neighboring cells. One of the main features of this scheme is its simplicity in that the admission decision can be made in real time and does not require much computational effort but this scheme cannot always guarantee the target call dropping probability.

Yu & Leung (1997) introduced a dynamic guard channel scheme in which each base station dynamically adapts the number of channel to be reserved based on the current estimates of the rate at which mobiles in the neighboring cells are likely to incur a handoff into this cell. The objective of the adaptation algorithm is to maintain a specified level of QoS for handoff calls despite of the temporal fluctuations in the traffic into the cell. The determination of the number of channels to be reserved is based on an analytical model which relates number of reserved channels to the dropping probability of handoff calls and the blocking probability of new calls.

In (Oliveria & Kim & Suda, 1998), the number of channels that must be reserved is estimated according to the requested bandwidth of all ongoing connections. Each base station keeps monitoring the dropping probability of handoff calls and the utilization of channels in its cell. Then base station according to this information adjusts the number of guard channels. Lee & Park (1998) proposed a call admission algorithm in which when a new or a handoff call arrives at the test cell, a number of channels in the neighboring cells is reserved. The number of channels to be reserved varies dynamically depending on the number current ongoing calls in the test cell and its neighboring cells.

Choi & Shin (1998) have proposed a scheme based on prediction of the probability that a call will be handed off to a certain neighboring cell from aggregate history of handovers in each cell and determines the number of reserved channels. In this scheme, each base station records the number of handoff failures and adjusts the reservation by changing the estimation window size. Boumerdassi & Beylot (1999) proposed a call admission algorithm for multi-rate personal communication networks in which the number of channels that must be reserved is determined periodically based on the estimated parameters, such as handoff rate. In the beginning of each period, the traffic parameters are estimated and it is assumed that for a given period, traffic parameters are fixed. In this scheme, when the number of occupied channels reaches the threshold T_1, the cell reserves a resource in the neighbors for which the probability of transition is high. If they have free channels, the reservation takes place immediately; otherwise, the algorithm waits for a free channel.

In (Ramanathan & Sivalingam & Agrawal & Kishore, 1999), two dynamic EASWR algorithms are given for wireless networks that support several types of traffic such as voice, data, and video applications, each with different channel requirements. The objective of these algorithms is to accept all handoff calls. Then the base station accepts new calls if and only if the additional channels need to accept all incoming handoff calls (the number of channels to be reserved) and this new call is available. The number of reserved channel is determined according to the estimation of the exact arrival time and channel requirements of future handoff calls. An extension of guard channel scheme is given in (Bozinovski & Popovski & Gavrilovska, 2000; Bozinovski & Popovski & Gavrilovska, 2000). This scheme operates same as the guard channel scheme when a new call arrives and $x < T_1$ or $x = C$; when $T_1 \leq< C$, the algorithm estimates the dropping probability of handoff calls during a period. Then the algorithm accepts new call if the estimated dropping probability of handoff calls is less than the predetermined QoS; otherwise reject the new call. A dynamic channel reservation algorithm, which is presented in (Rappaport & Purzynski 1996), the number of channel to be reserved in each cell is determined dynamically based on the number of ongoing calls in the neighboring cells. This scheme ensures that QoS is maintained in all cells.

Beigy and Meybodi (in press) proposed two learning automata based algorithms to determine the near optimal number of the guard channels when the parameters traffic parameters are unknown and possibly time varying. In these algorithms, learning automata are used to adapt the number of guard channels as the network operates. Let $g(t)$ be the number of guard channels at time instant t which takes values in interval $[g_{min}, g_{max}]$, (for $0 \leq g_{min} < g_{max} \leq C$). In these algorithms, each base station uses one learning automaton with action set $\underline{\alpha} = \{\alpha_1, \alpha_2, \ldots, \alpha_r\}$ alpha}, where $r = g_{\max} - g_{\min} + 1$. Selection of action α_i by learning automaton means that the base station uses $g(t) = g_{\min} + \alpha_i - 1$ guard channels.

The operation of these algorithms can be described as follows. These algorithms accept handoff calls as long as the cell has free channels. When a new call arrives at a given cell, the learning automaton associated to this cell chooses one of its actions, say α_i. If the cell has at least $g_{\min} + \alpha_i - 1$ free channels, then the call will be accepted; otherwise it will be blocked. Then the base station computes the current estimate of the dropping probability of handoff calls $\hat{B}_h$ and based on the result of comparison of this quantity with the specified level of QoS (p_h), the reinforcement signal will be produced and the action probability vector of the learning automaton will be updated using a learning algorithm. The differences between these algorithms are the way that they produce reinforcement signal for the learning automata and learning algorithm used to update the action probability vector.

The first algorithm uses a SL_{R-I} learning automaton in each cell and the reinforcement signal at time instant *n* is equal to $\psi\left(\left|\hat{B}_h - p_h\right|\right)$, where $\psi : R \longrightarrow [0,1]$ is a projection function. The projection function is considered to be a continuous, nondecreasing and nonnegative function that maps the set of real numbers into [0,1], for example $\psi(x) = x$ can be a projection function, which maps [0,1] into [0,1]. The continuity of the projection function is needed because the response produced by the environment is a real number in interval [0,1], its nonnegativity is needed in order to maintain the reward and penalty nature of updating, and the nondecreasing property is needed for preserving the relative strength of the reinforcement signal. It is obvious that when $\hat{B}_h$ is far from p_h, and then the reinforcement signal will be large, which causes the selected action of the learning automaton to be penalized. When $\hat{B}_h$ is near to p_h, the reinforcement signal will be small and near to zero which causes the selected action of the learning automaton to be rewarded. In other words, when $\hat{B}_h$ is greater than p_h, the chosen number of guard channels is too small and when $\hat{B}_h$ is smaller than p_h, the number of guard channels chosen by learning automaton is large. In other words, the reinforcement signal is an indicator of the relative distance of the dropping probability of handoff calls to the predefined level of QoS.

Simulation results showed that the blocking probability of new calls for the first algorithm is lower than the blocking probability of the guard channel algorithm, but it can not maintain the predefined level of QoS, as evidenced by the results of simulation. The second algorithm tries to minimize the blocking probability of new calls and at the same time to maintain the specified level of QoS. This algorithm uses a L_{R-I} learning automaton in each cell for determination of the number of guard channels. The selected action of learning automaton in a cell will be rewarded if the incoming new call is accepted and the current estimate of dropping probability of handoff calls

Figure 12. Blocking probabilities of new calls for learning automata based dynamic guard channel algorithms

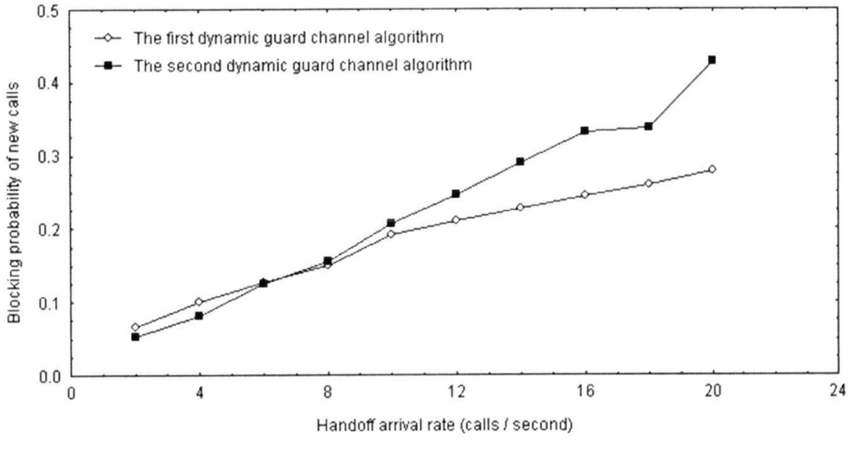

Figure 13. Dropping probabilities of handoff calls for learning automata based dynamic guard channel algorithms

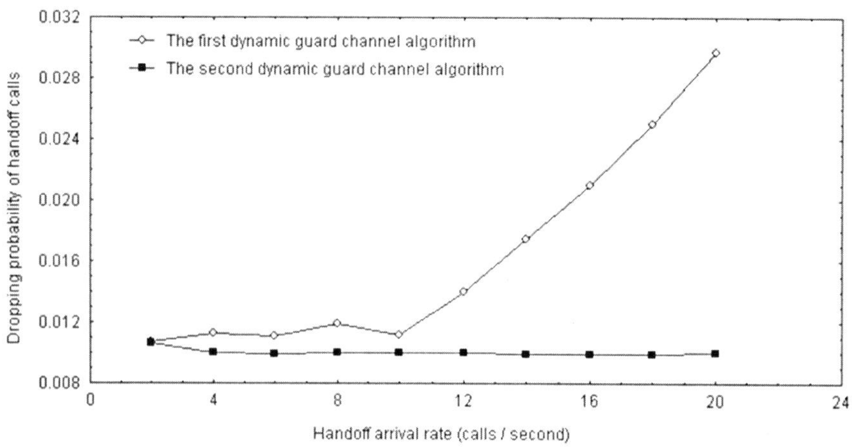

$\hat{B}_h$ is less than the specific level of QoS (p_h) or the incoming new call is rejected and the current estimate of dropping probability of handoff calls is greater than the specific level of QoS; the selected action neither rewarded nor penalized otherwise. Figures 12 and 13 show the performance of these algorithms.

Yu & Leung (1996) proposed a call admission algorithm is given, in which when a new or handoff call arrives at a neighboring cell, number of channels that must be reserved in the test cell is increased by a fraction amount and when a call is completed at or moved out of the neighboring cells, the number of reserved channels is decreased by the same fractional amount. Han & Nilsson (2000) proposed a *population-based channel reservation scheme*. This scheme dynamically adjusts the number of channels that must be reserved for handoff calls according to the amount of cellular traffic in its neighboring cells. Assume that cell i have n_i neighboring cells. Whenever a call which consumes b channels is accepted into cell j as either a newly call or a handoff call, the base station of the cell requests a fractional channel reservation for the amount of b / n_j to each of its n_j neighboring cells. Whenever this call is leaving the cell either by call completion or by handoff into one of its neighboring cells, the base station requests a fractional channel release for the same amount as requested for the reservation to each of its n_j neighboring cells, even to the cell into which this call is handed over. This step is to inform the neighborhood of appearance and disappearance of a potential handoff. Each base station network maintains a counter that records transactions for fractional channel reservation or release requests from its neighboring cells. Every time it receives a fractional channel reservation request or a release request, it increments or decrements the counter by the requested amounts, respectively.

Beigy and Meybodi (2005b) proposed an adaptive limited fractional guard channel algorithm for two classes of calls: new and handoff calls. The objective of this algorithm is to adapt parameter $T + p$ in such a way that minimizes the blocking probability of new calls subject to the constraint that the dropping probability of handoff calls be at most p_h. Since $T + p$ is a continuous parameter, the algorithm uses a continuous action-set learning automaton for adaptation of the value of parameter $T + p$. Let $x(n) = T(n) + p(n)$ be the parameter of the limited fractional guard channel algorithm at instant *n*, and *x(n)* takes values in the interval $\left[x_{\min}, x_{\max}\right]$, where $0 \leq x_{\min} < x_{\max} < C$. The action-set for learning automaton is the real line and it uses the Gaussian distribution, $N(\mu, \sigma)$, to choose its actions. This Gaussian distribution is updated using the reinforcement signal, which is emitted from the environment. Initially, the learning automaton chooses one of its actions with equal probability using a Gaussian distribution with a large variance. Since *x(n)* and *μ(n)* must be in the interval $\left[x_{\min}, x_{\max}\right]$, the above mentioned learning automaton cannot be used directly to adapt the value of $T + p$, and hence a projected version of learning automaton will be used. In the projected version of learning automaton, a constraint set $H = \left\{y \middle| x_{\min} \leq y \leq x_{\max}\right\}$ is used for updating μ as well as choosing actions of learning automaton. In the projected version, when the updated value of μ goes outside of the constraint set *H*, then μ is pushed into *H* and when the action chosen by the learning automaton does not belong to *H*, then the action is pushed into *H*. This algorithm can be described as follows. Each base station is equipped with a learning automaton for adapting $T + p$. When a new call arrives at a given cell, the learning automaton associated to that cell chooses one of its actions, say *x(n)*. Let $T(n) = \lfloor x(n) \rfloor$ and $p(n) = x(n) - \lfloor x(n) \rfloor$. If the number of busy channels of a cell is less than *T(n)*, then the incoming call will be accepted; when the cell has *T(n)* busy channels, then a call will be accepted with probability $p(n)$; otherwise the incoming call will be blocked. On the arrival of a new call the base station computes the current estimate of the dropping probability of the handoff calls and based on the result of the comparison of this quantity with the specified level of QoS, then it computes the reinforcement signal as $\left|\hat{B}_h - p_h\right|$ and the learning automaton updates its action probability distribution. Beigy and Meybodi (2005b) showed that this algorithm finds the optimal number of channels that must be reserved.

In (Beigy & Meybodi, 2002a; Beigy & Meybodi, 2002b), two adaptive limited fractional guard channel algorithm based upon continuous action-set learning automata are reported. These algorithms adjust the number of channels to be reserved in the cell according the traffic of the cell and the predefined QoS. The differences between these algorithms are the learning algorithm used for learning automata and the ways that the reinforcement signal will be produced. Salamah & Lababidi (Salamah & Lababidi, 2005) proposed an adaptive channel reservation scheme for cellular networks. In this algorithm, the base station measures the signal strength to predict the handoff. When there is no handoff in the new future, some of the reserved channels can be used for new calls.

Dynamic EASWR schemes based on mobility: The most salient feature of the mobile wireless network is the mobility, which can be used for adjusting the $T_w s$. Since the handoff occurs when the mobile users are moving during the call connection, thus good call admission control algorithms should consider the mobility pattern. Hence, in order to make a reservation schemes effectively adapt to the network traffic situations, the user mobility information must be deployed. In these schemes, each base station adjusts the reservation by employing the mobility information. The mobility pattern is influenced by many factors such as destinations of mobile users, the layout of the network, and the traffic condition in the network. Since it is not easy to specify the mobility pattern of each mobile user in detail, therefore the statistical mobility patterns of users are more useful. Based on the values of thresholds, $T_w s$, these schemes can reserve an integral or fractional number of channels.

Concept of *shadow cluster*, which is introduced by Levine & Akylidiz & Naghsineh (1997), estimates the future resource requirements based on the current movement pattern of the mobile users. The fundamental idea of the shadow cluster concept is that as an active user travels to other cell, the region of influence also moves. The base stations currently being influenced are said to form a shadow cluster, because the region of influence follows the movement of the active mobile terminal like a shadow. However, the strength of this scheme depends on the accuracy of the knowledge of users' movement patterns, such as trajectory of a mobile user, which is difficult to predict in real time systems. Hou & Fang (2001) proposed an *integral mobility based channel reservation scheme* in which mobile users are classified in two classes according to their velocities: high and low speed users. Thus the average cell dwell time of high speed users are shorter than that of the low speed users. Based on the velocity of each mobile user, the handoff probability of each class is predicted and the number of channels that must be reserved is determined. It is also noted that the better performance will be achieved if this scheme and a new call bounding scheme are combined. Hu & Sharma (2003) proposed a dynamic reservation scheme for multimedia cellular networks in which the handoff calls have a higher priority than the new calls. The prerequisite of this scheme is that base stations can estimate future trajectory of mobile computers with high degree of accuracy, which is possible in today's increasing improved position location techniques. This scheme uses the Kalman filter to predict the next cell for every mobile computer. Huang and et. Al. (Huang & Chuang & Yang, 2008) proposed a reservation based adaptive call admission algorithm in which a fuzzy logic system is used to estimate the number of channels to be reserved for handoff calls and particle swarm optimization (PSO) technique used to adjust the parameters of the membership functions in the fuzzy logic system. In (Martinez-Bauset & Gimenenz-Guzman & Pla, 2008) the problem of optimizing admission control policies in mobile multimedia cellular networks when predictive information regarding the movement of mobile terminals is available was studied. For the optimization process a reinforcement learning approach was used.

When a fractional number of channels are reserved, $T_w s$ are real numbers and the call admission controller have more control on both the dropping probability of handoff calls and the blocking probability of new calls because the rounding of $T_w 's$ lost some information. *Fractional mobility based channel reservation scheme* is given in (Hou & Fang, 2001) in which mobile users are classified in two classes according to their velocities: high and low speed users. Thus the average cell dwell time of high speed users are shorter than that of the low speed users. Based on the velocity of each mobile user, the handoff probability of each class is predicted and values of $T_w 's$ are determined.

4.2.2 Complete Partitioning Policies

Complete partitioning policies are subsets of reservation based call admission policies. In these policies, we have $S = S_{G_1} \cup S_{G_2} \cup \ldots \cup S_{G_W}$ Complete partitioning policies partition the channels among the different classes of calls by dedicating a certain number of channels to each class. This policy takes place when the threshold point for traffic class w is inside the state space, i.e. $T_w \in S$. These policies isolate each class of calls and the resulting process is simply the aggregation of N independent $M / M / T_w / T_w$ processes. Leong & Zhuang (2002) considered a cellular network that supports two traffic types of voice (constant-rate) and data (variable-rate). In this scheme, voice calls have a higher priority than the new calls. The channels in each cell are partitioned into two subsets, one for voice calls and the other for data calls. Each partition uses the standard LFG policy to accept/reject new calls in that class. Ahn & Kim (2003) proposed a dynamic channel allocation for multimedia cellular networks that uses the guard channel scheme for maintaining the level of QoS and works the same as the guard channel scheme when a new or handoff call arrives, but when a call is terminated or completed, it differs from the guard channel policy. If a call that uses a guard channel is terminated or completed, then that channel is reserved for future incoming handoff calls. On the other hand, if a call that uses an ordinary channel is terminated or completed, then the bandwidth adaptation, which is the allocation of freed bandwidth to the ongoing calls, is applied. In order to allocate the freed bandwidth to the ongoing calls, a Lagrangean relaxation procedure is used that leads to a sub-optimal solution. Kulavaratharash & Aghvami (1999) divided channels of a cell into two groups: ordinary and guard channel groups. The new calls are accepted if the ordinary channel group has free channel; otherwise the call will be blocked. For handoff calls, three different strategies are used: 1) first guard channel group and then the ordinary channel group is selected, 2) first ordinary channel group and then the guard channel group are selected, and 3) randomly one of the preceding strategies is selected. In order to improve the blocking probability of new calls without trading off the dropping probability of handoff calls, an algorithm is given in (Kulavaratharash & Aghvami, 1999). In this algorithm, if all channels in the ordinary channel group are occupied at the arrival time of a new call and there is at least one free channel in guard channel group, then any free guard channel can temporarily be lent to the ordinary channel group to prevent the new call to be blocked. Such transferring can only be carried out if the base station can predict that there are no handoff attempts from neighboring cell, while the borrowed channel is used for the new call. This prediction is done with the aid of power measurements. AlQahtani & Mahmoud (AlQahtani & Mahmoud, 2008) extended complete partitioning and the queuing priority call admission schemes for operation in 3G WCDMA networks. In their complete partitioning, each class of calls has its own queue and resource partition whereas in queuing priority, each call class has its own queue and all classes share the available resources. Then they develop an analytical model for the queuing priority algorithm to study the behavior of this algorithm.

4.3 Queuing Priority Schemes

These schemes reduce the blocking probability of new calls and the dropping probability of handoff calls by employing a queuing mechanism. In queuing priority schemes, calls of each class are accepted whenever there is a free channel for that class. When there is no free channels for a class, calls may be queued and calls of other classes are blocked and cleared from system. One key point of using queuing in call admission control algorithms is that the service differentiation could be managed by modifying the queuing discipline. For example, instead of FIFO queuing strategy, other prioritized queuing discipline can be used to maintain priority level in each service class. Another key point is the mobility of the users, which results difficulties in management of queue. These schemes consider two traffic classes, new calls and handoff calls. Based on the type of calls that is queued, these schemes are divided in three groups: *new call queuing schemes*, *handoff call queuing scheme* and *all call queuing schemes*. Some of the reported schemes are briefly described below.

4.3.1 New Call Queuing Schemes

In a new call queuing scheme, a certain number of channels is reserved in each cell exclusively for handoff calls. In new call queuing schemes, the new calls and the handoff calls are treated equally on a first-come first-served basis for channel allocation until the number of occupied channels in the cell becomes T_1. When the predetermined channel utilization threshold, T_1, is reached, new calls are queued and only handoff call requests are accepted. In other words, a new call is accepted if $c < C - T_1$, where $T_1 \geq 0$ is the number of channels reserved specifically for handoff (guard channels), that is,

$$u(x,w)=\begin{cases} accept & if\ x<C\ and\ w=handoff\ calls\\ accept & if\ x<T_1 and\ w=newcalls\\ queue & if\ x\geq T_1 and\ w=new\ calls\\ reject & if\ x=C \end{cases} \quad (33)$$

The only reported new call queuing scheme is given in (Guern, 1998). In this scheme, when the number of free channels is less than the number of guard channels, the new calls are queued. It is pointed out that the blocking probability of new calls can be drastically reduced by reserving some channels for handoff calls and using a queuing mechanism for new calls.

4.3.2 Handoff Calls Queuing Schemes

Handoff queuing schemes reserves a number of channels for use of handoff calls. In these schemes, the new calls are serviced as same as handoff calls until the number of free channels becomes less than the number of reserved channels $\left(C-T_1\right)$. When the number of occupied channels is greater than threshold T_1, new calls are blocked and handoff call requests are accepted. When all channels are occupied, the handoff calls are queued, that is

$$u(x,w)=\begin{cases} accept & if\ x<C\ and\ w=handoff\ calls\\ queue & if\ x=C\ and\ w=handoff\ calls\\ accept & if\ x<T_1\ and\ w=new\ calls\\ reject & if\ x\geq T_1\ and\ w=new\ calls \end{cases} \quad (34)$$

Hong and Rappaport analyzed handoff queuing scheme with an infinite buffer for handoff calls and this scheme with finite buffer is analyzed in (Yoon & Kwan, 1993). The extension of handoff queuing scheme with finite buffer size to multi-class of calls is proposed in (Tian & Ji, 2001). Agrawal & Anvekar & Naredran (1996) introduced a handoff call queuing scheme, which reserves no channels for handoff calls. In this scheme, when a new call arrives and all channels are busy, then the call will be blocked; when a handoff call arrives and all channels are busy, the call will be queued. Both types of calls will be accepted if there are any free channels. When a channel becomes free, then a handoff call from the queue, if queue is not empty, will be serviced. Agrawal & Anvekar & Naredran (1996) also proposed some queuing discipline such as first-in first out, most critical first. Cho & Ko & Kwang (1997) proposed a dynamic channel reservation scheme with handoff queuing. In this scheme, the number of channels to be reserved is adjusted based on the handoff traffic and the current number of reserved channels. Zheng & Lam (2002) introduced a dynamic channel reservation scheme with handoff queuing in which the number of channels to be reserved is adjusted based on the occupied channels in the neighboring cells. It must be pointed out that queuing of handoff calls is more sensitive to delay (time between request and the time for allocation of channels) in the service than queuing of new calls, because as mobile users move the signal strength decreases and the call may be dropped. However, this delay depends on the speed of the mobile user.

4.3.3 All Calls Queuing Schemes

These schemes wok as same as guard channel scheme when the number of occupied channels in the cell is less than T_1. When the number of occupied channels is equal or greater than T_1, new calls are queued and only handoff call requests are accepted. When all channels are occupied, the handoff calls are also queued, that is

$$u(x,w)=\begin{cases} accept & if\ x<C\ and\ w=handoff\ calls\\ queue & if\ x=C\ and\ w=new\ calls\\ accept & if\ x<T_1\ and\ w=new\ calls\\ queue & if\ x\geq T_1\ and\ w=new\ calls \end{cases} \quad (35)$$

Yoon & Kwan (1993) proposed a call admission scheme in which the value of T_1 is equal to C. In this scheme, the new calls are put after all handoff calls in the queue and the queue is serviced in the FIFO manner. When the queue is full, then all incoming calls will be blocked. Yoon & Kwan (1993) also used a rearranging mechanism in which when the queue is full, then the last new call is pushed out from the queue and the incoming handoff call will be placed after the last handoff call. Chang & Chang & Lo (1999) introduced a call admission scheme in which all calls are queued with certain rearrangements in the queue.

5. OPTIMAL CALL ADMISSION POLICIES

Let assign a cost to each blocked call, low cost for new calls and high cost for handoff calls, the optimal policy is the one that that finds $u(x,w)$ in such a way that the cost is minimized. In these policies, the call admission is formulated as as Markov decision process and actions of this Markov decision process are used as function $u(x,w)$. Saquib & Yates (1995) used value iteration algorithm of Markov decision process as a technique to search for the optimal policy, that is, the policy which minimizes a weighted blocking criterion. In (Kwon & Choi & Naghshineh (1998); Choi & Kwon & Choi & Naghshineh (2000)), a call admission control algorithm is given which focuses the forced termination probability (call dropping probability) as the main QoS requirement. In this approach the cellular system is modeled using semi-Markov decision process. The linear programming method for solving semi-Markov decision process is employed to find out the optimal call admission control decision in each state. Morley & Grover (2000) formulated the call admission problem in dual-mode cellular networks as a Markov decision process and the linear programming is used for finding the optimal call admission policy.

6. FUTURE RESEARCH DIRECTIONS

Voice telephony and short message services were two first applications that mobilized. Now mobile networks support many other services such as email, web browsing, and push to talk by introduction of packet based networks. Current 3G and 3.5G wireless networks are able to cope with several such applications and offer a sufficient bandwidth. Due to the rapid development and growth of mobile communications, there will be a rapid growth in demand for new wireless services in next-generation wireless networks. The next generation wireless networks such as UMTS long term evolution (LTE) and WiMAX will support a wide variety of multimedia services at higher bandwidths. These services have different traffic characteristics, bandwidth requirements, and quality of service requirements. To support such integrated services, call admission control algorithms become more important. Most of the call admission algorithms reported in the literature support only voice service.

One challenge is how to implement handoff in next generation networks with minimum packet loss and handoff latency. Call admission and handoff management in these networks are more complex, as they must cover both horizontal and vertical handoffs. Therefore, fast and seamless handover is a big challenge for these networks. Since these networks must support real-time multimedia applications that require small delay and high-rate data transmission, the future researches on call admission algorithms and handoff management will focus on algorithms for services that have different bandwidth requirements, different quality of service requirements, and delay and cross-layer call admission and handoff management is one of such area.

Table 1. Features of the proposed classification

Call admission algorithm	Ability to have control on blocking / dropping probabilities	Ability to track traffic variation	Usability for delay sensitive applications
Non-prioritized schemes	No	No	Yes
Prioritized schemes			
• Static equal access sharing with priority	Low	No	Yes
• Dynamic equal access sharing with priority	Low	Yes	Yes
• Static reservation based schemes			
o Fractional number of channels	High	No	Yes
o Integral number of channels	Medium	No	Yes
• Dynamic reservation based schemes			
o Fractional number of channels	High	High	Yes
o Integral number of channels	Medium	Medium	Yes
• Queuing schemes	Yes	No	No, it needs careful management of queues

7. CONCLUSION

In this chapter, we proposed a classification of user based call admission policies in mobile cellular networks. The proposed classification not only provides a coherent framework for comparative studies of existing approaches, but also helps future researches and developments of new call admission policies. Much of research has been done in reservation based call admission policies. One critical issue in all reservation based call admission control policies is how the reservation is made. In traditional guard channel policy, the number of guard channels is determined based on the priori knowledge of the cell traffic and the QoS requirements. Obviously, the performance will degrade if the cell traffic is not conformal to the priori knowledge; thus it will be better to use dynamic reservation schemes: adjusting the number of guard channels with the network traffic. In order to determine an optimal or near optimal value for number of guard channels one first answer the following question: when do reserve channels for incoming handoff calls? If the reservation is made at time when it is needed, the resulting scheme will definitely achieve the best performance. However, such timing will be very difficult, if it is not impossible, to acquire. Since the reservation is a waste of resources if it not used by handoff calls, the shorter the time the reservation is actually used (reservation time), the better performance will be achieved. Table 1 summarizes some features of algorithms in our classification.

REFERENCES

Acampora, A. S., & Naghshineh, M. (1994a). An Architecture and Methodology for Mobile-Executed Handoff in Cellular ATM Networks. *IEEE Journal on Selected Areas in Communications*, *12*, 1365–1374. doi:10.1109/49.329339

Acampora, A. S., & Naghshineh, M. (1994b). Control and Quality-of-Service Provisioning in High Speed Microcellular Networks. *IEEE Personal Communications*, 36-43.

Agrawal, P., Anvekar, D. K., & Naredran, B. (1996). Channel management policies for handovers in cellular networks. *Bell Labs Technical Journals*, *1*(2), 97–110. doi:10.1002/bltj.2018

Ahn, K., & Kim, S. (2003). Optimal bandwidth allocation for bandwidth adaptation in wireless multimedia networks. *Computers & Operations Research, 30*, 1917–1929. doi:10.1016/S0305-0548(02)00115-6

AlQahtani, S. A., & Mahmoud, A. S. (2008). Performance analysis of two throughput-based call admission control schemes for 3G WCDMA wireless networks supporting multi-services. *Computer Communications, 31*, 49–57. doi:10.1016/j.comcom.2007.10.029

Ayyagari, D., & Empremides, A. (1999). Admission control with priorities: approaches for multi-rate wireless systems. *ACM Mobile Networks and Applications, 4*, 209–218. doi:10.1023/A:1019103016426

Beigy, H., & Meybodi, M. R. (2002a). An adaptive limited fractional guard channel policy using continuous action learning automata. In *Proceedings of the 10th International Conference on Software, Telecommunications and Computer Networks (SoftCom 2002),* Croatia, Italy (pp. 536-540).

Beigy, H., & Meybodi, M. R. (2002b). A limited fractional guard channel scheme using learning automata. In *Proceedings of the First EurAsian Conference on Advances in Information and Communication Technology (EURASIA-ICT 2002),* Shriaz, Iran (pp. 343-347).

Beigy, H., & Meybodi, M. R. (2003a). Adaptive uniform fractional guard channel algorithm: the steady state analysis. In *Proceedings of the 9th International Symposium on Wireless Systems and Networks (ISWSN'03),* Dhahran, Saudi Arabia (pp. 177-181).

Beigy, H., & Meybodi, M. R. (2003b). An optimal prioritized channel assignment scheme for using in mobile transaction environments. In *Proceedings of the 8th Annual International Computer Society of Iran Computer Conference (*CISCC-2003), Iran (pp. 68-74).

Beigy, H., & Meybodi, M. R. (2003c). An optimal channel assignment scheme. In *Proceedings of the 11th Iranian Conference on Electrical Engineering (ICEE-2003),* Shiraz, Iran (Vol. 2, pp. 634-641).

Beigy, H., & Meybodi, M. R. (2003c). User based call admission control policies for cellular mobile systems: A survey. *The CSI Journal on Computer Science and Engineering, 1*(2b), 27–40.

Beigy, H., & Meybodi, M. R. (2004a). A new fractional guard channel policy. *Journal of High Speed Networks, 13*(1), 25–36.

Beigy, H., Meybodi, M. R. (2004b). Adaptive uniform fractional guard channel algorithms. *Iranian journal of Electrical and Computer Engineering, 3*(1), 47-53.

Beigy, H., & Meybodi, M. R. (2005a). A general call admission policy for next generation wireless networks. *Computer Communications, 28*, 1798–1813. doi:10.1016/j.comcom.2004.12.009

Beigy, H., & Meybodi, M. R. (2005b). An adaptive call admission algorithm for cellular networks. *Computers & Electrical Engineering, 31*, 132–151. doi:10.1016/j.compeleceng.2004.12.002

Beigy, H., & Meybodi, M. R. (in press). Learning automata based dynamic guard channel algorithms. *Journal of High Speed Networks.*

Boumerdassi, S., & Beylot, A. (1999). Adaptive channel allocation for wireless PCN. *ACM Mobile Networks and Applications, 4*, 111–146. doi:10.1023/A:1019186412792

Bozinovski, M., Popovski, P., & Gavrilovska, L. (2000). Novel strategy for call admission control in mobile cellular network. In *Proceedings of IEEE Conference on Vehicular Technology (VTCC 2000)* (Vol. 4, pp. 1597-1602).

Bozinovski, M., Popovski, P., & Gavrilovska, L. (2000). QoS-Based policy for call admission control in mobile cellular networks. In *Proceedings of IEEE Conference on Wireless Communications and Networking (WCNC 2000)* (Vol. 2, pp. 502-506).

Chang, C., Chang, C. J., & Lo, K. R. (1999). Analysis of hierarchical cellular systems with reneging and dropping for waiting new calls and handoff calls. *IEEE Transactions on Vehicular Technology, 48*(4), 1080–1091. doi:10.1109/25.775357

Chang, K., & Kim, D. (2001). Optimal prioritized channel allocation in cellular mobile systems. *Computers & Operations Research, 28*, 345–356. doi:10.1016/S0305-0548(99)00121-5

Chen, G. C., & Lee, S. Y. (2001). Modeling of static and dynamic guard channel schemes for mobile transactions. *IEICE Transactions on Information and Systems . E (Norwalk, Conn.), 84-D*, 87–99.

Cho, C., Ko, Y., & Kwang, H. L. (1997). Fuzzy adaptive guard channel assignment strategy for handoff in PCS system. *Proceedings of 6th IEEE International Conference on Fuzzy Systems, (pp.* 1511-1518).

Choi, J., Kwon, T., Choi, Y., & Naghshineh, M. (2000). Call admission control for multimedia services in mobile cellular networks: A Markov decision approach. In *Proceedings of 5th IEEE Symposium on Computers and Communications (ISCC 2000)* (pp. 594-599).

Choi, S., & Shin, K. G. (1998). Predictive and adaptive bandwidth reservation for handoffs in QoS-sensitive cellular networks. *ACM SIGCOMM Computer Communication Review, 28*(4), 155–166. doi:10.1145/285243.285278

Chung, S. P., & Chiu, C. L. (2002). Joint call admission control / congestion for wireless integrated voice/data Networks. *Computer Communications, 25*, 1653–1664. doi:10.1016/S0140-3664(02)00050-6

Cruz-Perez, A., & Ortigoza-Guerrero, L. (2007). Fractional resource reservation in mobile cellular systems. In Y. Zhang, H. Hu & M. Fujise (Ed.), *Resource, Mobility, and Security Management in Wireless Networks and Mobile Communications* (pp. 122-138). New York: Auerbach Publishing.

Das, S. K., Sen, S. K., & Jayaram, R. (1998). A novel of load balancing scheme for the tele-traffic hot spot problem in cellular networks. *ACM/Baltzer Journal on Wireless Networks*, 4(4), 325-340.

Fang, Y. (2003). Thinning schemes for call admission control in wireless networks. *IEEE Transactions on Computers, 52*(5), 686–688. doi:10.1109/TC.2003.1197136

Fang, Y., & Zhang, Y. (2002). Call admission control schemes and performance analysis in wireless mobile networks. *IEEE Transactions on Vehicular Technology, 51*, 371–382. doi:10.1109/25.994812

Ghaderi, M., & Boutaba, R. (2006). Call admission control in mobile cellular networks: A comprehensive survey. *Wireless Communications and Mobile Computing, 6*(1), 69–93. doi:10.1002/wcm.246

Guern, R. (1988). Queuing-blocking system with two arrival streams and guard channels. *IEEE Transactions on Communications, 36*, 153–163. doi:10.1109/26.2745

Han, M., & Nilsson, A. A. (2000). Population-based call admission control in wireless cellular networks. In *Proceedings of the 2000 IEEE International Conference on Communications* (pp. 1519-1523).

Haring, G., Marie, R., Puigjaner, R., & Trivedi, K. (2001). Loss formulas and their application to optimization for cellular networks. *IEEE Transactions on Vehicular Technology, 50*, 664–673. doi:10.1109/25.933303

Ho, C.-J., & Lea, C.-T. (1999). Improving call admission policies in wireless networks. *Wireless Networks, 5*, 257–265. doi:10.1023/A:1019107329380

Hong, D., & Rappaport, S. (1986). Traffic modeling and performance analysis for cellular mobile radio telephone systems with prioritized and non-prioritized handoffs procedure. *IEEE Transactions on Vehicular Technology, 35*, 77–92. doi:10.1109/T-VT.1986.24076

Hou, J., & Fang, Y. (2001). Mobility-based call admission control schemes for wireless mobile networks. *Wireless Communications and Mobile Computing, 1*, 269–282. doi:10.1002/wcm.18

Hu, F., & Sharma, N. K. (2003). Multimedia call admission control in mobile networks: A dynamical reservation-pool approach. *Computer Networks, 43*(3), 263–288. doi:10.1016/S1389-1286(03)00271-8

Huang, C., Chuang, Y., & Yang, D. (2008). Implementation of call admission control scheme in next generation mobile communication networks using particle swarm optimization and fuzzy logic systems. *Expert Systems with Applications, 35*, 1246–1251. doi:10.1016/j.eswa.2007.08.023

Ishikawa, Y., & Umeda, N. (1997). Capacity design and performance of call admission control in CDMA systems. *IEEE Journal on Selected Areas in Communications, 15*, 1627–1635. doi:10.1109/49.634800

Katzela, I., & Naghshineh, M. (1996). Channel assignment schemes for cellular mobile telecommunication systems: A comprehensive survey. *IEEE Personal Communications, 3*, 10–31. doi:10.1109/98.511762

Kulavaratharash, M. D., & Aghvami, A. H. (1999). Teletraffic performance evaluation of microcellular personal communication networks (PCN's) with prioritized handoff procedures. *IEEE Transactions on Vehicular Technology, 48*, 137–152. doi:10.1109/25.740076

Kwon, T., Choi, Y., & Naghshineh, M. (1998). Optimal distributed call admission control for multimedia services in mobile cellular networks. In *Proceedings of 5th International Workshop on Mobile Multi-Media Communications(MoMuc98).*

Lee, S., & Park, S. (1998). Handoff with dynamic guard channels in ATM-based mobile networks. In *Proc. of 12th International Conference on Information Networking (ICOIN-12)* (pp. 439-442).

Leong, C. W., & Zhuang, W. (2002). Call admission control for voice and data traffic in wireless communications. *Computer Communications, 25*, 972–979. doi:10.1016/S0140-3664(01)00434-0

Levine, D. A., Akylidiz, I. F., & Naghsineh, M. (1997). A resource estimation and call admission algorithm for wireless multi-media networks using the shadow cluster concept. *IEEE/ACM Transactions on Networking*, 5, 1-12.

Martinez-Bauset, J., Gimenenz-Guzman, J. M., & Pla, V. (2008). Optimal admission in multimedia mobile networks with handover prediction. *IEEE Wireless Communications*, 38-44.

Morley, C. D., & Grover, W. D. (2000). Strategies to Maximize Carried Traffic in Dual-Mode Cellular Systems. *IEEE Transactions on Vehicular Technology, 49*(2), 357–366. doi:10.1109/25.832966

Naghshineh, M., & Schwartz, M. (1996). Distributed Call Admission Control in Mobile/Wireless Networks. *IEEE Journal on Selected Areas in Communications, 12*, 711–717. doi:10.1109/49.490422

Oh, S., & Tcha, D. (1992). Prioritized channel assignment in a cellular radio network. *IEEE Transactions on Communications, 40*, 1259–1269. doi:10.1109/26.153371

Oliveria, C., Kim, J. B., & Suda, T. (1998). An adaptive bandwidth reservation scheme for high-speed multimedia wireless network. *IEEE Journal on Selected Areas in Communications, 16*(6), 858–874. doi:10.1109/49.709449

Peha, J. M., & Sutivong, A. (2001). Admission Control Algorithms for Cellular Systems. *Wireless Networks, 7*, 117–125. doi:10.1023/A:1016629421079

Ramanathan, P., Sivalingam, K. M., Agrawal, P., & Kishore, S. (1999). Dynamic resource allocation schemes during handoff for mobile multimedia wireless networks. *IEEE Journal on Selected Areas in Communications*, *17*, 1270–1283. doi:10.1109/49.778185

Ramjee, R., Towsley, D., & Nagarajan, R. (1997). On optimal call admission control in cellular networks. *Wireless Networks*, *3*, 29–41. doi:10.1023/A:1019172226345

Rappaport, S., & Purzynski, C. (1996). Prioritized resource assignment for mobile cellular communication systems with mixed services and platform types. *IEEE Transactions on Vehicular Technology*, *45*, 443–458. doi:10.1109/25.533760

Salamah, M., & Lababidi, H. (2005). Dynamically adaptive channel reservation scheme for cellular networks. *Computer Networks*, *49*, 787–796. doi:10.1016/j.comnet.2005.02.005

Saquib, M., & Yates, R. (1995). Optimal call admission to a mobile cellular network. *Proceedings of IEEE Vehicular Technology Conference* (pp. 190-194).

Tian, X., & Ji, C. (2001). Bounding the performance of dynamic channel allocation with QoS provisioning for distributed admission control in wireless networks. *IEEE Transactions on Vehicular Technology*, *50*, 388–397. doi:10.1109/25.923051

Tzeng, S. S., & Lu, H. Y. (2008). Dual-threshold admission control for non-real-time traffic in wireless data networks. *Computer Communications*, *31*, 2428–2434. doi:10.1016/j.comcom.2008.03.010

Vazquez-Avila, L. J., Cruz-Perez, F. A., & Ortigoza-Guerrero, L. (2006). Performance Analysis of Fractional Guard Channel Policies in Mobile Cellular Networks. *IEEE Transactions on Wireless Communications*, *5*(2), 301–305. doi:10.1109/TWC.2006.1611053

Wang, X., Fan, P., & Pan, Y. (2008). A more realistic thinning scheme for call admission control in multimedia wireless networks. *IEEE Transactions on Computers*, *57*(8), 1143–1148. doi:10.1109/TC.2008.29

Wu, S., Wong, K. Y. M., & Li, B. (2002). A dynamic call admission policy with precision QoS guarantee using stochastic control for mobile wireless networks. *IEEE Transactions on Networking*, *10*, 257–271. doi:10.1109/90.993306

Yavuz, E. A., & Leung, V. C. M. (2006). Computationally efficient method to evaluate the performance of guard channel based call admission control in cellular networks. *IEEE Transactions on Vehicular Technology*, *55*(4), 1412–1424. doi:10.1109/TVT.2006.877470

Yin, L., Li, B., Zhang, Z., & Lin, Y. (2000). Performance analysis of a dual-threshold reservation (DTR) scheme for voice/data integrated mobile wireless networks. In *Proceedings of the IEEE Wireless Communications and Networking Conference (WCNC-2000)* (pp. 258-262).

Yoon, C. H., & Kwan, C. (1993). Performance of personal portable radio telephone systems with and without guard channels. *IEEE Journal on Selected Areas in Communications*, *11*, 911–917. doi:10.1109/49.232300

Yu, O., & Leung, V. (1996). Self-tuning prioritized call handling mechanism with dynamic guard channels for mobile cellular systems. In *Proceedings of IEEE Vehicular Technology Conference*, (pp. 1520-1524).

Yu, O. T. W., & Leung, V. C. M. (1997). Adaptive Resource Allocation for Prioritized Call Admission over an ATM-Based Wireless PCN. *IEEE Journal on Selected Areas in Communications*, *15*, 1208–1225. doi:10.1109/49.622906

Zheng, Z. H., & Lam, W. H. (2002). Performance analysis of dynamic channel assignment with queuing and guard channel scheme combined scheme for handoff prioritization. *Electronics Letters*, *38*, 1728–1729. doi:10.1049/el:20021177

Chapter 9
Admission Control and Scheduling for QoS Provisioning in WiMAX Networks

Juliana Freitag Borin
University of Campinas, Brazil

Nelson L. S. da Fonseca
University of Campinas, Brazil

ABSTRACT

Although the IEEE 802.16 standard, popularly known as WiMAX, defines the framework to support real-time and bandwidth demanding applications, traffic control mechanisms, such as admission control and scheduling mechanisms, are left to be defined by proprietary solutions. In line with that, both industry and academia have been working on novel and efficient mechanisms for Quality of Service provisioning in 802.16 networks. This chapter provides the background necessary to understand the scheduling and the admission control problems in IEEE 802.16 networks. Moreover, it gives a comprehensive survey on recent developments on algorithms for these mechanisms as well as future research directions.

INTRODUCTION

The IEEE 802.16 (2004) standard, often referenced as WiMAX (Worldwide Interoperability for Microwave Access Forum), has been developed aiming at standardizing the broadband wireless technology. The standard defines the air interface and the medium access protocol for Wireless Metropolitan Area Networks (WMAN), providing high transmission rates for commercial and residential access to the Internet.

DOI: 10.4018/978-1-61520-680-3.ch009

In order to provide support to the big diversity of applications available on the Internet, such as voice, video and multimedia services as well as files transfer, the standard and its extension, IEEE 802.16e (2005), define signaling mechanisms between the base station and the subscriber stations and also five service levels: unsolicited grant service, real-time polling service, extended real-time polling service, non-real-time polling service and best-effort. In both directions, uplink (from the subscriber stations to the base station) and downlink (from the base station to the subscriber stations), the packets are associated with a service flow by the Medium Access Control

(MAC) layer. A set of Quality of Service (QoS) parameters, maximum latency and minimum rate among them, is associated with each flow.

Despite such services provide the basis for QoS provisioning, a complete solution requires traffic control mechanisms, such as admission control and scheduling, not defined in the standard. The traffic control mechanisms enable a balance between the utilization of the network resources and the Quality of Service provisioning. Conservative mechanisms can increase the level of QoS offered to the users but, on the other hand, can result in a low network utilization. An aggressive traffic control, in turn, can increase the network utilization on the account of Quality of Service degradation. This tradeoff between utilization and Quality of Service is of fundamental importance in WiMAX networks, which aggregate different types of traffic in a limited resources architecture.

The scheduling mechanism aims at guaranteeing the bandwidth required by the subscriber stations as well as enabling the efficient wireless link usage. In a WiMAX point-multipoint topology network, the downlink scheduling requires a single scheduler at the base station, whereas the uplink scheduling needs two components, one of them at the base station and the second one at the subscriber station. The base station scheduler allocates bandwidth for the subscriber stations and the subscriber station scheduler determines which packets will be sent in the received transmission opportunities. The admission control mechanism restricts the number of users simultaneously present in the network so as to avoid the wireless link saturation and, consequently, violation of QoS contracts.

Though admission control and scheduling are distinct mechanisms, investigation is essential on mechanisms operating in conjunction so that the WiMAX networks fulfill one of their main purposes: to provide high data rates with Quality of Service. The rest of this chapter discusses in more details the admission control and scheduling mechanisms in WiMAX networks and presents a survey of the solutions proposed in the literature.

BACKGROUND

The architecture of a network utilizing the IEEE 802.16 standard has two main elements: Base Station (BS) and Subscriber Station (SS). The BS makes the communication between the wireless network and the core network, whereas the SS provides the user access to the core network by establishing connections with the BS in a Point-Multipoint (PMP) topology. The standard also allows Mesh topologies (optional). The main difference between the PMP and Mesh topologies lies on the fact that in a PMP network the traffic flows only between the BS and the SSs, whereas in the Mesh mode, the traffic can be routed through the SSs and can occur directly between two SSs. In this chapter we will analyze PMP topology networks.

The physical layer operates in a frames format, which are subdivided in time intervals called physical slots. In each frame, the slots are organized in a downlink sub-frame and an uplink sub-frame. The downlink sub-frame is utilized by the BS for the transmission of data and control information to the SSs. The uplink sub-frame is shared among all SSs for transmissions addressed to the BS.

The IEEE 802.16 standard allows two physical medium access modes: Frequency Division Duplexing (FDD) and Time Division Duplexing (TDD). In the FDD mode the downlink and uplink channels operate simultaneously in different frequencies. In the TDD mode the uplink and downlink sub-frames share the same frequency, and so it is not possible to perform simultaneous transmissions in both directions. Each TDD frame has a downlink sub-frame followed by an uplink sub-frame.

The Medium Access Control (MAC) layer is connection oriented. Each connection is identified by a 16 bit identifier (Connection Identifier

- CID) and each SS has a unique MAC address which identifies it and is utilized to register it and authenticate it in the network. All the traffic, including the non-connection oriented one, is mapped for one connection. Besides the connections management, the MAC layer is responsible for the medium access control and also for the bandwidth allocation.

The allocation of resources for the SSs is performed on demand by a scheduler implemented at the BS. When a connection has data to be transmitted, the SS sends a bandwidth request message to the BS. A bandwidth request can be sent as an individual packet in a grant reserved for such purpose, or can be sent together with a data packet (Piggyback). The bandwidth request can be either incremental or aggregated. An incremental request indicates the additional bandwidth the SS needs, whereas an aggregated request indicates the total bandwidth requested by the SS. For the SS, the bandwidth requests always refer to a determined connection, whereas the grants allocated by the BS are addressed to an SS and not to a connection in particular. This way, the SS can utilize the grant received for a connection other than the one to which the request was made. The SSs have a scheduler which decides which packets originating from the upper layer will be sent in the received grants.

The allocation of grants for the sending of bandwidth requests, called polling, can occur in two ways:

- **Unicast:** The SS receives a grant whose size is enough for the sending of a bandwidth request;
- **Contention based:** Utilized when there is no available bandwidth for the BS to individually poll all the SSs. In this case, the BS allocates a grant for a group of SSs, which must compete for the opportunity to send a request message. In order to reduce the probability of collision, only the SSs needing bandwidth take part in the contention. For contention resolution, the SSs utilize the exponential backoff algorithm. The contention minimum window and maximum window size is BS controlled.

The MAC also provides mechanisms to deliver QoS to the uplink and downlink traffics. The main QoS provisioning mechanism consists in associating each connection with a service flow. The service flow is a MAC layer service which provides packets unidirectional transport. Several upper layer sections can operate over the same service flow in the MAC layer when their QoS requirements are identical. Each service flow must define its set of QoS parameters and deliver it during the establishment of the connection so that the BS can decide on the admission of the new connection. The standard and its IEEE 802.16e (2005) extension specify five types of service flow.

The Unsolicited Grant Service (UGS) supports real time flows generating periodically fixed size data packets, such as voice over IP. UGS connections receive fixed size periodic grants and must deliver the following QoS parameters: minimum reserved traffic rate, maximum latency, tolerated jitter, unsolicited grant interval, and request/transmission policy.

The second type of service is the real-time Polling Service (rtPS), designed for applications with real time requirement generating periodically variable size packets, as for example MPEG video applications. rtPS flows require bandwidth periodically through unicast polling and the QoS is guaranteed by satisfying the maximum latency and minimum rate requirements. An rtPS flow must specify the following QoS parameters: minimum reserved traffic rate, maximum sustained traffic rate, maximum latency, unsolicited polling interval, and request/transmission policy.

The extended real-time Polling Service (ertPS) is designed for variable rate real time traffic, as for example, voice over IP with silence suppression applications. This service uses a grant mechanism similar to the one used by the UGS connections. Nevertheless, the periodically allocated grants can be used to send bandwidth requests in order to inform the BS on the need of a new grant size. The BS does not change the grants size until it receives a bandwidth request from the SS. An ertPS flow must deliver the following QoS parameters: minimum reserved traffic rate, maximum sustained traffic rate, maximum latency, tolerated jitter, unsolicited grant interval, and request/transmission policy.

The non-real-time Polling Service (nrtPS) supports non-delay sensitive traffic which requires regularly variable size grants, such as FTP traffic. The service is similar to that offered by the rtPS service, however, it offers unicast polling less frequently and allows the SS to utilize the contention slots reserved for bandwidth request. nrtPS connections must specify the following QoS parameters: minimum reserved traffic rate, maximum sustained traffic rate, and request/transmission policy.

The Best Effort (BE) service supports best-effort traffic with no QoS guarantees. The SS can utilize both unicast slots and contention slots to request bandwidth.

SCHEDULING

Scheduling in IEEE 802.16 networks covers the downlink traffic scheduling, performed by the BS, and also the uplink traffic scheduling, performed by two schedulers, one at the BS and another one at the SSs. In order to carry out the resources allocation, the schedulers use information on the QoS requirements and the occupancy of the connections queues.

The downlink scheduler and the SSs uplink scheduler have direct access to the connections queues. The uplink scheduler located at the BS, in turn, depends on the bandwidth requests sent by the SSs in order to keep informed about each connection status. Such requests, besides incrementing the network load, may suffer delays, generated by the contention mechanism, for example, resulting in delivering outdated info.

In addition, unlike the other two schedulers, the BS uplink scheduler must allocate resources not only for data transmission, but also for sending bandwidth requests. In both cases, the resources allocation must be performed so that the QoS requirements of every connection are guaranteed.

This way, we can say that one of the greatest challenges in the MAC layer design for WiMAX networks equipments lies on the implementation of the BS uplink scheduler. On the other hand, by leaving the solution for this problem open, the standard provides an excellent chance for vendors to research innovative algorithms, which may differentiate their products.

Regardless of the scheduling policy adopted for the uplink traffic, the following requirements should be taken into account:

- The resources distribution must be based on the bandwidth requests sent by the SSs and also on each connection's QoS parameters; and distinct connections utilizing the same type of service may have different values for the QoS parameters;
- The bandwidth allocation must allow not only data transmission, but also the transmission of bandwidth requests according to the request mechanism defined for each type of service;
- All standard defined QoS parameters must be guaranteed.

In addition to the mentioned requirements, the scheduler is expected to efficiently use the available bandwidth so that a greater number of users can be admitted, resulting in high network utilization levels.

Although the scheduler is implemented at the MAC layer, the technology used in the physical layer (PHY) impacts the scheduler design. The IEEE 802.16 standard supports three physical layer technologies: Single Carrier (SC), Orthogonal Frequency Division Multiplexing (OFDM), and Orthogonal Frequency Division Multiple Access (OFDMA). In SC, there is only one carrier and the entire frequency is given to the SS. In OFDM, multiple subcarriers constitute a physical slot, but since they are transparent to the MAC layer, they can be seen as one logical channel from the scheduler point of view. However, each subchannel can be modulated differently; therefore, different users can achieve different transmission efficiencies, yielding to additional challenges to the scheduler design. Different from SC and OFDM, in which only time domain is considered, the OFDMA physical layer requires allocation in two dimensions (frequency and time). As a result, scheduling for OFDMA systems is the most complex one.

The next section surveys representative solutions presented in the literature for the uplink scheduling at the BS. Solutions are distinguished between PHY-unaware scheduling mechanisms and PHY-aware scheduling mechanisms. PHY-unaware schedulers do not consider the physical layer characteristics; their main goal is to assure the QoS requirements of each service flow. PHY-aware schedulers do take into account the technology used in the physical layer.

State of the Art of IEEE 802.16 Scheduling Research

PHY-Unaware Scheduling Mechanisms

The trend in the first solutions proposed in the literature (Hawa & Petr, 2002), (Wongthavarawat & Ganz, 2003) and also in some more recent proposals (Chen, Jiao, & Wang, 2005), (Tarchi, Fantacci, & Bardazzi, 2006) consists in adapting, for IEEE 802.16 networks, classic scheduling policies proposed for wired networks. Those works utilize a combination of policies, such as Strict Priority, Weighted Fair Queuing (WFQ) (Parekh & Gallager, 1993) and Earliest Deadline First (EDF), resulting in complex scheduling schemes. The most recent proposals (Sayenko, Alanen, & Hämäläinen, 2008), (Borin & Fonseca, 2008a), and (Borin & Fonseca, 2008b) adopt simpler implementation ideas. Given that the BS uplink scheduler is executed at each frame and, in determined configurations, there can be as many as 400 frames per second, simpler solutions become more attractive.

Hawa and Petr (2002) propose a mechanism implementing a priorities and Weighted Fair Queuing (WFQ) based scheduling combination. The UGS service has the highest priority for bandwidth allocation, the remaining of the flows is served by using WFQ with priority, i.e., if two data grants (from two different queues) have identical WFQ virtual finish times, then the one with the higher priority will be served first. A weight value is assigned to each queue based on the minimum rate reserved for the corresponding service flow. No result evincing the efficacy of the proposed scheduler is presented.

Wongthavarawat and Ganz (2003) propose a scheduling mechanism which utilizes a combination of the Strict Priority, Earliest Deadline First (EDF) and Weighted Fair Queuing (WFQ) disciplines. The bandwidth allocation occurs in two steps. In the first step, the scheduler distributes the bandwidth among the different types of service utilizing the Strict Priority discipline. In the second step, the bandwidth received by each service is distributed among the connections. At this stage the UGS service uses a fixed bandwidth policy, the rtPS service utilizes the EDF discipline, the nrtPS service utilizes the WFQ discipline and the BE service equally distributes the band among all the connections. The rtPS deadline information is determined by using the Arrival-Service curve concept (Cruz, 1991), while the WFQ weights are based on the ratio of a connection's average data rate to the total average data rate. The proposed mechanism is evaluated through simulation experiments, however, only rtPS and BE traffic is utilized.

Chen, Jiao and Wang (2005) utilize the same scheduler structure proposed by Wongthavarawat and Ganz. However, in the first step, the bandwidth allocation is performed with the Deficit Fair Priority Queue (DFPQ) technique. Simulation results show that the DFPQ algorithm is fairer than the Strict Priority algorithm and avoids starvation of best-effort traffic.

Tarchi, Fantacci and Bardazzi (2006) also propose a combination of scheduling disciplines. In the first step, the bandwidth is distributed among the UGS, rtPS, nrtPS and BE services utilizing a strict semi-preemptive priority technique. In the following step, the bandwidth is distributed among the connections utilizing the Packet Based Round Robin (PBRR) algorithm for the UGS service, the EDF discipline for the rtPS service and the WFQ discipline for the nrtPS and BE services.

Sayenko, Alanen and Hämäläinen (2008) present a Round Robin (RR) discipline based solution consisting in three stages. In the first stage, the BS calculates the minimum number of slots to be provided for each connection so that the QoS requirements are guaranteed. This computation utilizes differential equations for each type of service which take into account the bandwidth requests sent by the SSs as well as the QoS parameters provided by the connections. In the second stage, if slots are not allocated, the distribution of the slots among the rtPS, nrtPS and BE connections follows the emulation of the Round Robin discipline. However, unlike the standard RR algorithm, the proposed solution utilizes just one step, instead of multiple steps, and prevents the maximum rate requirement from being exceeded. Finally, in the third stage, the BS ranks the slots within the UL-MAP so that time requirements are guaranteed.

Borin and Fonseca (2008a) propose a scheduling algorithm utilizing three queues with different priorities. The low priority queue stores the bandwidth request sent by BE connections. The intermediate queue stores the bandwidth requests sent by rtPS and nrtPS connections. These requests can migrate to the high priority queue so that their QoS requirements are guaranteed. In addition to the requests from the intermediate queue, the high priority queue stores the periodic grants for the UGS and ertPS services and also the unicast request opportunities for the rtPS and nrtPS services which must be scheduled in the next frame. Results obtained from simulation experiments show that the proposed solution is capable of guaranteeing the maximum latency requirement for the UGS, ertPS and rtPS services, as well as the minimum rate requirement for the rtPS and nrtPS services.

In (Borin & Fonseca, 2008b), the authors extend the previous proposal so that the maximum sustained traffic rate and maximum traffic burst parameters are also considered by the scheduling mechanism. Additionally, in the first algorithm version, the scheduler tries to guarantee the maximum latency for every rtPS packet. This technique requires the employment of complex admission control mechanisms, such as Measurement Based Admission Control (MBAC) mechanisms, in order to guarantee the maximum latency requirement as well as high network utilization levels. The version presented in (Borin & Fonseca, 2008b), guarantees the maximum latency for the rtPS traffic not exceeding the minimum rate requirement. This technique, in addition to being in accordance with the standard, allows utilizing simpler bandwidth reserve based admission control mechanisms.

The performance of the scheduler proposed in (Borin & Fonseca, 2008b) was evaluated via simulation by using an ns-2 module for IEEE 802.16 networks (Borin & Fonseca, 2008c). The topology of the simulated network consisted of a BS, with the SSs uniformly distributed around it. The frame duration was set to 5 ms, and the capacity of the channel was assumed to be 40 Mbps with a 1:1 downlink-to-uplink TDD split. To eliminate the impact of packet scheduling at the SSs on uplink scheduling, each SS had only one service flow. Five different types of traffic were considered: voice (Brady, 1969), voice

Figure 1. Latency of UGS, ertPS, and rtPS connections as a function of the number of SSs

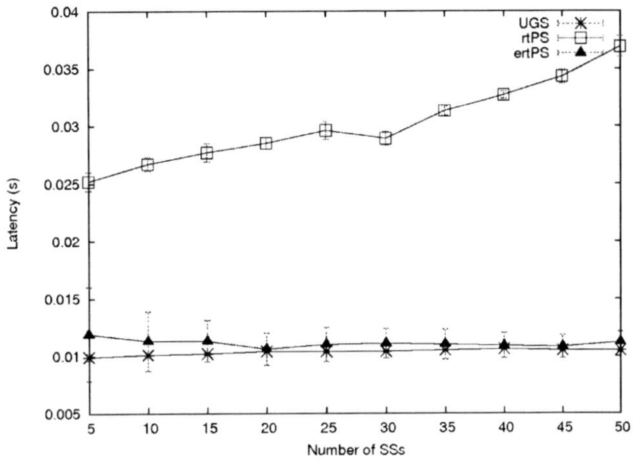

with silence suppression (3GPP2, 1999), video (Seeling, Reisslein, & Kulapala, 2004), FTP and WEB (Barford, Bestavros, Bradley, & Crovella, 1998), which were associated with UGS, ertPS, rtPS, nrtPS, and BE services, respectively.

The unsolicited grant interval for the UGS and for the ertPS services was 20 ms. The unsolicited polling interval of the rtPS service was 20 ms and the polling interval of the nrtPS service was 1 s. The maximum latency requirement of the rtPS service was 100 ms and each connection had its own minimum reserved traffic rate and maximum sustained traffic rate requirements (which varied according to the rate of the transmitted video). The nrtPS service had minimum reserved traffic rate requirement of 200 Kbps, and maximum sustained traffic rate requirement of 300 Kbps. The BE service did not have any QoS requirement.

In the simulation experiments, the number of SSs increased from 5 to 50 in steps of 5 units (one for each type of service). Results were produced by running the simulation ten times with different seeds for the random number generator. The mean values and the 95% confidence intervals are shown in the following figures.

Figure 1 shows the mean latencies of UGS, rtPS, and ertPS uplink connections as a function of the number of SSs. The latencies of UGS and ertPS connections were not affected by the load increase due to the increase of the number of SSs, which shows that the uplink scheduler is able to provide data grants at fixed intervals as required by these services. Conversely, the latency of rtPS connections increased with the offered load, however, it did not surpass the required value of 100 ms.

Figure 2 shows the average throughput of the rtPS connections transmitting the Lecture video (nine different video traces were used, for additional results see (Borin & Fonseca, 2008b)), while Figure 3 shows the average throughput of the nrtPS connections. In all the simulated scenarios, the average throughput values of the rtPS and of the nrtPS connections were in the range defined by the minimum and the maximum rates requirements as specified by the standard.

The throughput of the nrtPS connections, measured at the MAC layer, was a little higher than the offered load, measured at the transport layer, due to additional bits added by MAC headers. Although the nrtPS flows were configured to generate an average rate equals to the maximum sustained traffic rate, the joint effect of the maximum rate control done by the scheduler and the TCP congestion control resulted in an offered load lower than 300 Kbps. Consequently, all the nrtPS

Figure 2. Throughput of the rtPS connection transmitting the lecture video as a function of the number of SSs

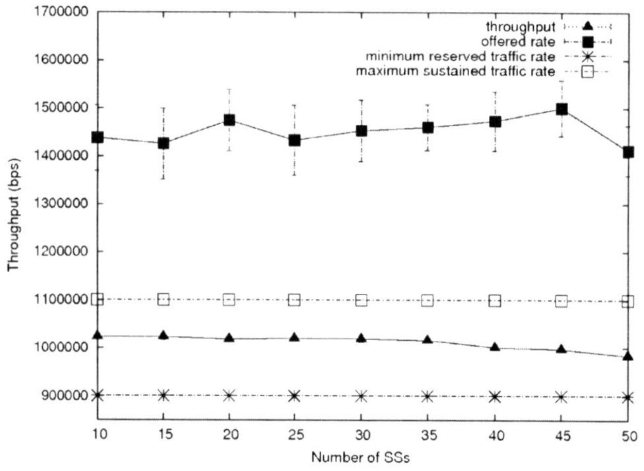

traffic generated by the upper layers was served at the MAC layer in all the simulated scenarios. The interaction between the scheduling mechanism and the TCP congestion control mechanism shall be investigated in the future to counteract resource underutilization side effect. Although not shown here, in all the simulated scenarios the BE connections were able to transmit in the slots not used by the higher priority service flows.

The scheduler proposed by Borin and Fonseca (2008b) uses a simple approach which provides maximum latency and minimum rate guarantees without violating the maximum sustained traffic rate and the maximum traffic burst values. Simulation results show that the proposed solution is able to provide QoS for the different types of service defined by the IEEE 802.16 standard, yet being standard-compliant.

In addition to the works discussed in this section, there are also proposals dealing with either just real time traffic (Lee, Kwon, & Cho, 2005), (Yang & Lu, 2006), (Mohammadi, Akl, & Beh-

Figure 3. Throughput of nrtPS connections as a function of the number of SSs

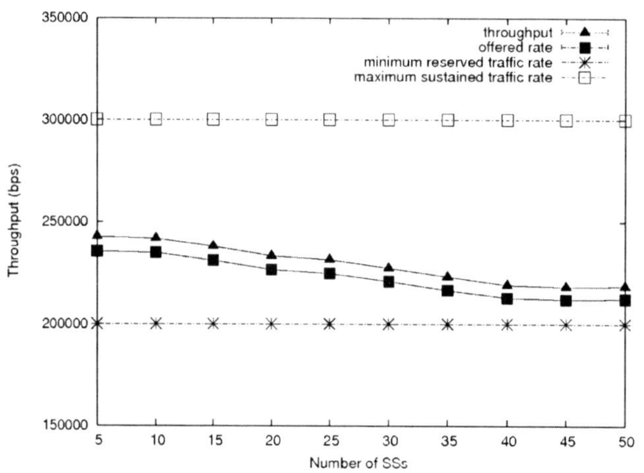

Table 1. Advantages and disadvantages of the discussed proposals

Study	Advantages	Disadvantages
(Hawa & Petr, 2002)	Guarantees latency for the UGS service and minimum rate for the remaining services.	Complex algorithm based on a hierarchy of schedulers; does not include the ertPS service; authors do not present performance analysis.
(Wongthavarawat & Ganz, 2003)	Guarantee of latency for the UGS and rtPS services.	Complex algorithm based on a hierarchy of schedulers; does not include the ertPS service; simulations consider just the nrtPS and BE services; it does not guarantee minimum rate.
(Chen, Jiao, & Wang, 2005)	Guarantees of latency, minimum rate and maximum rate.	Complex algorithm based on a hierarchy of schedulers; does not include the ertPS service; no information on the utilized simulator.
(Tarchi, Fantacci, & Bardazzi, 2006)	Guarantees of latency for the rtPS and UGS services.	Complex algorithm based on a schedulers hierarchy; does not include the ertPS service; does not provide minimum rate guarantees; simulations include just the UGS and nrtPS services.
(Sayenko, Alanen, & Hämäläinen, 2008)	RR policy based simple algorithm; includes the 5 types of service; guarantees minimum rate and maximum rate; simulations with the 5 types of service utilizing the ns-2 tool;	Provides no latency guarantees.
(Borin & Fonseca, 2008a)	3 priority queues based simple algorithm; includes the 5 types of service; provides latency and minimum rate guarantees; simulations with the 5 types of service utilizing the ns-2 tool.	BE is served by a FIFO policy, what can result in uneven resources distribution for this service; the latency guarantee requires the use of a complex admission control mechanism; does not consider the maximum rate requirement .
(Borin & Fonseca, 2008b)	3 priority queues based simple algorithm; includes the 5 types of service; provides latency, minimum rate and maximum rate guarantees; it performs maximum burst control; simulations with the 5 types of service utilizing the ns-2 tool.	BE is served by a FIFO policy, what can result in uneven resources distribution for this service.

namfar, 2008) or just best-effort traffic (Hou, She, Ho, & Shen, 2008), (Kim & Yeom, 2007). Table 1 summarizes the advantages and disadvantages of the solutions discussed in this section.

PHY-Aware Scheduling Mechanisms

Bai, Shami and Ye (2008) propose a set of mechanisms for QoS provisioning in IEEE 802.16 networks utilizing the single carrier technology in the physical layer. The solution includes a scheduling mechanism for the BS utilizing the cross-layer approach. Besides the bandwidth requests, the scheduler considers the type of modulation utilized by the SSs in the physical layer, for allocating modulated symbols rather than slots. In different modulations, the number of symbols necessary for sending the same amount of bits varies. The choice of the requests to be served follows a priority value defined by the SSs. Though this technique results in a less complex scheduler at the BS, it limits the interoperability among equipment from different vendors, given that all SSs should be capable of calculating the priority values. At the end of the allocation process, a UL-MAP message creation module converts the allocated symbols into number of slots. Results obtained from simulation experiments with the rtPS, nrtPS and BE services show that the proposed solution is capable of guaranteeing the QoS requirements of the connections.

The uplink scheduling problem for OFDM systems is analyzed in (Huang, Subramanian, Berry, & Agrawal, 2007). The authors approach this problem using a gradient-based scheduling framework. Bandwidth and power are allocated to maximize the projection onto the gradient of the total system utility function, which models the application-layer Quality of Service (QoS). An optimal solution as well as a family of lower complexity sub-optimal algorithms (SOAs) are introduced. The optimal algorithm determines the optimal carrier allocation by sorting the users on each tone. Given an optimal carrier allocation, the optimal power allocation is given by a per-user water-filling allocation. In each SOA, the same two phases are used but with some modifications to reduce the complexity of the computation of the optimal carrier allocation. Specifically, the algorithms begin with a carrier allocation phase which assigns each subcarrier to at most one user. Instead of using metrics given by optimal values, metrics based on a constant power allocation over all carriers assigned to a user are considered. Next, in the power allocation phase, power of each user is allocated across the assigned carriers using a water-filling allocation as in the optimal algorithm. As stated by the authors, the proposed optimal solution has prohibitively high computational complexity and was used to derive the SOA solutions. Simulation results in terms of fairness among users and the average number of users who receive positive rates within one scheduling interval are shown for the SOA algorithms. However, results for the QoS provision capability of the proposed solutions were not presented.

Singh and Sharma (2006) introduce a set of algorithms for the BS to allocate channels/slots to different SSs in an IEEE 802.16 OFDMA/TDD network. A heuristic algorithm is proposed to allocate different subchannels to the SSs one slot at a time aiming to maximize the throughput. However, the authors show that this algorithm is not fair since, in order to optimize system performance, it favors SSs with better channel conditions. To obtain fairer solutions, the heuristic algorithm is modified and four algorithms are presented. Different combinations of these algorithms are proposed for the uplink scheduler at the BS according to the load and channel characteristics of the network. The proposed scheduler tries to satisfy, first, the needs of the UGS connections. Next, the scheduler serves the rtPS connections followed by the nrtPS connections. nrtPS connections without minimum rate requirements and BE connections share the remaining resources. No evaluation of the proposed scheduler is provided.

An approach with fuzzy controls for the provisioning of fairness and QoS in WiMAX OFDMA networks is proposed in (Chen, Lee, Wu, & Kuo, 2009). The adopted FQFC scheduler assigns each connection a priority and a transmission oppor-

Table 2. Advantages and disadvantages of the discussed proposals

Study	Advantages	Disadvantages
(Bai, Shami, & Ye, 2008)	Utilizes a cross-layer approach which considers the different modulations utilized by the connections; provides an integrated uplink scheduling solution for the BS and the SSs.	Latency, minimum rate and maximum rate guarantees provided only by the SSs.
(Huang, Subramanian, Berry, & Agrawal, 2007)	Considers per-user power constraint.	Authors do not include results showing the scheduler capability to provide QoS; QoS provision does not include the IEEE 802.16 QoS parameters.
(Singh & Sharma 2006)	Attempts to satisfy QoS requirements while exploiting the channel-user diversity.	Best subchannels are allocated to UGS; authors do not present performance analysis.
(Chen, Lee, Wu, & Kuo, 2009)	Takes into account latency, minimum rate and maximum rate requirements while providing intraclass and interclass fairness.	The minimum reserved traffic rate requirement of the rtPS and the ertPS service is not considered.

tunity (TXOP) with values inferred by a fuzzy estimator. The scheduler first initializes the two variables based on the characteristics of connections and adjusts them to adapt to the dynamics of the system. The priority value is used to determine the transmission order of the connections and it is adjusted according to channel transmission quality, QoS requirements, and service classes. The TXOP value is set considering the transmission rate and the queue length difference between two consecutive transmissions at the MAC layer, which gives the maximum number of packets that a connection can transmit in a frame. For real-time connections, the priority value is further adjusted considering the packet loss rate such that all connections in the same class receive the same packet loss rate. Fairness among non-real-time connections is provided by guaranteeing that all nrtPS connections have the same ratio between average throughput and the minimum reserved rate. The scheduler also provides interclass fairness by guaranteeing that the connections average throughput does not exceed their maximum sustained traffic rate. Results obtained via simulation of different scenarios including diverse channel conditions show that the proposed solution is able to provide QoS as well as intraclass and interclass fairness.

ADMISSION CONTROL

While the scheduling guarantees that the required bandwidth is allocated to the connections so that the QoS requirements are supported, admission control limits the number of connections so that the network is not overloaded by a very high number of users. Whenever a user wishes to establish a new connection, a request is sent to the BS and the admission control mechanism decides if the new connection will be accepted. In order to make such decision, the admission control must check if there are sufficient resources to meet the QoS requirements of the new connection without degrading the QoS of the ongoing connections.

The choice of the admission control policy to be adopted in an IEEE 802.16 network is strongly associated with the utilized scheduling mechanism. For example, when adopting an admission control mechanism which estimates the available resources from the difference between the total capacity of the link and the sum of minimum rate requirements of the already admitted connections, one should make sure the scheduler does not allocate more than the minimum rate for a connection when other connections have not yet had their requirement provided. In addition, the integration of scheduling and admission control solutions may result in simpler mechanisms. When one of the mechanisms is capable of guaranteeing response to a QoS requirement, for example, the same guarantee does not need to be implemented by another mechanism.

The IEEE 802.16 standard defines no admission control policies, what has encouraged researchers from industry and academia to investigate solutions for that problem. The next section presents some of the already proposed solutions.

State of the Art of IEEE 802.16 Admission Control Research

Chen et al (2005) propose an admission control algorithm utilizing the minimum reserved traffic rate parameter to check if a new connection can be accepted without compromising the QoS provided for the traffic present in the network. The algorithm adds up the minimum rate requested by all the connections already admitted and subtracts this amount from the total capacity of the network, thus obtaining the available capacity (Ca). When a new connection comes or an already active connection requests changes in the QoS requirements, the algorithm checks if the $Ca > 0$ condition is met. If this condition is true, then the request is accepted. Nevertheless, it is important to remember that rtPS and nrtPS connections need bandwidth not only for data transmission, but also for sending bandwidth requests, so considering just the minimum rate requirement in order to calculate the available capacity may result in an inaccurate estimate.

In (Wang, Liu, Ju, & Ruangchaijatupon, 2007), the authors propose an admission control algorithm utilizing the minimum reserved traffic rate and maximum sustained traffic rate parameters in the decision process. UGS connections are accepted whenever the maximum sustained traffic rate parameter value is smaller then or equal to the available bandwidth. rtPS, ertPS and nrtPS connections are accepted when the minimum rate requirement can be met and BE connections are always accepted. When a connection is admitted, the BS reserves an amount of bandwidth for the connection. For a UGS connection the reserved bandwidth is equal to the maximum sustained rate requirement and for rtPS, ertPS or nrtPS connections the reserved bandwidth is equal to the minimum between the available bandwidth and the maximum sustained traffic rate requirement. The admission control mechanism also makes a bandwidth reservation for connections from other cells (handoff) with the purpose of reducing the ongoing connections blocking rate. In addition, the mechanism utilizes a bandwidth borrowing scheme, in which lower priority connections borrow bandwidth for the admission of connections with higher priority. The authors assume that the channel capacity is C, the available bandwidth amount is bl, and a percentage n of the channel capacity is reserved as a guard channel for handoff connections. When a new connection comes with a bandwidth requirement br, if $(bl - br) > C.n\%$, then the connection is admitted. For handoff connections, if $(bl - br) > 0$, then the connection is admitted. The authors do not define the way that C, bl and n variables must be estimated.

Wang et al (2007) propose an admission control mechanism which provides high priority to UGS connections and utilizes a bandwidth borrowing and degradation scheme aiming at maximizing the channel utilization. In this mechanism, UGS connections are always accepted when there is available bandwidth, i.e., when the total utilized bandwidth added to the bandwidth requested by the connection is smaller than or equal to the available bandwidth (B). An rtPS connection is accepted when the total utilized bandwidth added to the bandwidth requested by the connection is smaller than or equal to $B - U$, where U is the bandwidth reserved exclusively for UGS connections. If the condition is not met, then the algorithm tries to degrade the nrtPS connections bandwidth down to the minimum requested value and checks the admission condition once again. When an nrtPS connection requests admission, the algorithm checks if the total utilized bandwidth added to the bandwidth requested by the connection minus a degradation coefficient is smaller than or equal to $B - U$. If the condition is not met, the algorithm tries to degrade the other nrtPS connections bandwidth down to the minimum requested value. BE connections are always accepted. The disadvantage of this mechanism lies on the reservation of part of the band for UGS services. The authors justify the adoption of this strategy based on the idea that, from the user's perspective, blocking a new UGS flow, which usually serves voice traffic, brings more problems than blocking a non-UGS flow. However, reserving part of the bandwidth for a sole type of service may lead to network underutilization when the bandwidth requested by the UGS flows is smaller than the amount reserved for this service.

Niyato and Hossain (2007) propose a bandwidth allocation and admission control mechanism based on queues analysis and game theory. The modeling of the queues is used to calculate the throughput of the nrtPS and BE services and the delay of the real time services, in order for them to be used by the game theory model in the admission of new connections. The game theory formulation is based on the fact that the new connection accepts the service offered by the BS only if the latency and rate requirements can be guaranteed. When a new connection requests admission, it informs which type of service it desires, as well as the traffic parameters and QoS requirements. The BS establishes a set of strategies and calculates the expected payoff for each strategy. Next, the game is resolved in order to obtain the Nash balance, which is used to decide whether the new connection will be accepted. If the connection is accepted, then the amount of bandwidth to be reserved for the new connection is determined.

In the admission control mechanism proposed in (Chandra & Sahoo, 2007), requests for establishing new connections are classified in queues according to the type of service requested. The mechanism first serves the UGS queue, followed by the rtPS queue and then by the nrtPS queue, respectively. BE connections do not go through the admission control process since they need no type of QoS guarantee. At every scheduling interval, the admission control mechanism polls the requests queues for new connections and then decides which connections will be accepted. If the request is of UGS type, then the following condition should be met: the number of requested slots, according to the maximum rate and interval between grants requirements must be smaller than or equal to the total number of slots which can be accommodated within the maximum jitter tolerated by the connection. If this condition is met, then the mechanism further checks if the number of required slots can be met within a so called HyperInterval interval. For UGS connections, the HyperInterval value equals the minimum common multiple of the unsolicited grant interval parameter of all the connections, whereas for rtPS and nrtPS connections, it equals the minimum common multiple of the polling interval parameter of all connections. In this stage, the admission control mechanism utilizes the routine *search(no._of_slots, initial_slot, final_slot)*, which searches the number of required slots (*no._of_slots*) within the interval [*initial_slot, final_slot*]. If the request for a new connection is of rtPS type, then the number of required slots within a polling interval, according to the minimum rate requirement, must be smaller than or equal to the number of slots which can be accommodated by the available bandwidth within that same interval. Then the mechanism verifies the availability of slots, both for sending bandwidth request and for sending data, at every polling interval within the HyperInterval. The admission of nrtPS connections is similar to the admission of rtPS connections. Though this solution is interesting, since it considers the majority of the standard defined QoS parameters, as well as the rate required for the sending of bandwidth requests, the authors do not detail the algorithm used by the search routine, which plays an essential role on the mechanism.

The admission control mechanism proposed by Chang, Chen e Chou (2007) utilizes two levels of priority: one for the SSs and another one for the UGS, rtPS, nrtPS and BE services. The priority levels are used to establish the $RW_{r,k}$ parameter value, which defines the network revenue when admitting a connection from service *k* to a priority *r* node. The purpose of the admission control mechanism is to increase the network revenue, however the authors do not inform how to configure the $RW_{r,k}$ parameter value. The admission control mechanism also utilizes a cost function called COL (Competitive On-Line). When the BS receives a request for the establishment of a connection, the mechanism calculates a cost value based on the service type, on the minimum rate requested, on the link capacity and on a predefined Δ constant. The connection will be accepted if the obtained payoff is higher than the cost.

Guo et al (2007) point out two main issues in the admission control and resources reservation mechanisms design. One of them is to reserve resources for real time traffic so that an efficient use of the bandwidth is achieved. The second issue is reserving resources for connections from other cells (handoff) so as to reduce the probability of these connections blocking. In order to face these two problems, the authors propose the Dynamic Bandwidth Reservation Admission Control mechanism (DBRAC). For the rtPS service, the mechanism reserves an amount of bandwidth equal to the sum of *MinTR* and a value within the

[*0, MaxTR - MinTR*] range, where *MaxTR* and *MinTR* are the maximum rate and minimum rate requirements, respectively. The sum of the resources reserved for the rtPS connections is called *Rrt_vbr*. For handoff connections, the mechanism reserves an amount of bandwidth *Rhf* calculated through an unidimensional Markov chain where the state variable represents the number of handoff connections. The mechanism further utilizes a third variable, *R*, which is the maximum value between *Rrt_vbr* and *Rhf*. The DBRAC mechanism performs the admission decision as follows: if the available bandwidth (cell capacity minus the sum of the already admitted connections minimum rate requirements) is higher than the minimum rate required by the connection and higher than *R*, then the connection is accepted, otherwise it is rejected. Handoff connections are rejected only when the required minimum rate cannot be met.

Three of the admission control schemes surveyed in this section present interesting characteristics. The scheme proposed by Chandra and Sahoo (2007) takes into account not only the minimum reserved traffic rate solicited by the new connection, but also the rate used by the bandwidth request mechanism. Moreover, for connections requesting UGS service, the proposed scheme guarantees that the required number of data slots is available within the tolerated grant jitter for each nominal grant interval in a predefined period. These features show the authors attention to ensuring standard compliance which is of paramount importance when developing QoS mechanisms for WiMAX networks. In (Chang, Chen, & Chou, 2007) a scheme was proposed for revenue maximization at the admission control time scale. Since WiMAX is a multi-service network, it is expected that users of different types of services will pay different prices. From the service provider perspective, it is important to maximize the revenue while supporting Quality of Service. Therefore, a mechanism that integrates pricing and admission control is quite promissing. The IEEE 802.16e extension introduced mobility in WiMAX networks, therefore mechanisms for the admission of handoff connections are essential. The work in (Guo, Ma, Guo, & Hou, 2007) includes the admission of handoff connections by performing dynamic reservation of resources and yet reducing bandwidth waste. The combination of the main characteristics of these three mentioned solutions in one admission control mechanism is not trivial, nevertheless, it yields to a mechanism that fulfills three important expectations of WiMAX users and service providers: Quality of Service, mobility and profitability. To the best of our knowledge, such an admission control mechanism has not been proposed yet.

Despite the real time connections have maximum latency requirement, the majority of the discussed mechanisms uses just the minimum rate requirement of the connections in the decision process. Such approach is not a problem if we consider that, according to the standard, the maximum latency needs to be guaranteed only for the traffic not exceeding the connection's minimum rate requirement and that such guarantee can be implemented by the scheduler. Admission control algorithms which include latency guarantees for all real time traffic tend to be more complex, like the proposal presented in (Niyato & Hossain, 2007). Table 3 summarizes the advantages and disadvantages of the solutions presented in this section.

FUTURE RESEARCH DIRECTIONS

In order to WiMAX networks consolidate as broadband wireless Internet access technology, a number of Quality of Service provisioning related aspects still need to be investigated.

Despite several scheduling algorithms have been proposed, the great majority does not cover all IEEE 802.16 standard specifications. From the proposals presented in this chapter, for example, the majority does not include guarantees for all the standard defined QoS parameters and half of them

Table 3. Advantages and disadvantages of the discussed proposals

Study	Advantages	Disadvantages
(Chen, Jiao, & Wang, 2005)	Simple algorithm.	Does not include the ertPS service; does not consider the rate used by the bandwidth request mechanism.
(Wang, Liu, Ju, & Ruangchai-jatupon, 2007)	Includes the 5 types of service; considers handoff connections.	Does not consider the rate used by the bandwidth request mechanism; does not inform how important variables in the algorithm must be estimated; does not present throughput and latency results.
(Wang, He, & Agrawal, 2007)	Tries to maximize the channel utilization by using a bandwidth borrowing and degradation scheme.	Does not consider the rate used by the bandwidth request mechanism; reserves part of the bandwidth for UGS service.
(Niyato & Hossain, 2007)	Considers the latency and minimum rate requirements; performance analysis through the simulation of several scenarios.	Does not consider the ertPS service; complex algorithm.
(Chandra & Sahoo, 2007)	Considers the latency and minimum rate requirements, as well as the rate necessary for the bandwidth request mechanism.	Does not include the ertPS service; does not detail the algorithm utilized by the search routine, which is an important part of the proposed mechanism.
(Chang, Chen, & Chou, 2007)	Unlike the other proposals, it uses the cost and payoff idea, what facilitates the inclusion of price variable in the admission of the connections in order to maximize the provider's profit.	Assumes that the stations have different priorities, what is not in accordance with the standard; does not define how to configure the value of the payoff obtained by the network upon admitting a determined connection.
(Guo, Ma, Guo, & Hou, 2007)	Includes the 5 types of service; considers handoff connections; performs dynamic reservation of resources for the handoff connections reducing the bandwidth waste.	After guaranteeing the connections minimum rate requirement and reserving part of the capacity for handoff connections, the available capacity is shared among the rtPS connections, however, ertPS and nrtPS connections could also benefit from extra rate..

do not include the allocation of resources for the ertPS service. In addition, just a few works propose integrated scheduling and admission control solutions. So, future research focus must target the investigations of complete QoS solutions for the IEEE 802.16 standard, i.e., solutions integrating scheduling, admission control and the five types of service available in the standard together with their respective QoS parameters.

Other challenges that require attention are related to the bandwidth request mechanisms. For the rtPS service, it is necessary to understand how to find an optimum interval for connections polling, so as to meet the connections QoS requirements without wasting resources (Rath, Bhorkar, & Sharma, 2006). For the nrtPS and BE services, one must investigate dynamic adjustment mechanisms for the contention period which minimize the probability of collisions (Oh & Kim, 2005), (He, Guild, Yang, & Chen, 2007).

In order to offer end-to-end access solutions, WiMAX networks are expected to work in conjunction with other technologies, such as optical networks. This integration of different standards implies QoS management considerations which need to be investigated. For instance, the mapping of QoS requirements among networks having different standards so that the end-to-end Quality of Service can be guaranteed is still open for research.

CONCLUSION

This chapter has presented two fundamental mechanisms for provisioning Quality of Service in IEEE 802.16 standard based networks, the scheduling mechanism and the admission control mechanism. Though the standard includes these mechanisms in its QoS architecture, specific policies are not defined.

The downlink traffic scheduling is carried out by a scheduler implemented at the BS, whereas the uplink traffic is served by a scheduler at the BS and another one at the SSs. The BS uplink scheduler poses a bigger challenge than the other two schedulers as the BS has no direct access to the traffic storage queues. Thus, the analysis and bibliographic survey presented in this chapter has concentrated in this part of the problem. The uplink scheduling at the BS must be based on bandwidth requests sent by the SSs and also on each QoS requirement. Each connection is associated with one of the five types of service available in the standard, each service being characterized by a set of QoS parameters. Many of the scheduling mechanisms proposed in the literature do not consider all the possible QoS parameters. In addition, only a few proposals deal with the complexity of the mechanism, despite this is an essential aspect given the frequency at which the scheduler must be executed.

Similarly, when dealing with the admission control mechanism, focus was given to the admission of the connections serving the uplink traffic. Scheduling and admission control are complementary mechanisms in provisioning Quality of Service. In addition to the link capacity, the admission control mechanism must know the way the scheduler makes use of that capacity so that such resource, which is limited in wireless networks, can be used in efficiently. A big part of the proposed solutions for the admission control does not take into consideration the adopted scheduling policy. Mechanisms for provisioning QoS in WiMAX networks which deliver an integrated scheduling and admission control solution for the five IEEE 802.16 standard defined types of service need to be investigated.

REFERENCES

Air interface for fixed broadband wireless access systems - amendment for physical and medium access control layers for combined fixed and mobile operation in licensed bands. (2005). *IEEE Standard 802.16e*.

Bai, X., Shami, A., & Ye, Y. (2008). Robust QoS control for single carrier PMP mode IEEE 802.16 systems. *IEEE Transactions on Mobile Computing*, *7*(4), 416–429. doi:10.1109/TMC.2007.70738

Barford, P., Bestavros, A., Bradley, A., & Crovella, M. (1998). *Changes in web client access patterns: Characteristics and caching implications* (Tech. Rep. No. 1998-023). Boston, MA: Boston University.

Borin, J. F., & Fonseca, N. L. (2008a). Scheduler for IEEE 802.16 Networks. *IEEE Communications Letters*, *12*(4), 274–276. doi:10.1109/LCOMM.2008.072110

Borin, J. F., & Fonseca, N. L. (2008b). A standard-compliant scheduler for WiMAX networks. In *4th IEEE Broadband Wireless Workshop* (pp. 1-5).

Borin, J. F., & Fonseca, N. L. (2008c). Simulator for WiMAX networks. *Simulation Modelling Practice and Theory*, *16*(7), 817–833. doi:10.1016/j.simpat.2008.05.002

Brady, P. (1969). A model for generating On-Off speech patterns in Two-Way conversations. *The Bell System Technical Journal*, *48*, 2445–2472.

Chandra, S., & Sahoo, A. (2007). An efficient call admission control for IEEE 802.16 networks. In *15th IEEE Workshop on Local & Metropolitan Area Networks* (pp. 188-193).

Chang, B.-J., Chen, Y.-L., & Chou, C.-M. (2007). Adaptive hierarchical polling and cost-based call admission control in IEEE 802.16 WiMAX networks. In *Wireless Communications and Networking Conference* (pp. 1954-1958).

Chen, C.-L., Lee, J.-W., Wu, C.-Y., & Kuo, Y.-H. (2009). *Fairness and QoS Guarantees of WiMAX OFDMA Scheduling with Fuzzy Controls.* EURASIP Journal on Wireless Communications and Networking, Article ID 512507.

Chen, J., Jiao, W., & Wang, H. (2005). A service flow management strategy for IEEE 802.16 broadband wireless access systems in TDD mode. In *IEEE International Conference on Communications* (pp. 3422-3426).

Chu, G., Wang, D., & Mei, S. (2002). A QoS architecture for the MAC protocol of IEEE 802.16 BWA system. In *IEEE International Conference on Communications, Circuits and Systems and West Sino Expositions* (pp. 435-439).

Cruz, R. L. (1991). A Calculus for network delay, Part 1: Network elements in isolation. *IEEE Transactions on Information Theory, 37*(1), 114–121. doi:10.1109/18.61109

Einhaus, M., & Klein, O. (2006). Performance Evaluation of a Basic OFDMA Scheduling Algorithm for Packet Data Transmissions. In *11th IEEE Symposium on Computers and Communications* (pp. 695-702).

Favalli, L., Lanati, M., Monighini, S. L., & Savazzi, P. (2007). Resource assignment in multiservice 802.16e. In *The 18th Annual IEEE International Symposium on Personal, Indoor and Mobile Radio Communications* (pp. 1-5).

3GPP2 C.S0014-0 version 1.0. (1999). Enhanced Variable Rate Codec (EVRC).

Guo, X., Ma, W., Guo, Z., & Hou, Z. (2007). Dynamic bandwidth reservation admission control scheme for the IEEE 802.16e broadband wireless access systems. In *Wireless Communications and Networking Conference* (pp. 3420-3425).

Hawa, M., & Petr, D. D. (2002). Quality of Service scheduling in cable and broadband wireless access systems. In *10th International Workshop on Quality of Service* (pp. 247-255).

He, J., Guild, K., Yang, K., & Chen, H.-H. (2007). Modeling Contention Based Bandwidth Request Scheme for IEEE 802.16 Networks. *IEEE Communications Letters, 11*(7), 698–700.

Hou, F., She, J., Ho, P.-H., & Shen, X. (2008). Performance Analysis of Weighted Proportional Fairness Scheduling in IEEE 802.16 Networks. In *IEEE International Conference on Communications* (pp. 3452-3456).

Huang, J., Subramanian, V. G., Berry, R., & Agrawal, R. (2007). Joint Scheduling and Resource Allocation in Uplink OFDM Systems. In *41st Asilomar Conference on Signals, Systems and Computers* (pp. 265-269).

IEEE Standard for Local and Metropolitan Area Networks - Part 16: Air Interface for Fixed Broadband Wireless Access Systems. (2004). *IEEE Standard 802.16*.

Kim, S., & Yeom, I. (2007). TCP-Aware Uplink Scheduling for IEEE 802.16. *IEEE Communications Letters, 11*(2), 146–148. doi:10.1109/LCOMM.2007.061509

Lee, H., Kwon, T., & Cho, D.-H. (2005). An enhanced uplink scheduling algorithm based on voice activity for VoIP services in IEEE 802.16d/e system. *IEEE Communications Letters, 9*(8), 691–693. doi:10.1109/LCOMM.2005.1496584

Mohammadi, A., Akl, S. G., & Behnamfar, F. (2008). Optimal linear-time QoS-based scheduling for WiMAX. In *21stCanadian Conference on Electrical and Computer Engineering* (pp. 1811-1814).

Niyato, D., & Hossain, E. (2007). QoS-aware bandwidth allocation and admission control in IEEE 802.16 broadband wireless access networks: A non-cooperative game theoretic approach. *Computer Networks*, *55*(11), 3305–3321. doi:10.1016/j.comnet.2007.01.031

Oh, S.-M., & Kim, J.-H. (2005). The Analysis of the Optimal Contention Period for Broadband Wireless Access Network. In *Third IEEE International Conference on Pervasive Computing and Communications Workshops* (pp. 215-219).

Parekh, A. K., & Gallager, R. G. (1993). A generalized processor sharing approach to flow control in integrated services networks: the single-node case. *IEEE/ACM Transactions on Networking*, *1*(3), 344-357.

Rath, H. K., Bhorkar, A., & Sharma, V. (2006). An Opportunistic Uplink Scheduling Scheme to Achieve Bandwidth Fairness and Delay for Multiclass Traffic in Wi-Max (IEEE 802.16) Broadband Wireless Networks. In *IEEE Global Telecommunications Conference* (pp. 1-5).

Sayenko, A., Alanen, O., & Hämäläinen, T. (2008). Scheduling solution for the IEEE 802.16 base station. *Computer Networks*, *52*(1), 96–115. doi:10.1016/j.comnet.2007.09.021

Seeling, P., Reisslein, M., & Kulapala, B. (2004). Network Performance Evaluation Using Frame Size and Quality Traces of Single-Layer and Two-Layer Video: A Tutorial. *IEEE Communications Surveys and Tutorials*, *6*(2), 58–78. doi:10.1109/COMST.2004.5342293

Singh, V., & Sharma, V. (2006). Efficient and fair scheduling of uplink and downlink in IEEE 802.16 OFDMA networks. In *Wireless Communications and Networking Conference* (pp. 984-990).

Tarchi, D., Fantacci, R., & Bardazzi, M. (2006). Quality of Service management in IEEE 802.16 wireless metropolitan area networks. In *IEEE International Conference on Communications* (pp. 1789-1794).

Wang, H., He, B., & Agrawal, D. P. (2007). Above packet level admission control and bandwidth allocation for IEEE 802.16 wireless MAN. *Simulation Modelling Practice and Theory*, *15*(14), 366–382. doi:10.1016/j.simpat.2006.11.010

Wang, L., Liu, F., Ju, Y., & Ruangchaijatupon, N. (2007). Admission Control for Non-preprovisioned Service Flow in Wireless Metropolitan Area Networks. In *Fourth European Conference on Universal Multiservice Networks* (pp. 243-249).

Wang, L.-C., & Lin, W.-J. (2004). Throughput and fairness enhancement for OFDMA broadband wireless access systems using the maximum C/I scheduling. In *IEEE 60th Vehicular Technology Conference* (pp. 4696-4700).

Wongthavarawat, K., & Ganz, A. (2003). IEEE 802.16 based last mile broadband wireless military networks with Quality of Service support. In *IEEE MILCOM* (pp. 779-784).

Yang, O., & Lu, J. (2006). New scheduling and CAC scheme for real-time video application in fixed wireless networks. In *3rd IEEE Consumer Communications and Networking Conference* (pp. 303-307).

ADDITIONAL READING

Abu Ali, N., Dhrona, P., & Hassanein, H. (2009). A performance study of uplink scheduling algorithms in point-to-multipoint WiMAX networks. *Computer Communications*, *32*(3), 511–521. doi:10.1016/j.comcom.2008.09.015

Ahmed, M. H. (2005). Call admission control in wireless networks: a comprehensive survey. *IEEE Communications Surveys & Tutorials*, *7*(1), 49–68. doi:10.1109/COMST.2005.1423334

Belghith, A., & Nuaymi, L. (2008). WiMAX Capacity Estimations and Simulation Results. In *IEEE Vehicular Technology Conference* (pp. 1741-1745).

Chen, Y.-C., Hsia, J.-H., & Liao, Y.-J. (2009). Advanced seamless vertical handoff architecture for WiMAX and WiFi heterogeneous networks with QoS guarantees. *Computer Communications, 32*(2), 281–293. doi:10.1016/j.comcom.2008.10.014

Cho, D.-H., Song, J.-H., Kim, M.-S., & Han, K.-J. (2005). Performance Analysis of the IEEE 802.16 Wireless Metropolitan Area Network. In *First International Conference on Distributed Frameworks for Multimedia Applications* (pp. 130-137).

Chu, G., Wang, D., & Mei, S. (2002). A QoS architecture for the MAC protocol of IEEE 802.16 BWA system. In *IEEE International Conference on Communications, Circuits and Systems and West Sino Expositions* (pp. 435-439).

Cicconetti, C., Erta, A., Lenzini, L., & Mingozzi, E. (2007). Performance Evaluation of the IEEE 802.16 MAC for QoS Support. *IEEE Transactions on Mobile Computing, 6*(1), 26–38. doi:10.1109/TMC.2007.250669

Cicconetti, C., Lenzini, L., Mingozzi, E., & Eklund, C. (2006). Quality of service support in IEEE 802.16 networks. *IEEE Network, 20*(2), 50–55. doi:10.1109/MNET.2006.1607896

De Pellegrini, F., Miorandi, D., Salvadori, E., & Scalabrino, N. (2006). *QoS Support in WiMAX Networks: Issues and Experimental Measurements* (Tech. Rep. 200600009). CREATE-NET.

Eklund, C., Marks, R. B., Stanwood, K. L., & Wang, S. (2002). IEEE standard 802.16: a technical overview of the WirelessMANTM air interface for broadband wireless access. *IEEE Communications Magazine, 40*(6), 98–107. doi:10.1109/MCOM.2002.1007415

Gakhar, K., Achir, M., & Gravey, A. (2007). How Many Traffic Classes Do We Need In WiMAX? In *IEEE Wireless Communications and Networking Conference* (pp. 3703-3708).

Ghosh, D., Gupta, A., & Mohapatra, P. (2008). Scheduling in multihop WiMAX networks. *ACM SIGMOBILE Mobile Computing and Communications Review, 12*(2), 1–11. doi:10.1145/1394555.1394557

Israeli, A., Rawitz, D., & Sharon, O. (2008). On the complexity of sequential rectangle placement in IEEE 802.16/WiMAX systems. *Information and Computation, 206*(11), 1334–1345. doi:10.1016/j.ic.2008.07.002

Kim, E., Kim, J., & Kim, K. S. (2007). An Efficient Resource Allocation for TCP Services in IEEE 802.16 Wireless MANs. *Vehicular Technology Conference* (pp. 1513-1517).

Kim, S., Lee, M., & Yeom, I. (2009). Impact of bandwidth request schemes for Best-Effort traffic in IEEE 802.16 networks. *Computer Communications, 32*(2), 235–245. doi:10.1016/j.comcom.2008.10.006

Lakkakorpi, J., Sayenko, A., & Moilanen, J. (2008). Comparison of Different Scheduling Algorithms for WiMAX Base Station: Deficit Round-Robin vs. Proportional Fair vs. Weighted Deficit Round-Robin. *IEEE Wireless Communications and Networking Conference* (pp. 1991-1996).

Lera, A., Molinaro, A., & Pizzi, S. (2007). Channel-Aware Scheduling for QoS and Fairness Provisioning in IEEE 802.16/WiMAXBroadband Wireless Access Systems. *IEEE Network, 21*(5), 34–41. doi:10.1109/MNET.2007.4305171

Liu, C.-Y., & Chen, Y.-C. (2008). An Adaptive Bandwidth Request Scheme for QoS Support in WiMAX Polling Services. In *International Conference on Distributed Computing Systems* (pp. 60-65).

Liu, N., Li, X., Pei, C., & Yang, B. (2005). Delay Character of a Novel Architecture for IEEE 802.16 Systems. In *Sixth International Conference on Parallel and Distributed Computing Applications and Technologies* (pp. 293-296).

Niyato, D., & Hossain, E. (2007). A Hierarchical Model for Bandwidth Management and Admission Control in Integrated IEEE 802.16/802.11 Wireless Networks. In *IEEE Wireless Communications and Networking Conference* (pp. 3763-3767).

Nuaymi, L. (2007). *WiMAX: Technology for Broadband Wireless Access.* Hoboken, NJ: John Wiley & Sons.

Saffer, Z., & Andreev, S. (2008). Delay analysis of IEEE 802.16 wireless metropolitan area network. In *International Conference on Telecommunications* (pp. 1-5).

Sayenko, A., Alanen, O., & Hamalainen, T. (2007). Adaptive Contention Resolution for VoIP Services in the IEEE 802.16 Networks. In *IEEE International Symposium on a World of Wireless, Mobile and Multimedia Networks* (pp. 1-7).

So-In, C., Jain, R., & Tamimi, A.-K. (2009). Scheduling in IEEE 802.16e Mobile WiMAX Networks: Key Issues and a Survey. *IEEE Journal on Selected Areas in Communications, 27*(2), 156–171. doi:10.1109/JSAC.2009.090207

Triantafyllopoulou, D., Passas, N., Salkintzis, A. K., & Kaloxylos, A. (2007). A heuristic cross-layer mechanism for real-time traffic over IEEE 802.16 networks. *International Journal of Network Management, 17*(5), 347–361. doi:10.1002/nem.668

Xiao, Y. (2007). *WiMAX/MobileFi – Advanced Research and Technology.* Boca Raton, FL: Auerbach Publications.

Yu, J. T. (2007). Scheduling and Performance Analysis of QoS for IEEE 802.16 Broadband Wireless Access Network. In S. Ahson & M. Ilyas (Eds.), *WiMAX Standards and Security.* New York: CRC Press.

Chapter 10
Advancements on Packet Scheduling in Hybrid Satellite-Terrestrial Networks

Hongfei Du
Simon Fraser University, Canada

Jiangchuan Liu
Simon Fraser University, Canada

Jie Liang
Simon Fraser University, Canada

ABSTRACT

The past years have seen an explosion in the number of broadcasting network standards and a variety of multimedia services available to the mobile mass-market. Satellite communications has been gaining phenomenal growth and increasing interest over the last decade in its complementary but essential role for offering seamless broadband service coverage to potential users at every inch of the earth's surface. However, mobile satellite network often feature unidirectional and long-latency, a great deal of research effort has been attempted for this bottleneck. Given the absence of feasible power control mechanism and reliable feedback information, the role of packet scheduling in such a network with large delay-bandwidth product is extremely challenging. In fact, an optimized media access control (MAC) layer protocol is essential for cost-efficient satellite networks to compete with other terrestrial modalities. In particular, the integration and convergence between satellite network and conventional terrestrial backbone infrastructure offers promising solutions for next generation service provisioning. In this chapter, the authors give a survey on the state-of-the-art on packet scheduling in hybrid satellite-terrestrial networks (HSTN). A whole range of issues, from standardization, system to representative scheduling methodologies as well as their performance trade-offs have been envisioned. Moreover, the authors investigate viable solutions for effectively utilizing the limited/delayed feedbacks in resource management functions. They examine the flexibility and scalability for the alternative schemes proposed in this context, and analyze the performance gain achievable on essential QoS metrics, channel utilization, as well as fairness.

DOI: 10.4018/978-1-61520-680-3.ch010

INTRODUCTION

The proliferation of digital multimedia broadcasting (DMB) transmission technology and the increasing demand for resource-consuming rich-media and streaming video applications entail that the future generation wireless network should be capable of supporting heterogeneous multimedia provisioning over an extensive range of underlying network standards and protocols. Due to the unique broadcast nature and ubiquitous coverage of satellite communication system, the broadband satellite network, whilst concurrently integrating with terrestrial backbone infrastructures, has been gaining importance and provides immense brand new opportunities for delivering point-to-multipoint (p-t-mp) multimedia content to a large number of mobile audiences spreading over extensive geographical area. It is expected that the satellite component will play not only a complementary, but also essential role in delivering multimedia data to those areas where the terrestrial high-bandwidth communication infrastructures are, either economically prohibitive or technically unreachable.

A variety of multimedia broadcasting initiatives, such as the Multimedia Broadcast/Multicast Services (MBMS), Digital Video Broadcasting-Handheld (DVB-H), and terrestrial/satellite-DMB (T-/SDMB), Media Forward Link Only (MediaF-LO) and Digital Terrestrial/Television Multimedia Broadcasting (DTMB) have been envisioned as viable solutions to provide one-to-many content distribution to mobile/portable devices. The 3rd Generation Partnership Project (3GPP) within the MBMS framework (3GPP, 2008) defines a unidirectional point-to-multipoint mode for the provisioning of multimedia services and thereby optimizes the available capacity in cellular networks. DVB-H (ETSI, 2004), as initiated by the DVB forum implements additional features based on the DVB-T standard to address the specific constraints associated with mobile handheld terminals. At the same time, MediaFLO (TIA, 2006) developed by Qualcomm was recently approved by the Telecommunications Industry Association (TIA) as a new air interface standard for multicast delivery, aimed at delivering high-quality multimedia services to the U.S. mobile market. As the largest single digital communication market in the world, the Chinese government recently announced its national DTTB (Digital Terrestrial Television Broadcasting) standard and it has been widely expected that the massive deployment in China will begin in 2008 (GB, 2006). DTMB, the non-official acronym of the DTTB standard, is attracting a great deal of attention within the broadcasting community (Song, et al, 2007) as a cost-effective approach for delivering multimedia services over the Chinese market. Meanwhile, large number of research projects have been conducted to investigate viable solutions for facilitate the multimedia services provisioning via the so-called hybrid satellite-terrestrial network (HSTN), where terrestrial gap fillers are employed as the key functional element to provide the missing coverage when the direct LOS (line-of-sight) signals from satellite are temporarily unavailable. Mobile satellite networks often feature unidirectional and long-latency, which have posed challenging research barrier and attracted a great deal of research effort for this bottleneck. Given the absence of feasible power control mechanism and reliable feedback information, the role of packet scheduling is becoming a challenging task. In fact, an optimized design on media access control (MAC) layer protocols is essential for cost-efficient satellite networks to compete with other terrestrial modalities.

The rest of the chapter is organized as follows. We continue with an introduction of packet scheduling issues in HSTN. An overview of standardization issues and radio resource management (RRM) functions in multimedia broadcasting over HSTN are analyzed in the following section. In Section III, we review research and development efforts on packet scheduling schemes in satellite multimedia broadcasting drawn from existing literature; the challenging issues on the packet scheduling optimization and QoS concerns from diverse aspects are detailed. In Section IV, we address the advancements on the packet scheduling optimization technique in diverse aspects, e.g., the proportional delay-differentiation based scheme, cross-layer design, multi-dimensional optimization, usage of channel feedbacks and queue states, and etc. Adaptations on scheduling algorithms to the unique satellite broadcasting environments are studied via hierarchical distributed approach. The proceeding section discusses the performance trade-offs of the optimized packet scheduling schemes, in comparison with existing schemes. We conclude this chapter by raising some opportunities and challenges for moving the research agenda in this field forward.

1. OVERVIEW OF MULTIMEDIA BROADCASTING IN HSTN

3.1 Standardization and System

The continuing evolution of satellite delivered multimedia broadcasting lies in the integration and convergence between satellite networks and the terrestrial backbone infrastructures. Traditionally, satellite communication has been dedicated to long-distance intercontinental connectivity for audio/video (AV) transmission. From mid 1980s, very small aperture terminal (VSAT) (Pelton, 1989; ETSI, 1992) applications emerged as a promising solution and remained a niche market due to the cost of the transponders/terminals. The success of VSAT networks is mainly because they address a topology that appears to be ideally suited to satellite communication - point-to-multipoint, which greatly facilitates multimedia broadcast/ multicast provisioning. In the 1990s, the direct-to-home television broadcasting (Sandberg, 1995) business has been gaining great popularity and fostered the satellite industry a further growth in the fast growing personal satellite communication market. The explosive increase of Internet has pushed it as a driving force for personal satellite communications targeting at the Internet-based applications. Furthermore, being defined by 3GPP, the broadcast/multicast services is expected to play a fundamental role in the upcoming 4G mobile systems, whilst the satellite component becomes one of the most competitive solution for this mission. In Europe, the EU IST SATIN project has investigated the whole range of issues pertinent to the satellite UMTS (S-UMTS). In this context, the S-UMTS expands the reach of T-UMTS, in terms of geographical coverage, coverage completion, disaster-proof availability, dynamic traffic management as well as rapid service deployment.

As a representative HSTN system as well as a complementary alternative to 3G mobile networks, the SDMB system (Chuberre et al, 2004; Chuberre et al, 2005) is attracting a great deal of attention within the satellite community in Europe as a cost-effective approach for the delivery of MBMS to mobile users over satellite broadcasting networks. The system is constructed based on the S-UMTS concept and closely in line with the 3GPP standardizations in terms of service provisioning and QoS requirements. Based on its broadcast nature and point-to-multipoint service it provides, the SDMB system offers extensive coverage, low transmission cost for large numbers of terminals, as well as high QoS guarantees for multimedia provisioning. By employing the wideband code-division multiple access (WCDMA) with frequency division duplexing (FDD) (Holma & Toskala, 2002), the system can be closely integrated with existing mobile cellular networks and minimizes the potential cost impact on both 3G cellular terminals and network operators. Its network architecture enables the satellite system to be seamlessly integrated with the terrestrial 2G/3G mobile infrastructures by extending and adapting the 3GPP standards over a GEO satellite network. Its access layer uses new multiplexing scheme and packet scheduling algorithm to achieve high utility of the satellite bandwidth and optimized queue and buffer performance. The whole range of issues pertinent to the SDMB system, from system definition to demonstration and validation, are addressed in the following European IST projects: Mobile Digital broadcast Satellite (MoDiS) and Mobile Applications & sErvices Satellite & Terrestrial inteRwOrking (MAESTRO).

With the dominant role of Internet traffic over the satellite, digital television and rich multimedia content to home via satellite was originally envisaged in the European DVB-S and DVB-RCS standards, which empower interactive satellite communications with economical satellite terminals. Recently, the DVB-SH (Digital Video Broadcasting – Satellite services to Handhelds) specification (ETSI, 2007) is approved by the DVB Forum, to deliver IP-based media content and data to handheld terminals as a representative HSTN system, replacing current SDMB system in Europe. As the majority research work in the field have been devoted to the original SDMB system, in this chapter, we consider the SDMB system as a major representative HSTN system to study the whole range of issues relating to respective packet scheduling functions. Therefore, we use HSTN and SDMB interchangeable in the following context. Nonetheless, the scheduling algorithm itself is well-adapted to a wide range of WCDMA-based systems and therefore is not confined to SDMB.

The interoperability between different networks and their software is called "network convergence". This definition usually encompasses the terminals, the network operators and the service providers. In this vision, three different types of networks are considered. First of all are the broadcast-type networks used for the digital TV standards including T-DMB, DVB-H, DTMB, and MediaFLO. Besides, video and television content can also be transmitted via telecommunication networks such as UMTS. Fig. 1 defines typical hybrid satellite-terrestrial networks (HSTN), which is responsible for delivering multi-session multimedia content with diverse QoS requirements. The system relies on a hybrid broadcast infrastructure, which encompasses high-power, geostationary satellite and low-power terrestrial repeaters, both operating in a 30MHz spectrum between 2170-2200 MHz frequency band, i.e., IMT2000 band allocated to Mobile Satellite Systems. This frequency band is adjacent to the T-UMTS FDD downlink band (2110-2170MHz), which allows the system enjoy maximum reusing of technology and infrastructure and minimum potential cost impacts on both 3G cellular terminals and terrestrial repeaters. The hybrid system takes advantage of a broadcast capability to provide efficient delivery of MBMS contents to extensively mass mobile market. The user applies the standard 3G terminal enriched with satellite associated functions, which is responsible for measuring the associated CSI and end-to-end characteristics and feeding them back to the Sat-GW through the core network. The terrestrial gap-fillers, identified as intermediate module repeater (IMR), are co-located at the base stations to enhance signal reception quality and provide adequate coverage in urban/built-up areas. MBMS services are transmitted to the users in either a broadcast or multicast mode. In the latter case, service is only delivered to the users within a specific multicast group. Direct access via satellite is the preferred forward link, offering essential coverage over rural/remote communities; nevertheless, the communication will be retained via IMRs if the direct access path is temporarily unavailable. It is noteworthy that, there exist two transmission modes: unidirectional in baseline SDMB scenario and bi-directional transmission with a return channel via the terrestrial mobile networks, supporting interactive multimedia applications.

The multimedia services are delivered from the content provider through satellite gateway (Sat-GW), Geo-stationary satellite (GEO-Sat), to the user equipment (UE). The UE operates at either direct access or indirect access mode, depending on whether a direct LOS signal from

Figure 1. HSTN broadcast system architecture

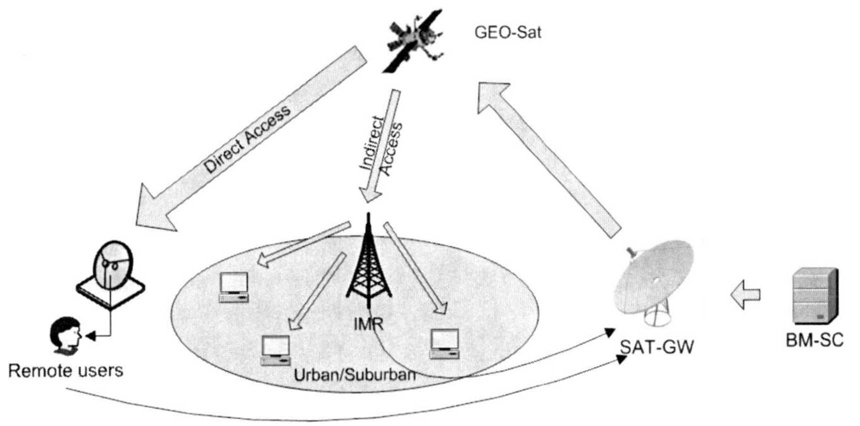

satellite is practically usable. Typically, the telecommunication networks will also provide the uplink channel including feedback information and interactive functions when the main media download is transmitted via a broadcast network. Finally, clients can access multimedia content via a direct internet connection including WiFi and WiMax, which provides an attractive solution for both indoor/outdoor receptions.

The hybrid architecture defines a hierarchical tree network where the available bandwidth from the GEO-Sat is distributed amongst the underlying nodes, i.e., the IMRs and the UEs. The functional components involved in this topology are described as follows:

- **Sat-GW:** It is connected to an interactive broadcast network service provider via Broadcast/Multicast Service Center (BM-SC) and terrestrial core networks.
- **GEO-Sat:** It is capable of provide outdoor and indoor coverage in rural areas under spot beams and with on-board processing capability. A single spot beam is typically 700-1000 km in diameter, providing national wide umbrella cell. GEO-Sat is controlled by a remote Network Operation Center (NOC) through a dedicated high-bandwidth channel. The GEO-Sat is capable of supporting high bandwidth for downlink (i.e. 90Mbps) and moderate bandwidth for uplink (i.e. 1.5 Mbps).
- **IMR:** The GEO-Sat is complemented by terrestrial gap fillers to address the indoor coverage in urban areas, where severe blockages of direct satellite signals may occur. It includes a full replacement with terrestrial core networks, e.g. UMTS, WLAN, WiFi/WiMAX, which can be used to complement the satellite unreachable/blocked coverage. A satellite blocked area is defined as the area within which the UEs have limited/no access to satellite signal. This event may arise from serious multipath impairments in built-up areas, or deep fading/shadowing effects with satellite associated link. To minimize cost and environmental impact, the IMRs are designed to be co-sited with 2G & 3G cellular base stations and share their antennas. Therefore, no new sites are required, resulting in a cost-efficient, rapid deployment of the infrastructure. The mode switching between direct access (DA) and indirect access (IA) is triggered by appropriate link quality measurements over available signal strength. Thus, the cooperation between the terrestrial and satellite system is of prime importance to ensure a

continuous and smooth coverage of the integrated system.

- **UE:** The UE used in the HSTN applies the standard return channel satellite terminals, embedded with built-in channel measurements and evaluation model. The terminal is capable of collecting channel state information from the detectable signals coming from both DA and IA links.

3.2 Radio Resource Management Functions

Radio resource management (RRM) functions in HSTN allocate and manage the radio resources to provide optimized network performance and guaranteed QoS level. Satellite communications are dynamic in nature; the dynamism arises from multiple dimensions, namely propagation conditions, traffic generation conditions, interference conditions, etc. Therefore, the dynamic network evolution calls for a RRM in a dynamic way, which is carried out by the RRM mechanism with series of associated parameters that need to be chosen, measured, analyzed and optimized. Since the RRM strategies are not subject to standardization activities, to improve overall system performance and reduce operator infrastructure cost, RRM functions can be implemented via many different algorithms. Such algorithms may be designed in a dynamic and adaptive manner to compensate with the instantaneous performance variations incurred from traffic source, network nodes and transmission media. Research and development in this important field are especially demanding and challenging. The main problems of designing appropriate RRM functions faced in satellite systems are explained in (Giambene, et al, 2007).

Given the unidirectional nature and the point-to-multipoint services it provides, aimed at maximizing spectrum efficiency and satisfying diverse QoS, the design of RRM functionality implemented at the satellite access layer is challenging. The packet scheduling algorithm, which is the single function performing short-term resource allocation, is the focus of efficient resource allocation. HSTN allows a user or an application to negotiate the characteristics of the network at the connection set up phase. The network may check whether sufficient resources are available, and returns the results to the application, which can accept or deny the connection request according to an admission control scheme. After admission of the connection request, the network should keep the performance of the connection as contracted by dynamically adjusting its network parameters. The above rules also apply for broadcast and multicast scenarios. By admitting the connection request the access network has to make a choice for the type of the radio access bearer taking into account several conditions such as attributes of the requested service, number of group members in the cell, current traffic, channel and load conditions and etc. In contrast to unicast, i.e., point-to-point (p-t-p) service delivery, the p-t-mp network has to select the type of the transport channel, namely if a common channel should be used or a dedicated channel is used instead. For instance, if there is small number of multicast members (1 or 2) in the cell, it is not worth using a common channel since a common channel needs additionally a return link dedicated channel to maintain the quality of the connection, i.e., measurement control/report, power control and the error correction due to its unidirectional nature. In other words, usage of a common channel is not always more effective than that of dedicated channels. Therefore, well defined criteria for selecting the transport channel type among others is necessary for optimally utilizing system capacity, e.g., the minimum number of members in the multicast group, momentary load condition, current/predictable channel condition, QoS constraints of the session, etc. Moreover, since the number of members in a multicast session can be dynamically changing, there should be another criterion for the appropriate timing when a *Radio Access Bearer* (RAB) re-assignment is necessary.

Figure 2. Interactions between RRM functions

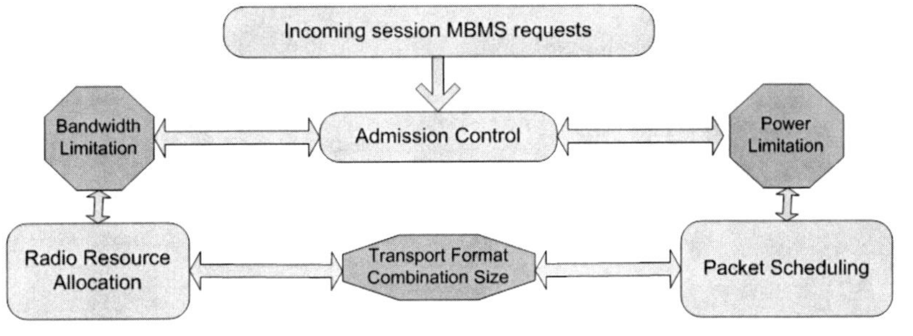

The RRM functionalities comprise three main parts: packet scheduling (PS), radio resource allocation (RRA) and admission control (AC). These functionalities cooperate interactively during the resource allocation procedures.

- **Packet scheduling:** Packet scheduling operates periodically in each transmission time interval (TTI) of the radio bearers. It time-multiplexes flows with diverse QoS into physical channels and adjusts the transmit power for physical channels on the basis of the packets to be served and the required reception quality of the service in terms of the target block error rate (BLER).
- **Radio resource allocation:** This entity is responsible for the radio bearer configuration at the beginning of each session, which includes the estimation of the required number of logical/transport/physical channels along with their mappings for each physical channel through the scheme layers/sub-layers.
- **Admission control:** The admittance decision of each incoming MBMS session is handled by the admission control function during the phase of service establishment/re-negotiation, aimed at preserving the required QoS while making efficient utilisation of resources.

As illustrated in Fig. 2, the interaction between RRM entities can be described as follows:

When there is a new session request, the AC will check the service QoS constraint and load/power of the system to determine whether to accept or reject the request. Also the AC needs to predict the long-term load of the network and interference level that may arise.

If the request is accepted, the RRA module is triggered to estimate the radio bearer configuration according to the traffic characteristics and specific scenarios (e.g. available rate of FACHs and S-CCPCHs). The radio bearer will be reconfigured whenever there is a new session request admitted by AC, or an existing session completes. During the RB reconfiguration, TFCS is derived for each S-CCPCH according to the service characteristics. The mappings of the MTCHs/FACHs to S-CCPCHs, as well as the TFCS available to S-CCPCHs, are passed to PS for the short-term selection of TF(C).Packet scheduler time-multiplexes service flows with different QoS requirements into physical channels, in such a way as to satisfy these requirements and adjusts the transmit power of the physical channel on the basis of the required reception quality of the service under the constraint that the total available power for all the physical channels within a beam is fixed. AC predicts the total system load based on the information regarding the current system status (number of admitted flows, requested QoS etc) and on the declared QoS requirements

Figure 3. Packet scheduler functional description

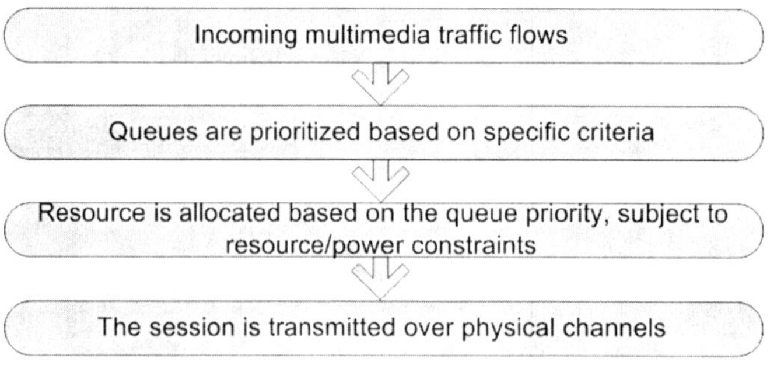

of the incoming traffic. PS calculates the actual system load resulting from the per-TTI scheduling decisions.

3.3 Packet Scheduling

The baseline HSTN system is defined as unidirectional transmission system, where the effective power control and reliable feedback information are unavailable. Therefore, efficient packet scheduling optimizations are critical for the overall capacity and performance of the network. Being the single function performing short-term resource allocation, the packet scheduler is implemented at the HSTN access layer and operates periodically in each transmission time interval (TTI) of the radio bearers. The main functionality of packet scheduling operates at the media access control (MAC) sublayer of the data link layer, aimed at coordinating the access among competing flows arrived at the queuing buffer in the radio link control (RLC) sublayer of the data link layer. The packet scheduler time-multiplexes competing flows into physical channel and adjusts the transmit powers settings on the basis of the required reception quality of the service in terms of the target transport block error rate (BLER), and under the limited total available power within a satellite beam.

As illustrated in Fig. 3, the packet scheduling strategy can be conceptualized into the following two steps:

- **Service prioritization:** The incoming service requests are prioritized according to priority criteria, considering performance variations and QoS guarantees.
- **Resource allocation:** The resource is allocated accordingly to the sessions, where the instantaneous data rate and transmission power are assigned within the specific resource allocation interval (i.e. one TTI).

The data rate assignment is essentially performed via the selection of the Transport Format Combinations (TFCs) (3GPP, 2008), which directly determines how much data from each transport channel is allowed to be forwarded to the physical layer in the particular TTI. In each TTI, the scheduler selects an appropriate Transport Format (TF) or TF set (TFS) from each transport channel. The combination of all the selected TFs/TFSs in all multiplexed transport channels within a Secondary Common Control Physical CHannel (S-CCPCH) forms a TFC. The exact TFC is selected for each active S-CCPCH from the Transport Format Combination Set (TFCS), which is derived during the session starts. The selection of TFC is of paramount importance since the capacity allocated to each service at the Sat-GW is strongly related with the QoS perceived by the end users. Therefore, the resource allocation has to be designed considering both the QoS demands and the system power/load constraints. The exact relationship between power constraints and transport block size (TBS) depends on the choice of modulation and coding schemes, the radio channel and the receiver architecture. In general, the bigger TBS, the more power that is consumed. Transmit power for each session is determined by the following parameters including: Thermal noise, Path loss, Eb/No requirements, required transmit power, code rate and Rate Matching ratio.

The decision of the packet scheduling is made in coordination with some specific criteria in terms of fairness and service requirements, which varies from one scheduling algorithm to another, effectively impacting the overall QoS guarantees and the network performance. Therefore, the packet scheduling shall be in compliance with the following objectives:

- Coordinate the serving order of contending multimedia traffic flows, aimed towards the highest possible degree of resource utilization and spectrum efficiency.
- Minimize the transmission power consumption so as to meet the system power constraints.
- Guarantee the application-specific QoS satisfaction in terms of different performance criteria, on the basis of the service prescribed requirements.
- Incorporate the instantaneous traffic dynamics of contending flows and thereby maintain a certain level of performance requirements.
- Consider the flexibility and scalability features.

2. REVIEW OF PACKET SCHEDULING SCHEMES

4.1 QoS Considerations

Quality of service (QoS) is broadly noted as an elusive term denotes the assurance of user perceived performance level in terms of bandwidth, packet loss, delay, and delay variation (jitter). QoS level is measured against its threshold levels which can be either statistically or dynamically determined based on the service QoS class and its instantaneous performance metrics. Providing such assurances in resource and power limited satellite network is challenging. What makes this challenge even worse is the long-latency and highly vibrating satellite link. Furthermore, there exist more stringent performance requirements on heterogeneous multimedia services, in terms of tolerance of loss, delay, and delay jitter. Whereas voice and video streaming applications have stringent requirements on transmission delays and delay jitter, and are error-tolerant, interactive applications like Web browsing are very sensitive to losses but can bear considerable delays. On the other hand, IP services and applications are dominating terrestrial networks, the space segment is challenged to be "QoS-aware" to seamlessly integrate with IP terrestrial networks to efficiently utilize resources and serve a maximum number of connections (Courville & Bischl, 2005). This entails the space segment not only be able to interpret the QoS parameters engineered by terrestrial networks, but also can actively perform QoS-based adaptations.

In HSTN, heterogeneous services are characterized by various applications with diverse QoS requirements. File download applications normally requires a critical error bound without tight timeliness demands. However, video broadcast applications pose interesting challenges, specifically, video broadcast impose stringent real-time performance requirements in terms of bandwidth and latency. In the following, we discuss some of the most important QoS performance metrics in heterogeneous multimedia communications.

- **Data rate:** Applications such as video streaming, media-cast distributions, telemedicine, two-way telephonic education, require rates ranging from a few hundred megabits to gigabits.

- **Delay:** The time for a packet to be transported from the sender to the receiver. It includes queuing delay, processing delay, propagation delay, etc. Real-time applications require a maximum delay of 400 ms and packet transfer delays for other classes of service are even more stringent.
- **Jitter:** Is the variation in end-to-end transit delay. Multimedia services have more stringent jitter demands than delay itself.
- **Bandwidth:** Is the maximal data transfer rate that can be sustained between two end points.
- **Packet Loss:** Is defined as the ratio of the number of undelivered packets to the total number of sent packets.
- **Reliability:** Is the percentage of network availability depending upon the various environmental parameters such as rain.
- **Scalability:** Normally considered as the complexity involved when the network increases its scale, namely the number of nodes or users.
- **Flexibility:** Refers to the ability of adapting the protocol design in response to some critical parameters, e.g., the network dynamics, channel variations and terminal heterogeneities.

To achieve an end-to-end QoS in HSTNs is a non-trivial problem. A successful end-to-end QoS model depends upon the various interfaces at each subsequent lower layer to the upper layers. In the HSTN system, the service types are categorized as: "streaming", "hot download", and "cold download":

- **Streaming:** Streaming service allows multimedia to be stored temporarily in the receiver buffer and displayed continuously even before the completion of transmission. Service in this category requires explicit upper bound on queuing delay/jitter.
- **Hot download:** The service in this category is to be stored at the receiver for their offline access. Compared with streaming, the hot download service has more tolerant demand on delay and jitter but more stringent demand on packet loss.
- **Cold download:** It requires the least demand on delay/jitter but the most stringent demand on packet loss, services in this category are often transmitted as individual file, such as software package, video/images, and text messages.

Given the unidirectional nature and long propagation delay, the baseline mobile satellite broadcasting system is incapable of effectively tracking real-time channel state information (CSI) from the mobile terminal side, which makes CSI-dependent scheduling a challenging task. In the following context, we first envisage possible optimizations on scheduling skills under typical unidirectional satellite system, thereafter, we investigate feasible solutions for utilizing the CSI-based information in such an environment, and seek to obtain better performance gain via novel approaches, e.g., channel-aware and hierarchical distributed scheduling.

4.2 Conventional Scheduling Schemes

In this section, we review classical packet scheduling schemes and discuss their pros and cons when applied to the mobile satellite systems. We consider Round Robin (RR) as one of the simplest scheduling algorithms, where queues are served recursively in their order in a non-preemptive manner. It is non-priority based and offers no differentiation between differentiated service classes. Therefore, the RR discipline is insensitive to packet size, where large packet size would be favored over other queues. As it does not consider any differentiation among users; the overall system throughput is fairly low. To ensure a minimum bandwidth allocation and distribution, the Weighted Round Robin (WRR) (Katevenis, Sidiropoulos & Courcoubetis, 1991) assigns a weight to each class. In proportion to the prescribed weights, the available bandwidth is allocated to each class in a round robin manner. Therein the weight assigned to each class can be regarded as a tunable parameter that can effectively provide QoS differentiation.

Previous studies (Karaliopoulos, Henrio, Narenthiran, Angelou & Evans, 2004) have systematically addressed the packet scheduling problems in a typical HSTN system, namely SDMB, via classical packet scheduling schemes, namely multi-level priority queuing (MLPQ) and weighted fair queuing (WFQ).

MLPQ-based scheduling is effectively the adaptation of the multi-level, non-pre-emptive priority discipline. MLPQ always processes packets starting from those non-empty queues having the highest priority first, with queues having the same priority served in a round robin fashion. Firstly, MLPQ employs a strict QoS-based prioritization scheme, in which a lower-priority service may suffer from considerably longer queuing delays. It always processes packets from those non-empty queues with the highest priority; as a result, packets waiting in the lower priority queues may suffer from considerably longer queuing delays. This scheme favors the high priority classes, assuring a delay bound for their packets, whilst it provides no guarantees for lower priority classes. Furthermore, it is generally agreed that background applications have no stringent delay constraint, and the only requirement for application in this category is that information should be delivered to the user essentially error free. In fact, background applications still need a delay constraint (at least an upper bound), since data can effectively be useless if it is received too late for practical purposes. Finally, MLPQ deals with queues having the same priority in a round-robin fashion. Consequently, there is no differentiation made between queues with the same QoS rank. However, this is not an efficient mechanism. Rather than prioritizing queues in a strict manner, other essential QoS metrics (e.g., delay tolerance and guaranteed data rate) should also be considered in the scheduling discipline design.

WFQ-based scheduling was motivated and developed in the SDMB system based on the well-known WFQ scheme (Demers, Keshav & Shenkar, 1990), being capable of guaranteeing a minimum bandwidth per bearer or per set of bearers grouped together for traffic handling purposes. The WFQ-based scheduler is more specifically based on the Virtual Spacing policy that uses the notion of Virtual Time (Zhang, 1991). The weights are primarily set according to the data rates of the multiplexed service flows rather than its priority. The weight distribution amongst flows can be adapted in response to new acceptances of a service or variation of channel mapping. The serving orders of the queues are computed depending on the time-stamp of the head packet of each queue, queues with the lowest time-stamp on its head packet will be served first. The non-priority nature of this scheduling policy leads to unacceptably long queuing delays in higher priority queues. The performance of WFQ is worse than that of MLPQ in terms of both delay and delay variation. Although MLPQ and WFQ have advantages in computational and implementation complexity, however, both of these schemes feature major weaknesses in QoS-differentiated multimedia services provisioning with respect to both efficiency and fairness.

4.3 Problems and Challenges

In satellite environments, the signal strength may strongly vary over time, making it necessary to perform instantaneous adaptations on the resource management functions in response to the dynamic network requirements. Given the precious usable resources at the satellite transponder, it is noted that the total available power for all the multiplexed physical channels within a satellite beam is strictly limited. Therefore, an efficient design of a packet scheduler replies on the intelligent usage of available power as well we minimizing the unnecessary power waste. On the other hand, the bandwidth constraints associated with a satellite link pose similar concerns as the power constraints.

Multimedia services are delivered to the end user in a continuous manner, enabling the contents can be displayed even before the termination of the session transmission. To support a variety of heterogeneous multimedia applications, the packet scheduler has to collect the application-specific QoS targets in terms of delay, jitter, throughput, packet loss, and etc, during the session initialization procedure. Consequently, the scheduler is desired to dynamically adjust its scheduling policies, according to the variations between instantaneous performance and their targets, thereby avoiding performance degradation and improving the QoS guarantees.

In satellite communication, the long latency inevitably exists in any satellite-associated link, 400-500 ms delay makes the Sat-GW difficult to effectively track the real-time channel state information (CSI). Therefore, another challenging issue is to develop a feasible solution for utilizing imperfect, i.e., limited or delayed, CSI feedback at the Sat-GW. Moreover, in order to effectively utilize the channel status, it is important to take into account the BC/MC nature of the HSTN system, where each session is expecting multiple simultaneous channel feedbacks from different receipts.

In the conventional unidirectional HSTN system, e.g., SDMB, the scheduler is physically located at the Sat-GW, where the CSI and the end-to-end performance are not obtainable, and the scheduler has to perform the resource allocation based on the queuing dynamics at the Sat-GW. However, in a BC/MC satellite network, the information reflecting the traffic and channel characteristics from the Sat-GW to the users plays a dominant role in determining the overall system performance and final delivered QoS.

The availability of a return link in bidirectional HSTN enables the scheduler to utilize the aforementioned essential user-associated information, such as channel status and network performance, into its scheduling decision. However, the packet scheduling problem becomes far more sophisticated than conventional unidirectional environment in that:

- The satellite is the only path that all the downlink traffic has to go through, where the long latency and the noisy propagation condition are inevitable.
- The feedback information from the user is overdue when received at the Sat-GW even through a terrestrial reverse link, thereby not accurately reflecting instantaneous channel and traffic status.
- Each session is received by multiple dedicated users experiencing different channel impairments; therefore the CSI and end-to-end performance for the session has to be exploited by packet scheduling considering the diverse feedback information from all clients belonging to the service group.
- Users in a BC/MC group span over a wide geographical area via either GEO-Sat or IMRs, it is therefore desired that the packet scheduling is capable of allocating bandwidth amongst users within the same BC/MC group in accordance with their traffic and link status.
- In the HSTN system, a return link from the UE to the Sat-GW is established via terrestrial mobile networks, supporting interactive communications. The presence of a return link provides feasibility for packet scheduler to utilize the user-related information, such as CSI and end-to-end performance into its scheduling decision. However, the packet scheduling problem becomes far more sophisticated than conventional unidirectional environment in that:
- The system features a GEO-Sat link, i.e. there is a substantial delay in forward link between the Sat-GW and UEs; the feedback

information from the UE is overdue when received at the Sat-GW and thereby not accurately reflecting current channel and traffic status. Therefore the scheduler has to exploit this overdue feedback information in a non-conventional manner.

- In point-to-multipoint delivery, each MBMS session is received by multiple dedicated UEs experiencing different channel impairments in a BC/MC service group. The scheduler has to exploit the feedback information from all members in the corresponding BC/MC group and derives an appropriate estimation for each session reflecting the average CSI and end-to-end performance.
- The UEs in a BC/MC group can span over a wide geographical area via either GEO-Sat or IMRs, it is therefore desired that the packet scheduling is capable of allocating bandwidth amongst UEs in the BC/MC group in accordance with their respective traffic and link status.

3. ADVANCED PACKET SCHEDULING SCHEMES

5.1 Proportional-Differentiation Based Schemes

To overcome the inherent deficiencies in MLPQ and WFQ, the challenge to the design of packet scheduling algorithms is to optimally utilize resources and efficiently schedule traffic whilst guaranteeing the prescribed QoS demands given the dynamic channel and network variations. Proportional differentiation based scheme assigns the priority of each queue proportionally based on some specific criteria, e.g., the QoS targets, queuing behaviors and etc.

Existing algorithms in this category like proportional fair (PF) (Jalalim, Padovani, & Pankai, 2000; Pandey, et al, 2002) packet scheduling is applied to wireless communication systems by scheduling the radio resource according to the pre-assigned priority associated to each user. This scheme provides better fairness than Max C/I and better throughput than Round Robin. However, the PF does not necessarily provide a good overall system throughput, e.g., it provides a poor delay profile compared to Max C/I (Abedi, 2005). It is also shown that the PF could provide a fair output for the wireless end-users as time elapses. Representative schemes in satellite networks, e.g., MLPQ and WFQ, perform poorly in terms of both fairness and delay. The main reason is because both of the algorithms based on single metric, i.e., either QoS class or date rate. As such, dynamic factors induced from network components and channels can not be effectively incorporated, which largely limit the efficiency and intelligence of the scheduling function. For this reasons, we suggest to perform proportional-differentiation based scheduling, taking into account diverse aspects from buffer status, queuing dynamics, channel variations, and etc.

Within this framework, novel scheduling schemes have been studied. Firstly, Buffer-Length Related Queuing (BLRQ) is introduced which considers the buffer status into the scheduling, queues with heavy queuing data will be given priority. When a finite length buffer size is assumed, it is essential to maintain a reasonable buffer status to prevent excess packet loss due to buffer overflow. In order to take account of buffer status during the packet scheduling procedure, BLRQ scheme is proposed aimed at balancing all the traffic flows with regard to their respective queue lengths. This approach is designed to reduce the probability of packet loss due to buffer overflow in the case of finite RLC buffer size. Since BLRQ can be regarded as a modified form of MLPQ, it is still a priority scheduling scheme, in which the packets in higher priority queues will be processed first. For those queues having same priority class, the queue with the longest packet queue in its buffer will be served first, instead of adopting the traditional round-robin approach. BLRQ is essentially an enhanced version of MLPQ; the scheduler operates exactly the same for traffic flows featuring different QoS rank. The difference between them is that BLRQ will provide service differentiation between the traffic flows within the same QoS rank.

In each TTI, the packet scheduler will scan all FACH queues and schedule packets accordingly. Firstly, the queue with highest QoS rank is served ahead of those with lower QoS rank. And then, once there is more than one queue within each QoS rank, the FACH queue with the longest queue length in its buffer will be served first. The mathematical presentation can be expressed as:

$$FACH\_selected = MAX\{queue\_length(i)\} \quad (1)$$

where $FACH\_selected$ is the ID of the FACH queue with the longest queue length; $queue\_length(i)$ is the queue length for the ith FACH within the same QoS rank at current TTI slot.

The scheduler will allocate resource according to both the QoS rank and the buffer status of each Forward Access CHannel (FACH) queue on a TTI-scale. In this respect, the differentiation is made available for those FACHs with same QoS rank but featuring discrepancies in their relative queue length, which can arise from the asymmetry of the network resource allocation (i.e. different number of FACHs mapped onto respective S-CCPCH), different traffic mixes under certain scenarios, traffic dynamics for each incoming traffic flow as well as the propagation and interference variations in the satellite system.

By considering the buffer status of individual queue, the BLRQ can effectively improve the queuing performance in terms of both buffer occupancy and packet drop rate. However, the BLRQ can only differentiate queues with same QoS rank according to the buffer length, there are other performance metrics should also be considered.

In order to achieve better packet scheduling performance in terms of both efficiency and fairness, inherited from the proportional delay differentiation (PDD) scheme (Dovrolis et al, 2002) in the context of differentiated service network. It assumes that there are QoS ratios between different QoS priority classes, offering improved performance in delay, jitter, and channel utilization.

In the PDD model, the hybrid delay consists of two separate parts: average queuing delay and head waiting time. The head waiting time is the waiting time of the packet at the head of each class. We modified this algorithm as follows: 1) The waiting time used in our algorithm is the average waiting time of all the packets in the queue of each class instead of the waiting time of the packet at the head of each class; 2) Instead of separating the average queuing delay and head waiting time, both queuing delay and waiting delay have been considered together in our algorithm and have also been assigned to the same weight in order to obtain the overall delay performance.

Delay differentiation queue (DDQ) (Fan, Du, Mudugamuwa, & Evans, 2006) was proposed for the delay differentiation services in a satellite environment, assuming there are QoS ratios between different traffic priority classes. For each resource allocation interval (e.g., TTI), the serving indices are obtained based on the average waiting delay for all packets currently in the queue, the average queuing delay for all the packets having left the queue, the packet arrival rate and QoS ratio. In this scheme, the instantaneous queuing delay is effectively considered for queues with the same QoS rank.

DDQ performs service prioritization dynamically depending on the QoS and the waiting time/queuing delay experienced by packets in each FACH. It assumes that each MBMS session maintains a separate FACH queue and that there are QoS ratios between different QoS priority classes. In each TTI, the serving indices are calculated for each queue. These serving indices are obtained based on the average waiting delay for all the packets currently in the queue, the average queuing delay for all the packets that have left the queue prior to this TTI, and the QoS priority ratio index.

The QoS factor α indicates the QoS priority of the MBMS services. In the SDMB system, there are three different service classes: streaming, hot download and cold download.

The fairness factor δ indicates the fairness among the MBMS services, and is expressed by the average waiting delay for all the packets currently in the queue and the average queuing delay for all the packets that have left the queue before the current TTI. Mathematical formulation of the DDQ can be expressed as follows.

Let $\delta_i(n)$ be the average queuing/waiting delay at current time slot n for each queue i. This measure describes the delay status of all packets passing through the respective queue, including both the packets which are currently in the queue and those packets which have already left the queue (been served). Delay index will be calculated for each queue i in each TTI as:

$$\delta_i(n) = \frac{\sum_{j=o}^{N_q} W_{i,j}^q(n) + \sum_{j=o}^{N_d} W_{i,j}^d(n)}{(N_q + N_d)} \quad (2)$$

where $\delta_i(n)$ is the fairness factor for queue i, N_q is the number of packets that are currently in the queue, $W_{i,j}^q$ is the waiting delay for packet j currently in the queue i, N_d the number of packets that have left the queue before the current TTI, $W_{i,j}^d$ is queuing delay for packet j, which has left the queue i before the current TTI.

Let α_i be the QoS priority factor for the service flow at the FACH queue i; the priority for queue i in TTI n can be defined as:

$$P_i(n) = \alpha_i \times \delta_i(n) \quad (3)$$

where α_i is the QoS class factor, which is essentially a time-independent parameter designated, for each queue i.

Consequently, the serving orders are calculated and assigned to each FACH by (3)**(3)** at the beginning of each TTI. Compared with WFQ, MLPQ and BLRQ, DDQ offers improved performance in delay, jitter, and channel utilization. However, DDQ experiences unbalanced performance among multiple QoS attributes, namely the gain achieved in one performance attribute leads to the performance degradation in other attributes. Furthermore, multimedia services feature differentiated delay constraints and applies the delay constraints for differentiated services in an equal way may lead to poor QoS guarantee for high priority queues. Therefore the delay profile has to be considered against the respective delay constraints (i.e., maximum acceptable delay) specified by the class of service. Finally, rather than scheduling competing flows in a static manner, to provide more flexible QoS provisioning and maintain optimal resource utilization, it is highly desired that the scheduler is capable of choosing the best scheduling policy according to diverse QoS preferences of the services and instantaneous performance dynamics.

In Combined Delay and Rate Differentiation (CDRD) (Du, Fan, & Evans, 2007) a joint judgment function (JJF) is developed to provide a dynamic intelligent scheduling task whilst considering several essential QoS factors that have crucial impact on system performance.

In each TTI, the scheduler will sort the FACH queues according to their priority index calculated from the JJF in descending order. The priority index is essentially determined by the difference between the instantaneous performance and the predefined performance threshold. The FACH queues with higher derived priorities will be served ahead of their lower priority counterparts.

The difference between them is that DDQ only focuses on delay differentiation and does not consider other QoS factors, but CDRD takes into account several other key performance parameters (i.e. required data rate and maximum acceptable delay for the specified service) for assuring satisfaction of service QoS requirements. In brief, CDRD aims to balance all service flows in order to achieve high-grade QoS satisfaction and improve overall system performance.

In the CDRD scheme, each of the QoS parameters is represented as a contributing factor in the JJF. For the service flow at FACH queue *i* at current time slot (i.e. TTI for UMTS) *n*, the JJF is defined as follows:

$$P_i(n) = \alpha_i \times \delta_i(n) \times \lambda_i(n) \times \gamma_i(n) \tag{4}$$

where $P_i(n)$ is the priority index for each queue *i* at current time slot *n*. *n* is the sequence number of the TTI at current time. α_i is QoS class factor. $\delta_i(n)$ is the delay serving index at current time slot *n* for queue *i*. $\lambda_i(n)$ represents the data rate factor for queue *i* at current time slot *n*. It is based on the ratio of the service data rate required against the average transmitted data rate.

The average transmitted data rate $\overline{\lambda_i(n)}$ for queue *i* at time slot *n* can be expressed as follows:

$$\overline{\lambda_i(n)} = \frac{\sum_{k=1}^{N_d} s_{i,k}}{(n-1) \times T_{tti}} \tag{5}$$

where $s_{i,k}$ is the packet size for k th packet in queue *i*; N_d is the number of packets that have left the queue prior to this TTI; T_{tti} is the value of TTI (i.e. 0.08 seconds in our case).

Therefore, the data rate factor $\lambda_i(n)$ is defined as follows:

$$\lambda_i(n) = \frac{\lambda_i^{req}}{\overline{\lambda_i(n)}} \tag{6}$$

where λ_i^{req} is the required/guaranteed data rate specified by the service QoS level. If the average transmitted data rate served by the scheduler is smaller than the required data rate targeted by the specific service, $\lambda_i(n)$ is larger than 1, thus the priority index for this queue is increased for better chance of being served; otherwise, it is smaller than 1, and the priority index is decreased for this over-satisfied queue. This factor is used to fine-tune the priority and leads the transmitted data rate to approach the guaranteed data rate. $\gamma_i(n)$ is the delay constraint factor for queue *i* at current time slot *n*, depending on the maximum queuing delay tolerated by a service. This factor can be expressed as follows:

$$\gamma_i(n) = \begin{cases} 2 \;, & \forall n: \dfrac{\sum_{j=1}^{N_q} W_{i,j}^q(n)}{N_q} \geq W_i^{threshold} \\ 1 \;, & \forall n: \dfrac{\sum_{j=1}^{N_q} W_{i,j}^q(n)}{N_q} < W_i^{threshold} \end{cases} \tag{7}$$

where $W_{i,j}^q(n)$ is the waiting delay for the jth packet currently in the queue *i*; N_q is the number of packets that are currently in queue *i*; $W_i^{threshold}$ is the delay threshold for the service queue *i*.

If the average queuing delay for queue *i* is larger than its delay threshold, the delay constraint factor $\gamma_i(n)$ is set to 2, which doubles the priority of this queue for improved chance to be processed; otherwise, it is set to 1. It is noted that delay threshold can be chosen as an adjustable parameter, which depends on the maximum tolerable delay of the corresponding service. $\gamma_i(n)$ is only in effect when the average queuing delay beyond the designated delay threshold, which provides a more efficient action to be taken for better QoS provisioning amongst differentiated traffic flows.

In each TTI, the scheduler will sort the FACH queues according to their priority index calculated from the JJF in descending order. The FACH queues with higher priorities will be served ahead of their lower priority counterparts.

5.2 Cross-Layer Design

Cross-layer design becomes popular topic in recent years, the methodology introduced for this type of scheduling uses cross-layer information for the scheduling decision. As proposed by Liu (2005), in order to achieve more efficient scheduling for diverse QoS guarantees, the interactive queuing behaviours induced by heterogeneous traffic and the dynamic variation of wireless channels are considered in the scheduler design. One of the most popular schemes in this area is to design an adaptive modulation and coding (AMC) scheme at the physical layer in conjunction with the packet scheduling procedure at the data link layer to guarantee the prescribed QoS and achieve efficient bandwidth utilization simultaneously (Liu et al., 2005). For example, Liu (2005) utilizes the CSI estimated at the receiver, and select the most appropriate modulation-coding pair, which is sent back to the transmitter through a feedback channel for updating the AMC mode. We exploit multimedia QoS requirements in accordance with packet scheduling decisions, the cross-layer scheme considers both the higher layer QoS targets as well as the lower layer dynamic queuing behaviors, aimed at achieving the highest possible degree of efficient resource allocation subject to resource/power constraints. In this context, a Cross-layer Joint Priority Queue (CJPQ) scheme is studied (Du, Fan, & Evans, 2007).

Fig. 4 illustrates the layer/sublayer interactions of the proposed cross-layer packet scheduling scheme. The RRM is mainly handled at the data link layer, which can be further divided into RRC, RLC and MAC sublayers. As seen from the proposed scheme, the cross-layer/sublayer correspondence is set up from both top-down and bottom-up directions to the packet scheduler at the MAC sublayer of the data link layer. Firstly, the MBMS sessions' prescribed QoS demands are retrieved at the RRA at the RRC sublayer of the data link layer at the beginning of each admitted session starts. During the radio bearer configuration, the RRA abstracts the prescribed QoS demands of admitted sessions and passes them to the joint priority function (JPF) as one set of priority criteria. The queuing dynamics in the RLC queuing buffer are monitored and passed to the JPF as another set of priority criteria. Upon receiving multiple performance metrics, the JPF derives the serving orders for competing flows for service prioritization. On the other hand, the resource allocation performs dynamic resource allocation based on the derived priority from the service prioritization and instantaneous data rate information from the rate matching function at the physical layer.

By utilizing the cross-layer correspondence through the layered protocol stacks, the proposed scheme exploits multimedia QoS requirements and seeks to provide better system performance by dynamically adapting to the queuing behaviors of each competing flow. To efficiently schedule wireless resources (such as bandwidth and power) and

Figure 4. Illustration of the layer interactions of the proposed cross-layer packet scheduling scheme

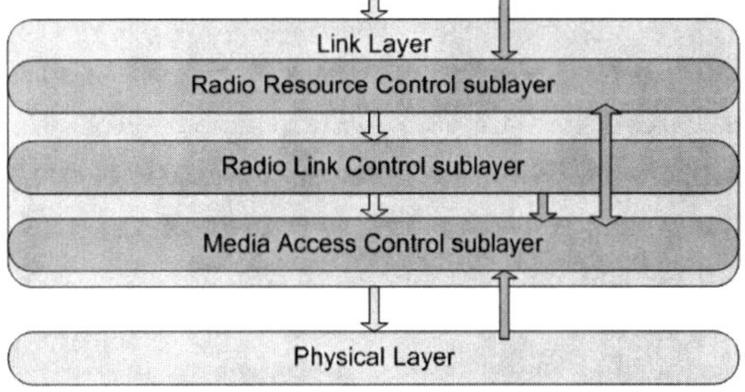

satisfy diverse QoS guarantees, the QoS demand at both application layer and transport layer as well as the instantaneous queuing behaviors induced by heterogeneous multimedia traffics are used in the design of the proposed scheduling scheme.

The contributing factors can be classified into two main sets, according to their frequency of variation. The first set of factors is set at the beginning of each session and kept constant during the session transmission. These factors include service type, required data rate, delay or packet loss constraints, etc. The second set of factors, which determine the average performance behaviors of the competing sessions achieved prior to the current TTI during the transmission, are reset at the beginning of each TTI and kept constant within a TTI. These factors include the average history queuing delay, instantaneous buffer length, etc. For multimedia delivery, the QoS rank of the specific session is the decisive factor in the JPF function, whilst other factors depend on the satisfaction of the above performance metrics by comparing the instantaneous performance metrics with their corresponding targets.

In the duration of each TTI, the JPF value is evaluated and assigned to each competing session. The session with the highest priority evaluated from the JPF function is scheduled first compared with their lower priority counterparts.

5.3 QoS-Based Multidimensional Adaptation

To provide better QoS guarantee whilst achieving more efficient resource utilization, an adaptive multidimensional QoS-based (AMQ) packet scheduling framework is developed for provisioning heterogeneous multimedia services (Du, Fan, & Evans, 2007). By taking into account essential aspects of QoS provisioning whilst preserving the system power/resource constraints, the AMQ packet scheduling scheme is capable of satisfying diverse QoS requirements and adaptively optimizing the resource utilization for satellite multimedia broadcasting. The proposed scheme is implemented by two cooperative algorithms: adaptive service prioritization (ASP) algorithm and adaptive resource allocation (ARA) algorithm. By taking into account multiple essential performance attributes, the former is capable of prioritizing contending sessions based on their QoS preferences and traffic dynamics, whilst the latter performs the resource allocation, in a dynamic and adaptive manner, according to the current QoS satisfaction degree of each session.

Compared with most existing packet scheduling algorithms used in satellite communication systems, the AMQ scheme is distinct in that it is capable of: 1) satisfying multiple essential QoS requirements, 2) adaptively tracking the queuing dynamics induced by heterogeneous traffics, 3) dynamically adapting itself to the most appropriate scheduling policy according to service QoS preferences and instantaneous performance variations, and 4) intelligently allocating the radio resources to contending sessions based on their degree of instantaneous QoS satisfaction.

A novel scheme, namely adaptive service prioritization (ASP) algorithm, is proposed at the MAC layer that considers multiple performance criteria across layers in order to adopt the most appropriate packet scheduling policy in response to diverse QoS demands and traffic dynamics. By taking into account the session's traffic priorities, QoS requirements at both application layer and transport layer, and the queuing dynamics induced by heterogeneous traffic at the RLC layer, the proposed ASP can satisfy multiple essential QoS requirements and provide efficient resource utilization. Moreover, we exploit the desired flexible feature of the ASP in dynamically adapting itself to the most appropriate scheduling policy according to service QoS preferences and instantaneous performance variations.

The traditional resource allocation procedure operates based on the existing static rate matching (SRM) technique, where the allocated data rate is based on the maximum data rate supportable for each physical channel. This strategy can only influence long term resource allocation, whilst the short term physical layer data rate variations can dramatically waste system capacity. Since the rate matching functionality is performed at the physical layer in accordance with other physical layer procedures, cross-layer interactions between physical layer and MAC layer are proven to be capable of obtaining performance gain in resource utilization. A novel dynamic rate matching (DRM) scheme has been proposed in (Du, Fan, Mudugamuwa, & Evans, 2007).The proposed DRM relies on instantaneous data rate instead of maximum data rate used in the SRM. The rate matching ratio is calculated for every TTI and corresponds to the instantaneous data rate of each Transport CHannel (TrCH). Based on the novel DRM technique, resource allocation is desired to be performed in conjunction with DRM for higher utilization efficiency. Therefore, a dynamic DRA scheme was proposed. This new resource allocation algorithm uses DRM to select the required transmission power for all physical channels according to their instantaneous data rate requirements. Therefore, it offers two main advantages: 1) it allows better discontinuous transmission (DTX) minimization, and 2) it requires less power when the instantaneous data rate is lower than the maximum data rate. However, compared with SRM technique, DRM technique involves more processing and memory.

Previously, resource allocation operated separately with the service prioritization procedure, which provides the serving orders upon scheduling the contending traffic sessions on a TTI-by-TTI basis. Based on the instantaneous supportable data rate derived from the DRM functions, the resources are allocated to the selected FACH queue in a strict-priority based manner, i.e., the tentative TF size is checked and assigned, from the maximum supported TF size to zero, in the highest priority FACH queue prior to the lower priority FACH queues. In this case, the high priority queues are always allocated with resources ahead of their low priority counterparts, TFs in lower priority queues are only checked when all the TFs in higher priority queue cannot be granted. In this scheme, high priority queues always obtain a high degree of QoS satisfaction, whilst the lower priority queues can only be allocated with resource at the expense of higher priority queues, which leads to inferior performance in terms of both delay and throughput.

To tackle this challenge, a more adaptive and dynamic resource allocation algorithm, namely adaptive resource allocation (ARA) algorithm is proposed in (Du, 2008). The introduction of this scheme will allow low priority queues to be allocated with more bandwidth by moderately utilizing the resources which should be assigned to those higher priority queues with enough QoS satisfaction. It is noted that to maintain the QoS satisfaction for high priority queues above their required level, only high priority queues with adequate QoS satisfaction performance at the particular resource allocation interval is eligible for sharing their resources with other lower priority queues with unsatisfied QoS performance. For each resource allocation interval, queues with either high-priority unsatisfied QoS or low-priority satisfied QoS are excluded from the adaptive resource sharing mechanism of the ARA algorithm. To this end, the proposed ARA scheme enables the maximum possible resource sharing between diverse QoS traffic classes at the minimum expense of the performance degradations on high QoS traffic classes.

The proposed AMQ scheme takes into account several key performance criteria simultaneously in order to assure comprehensive QoS satisfaction. On one hand, rather than differentiating the competing sessions with respect to their inherent traffic priorities (i.e. service types), the AMQ scheme considers the application prescribed QoS requirements as a combination of multiple attributes. On the other hand, the queuing dynamics of the competing flows at the RLC layer are monitored and evaluated in response to the fast-varying traffic dynamics. The proposed AMQ mechanism operates in the MAC sub-layer of the data link layer within the RRM scheme.

To consider both QoS criteria and queuing behaviors, we introduce an adaptive priority function (APF) for handling the contributing parameters from the aforementioned two modules. The parameters involved can be effectively sub-categorized into two main streams: Static priority factor (SPF) and dynamic priority factor (DPF).

- **SPF** refers to a set of prescribed QoS demands of each service class which is kept constant during the service session transmission. The parameters included in the SPF list are the QoS guarantees expressed in terms of application prescribed QoS rank, required data rate, upper bound on queuing delay/buffer occupancy, and target PLR and throughput.
- **DPF** refers to the performance criteria which keeps track of queuing status of each queue dynamically and updates on a TTI scale. The DPF parameters represent the dynamic queuing behaviour in terms of queuing delay, queue length, PLR and throughput.

Upon receiving the SPF and DPF parameters on either a per-session or a per-TTI scale, the APF function carries out the ranking and priority derivation process and comes up with a quantified priority associated with each FACH queue for the current TTI. The FACH queue with the highest priority traffic flow is served ahead of the other competing flows. The objective of the AMQ scheme can be identified as: to provide the high-level diverse QoS satisfaction among heterogeneous multimedia services, subject to the system resource and power constraints. The prioritized queues are then passed to resource allocation and are allocated the required resources.

To tackle this challenge, it is highly desired that the assignment of TFs can be performed more adaptively taking account of the instantaneous performance demands. We propose an innovative approach, namely adaptive resource allocation (ARA), which is capable of allocating the resource based on the current performance and QoS satisfactions of respective FACH queues.

The proposed methodology can be summarized as: the resource can be shared between high priority queues with over-satisfied QoS performance and those low priority queues with under-satisfied QoS performance, under the constraints that the QoS demands of high priority queues are guaranteed to be met, i.e. the sharing mechanism will not apply to those high priority queues with under-satisfied QoS performance. The amount of the resource to be shared is proportional to the QoS satisfaction factor of the high priority queues, which is derived on a TTI-scale; the better QoS satisfaction the high priority queue has, the more resources that could be shared with other demanding low priority queues.

5.4 CSI- and QSI- Based Schemes

Based on our discussions, channel and queuing status is important and are desired to be effectively tracked during the packet scheduling procedures. Notable existing algorithms in this category for wireless networks are discussed as follows.

To apply packet scheduling to wireless networks, compensation is used for offering differentiated treatments for different channel conditions (Bharghavan et al., 1999), namely, channel-state dependent scheduling. Priority is given to users who experience bad channel conditions during the scheduling decision period. This type of scheduling classifies the wireless channel into two states, namely BAD and GOOD states, representing the error and error-free channel conditions, respectively. One of the most popular models for emulating the channel state transition procedure is Finite-state Markov channel (FSMC) model (Wang & Moayeri, 1995) with specified error probability associated to wireless channel.

Most of the existing literature on wireless

fair queuing algorithms suggested using the well-known wire line fair queuing algorithms for their error-free service model. Representatives of the channel-state dependent scheduling are the Idealised Wireless Fair Queuing (IWFQ) (Lu, Bharghavan, & Srikant, 1998), the Channel Condition Independent Fair Queuing (CIF-Q) (Ng. T. S. E. et al., 1998), the Service Based Fairness Approach (Ramanathan & Agrawal, 1998) and the Wireless Fair Service scheduler. They all apply a compensation model on top of classical wireline queuing algorithms, for example, IWFQ uses WFQ or its variants to compute its error-free service, while the CIF-Q simulates the error-free service by applying a compensation model on top of the Start-time Fair Queuing (STFQ) (Goyal, Vin, & Chen, 1996), which can be regarded as an enhanced variation of WRR.

In Max C/I scheduling (Wong, et al, 2003), the wireless channel quality in terms of the carrier-to-interference ratios (C/I values) is estimated by the receiver and reported back to the transmitter via a feedback channel. A most proper modulation and coding scheme is derived for each user based on the reported C/I and system capacity specifications. Max C/I scheduling technique ranks the mobile users in terms of their respective channel quality, users with the best C/I value have the highest rank and resources are allocated to users according to some predefine criteria. This approach is easy to implement and capable of providing an upper bound on system capacity. However, the performance of Max C/I scheme depends on the distance between mobile users and base station, and the "starvation problem" is more severe for those users near the edge of cell. Therefore, it can be regarded as one of the most unfair schemes for wireless cellular networks.

One of the key difficulties experienced in wireless networks is that a multimedia session can experience location-dependent channel errors, which may have significant impact on the amount of data the session can effectively transmit. Representative contributions in this subject are Channel-condition Independent Fair (CIF) algorithms proposed in (Ng. T. S. E. et al., 1998), where delay and throughput are guaranteed for error-free sessions and both long-term and short-term fairness are considered for error sessions. A token bank fair queuing (TBFQ) scheduling is proposed in (Wong, et al, 2003) for broadband point-to-multipoint WLAN, considering both throughput and fairness under location-dependent channel error conditions.

Notably, the packet scheduling algorithms in the previous subsections are confined to the baseline unidirectional system, where return link is not envisaged and the scheduler located in the satellite gateway (Sat-GW) has to perform the fast resource allocation task without the knowledge of the state of individual channels, i.e. channel state information (CSI) dependent scheduling is not possible. Although the aforementioned packet scheduling schemes are capable of improving the performance in the Sat-GW, the lack of interaction between the user and the Sat-GW largely limits the efficiency and effectiveness of the resource management functions, and leads to inferior performance on the end-to-end behavior and QoS guarantee. As an attempt on investigating the packet scheduling in SDMB with a return channel via terrestrial mobile network, based on the concept proposed for wireless network and the research findings from our previous work, we address the major problems encountered in the current packet scheduling framework in unidirectional system. We propose a feasible solution, namely Proportional Channel-aware Packet Scheduling (PCPS), to exploit the performance gain obtainable on the packet scheduling from the establishment of a return link. The novelties of this scheme can be identified as: 1) considers the end-to-end QoS guarantees of multimedia services, 2) tracks the traffic dynamics in the queuing buffer of the Sat-GW, and 3) exploits the CSI associated with each mobile user within a broadcast/multicast (BC/MC) group, aimed at achieving the highest possible degree of efficient resource allocation in response to fast varying traffic and link status and subject to physical layer resource/power constraints.

Fig. 5 illustrates a simplified end-to-end architecture for implementing the PCPS framework in the SDMB system, where a single MBMS session is demonstrated for transmitting over multiple recipients in a BC/MC group. A channel-aware service differentiation (CSD) mechanism is introduced to compute the priority of each MBMS session, based on the following criteria:

The prescribed QoS requirements abstracted from the incoming MBMS sessions.

- The queuing dynamics tracked from the queuing buffer.
- The end-to-end performance and the CSI obtained from the terrestrial return channel.

Based on the QoS constraints and the feedback information, the CSD module performs priority-based service prioritization. It is noted that a Channel Estimation model is introduced inside the UEs, being responsible for estimating the received performance metrics and generating the feedback reports. The judgment and evaluation of the overall reception condition of a BC/MC group is performed at the Sat-GW, via CSD, based on the received feedback reports from all its prescribed group members.

The availability of feedback reports from the UEs to Sat-GW enables the dynamic adaptation on the packet scheduling mechanism at the transmitter, in response to the heterogeneity and variations induced from both the terminal and network domains. The feedback report is generated by the Channel Estimation at the individual UE, including the network performance and channel conditions. Upon receiving this feedback report, the CSD is able to evaluate the current reception status of the corresponding BC/MC group and perform the most appropriate differentiation mechanism based on the prescribed QoS constraint of each video session. In a BC/MC scenario, all

Figure 5. Simplified framework of the PCPS scheme over SDMB

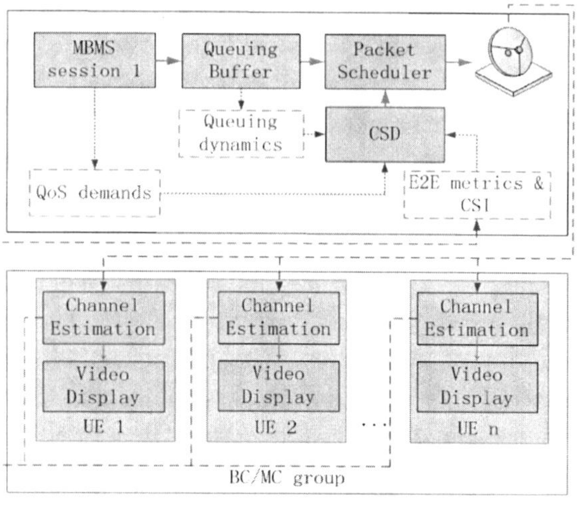

users within a BC/MC group receive the same content from a single video session; thereby the CSD can only perform the service prioritization based on an estimated overall status of both the user and the network.

To differentiate and schedule the multiplexed sessions with diverse QoS, queuing and link status, a proportional priority index (PPI) is defined and applied for each admitted session and updated dynamically in each TTI depending on multiple criteria. The instantaneous performance of respective users is defined in the form of a multi-dimensional metric taking into account multiple performance profiles as follows:

- **End-to-end network constraints:** Characterized by throughput, delay and packet loss rate (PLR):
- **Channel state information (CSI):** Characterized by the received signal to interference and noise ratio (SINR):
- **Queuing behaviour:** Queuing delay, buffer occupancy, buffer drop probability.

We assume the feedback report is perfectly generated at the UEs and reliably fed back to the CSD through the mobile network uplink without delay, reporting the current reception condition for respective UEs. Upon receiving the feedback reports from all members of the intended BC/MC group in a dynamic and periodical manner, i.e. in the scale of one TTI, along with the QoS demands and queuing behaviors for each session, the CSD compares the instantaneous performance for each user with its predefined performance thresholds and derives the PPI value for each session.

A greater value of PPI reflects a better performance and reception condition of associated UE. Subsequently, based on the individual performance of each member, the CSD further derives the overall performance index for each session associated with the entire BC/MC group. This measure essentially represents the percentage of UEs whose performance and reception level are above the predefined performance thresholds within the specific BC/MC group.

Based on the estimated PPI value for each session, the CSD performs dynamic network-aware prioritization mechanism accordingly. The priority will be given to those sessions with degraded network performance and channel conditions to ensure a minimum acceptable reception quality and balanced network performance.

The main concept of the CSD mechanism is to achieve the highest possible degree of the channel/bandwidth utilization and QoS satisfaction. In brief, the priority of each session is jointly determined by the QoS requirements, queuing behaviors and overall network/channel conditions of all users within the BC/MC group. By employing the CSD scheme, the scheduler is able to effectively scheduling and managing the limited radio resources under the system resource/power constraints.

5.5 Hierarchical Packet Scheduling

In (Du, Evans & Chlamtac, in press), a Hierarchical Packet Scheduling (HPS) scheme is proposed to adapt the satellite hierarchical topology. The HPS scheme employs an adaptive mechanism over essential end-to-end performance metrics:

- **Delay:** both queuing delay and end-to-end delay;
- **Packet loss over the link**, rather than the buffer overflow;
- **End-to-end throughput,** rather than buffer throughput.

The novelties of HPS can be identified as: 1) introduces a hierarchical packet scheduling framework for the satellite broadcasting network, 2) considers the service prescribed QoS demands as a multi-dimensional profile considering essential performance metrics, 3) tracks the traffic dynamics and channel variations from different parts of the network (i.e. Sat-GW, GEO-Sat, IMRs and UEs), and 4) performs adaptive resource allocation based on the current QoS satisfaction of each session. The goal is to perform efficient packet scheduling at multiple stages of the forward link, in response to fast varying traffic and link status at each stage and subject to service QoS guarantees and physical layer resource/power constraints.

In the hierarchical packet scheduling, the scheduling task is performed at multiple stages explicitly as:

- **Stage 1:** Adaptive packet scheduling (APS) at the Sat-GW- Performs the service prioritization and resource allocation procedures based on the traffic and link variations of the broadcast channels, while considering the buffer dynamics at the Sat-GW.
- **Stage 2:** Adaptive bandwidth allocation (ABA) on board of GEO-Sat- Allocates bandwidth amongst both users and IMRs based on their performance and link variations, while considering the queuing behaviours at the GEO-Sat.
- **Stage 3:** Adaptive bandwidth allocation (ABA) at the IMR- Allocates bandwidth amongst users within an IMR cell based on their performance and link variations, while considering the network status at the IMR.

To investigate the proposed HPS scheme in SDMB with return link, we employ the classical Finite-State Markov Chain (FSMC) (Wang & Moayeri, 1995; Zhang & Kassam, 1999) channel model to emulate the channel conditions, where slow fading is assumed in presence; thereby the state transition happens only between adjacent states, i.e., the probabilities of transitions exceeding two consecutive states are set zero. We assume the user only receives one session at a single TTI from either direct access via GEO-Sat or indirect access via IMRs.

Each session is assumed to retain an individual queue in the buffers at Sat-GW, GEO-Sat and IMRs before passing to the scheduler, where the queues are prioritized and allocated with resource/bandwidth according to their respective performance metrics. It is noted that the APS at the Sat-GW plays a dominant role in guaranteeing the overall performance and final delivered QoS.

During the radio bearer configuration, the radio resource allocation (RRA) abstracts the prescribed QoS demands from each session and passes them to the APS as the first set of criteria, namely, the service profile (SP). The queuing dynamics at the Sat-GW are tracked as the second set of criteria, i.e., the network profile (NP). Finally, and most importantly, the end-to-end metrics and CSI collected from the return link are defined as user profile (UP), which are effectively tracked when scheduling. We introduce a hybrid computing unit (HCU), to handle the above priority criteria and derive the priority index (PI) and QoS index (QI) in each TTI. The Forward Access CHannels (FACHs) are carried by S-CCPCH via transport channel multiplexing at the physical layer.

A bidirectional interactive satellite multimedia broadcasting typically consists of a satellite gateway (Sat-GW), a geostationary satellite (GEO-Sat), one or more terrestrial gap-fillers, i.e., intermediate module repeaters (IMRs), and a wide variety of users terminals (UEs) with different bandwidth/power constraints and fast-varying channel conditions. Given the severe channel conditions associated with the satellite links, the system employs the Forward Link (FL) via either GEO-Sat (FL-GS) or IMRs (FL-IMR), whilst the interactive activities are maintained by the Return Link (RL) via either Terrestrial Network (RL-TN) or GEO-Sat (RL-GS), depending on the instantaneous signal-to-noise-ratio (SNR) measured at the respective UEs. Due to the major discrepancies induced between GEO-Sat associated links and terrestrial link in terms of bandwidth, latency and packet loss, the preferred interactive communication link is defined as FL-GS with a RL-TN, nevertheless, in the presence of blocking or fading with the preferred links, the interactive activities will be maintained via other available links in order to adapt to the fast-varying channel conditions whilst securing an extensive geographical coverage. For instance, a UE will be set to RL-GS mode only when the received signal strength from the IMRs $SS_{i,j}^{IA}(n)$ is lower than the reception threshold $SS_{i,j}^{min}$, which is essentially determined by the terminal bandwidth/power constraints, targeting at guaranteeing the minimum acceptable displaying quality for the respective UEs. The hybrid architecture defines a hierarchical star network where the available bandwidth from the GEO-Sat is distributed amongst the underlying nodes, i.e., the IMRs and the UEs.

The scheduling area is associated with a single spot beam from a single Sat-GW, where the resource is allocated to the UEs in the area according to session QoS demands, and respective user conditions.

- **Two types of the reception signals can be identified at the UEs:** Signal from GEO-Sat via direct access (SSDA); signal from IMRs via indirect access (SSIA). Therefore, three types of receiver reception conditions can be identified as:
- **Type A:** $SS_{i,j}^{DA}(n) > SS_{i,j}^{min}$ & $SS_{i,j}^{IA}(n) < SS_{i,j}^{min}$: an example of this scenario can be remote users in far-flung geographical locations without the access of any terrestrial

wired/wireless infrastructures. In this case, the RL-GS is the only available feedback link.

- **Type B:** $SS_{i,j}^{DA}(n) > SS_{i,j}^{min}$ & $SS_{i,j}^{IA}(n) > SS_{i,j}^{min}$: this scenario applies to the users in unban/build-up area, where the UEs have excellent access to signals from both GEO-Sat and IMRs. It is noted that, the FL-GS and RL-TN are the default communication links; however, due to the fast vibrating nature of the wireless channel, the interactive communication will be maintained via FL-IMR and RL-GS if the default link is temporarily blocked or unavailable.
- **Type C:** $SS_{i,j}^{DA}(n) < SS_{i,j}^{min}$ & $SS_{i,j}^{IA}(n) > SS_{i,j}^{min}$: this scenario typically applies to the indoor/in-building users, where the satellite signal is currently blocked; the FL-IMR and RL-TN will be the only link available to maintain the interactive activities.

Based on the above discussions, the recipient member group associated with a typical broadcast scenario is formed as a combination of UEs with the reception conditions of Type A, B and C. While the recipient member group associated with a multicast scenario is formed as any possible subsets of a combination of UEs with the reception condition of Type A, B and C. With this heterogeneity considered in possible scenarios, the packet scheduling problem becomes a challenging task.

The reception evaluation process is illustrated in Fig. 6. As an essential part of the proposed integrated packet scheduling framework, the UE performs the measurements and the evaluations on the received channel quality, and generates a "reception status table (RST)", which includes the following user-associated performance metrics:

- The instantaneous SNR
- End-to-end delay and delay variation (jitter)
- End-to-end packet loss rate (PLR)
- End-to-end throughput

To effectively managing the radio resources and maximize the channel capacity. A novel return link adaptation (RLA) scheme is developed to adapt the scheduling policies in accordance with the diverse characteristics arisen from different return paths.

In heterogeneous satellite channel environments, the propagation channels associated with a satellite return link and a terrestrial return link feature major discrepancies.

- **RL-TN:** Low delay, low PLR, high SNR
- **RL-GS:** High delay, high PLR, low SNR

It is therefore desired that the reception conditions can be estimated at the receivers in a unified way for different return links associated with respective geographically dispersed users. A unified reception estimation (URE) is defined to effectively retrieve the reception conditions for a UE, based on the measurements on both SSDA and SSIA.

To increase the reliability and scalability of the overall scheduling performance, we propose to perform an intermediate evaluation at the IMRs, which conducts the measurements and assessments on all the BC/MC members in its IMR cell, and then reports the overall status of the respective IMR cell to the Sat-GW. The URE applies differentiated treatments on the UEs accordingly, based on whether RL-TN or RL-GS are used for the channel feedback.

4. DISCUSSIONS AND PERFORMANCE ANALYSIS

6.1 Flexibility

In the above context, we assume that all the contributing profiles behave and influence the

Figure 6. The effective reception evaluation process

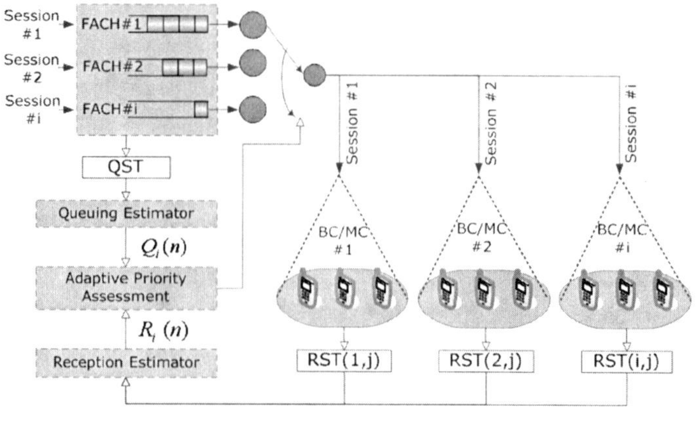

service prioritization in an equal way during the session transmission. However, fixed settings upon all performance criteria may not work well in provisioning multimedia data with different QoS demands and fast-varying traffic dynamics. The performance gain achieved in one profile may sacrifice other profiles, which may be even more important for the specific service. To offer more flexibility and enhance the system performance, tuning mechanism over essential performance profiles may be performed to further optimize the scheduling performance. By observing the QoS preferences specified by the service and the behaviors of queuing dynamics, the "tuning knobs" can be dynamically adjusted on a TTI-scale, e.g., queuing delay threshold, PLR threshold, throughput threshold and etc. By selecting an appropriate combination of the above threshold parameters for each FACH queue, the serving orders of competing flows can be effectively managed. According to

Figure 7. Comparison between different scheduling schemes

Scheme identifier	Mechanisms	Performance metrics	Cross-layer profiles	Incurred complexity
RR	Applies no QoS differentiation amongst queues, queues are served in turn.	N/A	N/A	O(1)
WFQ	Priority is set based on the guaranteed data rate, queues with the same rate are served based on the timestamp of the head packet	data rate	N/A	O(n)
MLPQ	Queues are served from the highest QoS class to the lowest QoS class, queues with the same QoS type are served in Round-robin fasion.	QoS class	N/A	O(n)
DDQ	The priority of queues are calcualted in porportion to QoS class and delay status.	QoS class, queuing delay	intralayer: RLC sublayer	O(n^2)
BLRQ	An enhanced version of MLPQ, queues with same QoS are served based on their instantaneous queue length	QoS class, queue length	intralayer: RLC sublayer	O(n^2)
CDRD	The priority of queues are determoned based on QoS class, queuing delay and data rate.	QoS class, QoS requesite, queuing behaviors	intralayer: RLC sublayer; interlayer: application layer	O(n^2)
CJPQ	The scheduling decision is made via utilizing interactive information sharing amongst layers	QoS class, QoS requesite, queuing behaviors	intralayer: RLC sublayer; interlayer: application layer, physical layer	O(n^2)
AMQ	The scheduler prioritizes traffic flows adaptively based on queuing vairation, QoS satisfaction,	QoS class, QoS requesite, queuing behaviors	intralayer: RLC sublayer; interlayer: application layer, physical layer	O(n^2)
PCPS	Prioritize and allocate radio resources in response to the end-to-end behaviors, queuing dynamics and channel variations associated with each user.	QoS class, QoS requesite, queuing behaviors, end-to-end performance, channel state information	intralayer: RLC sublayer; interlayer: application layer, physical layer	O(n^2)
HPS	The packet scheduling is performed in different part of the network, considering multiple performance metrics, including channel variations, queuing status, etc.	QoS class, QoS requesite, queuing behaviors, end-to-end performance, channel state information	intralayer: RLC sublayer; interlayer: application layer, physical layer	O(n^2)

Figure 8. Comparison between different scheduling schemes

return channel	priority	PD-based	CL-based	QS-based	CSI-based	layers/sub-layers	complexity	flexibility	scalability	robustness	fairness
RR: Applies no QoS differentiation amongst queues, queues are served in turn.											
no	static	X	X	X	X	MAC	O(1)	low	high	low	low
WFQ: Priority is set based on the guaranteed data rate, queues with the same rate are served based on the timestamp of the head packet											
no	static	X	X	X	X	MAC	$O(n^2)$	low	high	low	low
MLPQ: Queues are served from the highest QoS to the lowest QoS, queues with the same QoS type are served in Round-robin fasion.											
no	static	X	X	X	X	MAC	O(n)	low	high	low	high
BLRQ: The priority of queues are calcualted in porportion to QoS class and delay status.											
no	dynamic	√	X	√	X	RLC, MAC	$O(n^2)$	medium	medium	medium	medium
DDQ: An enhanced version of MLPQ, queues with same QoS are served based on their instantaneous queue length											
no	dynamic	√	X	√	X	RLC, MAC	$O(n^2)$	medium	medium	medium	medium
CDRD: The priority of queues are determoned based on QoS class, queuing delay and data rate.											
no	dynamic	√	√	√	X	APP, RLC, MAC	$O(n^2)$	medium	medium	medium	medium
CJPQ: The scheduling decision is made via utilizing interactive information sharing amongst layers											
no	dynamic	√	√	√	X	APP,RRC,RLC,MAC,PHY	$O(n^2)$	high	medium	medium	high
AMQ: prioritizes traffic flows adaptively based on queuing vairation, QoS satisfaction,											
no	dynamic	√	√	√	X	APP,RRC,RLC,MAC,PHY	$O(n^2)$	high	low	high	high
PCPS: in response to the end-to-end behaviors, queuing dynamics and channel variations associated with each user.											
TN	dynamic	√	√	√	√	APP,RLC,MAC,PHY	$O(n^2)$	medium	medium	high	medium
HPS: performed in different part of the network, considering multiple performance metrics, including channel variations, queuing status, etc.											
TN/GS	dynamic	√	√	√	√	APP,RRC,RLC,MAC,PHY	$O(n^2)$	high	medium	high	high

the sensitivity preferences of differentiated QoS traffic classes, through giving flexible importance to different profiles in terms of delay, PLR and throughput, it is therefore possible to adaptively select the best possible scheduling policy to allow for different treatments of diverse QoS demands and maintain optimal resource utilization. For example, the delay threshold is preferred to be set higher for delay-tolerant PLR-sensitive service, whilst preserving a target PLR and throughput. Some applications have stringent constraints on the throughput rather than PLR, thus the scheduler should apply a higher throughput whilst releasing the constraints on other profiles.

6.2 Scalability

From the implementation point of view, the complicated packet scheduling schemes may introduce extra computational complexity due to theirs nonlinear (with loop iterations for selection sort operations) and nondeterministic (with unpredictable variables) nature. In order to examine the scalability of the packet scheduling schemes, the *Big O notation* is employed for determining the involved computational complexity (Homer & Selman, 2000). It is assumed that there are *n* sessions to be transmitted to UEs in a number of multicast groups, located within multiple sectors of a satellite beam. We consider the computational complexity for a scheduling algorithm during one TTI period, with all the tunable thresholds already assigned for the current TTI. Derived from the worst case scenario, where the processing time is the most expensive among all possible scenarios, with the input size of n (i.e., total number of TrCHs), the involved computational time complexity (running time) required for RR and MLPQ are derived as $O(1)$ and $O(n)$ respectively, whilst the other schemes require similar computational complexity of $O(n^2)$, featuring typical quadratic statistics. Although the proposed schemes added more performance metrics and optimization mechanisms, their computational complexity remain the same, since only maximum complexity amongst all procedures is considered effective. This is because only sequence statements are involved, the deterministic factor cause quadratic level is the iterative loop for the selection sort process for different FACHs.

6.3 Performance Analysis

This section discuss the performance trade-offs in different scheduling schemes. Different priority decision functions may lead to unbalanced

Table 1. Radio bearer mapping configuration (kb/s)

S-CCPCH id	**1**			**2**		**3**
S-CCPCH bit rate	384			384		384
FACH id	1	2	3	4	5	6
Streaming	-	256	64	256	128	-
Hot Download	64	-	-	-	-	-
Cold Download	-	-	-	-	-	384

performance amongst competing queues. Human perception is highly intolerant of short-term delay variation, it is therefore paramount that the jitter is reduced as lower level as possible. For UMTS streaming QoS class, delay-variation of the end-to-end flow shall be limited in order to preserve the time relations (variation) between information entities (i.e. packets) of the stream (3GPP, 2007). Consequently, the unidirectional streaming service is quite sensitive to delay-variation, but less sensitive to delay itself, this result proves that the proposed packet scheduling provides a way to balance all queues in order to get minimum delay variation for streaming services. Although delay variation (jitter) is the most critical factor influencing the overall reception performance of the streaming service, delay, especially the queuing delay, must be controlled carefully to avoid packet drop. Furthermore, it is generally agreed that background applications do not have strict delay constraint, and the only requirement for applications in this category is that information should be delivered to the user error free. In fact, background applications still need a delay constraint as there will always be an upper limit for any service category to remain the service practical usable.

The performance of our proposed scheme was evaluated via simulations over a wide variety of traffic mix scenarios. In these scenarios, we consider individual MBMS session with diverse QoS profiles in terms of service type, data rate, and QoS constraints for broadcast transmission, each of which is carried by a single FACH queue. Multiple S-CCPCHs are used for carrying heterogeneous multimedia services and the considered radio bearer mapping scenarios are given in Table 1, where heterogeneous traffic types are carried by arbitrate S-CCPCHs. The task of the packet scheduler not only includes the differentiation of the session within a single S-CCPCH, but also embraces the traffic differentiation between FACHs which are carried by different S-CCPCHs.

Table 2 compares the queuing delay performance for different scheduling scheme, we found that each stage of our proposed schemes provides performance improvements against previous one. And HPS achieves the best performance amongst all the schemes. In Table 3we provide the performance comparison on the physical channel S-CCPCH utilization. WFQ achieves the highest utilization on channel with the highest rate, i.e., 384kbps FACH 6, while MLPQ performs better for FACH 1-3, which carry high priority streaming service. It is easy to find that the HPS achieves the most fair utilization amongst FACHs, i.e. significant gain is achievable on channel carrying download services with minor reduction on other channels.

DDQ is quite sensitive to queuing delay status, and thus keep maintaining a balanced queue length for each queue, which will ultimately leads to better buffer status as well as channel utilization. However, queuing status (e.g., delay or queue length) oriented scheme may pose unfairness issues, since flows with over-loaded traffic will block other queues and occupy significant excess bandwidth. More effective solution considers multiple metrics for determining the priority, the bursty traffic normal can be controlled gratefully by gradually release free bandwidth for the excess traffic rather than shock all the other queues.

The channel multiplexing in HSTN system follows 3GPP recommendations for the transport channel and physical channel mapping structure. We consider single-level channel multiplexing, where only multiple common transport channel

Table 2. Comparison of queuing delay for different scheduling schemes (seconds)

Mean queuing delay	FACH 1	FACH 2	FACH 3	FACH 4	FACH 5	FACH 6
RR	0.47	0.42	0.22	0.99	0.39	2.69
WFQ	38.11	55.82	33.49	0.18	61.49	0.21
MLPQ	3.28	0.40	0.24	0.50	0.40	7.58
DDQ	1.80	0.61	0.36	0.78	0.63	3.59
CDRD	1.33	0.49	0.294	0.66	0.46	3.15
CJPQ	0.80	0.50	0.30	0.63	0.47	1.34
AMQ	0.60	0.38	0.22	0.37	0.38	0.98
HPS	0.56	0.37	0.15	0.23	0.34	0.87

(i.e., FACH) are multiplexed simultaneously to a single common physical channel (i.e., S-CCPCH). However, to obtain further performance gain on the channel utilization, 2-level channel multiplexing scheme (Du, Fan, & Evans, 2006) deserves further investigation in line with packet scheduling functions. For instance, multiple logical channels can be multiplexed together to a single transport channel, where the capacity may be shared amongst logical channels within a transport channel.

It is noteworthy that the heterogeneity of traffic mixture in an S-CCPCH can have essential impact on the physical channel utilization. It is shown that the performance gain in the S-CCPCH channel utilization is higher in a more heterogeneous traffic mix scenario than those in other relatively simple scenarios. It therefore proves that the proposed algorithm is capable of offering higher resource utilization score when the traffic mix in the S-CCPCH becomes more heterogeneous.

As aforementioned, the delay threshold is an adjustable parameter upon balancing the system performance. Simulation results prove that more stringent delay threshold leads to better performance for the corresponding QoS traffic class and causes longer delays for the others. It is worth mentioning that the JPF function employs multiple performance criteria for determining the traffic serving priority, which effectively prevents queues with performance degradation on a single profile from gaining extra unnecessary resources. In the case that a flow experiences instantaneous extra burst traffic, the increase on its queue length may lead to its queuing delay out of profile. However, extra incoming traffic can also lead to instantaneous increase on the transmitted data rate, which effectively influences its serving priority and prevents the queue from obtaining unnecessary additional resources.

Besides, it is noticed that the impact of TTI variation on the performance of the proposed scheduling mechanism are two-fold. On one hand, the simulation results proved that a higher TTI setting improves the system performance on

Table 3. Comparison of physical channel utilization for different scheduling schemes

Mean S-CCPCH utilization	FACH 1	FACH 2	FACH 3	FACH 4	FACH 5	FACH 6
WFQ	18.00	18.00	18.00	62.33	62.33	95.33
MLPQ	96.74	96.74	96.74	92.75	92.75	68.05
DDQ	93.70	93.70	93.70	91.45	91.45	74.71
CDRD	93.46	93.46	93.46	91.07	91.07	74.32
CJPQ	92.84	92.84	92.84	90.36	90.36	85.97
AMQ	92.66	92.66	92.66	90.32	90.32	89.63
HPS	92.67	92.45	92.43	90.31	90.29	90.23

throughput and channel utilization. On the other hand, it is found that a lower TTI setting is capable of providing higher sensitivity for capturing the traffic dynamics.

5. CONCLUSION AND FUTURE WORKS

The past decade has witnessed the evolution of multimedia techniques and the growing number multimedia broadcasting standards. A great deal of research efforts has been devoted to seek ways for the efficient allocating resource to mobile users whilst guaranteeing the QoS demands from both user and application. In this chapter, we investigate the packet scheduling optimizations over satellite multimedia broadcasting system. We studied the existing packet scheduling schemes employed in such a system, and discussed their pros and cons with respect to QoS provisioning and fairness. Consequently, we present the state-of-the-art approaches for these optimization issues. These optimizations are categorized into different solutions. Namely, proportional differentiation, cross-layer design, adaptive multi-dimensional QoS-aware, and hierarchical packet scheduling scheme. We analyzed the performance tradeoffs of these proposals, highlighting performance gains achievable on multiple metrics. Scalability and flexibility issues are discussed for advanced scheduling scheme. It is proven that the overall scheduling performance depends not only on the respective scheduling decision algorithms, but also on system configurations, such as traffic mix, channel multiplexing, and TTI variations.

It is concluded that the design of a fair, efficient and adaptive packet scheduling algorithm entails comprehensive and simultaneous considerations upon different performance profiles as well as diverse demands from application, traffic, network and channel conditions. Dynamic adaptation on the scheduling algorithm is highly desired, especially for heterogeneous satellite broadcasting environments. Due to the inherent nature of wireless transmission, satellite communications suffer from strong variations of the received signal power caused by shadowing and multipath fading. Shadowing of the satellite signal is due to obstacles in the propagation path (buildings, trees, bridges, etc). Whereas for multipath, the fading occurs because the satellite signal is received not only via the direct LOS path but also being reflected from objects in the surrounding area (Du, 2007). The difference in propagation distances for the multipath signals may add destructively and lead to deep fades. Unlike its terrestrial counterpart, the design of the scheduling scheme in the satellite environment cannot rely on better utilisation of the instantaneous information reflecting frequent channel variations, since its long propagation delay for a GEO satellite makes it impossible to utilize the channel status from lower layer. Therefore, we suggest a cross-layer approach for utilizing information from higher layers of protocol stack, e.g., application layer and transport layer. One promising solution could be a TCP-driven MAC scheme. The transport layer is in charge of establishing end-to-end network connections and maintaining target transmission quality and reliability. For example, TCP will deem large delays or packet losses as a signal reflecting the wireless channel congestion status and thereby adjust its mechanism accordingly. However, in satellite communication system, large delays or packet loss event occur more frequently than terrestrial case, therefore, appropriate mechanism has to be designed to avoid TCP misunderstanding these indicative signals. MAC protocols play a fundamental role in guaranteeing good performance to higher-layer functions, by managing the arbitration of radio access. In fact, decisions made within the satellite RRM in MAC can significantly impact the end-to-end performance of TCP flows over a satellite network. By investigating the interactions between MAC and TCP functions, system performance is expected to be enhanced for both MAC resource utilization and TCP performance.

Another research challenge foreseen is that of supporting interactive applications over satellite broadcasting network, which can be regarded as a promising solution in delivering future advanced multimedia applications. Previously, the return link was not envisaged in the baseline system and the gateway has to perform the resource allocation without knowledge of CSI for different users. With the growing demand for supporting advanced multimedia applications, it is highly desired that the return link can be exploited in future systems for providing two-way interactive transmissions and supporting a variety of multimedia applications, such as interactive TV/video broadcasting, video/telephone conferences, disaster recovery and emergency broadcasting. This innovative concept of providing interactive services in advanced SDMB system will have major impact on the satellite broadcasting industry. When the return link via the terrestrial/satellite network infrastructure is adopted in the system, reliable transport protocol(TCP/RMTP) based applications, such as FTP (File Transfer Protocol), HTTP (HyperText Transfer Protocol - WWW), TELNET (TELetype NETwork), SMTP (Simple Mail Transfer Protocol) and etc., are expected to be supported. As a follow on from this direction, research can be conducted to investigate both the system infrastructure and the algorithm optimization for an efficient delivery of interactive multimedia content to mobile users with return links, on either terrestrial or satellite components.

Scheduling issues in HSTN are not standardized and remain open for research and industry communities to implement their respective algorithms in accordance with their system, service and economical factors. As both the network and service are increasing in their size and dimension, the design of scheduling algorithm itself becomes complicated and challenging optimization problem considering dynamics and heterogeneities involved, in this chapter, we aim to provide some basic key solutions for moving the research activities forward in the field.

REFERENCES

Abedi, S. (2005). Efficient Radio Resource Management for Wireless Multimedia Communications: A Multidimensional QoS-Based Packet Scheduler. *IEEE Transactions on Wireless Communications*, *4*(6), 2811–2822. doi:10.1109/TWC.2005.858032

Bharghavan, V. (1999). Fair Queuing in Wireless Networks: Issues and Approaches. *IEEE Personal Comm.*, *6*(1), 44–53. doi:10.1109/98.752787

Choi, Y.-J., & Bahk, S. (2006). Delay-Sensitive Packet Scheduling for a Wireless Access Link. *IEEE Transactions on Mobile Computing*, *5*(10), 1374–1383. doi:10.1109/TMC.2006.149

Chuberre, N., et al. (2004). Satellite Digital Multimedia Broadcasting for 3G and beyond 3G systems. In *13th IST Mobile &Wireless Communication Summit 2004*, Lyon, France.

Chuberre, N., et al. (2005). Relative Positioning of the European Satellite Digital Multimedia Broadcast (SDMB) Among Candidate Mobile Broadcast Solutions. In *IST Mobile & Wireless Communications Summit 2005*, Dresden, Germany.

Courville, N., & Bischl, H. (2005). Critical Issues of Onboard Switching in DVB-S/RCS Broadband Satellite Networks. *IEEE Wirelesss Communications*, *12*(5), 28–36. doi:10.1109/MWC.2005.1522101

Dovrolis, C. (2002). Proportional Differentiated Services: Delay Differentiation and Packet Scheduling. *IEEE Transactions on Networking*, *10*(1), 12–26. doi:10.1109/90.986503

Du, H. (2007). *Efficient radio resource management for satellite digital multimedia broadcasting*. Unpub. doctoral dissertation, University of Surrey.

Du, H. (2008). Adaptive QoS-aware Resource Allocation Scheme for Satellite Multimedia Broadcasting. In *Proc. IEEE WoWMoM'08*, Newport Beach, CA, USA, June 23-27.

Du, H., Evans, B. G., & Chlamtac, I. (in press). Hierarchical Packet Scheduling for Satellite Multimedia Broadcasting: An Adaptive QoS-aware Design . *IEEE Wireless Commun. Mag.*

Du, H., Fan, L., & Evans, B. G. (2006). Two-level Channel Multiplexing: A Novel Radio Resource Allocation Strategy for SDMB system. In *Proc. IEEE ICC'06*, Istanbul, Turkey (pp. 4445-4450).

Du, H., Fan, L., & Evans, B. G. (2007a). Adaptive Multi-dimensional QoS-based Packet Scheduling scheme for Multimedia Broadcasting over Geostationary Satellite Networks. In *Proc. IEEE GLOBECOM'07* (pp. 4613-4618), Washington DC, USA, November 26-30.

Du, H., Fan, L., & Evans, B. G. (2007b). Combined Delay and Rate Differentiation Packet Scheduling for Multimedia Content Delivery in Satellite Broadcast/Multicast Systems. In *Proc. IEEE ICC'07*, Glasgow, Scotland, UK (pp. 4444-4449).

Du, H., Fan, L., & Evans, B. G. (2007c). Cross-Layer Joint Priority Packet Scheduling Scheme for Multimedia QoS Provisioning in Satellite Broadcast/Multicast Systems. In *Proc. AIAA ICSSC'07*, Seoul . *Korea & World Affairs*, (April): 10–13.

Du, H., Fan, L., Mudugamuwa, U., & Evans, B. G. (2007). A Cross-Layer Packet Scheduling Scheme for Multimedia Broadcasting via Satellite Digital Multimedia Broadcasting System. *IEEE Communications Magazine*, *45*(8), 94–101. doi:10.1109/MCOM.2007.4290320

ETSI. (1992). *Satellite Earth Stations (SES), Transmit/receive VSATs used for data communications operating in the FSS 11/12/14-GHzbands*. (DE/SES-2002), ETS 300159.

ETSI EN 302 304. (2004). Digital Video Broadcasting: Transmission System for Handheld Terminals (DVB-H).

ETSI TS 102 585 V1.1.1. (2007). *System Specifications for Satellite services to Handheld devices (SH) below 3 GHz.*

EU FP5 IST MoDiS. (n.d.). Retrieved from http://www.ist-modis.org

EU FP6 IST MEASTRO. (n.d.). Retrieved from http://ist-maestro.dyndns.org

Fan, L., Du, H., Mudugamuwa, U., & Evans, B. G. (2006). Novel Radio Resource Management Strategy for Multimedia Content Delivery in SDMB system. In *Proc. AIAA ICSSC'06*, 2006-5476, San Diego, California, USA.

Farserotu, J., & Prasad, R. (2000). A Survey of Future Broadband Multimedia Satellite System, Issues and Trends. *IEEE Communications, 38*(6), 128–133. doi:10.1109/35.846084

GB 20600—2006. (2006). Framing Structure, Channel Coding and Modulation For Digital Television Terrestrial Broadcasting System, Chinese National Standard.

Giambene, G., et al. (2007). *Adaptive Resource Management and Optimization in Satellite Networks: Optimization and Cross-Layer Design.* New York: Springer.

Goyal, P., Vin, H. M., & Chen, H. (1997). Start-time Fair Queuing: A scheduling Algorithm for Integrated Service Packet Switching Networks. *IEEE/ACM Trans. Networking, 5*(5), 690–704. doi:10.1109/90.649569

3GPP. TS 23.107 v7.1.0. (2007). *Quality of Service (QoS) concept and architecture.*

3GPP. TS 25.301, v8.2.0. (2008). *Radio interface protocol architecture.* A. Demers, S. Keshav, S. Shenkar. (1990). *Analysis and Simulation of a Fair Queueing Algorithm, Internet. Res., & . Exper., 1*(1), 3–26.

3GPP. TS 23.246 v8.2.0. (2008). Multimedia Broadcast/Multicast Service, Architecture and functional description (Release 8).

Holma, H., & Toskala, A. (2002). *WCDMA for UMTS: Radio Access For Third Generation mobile Communications* (2nd Ed.). Hoboken, NJ: John Wiley & Sons Ltd.

Homer, S., & Selman, A. (2000). *Computability and Complexity Theory.* New York: Springer.

Huang, V., & Zhuang, W. (2004). QoS-Oriented Packet Scheduling for Wireless CDMA network. *IEEE Transactions on Mobile Computing, 3*(1), 73–85. doi:10.1109/TMC.2004.1261818

Jalalim, A., Padovani, R., & Pankai, R. (2000). Data throughput for CDMA HDR a high efficiency-high data rate personal communication wireless system. *IEEE International Vehicular Technology Conference* (pp. 1854-1858).

Karaliopoulos, M., Henrio, P., Narenthiran, K., Angelou, E., & Evans, B. G. (2004). Packet scheduling for the delivery of multicast and broadcast services over S-UMTS . *International Journal of Satellite Communication and Networking, 22*, 503–532. doi:10.1002/sat.797

Katevenis, M., Sidiropoulos, S., & Courcoubetis, C. (1991). Weighted Round-Dobin Cell Multiplexing in a General Purpose ATM Switch Chip. *IEEE Journal on Selected Areas in Communications, SAC-9*(8), 1265–1279. doi:10.1109/49.105173

Liu, Q. (2005a). Cross-Layer Scheduling with Prescribed QoS Guarantees in Adaptive Wireless Networks . *IEEE Journal on Selected Areas in Communications, 23*(5), 1056–1066. doi:10.1109/JSAC.2005.845430

Liu, Q. (2005b). Queuing with Adaptive Modulation and Coding over Wireless Links: Cross-Layer Analysis and Design. *IEEE Transactions on Wireless Communications, 4*(3), 1142–1153. doi:10.1109/TWC.2005.847005

Liu, Q. (2006). A Cross-Layer Scheduling Algorithm with QoS Support in Wireless Networks. *IEEE Transactions on Vehicular Technology*, *55*(3), 839–847. doi:10.1109/TVT.2006.873832

Lu, S., Bharghavan, V., & Srikant, R. (1998). Fair Scheduling in Wireless Packet Networks. *ACM SIGCOMM.*

Ng, T. S. E., et al. (1998). Packet Fair Queueing Algorithms for Wireless Networks with Location-Dependent Errors. In *IEEE INFOCOM* (pp. 1103-1111).

Pandey, A., et al. (2002). Application of MIMO and proportional fair scheduling to CDMA downlink packet data channels. In *IEEE International Vehicular Technology Conference* (pp. 1046-1050).

Pelton, J. N. (1989). International VSAT Applications and ISDN. *IEEE Communications Magazine*, *27*(5), 60–61.

Ramanathan, P., & Agrawal, P. (1998). Adapting Packet Fair Queuing Algorithms to Wireless Networks. *ACM MOBICOM*(pp. 1-9).

Sallent, O., Pérez-Romero, J., Agusti, R., & Casadevall, F. (2003). Provisioning multimedia wireless networks for better QoS: RRM strategies for 3G W-CDMA. *IEEE Communications Magazine*, *41*(2), 100–106. doi:10.1109/MCOM.2003.1179558

Sandberg, U. (1995). Building direct-to-home television and entertainment networks forEurope using the new digital video broadcast standards. In *International Broadcast Convention IBC* (pp. 175-177).

Song, J., Yang, Z., Yang, L., Gong, K., Pan, C., Wang, J., & Wu, Y. (2007). Technical Review on Chinese Digital Terrestrial Television Broadcasting Standard and Measurement on Some Working Modes. *IEEE Transactions on Broadcasting*, *53*(1), 1–7. doi:10.1109/TBC.2007.891835

Srivastava, V., & Motani, M. (2005). Cross-Layer Design: A Survey and the Road Ahead. *IEEE Communications Magazine*, *43*(12), 112–119. doi:10.1109/MCOM.2005.1561928

TIA-1099. (2006). Forward Link Only Air Interface Specification for Terrestrial Mobile Multimedia Multicast.

Wang, H. S., & Moayeri, N. (1995). Finite-state Markov channel – A useful model for radio communication channels. *IEEE Transactions on Vehicular Technology*, *44*(1), 163–171. doi:10.1109/25.350282

Wang, H. S., & Moayeri, N. (1995). Finite-state Markov channel – a useful model for radio communication channels. *IEEE Transactions on Vehicular Technology*, *44*(1), 163–171. doi:10.1109/25.350282

Wong, W. K., et al. (2003). Scheduling Algorithm in a Point-to-Multipoint Broadband Wireless Access Network. In *IEEE International Vehicular Technology Conference (VTC 2003)*(pp. 1593-1597).

Zhang, L. (1991). VirtualClock: A New Traffic Control Algorithm for Packet Switching Networks. *IEEE . ACM Transactions on Computer Systems*, *9*(2), 101–124. doi:10.1145/103720.103721

Zhang, Q., & Kassam, S. A. (1999). Finite-state Markov model for Reyleigh fading channels. *IEEE Transactions on Communications*, *47*(11), 1688–1692. doi:10.1109/26.803503

Section 3
Mobility

Chapter 11
Quality of Service Issues in Micro Mobility Enabled Wireless Access Networks

A. Dev Pragad
King's College London, UK

Vasilis Friderikos
King's College London, UK

A. Hamid Aghvami
King's College London, UK

ABSTRACT

Provision of Quality of Service (QoS) and Micro Mobility management is imperative to delivering content seamlessly and efficiently to the next generation of IP based mobile networks. Micro mobility management ensures that during handover the disruption caused to the live sessions are kept to a minimum. On the other hand, QoS mechanisms ensure that during a session the required level of service is maintained. Though many micro mobility and QoS mechanisms have been proposed to solve their respective aspects of network operation, they often have interaction with each other and can lead towards network performance degradation. This chapter focuses specifically on the issues of interaction between micro mobility and QoS mechanisms. Special focus is given to the relatively unexplored area of the impact Mobility Agents can have on the wireless access network. Mobility Agents play a central role in providing micro mobility support. However, their presence (location and number) can affect the routing as well as the handover delay. Through an example network this issue is highlighted. Following which an optimization framework is proposed to deploy Mobility Agents optimally within a micro mobility enabled wireless access network to minimise both the routing overhead as well as the handover delay. Results show considerable improvements in comparison to deploying the Mobility Agents arbitrarily.

DOI: 10.4018/978-1-61520-680-3.ch011

1. INTRODUCTION

Quality of Service (QoS) and micro mobility have become significant pillars for the successful deployment of the next generation of wireless IP based mobile communication networks. The next generation of IP based mobile networks such as WIMAX and LTE (advanced) are expected to support a wide array of services and mobile devices requiring strict QoS and micro mobility support. Providing seamless delivery of content to the numerous mobile devices on the move with appropriate QoS has become a prominent challenge to be met. In order to tackle this various IP level QoS mechanisms were proposed to ensure that sessions are provided appropriate QoS guarantees. To tackle the issue of moving Mobile Nodes (MN) Mobile IPv4 (Perkins, 2002) and then Mobile IPv6 (Johnson, Perkins, & Arkko, 2004) were proposed. However, Mobile IP can not be considered as an appropriate solution for optimal handover management i.e. to minimise handover latency. Mobile IP is associated with large handover delays as at each handover a location update needs to be sent to the HA through the Internet leading to large signalling overheads (A. T. Campbell & Gomez-Castellanos, 2000). These delays are predominantly due to the necessary registration signalling to the Home Agent and the establishment of the new tunnel. To counter the large handover latency of Mobile IP, various local mobility or micro mobility solutions were proposed to ensure a seamless handover performance by minimising the packet loss and delay during handovers, especially for time critical applications such as Voice over IP (VoIP). Moreover, micro-mobility can be thought as being inherently a QoS solution to address the degradation caused during Mobile IP handovers.

IP based networks such as the Internet in its original form does not provide any QoS nor Mobility support. As it stands the existing Internet cannot be used to deploy IP based mobile networks. The flexibility as well as other benefits of deploying IP based mobile networks has lead to numerous research activities in developing QoS and mobility mechanisms for the Internet. On the other hand there are strong incentives of mobile wireless networks to move towards IP technology. The most prevailing of them is to capitalize on the success of Internet applications but also to provide a common forwarding and management plane where convergence of the different wireless networks can be built (Wisely, 2009). In that integrated environment, provisioning the mobile Internet with QoS and mobility support will lead to the realization of ubiquitous communications (communication anytime and anywhere). Such a paradigm can bring forth numerous benefits both to end users by allowing them to use transparently the best available network and the network operators by reducing the cost of managing their infrastructure.

This chapter provides an overview of the recent QoS and micro mobility works as well as their interactions between them. In particular, the interaction between micro mobility and routing (QoS and normal routing) are considered. The impact of Mobility Agent (MA) based micro mobility is shown through examples followed by a proposed optimization framework that allows deploying Mobility Agents so that adverse effects of Mobility Agents on routing are minimized. Finally, avenues of future research work are also given towards the end of the chapter.

BACKGROUND

Future mobile access networks are expected to support a variety of mobile devices over IP. Hence, having efficient support of Mobility and QoS are of paramount importance for the successful deployment of IP based access networks. This section explores the major aspects of the QoS and mobility mechanisms and provides a background towards the main contribution of this chapter.

2.1 Overview of Quality of Service

IP level of Quality of Service consists of two branches namely QoS routing and QoS forwarding. Each plays an important role in efficient provisioning of QoS for a given session. QoS routing deals with finding the best available path with suitable QoS requirements for a session while QoS forwarding ensures that once a session starts the level of guaranteed QoS is maintained.

QoS Forwarding Architectures

The IETF proposes two types of internet QoS architectures namely Integrated Services (IntServ) (Braden, Clark, & Shenker, 1994) and Differentiated Services (DiffServ) (Blake et al., 1998) and their combination IntServ over DiffServ (Bernet et al., 2000).

Integrated Services

IntServ architecture is per flow based where every flow is treated independently. It provides individualized QoS for individual sessions. This involves maintaining individual states in each router through which the packet flows. The IntServ is similar to virtual circuit in nature where it reserves resources along the path that meets the required QoS. To achieve this reservation protocols such as the RSVP signalling protocol (Barzilai, Kandlur, Mehra, & Saha, 1998; Braden, Zhang, Berson, Herzog, & Jamin, 1997; Wroclawski, 1997a) is used, which is first transmitted from the transmitter to the receiver creating states in the intermediate routers and on its way from the receiver to the transmitter the actual reservation of resources is done. The IntServ model provides three types of services namely Controlled Load Service (CLS) (Wroclawski, 1997b), Guaranteed Service (GS) (Shenker, Partridge, & Guerin, 1997) and the Best Effort Service.

Differentiated Services

The DiffServ model on the other hand maps multiple flows into few classes of service; it is based on aggregate flows rather than per flow. DiffServ was designed keeping in mind scalability, flexibility and ability to work without any signalling such as RSVP. The packets are marked with DiffServ Code Point (DSCP) at the edge routers then based on this value the intermediate routers process the packet by giving it the appropriate priority. When packets transit between domains, Per-Domain Behaviours and Service Level Agreements (SLA) are used to provide end to end DiffServ QoS. The Per Hop Behaviour (PHB) plays an important role in prioritising the packets. The PHB can be described as a set of rules based on which a router decides how to schedule packets onto the output link. The two main PHB defined by IETF are Expedited Forwarding (EF) (Jacobson, Nichols, & Poduri, 1999; Davie et al., 2002) and Assured Forwarding (AF) (Heinanen, Baker, Weiss, & Wroclawski, 1999). For further information on QoS architectures, readers are directed towards Armitage (2000) and Wang (2001).

QoS Routing

QoS routing deals with primarily finding paths that support the required QoS parameters for a session. The QoS parameters can be bandwidth, delay, jitter, etc. The shortest path routing such as the OSPF (Coltun, Ferguson, Moy, & Lindem, 2008) have many insufficiencies such as inability to provide paths with necessary QoS for various sessions and disability to optimally utilise the network resource. A plethora of QoS routing solutions have been developed over the past decade (Chen & Nahrstedt, 1998; Apostolopoulos et al., 1999; Paul & Raghavan, 2002). One of the challenges of QoS routing is in finding the optimal paths in quick time to be practically implementable. The computations complexity of the QoS routing problem is very high. This has been one of the major focus on research on QoS routing – towards finding quicker and accurate route computation. However, all of the existing work aims to achieve the main objectives of QoS routing. According to Crawley, Nair, Rajagopalan, & Sandick (1998), the three main objectives of QoS routing are:

1. **Dynamic determination of feasible paths:** The QoS routing should be able to obtain a path capable of accommodating the required QoS for the session dynamically.
2. **Optimization of resource usage:** QoS routing can enable optimal utilization of the network resources in comparison to shortest path routing such as OSPF and improve the total network throughput.
3. **Graceful performance degradation:** QoS routing being aware of the network state can handle better under heavy network load in comparison to network state independent routing. This will lead to a more graceful degradation of network performance.

2.2 Overview of Micro Mobility Management

The need for micro mobility arises due to the inefficiency of macro-mobility protocols to provide seamless communication link within access networks. Mobile IP requires binding updates to be sent to the HA when a node moves into a new access point and acquires a new IP address. Therefore, each time the MN attaches itself to a new access point and acquires an IP address it will have to send a location update to the HA, this creates excessive signalling overhead and as a result longer disruption during handovers. In order to counter this negative effect of Mobile IP a profusion of micro mobility protocols have been developed to deal with such problems. However none have reached the point of full standardization (HMIPv6 at experimental stage). The protocols can broadly be classified into two categories (Eardley, Mihailovic, & Suihko, 2000) – Mobility Agent (MA) Architecture Schemes such as Hierarchical MIPv6 (Soliman, Castelluccia, & K. El Malk and, 2005) and Localised Enhanced-Routing Schemes such as Cellular IP (A. Campbell et al., 2000) and HAWAII (Ramjee et al., 2002). In depth evaluation of different types of Micro-mobility protocols are given in (Reinbold & Bonaventure, 2003; A. Campbell et al., 2002).

Localised Enhanced-Routing Schemes

These schemes use protocols which superimpose the normal IP routing by their own forwarding mechanisms within an access network. The per host forwarding schemes are a subset of this class of protocols and has their own forwarding entries at each router thereby superimposing the conventional IP routing. The protocols differ in the method of creating and maintaining the forwarding entries. Once the entries are created the gateway then uses these entries to forward the packets to the MN. Examples of LERS include Cellular IP (A. Campbell et al., 2000), and HAWAII (Ramjee et al., 2002). However, LERS have not been popular in comparison with Mobility Agents based schemes due to the requirement of installing additional per host routing entries within the access network.

Mobility Agents Architecture Schemes

This class of protocols use Agents which utilize tunnels to deliver the packets to the MN. When a MN handovers to a new access point it registers its new address with a mobile agent located within the access network and receives a care of address (CoA) at the MA. The mobile agent then tunnels all the packets addressed to the MN's CoA to the new address of the MN. In this way no signalling needs to go out of the access network. These protocols use IP routing within the access networks.

Hierarchical Mobile IPv6

The Hierarchical Mobile IPv6 (HMIPv6) introduces a Mobile IPv6 node (MA) called the Mobility Anchor Point (MAP) which can be located at any level in a hierarchical topology including Access Routers (AR). This primary function of the MAP is to reduce the signalling outside the local subnet or access network and thereby reduce the large delays which occur in normal Mobile IP handovers.

When a MN moves to a new subnet it obtains the on link Local Care of Address (LCoA) and receives the router advertisements which contains the information about the local MAPs and hence obtains the Regional Care of Address of the MAP domain. Then the MN sends a Local Binding Update to bind the LCoA with the MAP through the RCoA. The RCoA is registered with the HA and the CNs of the MN. All packets to the MN are sent with the RCoA. The MAP receives the packets addressed as RCoA and then tunnels the packets to the LCoA. A bidirectional tunnel is established between the MN and the MAP. All packets from the MN to the CNs are tunneled to the MAP and the MAP sends it to the CN with and all packets to the MN are sent to the MAP which tunnels it to the MN. As the node moves to a new access point it obtains the new LCoA and sends a local binding update to bind this new LCoA with the RCoA. As long as the MN stays within a MAP domain its RCoA doesn't change. This reduces the signalling overhead and the handover delay considerably compared to Mobile IP. Note that the MAP performs the role of MA.

Proxy Mobile IPv6

Proxy Mobile IPv6 (Gundavelli, Leung, Devarapalli, Chowdhury, & Patil, 2008) intends to provide network based mobility support for MNs without the need for direct participation of MNs. PMIPv6 is based on MIPv6 and uses many of the signalling of MIPv6 as well as HA functionalities. The primary difference between MIPv6 and PMIPv6 being that former is MN based while later in network based mobility management solution. The essential entities of PMIPv6 are the Local Mobility Anchor (LMA) and the Mobility Access Gateway (MAG). The LMA is charged with role of HA (and also MA) for the MN within the PMIPv6 domain and manages the MNs binding update state. The MAG functionality is implemented on the AR and is primarily in charge of managing the mobility related signalling on behalf of the MN. It is also responsible for detecting the movement of MNs and carrying out the handover process on behalf of the MN.

When a MN enters a PMIPv6 domain it will first attach itself with the MAG which will authenticate the MN for access to the network. Following which the MAG will be sending a Proxy Binding Update (PBU) to the LMA, upon validation of the binding update the LMA sending a Proxy Binding Acknowledgement (PBA) to the MAG with the home networks prefix option. Based on this information the MN obtains a home network address through an unicast router advertisement from the MAG. Any traffic originating form the MN is sent to the MAG which tunnels the data to the LMA, the LMA in turn sends the packet to the destination (CN). The same holds on the reverse direction, the data sent by the CN to the MN is sent to the LMA which then tunnels to the MAG where the packet is detunnelled and forwarded to the MN. Thus, for the MN it appears it is always located at the home network. When the MN moves, the MAG detects the movement and the new MAG sends a PBU to the LMA to indicate the new location of the MN. In this manner the MN is kept free of any mobility signalling. A thorough description of the PMIPv6 is provided in (Gundavelli et al., 2008).

2.3 Overview of Mobility and QoS Interactions

Mobility solutions such as Mobile IP were proposed to support the movement of IP enabled mobile devices. While, micro mobility solutions such as Proxy Mobile IPv6 and Hierarchical Mobile IPv6 were developed to provide seamless handover support to ongoing sessions. On the other hand, QoS mechanisms such as IntServ, DiffServ were developed to ensure a stable level of Quality of Service is maintained during a session and QoS routing to ensure the path with best available QoS resource is selected for a given session. The Mobility and QoS mechanisms were developed in isolation to address respective requirements. However, independent functioning of mobility and QoS mechanisms might not lead to the optimal performance.

To illustrate the above, let us consider the following scenario. During a handover the micro mobility and QoS mechanism will occur in respective sequence. Therefore the total delay before re-establishment of the session would be the signalling delay of micro mobility plus the signalling delay of the QoS forwarding plane (for example, in the case of IntServ). This can often lead larger delays and the session being dropped thus, invalidating the purpose of the micro mobility solution. The requirement of QoS re-establishing during a handover can significantly affect the desired objective of the micro mobility solution which is to minimise handover latency as much as possible. In order to address this, number of mobility aware QoS solutions (especially in the case of IntServ based networks) were proposed. The majority of the proposed solutions can be categorised into the following: Pre-Reservation Models (Talukdar, Badrinath, & Acharya, 2001; Tseng, Lee, Liu, & Wang, 2003), Dynamic RSVP (Kuo & Ko, 2000), RSVP Mobility Proxy (Paskalis, Kaloxylos, & Zervas, 2001; Paskalis, Kaloxylos, Zervas, & Merakos, 2002, 2003) and Fast Handover Trigger (Fu, Karl, & Kappler, 2002; Shin & Lee, 2004).

From the above discussions it is evident that although aimed at solving different aspects of network operations, both QoS routing and micro mobility protocols influence packet forwarding in the scope domain. Hence, applying different QoS (or even non QoS aware) routing schemes inside mobile network domains calls for an investigation of the cross issues with respect to the deployed micro mobility protocols. Friderikos, Mihailovic, & Aghvami (2004) showed that such cross issues can be so significant that routing decisions between the two mechanisms may contradict resulting in a sudden break of communication between the gateway of the scope domain and the MN. In the next section this impact of agent based micro mobility on the routing is investigated.

3. QOS AND MICRO MOBILITY ISSUES: IMPACT OF MOBILITY AGENTS

As previously mentioned, micro mobility solutions were proposed to minimise QoS disruptions caused to live sessions during handovers. However, the MA based schemes such as PMIPv6 and HMIPv6 can add potential overheads to the network such as tunnelling, processing and routing overhead. Because MA based schemes rely on IP in IP tunnelling to function, this can lead to reduction in the end to end throughput (Pack, Shen, Mark, & Pan, 2007). Not only this, the tunnelling header can consume valuable network bandwidth as it carries no useful information for the end user. The impact of tunnelling overhead on the performance of micro mobility protocol has been widely studied in the literature, the keen readers are directed to Pack & Cho, 2003; Pack, Nam, & Choi, 2004; Pack et al. 2007 for more detailed information. In addition to the above, there would also be an increase in the processing overhead at the MAs as each packet has to be tunnelled to and from the MN. This can add more complexity to the MA routers and potentially increasing the delay under heavy network load.

The location and number of MAs can affect the handover cost and the routing cost. In its simplest form, the routing cost is the number of hops a packet has to flow from the gateway of the access network to the AR to which the MN is associated with. Higher routing costs can increase the end to end content delivery delay. The handover signalling cost involves the number of hops the binding update has to travel to update the new address which is again dictated by the location of the MA. Higher handover costs leads to disruption of seamless content delivery. This section aims to tackle these two important costs with respect to the location of MAs. It becomes imperative that for seamless delivery of content to mobile devices accessing the Internet on the move, efficient management of micro mobility becomes important. Hence, minimising both of these costs will play an important role on the performance of the future mobile networks.

3.1 Interactions Between MA and Handover Cost

The MN handovers can be classified to layer 2 handover (represented by H1 in Figure 1), intra-MA handover (H2), inter-MA handover (H3) and inter-domain handover (H4). In MA micro mobility architecture the significant types of handovers are intra-MA handover and the inter-MA handover. Intra-MA handover deals with the handover that occurs within a MA domain and hence the location update is sent to the serving MA. In the case of inter-MA handovers, the MN handovers from one MA to another MA, this can occur when the MN moves between different MA domains. The BU for inter-MA handover has to be sent to the HA/CN. The MA domain is the number of ARs that a given MA can serve and is limited by the location of the MA. The higher the MA resides in the hierarchy of the topology the more number of ARs it is connected to, hence, it can cover a bigger geographical area. Figure 2 provides the binding update signalling for intra and inter MA handovers.

Intra MA Handovers: In an intra-MA handover the New AR (NAR) and the Previous AR (PAR) are both served by the same MA. This ensures that the mobility of the MN is kept local and the binding update is kept within the MA domain. This forms the fundamental functionality of Agent based Micro Mobility. The cost of the handover

Figure 1. Mobility architecture in future IP mobile networks

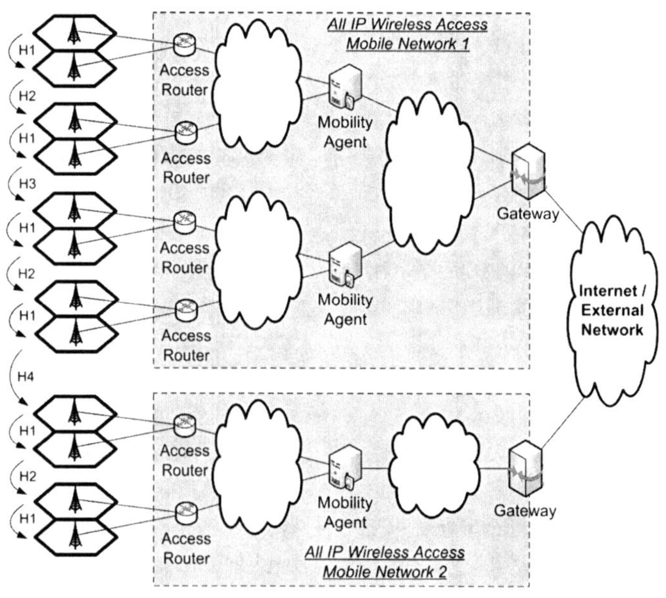

Figure 2. Intra and inter mobility agent handover signalling

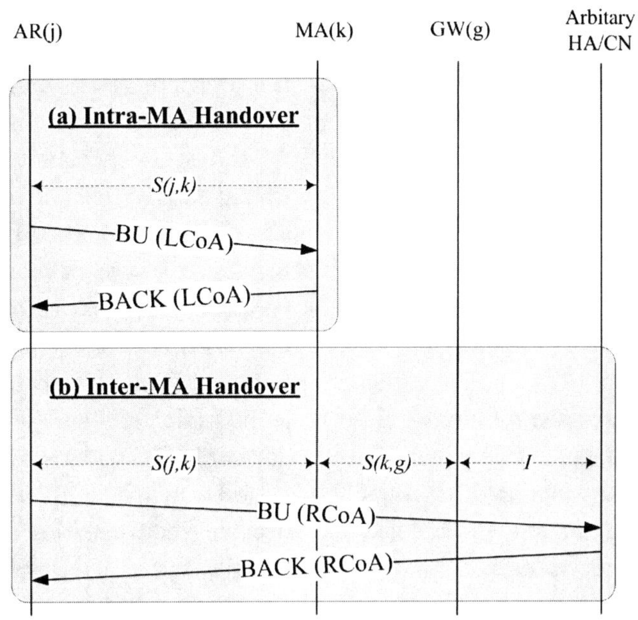

signalling can be defined as the cost of sending the BU from the MN to the MA and receiving a Binding Acknowledgement (BAck) from the MA. As a result, the cost is proportional to the number of hops that exists between the AR and the MA represented by S_{jk} (note: $S_{jk} = S_{kj}$). The closer the MA is to the AR the quicker the binding update can be performed resulting in smaller handover delay. From this argument it can be easily shown that intra-MA handover cost is directly proportional to the distance between AR and the MA, hence, it is directly related to the location of the MA in the topology. Thus the intra-MA handover cost is given by the number of hops taken by the BU from MN to MA (S_{jk})) and for the BAck from MA to MN S_{kj} as,

$$q_{ijk}^{in} = 2 \cdot S_{jk} \tag{1}$$

Inter MA Handovers: The inter-MA handover occur when the MN moves from one MA domain into another. This requires the binding update to be sent to the HA/CN to notify of the change in RCoA and to route the packets to the MN associated with the new MA. This involves the BU travelling over the Internet to reach the HA which can be very costly in certain cases where the HA is very far from the MN. The number of MAs has a direct impact on the inter-MA handover. A single MA supporting the whole network will ensure that no inter-MA handover occurs within that network. If the network is large and deploys more than one MA then inter-MA handover frequency will increase proportional to the number of MAs deployed. In general, the inter-MA handover cost for binding update and acknowledgement can be derived as,

$$q_{ijk}^{out} = 2 \cdot \left(S_{jk} + S_{kg} + I\right) \tag{2}$$

where, I represents the expected number of hops the BU will have to traverse to reach the HA through the Internet and the value of I will be relatively high compared to S_{jg}. The number of hops from the MN's AR to the gateway through the MA is given by $S_{jk} + S_{kg}$. Without loss of generality, it is assumed that the binding update and acknowledgement take the same route, hence the cost is doubled to obtain the total binding update and acknowledgement cost. To eliminate the BU through the Internet, a hierarchy of MAs can be implemented with the first level of MA located at the gateway and second hierarchy located within the network. In this paper we focus on a single hierarchy hence, we include the cost of BU through the Internet (***I***).

3.2 Interactions Between MA and Routing Cost

The location of MA can affect the routing within the network. First explored qualitatively by Friderikos et al. (2004), the presence of MA breaks the routing into two, from gateway to the MA and from the MA to the AR. In the absence of MA the routing would be directly from the gateway to the AR. This restriction on routing and its impact on the capacity of the network were analysed by Pragad, Friderikos, Pangalos & Aghvami (2007). It was shown that the location and the number of MAs affect the routing and can potentially reduce the capacity of the network especially in the case of mesh networks where multiple paths exists between source and destination but are not utilized. The packets going through the MA take more number of hops in comparison to the optimal route without the MA. The routing cost can be modelled as the number of hops the packet has to travel in the access network from the gateway to the MN's AR through the MA. The routing is assumed to take the shortest path from the source to destination and for a given network topology, the routing cost in the presence of the MAs is given as,

$$C_{ik} = S_{gk} + S_{kj} \tag{3}$$

We define C_{ik} to be the total shortest path routing cost from gateway node $g = 1$ to AR i through MA node k. Placing the MA at non optimal location can lead to very high routing cost.

3.3 Modelling of Handover Costs

In this section we derive handover costs to analyse the impact of MA location. The routing overhead cost is straight forward and is given by equation 3. The handover costs can be classified into intra-MA handover cost and inter-MA handover cost. The presence of MAs reduce the frequency of handovers where the binding update is sent to the HA. Thus, the MA is expected to be located in the network such that the frequency of handovers with binding update to the HA is reduce as much as possible. The handover probability between adjacent ARs is given by the H matrix.

In the matrix, h_{ij} is the probability of handover occurring between ARs i and j. The handover probabilities h_{ij} will affect both the intra and inter MA mobility overhead costs. The handover probability matrix can be obtained for a given network from network traces and statistics. To model the cost of each type of handover we derive the following handover costs functions based on the handover matrix H.

Intra-MA Handover Cost: The intra-handover cost is defined as the cost of sending a binding update from the new AR to the MA and for a binding acknowledgement to be sent from the MA to the MN plus some overheads such as processing and the L2 delay costs. Since processing and L2 delay cost are independent of the MA location we can ignore them without affecting the analysis in any manner. Given that a MN is located in AR i, the handover cost experienced by the MN as it moves to one of its neighbouring ARs is modelled as,

$$Q_{ik}^{in} = \sum_{j \in R} h_{ij} \cdot q_{ijk}^{in} \tag{4}$$

Where q_{ijk}^{in} is obtained from equation (1) and R is the set of ARs in the network and J be the set of MAs in the network. The total intra-MA handover cost for the network is obtained by summing over all ARs as follows,

$$Q^{in} = \sum_{k \in J} \sum_{i \in R} Q_{ik}^{in} \quad (5)$$

Inter-MA Handover Cost: The inter-MA handovers occur when a MN moves to a new AR which is served by a new MA, hence, has to initiate a global binding update to the HA to notify the change in MA. This is modelled as follows,

$$Q_{ik}^{out} = \sum_{j \in R} h_{ij} \cdot q_{ijk}^{out} \quad (6)$$

Where q_{ijk}^{out} is obtained from equation (2). The total inter-MA handover cost for the network is obtained by summing over all ARs as,

$$Q^{out} = \sum_{k \in J} \sum_{i \in R} Q_{ik}^{out} \quad (7)$$

3.5 Investigations on Impact of MA Deployment

To investigate the impact of MA location on the handover and routing costs we consider the topology given in Figure 3. It is assumed that for handovers between the base stations of a common AR, the AR acts an anchor point and provides the mobility support either using L2 techniques or alternatively following the MA functionality. This assumption allows us to consider all the four base stations belonging to each AR as a single location and investigate the handovers between the AR regions.

Figure 4 gives the routing cost, inter-MA and intra-MA handover costs for the given network for varying number and location of MAs. Deploying a single MA at the gateway results in the most optimal routing within the network since, the MA doesn't break the packet flow within the network. The number of routing hops per AR (destination) is 2.33 hops (14/6). The inter-MA handover cost is zero due to the lack of any inter-MA handover occurrence in this configuration. However, the intra-MA handover is at its maximum since for each handover the BU has to travel the farthest number of hops in the mobile access network (to the gateway). This configuration though results in optimal routing and inter-MA handover cost, produces the worst intra-MA handover cost which is the primary cost to be minimised in a micro mobility solution.

Having single MAs such as MA at router 3 and 5 leads to higher routing cost (18 and 20 units respectively) while MA at router 5 has lesser intra-MA handover cost (18) over the other (24). The intra-MA cost in comparison with having a single MA at gateway is smaller, hence, it may not be recommended to have the MA at GW as it is the furthest router from the ARs leading to higher

Figure 3. Example mobile network considered for the preliminary investigation

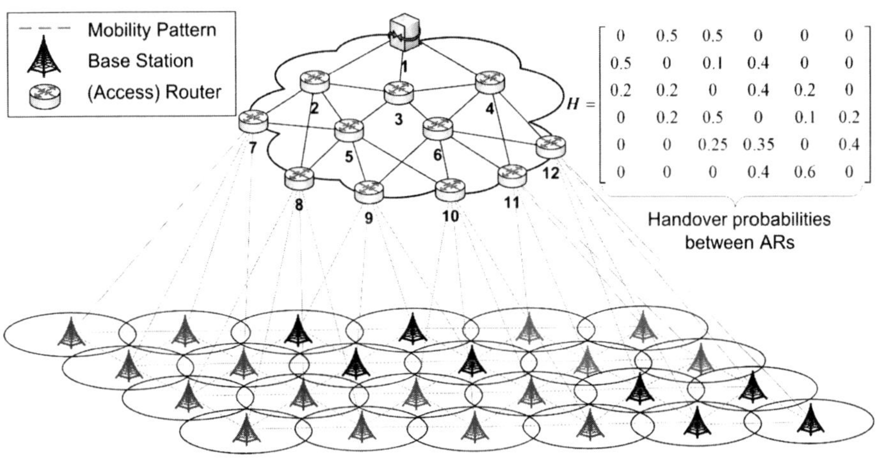

intra-MA handover costs. Moreover, real time applications that require very strict handover support might suffer degradation if the gateway is selected as the MA. It might be more appropriate for such sessions to use a nearer MA such as MA at router 5. The complexity of MA deployment is further illustrated in the case of network implementing two MAs. Deploying MA at routers 5 and 6 has the least intra-MA handover cost (compared to other two MAs location) at the expense of higher routing cost; for MA at router 2 and 4, the costs are inverted with higher intra-HA and least routing cost. Implementing three MAs (at routers 2, 3 and 4) is not ideal for a small network as the one considered here. This is primarily due to the very high inter-MA cost. A point worth nothing here is the effect of the handover probabilities between ARs. If the handover probability between two ARs is particularly high, it is more logical to have these two ARs attached to the same MA. Since the mobility patterns might change with time, it calls for a network management approach where two ARs that have different MAs are assigned to the same AR when the handover probability between them becomes high.

Figure 5 provides a spider/radar graph perspective of the three different costs when the network deploys one, two and three MAs. This figure clearly shows the tradeoffs that are involved in optimising the performance of micro mobility to ensure seamless delivery of content to mobile devices.

4. OPTIMAL DEPLOYMENT OF MOBILITY AGENTS WITH QOS REQUIREMENTS

In this section we formulate the MA location and domain problem as an integer linear program (ILP). Given the number of Mobility Agents (MAs), $\boldsymbol{K}$, that will be deployed, we aim to find the optimal location of the MAs and assign ARs to each of them (MA domain) so that the total routing and mobility handover costs in the access network is minimized. Let the network be modeled as a graph and the set of nodes in the network be given by V. By $\boldsymbol{R}$, and $\boldsymbol{J} = \boldsymbol{V/R}$, we express the set of ARs and potential MA location nodes in the network respectively. Without loss of generality we assume that node $\boldsymbol{g} = \boldsymbol{1} \in \boldsymbol{J}$ is the gateway node. As defined previously, $\boldsymbol{C}_{ik}$ is the total shortest path routing cost from gateway node 1 to AR $\boldsymbol{i}$ through MA node $\boldsymbol{k}$. Assuming that all-pairs shortest path costs (i.e., $\boldsymbol{S}_{ij}$ for all $\boldsymbol{i}, \boldsymbol{j} \in \mathrm{V}$) can be pre-calculated, the routing cost

Figure 4. Numerical investigations illustrating impact of mobility agents on various costs

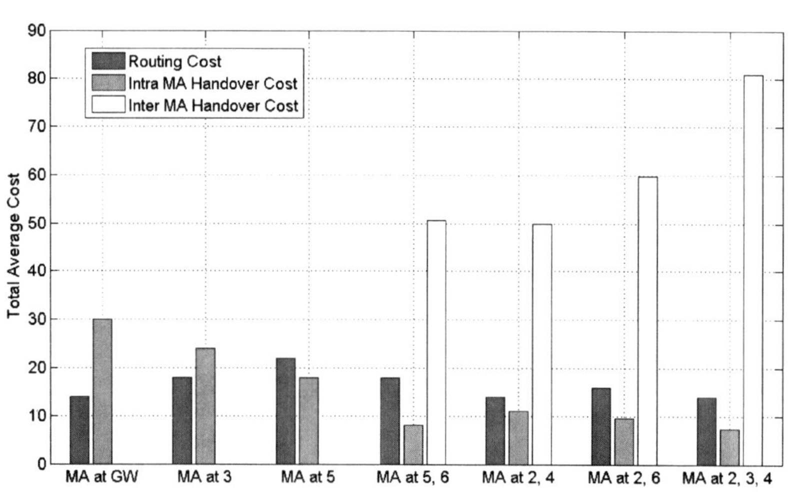

Figure 5. Spider diagram depicting the tradeoffs between various costs

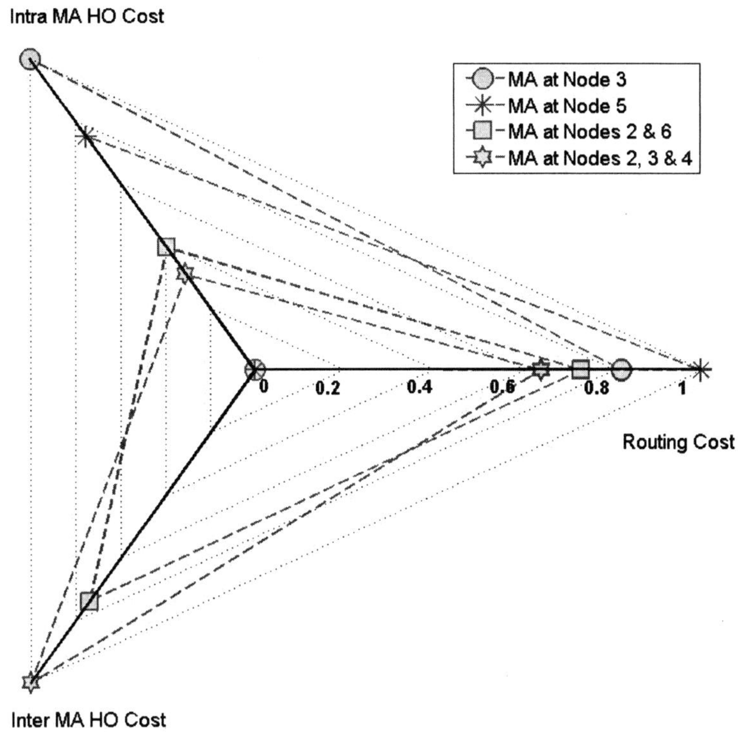

C_{ik} can be calculated according to equation (3). We also define by H the handover probability matrix between ARs, where h_{ij} is the probability of handovers occurring between ARs i and j. In order to capture the overhead due to mobility, i.e., handover between adjacent ARs, we define two distinctive cases: handover between ARs that are assigned to the same MA (intra-MA handover) and handover between ARs that are assigned to two different MAs (inter-MA handover). Based on the previous discussion the intra-MA handover cost between ARs i and j that are connected to the same MA node k is given according to equation (1). On the other hand, the inter-MA handover cost given by equation (2) where I encapsulates the cost for routing through the Internet to reach the HA. The value of I depends on the location of the HA for each of the MNs and is substantially greater than the distance between AR and gateway $I > S_{jg}$. Hence, the significance of the value of I is in capturing the essence that inter-MA handover cost is much higher than the intra-MA handover cost, and sending binding updates to the HA over the internet is undesirable and should be avoided where possible due to the large delays associated with it. Thus, the large value of I should penalize the inter-MA handovers and force the program to minimize the frequency of such handovers.

4.1 Uncapacitated Mobility Agent Location Problem

In order to be able to express the problem in a mathematical programming setting, we define the following boolean decision variables,

$$X_{ik} = \begin{cases} 1 & \text{If AR } i \text{ is assigned with MA } k \\ 0 & \text{Otherwise} \end{cases} \quad (8)$$

$$Y_k = \begin{cases} 1 & \text{If MA is located at node } k \\ 0 & \text{Otherwise} \end{cases} \quad (9)$$

Then the total routing cost can be expressed as follows,

$$\sum_{i\in R}\sum_{k\in J} C_{ik} X_{ik} \quad (10)$$

The intra-MA overhead cost should only be taken into account when the two AR $\boldsymbol{i}$ and $\boldsymbol{j}$ have the same MA node $\boldsymbol{k}$. In other words the total intra-MA handover cost can be written as follows,

$$\sum_{i\in R}\sum_{j\in R}\sum_{k\in J} h_{ij} q_{ijk}^{in} X_{ik} X_{jk} \quad (11)$$

The inter-MA handovers occur when the MN moves into an AR which has a different serving MA compared to the previous AR. Therefore, the inter-MA handover cost can be written as follows,

$$\sum_{i\in R}\sum_{j\in R}\sum_{k\in J} h_{ij} q_{ijk}^{out} X_{jk}(1 - X_{ik}) \quad (12)$$

The above two equations are non linear (Quadratic) in nature and in order to linearize the objective functions, we introduce a new Boolean variable $\boldsymbol{Z_{ijk}}$ such that $\boldsymbol{Z_{ijk}} = 1$ if and only if both $\boldsymbol{X_{ik}}$ and $\boldsymbol{X_{jk}}$ are equal to 1. Now, equation (11) can be rewritten as follows,

$$\sum_{i\in R}\sum_{j\in R}\sum_{k\in J} h_{ij} q_{ijk}^{in} Z_{ijk} \quad (13)$$

Similarly, equation (12) can be linearized to,

$$\sum_{i\in R}\sum_{j\in R}\sum_{k\in J} h_{ij} q_{ijk}^{out} (X_{jk} - Z_{ijk}) \quad (14)$$

Based on the above, the optimal location of MA and AR assignment can be formulated as an integer linear program as follows,

$$\min \sum_{i\in R}\sum_{k\in J}\left\{\sum_{j\in R} h_{ij}\left(\gamma\cdot Q_{ijk}^{in}\cdot Z_{ijk} + \sigma\cdot Q_{ijk}^{out}\cdot(X_{jk} - Z_{ijk})\right) + \delta\cdot C_{ik}\cdot X_{ik}\right\}$$

Subjected to,

$$\sum_{k\in J} X_{ik} = 1 \text{ for all } i \in R \quad (15)$$

$$\sum_{k\in J} Y_k = K \quad (16)$$

$$X_{ik} \le Y_k \text{ for all } i, j \in R, k \in J \quad (17)$$

$$Z_{ijk} \le X_{ik}, X_{jk} \text{ for all } i, j \in R, k \in J \quad (18)$$

$$X_{ik} + X_{jk} - Z_{ijk} \le 1 \text{ for all } i, j \in R, k \in J \quad (19)$$

$$X_{ik}, Y_k, Z_{ijk} \in \{0,1\} \text{ for all } i, j \in R, k \in J \quad (20)$$

The objective of the optimization problem is to minimize the total routing and mobility overhead due to handovers between AR that are assigned to the same or different MAs. Weights are assigned to each of the costs: $\boldsymbol{\gamma \in [0, 1]}$ represents the weight for intra MA handover cost while $\boldsymbol{\sigma \in [0, 1]}$ and $\boldsymbol{\delta \in [0, 1]}$ represents the inter MA handover cost and routing cost respectively. Constraints (15) ensure that each AR will be assigned to single MA only, while constraint (16) ensures that $\boldsymbol{K}$ MAs will be deployed. The binding constraints in (17) ensure that an AR will not be assigned to MA that hasn't been selected. Constraints (18) and (19) ensure that variable $\boldsymbol{Z_{ijk}}$ equals 1 only when both $\boldsymbol{X_{ik}}$ and $\boldsymbol{X_{jk}}$ are equal to 1. Note that this formulation allows the gateway node to be elected as a MA node.

4.2 Capacitated Mobility Agent Location Problem

We extend the previous formulation to take into account capacity constraints for each MA. Assuming, that each candidate MA $\boldsymbol{k}$ for the set of nodes $\boldsymbol{J}$ has capacity $\boldsymbol{W_k}$ on the maximum traffic flow that it can serve. And that the aggregate traffic demand for each AR is given by the vector D= [d1,d2,..., dr], the capacity constraint can be written as follows,

$$\sum_{i\in R} d_i X_{ik} \le W_k \cdot Y_k \text{ for all } k \in J \quad (21)$$

We can also introduce a lower bound of L_k on the aggregate demand a MA k must satisfy. To incorporate this requirement the following constraints need to be added,

$$\sum_{i \in R} d_i X_{ik} \geq L_k \cdot Y_k \text{ for all } k \in J \tag{22}$$

4.3 Numerical Investigation and Analysis

The optimization problem is solved for two Networks (A and B). For Network A the number of potential MA locations are $|J_A| = 6$ and the number of ARs are $|R_A| = 9$; for Network B, $|J_B| = 15$ and $|R_B| = 20$. Network A is the smaller and should provide the least combinatorial complexity; while, Network B is larger with a much larger potential MA node location set $|J_B|$ and ARs set $|R_B|$ which provides a more challenging combinatorial complexity. It should be noted that an AR can support number of Access Points (base stations) which are directly connected to a single AR as shown in Figure 3. As a result, if the AR belongs to a MA then the AP will also belong to that MA. If each AR supports five APs, then Network A and B can support up to 45and 100 APs / BSs respectively.

The formulated optimization program is executed in MATLAB using function *"bintprog"* with and without Tomlab/CPLEX solver for Network A and B, to obtain the optimal location of the MAs to be deployed to minimise handover latency as well and reduce routing overhead costs. The experiments were conducted in a Pentium Dual Core Processor with 2 GHz clock speed. For comparison purposes, a set of arbitrarily chosen MAs are selected and assigned to ARs in the shortest paths between gateway and AR on average. This will ensure that though the locations are selected arbitrarily the routing cost is kept to a minimum as much as possible. For Network A, the program is executed to obtain optimal locations for one, two and three MAs to be deployed. The results are shown in Figure 6. Upon closer examination it is evident that optimal handover cost is slightly higher than the arbitrary handover cost. This is due to the optimization program finding the tradeoffs between handover and routing cost. The routing cost on the other hand is considerably reduced (by almost 40%). When two MAs are to be deployed the program finds the location to be deployed such that the total cost in reduced by almost 22%. For the third case of deploying three MAs, the optimal handover cost is smaller than the arbitrarily selected MAs location. As the number of deployed MAs increases the program can find locations such that both the handover as well and the routing cost are minimised. The reduction in total cost by deploying three MAs optimally is 27%.

Figure 6. Routing and handover costs for optimal vs. arbitrary deployment of mobility agents for network A

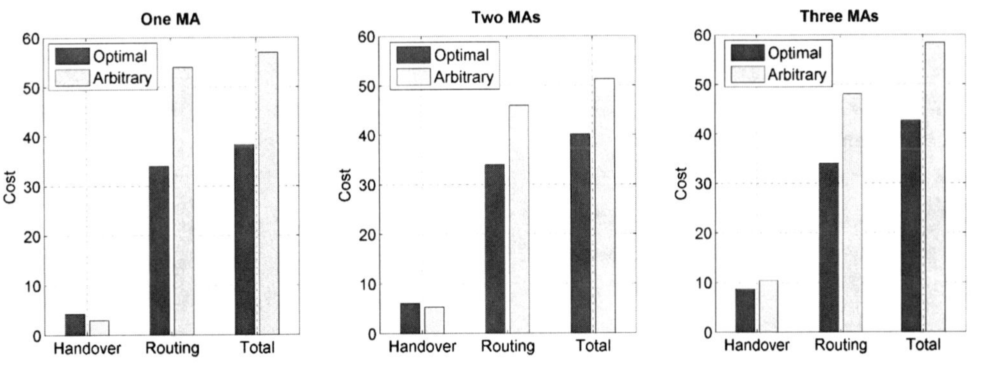

Figure 7. Routing and handover costs for optimal vs. arbitrary deployment of mobility agents for network B

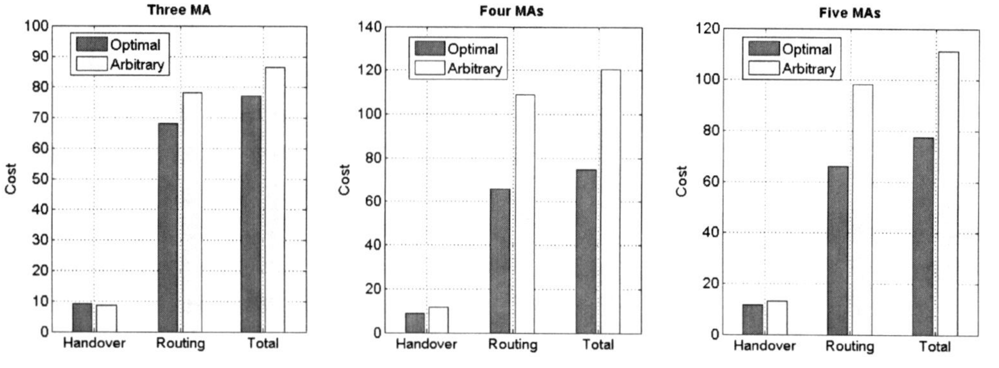

For the larger Network B,, the program is executed to obtain optimal locations for three, four and five MAs to be deployed. The results are shown in Figure 7. Similar pattern is observed here as in Network A for the handover cost. As more number of MAs are deployed the program can find the optimal location to minimise the handover delay in comparison with arbitrary deployment of MAs. For four and five MAs the optimal handover cost is lower than arbitrary handover cost. The reduction in total costs in optimally deploying three, four and five MAs are 12%, 38% and 31% respectively.

The run time complexity is reasonable considering the combinatorial complexity of the family of problems the formulated program belongs to. For Network A and B the approximate run time recorded in MATLAB was 88 and 1955 seconds respectively without Tomlab/CPLEX and 0.5 and 2 seconds respectively with Tomlab/CPLEX. The high runtime of Network B is due to the large possible MA location set (J_B) in comparison with Network A, which increases the combinatorial computational complexity drastically. Though the runtime of Network B appears to be large, it is reasonable enough for network planning and design purposes and falls within the range of runtime for other similar network planning problems (Toril & Wille, 2008). However, the binary integer program is not solvable in polynomial run time. The number of constraints for this program is given as $R + 3R^2K + 1$ and the total number of variables is $R^2K + RK+K$. For very large network two possible approaches can be followed. Where applicable the network can be divided into smaller network partitions and then solved locally for each of the partition. Alternatively, approximation algorithms can be developed to find quicker solutions. Such fast solutions can open up way for a more autonomic method of network management where the assignment of MAs to ARs can change depending on the traffic load, traffic pattern and mobility patterns.

5. FUTURE RESEARCH DIRECTIONS

There exist numerous avenues of future works on this topic. They can be broadly classified as autonomic joint network and mobility management and per flow based mobility management.

5.1 Autonomic Micro-Mobile Network Management

The work presented in this chapter focused on static location of MAs and assignment to AR within a network. However, it might be necessary for the network to be capable of adapting according to various network conditions. When the load over

Figure 8. Per flow based approach towards mobility management

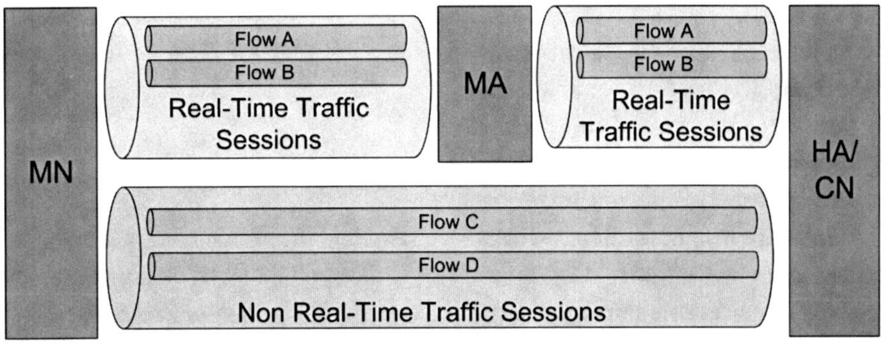

a particular MA increases, it might be necessary for some of the ARs to be switched to use different MA. To do this seamlessly without causing disruption to existing sessions is an open area of research.

Moreover, work can be carried out in obtaining fast approximation algorithms for the proposed mathematical program. This can enable deployment of MAs in a dynamic and autonomic fashion where by both the location as well as the ARs assigned to MAs can be changed dynamically. Factors such as varying mobility patterns can be considered as part of this open ended problem.

5.2 Per Flow Based Mobility Management

IETF working groups such as MONAMI and MEXT have carried out research work on efficient management of multiple care of addresses for a single MN. The flow based binding approach was considered in many of these works. The conventional mobile networks have considered the MNs to primarily use voice only. However, in future IP based mobile networks the MNs are expected to access a variety of data traffic such as Voice over IP, video (real time, streaming and download) and background data (web browsing, ftp, etc). This calls for an approach where each flow emanating from the MN are treated according to its merit rather than considering all traffic as same from a MN. DiffServ currently does this, however, when micro mobility is considered, all of these classes of traffic are provided the same level of mobility support. This can lead to over utilization of the MA by background traffic such as web browsing, ftp or download leading to higher blocking probability of real time sessions such as voice over IP and real time video streaming. Selection of MAs can also follow this approach whereby, individual sessions from a single MN can select the best available MA according to its QoS requirement and level of handover support. Figure 8 shows a figurative description of a per flow based mobility management rather than a per MN where all traffic emanating from the MN will have to use the MN.

6. CONCLUSION

Efficient micro mobility and QoS support has become two paramount requirements for the next generation of IP based mobile networks. Through efficient micro mobility and QoS mechanisms have been developed independently to meet different requirements, when deployed together they might not function optimally together. This chapter explored a variety of mobility and QoS interactions and covered the research activities over the recent past in this area. Following which, the impact MA based micro mobility solutions can have on the routing (QoS and traditional routing) of a network was explored. Specifically, the impact the exact MA location can have on the handover as well and the routing overhead was considered. It was shown that the location and number of MAs can significantly affect the performance of the network. To counter this, an optimization program was formulated to optimally deploy MAs by minimising handover as well as routing overheads. Results show considerable improvements by following an optimal approach in deploying MAs. New avenues of research works were also discussed with specific focus on dynamic and autonomic management of the mobile network when MAs are deployed. Moreover, a case for per flow based mobility management was also covered. It can be concluded that the next generation of network and mobility management require a rethink from the conventional approaches and that may include treating mobility from a per flow perspective rather than per MN perspective.

7. REFERENCES

Apostolopoulos, G., Kama, S., Williams, D., Guerin, R., Orda, A., & Przygienda, T. (1999, August). *RFC 2676: QoS Routing Mechanisms and OSPF Extensions* (experimental).

Armitage, G. (2000). *Quality of Service in IP Networks*. Indianapolis, IN: MacMillan Technical Publishing

Barzilai, T., Kandlur, D., Mehra, A., & Saha, D. (1998). Design and implementation of an RSVP-based quality of service architecture for an integrated services Internet. *IEEE Journal on Selected Areas in Communications*, *16*(3), 397–413. doi:10.1109/49.669047

Bernet, Y., Ford, P., Yavatkar, R., Baker, F., Zhang, L., Speer, M., et al. (2000, November). *RFC 2998: A Framework for Integrated Services Operation over Diffserv Networks* .

Blake, S., Black, D., Carlson, M., Davies, E., Wang, Z., & Weiss, W. (1998, December). *RFC 2475: An Architecture for Differentiated Service.*

Braden, R., Clark, D., & Shenker, S. (1994, July). *RFC 1633: Integrated Services in the Internet Architecture: an Overview* .

Braden, R., Zhang, L., Berson, S., Herzog, S., & Jamin, S. (1997, September). *RFC 2205: Resource ReSerVation Protocol (RSVP) – Version 1 Functional Specification.*

Campbell, A., Gomez, J., Kim, S., Valko, A., Wan, C.-Y., & Turanyi, Z. (2000). Design, Implementation, and Evaluation of Cellular IP. *IEEE Personal Communications*, *7*(4), 42–49. doi:10.1109/98.863995

Campbell, A., Gomez, J., Kim, S., Wan, C.-Y., Turanyi, Z., & Valko, A. (2002). Comparison of IP micromobility protocols. *IEEE Wireless Communications*, *9*(1), 72–82. doi:10.1109/MWC.2002.986462

Campbell, A. T., & Gomez-Castellanos, J. (2000). IP micro mobility protocols. *SIGMOBILE Mob. Computer. Communication. Rev.*, *4*(4), 45–53.

Chen, S., & Nahrstedt, K. (1998). An overview of Quality-of-Service routing for the Next Generation high-speed Networks: Problems and Solutions. *IEEE Network Magazine*.

Coltun, R., Ferguson, D., Moy, J., & Lindem, A. (2008, July). *RFC 5340: OSPF for IPv6.*

Crawley, E., Nair, R., Rajagopalan, B., & Sandick, H. (1998, August). *RFC 2386: A Framework for QoS-based Routing in the Internet.*

Davie, B., Charny, A., Bennet, J., Benson, K., Boudec, J. L., Courtney, W., et al. (2002, March). *RFC 3246: An Expedited Forwarding PHB (Per-Hop Behavior)* .

Eardley, P., Mihailovic, A., & Suihko, T. (2000). A Framework for the Evaluation of IP Mobility Protocols. In *The 11th IEEE International Symposium on Personal, indoor and mobile radio communications, 2000.* (Vol. 1, pp. 451–457).

Friderikos, V., Mihailovic, A., & Aghvami, H. (2004). Analysis of Cross Issues between QoS Routing and Micro-mobility protocols. *IEEE Proceedings Communications, 151*(3), 258–262. doi:10.1049/ip-com:20040574

Fu, X., Karl, H., & Kappler, C. (2002). QoS-Conditionalized Handoff for Mobile IPv6. In *Networking2002* (Vol. 2345, pp. 721–730).

Gundavelli, S., Leung, K., Devarapalli, V., Chowdhury, K., & Patil, B. (2008, August). *RFC 5213: Proxy Mobile IPv6.*

Heinanen, J., Baker, F., Weiss, W., & Wroclawski, J. (1999, June). *RFC 2597: Assured Forwarding PHB Group* .

Jacobson, V., Nichols, K., & Poduri, K. (1999, June). *RFC 2598: An Expedited Forwarding PHB.*

Johnson, D., Perkins, C., & Arkko, J. (2004, June). *RFC 3775: Mobility Support in IPv6.*

Kuo, G.-S., & Ko, P.-C. (2000). Dynamic RSVP for mobile IPv6 in wireless networks. In *IEEE 51st Vehicular technology conference, 2000* (Vol. 1, pp. 455–459).

Pack, S., & Cho, Y. (2003, 7-10 September). Performance analysis of hierarchical mobile ipv6 in IP-based cellular networks. In *14th IEEE proceedings of personal, indoor and mobile radio communications (PIMRC)* (Vol. 3, p. 2818-2822).

Pack, S., Nam, M., & Choi, Y. (2004). A study on optimal hierarchy in multi-level hierarchical mobile IPv6 networks. In *Proceedings of GLOBECOM.*

Pack, S., Shen, X. S., Mark, J. W., & Pan, J. (2007). Adaptive route optimization in hierarchical mobile IPv6 networks. *IEEE Transactions on Mobile Computing, 6*(8), 903–914. doi:10.1109/TMC.2007.1010

Paskalis, S., Kaloxylos, A., & Zervas, E. (2001). An Efficient QoS Scheme for Mobile Hosts. In *LCN'01: Proceedings of the 26th annual IEEE conference on local computer networks* (p. 630). Washington, DC: IEEE Computer Society.

Paskalis, S., Kaloxylos, A., Zervas, E., & Merakos, L. (2002). Evaluating the RSVP mobility proxy concept. In *the 13th IEEE international symposium on Personal, indoor and mobile radio communications, 2002.* (Vol. 1, pp. 270–274).

Paskalis, S., Kaloxylos, A., Zervas, E., & Merakos, L. (2003). An Efficient RSVP-Mobile IP Interworking Scheme. *Mobile Networks and Applications, 8*(3), 197–207. doi:10.1023/A:1023385429587

Paul, P., & Raghavan, S. V. (2002). Survey of QoS Routing. In *ICCC '02: Proceedings of the 15th international conference on computer communication* (pp. 50–75). Washington, DC: International Council for Computer Communication.

Perkins, C. (2002, January). *RFC 3220: IP Mobility Support for IPv4.*

Pragad, A. D., Friderikos, V., Pangalos, P., & Aghvami, A. H. (2007). The Impact of Mobility Agent based Micro-Mobility on the Capacity of Wireless Access Networks. *In Proceedings of Globecom 2007,* Washington DC.

Ramjee, R., Varadhan, K., Salgarelli, L., Thuel, S., Wang, S.-Y., & La Porta, T. (2002). HAWAII: a Domain-based Approach for supporting Mobility in Wide-Area Wireless Networks. *IEEE/ACM Transactions on Networking, 10*(3), 396–410.

Reinbold, P., & Bonaventure, O. (2003). IP Micro-mobility Protocols. *IEEE Communications Surveys and Tutorials, 5*(1).

Shenker, S., Partridge, C., & Guerin, R. (1997, September). *RFC 2212: Specification of Guaranteed Quality of Service*.

Shin, I.-H., & Lee, C.-W. (2004). A QoS Guaranteed Fast Handoff Algorithm for Wireless LAN. In *IEEE international conference on communications* (Vol. 7, pp. 3827–3832 Vol.7).

Soliman, H., Castelluccia, C., & El Malk, K. L. B. (2005, August). *RFC 4140: Hierarchical Mobile IPv6 Mobility Management (HMIPv6).*

Talukdar, A. K., Badrinath, B. R., & Acharya, A. (2001). MRSVP: a resource reservation protocol for an integrated services network with mobile hosts. *Wireless Networks*, *7*(1), 5–19. doi:10.1023/A:1009035929952

Toril, M., & Wille, V. (2008, June). Optimization of the Assignment of Base Station to Base Station Controllers in GERAN. *IEEE Communications Letters*, *12*(6), 477–479. doi:10.1109/LCOMM.2008.072044

Tseng, C.-C., Lee, G.-C., Liu, R.-S., & Wang, T.-P. (2003). HMRSVP: a hierarchical mobile RSVP protocol. *Wireless Networks*, *9*(2), 95–102. doi:10.1023/A:1021833430898

Wang, Z. (2001). *Internet QoS: Architectures and Mechanisms for Quality of Service (The Morgan Kaufmann Series in Networking)*. San Francisco: Morgan Kaufmann

Wisely, D. (2009). *IP for 4G*. New York: Wiley.

Wroclawski, J. (1997a, September). *RFC 2210: The Use of RSVP with IETF Integrated Services.*

Wroclawski, J. (1997b, September). *RFC 2211: Specification of the Controlled-Load Network Element Service.*

Chapter 12
Handover Analysis and Dynamic Mobility Management for Wireless Cellular Networks

Ramón M. Rodríguez-Dagnino
Tecnológico de Monterrey, México

Hideaki Takagi
University of Tsukuba, Japan

ABSTRACT

Dynamic location of mobile users aims to deliver incoming calls to destination users. Most location algorithms keep track of mobile users through a predefined location area. The design of these location algorithms is focused to minimize the generated signaling traffic. There are three basic approaches to design location algorithms, namely distance-based, time-based and movement-based. In this Chapter we focus only on the movement-based algorithm since it achieves a good compromise between complexity and performance. We minimize a cost function for this dynamic movement-based location algorithm in order to find an optimum threshold in the number of updates. Counting the number of wireless cell crossing during inter-call times is a fundamental issue for our analysis. We use renewal theory to capture the probabilistic structure of this model, and it is general enough to include a variety of probability distributions for modeling cell residence times (CRT) in exponentially distributed location areas and hyperexponentially distributed intercall times. We present numerical results regarding some important distributions.

1. INTRODUCTION

Counting the number of handovers (or wireless cell crossings) is an important problem in cellular wireless networks. In a typical cellular topology, the area to cover a city is designed as an irregular or regular layout having non-overlapping hexagon-shaped wireless cells. During a random duration call, mobile users will cross several cell boundaries spending a random time in each of the cells. The handover process is a complex function of many factors such as: size of wireless cells, user's mobility path, call patterns, (i.e., the number of renewals or handovers in random interval of duration T or CHT). This problem has been solved in several specific cases by Cox in his monograph. The CRTs are denoted by the sequence

DOI: 10.4018/978-1-61520-680-3.ch012

Figure 1. Cellular cell layout and cell residence time $X_i; i=1,2,\dots,k+1$.

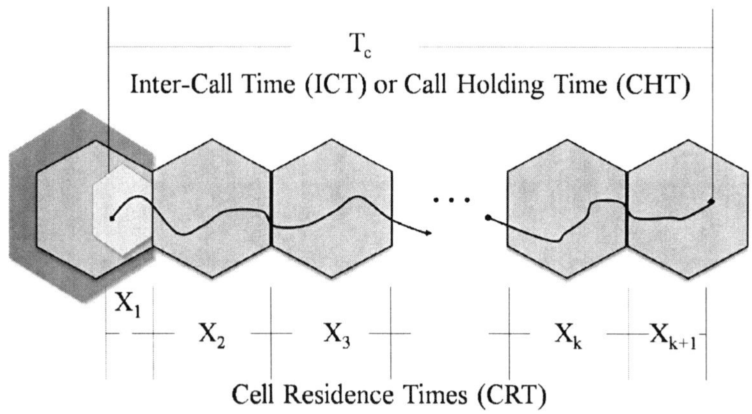

of random variables $X_1, X_2, \dots$ or renewal times, see Figure 1. Most of the results studied by Cox are based on the *ordinary* renewal process, i.e., all the random variables X_i; $i = 1,2,\dots,$ come from the same distribution with probability density function (pdf) $f_X(x)$. We denote $N(t)$ as the number of renewals in a fixed time interval (0, t], and the first renewal X_1 is started at time 0. We assume that T is independent of $\{X_1, X_2, \dots, X_i \dots\}$. Hence, $N(T)$ gives the counts of the number of renewals (handovers) in a random interval (0, *T*]. This basic model studied by Cox results restrictive in common cellular networks scenarios. It is common that a mobile user begins his call somewhere inside a wireless cell. Thus, we should consider the case in which only $X_i = X$; $i = 2,3,\dots$ have pdf $f_X(x)$ while X_1 may come from a different distribution. When X_1 is the residual life or forward recurrence-time of $X_2 = X$ (Cox, 1962, page 27), we have the *equilibrium* renewal process, which we have studied in (Rodríguez-Dagnino, Takagi, 2003). Another important situation occurs when X_1 has a different pdf from the remaining CRTs $X_2, X_3, \dots$ and it is called the *modified* or *delayed* renewal process. We have also studied a more general case where all the pdf's of the CRTs $X_1, X_2, \dots$ may be different. We call this case as the generalized renewal process (Rodríguez-Dagnino, Takagi, 2005) that is applicable to irregular layout typologies. In Figure 1 we show a basic layout where we emphasize the fact of a different pdf for the first CRT.

We have extended this basic approach in many directions, and we will discuss these counting handover methods in this Chapter and related results that can be found in (Yeung, 1997; Zonoozi, 1997; Orlik, 1998; Rodríguez-Dagnino, Leyva-Valenzuela, 1999; Rodríguez-Dagnino, Hernández-Lozano, Takagi, 2000; Rodríguez-Dagnino, Takagi, 2001; 2002).

Besides its importance in dimensioning wireless networks, counting the number of cell crossing boundaries is also important for location of mobile users in a specific location area.

The main goal in the location algorithms is to minimize the signaling cost resulting from the users updates in a database serving the location area. In spite of the fact that the user is not active in conversation, it is necessary to keep track of it by updating the database. This is a dynamic process and there have been several strategies to achieve this goal. The most studied strategies for this purpose are: Distance-based, time-based, and movement-based (Bar-Noy, Kessler, Sidi, 1994; Akyildiz, Ho, Lin, 1996).

Our analysis is aimed to find an optimal cost to reduce signaling traffic and database loads. In typical wireless networks the Mobile Switching

Centers (MSCs) are partitioned into Location Areas (LAs), for location purposes, and each LA has a set of wireless cells. There are several base transceiver stations in each LA, which communicate with a particular MSC through radio links or cable connections. The traffic patterns are highly dynamic in wireless environment. So, the static location algorithms are not efficient to dynamically update and store location information in some predefined databases. There are two relevant databases for this performance analysis: the Home Location Register (HLR) and the Visitor Location Register (VLR) databases. The VLR maintains temporary LA addresses of mobile users of a group of LAs, whereas a HLR maintains the permanent information to track the users. There is signaling information interchange between VLRs and one HLR through the Signaling Transfer Points (STP); see Figure 2. This signaling information obviously increases with the number of users in a location area and the minimization of this signaling traffic is desirable.

Most researchers agree that the movement-based scheme achieves a good tradeoff between algorithm complexity and effectiveness (Akyildiz, Ho, Lin, 1996; Fang, 2003). We will be focused on the movement-based scheme, and the relevant performance measure for signaling updates is the number of location updates for tracking non active mobile users. The movement-based update is a strategy where each mobile user counts the number of handovers incurred by its movements, and when this number exceeds certain threshold then it transmits an update message. The paging mechanism is also relevant to localize the destination user for call-delivery just after an incoming call arrives, and it should be included in the cost function. Our main goal in this location management system is to minimize the cost of these location updates and paging mechanisms.

Early works in this topic made the assumption of exponentially distributed ICT and any distribution for the cell residence times (Li, Kameda, Li, 2000), in a similar manner as in the handover problem. A generalization of these assumptions was done in (Rodríguez-Dagnino, Ruiz-Cedillo, Takagi, 2002) where the authors studied the case of circular cell configurations, and the residence time of first cell is different of the residence times of the remaining cells. Our mathematical modeling uses the delayed renewal process for solving this problem. Other generalizations of this scheme can be found in (Fang, 2003; Ma, Fang, 2004) See also (Rodríguez-Dagnino, Ruiz-Cedillo, Takagi, 2002; Li, Pan, Jia, 2002; Rodríguez-Dagnino, Takagi, 2007).

2. STATISTICAL MOMENTS OF HANDOVERS

2.1 The Mean Number of Handovers

The basic idea is to use well-known results of renewal theory on fixed intervals (0, t] and to extend them to random intervals $(0,T]$. Let $G_{N(T)}(z)$ be the probability generating function (pgf) for $N(T)$, the number of handovers in a *random* interval $(0,T]$. It is given by

$$G_{N(T)}(z)=\int_{t=0}^{\infty} G_{N(T)}(t,z)\, f_T(t)\, dt \qquad (2.1.1)$$

where $f_T(t)$ is the pdf of the random variable T, and

$$G_{N(T)}(t,z):=\mathrm{E}\left[z^{N(T)}|T=t\right]$$

$$=\sum_{j=0}^{\infty} P\left[N(T)=j\,|T=t\right]z^{j}$$

is the pgf of $N(t)$, the number of handovers in a *fixed* interval (0,t].

By taking the derivative with respect to z at both sides of (2.1.1), and evaluating at z=1 we find the following expression for the mean number of handovers:

Figure 2. Signaling traffic between the Visitor Location Registers (VLR) and the Home Location Registers (HLR) through Signaling Transfer Points (STP)

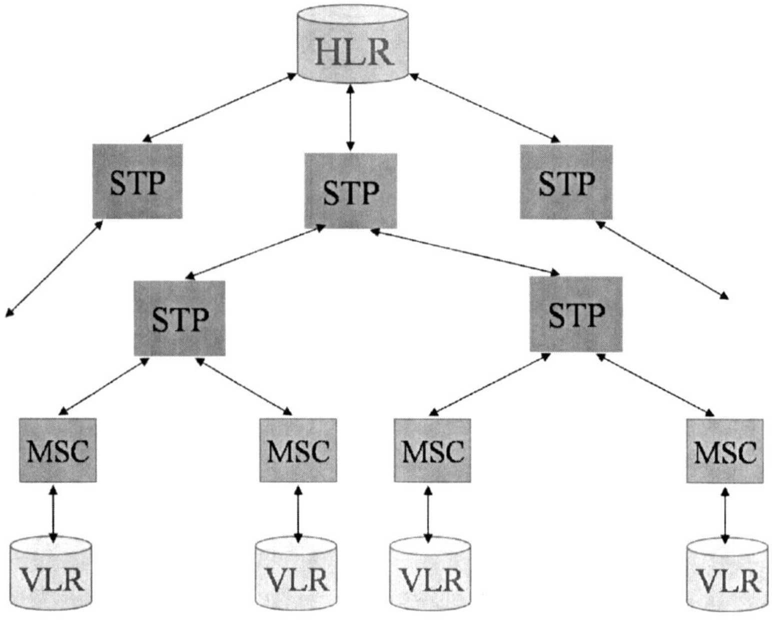

$$\mathrm{E}\left[N(T)\right] = \int_{t=0}^{\infty} \mathrm{E}\left[N(T)\,|\,T=t\right] f_T(t)\,dt$$

as shown by (Cox, 1962; p.46, eq. (3)), it holds for the equilibrium renewal process that

$$\mathrm{E}\left[N(T)\,|\,T=t\right] = \frac{t}{\mathrm{E}[X]}$$

for any distribution $f_X(x)$ having $\mathrm{E}[X]$. We will use the subindices *e*, *o*, and *m* to distinguish the equilibrium, ordinary and modified renewal process cases. Hence, we obtain

$$\mathrm{E}_e\left[N(T)\right] = \int_{t=0}^{\infty} \frac{t}{\mathrm{E}[X]} f_T(t)\,dt = \frac{\mathrm{E}[T]}{\mathrm{E}[X]} = \rho.$$

We call the parameter ρ as the *mobility ratio* parameter

$$\rho := \frac{\text{Expected value of CHT}}{\text{Expected value of CRT}} = \frac{\mathrm{E}[T]}{\mathrm{E}[X]},$$

which represents the average number of handovers a mobile users makes per call.

We remark that Nanda, 1993 suggested that the mean number of handovers is always equal to the mobility ratio, however we have shown that this result is valid only for the equilibrium renewal process and for the ordinary Poisson process (Rodríguez-Dagnino, Takagi, 2003). In general, it is difficult to obtain closed-form mathematical expressions for $\mathrm{E}\left[N(T)\,|\,T=t\right]$ in the ordinary and modified renewal processes cases. For CRT distributions having finite second-order moments, the following asymptotic result can be used for the ordinary renewal process

$$\mathrm{E}\left[N(T)\,|\,T=t\right] = \frac{t}{\mathrm{E}[X]} + \frac{\mathrm{Var}[X] - \mathrm{E}^2[X]}{2\,\mathrm{E}^2[X]} + o(1).$$

Hence,

$$\mathrm{E}_o[N(T)] = \rho + \frac{\mathrm{Var}[X] - \mathrm{E}^2[X]}{2\,\mathrm{E}^2[X]} + o(1).$$

Similarly, for the modified or delayed renewal process we have

$$E_m[N(T)] = \rho + \frac{\mathrm{Var}[X] - E^2[X]}{2E^2[X]} - \frac{E[X_1]}{E[X]} + o(1).$$

It is straightforward to extend these asymptotic results for CRT distributions having higher order moments.

2.2 Higher-Order Moments and Probability Distributions

Once $G_{N(T)}(z)$ is obtained by using Eq. (2.1.1), then the probability mass function (pmf) of $N(T)$ is given by

$$P[N(T)=j] = \frac{1}{j!}\frac{d^j}{dz^j}G_{N(T)}(z)\Big|_{z=0} \quad j=0,1,2,\ldots.$$

The lth binomial moment of $N(T)$ is given by

$$E\left[\binom{N(T)}{l}\right] = \frac{1}{l!}\frac{d^l}{dz^l}G_{N(T)}(z)\Big|_{z=1} \quad l=0,1,2,\ldots.$$

Now, let us define as $G^*_{N(T)}(s,z)$ the Laplace transform of $G_{N(T)}(t,z)$

$$G^*_{N(T)}(s,z) := \int_{t=0}^{\infty} e^{-st}G_{N(T)}(t,z)\,dt.$$

For the ordinary renewal process we have

$$G^*_{N(T),o}(s,z) = \frac{1-f^*_X(s)}{s[1-z f^*_X(s)]}.$$

Similarly, for the equilibrium renewal process we have

$$G^*_{N(T),e}(s,z) = \frac{1}{s} + \frac{(z-1)}{sE[X]}G^*_{N(T),o}(s,z),$$

and for the modified renewal process,

$$G^*_{N(T),m}(s,z) = \frac{1}{s} + \frac{(z-1)f^*_{X_1}(s)}{s[1-z f^*_X(s)]},$$

where $f^*_X(s)$ is the Laplace transform of the pdf $f_X(x)$ corresponding to the random variable X. See pages 37 and 38 of (Cox, 1962).

3. GENERAL DISTRIBUTED CELL RESIDENCE TIMES

3.1 Some Basic Results of Cox Approach

Let $N(t)$ be the number of renewals in a fixed interval $(0, t]$. Assume that T is a random variable with the pdf $f_T(t)$, then we can define $N(T)$ as the number of renewals in the random interval $(0,T]$. We can relate the pgf of $N(t)$ with that of $N(T)$ as follows

$$G_{N(T)}(z) = \int_{t=0}^{\infty} G_{N(T)}(t,z)f_T(t)\,dt.$$

An interesting relationship does occur when $f_T(t)$ is an Erlang pdf with parameters k and θ, i.e., with mean $E[T] = k/\theta$, so $\rho = k/(\theta E[T])$. In such a case

$$G_{N(T)}(z) = \int_{t=0}^{\infty} G_{N(T)}(t,z)\frac{\theta^k t^{k-1}}{(k-1)!}e^{-\theta t}\,dt$$

that is equivalent to

$$G_{N(T)}(z) = \frac{\theta^k}{(k-1)!}\left(-\frac{\partial}{\partial s}\right)^{k-1}\left\{G^*_{N(T)}(s,z)\right\}\Big|_{s=\theta} \quad (3.1.1)$$

where $G^*_{N(T)}(s,z)$ is the Laplace transform of $G_{N(T)}(t,z)$. Cox applied Eq. (3.1.1) to the case where $N(t)$ is an ordinary renewal process (Cox, 1962; p.43), and it can be applicable to the cases of equilibrium and delay renewal processes as well, as it is shown in the following subsection.

3.2 The Case of Modified Renewal Processes

Let us assume that the CHT can be fitted by a k-stage Erlang pdf, then for the *modified* renewal process we obtain

$$G_{N(T)}(z)=\frac{\theta^k}{(k-1)!}\left(\frac{\partial}{\partial s}\right)^{k-1}\left\{\frac{1}{s}+\frac{(z-1)f_{X_1}^*(s)}{s[1-z\,f_X^*(s)]}\right\}\Bigg|_{s=\theta}$$

$$=1+\frac{\theta^k(z-1)}{(k-1)!}\left(\frac{\partial}{\partial s}\right)^{k-1}\frac{f_{X_1}^*(s)}{s[1-z\,f_X^*(s)z}\Bigg|_{s=\theta}$$

From this we can express the pmf of $N(T)$ as

$$P[N(T)=j]=\begin{cases}1-\dfrac{\theta^k}{(k-1)!}\left(-\dfrac{\partial}{\partial s}\right)^{k-1}\dfrac{f_{X_1}^*(s)}{s}\Bigg|_{s=\theta} & ;j=0\\ \dfrac{\theta^k}{(k-1)!}\left(-\dfrac{\partial}{\partial s}\right)^{k-1}\dfrac{f_{X_1}^*(s)}{s}[1-f_X^*(s)][f_X^*(s)]^{j-1}\Bigg|_{s=\theta} & ;j=1,2,\ldots\end{cases}$$

We can also express the lth binomial moment of $N(T)$ as

$$\mathrm{E}\left[\binom{N(T)}{l}\right]=\frac{\theta^k}{(k-1)!}\left(-\frac{\partial}{\partial s}\right)^{k-1}\frac{f_{X_1}^*(s)[f_X^*(s)]^{l-1}}{s[1-f_X^*(s)]^l}\Bigg|_{s=\theta}\qquad l=1,2,\ldots$$

We note that if $f_{X_1}^*(s)\equiv f_X^*(s)$, we have an *ordinary* renewal process for the sequence of CRTs. On the other hand, if

$$f_{X_1}^*(s)\equiv\frac{1-f_X^*(s)}{s\mathrm{E}[X]},$$

we have the *equilibrium* renewal process.

Many examples can be found in Rodríguez-Dagnino, Takagi, 2003, for the equilibrium renewal process, and in Rodríguez-Dagnino, Takagi, 2005, for the modified renewal process.

4. GENERAL DISTRIBUTED CALL HOLDING TIMES

Now, we assume that the CRTs X is exponentially distributed with parameter μ. In this case, the sequence of CRTs is called a Poisson renewal process. Thus the pgf for $N(t)$, the number of renewals in the fixed interval $(0,t]$, is given by

$$G_{N(T)}(t,z)=\sum_{n=0}^{\infty}\frac{(\mu t)^n}{n!}e^{-\mu t}z^n=e^{-\mu(1-z)t}.$$

As a consequence of the memoryless property of the exponential distribution, this result is valid for both the ordinary and the equilibrium renewal process.

It can be observed that for any distribution for the CHT we can find that

$$G_{N(T)}(z)=\int_{t=0}^{\infty}e^{-\mu(1-z)t}f_T(t)\,dt=f_T^*[\mu(1-z)].$$

After finding the jth derivative of this pgf, we can obtain the following pmf for $N(T)$

$$P[N(T)=j]=\int_{t=0}^{\infty}\frac{(\mu t)^j}{j!}e^{-\mu t}f_T(t)\,dt=\frac{(-\mu)^j}{j!}f_T^{*(j)}(\mu);\quad j=0,1,2,\ldots$$

which is a Poisson mixture of the pdf $f_T(t)$.

A general expression for the binomial moments of $N(T)$ can also be obtained as

$$\mathrm{E}\left[\binom{N(T)}{l}\right]=\frac{\mu^l}{l!}\mathrm{E}[T^l];\quad l=1,2,\ldots.$$

Thus we get

$$\mathrm{E}[N(T)]=\mu\,\mathrm{E}[T]=\rho$$

and

$$\mathrm{Var}[N(T)]=\mu^2\,\mathrm{E}[T^2]+\rho(1-\rho).$$

Similarly, we consider a general pdf for the CHT and the delayed renewal process for the CRTs that are exponentially distributed with parameters μ_1 and μ.

The pgf is given by

$$G_{N(T)}(z)=\frac{(\mu-\mu_1)(z-1)}{\mu_1+\mu(z-1)}f_T^*(\mu_1)+\frac{\mu_1 z}{\mu_1+\mu(z-1)}f_T^*[\mu(1-z)],$$

which follows by substituting R = 2 into the more general case of all different CRTs considered in Section 4.1 of (Rodríguez-Dagnino, Takagi, 2005).

After finding the *j*th derivative of $G_{N(T)}(z)$ we can find the pmf and the moments of $N(T)$. The pmf of $N(T)$ is given by

$$P[N(T)=j]=\begin{cases} f_T^*(\mu_1) & ;j=0 \\ \dfrac{\mu_1\mu^{j-1}}{(\mu-\mu_1)^j}\left[f_T^*(\mu_1)-\sum_{i=0}^{j-1}\dfrac{(\mu_1\mu^i)}{i!}f_T^{*(i)}(\mu)\right] & ;j=1,2,\ldots \end{cases}, \quad (4.1)$$

where

$$f_T^{*(j)}(s):=\frac{d^j f_T^*(s)}{ds^j}=(-1)^j\int_{t=0}^{\infty}t^j e^{-st}f_T(t)\,dt;\quad j=1,2,\ldots.$$

The *l*th binomial moment of $N(T)$ is given by

$$\mathrm{E}\left[\binom{N(T)}{l}\right]=\left(1-\frac{\mu_1}{\mu}\right)\left(\frac{\mu}{\mu_1}\right)^l$$
$$\left\{(-1)^{l-1}f_T^*(\mu_1)+\sum_{i=0}^{l-1}\frac{(-1)^{i-1}}{i!}\mu_1^i\,\mathrm{E}[T^i]\right\}+\frac{\mu^l}{l!}\mathrm{E}[T^l];\quad l=1,2,\ldots.$$

For example, the mean is given by

$$\mathrm{E}[N(T)]=\frac{\mu-\mu_1}{\mu_1}[f_T^*(\mu_1)-1]+\mu\,\mathrm{E}[T].$$

4.1 Pareto Call Holding Times

Assuming a Pareto pdf

$$f_T(t)=\frac{\alpha\beta^\alpha}{t^{\alpha-1}};\quad \beta>0,\alpha>0,t>\beta,1<\alpha<2$$

with Laplace transform

$$f_T^*(s)=\alpha(s\beta)^{\frac{\alpha-1}{2}}e^{-\frac{s\beta}{2}}W_{-\frac{(\alpha+1)}{2},-\frac{\alpha}{2}}(s\beta);\quad s\beta>0$$

with *i*th derivative

$$f_T^{*(i)}(s)=(-1)^i\alpha s^{\frac{\alpha-1-i}{2}}\beta^{\frac{\alpha-1+i}{2}}e^{-\frac{s\beta}{2}}W_{-\frac{(\alpha+1-i)}{2},-\frac{(\alpha-i)}{2}}(s\beta);\quad s\beta>0,$$

where $W_{a,b}(x)$ is the Whittaker function (Rodríguez-Dagnino, Takagi, 2003).

Let us assume $\beta=0.3$, $\mu=1$, and we show in Figure 3 the pmf $P[N(T)=j]$ given in Eq. (5.1) for α = 1.1, 1.5 and 1.9, $\mu_1=0.1$, 1 and 10. The case of $\mu_1=\mu=1$ reduces to the equilibrium renewal process, see Rodríguez-Dagnino, Takagi, 2003.

By assuming the same parameters and equating the mean values of the CHT pdf, we show in Figure 4 the pmf $P[N(T)=j]$ when *T* is exponentially distributed with parameter *θ*, i.e.,

$$\frac{1}{\mathrm{E}[T]}=\theta=\frac{\alpha-1}{\alpha\beta}.$$

We can see that the pmf $P[N(T)=j]$ decays much slower when the CHT pdf is Pareto (heavy-tail) than when it is exponential (light-tail).

5. MOBILITY MANAGEMENT SCHEMES AND BASIC MODEL

5.1 Mobility Management Schemes

There are two basic tasks for location management: Location update (or registration) and call delivery to the mobile terminal. Both tasks are related to each other since the process of delivering calls with minimum delay requires the tracking of the location of each mobile terminal. To minimize the tracking process delay (or cost) several wireless cells are grouped to form a location area (LA). So, for tracking purposes the wireless typological layout of a city is divided into location areas, and each LA consists of a set of wireless cells, see Figure 5.

Figure 3. Probability mass function (log) for Pareto CHT and modified renewal process with exponential CRTs

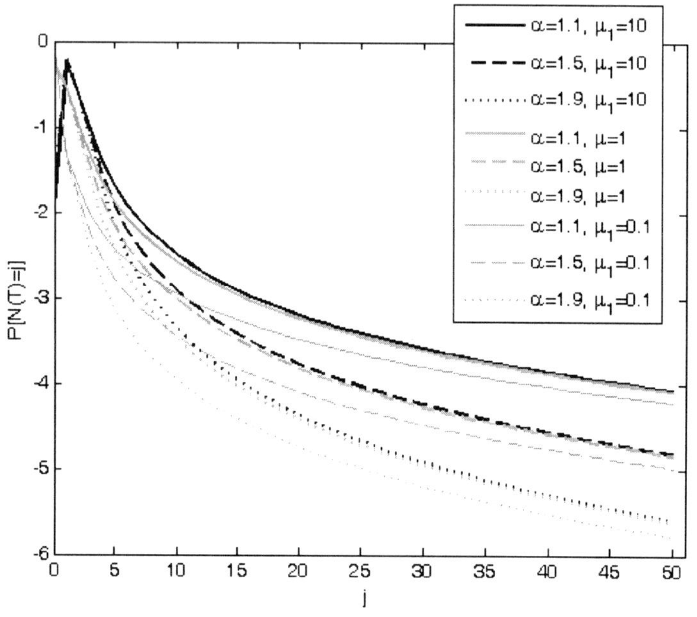

Figure 4. Probability mass function (log) for Exponential CHT and modified renewal process with exponential CRTs

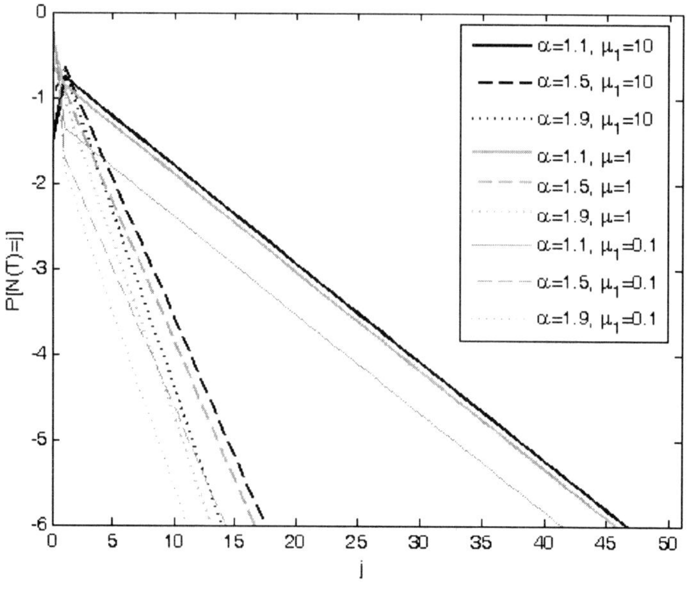

Figure 5. Dynamic movement-based location update with movement threshold d=5 and 7 location areas

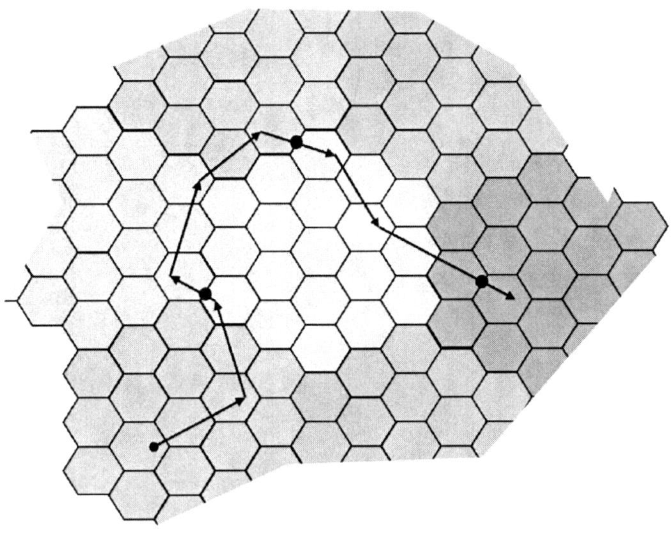

During the call establishment process the network begins paging to locate the base station serving the mobile terminal. Then, the tracking process consists of location update and paging procedures, and the corresponding incurred costs of updating C_u and searching C_p need to be minimized.

There are many proposed update strategies, and in one category of them the network takes the decision of when a mobile terminal must make the update. However, this static scheme suffers of serious drawbacks as it is discussed in (Kozatchok, Pierre, 2002). In most of the cases it is preferable a dynamic scheme where the mobile terminal takes the decision of when to update. There are three main dynamic schemes, namely time-based, distance-based, and movement-based schemes. Some regular or periodic events are used in these dynamic schemes. For instance, in the time-based procedure the updates are made at equally spaced intervals of time, in the distance-based procedure the updates are made at regular distances traveled by the mobile terminal, and in the movement-based the updates are carried out after a certain number of wireless cell crossing is achieved, or movement threshold. See Figure 5 for the movement–based scheme, where the updates are done after 5 cell crossings.

The distance-based and movement – based updating schemes are the most studied in the literature since they have a better performance than the time-based scheme, even though the time-based scheme seems to be simpler for implementation. The main difficulty with the distance-based scheme is in how to estimate in a reliable manner the travelled distance. This is a hard problem especially in irregular topologies. Moreover, many authors claim that the movement-based scheme is simpler for implementation (see Glisic, 2006 and references therein) and we will focus on it in the rest of this Chapter.

Most of the location management schemes are based in a two-level databases hierarchy. At the lower level is the Visitor Location Register (VLR) database involved in tracking a mobile terminal. There are several VLRs connected to the higher-level Home Location Register (HLR) database, see Figure 1. A mobile terminal is associated with a single HLR in the wireless network, where the permanent mobile profile is stored.

5.2 Mathematical Model

The basic structure relating the location area renewal process and its relationship with the cell residence times renewal process and the ICT is shown in Figure 1.

Let us denote the ICT random variable as T_c, i.e., the time between the previous and current phone calls, and we assume that T_c is a hyperexponential distributed random variable with pdf

$$f_{T_c}(t) = \sum_{n=1}^{N} q_n \mu_n e^{-\mu_n t},$$

where $\sum_{n=1}^{N} q_n = 1$. The mean value is given by

$$\mathrm{E}[T_c] = \sum_{n=1}^{N} \frac{q_n}{\mu_n}.$$

The cumulative distribution function is given by

$$F_{T_c}(t) = P\,[T_c \le t] = \int_0^t f_{T_c}(u)\,du = 1 - \sum_{n=1}^{N} q_n e^{-\mu_n t}.$$

Now, let us denote as T_L the random variables forming the renewal process of location areas $(LA_i,\ i = 0,1,\ldots,k)$ or Location Area Residence Time (LART). We assume that the pdf of T_L, say $f_{T_L}(t)$, is given by an exponential distribution as follows

$$T_{T_L}(t) = \theta_L e^{-\theta_L t},$$

where $L = \mathrm{E}[T_L] = \frac{1}{\theta_L}$. The Laplace transform of this LART pdf is given by

$$f_{T_L}^*(s) = \frac{\theta_L}{\theta_L + s}. \qquad (5.2.1)$$

The cell residence times (CRT) are denoted by the random variable X (see Figure 6), and we assume a general pdf $f_X(x)$ to model this renewal process.

We assume that LA boundary and cell boundary are independent in the model. In fact, since LA has exponential distribution, and CRT has general distribution, from the mathematical point of view

Figure 6. Time diagram relating all the important variables considered in the model

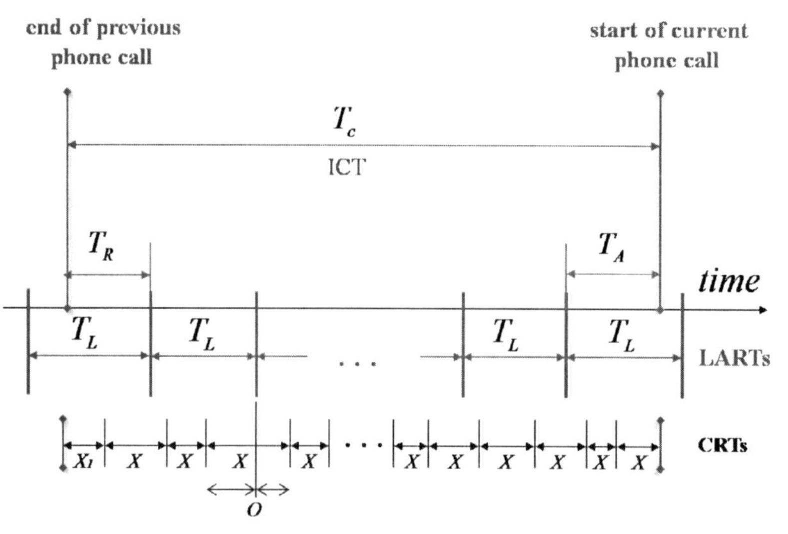

their boundaries coincide with probability zero. This is different from the real system where LA boundary coincides with cell boundary. A further consequence of this assumption is that if a CRT is divided into two LAs, see point O in Figure 6, there exists dependence between the residual (or excess) life and age of X. This dependence is neglected in our model.

In our model we reset the counter of the number of cell boundary crossings for VLR updates at every LA boundary. If this is not the case, we can just count the number of cell boundary crossings in a T_c, at once without bothering to count the number of cell boundary crossing for each LA.

The number of random variables T_L that occur during T_c constitutes an equilibrium renewal process. Thus, the excess life or residual life of T_L, say T_R, has a pdf given by

$$f_{T_R}(t) = \frac{1}{E[T_L]} \int_{\tau=t}^{\infty} f_{T_L}(\tau) d\tau = \frac{1}{L} e^{-\theta_L t} \quad (5.2.2)$$

By symmetry between T_R and T_A, as it can be seen in Figure 6, the pdf of the age T_A is $f_{T_A}(t) = f_{T_R}(t)$. As a consequence, the Laplace transform is given by

$$f_{T_R}^*(s) = f_{T_A}^*(s) = \frac{1 - f_{T_L}^*(s)}{sL} = \frac{\theta_L}{s + \theta_L}.$$

6. COUNTING THE NUMBER OF HLR AND VLR LOCATION UPDATES

6.1 Number of HLR Location Updates

Let $N(T_{cH})$ be the number of Home Location Registers (HLR) location updates in an ICT interval of duration T_c. Thus, its expected value can be expressed by

$$E[N(T_{cH})] = \sum_{k=1}^{\infty} k \, P[N(T_{cH}) = k],$$

where $P[N(T_{cH}) = k]$ is the probability of having k LA boundary crossings in T_c.

The number of random variables T_L occurring in an ICT constitutes an equilibrium renewal process. By applying the results in Section 4.1 of (Rodríguez-Dagnino, Takagi, 2003) we can readily obtain

$$P[N(T_{cH}) = k] = \begin{cases} 1 - \frac{1}{L} \sum_{n=1}^{N} \frac{q_n}{\mu_n} [1 - f_{T_L}^*(\mu_n)]; & k = 0 \\ \frac{1}{L} \sum_{n=1}^{N} \frac{q_n}{\mu_n} [1 - f_{T_L}^*(\mu_n)]^2 [1 - f_{T_L}^*(\mu_n)]^{k-1} & k = 1, 2, \ldots \end{cases} \quad (6.1.1)$$

Hence, by substituting Eq. (5.2.1) into Eq. (6.1.1) we find that $P[N(T_{cH}) = k]$ can be written as

$$P[N(T_{cH}) = k] = \begin{cases} 1 - \frac{1}{L} \sum_{n=1}^{N} \frac{q_n}{\mu_n} B_n & k = 0; \\ \frac{1}{L} \sum_{n=1}^{N} \frac{q_n}{\mu_n} B_n^2 A_n^{k-1} & k = 1, 2, \ldots \end{cases} \quad (6.1.2)$$

where

$$A_n = \frac{\theta_L}{\theta_L + \mu_n}, \qquad B_n = 1 - A_n = \frac{\mu_n}{\theta_L + \mu_n}.$$

From Eq. (6.1.2) $E[N(T_{cH})]$ can be easily obtained as:

$$E[N(T_{cH})] = \sum_{k-1}^{\infty} k \sum_{n=1}^{N} \frac{1}{L} \frac{q_n}{\mu_n} B_n^2 A_n^{k-1} = \frac{E[T_c]}{E[T_L]} = ,_L \sum_{n=1}^{N} \frac{q_n}{\mu_n}.$$

6.2 Number of VLR Location Updates

Let $N(T_{cV})$ be the number of Visitor Location Registers (VLR) location updates in a random time interval of duration T_c. Thus, its expected value is given by

$$E[N(T_{cV})] = \sum_{k=0}^{\infty} n_{v,k} \, P[N(T_{cH}) = k], \quad (6.2.1)$$

where $n_{v,k}$ is the average number of location updates in VLRs when a terminal receives the next phone call in the kth LA, $k = 0,1,2,\ldots$.

This average number of location updates can be calculated as (Li, Pan, Jia, 2000)

$$n_{v,k} = \begin{cases} \sum_{l=1}^{\infty} l \sum_{j=ld}^{(l+1)d-1} \varepsilon_1(j) & ;k=0 \\ \sum_{l=1}^{\infty} l \sum_{j=ld}^{(l+1)d-1} \{\varepsilon_2(j) + (k-1)\,\varepsilon_3(j) + \varepsilon_4(j)\} & ;k=1,2,\ldots \end{cases}$$

where d as the location update threshold distance, and the $\varepsilon_i(j)$'s allow us to describe the movement of a mobile user between both paging areas and location areas. We calculate these four probabilities in the following Subsections.

6.2.1 Calculation of $\varepsilon_1(j)$

This is the probability that there are j cell boundary crossings within the LA_0 between two consecutive calls. This is, both the previous call and the current call were received in LA_0 and, therefore, $k = 0$ (i.e. no LA boundary crossings). Therefore, $\varepsilon_1(j)$ is equal to the probability of number of CRT crossings during T_c such that $T_c < T_R$, i.e.,

$$P\,[T_c \le t \,|\, T_c \le T_R] = \frac{P\,[T_c \le t, T_c \le T_R]}{P\,[T_c \le T_R]} \qquad (6.2.2)$$

We have an exponentially distributed T_L with parameter θ_L.

We can calculate $P\ [T_c \le T_R]$ explicitly as

$$P\,[T_c \le T_R] = \int_0^{\infty} P[T_c \le t] f_{T_R}(t)\,dt$$

$$= \int_0^{\infty} \left(1 - \sum_{n=1}^{N} q_n e^{-\mu_n t}\right) \theta_L e^{-\theta_L t} dt = \sum_{n=1}^{N} q_n \frac{\mu_n}{\theta_L + \mu_n} \qquad (6.2.3)$$

The joint probability is given by

$$P\ [T_c \le t, T_c \le T_R] = \begin{cases} P[T_c \le t] & \text{if } t \le T_R \\ P[T_c \le T_R] & \text{if } t > T_R \end{cases}$$

Thus, we get

$$P[T_c \le t, T_c \le T_R]$$

$$= \int_0^{t} P[T_c \le x] f_{T_R}(x)\,dx + \int_0^{\infty} P[T_c \le t] f_{T_R}(x)\,dx$$

$$= \int_0^{t} \left(1 - \sum_{n=1}^{N} q_n e^{-\mu_n x}\right) \theta_L e^{-\theta_L x} dx$$

$$+ \left(1 - \sum_{n=1}^{N} q_n e^{-\mu_n t}\right) \int_t^{\infty} \theta_L e^{-\theta_L x} dx =$$

$$\sum_{n=1}^{N} q_n \frac{\mu_n}{\theta_L + \mu_n} (1 - e^{-(\theta_L + \mu_n)t}) \qquad (6.2.4)$$

By substituting Eq. (6.2.3) and (6.2.4) into Eq. (6.2.2) we obtain

$$P\,[T_c \le t \,|\, T_c \le T_R]$$

$$= 1 - \sum_{n=1}^{N} v_n e^{-(\theta_L + \mu_n)t}, \quad t \ge 0, \qquad (6.2.5)$$

where

$$v_n = \left[\sum_{j=1}^{N} q_n \frac{\mu_j}{\theta_L + \mu_j}\right]^{-1} \left(\frac{q_n \mu_n}{\theta_L + \mu_n}\right),$$

for all $1 \le n \le N$.

It is straightforward to see that $\sum_{n=1}^{N} v_n = 1$. Then, Eq. (6.2.5) is hyperexponential distributed with weights v_n.

Now,

$$\sum_{j=0}^{\infty} \varepsilon_1(j) z^j = \sum_{n=1}^{N} v_n (\theta_L + \mu_n) \left\{ G^*_{N(T)}(s,z) \right\} \Bigg|_{s=\theta_L + \mu_n},$$

where

$$G^*_{N(T)}(s,z) = \frac{1}{s} + \frac{(z-1)[1 - f^*_X(s)]}{s^2 \mathrm{E}[X][1 - z f^*_X(s)]} \qquad (6.2.6)$$

Hence,

$$\varepsilon_1(j)=\begin{cases} 1-\sum_{n=1}^{N} v_n \dfrac{1-f_X^*(\theta_L+\mu_n)}{(\theta_L+\mu_n)\mathrm{E}[X]}; & j=0 \\ \sum_{n=1}^{N} v_n \dfrac{[1-f_X^*(\theta_L+\mu_n)]^2[f_X^*(\theta_L+\mu_n)]^{j-1}}{(\theta_L+\mu_n)\mathrm{E}[X]}; & j=1,2,\ldots \end{cases} \quad (6.2.7)$$

6.2.2 Calculation of $\varepsilon_2(j)$ and $\varepsilon_4(j)$

$\varepsilon_2(j)$ is defined as the probability that there are j cell boundary crossings within the LA_0 (where the previous call arrived) when the mobile terminal enters LA_1 (where the current call arrived). For this case, $k=0$.

Let T_R be the forward recurrence time (excess or residual life) of T_L. Then $\varepsilon_2(j)=P[N(T_R)=j]$, where $N(T_R)$ is the number of CRT's during a random T_R duration time. In other words, we need to count the number of renewals in an equilibrium renewal process with a random duration time T_R.

The pdf for T_R is given by Eq. (5.2.2), and the pdf of $N(T_R)$ can be obtained in a similar manner as in (Rodríguez-Dagnino, Takagi, 2003)

$$G_{N(T_R)}(z)=\theta_L\left\{G^*_{N(T)}(s,z)\right\}\Big|_{s=\theta_L}$$

where $G^*_{N(T)}(s,z)$ is given by Eq. (6.2.6) for the equilibrium renewal process. Once $G_{N(T_R)}(z)$ is obtained, the pmf of $N(T_R)$ is given by

$$P[N(T_R)=j] = \begin{cases} 1-\dfrac{[1-f_X^*(\theta_L)]}{(\theta_L+\mu_n)\mathrm{E}[X]}; & j=0 \\ \dfrac{[1-f_X^*(\theta_L)]^2[f_X^*(\theta_L)]^{j-1}}{\theta_L\ \mathrm{E}[X]}; & j=1,2,\ldots \end{cases} \quad (6.2.8)$$

$\varepsilon_4(j)$ is the probability that there are j boundary crossings after entering the last LA_k until the current call arrives. Let T_A be the backward recurrence time (or age life) of the renewal time T_L.

Thus, according to the symmetry of the problem we conclude that

$$P\ [N(T_R)=j]=\varepsilon_2(j)=\varepsilon_4(j).$$

This equivalence is a direct consequence of the fact that the exponential distribution has the same memoryless distribution.

6.2.3 Calculation of $\varepsilon_3(j)$

$\varepsilon_3(j)$ is the probability that there are j cell boundary crossings in a time interval of duration T_L, i.e., we have to count the number of CRT's in a LART, or equivalently, $\varepsilon_3(j)=P[N(T_R)=j]$. In other words, we need to count the number of renewals in an ordinary renewal process in a random duration time T_L. However, since there is a probability zero for the coincidence of the LART and the CRT at the beginning of both renewal processes, then we should consider an equilibrium renewal process in this case as well.

Thus,

$$\sum_{j=0}^{\infty}\varepsilon_3(j)z^j=\theta_L\left\{G^*_{N(T)}(s,z)\right\}\Big|_{s=\theta_L}.$$

Hence,

$$\varepsilon_3(j)=\varepsilon_2(j) \quad (6.2.9)$$

This result is a consequence of considering exponentially distributed LARTs.

6.2.4 Average Number of VLR Location Updates

Finally, we are ready to calculate $E[N(T_{cV})]$. By substituting Eqs. (6.1.2), (6.2.7), (6.2.8), and (6.2.9) into Eq. (6.2.1), after some algebraic simplifications, we obtain

$$\mathrm{E}[N(T_{cV})] = \sum_{n=1}^{N} q_n \mu_n \frac{1-f_X^*(\theta_L+\mu_n)}{(\theta_L+\mu_n)^2\,\mathrm{E}[X]} \frac{[f_X^*(\theta_L+\mu_n)]^{d-1}}{1-[f_X^*(\theta_L+\mu_n)]^d} + \sum_{n=1}^{N} q_n \frac{(2-A_n)}{\mu_n\,\mathrm{E}[X]} \frac{1-f_X^*(\theta_L)[f_X^*(\theta_L)]^{d-1}}{1-[f_X^*(\theta_L)]^d} \quad (6.2.10)$$

6.3 Cost Calculation

We denote by TC_{dyn} the total cost of location updates and paging. Thus, TC_{dyn} is given by

$$TC_{dyn} = \delta_{HLR}\,\mathrm{E}[N(T_{cH})] + \delta_{VLR}\,\mathrm{E}[N(T_{cV})] + \delta_{poll}[1+3d(d-1)],$$

where δ_{HLR} is the cost of performing an HLR location update, δ_{VLR} is the cost of performing a VLR location update, δ_{poll} is the cost of polling a cell ($\delta_{poll} > 0$) or cost of paging, Li, Pan, Jia, 2002; Rodríguez-Dagnino, Takagi, 2007).

Similarly, the Total Cost for the static scheme is given by

$$TC_{static} = \delta_{HLR}\,\mathrm{E}[N(T_{cH})] + \delta_{VLR}\,\mathrm{E}[N(T_{cH})] + \delta_{poll}[1+3r(r-1)],$$

where r is the number of cells in a location area assuming that the cells are either hexagonal or circular.

The numbers δ_{HLR} and δ_{VLR} take into account the cost for wireless and wireline bandwidth utilization and the processing cost for making location updates in the databases HLR and VLR, respectively. These costs are generally much larger than the cost of paging δ_{poll}. Since these numbers depend of the actual cost in each wireless provider company, we are only interested on the relative importance of these numbers. However, our approach is general enough to be tailored to each company needs. These numbers are the same for both the static and dynamic schemes and $\mathrm{E}[N(T_{cV})]$ is the same as $\mathrm{E}[N(T_{cH})]$ for the static scheme. However, $\mathrm{E}[N(T_{cV})]$ and the paging area depend on d for the dynamic scheme.

7. PERFORMANCE EVALUATION OF MOBILITY MANAGEMENT

We should notice that the size of the location area, defined by r, should be larger or equal than the size of the paging area, defined by d (the movement – based scheme threshold). In addition, δ_{HLR} and δ_{VLR} should be larger than δ_{poll} since performing location updates in HLR and VLR usually spends more wireless or wireline bandwidth utilization than polling a cell.

The mean LART value must be larger than the mean CRT value. This is so since a location area is composed by tens of hundreds of cells. Similarly to the values studied in (Zonoozi, Dassanayake, 1997) we assume $\mathrm{E}[T_L] = 1800$ sec (30 min) for the LART mean value, whereas $\mathrm{E}[X] = 120$ sec (2 min) for the CRT mean value. In the same manner, we assume four ICT mean values, $\mathrm{E}[T_c]$; namely $\mathrm{E}[T_c] = 60$ sec, 600 sec, 6000 sec, 60000 sec.

We take as our performance measure for numerical comparisons, the ratio TC_{dyn} / TC_{static}, i.e., the total cost for the dynamic movement-based location update scheme versus the total cost for the static location update scheme. We are interested in observing how the minimum occurs for the performance measure.

Figures 7, 8 and 9 show the behavior of the ratio TC_{dyn} / TC_{static}, for ($\delta_{HLR} = 20$, $\delta_{VLR} = 20$, $\delta_{poll} = 1$ and $r=30$) and hyperexponential ICT's. We consider exponential LA's and CRTs with the mean values defined as above, and q_1=0.1, 0.5 and 0.9 for the ICT's. For the case q_1=0.1 we assume μ_1=1/150; μ_2=1/50 for E[T_c]=60; μ_1=1/1500 and μ_2=1/500 for E[T_c]=600; μ_1=1/15000 and μ_2=1/5000 for E[T_c]=6000; and μ_1=1/150000 and μ_2=1/50000 for E[T_c]=60000. For the case q_1=0.5 we assume μ_1=μ_2=1/60 for E[T_c]=60; μ_1=μ_2=1/600 for E[T_c]=600 and μ_1=μ_2=1/6000 for E[T_c]=6000; and μ_1=μ_2=1/60000 for E[T_c]=60000. For the case q_1 = 0.9, we assume μ_1=1/50 and μ_2=1/150 for E[T_c]=60; μ_1=1/500; μ_2=1/1500 for E[T_c]=600;

Figure 7. Comparison between the exponential case for LA, ICT, and CRT, and hyperexponential for ICT with N=2, q_1=0.1, μ_1=1/150, μ_2=1/50. $\delta_{HLR} = 20, \delta_{VLR} = 20, \delta_{poll} = 1,$ *and* $r = 30.$

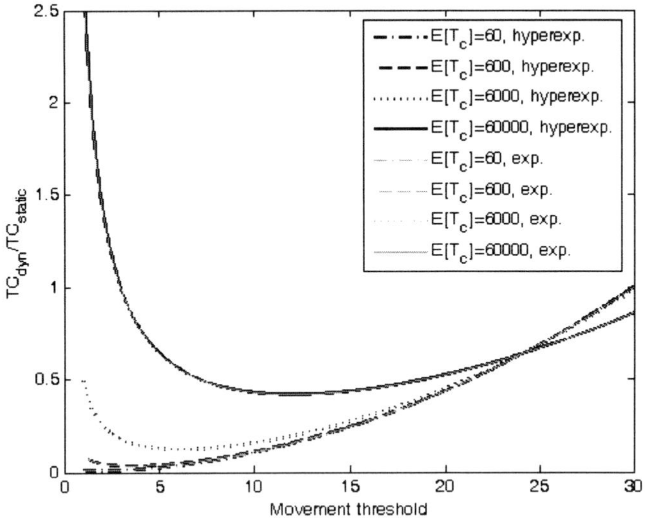

and μ_1=1/5000; μ_2=1/15000 for E[T_c]=6000; and μ_1=1/50000; μ_2=1/150000 for E[T_c]=60000. We have matched mean values, so the means for the exponential distribution for ICT are given by 60, 600, 6000 and 60000.

We can conclude from these plots that hyper-exponential distribution consideration has not a significant effect on the minimum values of TC_{dyn} / TC_{static}. Plots with similar parameters practically superpose each other even when $E[T_c]$ is increased. A similar conclusion can be drawn after looking at Figures 10, 11 and 12, where we have modified δ_{VLR} to be equal to 5.

As it is stated in (Orlik, Rappaport, 1998),

Figure 8. Comparison between the exponential case for LA, ICT, and CRT, and hyperexponential for ICT with N=2, q_1=0.5, μ_1=1/60, μ_2=1/60

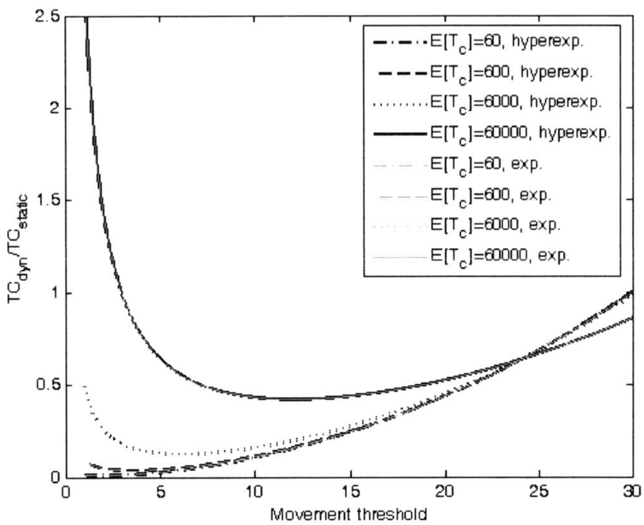

Figure 9. Comparison between the exponential case for LA, ICT, and CRT, and hyperexponential for ICT with N=2, q_1=0.9,μ_1=1/50,, μ_2=1/150

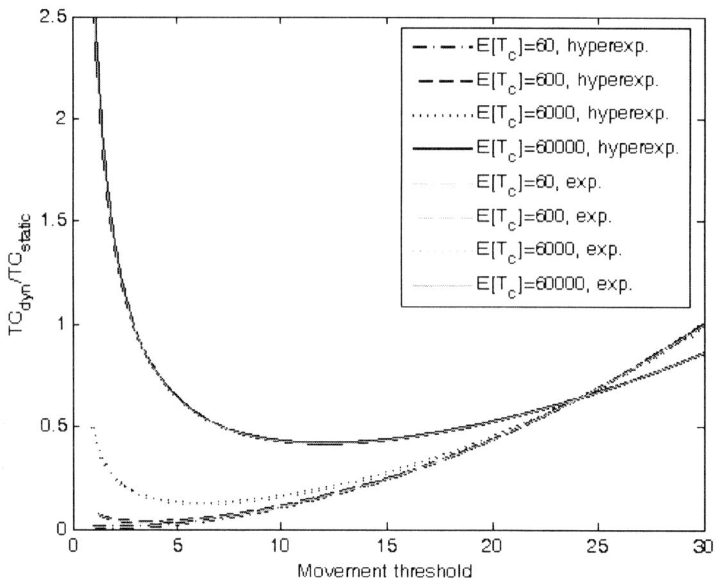

we can consider an equilibrium renewal process when the CRT are circular distributed. Let Z be the random variable for the distance traveled by a mobile in a straight line through a wireless cell. We assume circular wireless cells with radius R. Then, the pdf for the distance from an arbitrary point on the boundary of the circle where the mobile enters into a cell to another point on the boundary where the mobile exits from the cell in a straight line is exactly the same as the pdf for the length of a random chord of a circle with radius R in the sense of S-randomness (Rodríguez-Dagnino, Takagi, 2005; eq. (2.3.41), p. 198). This pdf is given by

$$f_Z(z) = \frac{2}{\pi\sqrt{4R^2 - z^2}}; \quad 0 \le z \le 2R.$$

Then the CRT for this cell is given by $X = Z / V$ where V is the velocity, and its kth moment is given by

$$\mathrm{E}[X^k] = \frac{\Gamma\left(\frac{k+1}{2}\right)}{\sqrt{\pi}\,\Gamma\left(\frac{k+2}{2}\right)}\left(\frac{\pi\,\mathrm{E}[X]}{2}\right)^k; \quad k = 1, 2, \ldots.$$

$$\delta_{HLR} = 20,\ \delta_{VLR} = 20,\ \delta_{poll} = 1 \ \ and \ \ r = 30.$$

where $\mathrm{E}[X] = \frac{2}{\pi}\left(\frac{2R}{V}\right)$, and $\Gamma(\alpha) = \int_0^\infty t^{\alpha-1} e^{-t} dt$ is the gamma function.

We should remember that if all moments of a random variable X exist, and it has a pdf $f_X(X)$, then its L.-S.T. $f_X^*(s)$ can be expanded as a function of the moments of X as follows:

$$f_X^*(s) = \sum_{k=0}^{\infty} \frac{(-s)^k}{k!} \mathrm{E}[X^k]. \tag{7.1}$$

For the numerical results we have chosen 40 terms in Eq. (7.1). The exponential case for LART, ICT, and CRT, versus circular distributed CRTs has

Figure 10. Comparison between the exponential case for LA, ICT, and CRT, and hyperexponential for ICT with N=2, q_1*=0.1,*μ_1*= 1/150,,* μ_2*= 1/50*

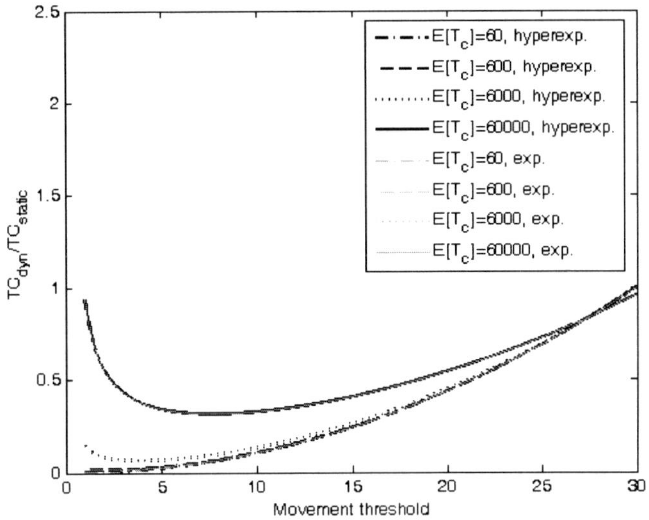

been considered in (Rodríguez-Dagnino, Takagi, 2007), and it was found that there is only a small effect when we consider the circular CRT instead of the exponential one.

We compare in Figures 13, 14 and 15, the cases hyperexponential ICT for N=2, exponential LART and exponential CRTs versus circular CRTs with the same mean value. We have fixed the parameters to $\delta_{HLR} = 20$, $\delta_{VLR} = 20$, $\delta_{poll} = 1$ and $r = 30$.

In Figure 13 we consider the case q_1*=0.1*, in Figure 14 we consider the case q_1*=0.5,* and in the same manner, in Figure 15 we consider the case q_1*=0.9*. We can conclude from these figures that minimums occur practically at the same value of d

Figure 11. Comparison between the exponential case for LA, ICT, and CRT, and hyperexponential for ICT with N=2, q_1*=0.5,* μ_1*= 1/60,* μ_2*= 1/60*

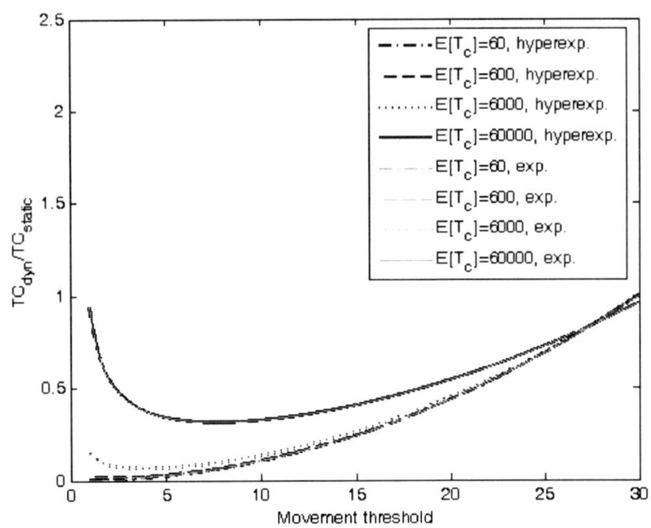

Figure 12. Comparison between the exponential case for LA, ICT, and CRT, and hyperexponential for ICT with N=2, q_1=0.9, μ_1=1/50, μ_2=1/150

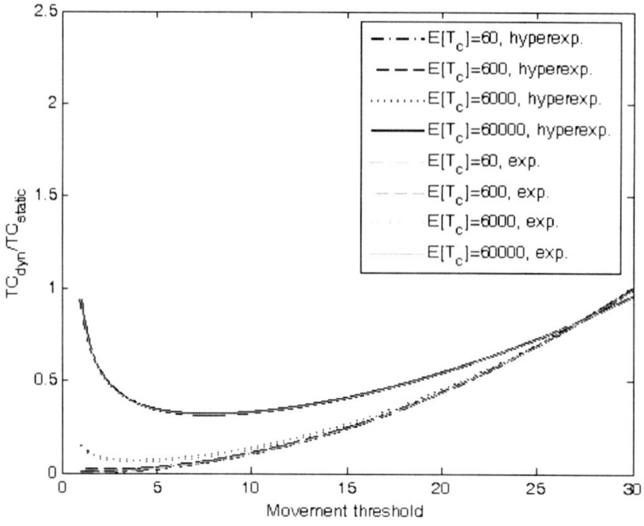

for the exponential CRT, and for the circular one, i.e., a few small changes occur by considering the circular CRT instead of the exponential one. A slight difference can be observed only when $\mathrm{E}[T_c]$ is increased up to 60000.

TC_{dyn} / TC_{static}, for $\delta_{HLR} = 20$, $\delta_{VLR} = 20$, $\delta_{poll} = 1$ *and* $r = 30$

TC_{dyn} / TC_{static}, for $\delta_{HLR} = 20$, $\delta_{VLR} = 5$, $\delta_{poll} = 1$ *and* $r = 30$

TC_{dyn} / TC_{static}, for $\delta_{HLR} = 20$, $\delta_{VLR} = 5$, $\delta_{poll} = 1$ *and* $r = 30$.

Figure 13. Comparison between exponential and circular CRT when ICT is hyperexponentially distributed

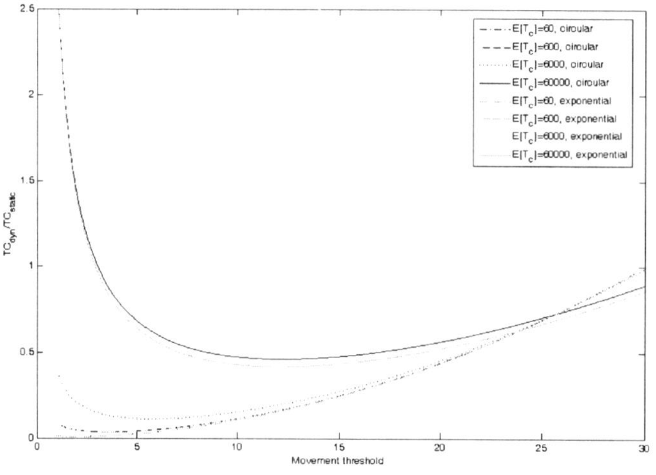

Figure 14. Comparison between exponential and circular CRT when ICT is hyperexponentially distributed

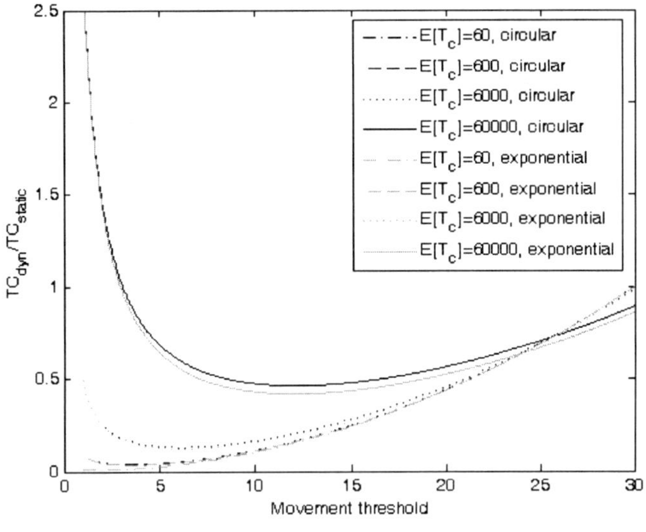

TC_{dyn} / TC_{static}, for
$\delta_{HLR} = 20,\ \delta_{VLR} = 5,\ \delta_{poll} = 1\ and\ r = 30.$

TC_{dyn} / TC_{static}, for
$\delta_{HLR} = 20, \delta_{VLR} = 20, \delta_{poll} = 1\, and\ r = 30.$

We consider N=2, q_1=0.1, μ_1= 1/150, μ_2= 1/50.

TC_{dyn} / TC_{static}, for
$\delta_{HLR} = 20, \delta_{VLR} = 20,\ \delta_{poll} = 1\ and\ r = 30.$

We consider N=2, q_1=0.5, μ_1= 1/60, μ_2= 1/60

Figure 15. Comparison between exponential and circular CRT when ICT is hyperexponentially distributed

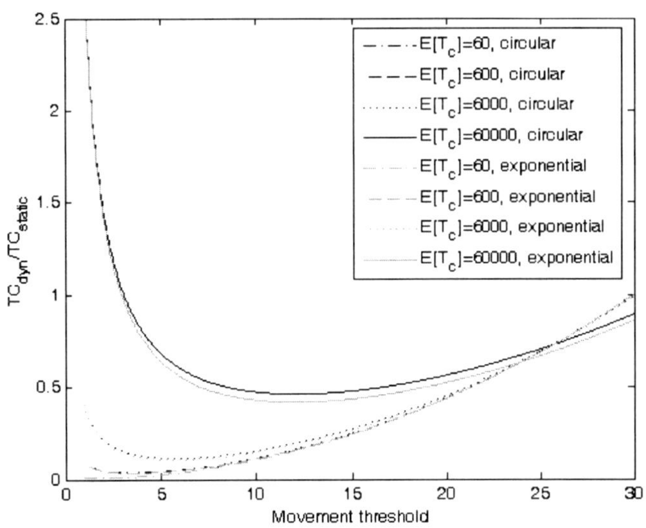

TC_{dyn} / TC_{static}, for

$\delta_{HLR} = 20, \delta_{VLR} = 20, \delta_{poll} = 1 \; and \; r = 30.$

We consider N=2, q_1=0.9, μ_1=1/50, μ_2=1/150

8. CONCLUSION

In the wireless literature context, by using different methods it was possible to find the pmf for general distributed CRTs, and exponentially distributed CHT for the equilibrium renewal process (Lin, Mohan, Noerpel, 1994), for distributions having rational Laplace transforms (Fang, Chlamtac, Lin, 1997), and for CHT having a PH distribution where the different exponential phases represent a stochastic number of subchannel-holding times (Christensen, Nielsen, Iversen, 2004); the PH distributed CHT allows for a new interpretation: as the time to absorption in a continuous time Markov process with two absorbing states, namely handover and terminated call. We realized that many of these results can be also found by using the methodology proposed by Cox in his monograph published in 1962, for mixture of Erlang distribution for T_c and general distributed X (Cox, 1962). Cox explained his methodology for simple cases related to the ordinary renewal process. We explored Cox's method in a more extensive manner for the equilibrium renewal process (Rodríguez-Dagnino, Takagi, 2003) and for the delayed renewal process (Rodríguez-Dagnino, Takagi, 2005) in a wireless environment context. We have also developed new formulae for general distributed CHT and exponentially distributed CRT, even for the case of all different distributed CRTs. For instance, we include examples of Pareto CHT in the modified renewal process, which cannot be analyzed by the other methodologies.

As it has been noted by Wang, Fan, Li, and Pan (2009) our approach for mobility management considers the more realistic situation of the interrelationships among LART, CRT, and ICT distributions and the associated cost.

In this Chapter we have modeled the behavior of a mobile relating the three important distributions, i.e., the exponential LART distribution, the hyperexponential ICT distribution, and general CRT distribution. We have studied the effect of having exponential and circular distributed CRTs.

The optimal performance measure TC_{dyn} / TC_{static}, for $\mathrm{E}[T_c] = 60$, 600, 6000 and 60000 seems to be always less than 1 for movement threshold values of $d \leq 15$. This means that the total cost of the dynamic scheme is smaller than the corresponding cost of the static scheme at the minimum of this performance measure. In fact, in many cases the minimum of this performance measure occurs for small values of d (location update threshold distance). On the other case, the minimums achieved are closer to 1 as $\mathrm{E}[T_c]$ is increased.

There are well-known published measurement studies regarding CRT and CHT and they cannot be assumed as exponential distributed in general. From this prospective our renewal process approach gives a general framework to consider distributions different from exponential. Unfortunately, there are no published measurement results for ICT and LART distributions, so there has not been comparison with real-world data as far as the authors are aware. There is an opportunity for research in gathering these measurements of actual wireless systems. Nonetheless, our approach can be adjusted to capture a large class of distributions for ICT, LART, and CRT, as we have shown in this Chapter. Our results have shown that the performance measure is only slightly affected by changing the ICT distribution from exponential to hyperexponential, and some small differences can be observed when we change the CRT distributions from exponential to circular and this effect is more noticeable for larger $\mathrm{E}[T_c]$ values.

So, the effects of a proper modeling of ICT and LART distributions may be important to make further justifications of the degree of sensitivity for movement threshold values after considering more probability distributions.

The insensitivity was also observed in our previous work (Rodríguez-Dagnino, and H. Takagi, 2007) regarding the situation of hyperexponential LART distribution, exponential ICT distribution, and general CRT distributions, and recently by Wang, Fan, Li, and Pan (2009) by considering distributions with rational Laplace transforms for LARTs and CRTs, and exponential interarrival distributions.

ACKNOWLEDGMENT

The first author thanks Tecnológico de Monterrey, for the support provided in the development of the work through the Research Chair of Telecommunications.

REFERENCES

Akyildiz, I. F., Ho, J. S. M., & Lin, Y.-B. (1996). Movement-based location update and selective paging for PCS networks. *IEEE Transactions on Networking*, *4*(4), 629–638. doi:10.1109/90.532871

Bar-Noy, A., Kessler, I., & Sidi, M. (1994). Mobile users: To update or not to update. In *IEEE Proc. INFOCOM*, Toronto, Ontario, Canada, June 14-16 (Vol. 1, pp. 570-576).

Barceló, F., & Jordán, J. (2000). Channel holding time distribution in public telephony systems (PARM and PCS). *IEEE Transactions on Vehicular Technology*, *49*(5), 1615–1625. doi:10.1109/25.892545

Christensen, T. K., Nielsen, B. F., & Iversen, V. B. (2004). Phase-type models of channel-holding times in cellular communication systems. *IEEE Transactions on Vehicular Technology*, *53*(3), 725–733. doi:10.1109/TVT.2004.825803

Cox, D. R. (1962). *Renewal Theory*. London: Methuen & Co.

Fang, Y. (2003). Movement-based mobility management and trade-off analysis for wireless mobile networks. *IEEE Transactions on Computers*, *52*(6), 791–803. doi:10.1109/TC.2003.1204834

Fang, Y., Chlamtac, I., & Lin, Y.-B. (1997). Modeling PCS networks under general call holding time and cell residence time distributions. *IEEE/ACM Transactions on Networking*, 5(6), 893-905.

Glisic, S. G. (2006). *Advanced Wireless Networks, 4G Technologies*. Hoboken, NJ: John Wiley & Sons, Ltd.

Jedrzycki, C., & Leung, V. C. M. (1996). Probability distribution of channel holding time in cellular telephony systems. In *Proc. Vehicular Technology Conf. (VTC'96)*, August 1996, (pp. 247-251).

Kozatchok, I., & Pierre, S. (2002). User tracking and mobility management algorithms for wireless networks. *The Computer Journal*, *45*(5), 525–539. doi:10.1093/comjnl/45.5.525

Li, J., Kameda, H., & Li, K. (2000). Optimal dynamic mobility management for PCS networks. *IEEE/ACM Transactions on Networking*, *8*(3), 319-327.

Li, J., Pan, Y., & Jia, X. (2002). Analysis of dynamic location management for PCS networks. *IEEE Transactions on Vehicular Technology*, *51*(5), 1109–1119. doi:10.1109/TVT.2002.800632

Lin, Y.-B., Mohan, S., & Noerpel, A. (1994). Queuing priority channel assignment strategies for handoff and initial access for a PCS network. *IEEE Transactions on Vehicular Technology*, *43*, 704–712. doi:10.1109/25.312778

Ma, W., & Fang, Y. (2004). Dynamic hierarchical mobility management strategy for mobile IP networks. *IEEE Journal on Selected Areas in Communications*, *22*(4), 664–676. doi:10.1109/JSAC.2004.825968

Nanda, S. (1993). Teletraffic models for urban and suburban microcells: Cell sizes and handoff rates. *IEEE Transactions on Vehicular Technology*, *42*, 673–682. doi:10.1109/25.260744

Orlik, P. V., & Rappaport, S. S. (1998). Traffic performance and mobility modeling of cellular communications with mixed platforms and highly variable mobilities. *Proceedings of the IEEE*, *86*, 1464–1479. doi:10.1109/5.681374

Park, K., & Willinger, W. (Eds.). (2000). *Self-similar network traffic and performance evaluation*. Hoboken, NJ: John Wiley & Sons, Inc.

Rodríguez-Dagnino, R. M. (1998). Handoff analysis in wireless multimedia networks. In *SPIE, ITCom Conference on Performance and Control of Network Systems II* (Vol. 3530, pp. 76-84).

Rodríguez-Dagnino, R. M., Hernández-Lozano, G., & Takagi, H. (2000). Wireless handover distributions in mixed platforms with multimedia services. In *SPIE, ITCom Conference on Internet Quality and Performance and Control of Network Systems* (Vol. 4211, pp. 59-69).

Rodríguez-Dagnino, R. M., & Leyva-Valenzuela, C. A. (1999). Performance analysis in broadband wireless networks. In *SPIE, ITCom Conference on Performance and Control of Network Systems III* (Vol. 3841, pp. 220-228).

Rodríguez-Dagnino, R. M., Ruiz-Cedillo, J. J., & Takagi, H. (2002). Dynamic mobility management for cellular networks: A delayed renewal process approach. *IEICE Transactions on Communications . E (Norwalk, Conn.)*, *85-B*(6), 1069–1074.

Rodríguez-Dagnino, R. M., Ruiz-Cedillo, J. J., & Takagi, H. (2005). Mobility management for general distributed location areas. In *SPIE, ITCom Conference on Performance, Quality of Service, and Control of Next-Generation Communication and Sensor Networks III* (Vol. 6011, pp. 1-13).

Rodríguez-Dagnino, R. M., & Takagi, H. (2001). Handover Distributions for wireless cellular networks with Pareto call holding times. In *SPIE, ITCom Conference on Internet Performance and Control of Network Systems II* (Vol. 4523, pp. 202-210).

Rodríguez-Dagnino, R. M., & Takagi, H. (2002). Counting handovers in a random number of talk-spurts and silence periods. In *SPIE, ITCom Conference on Internet Performance and Control of Network Systems III* (Vol. 4865, pp. 202-212).

Rodríguez-Dagnino, R. M., & Takagi, H. (2003). Counting handovers in a cellular mobile communication network: Equilibrium renewal process approach. *Performance Evaluation*, *52*(2-3), 153–174. doi:10.1016/S0166-5316(02)00178-5

Rodríguez-Dagnino, R. M., & Takagi, H. (2005). Distribution of the number of handovers in a cellular mobile communication network delayed renewal process approach. *Journal of the Operations Research Society of Japan*, *48*(3), 207–225.

Rodríguez-Dagnino, R. M., & Takagi, H. (2007). Movement-based location management for general cell residence times. *IEEE Transactions on Vehicular Technology*, *56*(5), 2713–2722. doi:10.1109/TVT.2007.900377

Rodríguez-Dagnino, R. M., & Takagi, H. (2007). Modeling of handover counting and location management for wireless mobile networks. In *SPIE, ITCom Conference on Next-Generation Communications and Sensor Networks* (Vol. 6773, pp. 1-13).

Takagi, H., & Rodríguez-Dagnino, R. M. (2007). Counting the number of renewals during a random interval in a discrete-time delayed renewal process. *Operations Research Letters*, *35*(1), 119–124. doi:10.1016/j.orl.2005.12.002

Wang, X., Fan, P., Li, J., & Pan, Y. (2009). Modeling and Cost Analysis of Movement-Based Location Management for PCS Networks with HLR/VLR Architecture and General Location Area and Cell Residence Time Distributions. *IEEE Transactions on Vehicular Technology*, *57*, 3815–3831. doi:10.1109/TVT.2008.919613

Yeung, K. L., & Nanda, S. (1996). Channel management in microcell/macrocell cellular radio systems. *IEEE Transactions on Vehicular Technology*, *45*, 601–612. doi:10.1109/25.543716

Zonoozi, M. M., & Dassanayake, P. (1997). User mobility modeling and characterization of mobility patterns. *IEEE Journal on Selected Areas in Communications*, *15*, 1239–1252. doi:10.1109/49.622908

Chapter 13
Supporting Multiple Quality-of-Service Classes in IEEE 802.16e Handoff

Melody Moh
San Jose State University, USA

Teng-Sheng Moh
San Jose State University, USA

Bhuvaneswari Chellappan
San Jose State University, USA

ABSTRACT

IEEE 802.16 WiMAX (Worldwide Interoperability for Microwave Access) is a major standard technology for Wireless Metropolitan Area Networks (Wireless MAN). Quality-of-service (QoS) scheduling class and mobility management are two main issues for supporting seamless high-speed data and media-stream communications. Previous works on WiMAX handoff however have mainly addressed a particular scenario or a single QoS class. This chapter first presents an overview of the QoS scheduling classes supported by the IEEE 802.16 standard, followed by a survey of major related works proposed to enhance 802.16e handoffs. Next, it will present a new context-sensitive handoff scheme that supports the five 802.16 QoS scheduling classes, and is energy-aware – it may switch to energy-saving mode during handoff. It will then illustrate performance evaluation, which will show that, compared to three existing methods, the proposed scheme successfully supports the five QoS classes in both layers 2 and 3 handoff, decreases end-to-end handoff delay, delay jitter, and service disruption time; it also increases throughput and energy efficiency. Finally, key implementation and cost issues are discussed. We believe that this chapter is a significant contribution for providing high-quality, seamless data and media streaming over 802.16 as well as LTE (Long-Term Evolution) cellular networks, and would be a valuable part of QoS architectures in the wireless networking domain.

DOI: 10.4018/978-1-61520-680-3.ch013

INTRODUCTION

The recent rapid progress in wireless networking has paved the way for ubiquitous and pervasive computing. Its fast advancement has motivated the evolution of next-generation wireless technologies to reach longer distance, higher data rate, and better QoS. This has culminated the widespread of academic and industry efforts in both broadband wireless networks and metropolitan area networks.

The IEEE standard 802.16, also called WiMAX, specifies the "Air-Interface for Fixed Broadband Wireless Access Systems" (IEEE 802.16 Working Group, 2004). The amendment 802.16e has been defined to support mobility and other extensions (IEEE 802.16 Working Group, 2005). Based on the 802.16e handoff (HO) mechanism, many new HO enhancements have been proposed (Chang, 2005; Chen & Hsieh, 2007; Cho et al., 2006; Choi et al., 2005; Das et al., 2006; Hu et al., 2007; Jang et al., 2007; Kim et al., 2005; Lee et al., 2006; Leung et al., 2005; Rouil & Golmie, 2006; Rouil & Golmie, 2007; Yang et al., 2007; Zhong et al., 2007). Most of them, however, try to address only a specific scenario or a particular QoS issue.

The IEEE 802.16d and 802.16e have specified five QoS scheduling classes (IEEE 802.16 Working Group, 2004;, IEEE 802.16 Working Group, 2005): (1) *Unsolicited Grant Services* (*UGS*) – for real-time uplink of fixed-size data packets generated periodically, such as voice over IP (VoIP). (2) *Real-Time Polling Services* (*rtPS*) – for real-time uplink of variable-sized data packets on a periodic basis, such as video streaming. (3) *Extended Real Time Polling Services* (*ErtPS*) – added in 802.16e and an enhancement of rtPS, in which the base station (BS) provides unicast grants in an unsolicited manner with dynamics allocations. (4) *Non-Real-Time Polling Services* (nrtPS) – for delay-tolerant, loss-sensitive data streams with variable-sized packets for which a minimum data rate is required, such as file transfer. (5) *Best Effort Services* (BE) – supporting data for which no minimum service level is required.

We observed that the QoS requirement of a mobile station (MS) may vary over time; a HO scheme that supports only one QoS class is neither complete nor practical. Further, many MSs are battery-powered, yet few of the existing schemes have addressed energy consumption issue.

This chapter proposes a QoS-aware HO scheme with the following major features:

1. *It is context-sensitive, supporting five QoS modes that correspond to 802.16 QoS scheduling classes* (Arunachalam, 1999; 11IEEE 802.16 Working Group, 2004; IEEE 802.16 Working Group, 2005).
2. *It is energy-sensitive and is designed to conserve energy during low-activity modes.*
3. *Depending on the particular handoff scenario, it may operate in either layer 3 or layer 2, and in either predictive or reactive modes (Jang et al., 2007).*

The chapter is organized as follows. Following this section, the next section presents background and related studies, including the five QoS scheduling classes, the basic IEEE 802.16 handoff scheme, and related works on the 802.16 handoff. Then, proposed context-sensitive, QoS-aware handoff scheme is described. This is followed by performance evaluation of the new scheme. Issues on implementation and costs are then discussed. Finally, a conclusion remark is presented, with suggested future directions.

BACKGROUND AND RELATED STUDIES

This section first illustrates the quality of services support in IEEE 802.16. Next, the basic handoff scheme of 802.16e is described. A brief survey of existing proposed HO mechanisms is then presented.

Quality of Services Support in IEEE 802.16

As mentioned in the introduction section, The IEEE 802.16d and 802.16e have specified five QoS scheduling classes (IEEE 802.16 Working Group, 2004; IEEE 802.16 Working Group, 2005). In this section, we provide an overview of the support of QoS in IEEE 802.16, and describe the five QoS scheduling classes in a greater detail.

The QoS support in the 802.16 standard is connection oriented. The standard specifies the offering of varying degrees of QoS for different type of transmissions. Initially, when a SS (Subscriber Station) is introduced into the network, three dedicated connections are set up for transfer of management messages. This reflects the fact that there are 3 levels of QoS for handling management traffic:

1. **QoS classification:** This function, performed at the Convergence Sub-layer (CS), maps the upper-layer data to the appropriate Connection Identifier (CID). By doing so, the proper QoS and data service parameters are set up at connection establishment, which enables the correct handling of data for the particular connection.
2. **QoS scheduling:** Scheduling data is performed by the MAC to ensure that data is handled appropriately. It offers ATM (Asynchronous Transfer Mode)-like data handling services. The QoS scheduling services will be described in details in the next subsection.
3. **Service flows:** The service flow in an additional level of QoS support. A service flow is a unidirectional flow of packets that is provided by a particular QoS support. To specify a service flow, the 802.16 standard uses five characteristics: the service flow ID (SFID), which uniquely identifies each service flow; the connection ID (CID), which maps to an SFID that exists only when the connection has an admitted or active service flow; the provisioned QoS parameter set (ProvisionedQoSParamSet), which is A QoS parameter set that is provided by means outside the standard of this network; the admitted QoS parameter set (AdmittedQoSParamSet), which defines a set of QoS parameters for which the nodes are reserving resources; and the authorization module, which is a logical function within BS that approves or denies changes to the QoS parameters. Finally, note that the active set is always a subset of the admitted set, which in turn is a subset of the authorized set.

QoS Scheduling Classes in IEEE 802.16

In this section, we describe the five QoS scheduling classes specified in the IEEE 802.16d and IEEE 802.16e standard. Most of them may be applied to the LTE (Long-Term Evolution) standard for cellular networks as well.

1. **Unsolicited grant service (UGS):** This class supports real-time data that generates fixed-size packets periodically. One important example is VoIP without silence suppression. In this class, the BS periodically provides data grant information elements to enable SS to transmit; SS does not need to participate in contention request opportunities. Furthermore, since constant data rates are supported, the overhead of bandwidth request by SS is avoided. The key parameters of this class include maximum sustained traffic rate, maximum latency, maximum jitter tolerance, and request/transmission policy. Note that this service is similar to the CBR (Constant Bit Rate) data service under ATM.
2. **Real-time polling service (rtPS):** This class supports real-time traffic that periodically

generates variable-sized packets. Examples include compressed audio and video streams such as MPEG (Motion Picture Expert Group) video. In order to improve transmission efficiency, the BS grants unicast request opportunities for the SS to specify its grant size dynamically according to the actual traffic needs. Similar to the UGS class, SS does not need to participate in contention request opportunities. Due to periodic requests of SS, the overhead of this class is more than that of UGS. Yet, it offers the flexibility by allowing efficient variable-sized data transmission. The key parameters include minimum reserved rate, maximum sustained rate, maximum latency, and request/transmission policy. Note that this service is similar to the rtVBR (real-time Variable Bit Rate) data service under ATM.

3. **Extended real-time polling service (ErtPS):** This class is later added to the IEEE 802.16e (with mobility support). It also supports real-time traffic that periodically generates variable-sized packets. Examples include VoIP with Activity Detection. This is a combination of UGS and rtPS. Like UGS, it supports periodic *unsolicited* grants, yet, like rtPS, the grant size may be dynamically changed by request. It however avoids the periodical granting of unicast request opportunities from BS to the SS. Thus, the overhead is less than that of rtPS. The key parameters include minimum reserved rate, maximum sustained rate, maximum latency, jitter tolerance, and request/transmission policy.
4. **Non-real-time polling service (nrtPS):** This class supports variable-traffic that tolerates delay, such as File Transfer Protocol (FTP). In this service, the BS periodically offers unicast request opportunities for connections, even in times of congestion, and SS participates in contention request mechanisms. The key parameters include minimum reserved rate, maximum sustained rate, and request/transmission policy. Note that this service is similar to the nrtVBR (non-real-time Variable Bit Rate) data service under ATM.
5. **Best effort (BE):** This class supports data that have no specific QoS (data rate or latency) needs, but are to be transmitted when there is available bandwidth. In this service, SS use contention request and unicast request opportunities to gain network access. The key parameters include maximum sustained rate and request/transmission policy. Note that depending on the applicability, this service may be similar to UBR (Unspecific Bit Rate), ABR (Available Bit Rate), and GFR (Guaranteed Frame Rate) data services under ATM.

IEEE 802.16e Handoff

The basic operation of 802.16e HO process is in layer 2, and may be depicted as in figure 1, and described below.

Before performing the HO, the MS acquires information about neighboring BS through the MOB_NBR-ADV (mobile neighbor advertisement) message broadcast by the serving BS. When the MS decides to perform a handoff, it sends MOB_MSHO-REQ (mobile MS HO request) message to its serving BS. The BS responds with a MOB_BSHO-RSP (mobile BS HO response) message, in which it will include the recommended target BS to which the MS can move. The MS finally sends the MOB_HO-IND (mobile HO indication) message before attaching to the target BS. Alternatively, the handoff may also be initiated by the serving BS by sending a MOB_BSHO-REQ (mobile BS HO request) message.

After sending out MOB_HO-IND, the MS synchronizes with the target BS's downlink. The MS obtains the DL-MAP (downlink map), DCD (downlink channel descriptor), UL-MAP (uplink map), and UCD (uplink channel descriptor) through the synchronized downlink. Finally, the

Figure 1. IEEE 802.16e Handoff (HO)

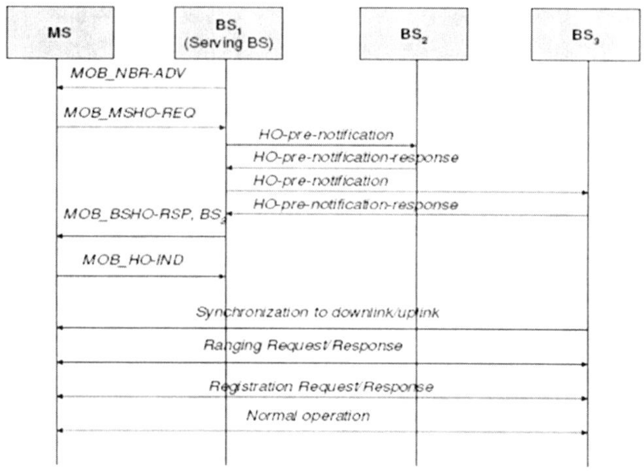

MS performs ranging and completes the network re-entry process with the target BS. After this, the MS and its new serving BS engage in normal operation.

Related Studies in IEEE 802.16 Handoff

Many handoff schemes have been proposed, they may be broadly categorized as layer-2 handoff (L2HO) and layer-3 handoff (L3HO) (Hu et al., 2007). In the following, several major ones are briefly summarized.

Layer-2 Schemes

Leung et al. aimed to achieve mobility with the 802.16d fixed WiMAX systems (2005). Choi et al. reduced service disruption during real-time data reception in the downlink during handoff (Choi et al., 2005); the target BS transmits data in the downlink before the MS completes the entire L2HO. This approach has been adopted in the proposed scheme presented in this chapter. Lee et al aimed to reduce handoff delay by avoiding unnecessary scanning of neighboring BSs (Lee et al., 2006). Rouil and Golmie focused on the scanning phase of handoff to minimize the impact of scanning on the QoS experienced (2006). Cho et al based their handoff decision on both uplink and downlink signals, aided by measurements from neighboring BS and MS (Cho et al., 2006).

In the following, we describe the fast handover scheme for real-time downlink services proposed by Choi; this scheme has been adopted in one handoff sub-scheme in the proposed propotocol. This scheme, suggested improvements so that real-time data reception is not affected by handoff latency and packet loss (Choi et al., 2005). The main idea of the scheme revolved around the idea of synchronization downlink before the entire handoff was complete. Therefore, real-time data reception on the downlink could resume before the entire handoff is complete. Since real-time data beyond a certain delay (playout delay) would be discarded, the scheme claimed to reduce the packet loss probability.

This scheme considered the hard handoff mode. When the MS sends the handoff request (MOB_MSSHO_REQ), the serving BS communicates with neighboring BSs and determines a suitable target BS for MS. The serving BS then sends a handoff response (MOB_HO_RSP) to MS containing information about the suitable target BS. Then, the MS sends handoff indication. The serving BS transfers the security related information of the MS to the target BS and starts forwarding the real-time data destined for the

MS to the target BS (MSS-Data-Forwarding). Upon completion of downlink synchronization, the MS will be able to receive the real-time data in the downlink through the reception of a new message, Fast_DL_MAP_IE. The remaining steps in 802.16e handoff are performed (uplink synchronization, ranging, registration etc.) simultaneously. After handoff is complete, normal operation proceeds, and the old serving BS stops forwarding data.

Layer-3 Schemes

Jang et al. (2007) proposed a L3 solution for the 802.16e network based on FHMIPv6 (Fast Handovers for Mobile IPv6) (Koodli, 2005). It aimed to reduce handoff latency by preparing L3 handoff simultaneously while L2 handoff is in process; it included two modes, predictive and reactive modes. This scheme has been adopted in our proposed scheme as a base framework. Chen and Hsieh improved the previous scheme by coinciding L2 and L3 HO through the combination of several control messages (Chen & Hsieh, 2007). Kim et al proposed a micro-mobility solution called LPM (Last Packet Marking), which used "bicasting" such that data for the MS is bicasted to both serving and target BS (Kim et al., 2005). Das et al. proposed to establish tunnels to potential target ASN-GW (access service network gateway) which is the L3 entity that the BS is attached (2006). Chang proposed to use HMIP (Hierarchical Mobile-IP) to avoid unnecessary home agent registrations for fast handoff (Chang, 2005).

In the following, the Mobile IPv6 fast handover scheme proposed by Jang et al. (2007) is described, since its predictive mode has been adopted in our proposed scheme as a base framework. This scheme aimed to reduce handoff latency by initiating steps for an impending handoff in advance. It is based on Mobile IPv6 Fast Handover protocol (FMIPv6), an IETF draft. Under this scheme, the handoff procedure of FMIPv6 has been reused to suit the 802.16 link layer technology. It uses the primitives proposed by IEEE 802.21 for performing a handoff. In this scheme, there are two modes of handoff, called predictive mode and reactive mode. The distinction between these two modes will be described later .We will initially describe the predictive mode of handoff.

After the 802.16e handoff mechanism, the L2 (layer 2) of MN notifies its L3 (layer 3) that a new link is detected through the LD (Link_Detected) primitive. In order to find those ARs to which potential target BSs are attached, the MN and PAR (previous access router) exchange RtSolPr (router solicitation for proxy advertisement) and PrRtAdv (proxy router advertisement) messages.

When L2 decides to perform a handoff (through the exchange of handoff request-response mechanism), it notifies L3 through the LHI (Link_Handover_Imminent) primitive. Then, the L3 of MN sends FBU (fast binding update) to the PAR. The PAR then sets up a tunnel with NAR (new access router) through the exchange of HI (Handover Initiate) and HACK (Handover Acknowledge) messages. Then the PAR sends an FBACK (fast binding acknowledgement) to the MN. Note that if the MN receives FBACK before sending MOB_HO-IND, it runs in Reactive mode.

PROPOSED SCHEME

This section first describes the basic features and QoS modes of the proposed scheme. Next, the handoff sub-schemes are illustrated; this is following by a description of three additional features supported by the proposed scheme.

Basic Features and QoS Modes

Five handoff QoS modes are defined for the proposed scheme, whose names adopted the names proposed very early in the 802.16 Working Group (Arunachalam, 1999). Each mode supports one or more QoS scheduling services defined in the 802.16 standard (IEEE 802.16 Working Group,

Figure 2. Mode 1 - Conversational Mode (UGS)

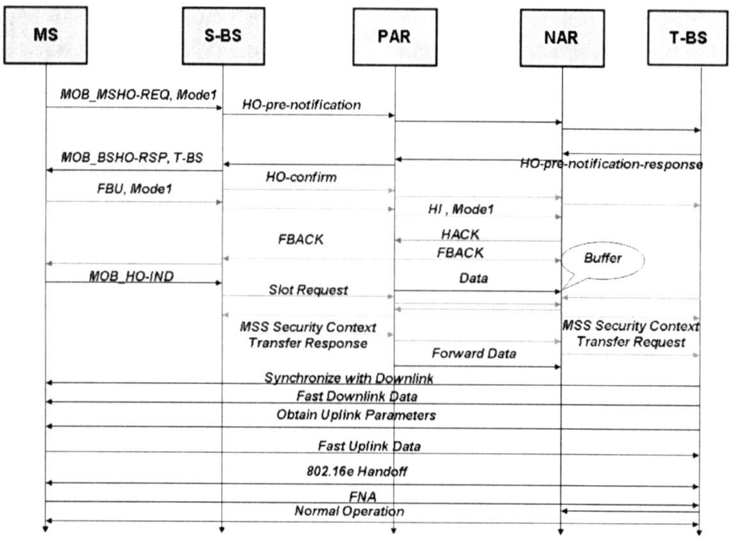

2004; IEEE 802.16 Working Group, 2005), described below:

- **Mode 1**: Conversational mode (for UGS)
- **Mode 2:** Streaming mode (for rtPS and ErtPS)
- **Mode 3:** Interactive mode (for nrtPS)
- **Mode 4:** Background mode (for nrtPS and BE)
- **Mode 5:** Standby mode (No traffic)

A particular handoff mode will be selected by the MS based on its current context (i.e., its QoS requirements). Once a particular mode is selected, its corresponding sub-schemes will be executed (to be described in Section 3.2). The proposed scheme is extended from two existing methods (Choi et al., 2005; Jang et al., 2007) with substantial enhancement. The main technical features are as follows:

1. A fast uplink service is designed for Mode 1 – Conversational mode.
2. A fast downlink service is adopted (Choi et al., 2005) for Modes 1 and 2 (Conversational and Streaming modes).
3. Two low-power handoff operations are designed for Modes 4 and 5 (Background and Standby modes).
4. *A basic L3HO framework (Jang et al., 2007) is adopted, with extension for concurrent L2HO support, in both predictive and reactive modes.*

Handoff Sub-Schemes

The proposal consists of five sub-schemes, each corresponding to a QoS mode. Once a mode is selected at the MS, it may specify the mode in the MOB_MSHO-REQ to the serving BS, and/or in the FBU (Fast Binding Update) to the PAR (Previous Access Router). The PAR may then inform the NAR (Next AR) through the Handoff Initiate (HI) message. The sub-scheme that corresponds to the selected mode would then be executed.

Mode 1: Conversational Mode (UGS)

The most important feature of conversational mode is its involvement of two-way exchange of real-time traffic (such as VoIP). This sub-scheme therefore employs a new design of *fast uplink* data transmission and adopts a fast downlink data reception (Choi et al., 2005), described below (refer to Fig. 2):

When the MS sends a handoff (HO) request (along with the HO mode), the serving BS notifies its neighbor BSs, receives their responses and eventually sends a HO response to the MS, with the target BS information. To support fast downlink data reception, the serving and target BSs exchange security information of the MS, (shown in red in Fig. 2) (Choi et al., 2005). On reception of a HO response, the MS transmits FBU to establish the tunnel (shown in blue) (Jang et al., 2007). After the tunnel is established, the MS sends HO indication and proceeds with standard 802.16e HO with the target BS.

After the serving BS receives a HO indication, it sends an uplink slot request to the target BS (depicted in green) to support *fast uplink* transmission. As the MS proceeds with HO, the data destined for the MS is tunneled by PAR to NAR, where the data is buffered. On the completion of downlink synchronization, the MS will be able to receive data transmitted by the target BS faster in the downlink. When the MS obtains uplink parameters, it will find its allocated slots in the uplink, and hence, will be able to transmit uplink data. The MS simultaneously completes 802.16e HO. Finally, it sends FNA (Fast Neighbor Advertisement) and proceeds with normal operation.

Mode 2: Streaming Mode (rtPS, ErtPS)

This mode (shown in Figure 3) supports audio and video download. It involves one-way reception of real-time data in the downlink, but does not require faster uplink transmission. It therefore does not need, as in mode 1, the request of slots by the serving BS to the target BS (note that the green arrow in Fig. 2 has been removed from Fig. 3).

Mode 3: Interactive Mode (nrtPS)

The interactive mode (Figure 4) involves applications that are loss-sensitive (such as FTP), but can tolerate longer delay. To avoid packet loss, the proposed scheme involves tunnel establishment and buffering by NAR (Jang et al., 2007). It however does not need fast uplink or downlink communication (the red and green arrows in Figure 2 are removed from Fig. 4).

Figure 3. Mode 2 - Streaming Mode (rtPS, ErtPS)

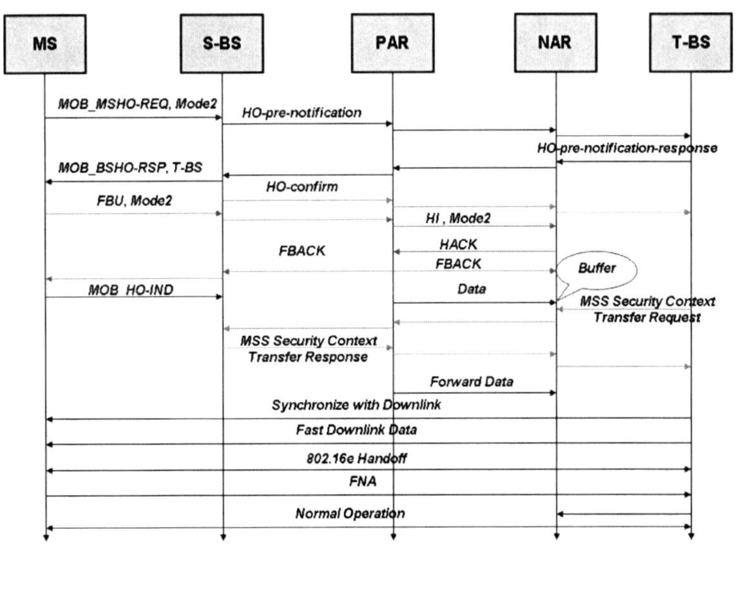

Mode 4: Background Mode (nrtPS, BE)

This mode is similar to the interactive mode except that it supports applications with a greater delay tolerance. The proposed scheme saves energy by requiring the BS to send an unsolicited sleep response to the MS. This allows a sleep interval before resuming HO (Figure 5).

Mode 5: Standby Mode (No traffic)

In the standby mode (Figure 6), the MS is idle. Tunnel establishment is therefore not needed. In addition, a sleep interval for the MS is added.

Additional Features

In the following, we discuss three additional features of the new proposed handoff scheme.

Figure 4. Mode 3 - Interactive Mode (nrtPS)

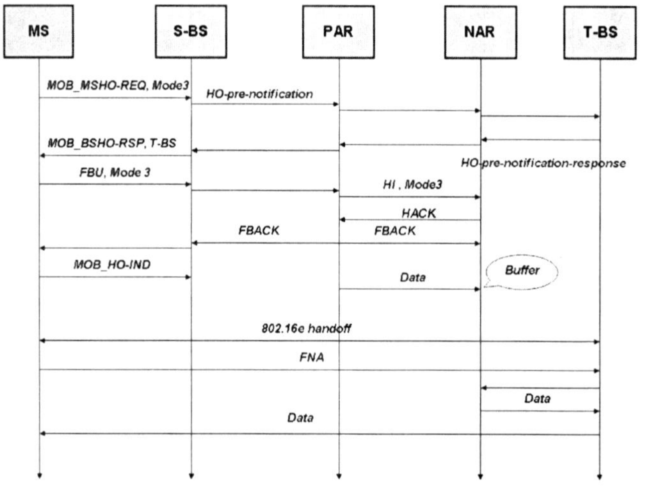

Figure 5. Mode 4 - Background Mode (nrtPS, BE)

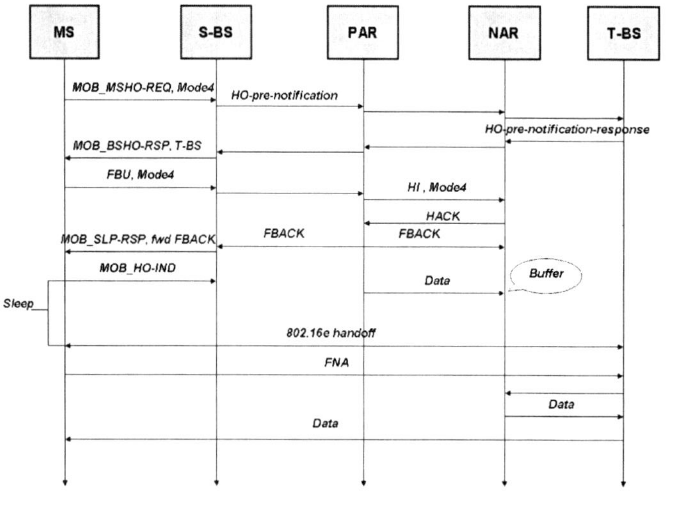

Figure 6. Mode 5 - Standby Mode (No traffic)

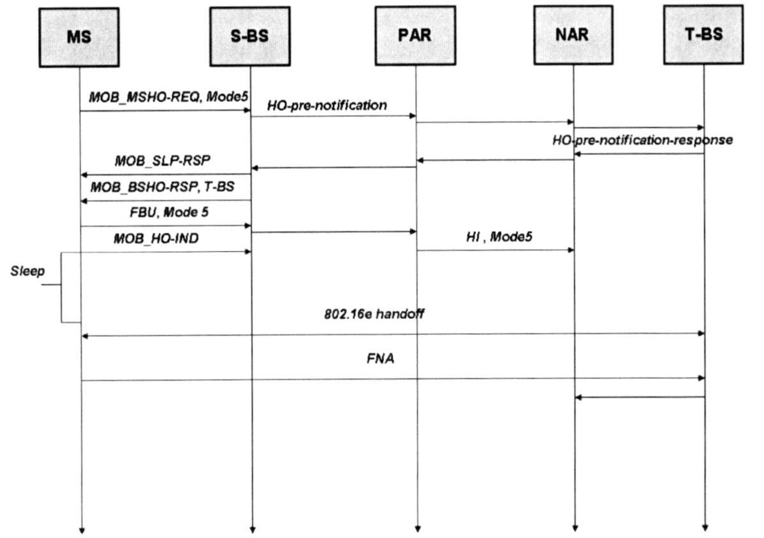

Figure 7. Topology

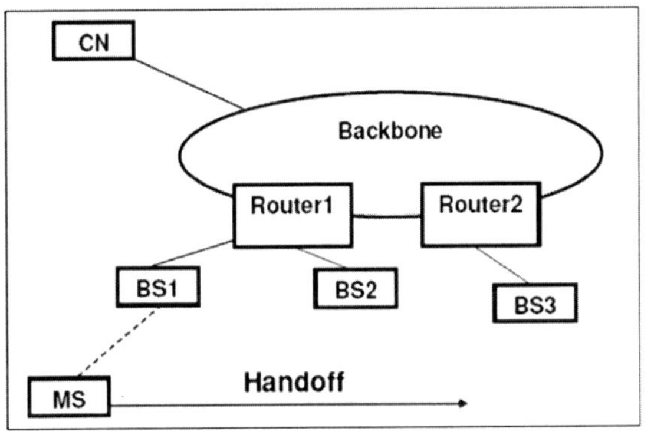

Supporting Predictive and Reactive Modes

The design has been presented using predictive mode (Jang et al., 2007). That is, FBU need to be sent in advance so that FBACK (Fast Binding Acknowledgement) is received by the MS before sending out a HO indication. This mode is vital in the first two (conversational and streaming) modes and should be supported. In the other three modes, this requirement may be relaxed, allowing the simpler, reactive mode to be followed.

Supporting Macro- and Micro-Mobility

The schemes described above mainly deal with macro-mobility. Micro-mobility may be supported by requiring the PAR to check if the target BS is attached. If so, L3 mobility and certain associated messages (HI, HACK, etc) may be avoided and only L2 mobility is performed.

Supporting Mode Selection

The MS may be simultaneously handling a combination of traffic types. Conventionally, the highest priority mode (mode 1 being the highest) is chosen. Alternatively, a user may be allowed to select a priority order according to a preference, a service policy, or a contract.

PERFORMANCE EVALUATION

This section first describes the simulation setting. Next, simulation results are presented with five scenarios, each illustrates the effect of the correspondence QoS mode.

Figure 8. Mode 1, L3HO: Throughput

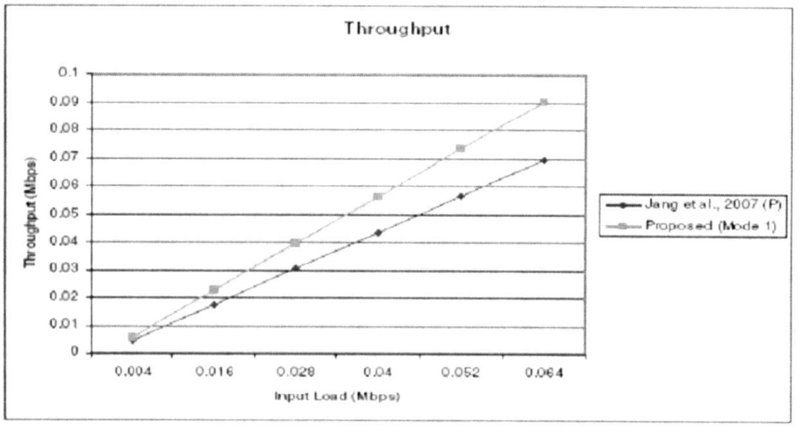

Figure 9. Mode 1, L3HO: Service disruption

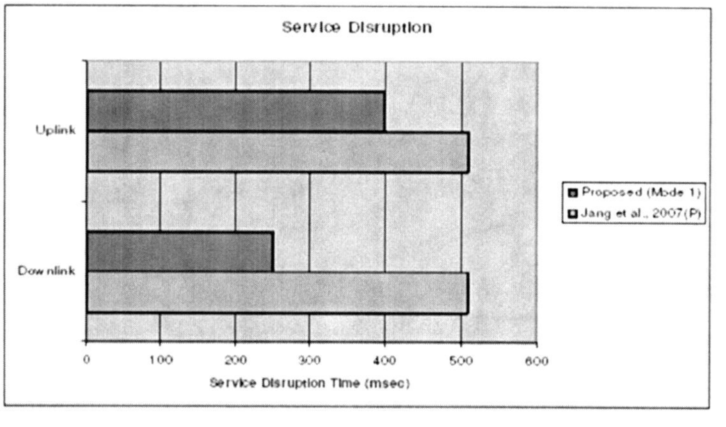

A network simulator has been developed using the JR programming language (Olsson & Keen, 2004). The topology of the network is depicted in figure 7. The proposed scheme is compared with a predictive baseline L3 scheme (Jang et al., 2007), denoted by Jang et al., 2007 (P) in the figures, for assessing the efficiency of macro-mobility. It is further compared with two L2 schemes (IEEE 802.16e and one by Choi et al., 2005) for assessing the efficiency of micro-mobility.

Table 1. Traffic characteristics

Parameters/Scenarios	Application	Input Load (Mbps)	Packet Size (bytes)	Avg PacketSize (bytes)
1.Conversational Mode	VoIP	0.004 - 0.064	120	NA
2. Streaming Mode	Video streaming	0.5 -1.5	1300	NA
3. Interactive Mode	FTP	0.5 - 1	NA	1200
4. Background Mode	Offline database	1.5	NA	1100
5. Standby Mode	No traffic	NA	NA	NA

Simulation Setting

The traffic characteristics used for the simulation experiments are given in Table 1.

The downlink and uplink synchronization delays are set at 250 ms and 150 ms, respectively. The ranging request/ response delay and registration request/response delay are all set at 25 ms each. The delay between MS to BS is set for a 4-mile distance. The link delays of correspondent node (CN) to router, inter-router, and router to BS are set as 30 ms, 20 ms, and 5 or 10 ms, respectively (Rouil & Golmie, 2007).

Simulation Results

In the simulation experiments, the MS performs HO periodically. Some preliminary results have appeared in an earlier conference paper publication (Chellappan et al., 2009).

Scenario 1: MS is Engaged in Two-Way Exchange of Real-Time Traffic (VoIP) with CN

The first experiment involves exchange of conversational data between CN and MS (VoIP). The play-out delay is set at 70 ms. Throughput and

Figure 10. Mode 1, L2HO: Throughput

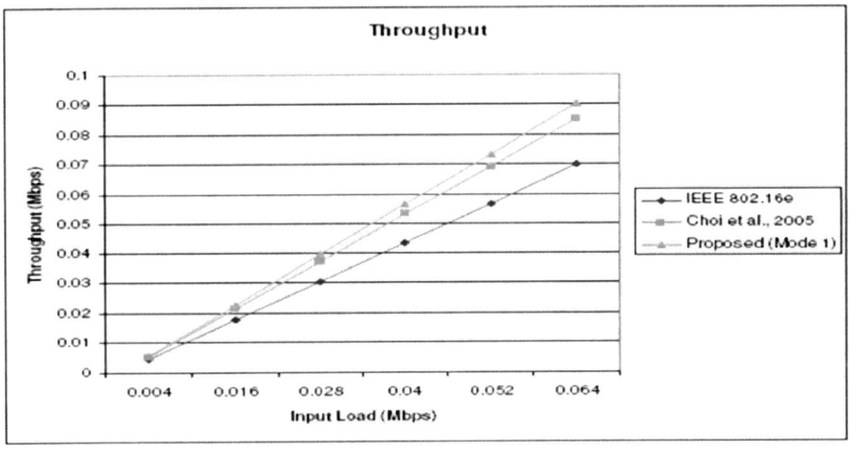

Figure 11. Mode 1, L2HO: Average delay deviation

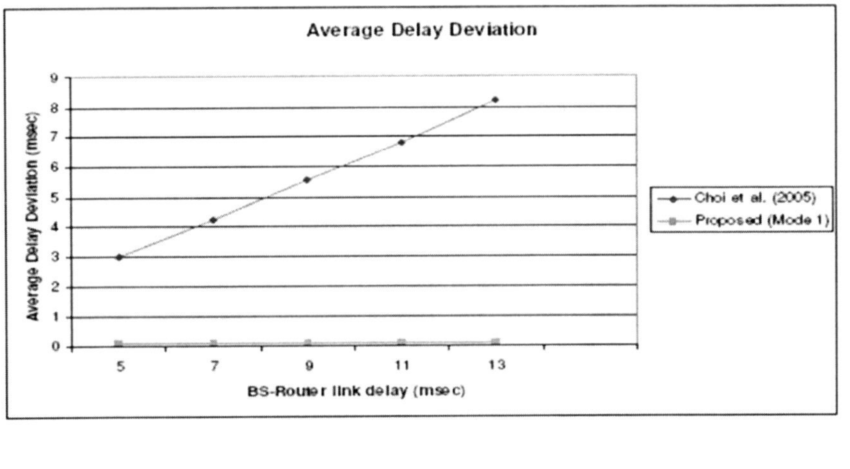

Figure 12. Mode 2, L3HO: Packets received (Jang et al., 2007)

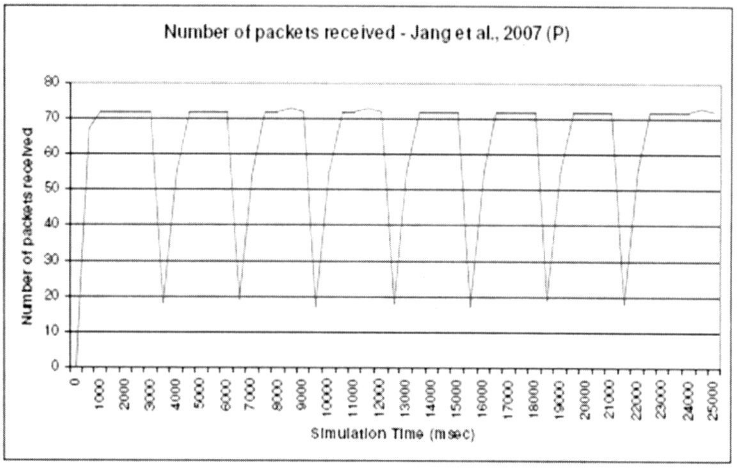

service disruption time in L3 HO (fig 8 and 9), and throughput and delay jitter in L2 HO (fig 10 and 11) are presented. As the proposed scheme implements both faster uplink and downlink accesses, it achieves higher throughput, smaller service disruption, and smaller delay jitter.

Real-time conversational data is extremely delay-sensitive. Hence, service disruption time should be minimized. Figure 9 shows that the proposed scheme achieved a smaller disruption time due to faster uplink and downlink services. Throughput (fig. 10) and service disruption (not shown) were also measured for micro-mobility scenario and results show that the proposed scheme performs better than the two existing methods. Furthermore, average delay jitter (deviation) is much smaller in the proposed method due to its faster uplink service, which is vital for real-time services.

Scenario 2: MS is Viewing a Video File (Video Streaming) from CN

To assess the efficiency of Mode 2 (streaming), the CN transmits a video file to the MS. In the macro-mobility scenario, the proposed scheme has

Figure 13. Mode 2, L3HO: Packets received (Proposed)

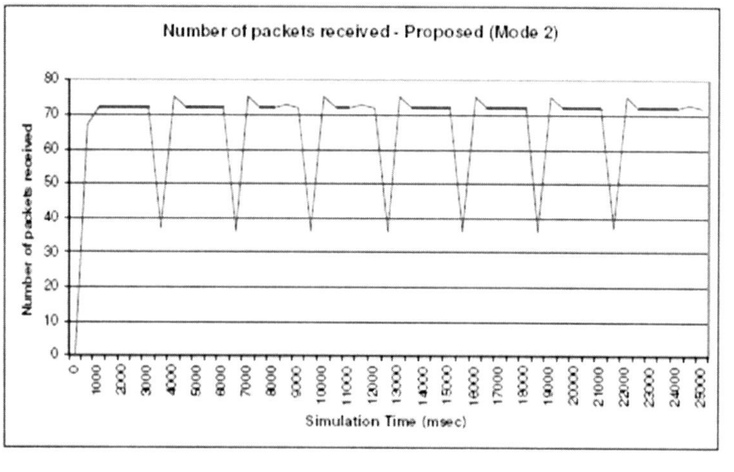

Figure 14. Mode 2, L3HO: End-to-end delay and average delay (Jang et al., 2007)

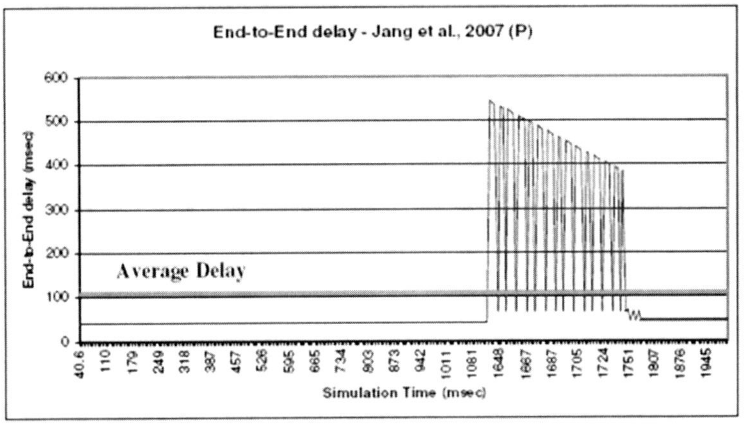

Figure 15. Mode 2, L3HO: End-to-end delay and average delay (Proposed)

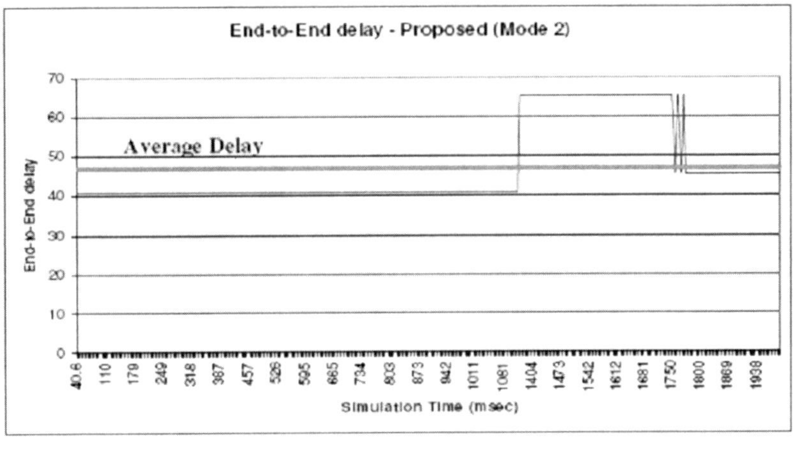

achieved to a higher throughput, as in Scenario 1. To examine more closely, figures 12 and 13 show the number of packets received at the MS where a HO takes place every 3,000 msec and causes the number to drop. The existing scheme by Jang el al. (2007) observes a drop of 40 packets while the existing scheme only drops 20 packets. The new scheme also results in a much smaller end-to-end delay during HO, as shown in figures 14 and 15 where a HO occurred from 1081 to 1750 msec; comparing up to 500 msec to only 67 msec.

Scenario 3: MS is Downloading a Data or Image File (FTP) from CN

To assess the efficiency of the scheme in Mode 3 (interactive mode), the experiment involved the MS downloading a data file from the CN. As the new scheme proactively tunnels the data, it has a higher throughput (figure 16) and lower loss rate (figure 17) than two other existing schemes, especially when using a larger BS buffer size of 0.6 Mbits.

Figure 16. Mode 3, L2HO: Throughput

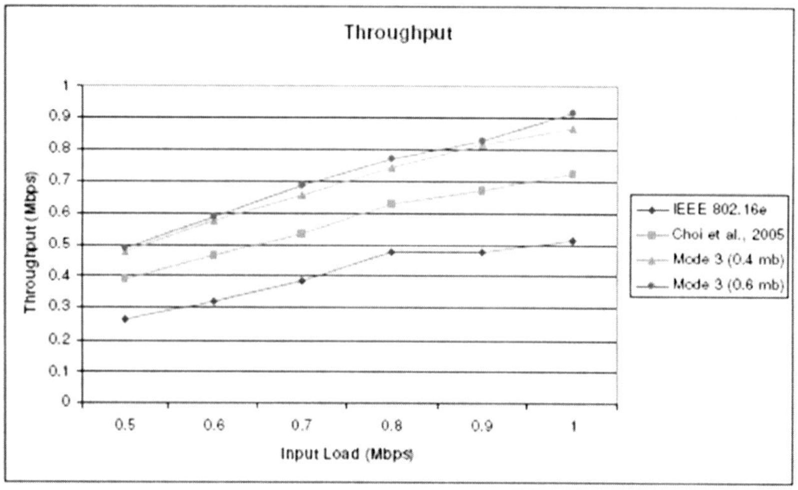

Scenario 4: MS is Performing a Backup of Data from the CN in the Background (Offline Database Operation)

Experiments have been carried out for Mode 4 in the macro-mobility scenario. As throughput measurements have shown comparable results, we focused on energy consumption. The sleep delay is set at 20 ms (twice the initial sleep delay (Han & Choi, 2006)); power consumed in idle and sleep states are taken as 170 mW and 50 mW, respectively (Han & Choi, 2006; Krashinsky & Balakrishnan, 2005). Fig. 18 shows that the proposed scheme has achieved significant energy savings.

Scenario 5: MS is Idle with No Active Connections

In Mode 5 (standby mode), the MS is idle and no tunnel establishment is needed. Thus, the apart from energy savings from standing by (not shown, but refer to Figure 18), the control overhead is reduced (figure 19).

Figure 17. Mode 3, L2HO: Loss Rate

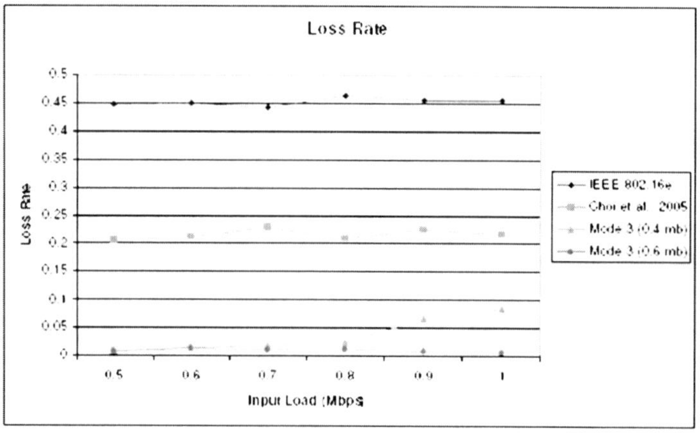

Figure 18. Mode 4: Energy savings

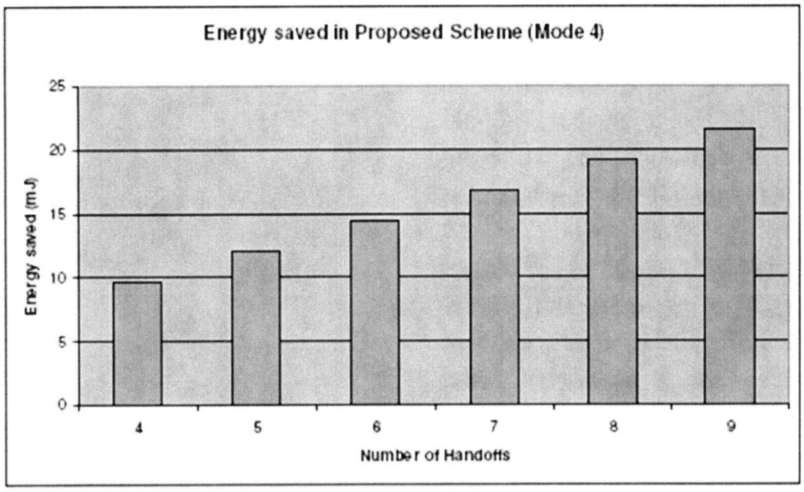

Figure 19. Mode 5: Control Packet Overhead

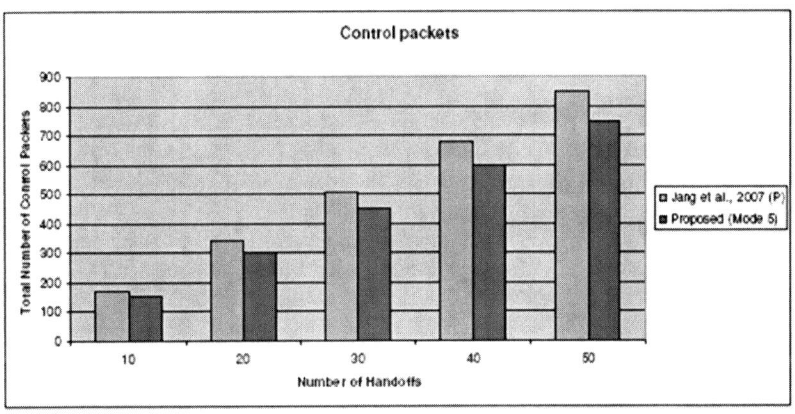

IMPLEMENTING ISSUES

In this section, we first discuss how multiple modes may be simultaneously supported in a single IEEE 802.16e mobile node. Next, we present a quantitative analysis of the costs involved in the proposed implementation.

Cost Involved

The proposed scheme uses different handoff modes based on the QoS needs of the MS. As each mode is based on a different approach, the resources used and the cost involved differs for each mode. The various costs involved include the following.

Mode Selection

This overhead is involved in the MS. The MS should be able to specify the mode of handoff. Including this overhead in the MS is reasonable, as the MS will be the best judge of the current traffic it is handling. A detailed discussion of mode selection is included in section 6.2.

Buffering

When MS performs a handoff, the PAR establishes a tunnel with the NAR. It forwards packets through this tunnel to the NAR. Hence, the NAR should be capable of maintaining data buffers. Buffering cost is involved in all handoff modes except the standby mode.

The size of the buffer needed varies between the different modes. In the streaming and conversational mode, the buffer size is small compared to that of interactive mode. In interactive mode, the NAR buffers data until the MS completes the entire L2 handoff compared to streaming and conversational mode, in which the NAR buffers data only till the target BS is ready to transmit data faster in the downlink. Once the target BS gets the information related to the MS from the serving BS, the NAR starts flushing the buffered data. Hence, the NAR does not buffer data for the entire period of L2 handoff. The buffer size is the biggest in the background mode. Apart from buffering data during entire L2 handoff, the NAR buffers data during the time the MS sleeps.

Tunneling

Tunneling is needed for reducing the delay (for streaming and conversational mode) and for avoiding packet losses (for interactive and background mode). This ensures that the packets destined for the MS are in transit as the MS performs handoff. This overhead is avoided in standby mode.

Fast Downlink Data Transmission

This refers to the data transmission in the downlink by the target BS even before handoff completes. This is applicable only to real-time data transmission, which is delay intolerant, and hence restricted only to the streaming and conversational modes. In the remaining cases, target BS forwards data only after L2 handoff is completed by the MS.

Table 2. Overhead involved in different modes

Overhead	*Mode 1: Conversational*	*Mode 2: Streaming*	*Mode 3: Interactive*	*Mode 4: Background*	*Mode 5: Standby*	*Jang et al. (2007)*	*Choi et al. (2007)*
Mode Selection	X	X	X	X	X		
Buffering	X	X	X	X		X	
Tunneling	X	X	X	X		X	
Fast Downlink Access	X	X					X
Fast Uplink Access	X						
Bicasting							

Fast Uplink Access

This overhead is incurred only in the conversational mode in which the serving BS sends a request for uplink slot allocation (on behalf of the MS) to the target BS.

Bicasting

This overhead is not incurred in any of the modes. By using a combination of tunneling and buffering techniques, bicasting is avoided.

Summary of Costs

This section summarizes the costs involved in all handoff modes in table 10. By varying the handoff method according to the current context of the MS, we are able to obtain varying degrees of cost, hence resulting in a good tradeoff between the cost involved and the QoS obtained.

Implementation of Mode Selection

In the proposed scheme, the MS will perform mode selection. The rationale behind this decision is that the MS has the most accurate knowledge of current applications or connections it maintains. It would be the best judge of its QoS requirements. This design decision requires a suitable methodology for the MS to perform mode selection. Mode selection could be implemented using the following strategy

- Service providers could provide policies based on the handoff modes. Each policy may contain a single mode or a combination of modes.
- The policies will be priced based on the overhead involved in its modes. For example, for real-time conversational mode, the cost involved includes mode selection, buffering, tunneling, fast downlink access and fast uplink access. Compare this to
  - The interactive mode, which includes only modes selection, buffering and tunneling.
  - The standby mode, which requires only mode selection.
- In this framework, the MS could sign for a particular policy while signing the agreement with the service provider. When the MS performs a handoff, the mode/modes included in the policy will be implemented according to priority rules described in section 3.2.13.
- To implement this scheme, the MS should implement the logic for mode selection. This can be implemented with ease, as different handoff modes are designed based on QoS scheduling services available in IEEE 802.16. Hence, depending on the scheduling service used by MS, the appropriate handoff mode would be specified by the MS.
- Apart from programming the MS to perform dynamic mode selection, a user-interface to control modes could also be provided. This will enable users to have additional means of controlling the context. For example, if the MS equipment is very low on power, the user could prioritize energy-saving by selection of standby mode, even if he is receiving video data. In this case, the user sacrifices application quality for energy savings.
- This framework will scale well to target different markets. As an example scenario, service providers can provide the following policies
  - *Premium Policy*: This policy may be targeted at enterprise and corporate users, who will be willing to pay huge fees for high quality services. Such a policy could include all five modes.
  - *Business Policy*: This policy could cater to the medium-sized and small businesses, in which they specify

what combination of modes they want in their policy.
  - *Economic Policy*: This policy could cater to customers seeking lowest price for minimum services. This policy could include just the standby mode.
- The target market for this framework would include
  - Online movie viewers who can sign up for high priced streaming mode.
  - Gamers, who can sign up for high priced conversational mode.
  - Software engineers (constantly checking emails or offline database operations) who can sign up for medium priced interactive or background mode.
  - Students, who can sign up for best effort or standby modes, with the lowest price.

CONCLUSION

IEEE 802.16 is a major long-distance, high-speed wireless technology that promises to provide both static and mobile stations with high data-rate wireless Internet access. It is therefore vital for supporting ubiquitous, pervasive computing. Two major areas of research works in the IEEE 802.16 networks are QoS and fast mobility support. Much work have devoted to designing scheduling algorithms for a particular QoS class, or proposing fast handoff methods for some specific scenarios. The work presented in this chapter serves as a bridge between these two important research areas. A context-sensitive handoff scheme has been proposed. Its five QoS modes are able to successfully support the corresponding five IEEE 802.16 QoS scheduling classes (IEEE 802.16 Working Group, 2004; IEEE 802.16 Working Group, 2005). Ongoing works consist of applying the proposed scheme to high-speed vehicular networks (Moh et al., 2010), developing enhanced cross-layer HO schemes for supporting real-time applications (Huynh et al., 2010). Future work may include strengthening the energy-saving mechanism by using a dynamic sleep interval that adapts to communication channel condition and energy levels, careful consideration of applying the proposed scheme onto the LTE cellular standard, and detailed conformance of the proposed scheme to the WiMAX stages 2 and 3 end-to-end architecture defined by the WiMAX Forum.

REFERENCES

Arunachalam, A. V. (1999). *Quality of Service (QoS) classes for BWA*. A contribution to the IEEE 802.16 Broadband Wireless Access Working Group.

Chang, C. (2005). A Mobile-IP Based Mobility System for Wireless Metropolitan Area Networks. In *Int. Conf. on Parallel Processing Workshops*, (pp. 429-435).

Chellappan, B., Moh, T.-S., & Moh, M. (2009). On Supporting Multiple Quality-of-Services Classes in Mobile WiMAX Handoff. Proceedings of *IEEE International Conference on Computing in Engineering, Science, and Information*, Fullerton, CA, Apr 2009.

Chen, Y., & Hsieh, F. (2007). A Cross Layer Design for Handoff in 802.16e Network with IPv6 Mobility. *Wireless Communications and Networking Conference*, 2007, 3844 – 3849.

Cho, S., Kwun, J., Park, C., Cheon, J., Lee, O., & Kim, K. (2006). Hard Handoff Scheme Exploiting Uplink and Downlink Signals in IEEE 802.16e Systems. *IEEE Vehicular Tec. Conference 2006, 3*, 1236 – 1240.

Choi, S., Hwang, G., Kwon, T., Lim, A., & Cho, D. (2005). Fast Handover Scheme for Real-Time Downlink Services in IEEE 802.16e BWA System. *IEEE 61st Vehicular Tech. Conf., 3*, 2028–2032.

Das, S., Klein, T., Rajkumar, A., Rangarajan, S., Turner, M., & Viswanathan, H. (2006). System Aspects and Handover Management for IEEE 802.16e. *Bell Labs Technical Journal, 11*(1), 123–142. doi:10.1002/bltj.20148

Han, K., & Choi, S. (2006). Performance Analysis of Sleep Mode Operation in IEEE 802.16e Mobile Broadband Wireless Access Systems. *Vehicular Technology Conference, 3*, 1141-1145.

Hu, R. Q., Paranchych, D., Fong, M.-H., & Wu, G. (2007). On the evolution of handoff management and network architecture in WiMAX. In *Proc. of IEEE Mobile WiMAX Symposium*, 2007.

Huynh, P.-Q., Jangyodsuk, P., & Moh, M. (2010). Supporting Video Streaming and VoIP over Mobile WiMAX Networks by enhanced FMIPv6-based Handover. Accepted to be presented at *the Fourth International Conference on Information Systems, Technology and Management (ICISTM-10)*, to be held in Bangkok, Thailand, March 2010.

IEEE 802.16 Working Group on Broadband Wireless Access. (2004). *Part 16: Air Interface for Fixed Broadband Wireless Access Systems.*

IEEE 802.16 Working Group on Broadband Wireless Access. (2005). *Part 16: Air Interface for Fixed and Mobile Broadband Wireless Access Systems Amendment 2: Physical and Medium Access Control Layers for Combined Fixed and Mobile Operation in Licensed Bands and Corrigendum 1.*

Intel. (2006). *Wireless Broadband EDUCAUSE 2006.*

Jang, H., Jee, J., Han, Y., Park, S. D., & Cha, J. (2007). *Mobile IPv6 Fast Handovers over IEEE 802.16e Networks*. IETF Internet Draft, March 2008 (replaces a 2007 version).

Kim, K., Kim, C., & Kim, T. (2005). A Seamless Handover Mechanism for IEEE 802.16e Broadband Wireless Access. In *Proceedings of 5th International Conference on Computational Science - ICCS 2005* (LNCS 3515(II), pp. 527-534).

Koodli, R. (2005). Fast Handovers for Mobile IPv6. *IETF RFC 2068.*

Krashinsky, R., & Balakrishnan, H. (2005). Minimizing Energy for Wireless Web Access with Bounded Slowdown. *Wireless Networks, 11*(1-2), 135–148. doi:10.1007/s11276-004-4751-z

Lee, D. H., Kyamakya, K., & Umondi, J.P. (2006). Fast Handover Algorithm for IEEE 802.16e Broadband Wireless Access System. *1st Int. Symp. on Wireless Pervasive Computing*, 2006

Leung, K. K., Mukherjee, S., & Rittenhouse, G. E. (2005). Mobility Support for IEEE 802.16d Wireless Networks. *2005 IEEE Wireless Comm., & . Networking Conf., 3*, 1446–1452.

Moh, M., Chellappan, B., Moh, T.-S., & Vanesgopal, S. (2010). Handoff Mechanisms for IEEE 802.16 Networks Supporting Intelligent Transportation Systems. to appear in M.-T. Zhou, Y. Zhang, and L. Yang (Eds.), *Wireless Technologies for Intelligent Transportation Systems*. Hauppauge, NY: Nova Science Publisher Inc.

Olsson, R. A., & Keen, A. W. (2004). *The JR Programming Language Concurrent Programming in an Extended Java*. New York: Kluwer Academic Publishers.

Rouil, R., & Golmie, N. (2006). Adaptive Channel Scanning for IEEE 802.16e. *IEEE Military Comm. Conf, 2006*, 1–6.

Rouil, R., & Golmie, N. (2007). Seamless Mobility in WiMAX. *WiMAX Forum Conference*.

Yang, K., Ou, S., Chen, H., & He, J. (2007). A multi-hop peer- communication protocol with fairness guarantee for IEEE 802.16-based vehicular networks. In *IEEE Trans. On Vehicular Technology* (pp. 3358-3370).

Zhong, L., Liu, F., Wang, X., & Ji, Y. (2007). Fast Handover Scheme for Supporting Network Mobility in IEEE 802.16e BWA System. In *Int. Conf. on Wireless Communications, Networking, and Mobile Computing*, 21-25 Sept., (pp. 1757-1760).

Chapter 14
QoS in Vehicular Communication Networks

Robil Daher
Rostock University, Germany

Djamshid Tavangarian
Rostock University, Germany

ABSTRACT

Vehicular communication networks (VCNs) have emerged as a key technology for next-generation wireless networking. DSRC/WAVE as a leading technology for VCN provides a platform for Intelligent Transportation System (ITS) services, as well as multimedia and data services. Some of these services such as active safety and multimedia services have special requirements for QoS provision. However, when providing QoS, the VCN characteristics are the cause for several new issues and challenges, especially when vehicles travel at high speeds of up to 200 km/h. These issues are addressed in the context of roadside networks and vehicular ad hoc (unplanned) networks (VANETs), including vehicle-to-vehicle (V2V) and vehicle-to-roadside (V2R) communications. Accordingly, plenty of solutions for provisioning QoS in VCNs have been classified in regards to VANETs and roadside networks, on the one hand, and to layer-2 and layer-3, on the other hand. Consequently, several QoS solutions, including medium access and routing protocols, are presented and discussed. Additionally, open research issues are discussed, with an objective to spark new research interests in the presented field.

INTRODUCTION

Vehicular communication networks (VCNs) have emerged as a key technology for next-generation wireless networking. Several national and international organizations (IEEE, ASTM, ISO, etc.), public transport authorities (US Department of Transportation and equivalent transport authorities in Europe and Japan, etc.), and vehicle manufacturers (DaimlerChrysler, BMW, GM – General Motors, Renault, Toyota, etc.) have been corporately working on the development of standards for VCNs. Accordingly, various projects are underway (AKTIV, COOPERS, etc.) or were completed just recently (FleetNet, ASV 4, VSC, etc.). Several consortia (C2C-CC, VSCC, etc.) were set up to explore the potential of VCNs

DOI: 10.4018/978-1-61520-680-3.ch014

(Hartenstein & Laberteaux, 2008). These projects are funded substantially by national governments and involve several constituencies, including the automotive industry, the road operators, tolling agencies, and other service providers.

While the adoption of vehicle manufactures for information technology address safety issues mainly, environmental and comfort issues of vehicles have accelerated the development process of VCNs (Hartenstein & Laberteaux, 2008). The increasing demand for broadband wireless services for Intelligent Transportation Systems (ITS) technologies, besides the wide use of IEEE 802.11, has led to adoption of the DSRC/WAVE - Dedicated Short Range Communications (DSRC) in accordance with IEEE 802.11p and IEEE 1609 Wireless Access in Vehicular Environments (WAVE) standards (Federal Communications Commission [FCC], 2004; DOT-HS810591, 2006). Furthermore, several national and regional governments contribute licensed spectrum for vehicular wireless communication, generally in the 5.8/5.9 GHz band and at least in Japan, the 700 MHz band (Stibor, Zang, & Reumerman, 2007). For instance, the U.S. Federal Communications Commission (FCC) has allocated 75 MHz of Radio Spectrum at 5.9 GHz band for DSRC (FCC, 2004).

Since VCNs form the basis for supporting not only the ITS-services, especially public-safety-related applications, but also a wide range of future multimedia and data applications, such as audio/video as well as e-maps and road/vehicle-related services (Su & Zhang, 2007), vehicles are envisaged to become a part of the Internet in the near future, either as mobile endpoints, as mobile backbone routers, or as mobile sensors (Kutzner et al., 2003). Thus, VCNs must support QoS for a number of applications and services, especially real-time applications addressing safety and VoIP services (Bossom et al., 2008). However, the dynamic architecture of the VCN, especially within the access level network, in which vehicles can move with speed of even more than 200 km/h, forms the main challenge for adopting the already existing QoS-models and solutions. To investigate related issues and solutions, this chapter deals with QoS-mechanisms, protocols and models for VCNs and concentrates on DSRC/WAVE as the leading technology for vehicular communications (Berger, 2007).

This chapter presents a detailed investigation of the current state-of-the-art of QoS-mechanisms, protocols and models for DSRC/WAVE-based VCNs and related solutions. Furthermore, open research issues in all protocol layers are also discussed, with an objective to spark new research interests in this field. The rest of this chapter is organized as follows. First of all, we briefly describe the specifications of DSRC/WAVE and related network architectures, as well as real-time applications and their QoS requirement. Then, the main issues and challenges for adopting QoS in VCNs are addressed. After that is done, we will classify the considered solutions in accordance with roadside networks and vehicular ad hoc networks, on the one hand, and with layer-2 and layer-3, on the other hand. Consequently, we present several QoS solutions, including medium access and routing protocols, to reflect the state of the art in this field. Finally, we summarize our chapter in the last section.

DSRC/WAVE: Background and Network Architecture

The DSRC/WAVE specifies using IEEE 802.11p for physical and MAC layers, while using IEEE 1609 for the upper layers. The IEEE 1609 family of WAVE defines the architecture, communications model, management structure, security mechanisms and network access for wireless communications in the vehicular environment. Additionally, IEEE 1609 family consists of four (trial-use) standards: 1609.1 for specifying the services and interfaces of resource manager applications; 1609.2 for defining security services for applications and management messages, including message format and processing; 1609.3 for defining networking services, including network and transport layer services; Finally, 1609.4 for providing multi-channel operations as enhancements for 802.11 for supporting WAVE. These trial-use standards are expected to be full-use standards in 2009/2010. Moreover, new IEEE 1609 members "1609.5 and 1609.11" are still in process: P1609.5 enhances the communication management services and is expected to be published in 2012 (Kurihara 2009); P1609.11 defines the services and secure message formats necessary to support ITS interoperability with secure electronic payments and is expected to be published in 2013 (Kurihara 2009). The IEEE 802.11p is an extension of IEEE 802.11 and IEEE 802.11a for vehicles travelling at high speeds and it introduces changes to parameters such as adjacent/non-adjacent channel rejection, receiver minimum input sensitivity, etc. Moreover, IEEE 802.11p supports two different stacks: IPv6 for service channels and WAVE Short Message Protocol (WSMP) on any channel. Figure 1.a shows the protocol suite DSRC/WAVE.

The DSRC defines two main components for a vehicular communications system: On-board Unit (OBU) and Roadside Unit (RSU). The former is installed in the vehicle, whereas the latter is established on the roadside as part of the access network. DSRC generally enables communications over line-of-sight with distances of up to 1000 meters between RSUs (RSU class D and max. output power of 28.8 dBm) and mostly high speed (up to 200 km/h), occasionally also for stopped and slow moving OBUs (FCC, 2004). In accordance with DSRC, a vehicular communications system specifies two levels of communications network in its infrastructure (Zhang, Su, & Chen, 2006; Hartenstein & Laberteaux, 2008; Stibor et al., 2007), as revealed in Figure 1.b:

1. **Vehicular ad-hoc network (VANET):** Provides direct communication between OBUs, Vehicle-to-Vehicle (V2V), including to and from RSUs, Vehicle-to-Roadside (V2R) (Hartenstein & Laberteaux, 2008)
2. **Roadside network:** Consists of twofold: (a) Roadside Access Network (RAN), which comprises the RSUs and enables the V2R-communications through appropriate connections to the backbone; (b) Roadside Backbone Network (RBN), which represents the backbone network of RSUs, and in which RSUs communicate with each other and with Internet (Kutzner et al., 2003)

Different frequency bands are used for DSRC in different countries. While the FCC has allocated 75 MHz of Radio Spectrum at 5.9 GHz (5.850-5.925 GHz) band in the USA (FCC, 2004), the European Electronic Communications Committee (ECC) has assigned 50 MHz (5875-5925 MHz) to DSRC in Europe (Electronic Communications Committee [ECC], 2008a). Besides that, the ECC has recommended additional 20 MHz (5855-5875 MHz) for DSRC in Europe (ECC, 2008b). In Japan 80 MHz (5770-5850 MHz) were allocated to DSRC (ECC, 2008a), whereas other countries worldwide have been considering the 5.9 GHz band for DSRC (ECC, 2008a). Table 1 summarizes the properties of channel assignment in regard with US FCC 90.377, where we can distinguish two types of channels: Service Channel (SCH) and Control Channel (CCH).

In accordance with 802.11a and due to the channel bandwidth, DSRC/WAVE provides transmission bitrates of 3, 4.5, 6, 9, 12, 18, 24 and 27 Mbps for 10 MHz-channels, whereas expecting to provide bitrates of 6, 9, 12, 18, 24, 36, 48 and 54 Mbps for 20 MHz channels. The selected transmission bitrate depends on several parameters, especially the link quality (for instance between OBUs in V2V communications). However, since the traffic safety applications require a high reliability, a low rate (e.g. 6 Mbps) will most likely be chosen for such cases (Bossom et al., 2008).

Table 1. DSRC channel assignment in USA (Adapted from FCC, 2004)

Channel No.	Freq. Range (MHz)	Max. EIRP of RSU (dBm)	Channel Use	Band width	Bitrate [Mbps]
170	5850-5855	-	Reserved	5 MHz	-
172	5855-5865	33	SCH [1]	10 MHz	3 – 27
174	5865-5875	33	SCH	10 MHz	3 – 27
175	5865-5885	23	SCH [2]	20 MHz	6 – 54
176	5875-5885	33	SCH	10 MHz	3 – 27
178	5885-5895	33 / 44.8	CCH	10 MHz	3 – 27
180	5895-5905	23	SCH	10 MHz	3 – 27
181	5895-5915	23	SCH [3]	20 MHz	6 – 54
182	5905-5915	23	SCH	10 MHz	3 – 27
184	5915-5925	33 / 40	SCH [4]	10 MHz	3 – 27

[1] specified for safety in V2V-communications [2] combined of channels 174 & 176 [3] combined of channels 180 & 182
[4] high power public safety and non-public safety DSRC operations

Real-Time Applications in VCNs

The DSRC/WAVE-based VCNs provide a solid ground for a wide range of applications and services in the field of ITS and Internet-related services. Schoch et al. (2008) has classified such applications in four categories:

1. Active safety for situations, such as: dangerous road features, abnormal traffic and road conditions, danger of collision (crash), imminent crashing, and occurred incident, e.g., low bridge warning, road condition warning, lane change warning, pre-crash sensing, and breakdown warning, respectively.
2. Public service for the purpose of emergency response and support for authorities, e.g., approaching emergency vehicle warning and electronic drivers license, respectively.

Figure 1. (a) DSRC/WAVE protocol suite. (Adapted from IEEE P1609.0 D0.2 (2007)); (b) VCN-architecture

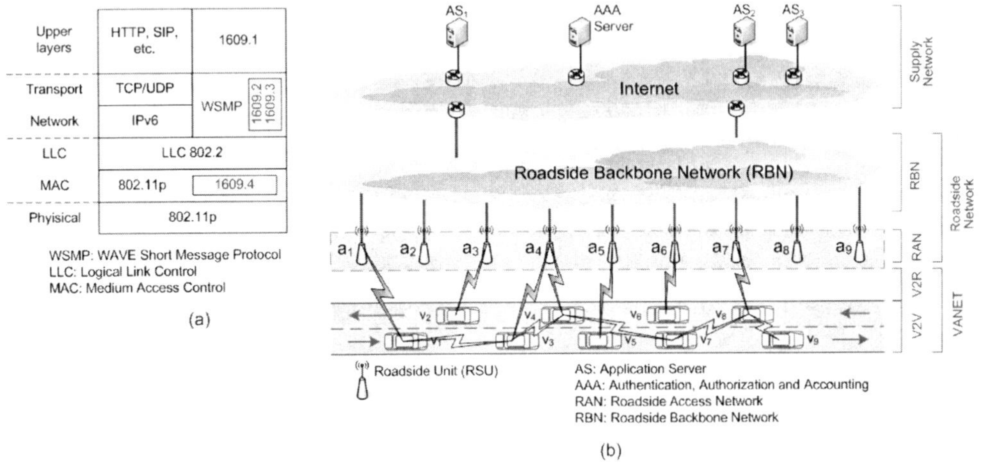

3. Improved driving for the purpose of enhanced driving and traffic efficiency, e.g., left turn assistant and enhanced route guidance and navigation, respectively.
4. Business and entertainment for the purpose of vehicle maintenance, mobile services, enterprise solutions, and e-payment, e.g., just-in-time repair notification, VoIP telephony, fleet management, and gas payment, respectively.

Among these applications, only the active safety features, as mentioned in the 1st category, can be classified as hard real-time applications, where a strict QoS guarantee is required for each kind of these applications, especially the end-to-end (E2E) delays. Indeed, the public safety message delivery applications can tolerate only up to 50 ms, particularly in case of V2V communications (Xu, Mak, Ko, & Sengupta, 2007). Accordingly, if this delay is exceeded in case of V2V communication, a vehicle's crash could be unavoidable.

Otherwise, applications of 2nd and 3rd categories, as well as real-time applications of the 4th category, such as voice and video-related services, can be classified as soft real-time applications, which can tolerate to a certain extent degradation of the promised QoS (Murthy & Manoj, 2004). For instance, the VoIP-applications can tolerate an E2E delay of up to 150 ms in case of using G.711 codec (ITU-T Recommendation G.114 [G114], 2003).

QoS-Parameters in VCNs

The diversity of applications supported by VCNs reflects the diversity of QoS requirements; therefore, the associated QoS parameters differ from one application to another (Murthy & Manoj, 2004). The main parameters used to describe the QoS requirements of an application are the E2E-delay, jitter, packet loss, and bandwidth. Table 2 shows the differences between QoS requirements of hard and soft real-time applications in a vehicular environment, where safety applications have stringent QoS requirements, especially regarding E2E delay and packet loss. For instance, broadcasting a safety message every 500 ms is probably too slow, because the driver's reaction time can be as small as 500 ms (Olson, 2006). Thus, if information is delayed by 500 ms, the driver may detect the threat before the onboard active safety system reacts (Xu et al., 2007). Based on that, Xu et al. (2007) assumes that the safety messages must certainly be generated by each vehicle with a higher rate than 1/500 ms (one message per 500ms). Although another study (VSC, 2005) estimates the transmission rate of safety messages to be between 1 and 10 Hz, i.e., up to 1/100 ms, Xu et al. (2007) ensures that a transmission rate of 1/50 ms of safety messages is sufficient for collision-warning applications, where a receiving rate greater than 1/50 ms is not required (Xu et al., 2007). In relation to the vehicle's speed, a certain extent of packet loss could be tolerated, in a sense where the slower the vehicle is, the higher the tolerated packet loss would be. In a worst case scenario, when two vehicles are moving with high speed, e.g. 200 km/h, towards each other on the highway, the packet loss of safety messages must be as small as possible; a value of zero is preferred in this case in order to guarantee that a crash can certainly be avoided. Xu et al. (2007) finds that "a probability of reception failure of 1/100 is a higher loss rate than accepted in many networks". However, the higher the transmission rate of safety messages is, the higher the tolerated probability of reception failure is allowed. Hence, more research is required in this field in order to verify if such a loss rate is of acceptable value (Xu et al., 2007)

The required bandwidth for active safety applications is relatively small and depends on the used safety service (Xu, Mak, Ko, & Sengupta, 2007), e.g., the Society of Automotive Engineers J1746 standard encodes vehicle location besides other data in less than a hundred bytes (SAE Std. J1746, 2001). The National Transportation

Table 2. QoS-parameters for active safety and VoIP-applications

Real-time application	Bandwidth (kbps)	E2E delay (ms)	Jitter (ms)	Packet loss (%)	QoS requirements
Active safety (message delivery)	< 16 [1]	≤ 50	< 25 [1]	<< 1[3]	hard
VoIP / G.711 (30 ms, 64 kbps)	69.33 [2]	≤ 150	≤ 60	≤ 1	soft

[1] estimated for safety application in case of 100 byte packet size and 50 ms transmit rate
[2] codec bitrate over IP, i.e., including RTP and UDP header
[3] estimated by Xu et al., (2007)

Communications for ITS protocol hazard codes also encode vehicle hazard information in only five bytes (Institute of Transportation Engineers [ITE], 2004). Therefore, a bandwidth of less than 16 kbps on average could be expected for safety application in case of 100 byte packet size and 50 ms transmit rate.

ISSUES AND CHALLENGES FOR PROVIDING QOS IN VCNS

VCNs define two types of operation mode in VANETs: ad hoc mode (Point-To-Point - P2P) for V2V communications and cell-based mode (Point-To-Multipoint - P2MP) for V2R communications. The QoS requirements of those modes are different due to different communication characteristics. For instance, vehicles travelling at a high speed of 200 km/h in the same direction can communicate with each other over V2V-VANET without any impact of this speed on the communication functionality itself or QoS support. This is the case, because the physical distance between the two vehicles is assumed to be relatively constant. On the contrary, keeping the real-time sessions between one of such vehicles and the roadside network open forms an essential challenge for the network performance due to frequent vehicle handoffs between RSUs. Additionally, there are difficulties for real-time packets routing over RBN to reach the appropriate targeted RSU at a suitable time. Mathematically, a vehicle travelling at 200 km/h remains approximately 18 s associated with each RSU (with 1000 m-range) along the related road. In this example, the vehicle changes six RSUs a minute in case of V2R communication. However, similar problems occur in case of V2V communications between vehicles travelling at different speeds, e.g., the first vehicle travels at 200 km/h, while the other one is stopped or moving much slower than the first one.

Consequently, several issues and challenges for providing QoS in VCNs have to be addressed for both VANETs and roadside networks, including wired and wireless roadside backbone networks (Daher et al., 2008).

Issues and Challenges for Providing QoS in VANETs

The unique characteristics of VANETs cause several difficulties for provisioning QoS. The characteristics of vehicles' mobility form the most important aspects of supporting QoS in VANETs. The vehicles' velocity, vehicles' movement patterns, and vehicles' density are the characteristics of mobility in VANETs (Schoch et al., 2008). First of all, vehicles' velocity may range from zero, when vehicles are stopped or stuck in a traffic jam, to over 200 km/h on highways. During high velocity, the wireless communication window between an OBU and RSU is very short because of the relatively small association time caused by a transmission range of several hundred meters (Schoch et al., 2008), as mentioned in the example above. Secondly, different types of road systems (city and rural roads and highways) have different vehicle movement patterns and thus different QoS requirements. That is, while the roads inside cities have straight streets and a relatively high density of traffic, the rural roads have a lower density of traffic and more curves. However, the traffic in both cities and rural roads is mostly unordered compared to highways, where vehicle movement is approximately one-dimensional. These different patterns pose special challenges, especially for routing (Schoch et al., 2008; Blum, Eskandarian, & Hoffman, 2004). Finally, the variation of vehicle density along the roads inside cities, on rural roads or highways form a key challenge for resource management of VANET, as well as roadside networking. Thus, the required bandwidth in some segments of the highways, e.g. in case of a traffic, could be very high in comparison to the rest of the highway. Moreover, a network overloading in such cases may not be avoidable, which degrades the provided QoS drastically for all participating RSUs and OBUs.

As a P2P network, the V2V-VANETs have similar QoS issues to those of Mobile Ad-Hoc Networks (MANETs). Except the high-speed nodes, issues such as dynamically varying network topology, imprecise-state information, error-prone shared-radio channel, and limited-resource availability (Murthy & Manoj, 2004) are existent in both networks. In relation with these issues, two main challenges for QoS provisioning could be addressed in V2V-VANETs:

1. **QoS-oriented routing:** Due to the rapidly varying network topology and related channel characteristics, maintaining a QoS-stable path over V2V communications from source to destination forms a huge challenge for routing protocols. The higher the speed differences among vehicles that build the V2V-VANET is, the more difficult it is to provide a guarantee for QoS between source and destination. Difficulties, such as state information updates, topology discovery, and best-route employment, are essential for a QoS guarantee in such networks. To compensate influence of these difficulties on the provided QoS, a relatively high signaling overhead of used routing protocol and related mechanisms could be required, which may drastically decrease the whole resource availability on used channels. Therefore, a trade-off between QoS requirements and resource utilization reserved for control traffic is required.
2. **QoS-based channel utilization:** DSRC provides one CCH reserved for only public safety services and six SCH for safety and non-safety services. While safety services are prioritized over other services independent of channel type, the SCHs are available on a shared basis for non-safety services (FCC, 2004). That is, while the QoS can be guaranteed for safety services, only soft QoS can be guaranteed for other types of real-time services, such as VoIP services.

The V2R-VANETs have similar characteristics to that of the infrastructure mode of 802.11, except the high speed of mobile stations, such as central point-of-access through the RSU, shared radio channel, etc. However, the main issue for maintaining QoS of vehicles travelling at high speeds is to enable a seamless layer-2 and layer-3 handoff between RSUs, especially in relatively very short times (≤50 ms). Not only the optimization of association mechanisms, but also the cooperation between RSUs, is required to achieve a seamless handoff. Subsequently, the used roadside network plays a key role in this issue, as outlined below.

Issues and Challenges for Providing QoS IN Roadside Network

The RBN addresses three types of backbone infrastructures in regards to communication mediums: wired, wireless, and mixed wired/wireless. Each

Figure 2. RBN cluster topologies; (a) bus cluster; (b) star cluster; (c) chain cluster; (d) umbrella cluster, mixed between star and bus model

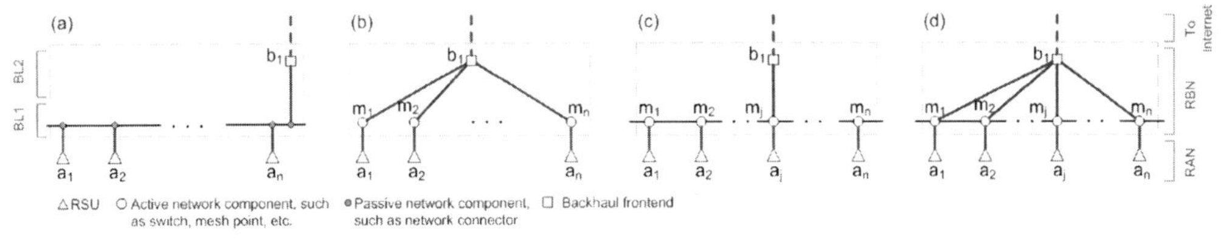

RBN type has different characteristics and thus causes different issues for providing QoS in roadside networks. To simplify the inspection of QoS issues and challenges in accordance with RBNs types, we have classified the RBNs into four main topologies (Daher, 2008), as shown in Figure 2: bus, star, chain, and umbrella topology, in which two roadside backbone levels (BL) are defined. As each certain set of RSUs can be connected to a single backhaul front-end b_l, which acts as a gateway to the supply network (Internet), we use the term *cluster* to indicate each one of these sets. In this respect, each RBN-architecture can be considered as a construction for a number of clusters, which are connected to each other in an appropriate manner to build flat, hierarchical, or other types of network architectures. The number of RBN components per cluster, number of clusters in RBN, and technologies used within clusters, are a matter of network design, which is beyond the scope of this chapter. Consequently, the wired type of RBN means, that communications among RBN components occur over wired connections, e.g., each RSU a_i (in case of chain topology) is connected through its switch/router m_i over one or more wired hops of BL1 to the central switch/router m_j, which in this case also acts as a gateway to the related backhaul front-end b_l. In similar manner, other types of RBN, wireless and mix wired/wireless types can be explained. For instance, we can do so through a combination of the star cluster and WiMAX. In case of wireless RBN, the BL1 components of Figure 2.b meet subscriber stations, while the backhaul front-end indicates a base station. However, on the contrary to other topologies of RBN, the bus topology does not exist for wireless RBNs.

The main challenge for roadside network performance in respect to QoS requirements is to be adaptive to the rapidly varying architecture of related VANET, especially in case of V2R-VANET, with minimum effect on the provided QoS within the access networks. Accordingly, two essential issues can be addressed in roadside networks independent of RBN type: seamless layer-2 and layer-3 handoff and QoS-oriented routing. In that respect, while the RAN must enable low latency layer-2 and layer-3 handoff mechanisms between RSUs, the related RBN must provide efficient low latency switching/routing between the RSUs, on the one side, and between RSUs and supply network (Internet) on the other side.

Different types of roadside networks have different issues and challenges for QoS provisioning. A well-designed wired RBN provides higher bandwidth and more reliability than that of usual wireless RBNs, and as such has relatively low routing and handoff latencies. However, the backbone architecture in combination with the used network as well as cable technology, such as with fiber optic, is essential for providing QoS. For instance, in case of bus topology, the uplink to backhaul front-end., the supply network must have adequate bandwidth compared to connected RSUs in related clusters, i.e., we need sufficient bandwidth for a maximum load of approximately "n*27 Mbps"- where "n" is the number of RSUs in the cluster. None considering these requirements could lead to degradation of provided QoS in heavy loaded roadside networks and V2R-VANETs.

On the contrary, using wireless RBN provides higher scalability and flexibility by network design and installation, but lower bandwidth and higher latencies. Therefore, we must expect a higher complexity for handoff and routing mechanisms. Protocols are expected to provide QoS in wireless RBN solutions (Daher et al., 2008). Similar to that of wired RBNs, the difficulties of wireless RBNs is to concentrate on the used network topology and wireless technology. In that respect, networks of WMAN, WWAN and WGAN, such as WiMAX 802.16d, UMTS and satellite network, can be used to bridge the relatively long distances among wireless RBN components, and also between wireless RBN components and components of the supply network (Internet). However, the bandwidth constrains, as they exist for WiMAX and UMTS, as well as the relatively high latencies for satellite communication, are essential issues for adopting wireless solutions for RBN. Moreover, the used operation mode (P2P and P2MP) for the wireless network affects the capability of such solutions to support QoS. For instance, the use of P2P-based wireless RBN (wireless mesh backbone) creates new challenges for adopting QoS models, due to lack of central coordination, error-prone shared-radio channel, limited-resource availability, etc. That is, the mesh nature of the network degrades the available bandwidth in comparison with P2MP-based solution, but provides more flexibility for network installation and configuration, including self-configuration and self-healing

Thus, different types of roadside networks have different types of issues and challenges for providing QoS. While wireless RBNs provide higher scalability and flexibility for network design and installation, wired RBNs offer higher bandwidth and more reliability than wireless RBNs would. Therefore, more complexity by handoff mechanisms and routing protocols is expected for providing QoS in wireless solutions of RBN (Daher et al., 2008).

Furthermore, the network design on layer-3 has a direct influence on the real-time capabilities of roadside networks. That is, the use of different IP-subnets within a roadside network may increase the layer-3 handoff latency drastically and as a result degrade the whole provided QoS. In addition, keeping the guaranteed level of QoS while mapping QoS between RAN and RBN, on the one hand, and between RBN and Internet, on the other hand, forms a real challenge for known solutions in all layers.

CLASSIFICATIONS OF QOS-SOLUTIONS IN VCNS

To simplify studying the considered QoS-solutions in VCNs, the classification of these solutions is very essential to address the differences in performance and reliability among these solutions in different VCN-levels, as well as different ISO OSI layers. Therefore, this chapter deals only with the VCN-level-wise and layer-wise classification schemes to classify the considered QoS solutions. In addition, only VCN-oriented QoS solutions are considered. Other solutions, especially those developed in the context of similar technologies, such as MANET solutions in respect with VANET, are not considered.

VCN-Level-Wise Classifications of Existing QoS Solutions

Due to the different network and operation characteristics of VANET and roadside networks, different QoS models and solutions have been developed for each of these networks. Thus, as a first classification level of QoS solutions, we classify the considered QoS solutions in accordance with VCN-level and the related network properties, as revealed in Figure 3.a. The dashed arrows between RAN and other boxes indicate that QoS solutions for V2R-communications are also a part of RAN. The most common QoS solutions of VANET are for V2V and V2R communications. Only a single study could be discovered, which addressed provision of QoS for only V2R–VANET.

Layer-Wise Classifications of Existing QoS-Solutions in VCNs

The layer-wise classification scheme helps understanding which layers of the network protocol stack are engaged into related QoS solutions. Although several QoS solutions operate in a single layer of the network protocol stack, a cross-layer interaction, especially between physical, MAC and network layer, is strongly required for provisioning guaranteed QoS. Figure 3.b shows layer-wise classification of considered solutions. Unfortunately, only few solutions about provisioning QoS over MAC and network layers could be found for VANET, while only two QoS solutions in the category "VANET/cross-layer solutions" could be addressed. This fact indicates the demand on research and development in this field.

QOS MODELS AND SOLUTIONS FOR VANETS

The QoS solutions for VANETs must deal with issues caused by the high speed of its clients during V2V or V2R communications. Although MANET and VANET have a similar architecture and mechanisms, but due to the relatively low speed of MANET clients, the MANET-based QoS-solutions cannot easily be integrated into VANETs without additional modifications or even essential change of mobility concepts, especially when considering the resulting challenges, such as layer-2 and layer-3 handoff and IP routing (Su & Zhang, 2007; Franz et al., 2005).

There are two main communication patterns that could be considered by real-time applications in VANETs: message dissemination and packet flow. The message dissemination is mostly used by safety applications to deliver a safety service-dependent message to other vehicles in a certain range and at a certain time (Xu et al., 2007). Since message dissemination is based on per-hop relaying, MAC layer-based QoS solutions could be sufficient for corresponding services and applications such as by Xu et al. (2007), Su & Zhang (2007), and IEEE 802.11e. On the contrary, the packet flow is used by multimedia applications such as VoIP and video streaming. Because sufficient resources between source and destination must be reserved to guarantee QoS, the use of only MAC layer-based QoS solutions could not be adequate to guarantee E2E QoS for such kind of real-time applications. Therefore, the IP-layer-based solutions, especially with an appropriate routing protocol, besides using MAC layer-based QoS solutions could drastically improve the network performance and the QoS provisioning for such applications. This section addresses the possible layer-2 and layer-3 models and solutions of QoS for VANETs and presents some of these solutions.

MAC Layer Solutions

The most important MAC layer QoS solution for VANET is the IEEE 802.11e enhanced with WAVE MAC, since this standard combination currently provides the only standard QoS solution for VANET. In general, only few studies could be found about MAC layer based QoS provisioning for VANETs. Beside the IEEE 802.11e, two other VANET specific QoS solutions were introduced by Su & Zhang (2007) and Xu et al. (2007). In this subsection we concentrate only on the WAVE MAC-related 802.11e.

IEEE 802.11e: WAVE and 802.11p QoS

The IEEE 802.11e is an amendment to the standard 802.11 for integrating QoS capability into the 802.11 MAC protocol. IEEE 802.11e defines a new medium access procedure, called the Hybrid Coordination Function (HCF), in order to enhance the contention-based, as well as the contention-free accesses by providing a priority mechanism as basis for QoS. The HCF provides two access methods: Enhanced Distributed Channel Access

Figure 3. Classification of QoS solutions in VCNs

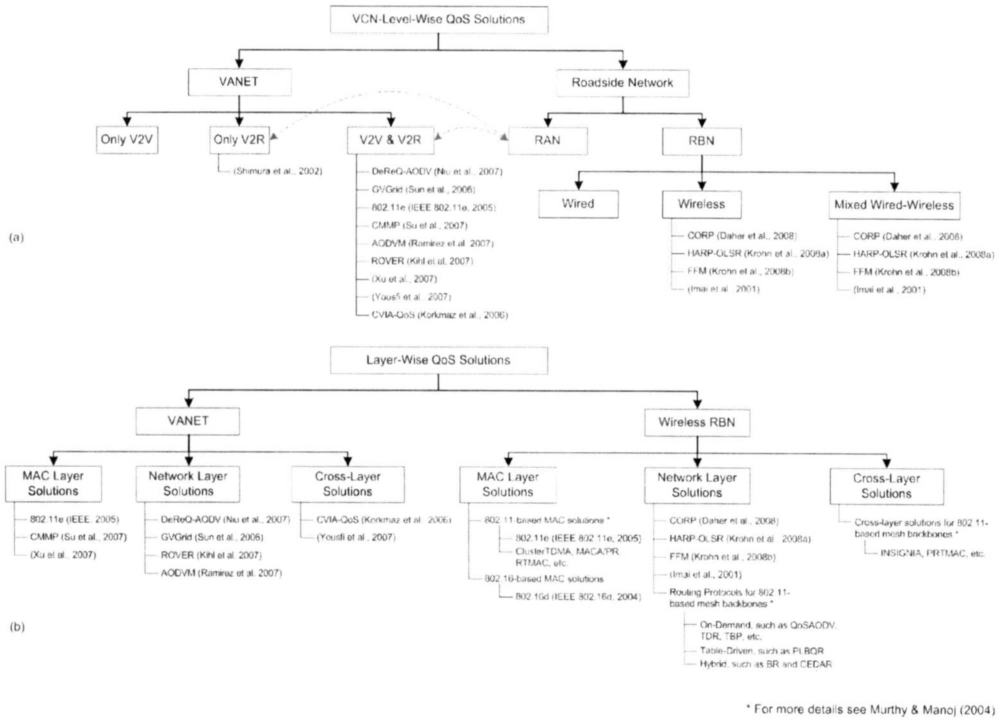

(EDCA) and HCF Controlled Channel Access (HCCA). However, the WAVE and 802.11p follow the 802.11e EDCA access method.

EDCA provides a differentiated and distributed access to the wireless medium, where the infrastructure and ad hoc operation modes are supported. IEEE 802.11e defines eight different User Priorities (UPs), which can be directly transferred from IEEE 802.1D standard capable MAC devices. However, the received frames will be mapped by MAC layer to the appropriate Access Categories (ACs). IEEE 802.11e defines four different ACs, where each AC has a different priority of access to the wireless medium. Figure 4a shows the mapping between UPs and ACs in accordance with IEEE 802.11D.

The WAVE enhances the 802.11e EDCA architecture to adapt the appropriate functionality in relation with IPv6 and WSMP. Based on 802.11e, WAVE MAC implements access category queues on a per-channel basis and in cooperation with channel coordination (Zhang, Yang, & Ma, 2008)., as shown in Figure 4b.

The channel selector is responsible for several actions, such as deciding when to monitor a channel, how the WAVE device utilizes a specific channel, what set of legal channels at a specific time point, and other aspects (Zhang et al., 2008). The channel router acts as a distributer of datagrams arriving from LLC layer on the appropriate channels. In that respect, the channel router forwards the WSMP datagrams arriving from LLC to the appropriate queue in accordance with the packet priority and channel identified in the WSMP header. Similarly, the channel router transfers the IP datagram to the data buffer related to the corresponded SCH, which is specified in the related transmitter profile registered by the corresponded IP Application. The appropriate priority queue is then selected by mapping UPs, in accordance with the table in Figure 4a, to an Access Category Index (ACI). Accordingly, the

Figure 4. (a) Mapping of UPs to ACs (Adapted from IEEE Std. 802.11-2007); WAVE MAC architecture (Adapted from IEEE 802.11p and 1609.4)

UP (Same as 802.1D user priority)	802.1D designation	AC	Designation (informative
1	BK	AC_BK	Background
2	—	AC_BK	Background
0	BE	AC_BE	Best Effort
3	EE	AC_BE	Best Effort
4	CL	AC_VI	Video
5	VI	AC_VI	Video
6	VO	AC_VO	Voice
7	NC	AC_VO	Voice

High ← priority — Low

(a)

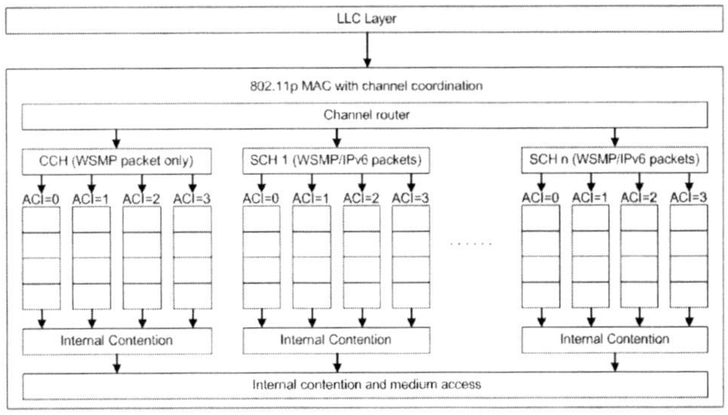

(b)

channel selector schedules the queued data for an external contention by de-queuing the queues based on their ACI (Zhang et al., 2008).

Network Layer Solutions

The existing network layer solutions for provisioning QoS in VANETs focus on QoS-oriented routing, where only four VANET-specific and QoS-oriented routing protocols could be addressed in the literature. Only one protocol uses multipath routing functionality to provide QoS for vehicular and intelligent transportation systems (Ramirez & Fernandez, 2007). The rest of these protocols employs the geographic and position information to enhance routing functionalities, such as presented by Niu, Yao, Ni, & Song, (2007), Sun, Yamaguchi, Yukimasa, & Kusumoto (2006), and Kihl, Sichitiu, Ekeroth, & Rozenberg, (2007). This section presents two of these solutions, namely the solutions of Niu et al. (2007) and Sun et al. (2006).

Delay and Reliability Constrained QoS Routing Algorithm (DeReQ)

Niu et al. (2007) developed a link Delay and Reliability constrained QoS routing algorithm (DeReQ) to support multimedia communications in VANETs. The main purpose behind developing the DeReQ algorithm is to find a best route in relation to reliability and QoS requirements, especially delays. The essence of this solution lies in the following three main aspects: (1) considering reliability estimation on active links, excluding alternative methods that did not consider the influence of node mobility patterns on the link reliability models; (2) considering two key QoS metrics satisfactions - link reliability and link delay; (3) reducing flooding overhead through setting up a time-to-live factor in terms of number of hops for each routing message.

As an algorithm, DeReQ does not include developing any routing protocol, but it could be integrated into other known MANET routing protocols. Thus, several MANET routing protocols can be extended by the DeReQ algorithm to provide QoS routing support. However, Niu et al. (2007) have used the AODV routing protocol as a basis to test their algorithm, which is added into the routing discovery process of AODV. Accordingly, the AODV routing table has to be extended in order to store the information required by DeReQ algorithm, such as link reliability, link delay, and node's position and speed.

The DeReQ algorithm supposes that links in VANET are symmetric. Accordingly, it describes each link through two main attributes: link reliability and link delay, which are used as QoS metrics. Since finding an optimal route in VANET under considering two QoS metrics is a NP-complete problem, Niu et al. (2007) simplified the "route search problem into a problem of finding the path that has the best link reliability while the link delay is under a desired time-delay bound". Three main steps are defined in the DeReQ algorithm, in order to achieve selection of the proposed best route. In the first step, the algorithm searches for the route "p" with the maximum link reliability among all the available routes, then it identifies the set "s" of routes whose link-delay satisfies the desired delay bound, and whose related broadcasting hops are suitable to the selected time-to-live metric. Thus, if p belongs to s, then p represents the final search results of DeReQ algorithm. If this is not the case, we move on to step 2. In the second step, the maximal link reliability requirement will be reduced with a certain factor, and the new resulting p will be checked if it belongs to s. Should this not be the case, then the link reliability requirement will be reduced again, and this process will be repeated until we only have a single route left, which will be selected in the third step as a final result. In that respect, Niu et al., (2007) suppose that the final result of the algorithm will be the most acceptable path that has an acceptable time-delay, best link reliability, and possibly minimum hop number.

The simulation results show that DeReQ-extended AODV (DeReQ-AODV) found the routes with the maximum link reliability for all simulated cases. DeReQ-AODV outperformed the original AODV and maintained link reliability at no less than 60%, even at high speed of more than 200 km/h. Also, DeReQ-AODV satisfied the QoS requirement of end-to-end delay limit of 40ms and achieved a comparable performance to that of AODV. Due to rapidly varying network topology, DeReQ-AODV achieved a route success ratio of 80%. DeReQ provides the first step towards adopting QoS in VANETs; however, the discussed study does not clearly state how to reduce the influence of other QoS parameters, such as packet loss in a QoS provision scenario in VANETs. Also, the characteristics for safety services and message dissemination were not considered in the concept.

GVGrid: A QoS Routing Protocol for VANETs

Sun et al. (2006) presented a QoS routing protocol named GVGrid, which is used for VANETs. GVGrid is an on-demand and position-based routing protocol, which initiates a route from a source as fixed node to vehicles existing in a destination region. The idea behind developing GVGrid stems from the fact that vehicles driving in the same direction at approx. same speed are assumed to remain at a relatively stable inter-vehicle distance, thus actually allowing relatively stable wireless connections to be initiated. Therefore, these stable links can be used for the purpose of QoS-oriented routing between vehicles in general, and in particular for building a V2V communication-based backbone for RSU-to-RSU communications.

In GVGrid it is assumed that each node is equipped with the same ranged wireless device, such as IEEE 802.11, and Car Navigator (GPS and digital map) for accurate geographic information as well as roads network and vehicles' driving direction information. GVGrid partitions the geographic region into squares of equal-size, which are named grids (Sun et al., 2007). The grid size "w" is selected according to cell radius "r", so that nodes can communicate with other nodes of neighboring grids. In addition, nodes in GVGrid exchange information concerning vehicle's position, driving direction and ID over hello messages, which will be generated periodically.

The route discovery process used in GVGrid concentrates on finding all route candidates that follow driving routes from a source that is located in the request zone to the destination region. The source node forwards a Route Request (RREQ) to a selected node of each neighboring grid of the requested zone. Each of such nodes forwards the RREQ in a similar way. The forwarding node adds road and node information to each forwarded RREQ. In that way, GVGrid provides a kind of selective forwarding similar to that of GPCR routing protocol (Lochert, Mauve, Fusler, & Hartenstein, 2005) to avoid route discovery flooding. On the other hand, in order to initiate a route discovery from the source node to the destination region, a node "d" of destination region "D" must confirm the received RREQ. The node d with the smallest ID in grid D becomes the "leader" node, which calculates the best route from the information included in RREQs by estimating route lifetime. Node d estimates the route lifetime by calculating the number of occured disconnections, using the information in RREQs. Accordingly, node d confirms building the route through transferring a Route Response (RREP) to the source node via the selected route.

Furthermore, GVGrid provides a route maintenance mechanism that restores the original route when the route breaks down. Since the original route is considered to be the best route according to the estimated route lifetime, the grids participant in the original route, will be saved by all nodes which were engaged in the respective route. Thus, when the route breaks down, only nodes that belong to the original route will be considered in the new attempt to construct an alternative route. Otherwise, if all nodes belonging to the original route broke down, alternative nodes from the front grid will be considered for selection.

Sun et al. (2007) used traffic simulator NETSTREAM (Toyota Central R&D Labs) as platform for their simulation; they additionally implemented the GPCR (Lochert et al., 2005) in order to compare the results. To evaluate the quality of network routes, two performance parameters are considered: the lifetime and the packet arrival ratio. In comparison to GPCR, GVGrid could achieve the longest route lifetime in all simulated cases, where the route lifetime in GVGrid was more than 30% better than that of GPCR. Also, in case of the packet arrival ratio, GVGrid provided better ratios in the used traffic density (720/km^2 with 3~6/grid and 240/km^2 with 1~2/grid) and sparse densities, where more than 10% performance gain, compared to GPCR, was achieved. However, although GVGrid causes lower packet loss rate in comparison to GPCR, GVGrid has a higher delay time. As a consequence, GVGrid can be considered to have several advantages for relatively high quality communication data transfers compared to existing methods. As another consequence, GVGrid still has to deal with critical issues, especially the combination of geographic and link state based routing. Also, we assume another performance comparison between GVGrid and other QoS-oriented routing protocols for VANET may give more details about GVGrid capabilities.

Open Research QoS-Issues in VANETs

Soon in the future VANETs will provide a wireless platform for several types of communications and services, such as active safety or multimedia services, for users as well as for vehicles. The most known studies about VANETs concentrate on the functionality of VANET, but less so on its reliability for real-time applications. Studies that deal with QoS solutions for VANET mostly concentrate on ITS services and active safety applications, while less focus is put on multimedia services over VANETs. In spite of the importance of QoS for safety applications, only few studies could be addressed about VANET-specific QoS solutions. Indeed, there is a general lack of research in the field of VANET-specific QoS solutions, as could be deduced by study of related literature. Several topics, such as frame/packet prioritization and forwarding, end-to-end QoS, seamless layer-2 and layer-3 handoff, QoS-oriented routing, and several other points, are still open to be addressed.

QOS MODELS AND SOLUTIONS FOR ROADSIDE NETWORKS

The QoS solutions for roadside networks usually depend on the used-network architecture and related-communication technologies. Due to the diversity of technologies that could be used for constructing roadside networks, we will focus in this section only on QoS solutions in respect to backbone infrastructures, which were developed especially for VCNs. In the rest of this section, only RBNs will be considered, since the RANs are adequately covered in VANETs' QoS solutions.

The network architecture and network technologies are very important factors in deciding which QoS models and protocols should be adopted for which VCN. That means, QoS solutions can be considered only in accordance with the used RBN infrastructure. In other words, the QoS solution in a RBN can be efficiently analyzed, if and only if sufficient knowledge about the related RBN infrastructure is available.

Worldwide, only several projects, such as FleetNet, ASV, VSC and lately COOPERS, VSC 2 and CIVS, have attempted to explore the potentials of VCNs and its ability to address related challenges and issues in order to develop solutions for adopting such networks as platforms for ITS and multimedia-related services (Hartenstein & Laberteaux, 2008). However, the main focus was/is the design and development of solutions only for VANETs. To the best of our knowledge, none of these solutions has quite obviously dealt with the challenges and issues of design and development of the roadside backbone network, especially for guaranteeing QoS provision in VANETs. Thus, these solutions suppose that a backbone infrastructure of the roadside network is already well established and providing the required resources and capabilities for vehicular communication (Franz, Hartenstein, & Mausve, 2005). Due to this gap in the research and development of RBNs, only very few VCN-specific QoS solutions could be considered for RBNs.

This section attempts to propose the possible layer-2 and layer-3 models and solutions of QoS for roadside networks in accordance with the wired and wireless backbone infrastructures. The category for mixed wired/wireless RBNs could not address any solutions.

QoS Solutions for Wired Roadside Backbone Network

The well-designed wired RBNs provide higher QoS satisfaction, as well as reliability compared to wireless RBNs. We believe that the already existing VANETs employ conventional wired RBN solutions for their test environments and system analysis, where the used RBN is/was only a means to an end. Therefore, conventional QoS solutions, such as those based on a combination of IEEE 802.1D, DiffServ, IntServ, MPLS, etc. are expected to be in use for wired RBNs.

Since the most known VCNs solutions and studies have provided marginal details, if any, about their used RBNs, a wide view on state of the art of QoS in wired RBNs could not be proposed for any layer of the network protocol stack. Instead, we will briefly outline two studies, in which a few details about the used RBN is discovered.

The known IEEE 802.11-based solutions for V2V and V2R present a flat architecture (Franz, Hartenstein, & Mausve, 2005; Wan, Tang, & Wolff, 2008), in which the access points (APs) and/or relays, as RSUs, are directly connected to a non-specified RBN through DSL or LAN connections. The internet is accessed through appropriate gateways. However, a concept for RBN or related QoS solutions was not mentioned, since these solutions were based on the existing wired infrastructure within towns and cities.

Okanishi et al. (2008) proposed one of the few studies on using an IP-based backbone-network infrastructure for highways. This study discusses the possibility of replacing currently existing dedicated communication network systems of roads' administrators on highways with an "Integrated

IP/optical network" that could also be used as a backbone-network platform for "road-to-vehicle" communications as well as for Internet services. An optical fiber with a bandwidth of at least 1 Gbps has been recommended, whereas up to 30 Gbps are foreseen for future applications. Okanishi et al. (2008) mentioned that the "integrated IP/optical network provides comfortable network access (quality of service) seamlessly for the core network, local networks, and roadside network terminals" (Okanishi, Kon, Chiku, Sugiyama, & Sakurai, 2008). However, Okanishi et al. (2008) did not give any details about the kinds of QoS mechanisms and protocols that could be used for their solution.

QoS Solutions for Wireless Roadside Backbone Network

Though some limitations of wireless RBNs arise, such as low reliability, bandwidth restrictions, and other factors, the wireless infrastructure of such RBNs has promising essential features, such as high flexibility, low decision-to-installation time, and reduced costs, in comparison to wired RBNs. The importance of wireless RBNs was addressed earlier. For instance, in the U.S.A., Lamm & Schneider (2001) reported that "Several state Departments of Transportation (DOTs) have expressed an interest in deploying wireless communications' infrastructures along the roadsides of their states to support traffic management and maintenance services." (Lamm & Schneider, 2001). However, the focus of such studies and/or researches can be divided into two main categories: (1) the use of wireless network as an alternative solution for RBN, especially on-road systems where no RBN has already been installed (Lamm & Schneider, 2001; Daher, Krohn, Gladisch, Arndt, & Tavangarian, 2008a); (2) the use of wireless RBN, especially those that are based on IEEE 802.11 as access technology, as a platform for providing broadband wireless Internet on highways as well as railways and other road systems (Lamm & Schneider, 2001; Bengsch, Kopp, Petry, Daher, & Tavangarian, 2004; Krohn, Unger, & Tavangarian, 2007; Daher et al., 2008a). Most of these studies concentrate on the challenges and issues faced in conception and network design, while subjects of QoS were not in the focus.

The attempts to develop QoS solutions for the wireless RBNs so far concentrated on layer-3, especially QoS-oriented routing protocols and packets pre-fetching mechanisms, as well as on providing an efficient platform for enabling seamless handoff mechanisms in RANs in the case of V2R communication. Specific layer-2-oriented QoS solutions for RBNs could not be discovered in available literature. This was an expected result, as the efforts and costs to develop layer-2 QoS solutions for wireless RBNs are currently unprofitable in a technical sense. In other words, the need for layer-2-specific QoS solutions for RBNs currently has a very low priority in the research and development facilities around the world. In the rest of this section, we propose the layer-3 based QoS solutions in relation to related wireless-network architectures.

Cluster-Oriented Routing Protocol (CORP)

The Cluster-Oriented Routing Protocol (CORP) provides a QoS-oriented routing protocol and is basically designed for hierarchical backbone infrastructures of roadside networks (Daher, Krohn, Gladisch, Arndt, & Tavangarian, 2008b). However, CORP is specifically developed for the hierarchical multi-layer backbone infrastructure that proposed by Daher et al. (2008a) as a part of the planned project "Wi-Roads" at the University of Rostock in Germany – Wi-Roads refers to "**Wi**reless Roadside Infrastructure for High-speed **Roads**". Figure 5 shows the related network architecture and its topology. Before presenting CORP we will briefly introduce the related wireless RBN, referred to as Wi-Roads RBN in the rest of this chapter, and its features. This way, we

Figure 5. Network topology of multi-layer backbone infrastructure

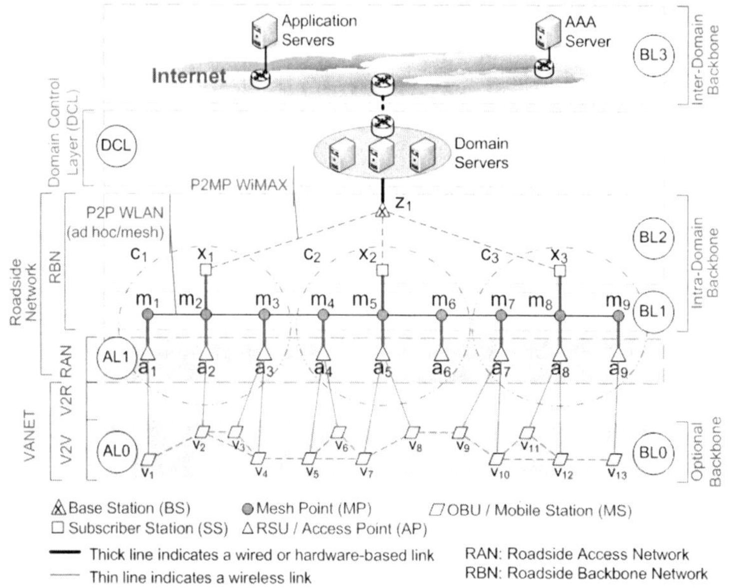

can simplify explaining CORP characteristics and functionalities in the rest of this section.

The architecture of Wi-Roads RBN forms a multiple-domain architecture, in which several components are defined: (1) Mesh Point (MP), which is connected to at least one RSU; (2) Cluster (chain topology), which comprises several MPs, three MP in this example, and a single Backhaul Frontend (BFE); (3) Domain, which consists of many clusters that are connected to a single Backhaul Backend (BBE). The MPs are based on IEEE 802.11 technology and configured in P2P mode as mesh routers. The BFE and BBE are based on IEEE 802.16d technology and configured in P2MP mode; thus, a WiMAX Subscriber Station (SS) and Base Station (BS) are used as BFE and BBE, respectively. Accordingly, the Wi-Roads RBN consists of the following layers, as shown in Figure 5:

1. **Inter-domain backbone layer,** considered as backbone layer-3 (BL3), refers to the network between domains and comprises the supply network / Internet
2. **Domain control layer (DCL)** indicates a network, in which the domain-related management servers bridge the BBE to the Internet. These servers act as a central point of connection between the intra-domain and the inter-domain backbones. Indeed, DCL has the domain intelligence and is responsible for mobility management, including dynamic addressing, intelligent routing, load balancing, etc.
3. **Intra-domain backbone,** considered as backbone layer-2 and 1 (BL2 and BL1), refers to the clusters within a domain, and presents the actual wireless infrastructure for RBN. Here, only WLAN and WiMAX technologies are used

The RAN is considered to be access layer 1 (AL1), while VANETs might be considered to be access layer 0 (AL0) for the cause of data exchange between vehicles, or as a backbone layer (BL0) when transferring backbone related traffic over V2V/V2R communications. While BL1 and BL2 provide a wireless-communication platform for RBN, DCL supplies the required management and control mechanisms, especially mobility management and traffic-congestion control (both of which are essential for provisioning QoS in such networks). The Wi-Roads RBN represents a three-layer physical hierarchy, whose network topology defines three types of nodes: MP, Cluster Head (CH) as BFE, and Domain Head (DH) as BBE, as shown in Figure 5.

However, the characteristics of Wi-Roads RBN in relation with VANETs cause several issues. On the one hand, the main problem of the presented wireless RBN is the *bandwidth-wall problem*, as we like to call it, which is explained as following: although the WiMAX technology used in the wireless RBN (Figure 5) successfully bridges the long distances between BFEs and related BBE, it forms a bandwidth bottleneck between BL1 and Internet due to the bandwidth restrictions of WiMAX (in comparison to DSRC/WAVE). That means, the wireless uplinks between SSs as BFEs and BS as BBE have lower bitrates than that of access links. For example, an RSU provides up to 27 Mbps, while a SS as a BFE has up to 10 Mbps. Here, the WiMAX bitrate of 75 Mbps cannot be reached, due to the long distance between SS and BS, as well as the channel sharing with other SSs. On the other hand, the Wi-Roads RBN provides QoS support on MAC layer of all backbone levels, because IEEE 802.11e is provided on BL1 and IEEE 802.16d already includes QoS mechanisms. However, the QoS solution on layer 2, especially with 802.11e, is based on per-packet and per-hop QoS. Thus, an end-to-end QoS could only be provisioned when layer-3 solutions are also integrated as end-to-end resource reservation and admission control (Daher & Tavangarian, 2006a), load balancing (Daher & Tavangarian, 2006b), and QoS-oriented routing. Due to the rapidly varying VANET topology, the use of reactive routing protocols, such as AODV over Wi-Roads RBN, causes frequent route-breakdowns, since the route must be re-established for each client/vehicle after each handoff to new RSU. Similarly, the rapidly varying VANET topology causes proactive routing protocols, such as OLSR, to generate relatively high signalling overhead, especially for updating IP tables after each client/vehicle handoff. Also, the hierarchical routing protocols such as HSR (Iwata, Chiang, Pei, Gerla, & Chen, 1999) build a hierarchical topology over usually flat network architecture and provide no possibility to bind the hierarchical topology with the physical hierarchy. In other words, the known hierarchical routing protocols for mesh networks do not profit from the already existing physical hierarchies in the Wi-Roads RBN.

Consequently, to benefit of the physical hierarchy of the Wi-Roads RBN and to deal with routing challenges related to the rapidly varying VANET topology, as well as bandwidth wall problem Daher et al. (2008b) developed the routing concept of CORP. CORP provides a QoS-oriented routing specified for the Wi-Roads RBN; routing within VANET (AL0/BL0 and AL1) does not belong to CORP. CORP supposes that each domain is configured in a single IPv6 subnets, in order to avoid high latency for layer-3 handoffs (in case of intra-domain communications). Accordingly, two scopes of routing could be defined in CORP: intra-domain routing and inter-domain routing. In this respect, a layer-3 handoff is expected only in case of inter-domain routing. In accordance with the physical hierarchy of related backbones, CORP distinguishes between two main types of routing: vertical routing and horizontal routing. Firstly, the vertical routing consists of downwards routing, from DH to MP in case of downloads, and upwards routing from MP to DH in case of uploads. Secondly, the horizontal routing is between MPs in case of inter-vehicle communications over roadside network.

The CORP-specific route discovery mechanism is based on the identification of each node within the topology through the 3-tuple (domin ID: cluster ID: MP ID). Accordingly, CORP specifies two types of routing mechanisms: relayed routing and QoS-oriented routing. Firstly, the relayed routing provides routing on the basis of per-hop decisions, i.e., each node forwards the packets depending on the related destination node ID, so that the forwarding node can determine the direction and distance to the destination node through comparing the destination node ID with its own ID. To deal with the issues of rapidly varying topology, CORP specifies a hierarchical forwarding mechanism and targeted home-MP forwarding mechanism for vertical and horizontal routing, respectively – a targeted home-MP indicates a MP to which the related RSU is associated with the destination client (vehicle). Secondly, the QoS-oriented routing provides an end-to-end QoS through additional mechanism that enable resource reservation and admission control (Daher & Tavangarian, 2006b). The QoS-oriented CORP also uses the forwarding mechanisms of relayed routing, but only in relation to a route reservation mechanism, which establishes and maintains the route until the end of the related data session. This route can be modified dynamically in accordance with the load variation in the network, e.g., for the purpose of achieving shorter routes or to follow the topology variation of related client (vehicle) in VANET. To guarantee E2E QoS via CORP, a cross-layer interaction with layer-2 is required in order to address the link state of each hop. Therefore, a load observation model is developed to provide the CORP-node platform with appropriate load and link state information of each related interface.

Due to the relatively stable architecture of Wi-Roads RBN, CORP initiates a mechanism of CORP-specific topology discovery when starting the protocol for the first time. However, when a node is broken down, or when a new node, such as MP, is connected to the network, the topology discovery mechanism will detect the new change and inform the related nodes to update their neighborhood graph. This functionality forms the basis for supporting self-configuration and self-healing mechanisms. In general, only the domain head must have the knowledge about the whole network graph, while the other nods only have a sub-graph of the whole network graph. The size of this sub-graph depends on the configured Neighborhood Degree (ND) that determines the node's sight-distance, based on hops, into the network graph, e.g., when cluster ND=1, the cluster knows about all clusters being on one hop distance. Only the domain head can change the ND in the network. The restriction of nodes' knowledge about the network graph leads to additional reduction of the required control and management signalling functionality of CORP. Also, in order to reduce the control and management signalling of CORP, without degrading the routing performance, a hierarchical IP address-to-node ID resolution is integrated. Each node has an IP table of all associated nodes and clients of the lower level, i.e., the MP IP table contains all IP addresses of the related RSU and its associated clients, while the CH IP table comprises IP tables of all member MPs. In this scenario, the DH has an IP table of all nodes and clients in its related domain. The IP table of MP, CH and DH will be updated periodically in long intervals and after a critical change, for example after a handoff to a new RSU. The update process also reflects the hierarchical property of CORP, e.g., the DH IP table will be updated only in case a vehicle has left a cluster, entered a new cluster, or associated/disassociated to a cluster-related RSU.

The most important characteristics of QoS-oriented CORP is its main focus on fairness-oriented balanced routing, as well as address-tracking forward mechanisms. Firstly, the fairness-oriented balanced routing mechanism is developed to compensate the influence of the *bandwidth-wall problem* on the whole performance of related RBN. In accordance with the used ND, which has the option to be changed dynamically, each node sends its load-state information to other nodes in an ND hops distance in RBN; load-state information is not transported to the RAN or to out-of-a domain, except in case of inter-domain routing. Since each node has knowledge about ND-determined sub-graph of the whole network, the node (in relation to the load-state information of some other nodes) can decide which path should be selected for related data traffic. The decision's complexity of a route selection is very low, since the number of routes is limited to the number of CHs in case of vertical routing and to a single route if zigzag (BL1 to BL2 and back to BL1) routing is not allowed (in the case of horizontal routing). However, since the size of the sub-graph determines the number of possible path options, the greater ND is, the higher the possibility of finding the best unloaded and real-time capable path is, as well. Secondly, the address-tracking forward mechanism is a result of the hierarchical feature of CORP. The DH offers only IP-to-cluster ID resolutions to the ID Resolution Requests that are received from MPs. Thus, the data packets will be routed to clusters and not directly to MPs. Since each MP saves a list of vehicles that are lastly handed off to RSUs of other MPs for a certain time, the arrival packets can still be forwarded in the right direction to the right MP (by the addressed vehicle). In other words, if the target vehicle changed its MP while the packet is underway, the destination MP will automatically forward that packet towards the new MP, to which the vehicle is recently associated.

Although CORP is expected to support routing in all possible hierarchical networks, the development of CORP is currently concentrated on the adoption of the cluster-oriented wireless RBNs (Daher et al., 2008b). However, the development of the CORP protocol is still a work in progress, and we expect a first complete specification and implementation of CORP in the third quarter of 2010 (Daher et al., 2008b).

Host Abstraction Routing Platform (HARP)

The Host Abstraction Routing Platform (HARP) presents another routing concept for providing a QoS-oriented routing in the wireless RBNs (Krohn, Daher, Gladisch, Arndt, & Tavangarian, 2008a). Similar to CORP, HARP is also specifically developed for the hierarchical multi-layer backbone infrastructure (Figure 5) developed by Daher et al. (2008a).

The main concept of HARP is based on providing an intermediate layer between the Wi-Roads RBN and routing protocols in order to enable addressing clients (vehicles) in relation to their associated RSUs - MPs from RBN point of view. In this respect, routing occurs between DH and MPs, in case of vertical routing, or among MPs in case of horizontal routing, without integrating the clients directly into the routing process. In other words, a virtual tunnel will be initiated between source and destination over the RBN, as the following example of downward routing explains: the DH marks the received IP packet from Internet with his IP as source and as destination with the IP address of the MP over which the destination client is reachable. Then, any kind of routing protocol, especially QoS-based routing protocols, could be used to route this packet from DH to the target MP, which decodes the packet and forwards the data packet to the destination client over the related RSU. Thus, converting the routing problem of rapidly varying network topology into a routing in a quasi stationary network environment, i.e., the negative influence of rapidly varying network topology of VANET on routing in RBNs should be reduced drastically. In contrast to CORP, there is no direct benefit from the hierarchical infrastructure. Instead, HARP should enable using other routing protocols, especially those that support QoS, where a domain-wide load balanced routing could be supported when using an appropriate routing protocol. Furthermore, to enable routing via HARP, some other mechanisms are still required in order to adapt the developed routing concepts to the wireless infrastructures. Several CORP's mechanisms, such as that for MP-to-client resolution, IP tables update, and node/link load observation and control, can be integrated directly into HARP. However, the portability, as well as the adaptability to the selected routing protocols, must be guaranteed.

Currently, Krohn et al. (2008a) did not present any list of routing protocols that can be used with HARP; however, Krohn et al. (2008a) preferred using a proactive routing protocol, with special emphasis on OLSR, for the proposed HARP (HARP-OLSR), since OLSR is expected to provide higher reliability and real-time capability in comparison to other considered protocols in conjunction with the used wireless RBN (Krohn et al., 2008a). The concept and development of HARP is still a work in progress and we expect a first complete specification and implementation of HARP in the second half of 2010 (Krohn et al., 2008a).

Packets Pre-Fetching Mechanisms

The Packets Pre-fetching Mechanisms (PPFMs) deal only with downwards packet forwarding. Since we have frequent handoffs, due to the relatively small RSU cells and high speed driving vehicles, we can observe degradation of QoS provided for vehicles in V2R communications. The main idea behind the PPFMs is to reduce the influence of frequent handoffs through forwarding data packets down to the next expected RSU before or during the handoff process. To achieve that, the PPFM must be able to get the vehicle's position at certain time intervals, and can accordingly estimate the vehicle's position for the coming time with certain accuracy. In this respect, PPFM can early react on the driving behavior of related vehicle - and thus forward the data packets to the appropriate RSUs.

In this respect, Krohn, Unger, and Tavangarian (2008b) proposed a novel mechanism, called Feed Forward Mechanism (FFM), for improving the packet delivery at the RSU over a wireless RBN through tracking and localizing the moving vehicles at the level of access networks. The FFM is proposed in accordance with a specific wireless RBN, based on a similar architecture to that developed for CORP. The main difference lies in the domain structure and the use of DCL. Also, the presented RBN is strongly dependent on the used RAN, since only IEEE 802.11 WLAN was considered for V2R communications, as shown in Figure 6. This network however was foreseen for wireless Internet on highways.

To support the presented FFM, a two-layer proxy system was developed, in which each node in the network is provided with a proxy. Here, the local proxy for each AP and a central proxy act as a gateway to the internet (Krohn, Unger, & Tavangarian, 2008b). In FFM, the local proxy provides the central gateway with information about velocity and location of related vehicles. To provide QoS in such a solution, the central proxy prioritizes the real-time related packets, such as that of VoIP traffic, over data packets, so that some downloaded data packets could be buffered in the central proxy for a certain time before they are forwarded to the appropriate local proxies that are responsible for the related desti-

Figure 6. Distributed proxy system for feed forward mechanism

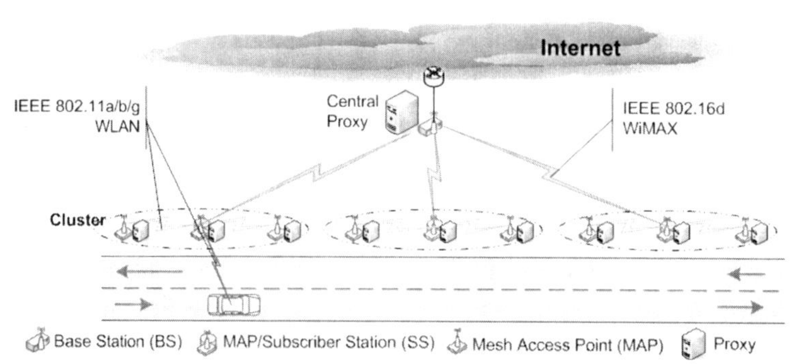

nations. The central proxy uses the location and velocity information about each client's vehicle, delivered by the local proxies, in order to calculate the best possible local proxy for delivering the related downloaded packets. In this solution, the cooperation between RBN and RAN is necessary for tracking the vehicles. In accordance with FFM concept, a distributed proxy system, algorithms, and a protocol for an efficient packet transfer between applications servers and vehicles was developed.

Another PPFM for hot-spotted networks (Imai, Morikawa, & Aoyama, 2001) was presented earlier, where a pre-fetching mechanism was discussed for IEEE 802.11 WLAN as a platform for providing wireless Internet for vehicles, especially for data download scenarios, in cooperation with other cellular networks. However, there was no focus on the QoS requirements in this study.

Open Research QoS-Issues in Roadside Backbone Network

Due to the lack of studies in the field of roadside backbone infrastructures of vehicular communication networks (VCNs), plenty of issues and challenges for QoS provision in VCNs remain, and could only be addressed partially. We believe that provisioning QoS in VANETs, without considering the roadside backbone network, cannot offer a reliable guarantee for QoS in the whole VCN. Therefore, issues such as providing an efficient platform for QoS-oriented routing, as well as seamless layer-2 and layer-3 handoff mechanisms for V2R communications, are very hot topics for research. Also challenges for designing reliable, modular and real-time capable roadside backbone infrastructures should lead the research topics in the future. Moreover, we believe that the research in the field of wireless roadside backbones and related routing protocols will become a very promising research field, especially in accordance with the rapidly-evolving wireless-mesh technologies.

SUMMARY

This chapter introduced vehicular communication networks (VCNs) and presented their importance for Intelligent Transportation System (ITS) services, as well as multimedia and data services. An overview about the DSRC/WAVE as a leading technology for VCNs was presented and the QoS requirements for real-time applications of ITS and multimedia services has been discussed. Only few VCN-specific QoS solutions could be found in the literature. However, the found VCN-specific QoS solutions are classified in accordance to VCN level, as well as layer-2 and layer-3. Some of these solutions are introduced. For instance, we briefly presented WAVE MAC-related IEEE 802.11e as a layer-2 QoS solution for VANETs, while we proposed CORP protocol as a layer-3 QoS solution that suited wireless backbone network specifications. Finally, we presented a short description about the open research QoS-issues in VANET and roadside backbone networks.

REFERENCES

Bengsch, A., Kopp, H., Petry, A., Daher, R., & Tavangarian, D. (2004). Efficiency of Wireless Local Area Networks in High Speed Moving Clients. In *Proc. Innovative Internet Community Systems (I2CS)*, Guadalajara, Mexico.

Berger, I. (2007). Standards for Car Talk. *IEEE – The Institute*, Vol. 31, No. (1).

Bossom, R., Brignolo, R., Ernst, T., Evensen, K., Frötscher, A., Höfs, W., et al. (2008). Communication for eSafety (COMeSafety). Overall Framework - Proof of Concept Implementation, Version 2.0. European ITS Communication Architecture.

Daher, R., Krohn, M., Gladisch, A., Arndt, M., & Tavangarian, D. (2008a). *Vehicular Communication Networks: Novel Concept for Wireless Backbone Infrastructure of Roadside Networks*. Internal Report, University of Rostock, Rostock, Germany.

Daher, R., Krohn, M., Gladisch, A., Arndt, M., & Tavangarian, D. (2008b). *Cluster-Oriented Routing Protocol (CORP) for Hierarchical Roadside-Backbone Infrastructures of Vehicular Communication Networks*. Internal Report, University of Rostock, Rostock, Germany.

Daher, R., & Tavangarian, D. (2006a). Resource Reservation and Admission Control in IEEE 802.11 WLANs. In *Proc. Third International Conference on Quality of Service in Heterogeneous Wired/Wireless Networks (QShine 2006)*, Waterloo, Canada.

Daher, R., & Tavangarian, D. (2006b). QoS-oriented Load Balancing for WLANs. In *Proc. First International Workshop on Operator-assisted (Wireless Mesh) Community Networks 2006 (OpComm'06)*, Berlin, Germany.

Electronic Communications Committee. (2008a). *ECC/DEC/(08)01. ECC Decision of 14 March 2008 on the harmonised use of the 5875-5925 MHz frequency band for Intelligent Transport Systems (ITS).*

Electronic Communications Committee. (2008b). *ECC/REC (08)01. ECC Recommendation (08)01 Use of the Band 5855-5875 MHz for Intelligent Transport Systems (ITS).*

Federal Communications Commission. (2004). *FCC 03-324 A1. Amend Rules Regarding Dedicated Short Range Communications Services and rules for Mobile Service for Dedicated Short Range Communications of Intelligent Transportation Services.* FCC 03-324 A1.

Franz, W., Hartenstein, H., & Mausve, M. (2005). *Inter-Vehicle-Communications Based on Ad-hoc Networking Principles.* The FleetNet Project, University of Karlsruhe.

Hartenstein, H., & Laberteaux, K. P. (2008). A Tutorial Survey on Vehicular Ad Hoc Networks. *IEEE Communications Magazine, vol. 41*, no. (6).

IEEE P1609.0 D0.2 (2007). *IEEE Trial Use Standard for Wireless Access in Vehicular Environments (WAVE) – Architecture.*

Imai, N., Morikawa, H., & Aoyama, T. (2001). Prefetching Architecture for Hot-Spotted Networks. ICC 2001. In *IEEE International Conference on Communications (ICC 2001).*, Helsinki, Finland.

ITE. (2004). *Institute of Transportation Engineers, Traffic Management Data Dictionary (TMDD) and Message Sets for External Traffic Management Center Communications (MS/ETMCC).* [Online]. Available:Retrieved from http://www.ite.org/tmdd

ITU-T Recommendation G.114 (2003). *One-way transmission time.*

Iwata, A., Chiang, C. C., Pei, G., Gerla, M., Chen, T. W. (1999). Scalable routing strategies for ad hoc wireless networks. *IEEE Journal on Selected Areas in Communications, vol. 17*, no. 8(8), pp. 1369-1379.

J. J. Blum, J. J., Eskandarian, A., & Hoffman, L. J. (2004). Challenges of Intervehicle Ad Hoc Networks. *IEEE Transactions. on Intelligent Transport.t. Systems,* vol. 5, no. (4), pp. 347–51.

Kihl, M., Sichitiu, M., Ekeroth, T., & Rozenberg, M. (2007). Reliable Geographical Multicast Routing in Vehicular Ad-Hoc Networks. In *Proceedings of the 5th international Conference on Wired/Wireless internet Communications. (Lecture Notes in Computer Science, vol. 4517).* Berlin, Germany: Springer-Verlag, Berlin.

Krohn, M., Daher, R., Gladisch, A., Arndt, M., & Tavangarian, D. (2008a). *Host Abstraction Routing Platform (HARP) for Hierarchical Roadside-Backbone Infrastructures of Vehicular Communication Networks.* Internal Report, University of Rostock, Rostock, Germany.

Krohn, M., Unger, H., & Tavangarian, D. (2007). Advanced Wireless Network Architectures for Highways. In *Proceedings of Wireless Congress*, Munich, Germany.

Krohn, M., Unger, H., & Tavangarian, D. (2008b). A Distributed Proxy System for High Speed Clients. In *Proceedings of 13th Joint International and National CSI Computer Conference*, Kish Island, Iran.

Kurihara, T. M. 2009. IEEE 1609 Working Group – Project Status Report (2009-05-11), doc: IEEE 802.11-09-0093-02-000p.

Kutzner, K., Tchouto, J.-J., Bechler, M., Wolf, L., Bochow, B., & Luckenbach, T. (2003). Connecting Vehicle Scatternets by Internet-Connected Gateways. In *Proceedings of Multiradio Multimedia Communications (MMC) - Communication Technology for Vehicles.*, Dortmund, Germany.

Lamm, R., & Schneider, S. J. (2001). *Investigation into the Development of a Wireless Ethernet Backbone, 10-9233 – Internal Report.* Southwest Research Institute. Retrieved from http://www.swri.org/3pubs/ IRD2001/10-9233.htm

Lochert, C., Mauve, M., Fusler, H., & Hartenstein, H. (2005). Geographic Routing in City Scenarios. In *ACM SIGMOBILE Mobile Computing and Communications Review*, (pp. pages 69–72).

Murthy, S. R. C., & Manoj, B. S. (2004). *Ad hoc Wireless Networks*. Upper Saddle River, NJ: PrinticePrentice Hall Inc.

Niu, Z., Yao, W., Ni, Q., & Song, Y. (2007). DeReQ: a QoS routing algorithm for multimedia communications in vehicular ad hoc networks. In *Proceedings of the 2007 international Conference on Wireless Communications and Mobile Computing (IWCMC '07),* Honolulu, Hawaii, USA.

Okanishi, S., Kon, M., Chiku, S., Sugiyama, M., Sakurai H. (2008). Traffic Network System. *NEC Technical Journal (special iss.: (ITS), vol.3* no.(1).

Olson, P. (2006). Perception-Response Time to Unexpected Roadway Hazards. *Hum. Factors, vol. 28*, no. (1), pp. 91–96.

Ramirez, C. L., & Fernandez, V. M. (2007). QoS in Vehicular and Intelligent Transport Networks Using Multipath Routing. ISIE 2007 In. *IEEE International Symposium on Industrial Electronics (ISIE 2007).* (pp. 2556-2561).

Schoch, E., Kargl, F., Weber, M. & Leinmüller, T. (2008). Communication Patterns in VANETs. *IEEE Communications Magazine, vol. 46*, no. (11).

Shimura, A., Aizono, T., & Sakaibara, T. (2002). A proposal of Quality of Service (QoS) Control Method to Realize Highly-Reliable Communications for Roadside System. In *IEEE Intelligent Vehicle Symposium,* IEEE.(vol.2, pp. 370- 377).

IEEE Std 802.11-2007 (2007). *IEEE Standard for Information technology - Telecommunications and information exchange between systems - Local and metropolitan area networks - Specific requirements Part 11: Wireless LAN Medium Access Control (MAC) and Physical Layer (PHY) Specifications.* (Revision of IEEE Std 802.11-1999)

Std, S. A. E. J1746. (2001). *ISP-Vehicle Location Referencing Standard.*

Stibor, L., Zang, Y., & Reumerman, H.-J. (2007). Neighborhood Evaluation of Vehicular Ad-Hoc Network Using IEEE 802.11p. In *13th European Wireless Conference*, Paris, France.

Su, H., & Zhang, X. (2007). Clustering-Based Multichannel MAC Protocols for QoS Provisioning over Vehicular Ad Hoc Networks. *IEEE Transactions on Vehicular Technology, vol. 56*, no. (6), part 1.

Sun, W., Yamaguchi, H., Yukimasa, K., and & Kusumoto, S. (2006). GVGrid: A QoS Routing Protocol for Vehicular Ad Hoc Networks. In *Fourteenth IEEE International Workshop on Quality of Service (IWQoS 2006)*. New Haven, CT: Yale University, New Haven, CT, USA.

VSC. (2005). Vehicle Safety Communications Project: Final Rep. Submitted to NHTSA and FHWA in response to cooperative agreement DTFH61-01-X-001.

Wan, S., Tang, J., & Wolff, R. S. (2008). Reliable Routing for Roadside to Vehicle Communications in Rural Areas. ICC '08. In *IEEE International Conference on Communications (ICC '08)*.

Xu, Q., Mak, T., Ko, J., & Sengupta, R. (2007). Medium Access Control Protocol Design for Vehicle–Vehicle Safety Messages. *IEEE Transactions on Vehicular Technology, vol. 56*, no. (22).

Yousefi, S., Fathy, M., & Benslimane, A. (2007). Performance of beacon safety message Dissemination in Vehicular Ad hoc NETworks (VANETs). *Journal of Zhejiang University – Science A., Vol. 8*, no. (12). Zhejiang, China: Zhejiang University Press.

Zhang, X., Su, H., & Chen, H. (2006). Cluster-Based Multi-Channel Communications Protocols in Vehicle Ad Hoc Networks. *IEEE Wireless Communications, vol. 13*, no. (5), pp. 44-51.

Zhang, Y., Yang, L. T., & Ma, J. (2008). Unlicensed Mobile Access Technology: Protocols, Architectures, Security, Standards and Applications. Published byBoca Raton, FL: CRC Press.

KEY TERMS AND DEFINITIONS

AKTIVE: "Adaptive und Kooperative Technologien für den Intelligenten Verkehrs", which means in English: Adaptive and Cooperative Technologies for the Intelligent Traffic, German R&D project, www.aktiv-online.org.

ASV: Advanced Safety Vehicle Program.

C2C-CC: Car-to-Car Communication Consortium.

COOPERS: CO-Operative SystEms for Intelligent Road Safety, European R&D project, www.coopers-ip.eu.

VII: Vehicle Infrastructure Integration.

VSC: Vehicular Safety Communication project.

VSCC: Vehicular Safety Communication Consortium.

Wi-Roads: (**Wi**reless Roadside Infrastructure for High-speed **Roads**), project in planning, by Chair for Computer Architecture, Faculty of Computer Science and Electrical Engineering, University of Rostock, Germany.

Section 4
Multimedia

Chapter 15
Correlating Quality of Experience and Quality of Service for Network Applications

Mihai Ivanovici
Transilvania University of Braşov, România

Răzvan Beuran
National Institute of Information and Communications Technology, Japan & Japan Advanced Institute of Science and Technology, Japan

ABSTRACT

There is a significant difference between what a network application experiences as quality at network level, and what the user perceives as quality at application level. From the network point of view, applications require certain delay, bandwidth and packet loss bounds to be met – ideally zero delay and zero loss. However, users should not be directly concerned with network conditions, and furthermore they are usually neither able to measure nor predict them. Users only expect good application performance, i.e., a fast and reliable file transfer, high quality for voice or video transmission, and so on, depending on the application being used. This is true both in wired as well as wireless networks. In order to understand network application behavior, as well as the interaction between the application and the network, one must perform a delicate task – the one of correlating the Quality of Service (QoS), i.e., the degradation induced at network level (as a measure of what the application experiences), with the Quality of Experience (QoE), i.e., the degradation perceived by the user at application level (as a measure of the user-perceived quality) (Ivanovici, 2006). This is done by simultaneously measuring the QoS degradation and the application QoE on an end-to-end basis. These measures must be then correlated by taking into account their temporal relationship. Assessing the correlation between QoE and QoS makes it possible to predict application performance given a known QoS degradation level, and to determine the QoS bounds that are required in order to attain a desired QoE level.

DOI: 10.4018/978-1-61520-680-3.ch015

APPLICATION REQUIREMENTS AND QUALITY OF EXPERIENCE

Applications drive the development of networks. The need to transfer huge amounts of data across long-haul connections drives the increase of network bandwidth. The need for seamless connectivity drives the development of wireless networks. All network applications, including the now ubiquitous e-mail or browsing, require the continuous refining of network technologies. New standards and protocols allow for more reliable and faster data handling. However, the user is the only one who can say whether data is transferred fast enough, or whether the application behaves the way it should.

Consider web browsing. Users may wish that pages are loaded as fast as possible, maybe even instantaneously. This is an *expectation* that depends on user experience, the type of data to be downloaded, etc. A *requirement* in this case is an expectation which is expressed with a time constraint. When browsing the Internet it is desirable that pages are loaded in a couple of seconds. If it takes longer than 10 seconds, the page may no longer be of interest. Therefore, for web browsing, a requirement may be that a web page is loaded in less than 10 seconds.

From the network point of view, each application requires certain delay, bandwidth and packet loss bounds to be met in order to provide a satisfactory performance to users. However, performance evaluation can be done using various metrics, and user satisfaction can have several levels. For example, a user of voice communication can say quality has been "excellent", "good", "fair", "poor" or "bad", according to a widely used Mean Opinion Score (MOS) as defined in the ITU-T P.800 recommendation (ITU-T, 1996). Usually numbers are associated to these quality levels, on the scale from 5 (excellent quality) to 1 (poor quality) for ITU-T P.800 recommendation. Objective metrics, such as ITU-T P.861 (ITU-T, 1998) or P.862 (ITU-T, 2001), use quality scales as well, but in this case the score will be computed by an algorithm instead of the subjective MOS that is assigned through trials by human observers. For each of the satisfaction levels, an associated set of Quality of Service (QoS) degradation bounds can be determined, and they will represent the requirements of the application under study in order to provide a desired Quality of Experience (QoE) level.

A network application typically implies a data transfer between two end points of a network; data can represent either text and static images in the case of HTTP transfers related to web browsing, binary files in the case of file transfers by the FTP protocol, or video and/or sound for video and voice conferencing.

Based on the time requirements of network applications, two main distinct classes are identified in (Fluckiger, 1995):

- **Real-time or time-critical applications,** that have strict time constraints, such as video or voice conferencing
- **Non-time-critical or asynchronous applications,** for which time constraints are more relaxed, such as file transfers

Note that even in the case of non-real-time applications, there are still some time constraints; for example, if web page loading experiences large delays, the user degree of satisfaction will decrease, therefore delay needs to be taken into account when considering the QoE for such an application.

Based on the type of traffic pattern generated by the application, (Beuran, 2004b) distinguishes between:

- **Elastic traffic applications,** for which the traffic adapts to network conditions (usually this traffic is generated by applications that use TCP/IP as transport protocol)
- **Inelastic traffic applications,** for which the traffic doesn't adapt to network

conditions (usually this traffic is generated by UDP-based applications)

The elastic traffic applications try to optimize performance by detecting network conditions and adapting themselves to these conditions. This is especially useful in wireless networks, where conditions have a high variability. Although the user or the application itself do not have direct control and cannot impose limitations on the instantaneous throughput it generates, some control can be achieved by tuning the TCP parameters (Tierney, 2005). The inelastic traffic applications do not take into account any feedback from the network. This class of applications is represented by most video streaming applications – the server will stream video data to all the clients at the same data rate, as required by the volume of information in the video stream, regardless of the network conditions.

A taxonomy of multimedia applications based on the type of user interaction can be found in (Fluckiger, 1995). The author distinguishes between *people-to-people* and *people-to-systems* or *people-to-information servers* applications. For any network application involving interaction with the human being, QoE metrics based on human perception should be defined and used.

A survey on the QoS requirements of network applications was published by the Internet2 QoS Working Group (Miras, 2002), but their approach relies on subjective measurements and the conclusions are vague. TF-STREAM reported on best-practice guidelines for deploying real-time multimedia applications (Cavalli, 2002); these guidelines are nevertheless generic and do not guarantee satisfactory results for any combination of network conditions and codec that may be encountered. ITU-T defined network performance objectives for IP-based services in ITU-T Y.1541 recommendation (ITU-T, 2001), again in a generic context. HEAnet reviewed several aspects of perceived quantitative quality of applications (Reijs, 2002), and in this sense their approach is closest to the goal of going beyond qualitative evaluations, and creating a quantitative representation of QoE that can be related to QoS parameters for practical purposes.

The objective and pertinent quantification of QoE for applications can be only performed empirically, by experimentally assessing the user satisfaction. There are studies that try to use mathematical expressions in order to quantify the dependency of QoE on QoS: either a negative exponential function (Hoβfeld, 2007) or a logarithmic one (Richards, 1998), are employed to determine intervals or curves of *satisfaction*. The reason for using a mathematical function like the exponential is natural: the higher the experienced quality, the higher its sensitivity to a small variation. If the QoE is already low, a large variation of QoS will not be perceived as significant. This relationship can be compared to the QoE for restaurants: "If we dined in a five-star restaurant, a single spot on the clean white table cloth strongly disturbs the atmosphere. The same incident appears much less severe in a beer tavern." (Hoβfeld, 2007, pp. 367).

File Transfer

File transfer is one of the basic applications running over today's networks. It is largely used for the simple purpose of transferring data between two points using FTP (File Transfer Protocol). File transfer via FTP is an *elastic* TCP-based application. TCP tries to occupy as much of the available bandwidth as it can handle. It also adapts its transmission rate to prevailing network conditions – with high loss rates it backs off to a slower transmission rate. It also provides reliable data transfer by means of its loss recovery mechanisms.

TCP behavior has been analyzed extensively. Some researchers take an analytical approach (Mathis, 1997; Padhye, 1998, 2000). Another path is that of simulation (Fall, 1996; Breslau, 2000). There exists also the possibility to do experimental work in real networks, so as to assess raw network performance (Korcyl, 2004), or to collect traffic traces (The Internet Traffic Archive). Each of these methods has certain advantages and disadvantages related to their accuracy and the range of conditions that are analyzed.

Note that since TCP was designed in the 70's-80's (Jacobson, 1988), networks have changed considerably by a huge increase in bandwidth, as well as the spreading of wireless networks. As an alternative to TCP/IP, a new protocol was proposed in 2000, named SCTP (Stream Control Transmission Protocol) (Stewart, 2007). Although SCTP is supposed to offer superior performance compared to TCP, the latter is still the most used protocol for data transfer since it is included by default with all operating systems.

Web Browsing

Web browsing is another form of file transfer, however in this case the files are not transferred explicitly, but the transfers are initiated by web browsers on behalf of the user. Each web page that a user accesses triggers a series of file transfers. Web pages used to consist mainly of text files, however in recent days the multimedia content increased significantly, and most web pages include images, and even richer multimedia content, such as music and sounds, or video sequences. For mobile devices, such as cell phones, web page content is usually simplified to minimize loading time.

Web browsing uses the HTTP (Hyper Text Transfer Protocol) for transferring web page content. HTTP is also based on TCP, just like FTP. For some results related to application performance in the case of HTTP see for example (Padhye, 2001).

Voice Over IP

As Voice over IP (VoIP) became sufficiently robust, it started becoming a real contender to the classical Public Switched Telephone Network (PSTN). There is at the moment a plethora of Voice over IP applications – Skype, Yahoo messenger, NetMeeting. However, when many users compete for network and server resources, the quality is sometimes disappointing.

In the case of VoIP applications, there are two QoS metrics that are important to evaluate communication quality: the mouth-to-ear delay and the jitter. According to some studies, the mouth-to-ear delay should not exceed 400ms (ITU-T Y.1541, 2001), while the jitter should be less than 40ms (Beuran, 2004c). However, these general indications do not allow any detailed analysis; for this purpose appropriate QoE metrics for VoIP must be used.

Video Streaming or Video Over IP

Video streaming applications are widely-used in nowadays Internet. Such applications are very demanding from the point of view of QoS requirements, and their real-time characteristics make video streaming applications very sensitive to network quality degradation, in particular to packet loss and jitter. In terms of bandwidth, the requirements of video streaming depend on the codec used and consequently on the compression rate, which is inversely proportional to the quality of the video signal.

Streaming applications usually use RTP (Real-Time Protocol) over UDP, therefore the traffic generated by such an application is inelastic and doesn't adapt to the network conditions. In addition, neither UDP itself nor the video streaming application implement a retransmission mechanism. Therefore, the video streaming applications are very sensitive to packet loss: any lost packet in the network will cause missing information in the video stream.

Given that losses can deeply affect the QoE, video streaming requires high reliability for the data transfer between the streaming server and the client. Today's networks, both wired and wireless, including the 3G cell phone networks do provide in theory sufficient quality to ensure satisfactory video streaming and communication.

Regarding jitter, it is necessary to control its boundaries, so that an appropriate dejittering buffer can be used. These boundaries must be sufficiently small for interactive video applications, but in the case of video streaming jitter values can have larger values, as long as they are bounded. These bounds determine the delay the user perceives before playback can start; therefore they should have reasonable values of the order of a couple of seconds.

The codecs that implement the MPEG standards are the most frequently used for video signal compression. One of them is the MPEG-4 standard, which is capable to compress the video signal at an extremely low bit rate and still preserve a relatively good quality. The high compression rate it achieves reduces the required bandwidth for a video streaming application with respect to that of previous standards.

NETWORK DEGRADATION AND QUALITY OF SERVICE

Network Degradation

The term network degradation is used to refer to the totality of network effects (bandwidth limitation, packet loss and reordering, delay and jitter) that perturb any network traffic. Ideally one may desire a zero-loss instantaneous transfer of application traffic, but in reality the network degradation will cause a certain delay between sending and receiving, with potential loss of information during transfer.

This degradation occurs because the information is transferred using communication channels that make use of different network elements to provide end-to-end connectivity. Such network elements can be very basic, such as network cables, optical fibers and hubs, or can be more complex devices, such as switches or routers. In the case of wireless networks, the cables are replaced by a transmission media such as air, which has considerably different properties compared to the network cables. Another important element in the case of many types of wireless networks are the base stations, which have a role similar to switches since they concentrate the wireless traffic, and may serve as gateways to the wired network.

From a behavioral point of view, one can identify two basic elements that are the building blocks of any wired network system: the *wire* and the *queue* (Ivanovici, 2006). The *wire* represents the transmission media, which can be considered, in a first approximation, error free. Therefore its main characteristic is the constant propagation delay. The *queue* is characterized by its length and service rate. It introduces variable delay and loss. This degradation is introduced by the *intra-stream* and *inter-stream* competition for resources: a packet competes both with other packets from the same traffic flow, as well as with packets from other streams. We can conclude that delay and loss have both a constant and a variable component, and we have to take into account their variations (hence, their instantaneous values). The constant component of the delay, for instance, is mainly the consequence of the transmission and propagation delays in networks, and therefore in general doesn't change for a certain route and traffic type. The variable component is caused by the varying queue occupancy in all the network elements along the route. This component depends on the other traffic flows in the network and the congestion level at each moment of time. Stochastic processes, such as the Poisson, Birth-and-Death or Markov chains, are used to model the network degradation (Allen, 1990; Papoulis, 2002). Therefore, the delay – or any other QoS parameter – is usually characterized by a distribution, i.e. a probability density function.

Wireless networks use electromagnetic waves for data communication, therefore the transmission media cannot be assumed error-free anymore.

The transmission delay itself has a constant component, the propagation delay, and a variable component that accounts for the way in which the wireless network protocol copes with errors. For example, for IEEE 802.11, the MAC protocol includes a certain number of frame retransmissions as a mechanism of sending data over lossy channels. The delay between retransmissions as specified by the standard causes an exponential increase in the overall delay perceived at packet level. However, there is a maximum number of retransmissions allowed, which if exceeded leads to packet loss. Hence, in wireless networks both delay and packet loss have a higher variability than in wired networks.

Degradation can also occur as a consequence of other network mechanisms, such as switching (Beuran, 2004), routing, address resolution (ARP), scheduling (Chao, 2002) or encryption in wireless networks. All such mechanisms may lead to increased delay, bandwidth utilization and packet loss.

QoS Metrics

The metrics for network degradation are widely accepted and are generally known under the name of QoS metrics. They are used to quantify the network degradation or the IP performance (Paxson, 1998). These metrics include:

- **Throughput:** The amount of information transferred per unit of time;
- **Delay:** The time interval needed to transfer information (Almes, 1999);
- **Jitter:** The variation of the delay;
- **Packet loss rate:** The proportion of the data that is lost during transfer (Almes, 1999);
- **Inter-packet arrival time** or inter-packet gap.

While for most of the metrics the definition is simple and their computation is straightforward, for jitter there are several definitions starting from the same generic formula, $j = \left|D_{ref} - D_i\right|$, where D_i is the delay of the packet for which one computes the instantaneous jitter. D_{ref} is the delay of reference, which can be the delay of the first packet (ITU-T I.380, 1999), the average delay, or the delay of the previous packet (Demichelis, 2002).

For each metric, averages or instantaneous values can be measured, as well as distributions. Theoretically, loss and throughput are both on/off functions; packet loss either occurs or not, and at the finest resolution data is sent or not over a wired or wireless channel. Averages can lead to false interpretation: imagine the following example – throughput and loss are measured in parallel as average on an interval of 60 seconds for a Fast Ethernet link. Suppose the average throughput was 50% of the line speed, and loss was 10%. One may wonder how come such a high percentage of loss if only half of the bandwidth was used. The answer is: averaging hides the bursts that were transmitted over the line and caused packet loss over very short periods of time.

Instantaneous values are most relevant for the characterization of degradation in wireless networks because network conditions change rapidly and significantly in such networks. The distributions are interesting from a statistical point of view, to characterize the behavior of the network in order to predict or estimate how the network will react to a certain application.

All the instantaneous parameters that characterize QoS are not independent, but intimately correlated – the variation of one of them implies the variation of the others. For instance, if the available bandwidth diminishes due to other traffic flows, the instantaneous throughput will decrease as well; the consequence may be either an increased instantaneous delay – if packets are buffered someplace – or an increased loss rate – if packets are dropped due to insufficient resources. On the other hand, if loss increases, the throughput becomes smaller, and so on.

Assessing the QoS

There are two ways of measuring the QoS parameters: *active* or *passive*. In the first approach, the measurement system injects artificial traffic into the network and assesses the QoS parameters of the network. This is unrealistic, since network conditions can change between the time of quantifying the QoS parameters for the artificial traffic and the time when the real application is run. Therefore, the application could experience some new network conditions. The second approach, passive measurement, is more realistic, since the measurement of the QoS parameters is performed in a non-invasive manner for the real traffic of the application under study.

There are also hybrid techniques, when a specially designed ICMP-based application is run in parallel with the application under study (Jiang, 1999). One drawback of this method is the fact that the measurements are not made for the traffic of the application under study. This may lead to differences in results caused by differences in packet size, mechanisms that depend on packet type, such as scheduling priority, etc. Moreover, the network state and the application under study may be perturbed by the injected measurement traffic packets. In particular, the assumption that the one-way delay equals half of the Round Trip Time (RTT) does not hold in practice in general, not even for symmetric paths between two end-nodes.

QoS parameters can be quantified for each and every packet, or only for certain packets, selected, for instance, by sampling at regular intervals. Given the variation of network parameters from a moment to another, it is desirable to build a system capable of assessing the QoS parameters for every packet of the application traffic flow. This would be the finest level of detail, and of course the most precious for a pertinent analysis and correlation with the QoE parameters. Sampling, similar to averaging, may lead to a false interpretation of the QoS since one may not know in advance what kind of sampling is suited to capture all the characteristics of the sampled data.

Wired vs. Wireless QoS

Although in terms of general properties and metrics the network degradation in wired and wireless networks are very similar, there are a series of differences that we review below:

1. For point to point transfers using network cables or fibers, wired networks can be considered to have 100% reliability, however point to point transfers in wireless networks are subjected to external interferences (noise, obstacles) that may significantly change how degradation occurs.
2. Using a hub in a wired network introduces some unpredictability, since the number of senders that share the same connection is increased. However, this quality degradation can be bounded since the maximum number of senders is determined by the architecture of the hub. On the other hand, in a wireless network the transmission media can be shared by a potentially infinite number of users, since no physical connection is required between a sender and a receiver. This introduces a potentially infinite network degradation.
3. A switch or router in a wired network has a certain number of input ports that can be even of the order of hundreds. By a careful design (fully meshed architectures, sufficient memory, etc.) the switching element can ensure a lossless transfer between its ports as long as there are no conflicts in the traffic itself (for example, two transmitters sending to one destination). However, a base station (i.e., access point) in a wireless network has a very limited number of input channels, usually ranging between 1 and 4. This means that although there may be no conflicts in the traffic itself, the access point

input channels must be shared between the wireless transmitters, and thus the probability that the switching element becomes a bottleneck is significantly higher.

4. In wired networks switches are usually connected to each other by connections that usually have higher bandwidth that their normal ports (for example, 1 Gbps switches may use 10 Gbps to connect to each other). Although this is true to some extent in wireless networks such as the mesh networks (for example the backbone access points may use 54 Mbps 802.11a links to connect to each other, while 802.11b/g connections are used for end nodes), the differences in maximum bandwidth are not so high since 802.11b has a 11 Mbps maximum bandwidth, whereas 802.11g has the same 54 Mbps maximum bandwidth with 802.11a. As a consequence the backbone connection can more easily become a bottleneck in wireless networks.

For all the reasons we have given above, in wireless networks one expects to see a larger and more variable network degradation that in wired networks (Nguyen, 2007; Beuran, 2007). This has lead researchers to try to adapt the application traffic to the wireless environment in order to maximize quality. An example in this sense is the wireless adaptation of TCP (Tian, 2005).

QUALITY OF EXPERIENCE AND QUALITY OF SERVICE MEASUREMENT

There are three steps to take in order to assess application performance: (i) observe the application behavior on an end-to-end basis; (ii) accurately measure the quality degradation experienced by the application traffic, and (iii) correlate the above. Scientific method requires the use of objective metrics to perform both the network and application level performance assessments.

First of all, one must observe the application outcome. A human observer could judge if the application behaved as expected, or objective metrics can be used (ITU-T P.862, 2001; Beuran, 2004). At application level the user is unaware of what is happening at network level. Moreover, the user should not care about the underlying mechanisms. However, the performance of the application strongly depends on network performance. Hence, it is mandatory to observe the network conditions (quality degradation) between the two end points.

However, observing is not enough. Accurate quantification of both the application performance and network conditions should be performed. The following step is to correlate the two results, thus experimentally determining the relationship between the application performance and the quality degradation at network level. This implies defining the application outcome and a metric to allow quantifying the application performance. The metric can be either objective or subjective. Next, the appropriate method to measure the application outcome should be chosen. Once the method and the outcomes are well defined, the application outcome can and must be accurately quantified.

There are three traditional methods for testing and validating network devices, protocols and applications: (i) *analytical*, i.e. mathematical modeling; (ii) *simulation*, i.e., running a model or a representation of the application's code in a completely synthetic environment; and (iii) *real network testing*, i.e., running the real application in a real environment.

Network emulation is a hybrid technique between simulation and real network testing, that allows the study of real network applications in a laboratory setup. Through emulation application behavior can be studied in a wide range of controllable and reproducible network conditions. This hybrid technique combines the advantages of network simulation with those of tests in real networks (Ivanovici, 2006) enabling

Figure 1. Application performance assessment setup

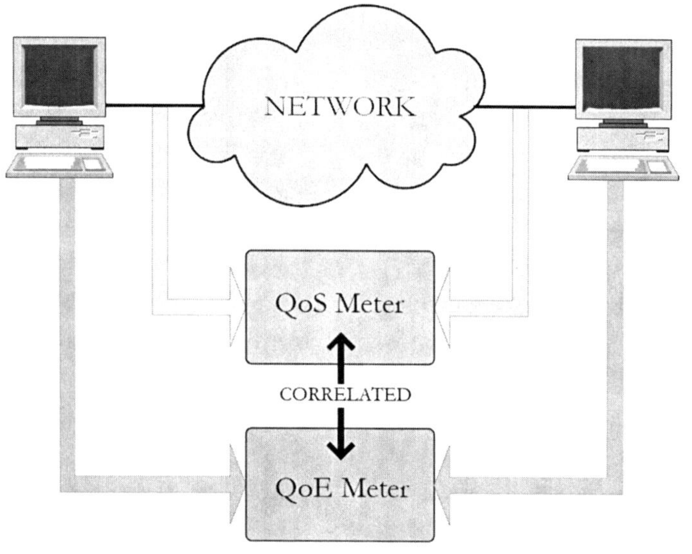

controlled and reproducible experiments. The same approach was used in (Hoβfeld, 2008) to analyze the behavior of VoIP in UMTS. Most of the existing network emulators are implemented in software; therefore the quality degradation they introduce is imprecise and irreproducible. Current hardware (Simena; Anue; Empirix; Shunra, 2004) and software (NISTNet; Rizzom; Yeom, 1998) approaches exhibit an additional important drawback: they all introduce unrealistic degradation. The reason is twofold: packets in a flow are treated independently, and quality degradation effects are not correlated (e.g., packet loss and delay are independent).

The authors of (Ivanovici, 2005, 2006; Ciobotaru, 2005) designed and implemented a dedicated hardware emulator in order to perform QoS and QoE experiments, which is an integral component of their approach. The authors used the generic setup shown in Figure 1 to assess application performance. This setup allows correlating the QoS and QoE; it comprises a QoS meter and a QoE meter, and will be detailed in the following section.

The QoS / QoE Measurement System

The QoS / QoE measurement system depicted above, and shown in more detail in Figure 2, is able to measure non-intrusively the network QoS parameters (Ivanovici, 2006). It is composed of commodity articles – FastEthernet taps, programmable network interface cards (NICs) – together with custom design clock cards for time synchronization. Using these components, the reported latency measurement accuracy is of 1 μs, for any size packets, up to loads of 100 Mbps.

In parallel with monitoring network traffic for computing the QoS parameters, the system can measure the elapsed time for a file transfer, or record the video or voice signal, depending on the type of application under study. Then, the perceived quality (QoE) is quantified based on specifically defined metrics. Thus it is possible to correlate network conditions with the QoE for these applications; this allows the performance of two main tasks:

1. **Predicting** the expected QoE for an application running over a given network knowing

Figure 2. The QoS / QoE measurement system in a typical test setup

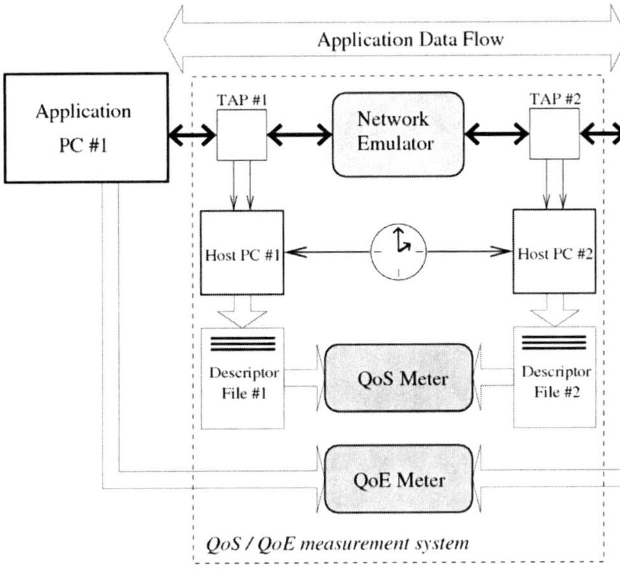

the corresponding measured QoS parameters; understanding the causes of application failure by defining minimum requirements that must be met by the network;

2. **Designing** or configuring a network to provide the necessary QoS conditions for an application to run at the desired QoE level.

One of the major advantages of this system is its versatility. It can be used to test network devices, small local networks and even large local or wide-area networks (in this case GPS cards could be used in order to have a global time reference). Another advantage is the fact that it can be reprogrammed as required for future work. The system is capable of measuring one-way latency, which is more relevant than Round Trip Time (RTT) measurement given the usual asymmetry of networks.

In addition, all the measurements are non-intrusive. After placing the taps in the test network, traffic flows unaffected and one can observe the behavior of real network applications. Taps are passive network devices that can be used to monitor a full-duplex link, in our case a FastEthernet connection. The setup described includes two FastEthernet taps manufactured by NetOptics that mirror the traffic flowing in both directions and feed it to the programmable NICs. Wireless networks are even easier to monitor, since the transmission media is open space. For WLANs, for example, any laptop equipped with one of several types of off-the-shelf WLAN NICs can be made to capture traffic, if a special driver and software for WLAN analysis are used. In what follows, we shall discuss the wired network setup in (Ivanovici, 2006). From an application point of view it is not important whether the application runs over a wired or wireless network, but which is the amount of QoS degradation in that particular network.

One important component of the QoS / QoE measurement system described in (Ivanovici, 2006) are the programmable Alteon Fast/Gigabit Ethernet NICs. The host PC communicates with the NIC through a shared memory segment and control structures. The NIC performs all the necessary Ethernet MAC and PHY layer processing and has a 1 MB memory, which is used to store the running software and packet descriptors extracted from the mirrored traffic. In addition, these NICs were programmed to monitor Ethernet connections and collect data about the packets on the tapped links.

One NIC is needed for each traffic direction; hence a total of four NICs are required in order to monitor the two full-duplex links in the experimental setup in Figure 2. These NICs produce for each packet a descriptor with the following fields: timestamp (32 bits), packet identifier (32 bits), packet size (16 bits), protocol number (8 bits). *Timestamp* represents the packet arrival time, including the time needed to store the packet in the receiving buffer. Synchronization between NICs is achieved by using a custom global clock system, formed of a master clock card and slave clock cards. Packet timestamps are obtained by transforming the local clock value to a global one, using conversion tables generated 128 times per second. The overall latency measurement error is bounded to 900 ns. *Packet identifier* is a value that uniquely identifies the packet. This value is obtained based on information from packets, such as sequence numbers from RTP and TCP headers, checksums, etc. *Packet size* contains the dimension of the packet expressed in bytes, including the four-byte CRC. *Protocol number* allows us to distinguish between different protocols and filter the packets of interest.

Based on the data collected by the QoS measurement system, the following QoS parameters are computed off-line: average latency and jitter, average throughput and packet loss (ITU-T I.380, 1999; Demichelis & Chimento, 2002). The average jitter an application would experience is given by the jitter determined with respect to the latency of the previous packet which (Beuran, 2003) considers as the most relevant from an application-oriented perspective. Packet loss is determined using the packet identifier from descriptors. A packet is considered lost if its identifier, which appears in the descriptor file at the first measurement point, doesn't appear in the descriptor file at the second measurement point. The system in (Ivanovici, 2006) can also compute instantaneous values (e.g., for throughput) and various histograms (e.g. inter-packet arrival time histograms).

Experiment Methodology

The measurement methodology proposed in (Ivanovici, 2006) is a set of sequential actions. For making the experiment itself (step 4 in the measurement procedure), each of the corresponding steps is performed automatically, according to a script that communicates with all the components of the test setup. The Python programming language (Python) was used to implement the script due to its libraries that allow implementing clients and servers easily.

The measurement procedure is the following:

1. Choose the QoS parameters to which the application is sensitive
2. Choose the QoE metrics of interest at application and user level
3. Define a set of network conditions for the experiments
4. Run one test and perform the measurements
    a. Configure the network emulator
    b. Start the monitoring system
    c. Start the network application
    d. Wait for a certain interval of time
    e. Stop the network application
    f. Stop the monitoring system
5. Correlate the QoS parameters with the QoE parameters, measured in parallel
6. Plot the QoE versus QoS dependency

The steps indicated above can be used to assess the performance of an application given a set of known conditions representing one state of the network. Therefore this approach can be used to assess application performance as if it would run in any wired or wireless network environment, depending on the QoS degradation conditions that are recreated. The main point of interest of the authors is to characterize application performance in a wide range of conditions. This can be done by repeating the above measurement while changing the network conditions (for example, vary packet loss). Another possibility is to vary network conditions during an experiment. This makes it possible to study application behaviour in dynamic conditions (e.g., how does application react to a sudden increase of packet loss due to congestion, as in real networks). Emulating dynamic conditions is an essential element in studying application performance over wireless networks, where conditions change continuously due to the varying environment properties and the motion of the mobile nodes.

CORRELATING QUALITY OF EXPERIENCE AND QUALITY OF SERVICE

To emphasize the usefulness of objectively measuring and correlating QoE and QoS for network planning (computing what network conditions are required to achieve a certain application performance level), as well as for network suitability decisions (determining whether a network is appropriate for a certain application, and compute the estimated application performance level), the following applications are presented as case studies: file transfer, web browsing, VoIP and video streaming.

1st Case Study: File Transfer

A very important aspect of this chapter is the definition and quantification of application-specific QoE metrics. The two QoE metrics proposed in (Beuran, 2003) for file transfer applications, *goodput* and *transfer time performance*, allow the assessment of user-perceived quality for this particular application.

Goodput (G) quantifies the network efficiency of the file transfer. It is computed as follows:

$$G = \frac{B_{\min}[bytes]}{B[bytes]},$$

where B_{min} is the minimum number of bytes required for that file transfer (including protocol overhead for Ethernet, IP, TCP and FTP) and B is the count of the actually transmitted bytes. Goodput values are on a scale from 0 to 1, where 1 means maximum efficiency of the file transfer. Goodput decreases due to packet retransmission when loss occurs. Given its definition, G doesn't depend on any time parameter related to the transfer (e.g. transfer duration or round-trip time), but only on the amount of bytes being effectively transmitted. Therefore an additional metric is required to take this aspect into account.

Transfer time performance (*TTP*) allows the evaluation of the time efficiency for a file transfer:

$$TTP = \frac{T_{th}[s]}{T[s]} = \frac{B_{\min}[bytes]}{L[bps] \cdot T[s]},$$

where T_{th} is the theoretical transfer duration, and T is the measured transfer duration. The theoretical transfer duration is the ratio of the minimum number of transmitted bytes required for that transfer, B_{min}, to the line speed, L (for instance, 100 Mbps). T is computed as the difference between the time when the last packet from a transfer was received and the time when the first packet was sent. *TTP* values are also on a scale from 0 to 1, with 1 meaning the ideal, optimum performance. Packet retransmission delays make *TTP* values decrease. *TTP* depends indirectly on all parameters that influence transfer duration, such as RTT, TCP window size etc.

In the experiments performed, the authors introduced packet loss up to 25% in both traffic directions and ran tests with different transferred file sizes. The conditions for file transfer tests were

Table 1. TTP for various file sizes and RTT values

File size		10 kB	100 kB	1MB	10MB
TTP	*0.8 ms RTT*	0.0219	0.1650	0.8696	0.8919
	60 ms RTT	0.0029	0.0141	0.0559	0.0791

the following: FTP client with Linux kernel 2.4.6 (64 kB maximum TCP window), ftp-0.17-7; FTP server with Linux kernel 2.4.9 (64 kB maximum TCP window), wu-ftpd-2.6.1-20. The values presented in what follows were obtained by averaging over 100 experiments for each intended loss rate. There are two series of tests, one with an RTT of 0.8 ms (emulating a local network scenario) and the other with a RTT of 60 ms (emulating a wide area network).

Table 1 shows the *TTP* values obtained in zero loss conditions for two different RTTs and several transferred file sizes. It can be seen that the time efficiency increases with file size, since the overhead of connection establishment and termination becomes less significant compared to the file transfer time itself. The variation of *TTP* between the two RTTs is of an order of magnitude.

TCP window size is an important parameter regarding TCP performance. The optimal window size, $W_{optimal}$, is given by the bandwidth-delay product:

$$W_{optimal} = BW \cdot RTT,$$

where BW is the bottleneck bandwidth of the connection (for these experiments, 100 Mbps).

Considering a 0.8 ms RTT gives $W_{0.8} = 10$ kB. For the 60 ms RTT it results $W_{60} = 750$ kB. Given that the default maximum window size was 64 kB, this doesn't represent a limitation for the 0.8 ms RTT, but it limits the traffic for 60 ms RTT, and the performance is one order of magnitude lower, exactly as observed in Table 1.

The results presented below were obtained for a 10 kB file, which is the typical file size for Internet traffic (Arlitt, 1996). For larger file sizes, the graphs of goodput and *TTP* have a similar shape. *TTP* values approach 1 for large files and small RTTs (see Table 1), which shows that it is more efficient to send the same amount of data in one large transfer than in multiple short ones.

Goodput (see Figure 3a) decreases almost linearly with packet loss, showing the diminution of link utilization efficiency. As expected, RTT doesn't have any influence on goodput, since G is not time dependent. Therefore goodput is not a stand-alone indicator of file transfer QoE, and must be correlated with *TTP*.

Transfer time performance (see Figure 3b) shows the significant dependency of transfer time on packet loss. The maximum value of *TTP* equals 0.0219 due to the additional durations of connection establishment and termination, which represent approximately 96% of the transfer time for 10 kB files. For 0.8 ms RTT, *TTP* value decreases 20 times for packet loss rates of 5% compared to the value obtained at zero loss. This is equivalent with an increase of 20 times of the transfer duration, which means a significant degradation of the QoE. For loss rates of 10% and higher, performance degrades hundreds of times. For 60 ms RTT *TTP* is smaller than for 0.8 ms RTT and loss has a less dramatic influence on it.

The influence of packet loss on TCP performance depends on the type of the lost packets: losing a data packet is easily hidden by the retransmission mechanism, whereas losing a TCP connection establishment or termination packet has a more important effect due to the relatively large timeouts. For 10 kB files, transfer duration has increased by an order of magnitude in such cases.

Figure 3. Goodput (a) and transfer time performance (b) versus packet loss for file transfer tests (10 kB file)

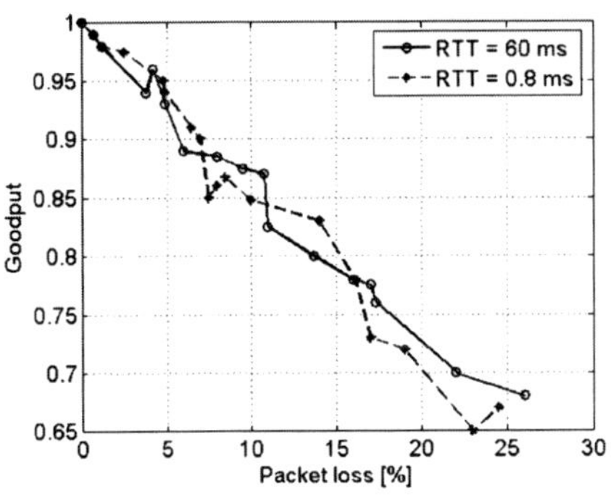

a

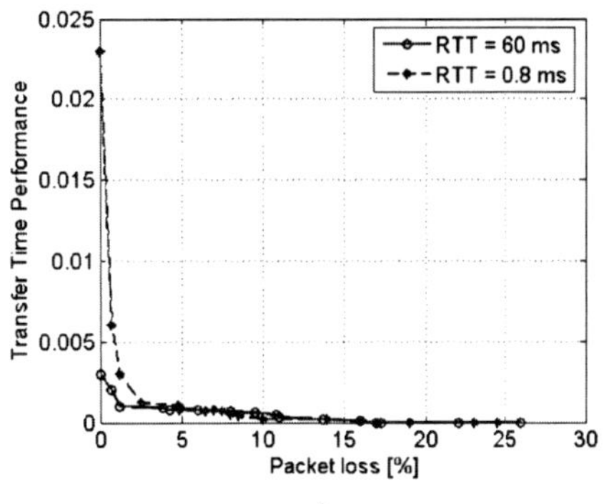

b

Goodput diminishes with packet loss, as expected. The dependency is linear, and goodput decrease is not very large in the range of 0 to 5% packet loss. Setting the value of 0.9 as the threshold of acceptability for network utilization efficiency, we determine that packet loss should not exceed 5%. For loss rates above 20%, goodput indicates a transfer efficiency lower than 0.7. This approaches 0.5 for loss rates close to 40%.

The transfer time performance graph has a negative exponential shape, showing that the time needed to transfer a file increases significantly with packet loss. For loss rates around 5% and low RTTs, the *TTP* is one order of magnitude smaller than the value obtained at zero packet loss. The degradation observed is less significant for the 60 ms RTT than for the 0.8 ms RTT. At 25% loss rate, the time to transfer has become several hundred times larger than in the case when the loss rate is smaller than 5%. This renders the connection practically unusable for file transfer.

By combining all the previous considerations, the conclusions are summarized in Table 2. File transfer applications require packet loss not to exceed 5% in order to keep the network utilization efficiency above 0.9, and in order not to have an increase of the transfer time larger by more than an order of magnitude with respect to no loss conditions. Excellent performance requires even tighter bounds: packet loss should not exceed 1% in order to reach a network utilization efficiency around 0.99, and a transfer time not larger than three times with respect to no loss conditions.

2nd Case Study: Web Browsing

Web browsing is an HTTP-based application that is characterized by short-lived TCP transfers. The performance of such an application strongly depends on packet loss, hence we chose to present the results obtained for such a case. The traffic of interest (HTTP) competes with the background traffic to occupy queue space – which induces loss, and for being serviced – which induces delay. Two scenarios of interest are compared in (Ivanovici, 2005): the case when the background traffic source has a CBR pattern, and that of a Poisson pattern. For all the tests the emulator was configured to introduce a fixed delay of 12.5

Table 2. Summary of the effects of packet loss on file transfer quality

Loss [%]	File Transfer Quality
0 – 1	Excellent
1 – 5	Good / Acceptable
> 5	Bad / Unacceptable

ms (equivalent to 25 ms RTT) and the available bandwidth was limited to 10 Mb/s. The end PCs ran Linux with kernel 2.4.21, the HTTP server was Apache 2.0 (httpd-2.0.46), and the client was wget (wget-1.8.2), a non-interactive network retriever that allows for the automation of tests. The interconnect employed was Fast Ethernet. For the Apache server all the parameters had default values, including the *Timeout* of 300s. *KeepAlive* was set to "on" and "off" in turn. When KeepAlive is "off", a new TCP connection is opened and closed for each file transfer. This represents the most inefficient case. When KeepAlive is "on", the same TCP connection is reused for up to *MaxKeepAliveRequests* = 100 transfers, if separated by no more than *KeepAliveTimeout* = 15 s.

A representative web-page structure that contains both images and text was selected for the experiments. The site consists of 499 files, with a total size of 1.6 MB. The average file size is approximately 3 kB, close to the average value of file sizes on the Web (Arlitt, 1996). The results in Figure 4 show the dependency of site download duration on the offered background traffic load, for *KeepAlive* "off" and "on", respectively. The site download duration is a measure of the QoE for web-browsing applications. The reference value is that obtained when the application has an exclusive use of the network, i.e., when there is no background traffic. The offered background-traffic load varies from 0 to 100%, being a measure of the congestion induced by the emulator, and implicitly a measure of QoS.

Note that for low loads CBR background traffic has almost no influence on the performance (Figure 4a), since this case is equivalent to a constant diminution of the bandwidth available for the application. A constant amount of available bandwidth leads to a steady performance of TCP. Since web browsing only implies transfers of relatively small amount of data, the available bandwidth can be low without a significant impact on performance. When the background traffic load approaches 100%, the available bandwidth becomes insufficient. Subsequently there is a steep increase of the download duration, followed by denial of service and leading to complete application failure.

When the background traffic is Poisson (and therefore more realistic) noticeable performance degradation starts occurring from loads of 60%. At loads larger than 80%, degradation becomes significant and reaches values with more than one order of magnitude higher compared to the CBR case. The intrinsic burstiness of the Poisson traffic determines the larger deviations of the results. One can observe in Figure 4b an improvement of the worst-case behavior of one order of magnitude

Figure 4. Site download duration versus offered background-traffic load when KeepAlive was (a) "off" and (b) "on"

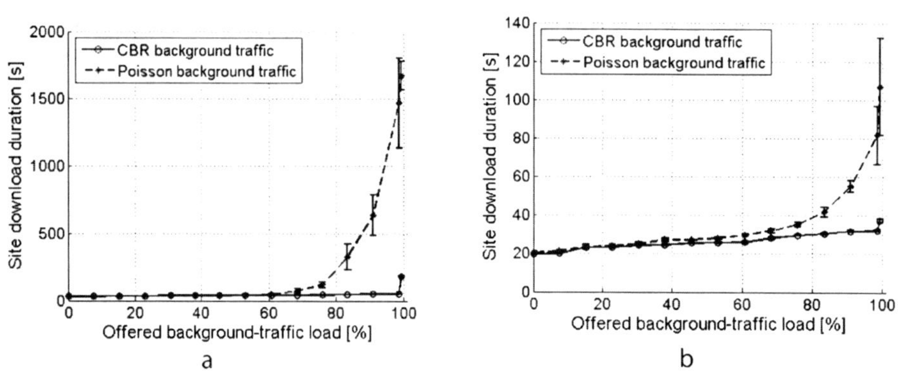

when *KeepAlive* is "on", due to the reutilization of the same TCP connection for multiple transfers. This reduces the probability of losing connection establishment and termination packets; such loss is the main culprit for the performance drop of TCP-based applications in these experiments.

Table 3 summarizes the conclusions related to the web browsing. The available bandwidth is the difference between the line speed and the background-traffic load.

3rd Case Study: Voice Over IP

VoIP is a widely-used interactive network application. The bandwidth requirements of speech transmission are low (64 kb/s voice data maximum), but interactivity implies high sensitivity to delay and jitter. The influence of one-way delay on VoIP UPQ is not considered here, since these requirements are generally known (ITU Y.1541, 2001; Reijs 2002): a mouth-to-ear delay of up to 150 ms gives good interactivity, a delay between 150 and 400 ms is acceptable, and delays higher than 400 ms are unacceptable. The work in (Beuran, 2003) is a study of uni-directional traffic, focusing on the perceived quality of the speech itself depending on packet loss and jitter. The QoE was determined by using the PESQ score (ITU-T P.862, 2001) in an implementation supplied by Malden Electronics Ltd.

For experiments with VoIP, a freeware application was used, namely Speak Freely v7.6a (Wiles). The application doesn't do any of the following: silence suppression, reordering of out-of-order packets, packet loss concealment, but it does uses a de-jittering buffer (default size is 80 ms). The study focused on a region with loss rates between 0 and 15%, and average jitter values ranging from 0 to 75 ms, since quality becomes unacceptable within these boundaries already. A detailed description of the test conditions is available in the technical report (Beuran, 2004c).

The four codecs analyzed in (Beuran, 2004c) were: G.711, G.726, GSM and G.729. The G.711 codec (ITU-T, 1993) sends data at 8 kHz with 8 bits per sample, resulting in a data rate of 64 kb/s. The sound is in PCM format, encoded using the μ-law. The G.726 codec (ITU-T, 1990) converts a 64 kb/s μ-law or A-law PCM channel to and from 40, 32, 24 or 16 kb/s channels. In our application only the 32 kb/s encoding is available. The GSM (Global System for Mobile telecommunications) codec (Rahnema, 1993) uses linear predictive coding (LPC) to compress speech data down to 13 kb/s. The G.729 codec (ITU-T, 1996) is frequently used for VoIP communication. It sends

Table 3. Summary of the effects of background traffic on web browsing

Background-traffic load [%]	Available bandwidth [%]	Web Browsing Quality
0 – 60	40 – 100	Excellent
60 – 90	10 – 40	Good / acceptable
90 – 100	0 – 10	Bad / unacceptable

Table 4. Codec characteristics

Codec	Data rate [kb/s]	Packet size [B]	Effective rate [kb/s]	Packet rate [packets/s]
G.711	64	378	75.6	25
G.726	32	382	38.2	12.5
GSM	13	190	19	12.5
G.729	8	170	17	12.5

Figure 5. VoIP results for G.711

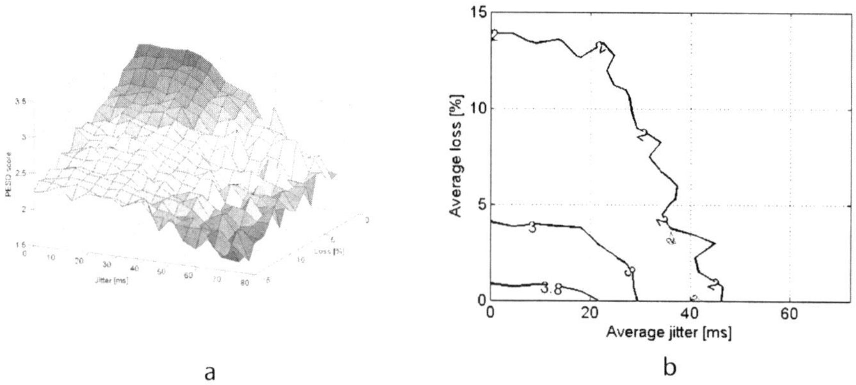

data at 8 kb/s using conjugate-structure algebraic code-excited linear-prediction (CSACELP). The basic characteristics of the analyzed codecs are summarized in Table 4 (RTP was used as a transport protocol).

Five series of tests were run for each codec in order to collect the data used for the results shown in Figure 5a – the dependency of the PESQ score on jitter and loss. According to (Servis, 2001) the relationship between PESQ scores and audio quality is the following: (i) PESQ scores between 3 and 4.5 mean acceptable perceived quality, with 3.8 being the PSTN1 threshold – this will be termed as *good quality*; (ii) values between 2 and 3 indicate that effort is required for understanding the meaning of the voice signal – this will be named *low quality*; (iii) scores less than 2 signify that the degradation rendered the communication impossible, therefore the quality is *unacceptable*.

Based on this information the Figure 5b shows the boundaries on QoS parameters that must be enforced in order to attain a certain quality level, for the codec G.711. One can notice that G.711 provides good quality as long as loss rate is below 4% and average jitter doesn't exceed 30 ms. The same codec will provide low but acceptable quality if loss rates are roughly between 4 and 14% and jitter is between 30 and 45 ms. Outside these bounds the quality will be unacceptable. Note that G.711 is the only codec amongst those tested that also provides very good (PSTN equivalent) quality.

The results regarding the mapping between QoE and QoS for G.711 shown in Figure 5 can be summarized roughly as follows (Table 5):

Table 6 shows a codec comparison from the point of view of PESQ score thresholds (Beuran, 2004c) for all four codecs used in these experiments. The codecs were classified based on the coverage of the area corresponding to a certain quality level with respect to the area of the studied

Table 5. Summary of the effects of packet loss and jitter on VoIP communication using G.711

Jitter [ms] / Loss [%]	*0 – 20*	*20 – 30*	*30 – 45*
0 – 1	Excellent	Good	Bad
1 – 4	Good	Good	Bad
4 – 14	Bad	Bad	Bad

Table 6. Codec classification based on good and low quality coverage

Codec	Good quality coverage [%]	Low quality coverage [%]
G.729	10.48	88.16
G.726	9.62	60.33
G.711	9.00	41.70
GSM	5.06	51.25

loss-jitter space. In what follows we present two such classifications, one for the area of *at least good quality* (PESQ scores larger or equal to 3) and one for the area of *at least low quality* (PESQ scores larger or equal to 2). Note however that this classification doesn't take into account the bit-rates of each codec, which are also important when making the trade-off between perceived quality and network utilization efficiency.

For demanding users that require at least good quality of the speech signal, one can choose the codec based on the column "Good quality coverage" in Table 6. Less demanding users, for which low quality is sufficient, can use VoIP in a wider range of network conditions, by choosing the appropriate codec by consulting the column "Low quality coverage" in the same table. Note that the codec G.729 is on the first position in both columns, indicating that it performs best in our study. Given that it is also the codec with the lowest bit-rate, we consider it as the codec of choice from those that were studied for almost any situation.

4th Case Study: Video Streaming

What does a user expect from a video application? *Good* or *excellent* video quality. In other words a clear picture, not affected by impairments or gaps (missing video information during playback). In order to objectively analyze the performance of a video application from a user perspective, we must identify its requirements, i.e. find the appropriate metrics and determine the bounds of acceptance.

The QoE metrics for video and voice applications are classified in two major categories:

- **Reference-Based Metrics,** when both the video sequence / voice signal at the transmitter and the video sequence / voice signal at the receiver are available; the sequence at receiver will be compared to the original sequence at transmitter;
- **No-Reference Metrics,** when the video sequence / voice signal at the transmitter is not available, therefore only the video sequence / voice signal at the receiver is being analyzed.

Another classification of the QoE metrics divides them in subjective and objective metrics. Subjective video quality measurements are time consuming and must meet complex requirements (see the ITU-R and ITU-T recommendations BT.500, P.910, J.140, J.143) regarding the conditions of the experiments, such as viewing distance and room lighting. The objective metrics are usually preferred, because they can be implemented as algorithms, and are human-error free. The most complex objective metrics are based on models of the human-vision system, but the most widely used are distance measures, such as the Root Mean Square Error (RMSE) or the Peak Signal-to-Noise Ratio (PSNR). These simple measures are unable to capture the degradation of the video signal from a user perspective. For a more realistic quantification of the user-perceived degradation, image attributes like sharpness and colorfulness are used in (Winkler, 1999; 2001).

The Video Quality Experts Group (VQEG) is the main organization concerned by the perceptual quality of the video signal, and they reported on the existing metrics and measurement algorithms (VQEG, ****). A survey of video-quality metrics based on models of the human vision system can be found in (Branden, 1997), and several no-reference blockiness metrics are studied and compared in (Winkler, Sharma, 2001). OPTICOM is the author of the latest metric for video quality evaluation, called Perceptual Evaluation of Video Quality (PEVQ), which seems to be generally accepted as the *de facto* standard. This reference-based metric is used to measure the quality degradation in case of any video application running in mobile or IP-based networks. The PEVQ Analyzer (OPTICOM) measures several parameters in order to characterize the degradation: brightness, contrast, PSNR, jerkiness, blur, blockiness, etc.

Most of the existing metrics for video quality quantify the degradation introduced by the compression algorithm itself or due to the frame rate that is used. There are very few studies that objectively assess the degradation in video quality caused by the packet loss at network level, such as (Malkowski, 2007).

A set of QoE metrics for video applications that take into account this aspect in particular is presented in (Ivanovici, 2006). Given the way the video signal is degraded in experiments, the authors identified two kinds of degradation: (i) the degradation that affects the sequence, i.e. the temporal component of the signal, and (ii) the degradation that affects the frames, i.e. the spatial component. For the quantification of the first type of degradation, the authors proposed three objective reference-based metrics: (i) the number of dropped video frames (NDF); (ii) the number of altered video frames (NAF) (Ivanovici, 2005), and (iii) the average signal unavailability (ASU). NDF indicates how many frames were skipped (not rendered) because of the missing bits in the MPEG video stream, and is computed as the difference between the number of frames in the original video sequence at transmitter and the number of video frames that are effectively rendered at receiver. NAF indicates how many frames from the ones received and rendered are affected by impairments. NAF could be computed based only on the received video signal, by using an appropriate algorithm to detect the degradation in the video frame. The authors chose to compute NAF by comparing each received video frames with the ones that were transmitted, therefore NAF is a reference-based metric. By putting together the two metrics, one can plot the total number of affected frames (TNAF), both dropped and altered, as a function of packet loss at network level. ASU is computed as the average of the intervals u_i when the video signal is degraded, therefore the video information is unavailable. The duration of the intervals can be expressed either in number of frames or in *ms*.

Given the way the majority of the video frames are degraded (see Figure 6), the most useful metric would be the blockiness, which objectively quantifies the impairments. To quantify the degradation of a single video frame, one could simply measure the affected area in number of pixels or in number of 8x8 blocks, or use an appropriate perceptual metric able to quantify the degradation from a human perspective. Apart from severe blockiness, many degraded frames are "dirty", i.e. have many blocks containing other information than they should, or even other colors. Therefore (Ivanovici, 2006) consider that metrics like blur, contrast, brightness lose their meaning, and are not able to accurately reflect the perceptual degradation. QoE metrics able to quantify the *dirtiness*, as well as the shift in colors or the amount of new colors, would be more appropriate and useful.

Figure 6. Example of two degraded video frames

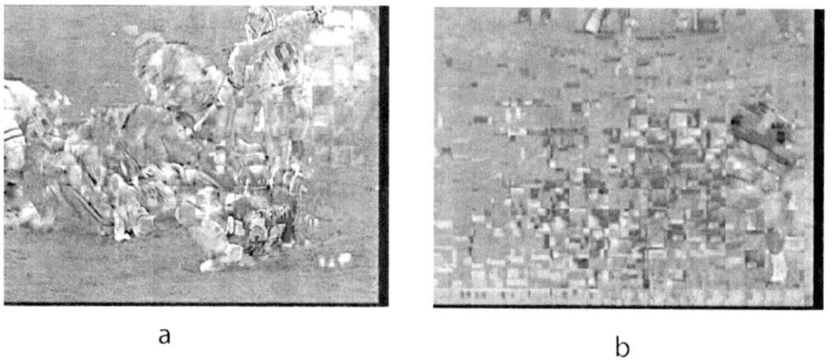

a b

Figure 7. The total number of affected frames (a), and the average signal unavailability (b) as a function of packet loss

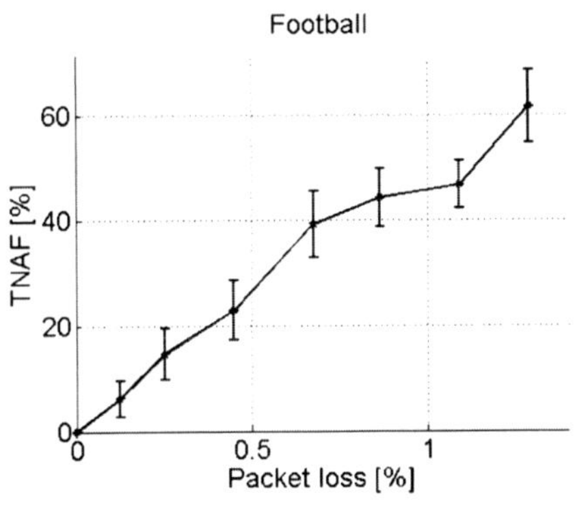

a

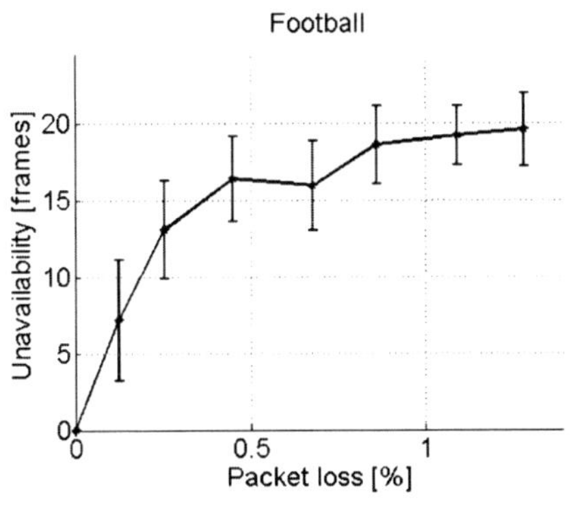

b

Given that packet loss is the major issue for an MPEG-4 video streaming application, the authors programmed the network emulator to degrade the performance of the emulated network connection by introducing packet loss. The induced loss percentage was from 0 to 1.3%. Above this threshold, the application can not longer function (i.e., the connection established between the client and the server breaks), and tests could not be performed. The MPEG-4 streaming server used was the Helix streaming server from Real Networks, and the MPEG-4 client was mpeg4ip. (Ivanovici, 2006) modified the source code of the client so as to record the received video signal as individual frames in bitmap format. The tests were run using several widely used video sequences ("football", "train", etc.), MPEG-4 coded. The video sequences are 10 seconds long, with 250 frames, each of 320 x 240 pixels. The average transmission rate was approximately 1 Mb/s, a constraint of the trial version of the MPEG-4 video streaming server used.

In Figure 7a the average and the standard deviation of the total number of affected frames (TNAF), both dropped and altered, is presented as a function of packet loss at network level, for the "football" video sequence. One can observe the monotonic increase of TNAF which is almost linear. A 1% packet loss causes approximately 45% of the frames to be affected. Other results, and the NDF and NAF represented as functions of packet loss at network level can be found in (Ivanovici, 2005). The authors also investigated the periods of time when the video signal is practically unavailable because of the degraded frames, and experimentally determined the dependency of the average video signal unavailability on packet loss, which is depicted in Figure 4b for the same video sequence. One can observe that, on average, the intervals of unavailability have a dependency on packet loss in the shape of a logarithmic function.

For the packet loss interval that was studied, as loss increases, the intervals of unavailability tend to remain constant at about 20 frames. One possible explanation can be found by investigating how the MPEG-4 data flow is encoded (Curet, 2002; Meer, 2002). If the information of an I (intra) frame from the MPEG-4 stream is missing, then all the following P (predictive) or B (bi-directional predictive) frames will be degraded. One missing P frame implies only the degradation of another adjacent frame. As a consequence, for higher packet loss values the intervals of unavailability may remain the same, depending on the loss pattern. More precisely, if one packet containing one

Table 7. Proposed mapping between TNAF and video signal quality

TNAF [%]	Video quality
0 – 10	Good
10 – 30	Acceptable
30 – 70	Bad

Table 8. Summary of the effects of packet loss rate on video streaming

Packet loss [%]	Video quality
< 0.2	Good
0.2 – 0.5	Acceptable
0.5 – 1.3	Bad
> 1.3	N.A.

I frame is lost, then other consecutive packets that follow can be lost as well, and the degradation level will not change, i.e. the interval of unavailability will be the same.

Given the results presented, one may wonder what are the bounds for good, acceptable and bad quality from a user perspective, and what are the requirements from the point of view of the video application. For example, one can arbitrarily choose the mapping presented in Table 7 (Ivanovici, 2006):

From the above mapping, and from the TNAF dependency on packet loss percentage depicted in Figure 4a, the application requirements will read as follows: in order to deliver good video signal quality, the loss percentage experienced by the application should not exceed 0.2%. For acceptable video quality, the loss should be between 0.2% and 0.5% and if the loss exceeds this last threshold then the video quality will be bad. Another requirement is definitely the following: packet loss must not exceed 1.3%, otherwise the application will completely fail. This relationship is summarized in Table 8:

FUTURE RESEARCH DIRECTIONS

In order to get a deeper understanding of the relationship between QoS and QoE, the authors plan to perform subjective tests both for video and VoIP applications. Since the authors believe that the area is not sufficiently standardized, they also intend to continue developing objective metrics for the evaluation of the video quality, especially by taking into account the human perception. For VoIP the authors shall study a larger number of wireless scenarios, given the high variability that characterizes that particular kind of networks.

Further application performance assessment would include the study of other network applications. A class of applications largely used in today's Internet is peer-to-peer. These applications are mainly used for file sharing, and use TCP to ship data between users. The study of TCP-based applications could continue with a closer look to the TCP's transient behavior. The intrinsic burstiness of the TCP traffic has a great impact on application performance. The different TCP flavors that exist, like FAST (CalTech, 2002) or High-Speed TCP (Floyd, 2002), are more "aggressive" because they try to recover faster than the classical TCP. This may lead to a rapid occupancy of all available resources, and consequently to congestion, which will affect other application streams.

CONCLUSION

For a long while researchers focused on QoS, the quality of the service delivered by a network in terms of bandwidth, packet loss, delay and jitter, as the most important set of metrics for network health status. Moreover, since large file transfers represented an overwhelming majority of network traffic, bandwidth used to be the most important QoS metric; whenever there were network complaints, network administrators used to just "throw bandwidth at the problem".

As the types of applications used in networks increased, and multimedia applications started spreading, other QoS metrics, such as delay and jitter grew in importance. Nevertheless, as the utilization of networks by ordinary people increased significantly, metrics that estimate their satisfaction had to be introduced. The reason is that for ordinary users the values of packet loss rate and delay have no meaning. In addition, application service providers needed to have ways to measure the quality from the point of view of users, so as to be able to take decisions about networks in an informed and objective fashion.

The authors were some of the first to talk in 2003 about the issues related to users' perspective on quality using the term UPQ, User-Perceived Quality. The concept spread subsequently, and it is now more widely encountered under the name of QoE, Quality of Experience. As both manufacturers and users recognized the importance of QoE, it has become an important research topic ever since.

Of course, QoE by itself cannot say much about what one can do to fix a user satisfaction problem. That is why correlating the QoS – the degradation induced at network level, as a measure of what the application experiences – with the QoE – the degradation perceived by the user at application level, as a measure of the user-perceived quality – is essential. This can be done by simultaneously measuring the QoS degradation and the application QoE on an end-to-end basis. These measures must then be correlated by taking into account their temporal relationship.

To perform such measurements in a wide range of network conditions in a laboratory setup, the authors employed the technique of emulation. Using very similar setups, either using a software network emulator, or a hardware network emulator, it was possible to establish the relationship between QoE and QoS for most of the typically used applications today's networks: file transfer, web browsing, voice over IP and video streaming. Assessing the correlation between QoS and QoE makes it possible to predict application performance given a known QoS degradation level, or to determine the QoS bounds that are required in order to attain a desired QoE level. For example, a network administrator will know that if he maintains the QoS within certain bounds (loss rate below 1% and jitter below 20 ms) the VoIP users in his network domain will be content. Reversely, knowing the values of QoS parameters in a network (e.g., packet loss 3%, jitter 25 ms), an administrator can determine that VoIP users will experience good, but not excellent quality. Other applications of this approach can be envisaged, such as those related to Authentication Authorization and Accounting (AAA) services. For all these tasks the key element is and remains the correlation between the objectively assessed QoS and QoE.

REFERENCES

Allen, A. O. (1990). *Probability, Statistics, and Queueing Theory with Computer Science Applications* (2nd ed.). New York: Academic Press.

Almes, G., Kalidindi, S., & Zekauskas, M. (1999). *A One-way Delay Metric for IPPM.* IETF RFC 2679.

Almes, G., Kalidindi, S., & Zekauskas, M. (1999). *A One-way Packet Loss Metric for IPPM.* IETF RFC 2680. Anue Systems. (n.d.). *Network Emulation.* Retrieved from http://www.anuesystems.com

Arlitt, M. F., & Williamson, C. L. (1996). *Web Server Workload Characterization: The Search for Invariants.* In *Proc. SIGMETRICS*, Philadelphia, PA, USA.

Beuran, R. (2004a). *Mesure de la Qualite dans les Reseaux Informatiques.* Ph. D. Thesis, Politehnica University of Bucharest, România & Jean Monet University of St. Etienne, France.

Beuran, R., & Ivanovici, M. (2004c). *User-Perceived Quality Assessment for VoIP Applications*. Technical report (delivered to U4EA Technologies), *CERN-OPEN-2004-007*, CERN, Geneva, Switzerland.

Beuran, R., Ivanovici, M., & Buzuloiu, V. (2004b). File Transfer Performance Evaluation. *Scientific Bulletin of Politehnica University Bucharest, C Series (. Electrical Engineering*, *66*(2-4), 3–14.

Beuran, R., Ivanovici, M., Davies, N., & Dobinson, B. (2004). Evaluation of the Delivery QoS Characteristics of Gigabit Ethernet Switches. In *CERNOPEN-2005-002*, CERN, Geneva, Switzerland.

Beuran, R., Ivanovici, M., Dobinson, B., Davies, N., & Thompson, P. (2003). Network Quality of Service Measurement System for Application Requirements Evaluation. In *International Symposium on Performance Evaluation of Computer and Telecommunication Systems*, July 20-24, 2003, Montreal, Canada (pp. 380-387).

Beuran, R., Nakata, J., Okada, T., Miyachi, T., Chinen, K., Tan, Y., & Shinoda, Y. (2007). Performance Assessment of Ubiquitous Networked Systems. In *5th International Conference on Smart Homes and Health Telematics (ICOST2007)*, Nara, Japan (pp. 19-26S).

Beuran, R., Nguyen, L. T., Latt, K. T., Nakata, J., & Shinoda, Y. (2007). QOMET: A Versatile WLAN Emulator. In *21st IEEE International Conference on Advanced Information Networking and Applications (AINA-07)*, Niagara Falls, Ontario, Canada (pp. 348-353).

Breslau, L., Estrin, D., Fall, K., Floyd, S., Heidemann, J., & Helmy, A. (2000). Advances in Network Simulation. *IEEE Computer*, *33*(5), 59–67.

CalTech. (2002). *FAST Kernel for Large Data Transfers*. California Institute of Technology. Retrieved from http://netlab.caltech.edu/FAST/

Cavalli, V., & Verharen, E. (2002). *TF-STREAM Real Time Multimedia Applications*. TERENA Technical Report.

Chao, H. J., & Guo, X. (2002). *Quality of Service in High-Speed Networks*. Hoboken, NJ: John Wiley & Sons, Inc.

Ciobotaru, M., Ivanovici, M., Beuran, R., & Stancu, S. (2005). Versatile FPGA based Hardware Platform for Gigabit Ethernet Applications. In *6th Annual Postgraduate Symposium*, Liverpool, UK.

Ciobotaru, M., Stancu, S., LeVine, M., & Martin, B. (2005) GETB - A Gigabit Ethernet Application Platform: its Use in the ATLAS TDAQ Network. In *Real Time 2005*, Stockholm, Sweden.

Curet, D., Gouleau, E., Relier, S., Roux, C., Clement, P., & Cherry, G. (2002). *RTP Payload Format for MPEG-4 FlexMultiplexed Streams* [IETF draft].

Demichelis, C., & Chimento, P. (2002). *Instantaneous Packet Delay Variation Metric for IPPM.* IETF RFC 3393.

Demichelis, C., & Chimento, P. (2002). *Instantaneous Packet Delay Variation Metric for IPPM* [Internet draft]. *Empirix*. (n.d.). Retrieved from http://www.empirix.com

Fall, K., & Floyd, S. (1996). Simulation-based Comparisons of Tahoe, Reno, and SACK TCP. *Computer Communication Review*, *26*(3), 5–21. doi:10.1145/235160.235162

Floyd, S. (2002). *High-Speed TCP for Large Congestion Windows* [Internet draft].

Fluckiger, F. (1995). *Understanding Network Multimedia - Applications and Technologies*. Upper Saddle River, NJ: Prentice Hall.

Hoßfeld, T., & Binzenhöfer, A. (2008). Analysis of Skype VoIP traffic in UMTS: End-to-end QoS and QoE measurements. *Computer Networks*, *52*, 650–666. doi:10.1016/j.comnet.2007.10.008

Hoßfeld, T., Tran-Gia, P., & Fiedler, M. (2007). Quantification of Quality of Experience for Edge-Based Applications. In *LNCS 4516* (pp. 361-373).

ITU-R. (1998). *Recommendation BT.500, Subjective quality assessment methods of television pictures*. International Telecommunication Union.

ITU-R. (1998). *Recommendation J.140, Subjective assessment of picture quality in digital cable television systems*. International Telecommunication Union.

ITU-T. (1990). *Recommendation G.726, 40, 32, 24, 16 kbit/s Adaptive Differential Pulse Code Modulation (ADPCM)*. International Telecommunication Union.

ITU-T. (1993). *Recommendation G.711. Pulse Code Modulation (PCM) of voice frequencies*. International Telecommunication Union.

ITU-T. (1996). *Recommendation G.729, Coding of speech at 8 kbit/s using Conjugate-Structure Algebraic-Code-Excited Linear-Prediction (CSACELP)*. International Telecommunication Union.

ITU-T. (1996). *Recommendation P.800, Methods for subjective determination of transmission quality*, International Telecommunication Union.

ITU-T. (1996). *Recommendation P.910, Subjective Video Quality Assessment Methods for Multimedia Applications*. International Telecommunication Union.

ITU-T. (1998). *Recommendation P.861, Objective quality measurement of telephone-band (300-3400 Hz) speech codecs*. International Telecommunication Union.

ITU-T. (1999). [*Internet Protocol][IP][Data Communication Service - IP Packet Transfer and Availability Performance Parameters*. International Telecommunication Union.]. *Recommendation, I*, 380.

ITU-T. (2000). *Recommendation J.143, User requirements for objective perceptual video quality measurements in digital cable television*. International Telecommunication Union.

ITU-T. (2001). *Recommendation P.862, Perceptual evaluation of speech quality (PESQ), an objective method for end to end speech quality assessment of narrow-band telephone networks and codecs*. International Telecommunication Union.

ITU-T. (2001). *Recommendation Y.1541. Network Performance Objectives for IP-Based Services*. International Telecommunication Union draft.

Ivanovici, M. (2005). Objective Performance Evaluation for MPEG-4 Video Streaming Applications. In *Scientific Bulletin of Politehnica University Bucharest, C Series (. Electrical Engineering*, *67*(3), 55–64.

Ivanovici, M. (2006). *Network Quality Degradation Emulation—an FPGA-based approach to application performance assessment.* Ph. D. thesis, Politehnica University of Bucharest, România.

Ivanovici, M., & Beuran, R. (2006). *User-perceived Quality Assessment for Multimedia Applications*. In *The 10th International Conference on Optimization of Electrical and Electronic Equipment (OPTIM'06),* Brasov, Romania, May 18-19 (vol. 4, pp. 55-60).

Ivanovici, M., Beuran, R., & Davies, N. (2005). Assessing Application Performance in Degraded Network Environments - an FPGA-based approach. In *Communicating Process Architectures*, Eindhoven, Netherlands.

Jacobson, V. (1988). Congestion Avoidance and Control. In *ACM Computer Communication Review, SIGCOMM '88 Symposium,* Stanford, CA, USA.

Jiang, W., & Schulzrinne, H. (1999). *QoS Measurement of Internet Real-Time Multimedia Services*. Technical Report, Department of Computer Science, Columbia University.

Korcyl, K., Sladowski, G., Beuran, R., Dobinson, R. W., Meirosu, C., Ivanovici, M., & Maia, M. L. (2004). Network performance measurements as part of feasibility studies on moving part of the ATLAS Event Filter to off-site Institutes. In *Grid Computing: First European Across Grids Conference*, Santiago de Compostela, Spain (pp. 206 - 213).

Malden Electronics Ltd. (n.d.). *Speech Performance Analyser*. Retrieved from http://www.malden.co.uk

Malkowski, M., & Claßen, D. (2007). Performance of Video Telephony Services in UMTS using Live Measurements and Network Emulation. *Wireless Personal Communications Journal*.

Mathis, M., Semske, M., Mahdavi, J., & Ott, T. (1997). The macroscopic behavior of the TCP congestion avoidance algorithm. *Computer Communication Review, 27*(3).

Meer, J. van der, Mackie, D., Swaminathan, V., Singer, D., & Gentric P. (2002). *Transport of MPEG-4 Elementary Streams* [IETF draft].

Miras, D. (2002). *A Survey on Network QoS Needs of Advanced Internet Applications* [working document]. Internet2 QoS Working Group.

MPEG4IP. (n.d.). *MPEG4IP Project*. Retrieved from http://mpeg4ip.sourceforge.net

MPEG. (n.d.). *Motion Picture Experts Group*. Retrieved from http://www.mpeg.org/

Networks, R. (n.d.). *Helix Server*. Retrieved from http://www.realnetworks.com/ & https://helixcommunity.org/

Nguyen, L. T., Beuran, R., & Shinoda, Y. (2007). Performance Analysis of IEEE 802.11 in Multi-hop Wireless Networks. In *3rd International Conference on Mobile Ad-hoc and Sensor Networks (MSN 2007)*, Beijing, China (LNCS 4864, pp. 326-337). Berlin, Germany: Springer-Verlag.

NIST. (n.d.). *NISTNet network emulator*. National Institute of Standards and Technology. Retrieved from http://www-x.antd.nist.gov/nistnet/

OPTICOM GmbH. Germany. (n.d.). *PEVQ – Advanced Perceptual Evaluation of Video Quality* [white paper]. Retrieved from http://www.opticom.de

Padhye, J., Firoiu, V., Towsley, D., & Kurose, J. (1998). Modeling TCP throughput: a simple model and its empirical validation. *Computer Communication Review, 28*(4), 303–314. doi:10.1145/285243.285291

Padhye, J, Firoiu, V., Towsley, D.F., & Kurose, J. (2000). Modeling TCP Reno Performance: A Simple Model and Its Empirical Validation. *IEEE/ACM Transactions on Networking, 8*(2), 133-45.

Padhye, J., & Floyd, S. (2001). Identifying the TCP Behavior of Web Servers. In *SIGCOMM 2001*.

Papoulis, A., & Pillai, S. U. (2002). *Probability, Random Variables and Stochastic Processes*, fourth edition, McGrawHill.

Paxson, V., Almes, G., Mahdavi, J., & Mathis, M. (1998). *Framework for IP Performance Metrics*. IETF RFC 2330.

Python. (n.d.). *Python Programming Language*. Retrieved from http://www.python.org

Rahnema, M. (1993). Overview of the GSM system and protocol architecture. *IEEE Communications Magazine*.

Reijs, V. (2002). *Perceived Quantitative Quality of Applications*. HEAnet. Retrieved from http://www.heanet.ie

Richards, A., Antoniades, M., Witana, V., & Rogers, G. (1998). Mapping User Level QoS from a Single Parameter. In *Proc. Second IFIP/IEEE International Conference on Management of Multimedia Networks and Services (MMNS'98)*, November 1998, Versailles, France.

Rizzom, L. (n.d.). *Dummynet: a simple approach to the evaluation of network protocols.*

Servis, V. (2001). *Measuring speech quality over VoIP networks.* The TOLLY Group.

Shunra. (2004). *The Virtual Enterprise: Eliminating the risk of delivering distributed IT services* [white paper]. Retrieved from http://www.shunra.com

Simena. (n.d.). *Network Emulators.* Retrieved from http://www.simena.net

Stewart, R. (Ed.). (2007). *Stream Control Transmission Protocol.* IETF RFC 4960.

The Internet Traffic Archive. (n.d.). Retrieved from http://ita.ee.lbl.gov/index.html

Tian, Y., Xu, K., & Ansari, N. (2005, March). TCP in wireless environments: problems and solutions. *IEEE Communications Magazine, 43*(3), 27–32. doi:10.1109/MCOM.2005.1404595

Tierney, B. (2005). *TCP Tuning and Network Troubleshooting.* Retrieved from http://www.ONLamp.com

van den Branden Lambrecht, C. J. (1997). *Survey of Image and Video Quality Metrics based on Vision Models.* Presentation.

Video Quality Experts Group (VQEG). (n.d.). *Final report from the Video Quality Experts Group on the validation of objective models of video quality assessment.* Retrieved from http://www.vqeg.org

Wiles, B. C., & Walker, J. (n.d.). *Speak Freely.* Retrieved from http://www.speakfreely.org

Winkler, S. (1999). Issues in Vision Modelling for Perceptual Video Quality Assessment. *Signal Processing, 78*(2), 231–252. doi:10.1016/S0165-1684(99)00062-6

Winkler, S. (2001). Visual Fidelity and Perceived Quality: Towards Comprehensive Metrics. In *Proc. SPIE Human Vision and Electronic Imaging, San Jose, California* (vol. 4299, pp. 114-125).

Winkler, S., Sharma, A., & McNally, D. (2001). Perceptual Video Quality and Blockiness Metrics for Multimedia Streaming Applications. In *Proc. 4th International Symposium on Wireless Personal Multimedia Communications*, Aalbord, Denmark (pp. 553-556).

Yeom, I., & Narasimha Reddy, A. L. (1998). ENDE: An End-to-end Network Delay Emulator. *Journal on Multimedia Tools and Applications.*

Chapter 16
Quality of Experience vs. QoS in Video Transmission

André F. Marquet
WIT-Software, Portugal

Jânio M. Monteiro
University of Algarve/ INESC-ID, Portugal

Nuno J. Martins
Nokia Siemens Networks, Portugal

Mario S. Nunes
IST/INESC-ID, Portugal

ABSTRACT

In legacy television services, user centric metrics have been used for more than twenty years to evaluate video quality. These subjective assessment metrics are usually obtained using a panel of human evaluators in standard defined methods to measure the impairments caused by a diversity of factors of the Human Visual System (HVS), constituting what is also called Quality of Experience (QoE) metrics. As video services move to IP networks, the supporting distribution platforms and the type of receiving terminals is getting more heterogeneous, when compared with classical video distributions. The flexibility introduced by these new architectures is, at the same time, enabling an increment of the transmitted video quality to higher definitions and is supporting the transmission of video to lower capability terminals, like mobile terminals. In IP Networks, while Quality of Service (QoS) metrics have been consistently used for evaluating the quality of a transmission and provide an objective way to measure the reliability of communication networks for various purposes, QoE metrics are emerging as a solution to address the limitations of conventional QoS measuring when evaluating quality from the service and user point of view. In terms of media, compressed video usually constitutes a very interdependent structure degrading in a non-graceful manner when exposed to Binary Erasure Channels (BEC), like the Internet or wireless networks. Accordingly, not only the type of encoder and its major encoding parameters (e.g. transmission rate, image definition or frame rate) contribute to the quality of a received video, but also

DOI: 10.4018/978-1-61520-680-3.ch016

QoS parameters are usually a cause for different types of decoding artifacts. As a result of this, several worldwide standard entities have been evaluating new metrics for the subjective assessment of video transmission over IP networks. In this chapter we are especially interested in explaining some of the best practices available to monitor, evaluate and assure good levels of QoE in packet oriented networks for rich media applications like high quality video streaming. For such applications, service requirements are relatively loose or difficult to quantify and therefore specific techniques have to be clearly understood and evaluated. By the mid of the chapter the reader should have understood why even networks with excellent QoS parameters might have QoE issues, as QoE is a systemic approach that does not relate solely to QoS but to the ensemble of components composing the communication system.

INTRODUCTION

Monitoring and improving video experience is gaining particular interest in Internet Protocol Television (IPTV), and Mobile TV as means of delivering TV broadcasts inside restricted network infrastructure, swayed by the fact that the main issue is no longer how to make video distribution a reality but rather how to improve the quality of the video stream delivered to the end device and ensure the best user experience, so that this can also be used as a value adding proposition to any solution available to an end consumer.

The usage of video encoding tools and optimization of the required bit rate for video transmission brings new multimedia opportunities for the service providers, e.g. delivering more TV services and the deployment of High Definition (HD) content distribution (Wiegand et al, 2003). While offering new services is important, it is also necessary to assure the quality of them so that the service level content service provider or carrier's brand is not diluted. Nevertheless, assessing the quality of the contents delivered to the end devices is still a huge challenge, but is fundamental for the eventual establishment of Service Level Agreements (SLA) between whoever provides the service and who consumes it.

From a technical point of view the Quality of Experience (QoE), when delivering video to an end device, can be seen as the quality remaining in the user's device after the whole encoding and delivery process, that means the distortion introduced to the raw content in every step until the content reaches the decoder at the end device. There are several elements involved in the video delivery chain, as depicted in Figure 1, and some of them introduce distortion. The ones marked in solid line are the

Figure 1. Overview of the video delivery chain

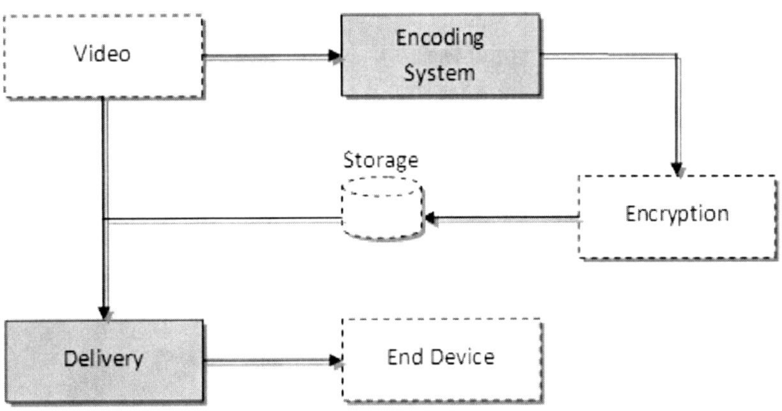

ones that may contribute to the overall distortion in the downstream video delivery. One approach to have a representation of the distortion and the resulting video quality in the last chain element, the end device, is quantifying the distortion of the video sequence when ingested to the several systems in the chain. Adopting such an approach requires modeling all the systems involved in the distortion process, namely, the encoding and the delivery process.

This chapter is especially centered on IPTV and Mobile TV QoE issues. Thus, the focus of this chapter is in connectionless IP networks as they are now prevalent and widely deployed for a variety of services. Audio and voice QoE aspects are not the focus of this chapter and they would certainly deserve a separate analysis per se, so they will not be covered in detail, however they might share some of the application level requirements of distribution and conference video applications, so some of the concepts presented can be transposed to those cases, while references to literature are given for particular cases.

We start by addressing what characterizes and causes video quality problems, afterwards explaining what is QoE and how it can be measured through different approaches and from there describe several models that are being defined to evaluate the quality of video distribution paths from the QoE point of view.

After QoE concepts are well understood we move to present some techniques developed to help assure QoE levels that are desired for IPTV services and we also show the effectiveness of currently used mitigation techniques. From these theoretical bases we present some techniques that can be integrated into generic IPTV and Mobile TV deployments towards monitoring and repairing of video, in view of attaining targeted QoE levels.

Finally we give a brief presentation of areas that present the most promises for future research regarding the development of QoE related techniques and applications.

BACKGROUND

The ITU-T in its Recommendation ITU-T P.10/G.100 of 2007 defines QoE as the overall acceptability of an application or service, as perceived subjectively by the end-user. The QoE includes the complete end-to-end system effects, the client, terminal, network, services infrastructure, etc. However, overall acceptability may be influenced by user expectations and context, and it is exactly that subjectivity that makes defining QoE such a difficult task

Standardization Bodies

There is an ongoing effort regarding the standardization of QoE and the adoption of better practices dealing with the ensemble of the QoE aspects in various standardization organizations like Broadband Forum, Digital Video Broadcasting (DVB), Open IPTV Forum, Alliance for Telecommunications Industry Solutions (ATIS) and International Telecommunication Union (ITU).

The Broadband Forum addresses the QoE by means of the release of TR-126 (Broadband Forum TR-126, 2006). The purpose of this Technical Report is to present the recommended minimum end to end QoE requirements that should be adopted while engineering triple play applications delivered through a broadband infrastructure. The release of QoE requirements reflects the concern in QoE-based engineering when designing a network to deliver triple-play services, more specifically to deliver multimedia services, as described in (Broadband Forum TR-059, 2003). These considerations include an engineering approach to have special attention to the whole encoding and delivery chain and is particularly relevant when defining the application layer QoE requirements but most importantly being able to translate the subjective QoE requirements into objective service metrics that can be used, for instance, to ensure that Servicel Level Agreements (SLAs) are meet.

The most relevant input work done by the DVB commercial group can be found in (DVB Project DVB-HN CR, 2008) and the DVB Home Network reference model (DVB Project DVB-HN A109, 2007). Treating the QoE as a QoS related topic, the DVB project attempts to develop new engineering approaches to improve QoS in the system and consequently improving QoE in the home network. The main requirements the DVB project are pursuing are focused on the quality aspects of home networks broadening to end to end delivery of a service up to the end device (DVB Project HLTR, 2008).

The Open IPTV Forum has not yet engaged on the QoE topic, being more concerned in establishing an architectural framework for the delivery of IPTV services and defining a blueprint for the control services (services and functions).

The ATIS is one of the most active standardization organizations in what refers to QoE related topics and has developed some work in order to define a conceptual framework which includes definitions of QoE under the scope of the ATIS IPTV Interoperability Forum (IIF) Quality of Service Metrics (QoSM) Task Force and types of metrics and measurements (ATIS, 2006). When developing a model for service delivery based on QoE, ATIS is giving particular attention to performance metrics that could correlate well with user opinion (mean opinion scores). These kind of subjective evaluations are considered extremely import, both when engineering the service as well as when performing the service level assessment. However, such assessment is still a challenge in real time monitoring. Nevertheless, this can be obtained from actual involvement of the end user (in IPTV this can be considered technically easy since the communication is bi-directional) or through the adoption of modeling techniques accepted as correlating well with human perception. Within the context of the work being developed by IFF QoSM Task Force, focus is being given on non-reference techniques which could be deployed on several points in the network, e.g. end equipment or Digital Subscriber Line Access Multiplexer (DSLAM). With these non-reference models estimations can be done to assess the quality score, e.g. Mean Opinion Score (MOS), Peak Signal-to-Noise Ratio (PSNR) of contents delivered to end equipments.

The ITU is a body of the United Nations and is currently addressing QoE related topics through the specific ITU-T Study Group 12. While ITU-R Recommendation BT.500-11 (ITU-R, 2002) has been the reference subjective evaluation procedure for broadcast quality content for more than 20 years, novel subjective evaluation procedures are being considered, for instance the Subjective Quality of Internet Video Codecs (SAMVIQ) (Kozamernik et al, 2005) has emerged as a strong candidate for normalization of standard subjective tests procedure within ITU-R.ITU-R for mobile TV and Internet TV class of applications. In June 2008, ITU-T Study Group 12, announced Recommendation G.1080 (formerly known as G.IPTV-QoE) which defines QoE requirements for video, audio, text, graphics, control functions and meta-data from an end user perspective. And also Recommendation G.1081 for IPTV performance monitoring, targeting higher QoS and QoE levels to individual customers by means of software, hardware or hybrid architectures that allow the operator to track network monitoring parameters of the IPTV solution and the way they impact on the end user.

The Video Quality Experts Group (VQEG) is a group of experts from various backgrounds and affiliations, including participants from several internationally recognized organizations, working in the field of video quality assessment. The group was formed in October of 1997 at a meeting of video quality experts. The majority of participants are active in both the ITU and VQEG, combining the expertise and resources found in several ITU Study Groups to work towards a common goal.

The Moving Picture Experts Group (MPEG) is a working group of ISO/IEC in charge of the development of standards for coded representation of digital audio and video. Established in 1988, the group has produced MPEG-1, the standard on which Video CD and MPEG-1 audio layer 3 (MP3) are based, MPEG-2, the standard on which products such as Digital Television set top boxes and DVD are based, MPEG-4, the *de facto* standard for multimedia for the fixed and mobile web, MPEG-7, the standard for description and search of audio and visual content and MPEG-21, the Multimedia Framework. MPEG has recently started a number of new standard lines: MPEG-A "Multimedia Application Format" provides application-specific standards by integrating multiple MPEG technologies. MPEG-B, MPEG-C and MPEG-D provide Systems, Video and Audio specific standards, respectively, and MPEG-E "MPEG Multimedia Middleware" (M3W) is the latest standard under development that will support download and execution of multimedia applications. Some MPEG standards are publicly available (including reference software).

Video Quality

Objective video quality metrics can be classified according to the availability of the original image that can be used as a reference to compare a distorted image or video signal against. Most of the proposed objective quality metrics in the literature assumes that the undistorted reference image or video is fully available, the reason why it is known as full-reference image and video quality assessment. In most IPTV service applications, where the main significance is in making a quality assessment after the delivery process, the reference images or video sequences are often not accessible. Therefore, it is highly desirable to develop measurement approaches that can evaluate image and video quality blindly. Blind or no-reference image and video quality assessment turns out to be a very difficult task, although human observers usually can effectively and reliably assess the quality of distorted image or video without using any reference. Another type of image quality assessment method exists, in which the original image or video signal is not fully available, instead only certain features are extracted from the original signal and transmitted to the quality assessment system as meta-information to help evaluate the quality of the distorted image or video. This is usually referred to as reduced-reference (RR) image and video quality assessment.

Characterization and Causes of Video Quality Problems

Video quality problems can be characterized according to the visual impact. Some of the most common video quality problems include video jerkiness, that means video being shown in a non fluid manner, which is a visual impairment not reproducible as a still image, however video jerkiness and video freezing are some of the most relevant, common and impairing QoE phenomena, possible causes for this problem are due to issues in the encoder, network loss/jitter, bad system clock synchronization, loss of synchronization or more rarely to a bad scene cut (in case of pre-edited video materials). The left side picture of figure 2 represents the original image of the video signal of a still camera and synthetic background with a news presenter and has no perceptible distortions, the picture on the right presents the same image with video blur, possible causes for this problem are due to issues on the camera, source (focus, motion), the encoder, or the decoding equipment.

Below, in figure 3, the image on the left hand side shows a video problem denominated of video noise, and possible causes are at the camera, source, encoder, and decoder or transcoding systems. The image on the right hand side shows an example of drastic video problem that is commonly called as video blackout (type 1) and possible causes include network related problems like information (packet) loss, lack of bandwidth but also others like problems at the encoder or the transcoder. Video blackout (type 2) means that no

Figure 2. On the left, the original video source, on the right the same image with video blur

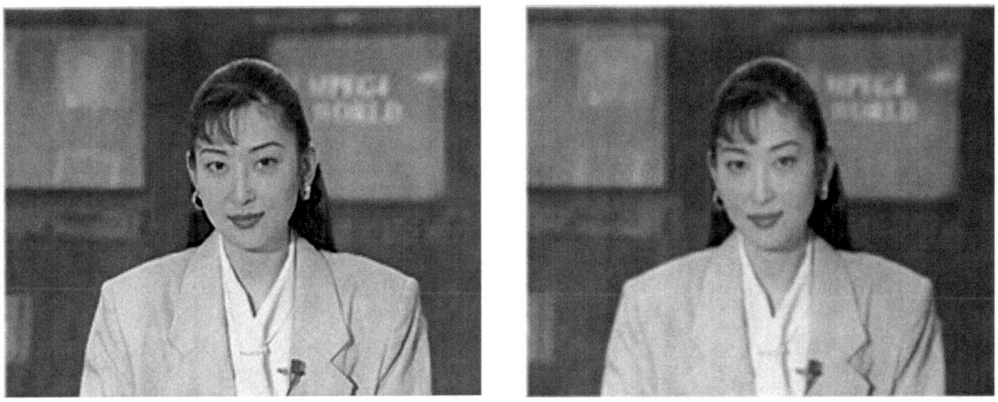

Figure 3. On the left, image with vide noise, on the right with video blackout (type1)

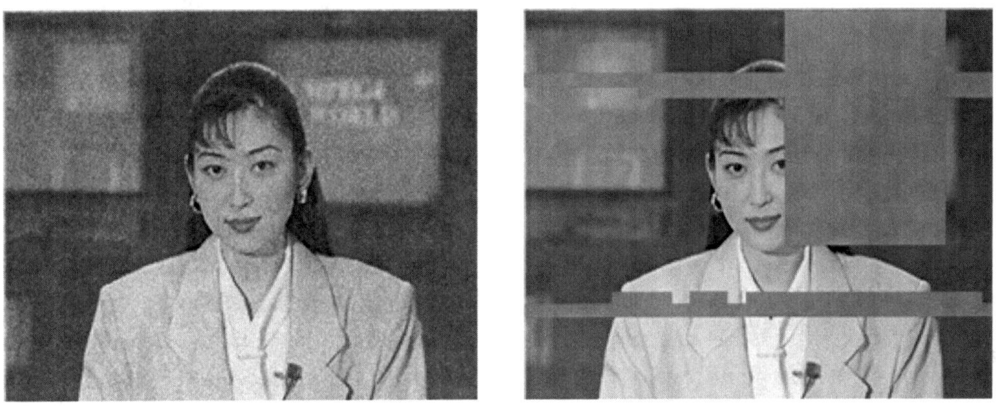

image is displayed at the receiver equipment and possible causes include no video signal at source, massive network losses or lack of bandwidth.

Figure 4 presents two problems that are quite common on IP video distribution networks and greatly impact the QoE. The image on the left hand side presents video blockiness of type 2, where the image can be seen with a rather unpleased effect of visible blocks. This problem is typically caused at the encoder level due to restrictive encoding video bit rates but it can also be caused by poor-transcoding, network loss or lack of bandwidth, depending on the video coding tool and transport protocol. The image on the right hand side presents a rather different class of problems, typified as video distortion, in this case of the video distortion of type 1, where the image is presented segmented, the cause of this problem can be identified at the encoder, transcoder, or because of network loss or lack of bandwidth.

In IP based networks, the loss of data packets due to network loss or lack of bandwidth is one of the most common and at the same time visually devastating problems, as a single loss of a data packet can cause a significant degradation on the QoE. Figure 5 depicts a clear example of two network caused video problems, the picture on the left shows a segmented image caused by a single B-frame IP packet loss and the picture on the right shows undecoded strips caused by a single I-frame IP packet loss.

Figure 4. On the left, image with video blockiness of type 2, on the right image with video distortion of type 1

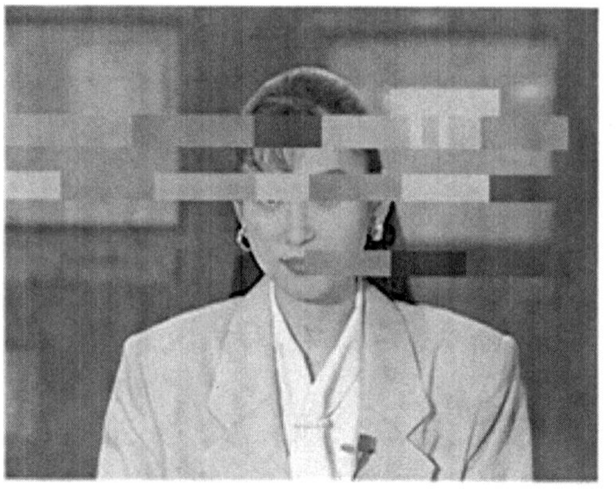

Subjective Metrics

Video quality using a simple five grade scale similar to the one used in MOS and has been commonly used to evaluate video quality in television services. At the basis of subjective assessment are a number of human observers, who rate the quality of video sequences. However, it is unfeasible to use these methods for the in-service continuous evaluation of video quality. Moreover, subjective quality assessments tend to be disregarded in favor of objective ones, because even if subjective measures are the ultimate video quality analysis procedure, they are considered too time consuming to be worth the effort of the rigorous set-up required by standard recommendations.

Conducting Subjective Evaluations

Video quality subjective assessments can be conducted following the lines of the BT.500 recommendation (ITU-R, 2002) and the SAMVIQ draft recommendation (Kozamernik, 2005). The norms advisory procedures should be followed including a) the selection of subjects, b) screen size, c) viewing distance ratio and d) stimulation method. The selection of the subjects can be done by applying an extensive questionnaire in order to validate that the set of individuals that are to be submitted to the video evaluation test are representative of the considered universe. The subjects should be nonexperts on video coding technologies and have a normal psychological and physiological profile.

Figure 5. Both pictures depict video distortion of type 1

BT.500 Recommendation

Subjective assessments conducted according to ITU Recommendation BT.500 are specially indicated to broadcast signals. The test sequences, typically Standard Definition (SD) or High Definition (HD) signals, are typically shown in random order, according to the Single Stimulus (SS) method, as described in BT-500 recommendation and depicted below in figure 6. In Double Stimulus (DS) methods, assessors are cyclically presented with an unimpaired reference of the video sequence followed by the impaired version of the same picture, which they must grade when comparing it against the unimpaired sequence.

In the test room, a supervisor, with no relation with the subjects should be present to assure that the tests go according the planed procedure.. Subjects are expected to grade on a linear MOS scale, which goes from 1, standing for very annoying impairments (bad quality), to 5 for imperceptible impairments (excellent quality). Although this recommendation outlines the test procedures and statistical analysis of results it does not impose a normative approach for the analysis of the results.

Subjective Quality of Internet Video Codecs Methodology

The SAMVIQ Subjective Quality of Internet Video Codecs Methodology most commonly called SAMVIQ focuses particularly on mobile terminals and reduced display devices oriented towards PCs and mobile screen displays. SAMVIQ methodology requires a especially designed software that implements the assessment procedure. Figure 7 shows a screenshot of a special design program for the purpose of running deploying the SAMVIQ. According to this methodology, several test sessions are organized in scenes in such a way that one scene follows the other according to the assessor selection. No more than four scenes should be displayed per session and in each session it is possible to play and grade any sequence in any order since each assessor has the complete control over the application. Accordingly, each sequence can be played and assessed as many times as the subject wants, as long as each video sequence is played out and viewed completely. The last grade always remains recorded. From one scene to the next, the sequences are randomized, preventing the "less collaborative" subjects from attempting to grade in an identical way according to a pre-established order.

Subjects are asked to assess the overall picture quality of each presentation by inserting the slider mark on the continuous scale on the right side of the user interface in a scale between 0 and 100%. They should also be instructed to consider a specific scenario of video distribution to mobile terminals, since this correctly biases the subjects to the intended application. The first sequence is always a pre-established reference video, so to stabilize the subjects' quality perception.

Figure 6. BT.500 recommendation single stimulus method

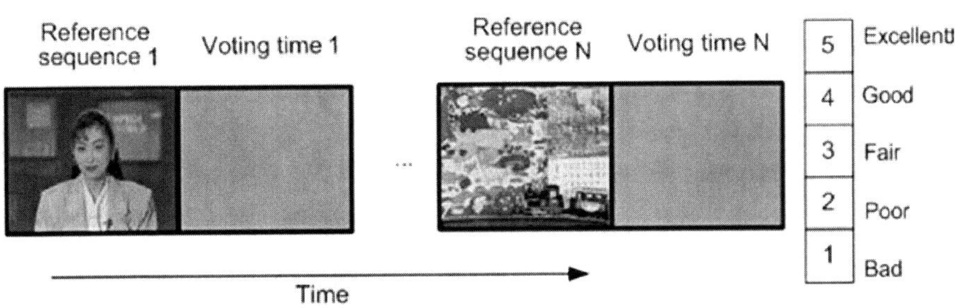

Figure 7. SAMVIQ reference software

Considerations on the Statistical Treatment of Collected Data

The collected data from the exposed subjective tests should be analyzed considering a confidence interval of 95% to compensate MOS evaluation error. The data tendency should be given by a trend line within the MOS bounded by the confidence intervals. Regressive analysis can then be applied to the data points of interest in order to attain the MOS results for the several sequences. The method yields the results of subjective evaluation scores and calculates the MOS and the standard deviation for each video presentation, allowing presenting the results with a confidence interval of 95% associated to each mean score.

Objective Metrics

Subjective video monitoring methods are time-consuming and complex and, as a consequence of that, several objective algorithms were developed to try to assess quality according to an end user's perception in a day-to-day performance evaluation and monitoring.

When the access to the media signal in an uncompressed form is somehow possible, three major classes of algorithms are used: full reference, reduced reference or no reference metrics.

Full reference (FR) metrics compare the undistorted signal (or reference) with a processed or distorted signal. These methods require that both signals are available, so they can be compared towards the computation of an objective metric, normally in an uncompressed form and they usually also require a high degree of spatial and temporal synchronization. Most of the objective quality metrics in the literature assume that the undistorted reference image or video is fully available, and that is the reason why it is known as full reference image and video quality assessment. In this context, the signal is the original message and the noise is the error added in the reconstruction. Currently, the most widely used and accepted FR objective image and video distortion (noise)/quality metrics (Winkler, 2005, pp. 54-55) are the Sum of Absolute Distortions (SAD), the Mean Square Error (MSE) and Peak Signal to Noise Ratio (PSNR):

- **SAD** measures the cumulative linear distortions between reference and reproduction picture, in this case if the SAD is 0 then there is no measurable distortion.

- **MSE** measures the cumulative square error between reference and reproduction picture.
- **PSNR** measures the distortion of the image in relation to the peak signal, in a logarithmic scale.

The MSE and PSNR are widely used because they are simple to calculate and have a clear physical meaning. However, they are not widely accepted as correlating well with perceived subjective quality measurements, specially what concerns noise introduced by the so-called block effect, which is typical of the DCT family of energy compacting macro-block based video coding tools. Other full-reference methods specified in ITU-T Recommendation J.144 (ITU, 2004) can be used to estimate the perceptual ratings of human subjects.

Reduced reference (RR) metrics extract partial evaluation features from the original signal, which need to be present in the measurement point along with the video signal. At the evaluation point, the same extraction process is performed from the received signal and the quality evaluation is performed comparing both original and received features. This method requires a time alignment between original and received evaluation features. Reduced reference methods can be used for quality evaluation in video transmission, however a reliable communication channel should be used to make the evaluation features available at the receiver end.

No reference (NR) methods use solely information retrieved from the received video in the process of quality assessment. The evaluation process usually uses frame blockiness analysis to obtain the impairment estimation. When evaluating video quality in transmission networks, these metrics are globally classified as Media Layer models, because they require access to the media decoded signal to perform quality estimation.

In recent years, a great deal of effort has been dedicated to develop objective image and video quality assessment methods (mostly for FR quality assessment), which incorporate perceptual quality measures by considering HVS characteristics, but they have failed in becoming largely accepted by the video quality community, hence PSNR maintains its popularity.

In fact, only limited success has been reported from evaluations of sophisticated HVS-based objective FR quality assessment models under strict testing conditions and a broad range of distortion and image types. The VQEG under the aegis of the ITU has recommended objective assessment methods for video quality most notably the Video Quality Metric (VQM) (ITU, 2004), which is applicable to the evaluation of macro-block DCT based coded video in broadcasting applications like MPEG-2, that still constitutes an important part of the broadcast industry but is being rapidly substituted by MPEG-4 Part 10 coded video, specially for HD services. VQM metric is considered to better match the observer's impression of spatial, temporal and color continuity (Xiao, 2007), thus correlating better with subjective metrics like MOS than PSNR or MSE. For instance, it is necessary to consider different format displays e.g. PC monitors or PDA screens, as IPTV and Mobile TV solutions are embracing new display mediums. Moreover, multimedia applications often suffer from distortion due to packet loss, which produces perceptual effects considerably different from those of coding distortion (see previous chapter Characterization and Causes of Video Quality Problems), namely propagation errors due to errors on I frames. Consequently, current standards for the evaluation of video quality are not directly applicable to multimedia communications and therefore the justification for the announcement of specific IPTV QoE recommendations. Okamoto et al (2004), proposed an objective video quality assessment method that is applicable to arbitrary video sequences in telecommunications environments, with a wide range of bit rates coding and viewing on PC monitors.

METRICS, MODELS AND ARCHITECTURES FOR A QOE ORIENTED VIDEO TRANSMISSION

QoS Metrics Affecting QoE

Although many parameters are defined for the characterization of QoS metrics at the IP layer, in a video distribution framework, the most important parameters include the IP packet delay variation (IPDV), packet error rate (IPER), packet loss rate (PLR), packet transfer delay (IPTD) and the per-link or end-to-end bandwidth. From these, packet loss is the parameter that mostly affects the decoding quality. Although losses are usually quantified using the PLR, they can also be characterized in terms of loss period, loss distance, loss noticeable rate, loss period length and inter loss period length (Koodli & Ravikanth, 2002). Other forms of packet loss descriptions like a loss run length distribution computed during a period of time, may be used to describe sparse bursts due to congestion and queue managing solutions like Random Early Discard (RED). The propagation effects of a lost packet can be significant, not only because it corresponds to the complete loss of a P or B frame or the partial loss of an I frame, but also, as a result of video compression and frame interdependency many arrived packets may be useless for decoding purposes. Another important QoS parameter that affects channel change times in multicast video distribution systems is the delay between a Join/Leave Internet Group Management Protocol (IGMP) messages (Cain, et al. 2002) and the corresponding packet arrival or withdraw.

Typically, in a well dimensioned and managed wired IPTV network, these QoS parameters are relatively constant in time and accordingly a reasonable level of quality assurance can be achieved through a proper QoS dimensioning or even over provisioning, which would be capable of guarantying high levels of user's QoE. As the challenge in wired IPTV networks, with a typical architecture depicted in figure 8, is moving to high definitions and 3D content, mobile networks are still trying to support good levels of QoE for lower resolution video contents, which typically range from QCIF, QVG and CIF to VGA. The demand for a deeper understanding of QoE issues and assurance is therefore critical in both of these networks.

As home network terminals move to IEEE 802.11 wireless, IPTV Set-Top-Boxes (STB) need to compete for bandwidth and channel access with other IP terminals and services. Other wireless terminals not only introduce bandwidth limitation, but also packet discards due to congestion, loss due to collisions and fluctuating service quality. In this scenario the wireless LAN gateway is an important element to extend the wired QoS quality to the wireless environment. However QoS levels in wireless

Figure 8. All-IP video distribution network

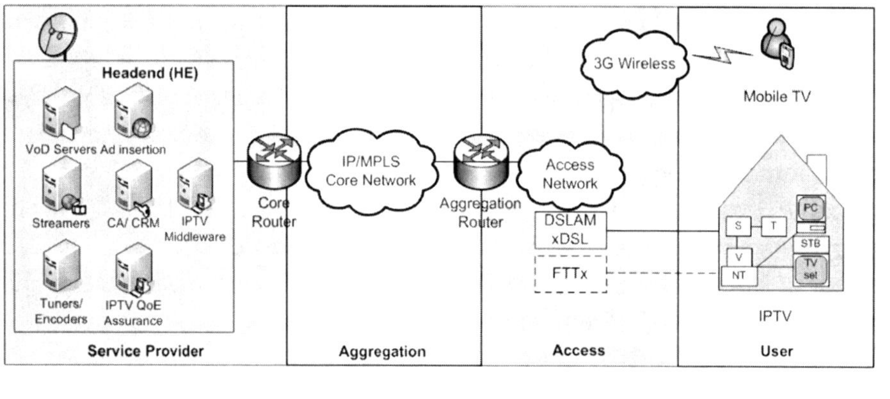

and mobile are more difficult to guaranty due to several factor like terminal mobility, channel changes, interference and path loss and also as a consequence of medium access (MAC) mechanisms that were not primarily designed for multimedia transmission. Although bandwidth over-provisioning can be seen as a solution to compensate for the lack of appropriate MAC mechanisms, field tests have shown (Lee et al, 2008) that when a 2 Mbps video stream is transmitted with concurrent traffic in a 802.11g wireless network, appropriate levels of user QoE could only be achieved if QoS prioritization of video is used.

As the default IEEE 802.11 MAC mechanisms didn't consider traffic prioritization or bandwidth reservation they were updated with QoS extensions in the IEEE Std 802.11e (2005), with two mechanisms: Enhanced Distributed Channel Access (EDCA) which enables traffic prioritization and HCF (Hybrid Coordinator Function) Controlled Channel Access (HCCA) which enables admission control and bandwidth reservation functions. From these, only EDCA mechanisms were implemented in Wi-Fi Multimedia (WMM) capable Access Points and cards, considering four QoS Access Categories (ACs): voice, video, best effort and background traffic.

In (Lee et al., 2008) the authors have used the Video Quality Metrics (VQM) (ITU-T Recommendation J.144, 2004) method to estimate subjective quality in a mixed wired and 802.11g/e IPTV distribution. They have found that the packet loss rate should be limited to a maximum of 3×10^{-3} in order to achieve good QoE levels.

Unfortunately none of these mechanisms can be used to differentiate intra-stream priorities, as required by I, P and B video frames. Additionally the lack of support for intra-stream priorities does not only occur at the MAC layer, but also extends to other transmission layers.

The EDCA's transmission opportunity (TXOP) is an 802.11e mechanism that is particularly suited to video streaming applications. The TXOP is the interval of time during which a QoS enabled station has the right to send multiple MAC Protocol Data Unit (MPDU) without having to re-contend for access after having succeeded in sending the first frame. Studies like (Cranley et al., 2007) have tested the performance of streaming video in the case where I, P, and B frames are transmitted through different access categories and the TXOP limit is dimensioned according to the mean video frame size. In (Suzuki et al., 2008) the authors assess QoE from MAC-level QoS for audio-video transmission over an IEEE 802.11e EDCA wireless LAN. They have verified that to achieve high QoE results the value of the transmission opportunity (TXOP) limit should be adapted according to the content types.

Finally the heterogeneous dimensions of mobile terminals require an adaptation of content to the definition, image size and user distance from the screen. In terms of definition, Scalable Video Coding tools already offer the possibility of transmitting different image resolutions adapted to the receiver's capabilities. However encoding of the same video sequence using different sizes may not guaranty a proper user satisfaction. In

Figure 9. Context based QoE in heterogeneous devices

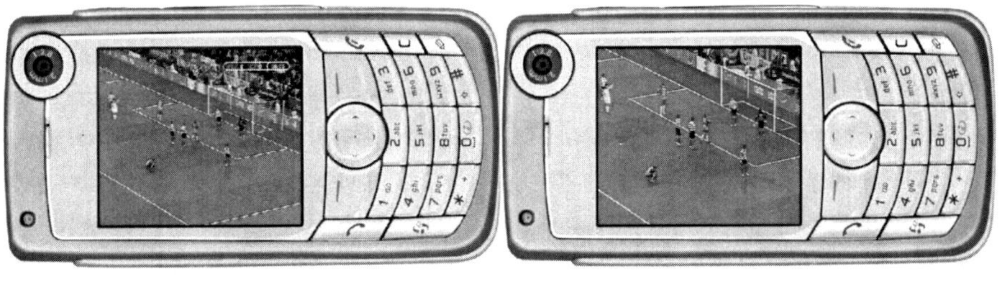

fact, a user trying to view two different images on a small display may have different satisfactory experiences according to image context, as exemplified by left and right images of figure 9. In these cases a specific context based analysis should be made, like the one proposed in (Bae et al., 2006), which, according to the target size of a LCD display selects which region of interest should be transmitted. If the resulting region is smaller than the original one, then it should be enlarged to meet the display's definition.

QoS and QoE Measurement and Assessment in Communication Networks

Different communication layers are used to measure, convey and adjust QoS and QoE measures. Traditionally, in media streaming, end-to-end QoS monitoring is commonly performed at the transport layer, using both the Real-time Transport Protocol (RTP) and RTP Control Protocol (RTCP) (Schulzrinne et al., 2003). RTP is a protocol designed for the transfer of real-time data over IP networks. Important features of RTP include sequence numbering, time stamping, inter-media synchronization and payload identification. Although other protocols could be used in unicast scenarios, RTP usually is transported over UDP, which makes it prone to packet loss. Nevertheless, RTP specification by itself does not consider any retransmission or flow control mechanism, leaving these tasks to either upper or lower layer protocols. RTCP is an accompanying protocol that can be used to monitor the data transfer quality, conveying quality parameters between both communication ends. RTCP Sender and Receiver reports enable the computation and retrieve of the reception packet rate, number of lost packets and inter-arrival jitter.

At the IP layer, service providers typically define QoS requirements using Service Level Agreements (SLAs). Network delay, jitter bounds and loss rates are usually defined through these SLAs and guaranteed using the Differentiated Services (DiffServ) QoS architecture. However, the effect of each of these requirements does not directly correspond to QoE measures. Instead, they depend on the service and media types being carried. For instance, in IPTV video streaming applications, the delay of a channel change procedure (defined as the time between a remote control button push until stable channel is displayed) should be limited to a maximum of 2 seconds (Broadband Forum TR-126, 2006). This overall delay however, also includes other delays like reception buffering and video decoding. For the network part, operators usually limit one-way network delays of video traffic to a maximum of 100ms. Jitter, is compensated using a reception de-jitter buffer. The delay introduced by this buffer typically adds less than 100ms to the network jitter value.

The PLR may be caused by several factors: congestion losses, bit error losses, link or equipment failure. From these, congestion losses can be limited with the appropriate capacity planning and using the SLAs and QoS solutions. However, and especially in wireless networks, packet losses that result from bit error losses might require other quality assuring solutions in order to achieve high levels of reliability. Packet loss is the parameter that most directly causes visual impairments in the decoding video. These impairments also depend on the encoding structure of the video, which can be made more resilient, as for instance happens with data partitioning features of H.264 (ITU-T Recommendation H.264 and ISO/IEC 14496-10, 2007).

At the application layer and in terms of QoE monitoring, an objective quality evaluation function can be implemented at the receiver terminals that reports back the perceived reception quality to the service managing or video streaming servers. A higher layer communication protocol must be used that transmits these QoE scores, together with other information concerning end-user's quality, to the service provider network. That protocol can also be used to transmit feature information in the

Figure 10. Architecture and quality measurement points (QMP) in an IPTV distribution framework

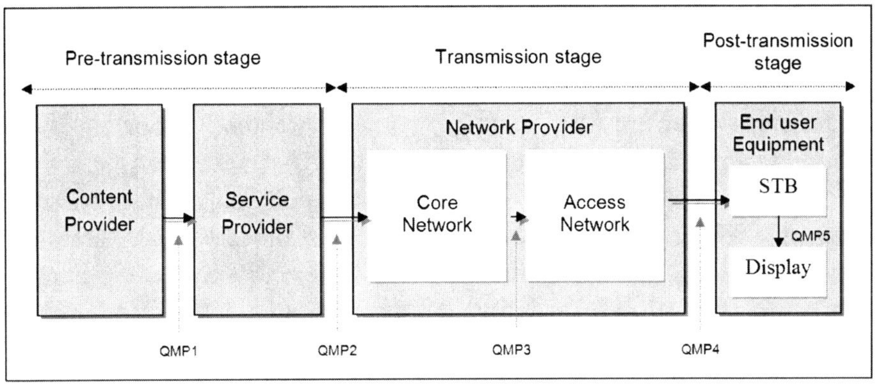

opposite (downstream) direction, as happens when using a reduced reference quality evaluation. It should be noted that the availability of end user's QoE metrics at the service provider may serve to adjust the encoding parameters, according to the monitored network conditions, or to adjust loss resilience mechanisms. They can also be used by a network management unit to adapt QoS parameters.

Determining QoE in Transmission Networks

Much of the effort related with video quality metrics, has been focused in evaluating artifacts related with compression. However, as current compressed video is usually transmitted in packet oriented networks, which are prone to the binary erasure of its content, the interest in the definition of monitoring models that evaluate end user's quality according to these artifacts is growing.

With the advent of new mobile TV, Internet TV or video streaming and IPTV video services, appropriate and standard measurement tools are needed by service and network providers to make a performance testing according to an end user's perception of quality. That information is also essential to make a proper decision of the encoding methods and parameters, with additional reflections on the quality assurance capability of core and/or access networks.

Traditionally, network operators have used QoS parameters like PLR or Jitter to evaluate network quality. However, for video content transmission, these metrics do not present a straightforward mapping with a user perceived quality because they do not consider the media content being carried.

Quality Measurement Points in a Video Distribution Framework

A video distribution framework can be partitioned in three main stages (Takahashi et al., 2008): a pre-transmission stage which includes video encoding; a network transmission stage which causes partial loss of media information; and finally a post-transmission stage where content is decoded and error concealment is performed according to a terminal's performance. The transmission stage is usually further sub-divided in a core, access and home network segments.

Between these transmission stages and inside some of them there are several Quality Measurement Points (QMPs) as presented in Figure 10.

In this architecture, the maximum level of video and audio quality is typically achieved at the input of the encoder and depends from several factors, including frame rate, image definition and Signal-to-Noise Ratio (SNR). As the encoding process typically reduces QoE quality, Quality Monitoring Point 1 (QMP 1) represents the place where the highest quality of the encoded signal is achieved, in the transmission path. The reduction in the signal bit rate at the cost of frame rate, image definition, quantization parameters and Group of Pictures (GOP) structure are the most important factors to contribute to the quality degradation in this first stage.

In the following step, service providers can directly forward the received signal, or transcode it to adapt its content according to receiver's capability or bandwidth. In intermediate points (QMPs 2, 3 and 4) the media signal is usually encapsulated in transport and network layers. This encapsulation hides some of the encoding structures and places some challenges to the evaluation of the quality using legacy methods that require access to a decoded version of the media signal.

There is an ongoing effort to develop different types of quality assessment methods for each of these quality measurement points to evaluate from an end-user point-of-view the media distribution path. Broadly, these models usually fall in the following four categories: Media Layer Models, Parametric Models, Bit-stream Models or a combination of these models known as Hybrid Models. Figure 11 compares them in terms of the required input information. In the following we will analyze each of these models.

Media Layer Models

In the Media Layer Models category, subjective quality is assessed using information taken from the media (audio, voice or video). The access to the media signal make these methods appropriate for pre-transmission and post-transmission stages.

These models usually are not dependent on the system that caused the impairment because they estimate subjective quality directly from the media data. As previously explained, FR, RR and NR metrics belong to this group of models. An example of a NR Media Layer Model can be found in Qiu et al. (2006).

Exiting Standards based in this type of metrics include ITU-T P.862 (2001) for speech quality evaluation, ITU-R BS.1387 (1998) for Audio, ITU-T J.144 (2004) for standard TV quality evaluation and ITU-T J.148 (2003) for quality evaluation of cross–modal influences between audio and video.

In terms of network and post-transmission stages, full-reference metrics can be computationally demanding and as it was mentioned they require the access to the original data which is usually not available in intermediate transmission networks and in end user's terminals. Similarly, reduced reference metrics require that information taken from the original signal should be transmitted in parallel with the encoded media signal. For these reasons, media layer models based in no reference metrics are preferred for

Figure 11. Comparison between media layer, parametric packet-layer, bit-stream and hybrid models in terms of model inputs

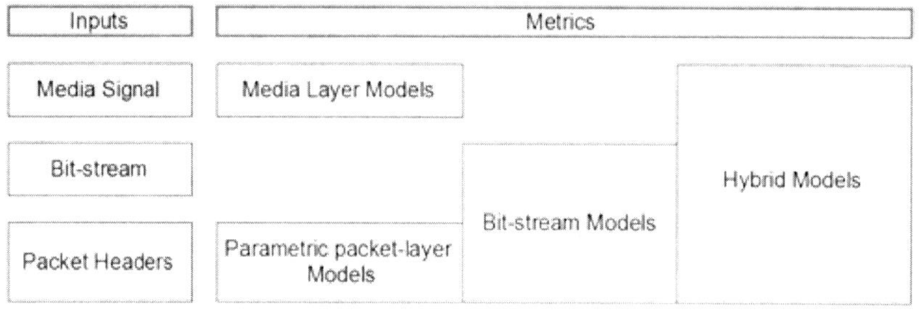

continuous quality evaluation and assurance of post-transmission stages.

Parametric Packet-Layer and Planning Models

Parametric models assess QoE using parameters taken from networks or terminals, like packet loss, jitter or delay. However, the resulting quality is dependent on the type of service being transmitted and in the codec being used. These models are usually subdivided in two categories: parametric packet-layer models and parametric planning models.

Parametric packet-layer models use the information carried in packet headers to obtain parameters like the packet loss rate and packet loss frequency. Those metrics are afterwards used with distortion measures for a specific encoder to estimate subjective quality. Examples of these models can be found in Yamagishi et al. (2008) where a model is proposed for IPTV quality assessment and in the ITU-T Recommendation P.564 (2007) that generates quality metrics for individual VoIP calls using network performance measurements from RTP, UDP or IP packet headers. Video information from the Transport Stream (TS) can also be used. These models are more appropriate for measurement points QMP2, QMP3 and QMP4, as presented in figure 10.

Parametric planning models use QoS parameters from receivers and network planning as input to assess QoE. They are dependent of a specific system to be modeled, assessing subjective quality based in impairments like encoding method, end-to-end delay and equipment. The E-model for VoIP (ITU-T Recommendation G.107, 2008) and the opinion model for video-telephony applications (ITU-T Recommendation G.1070, 2007) constitute two examples of these models. They are most appropriate for the planning of core, access and home networks.

The ITU-T study group on quality of service and quality of experience (SG12) is the standardization body that defined these models. Concerning IPTV, and in the field of parametric models, this group is currently working in an opinion model for video streaming applications (G.OMVAS) and in a non-intrusive parametric model for the assessment of performance of multimedia streaming (P.NAMS).

The advantage of parametric models in QoE evaluation is that they can be more easily implemented in communication networks since they are lightweight when compared with other models. However, they compute QoE as an average quality in time and cannot differentiate the loss effects of specific bit stream structures.

Bit-Stream Models

As the quality estimated by parametric models is obtained as time interval average, it cannot differentiate losses between specific encoded media. For instance, the impact of a single lost IP packet on an Elementary Stream (ES) video frame, typically not only causes a spatial loss propagation within the same I, P or B frame, but usually affects other dependent P or B structures causing a temporal loss propagation. Accordingly, the same packet loss ratio can result in considerably different visual effects depending in which parts of the bit-stream are affected.

This problem could be ultimately solved using Media Layer models, which analyze the effect in QoE of the fully decoded audio and video content. However, since these models present the drawback of computational complexity, especially for quality assessment in network transmission and post-transmission stages, other models defined as Bit-stream models are recently being developed.

As presented in figure 12, bit-stream models utilize the coded bit-stream information combined with the packet-layer information as used in parametric packet-layer models, to assess QoE without needing video decoding. Using that information, these models are capable of considering

Figure 12. Bitstream layer model - Protocol stack

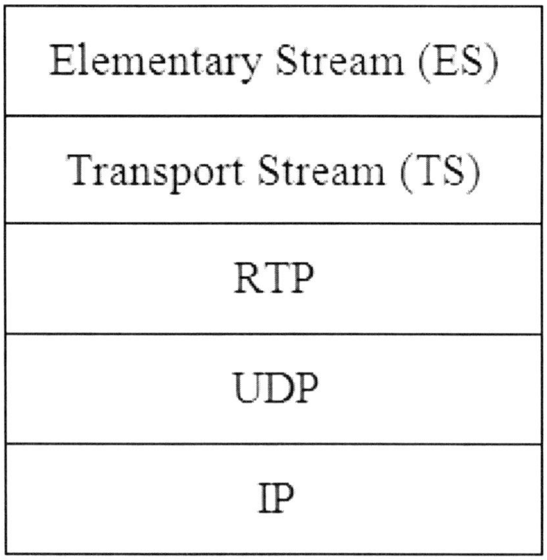

the interdependency between intra-coded video (or audio) structures, resulting in more precise metrics when compared with parametric models. Bit-stream models are most appropriate for post-transmission stages. In this field, the ITU-T SG12 is defining a non-intrusive bit-stream model for the assessment of performance of multimedia streaming (P.NBAMS).

Hybrid Models

Hybrid models are a combination of any of the previously mentioned solutions. As represented in figure 11, these models analyze the media signal, the bit-stream information and/or packet headers to assess subjective quality. An example of one of these models can be found in (Winkler & Mohandas, 2008). These models are most appropriate for pre-transmission and post-transmission stages. The ITU-T SG9 is developing in ITU-T J.bitvqm several hybrid models which use bit-stream data to assess QoE. Next table summarizes existing QoE models and refers some of the existing ITU standards and ongoing projects.

QoE Assurance Solutions

As previously analyzed, the quality of compressed media significantly degrades if the decoders are exposed to transmission impairments such as packet loss. Especially in IPTV services, users expect high levels of QoS, evaluated through QoE. When network based QoS mechanisms are available, they can be used to increase the reception quality by giving higher priority to more important data. These solutions however may need to be complemented by other mechanisms to increase the end-to-end reliability of highly sensitive data, when transmitted over IP based networks. In (ITU-T Recommendation Y.1541, 2008) network performance objectives for IP-based services are defined together with several QoS classes, numbered from 0 to 7, as presented in Table 2. QoS classes are classified in terms of packet transfer delay, delay variation, packet loss rate and error rate. From these, classes 6 and 7 are considered as an upper bound of quality, being more stringent than classes 0 to 4 and were provisionally defined in (ITU-T Recommendation

Table 1. Assessment models for subjective quality estimation

	Main Application	**IPTV QoE Standards**
Media-layer models	Encoding and decoded quality monitoring	ITU-T J.144
Parametric packet-layer models	Network in-service and nonintrusive quality monitoring	ITU-T P.NAMS
Parametric planning models	Network planning and terminal specification	ITU-T G.OMVAS
Bitstream layer models	Terminal in-service and nonintrusive quality monitoring	ITU-T P.NBAMS
Hybrid models	In-service nonintrusive quality monitoring	ITU-T J.bitvqm

Table 2. Network QoS requirements for multimedia services

QoS class	Packet Transfer Delay	Packet Delay Variation	Packet Loss Ratio	Packet Error Ratio	Packet Reorder Ratio	Typical Applications
0	100 ms	50 ms	0.1%	0.01%	Undefined	Streaming of live content including TV, speech and low resolution video content.
1	400 ms	50 ms	0.1%	0.01%	Undefined	Streaming of video and audio content.
2	100 ms	Undefined	0.1%	0.01%	Undefined	Streaming control.
3	400 ms	Undefined	0.1%	0.01%	Undefined	Access to web pages, interactive message exchange, payment transactions.
4	1 s	Undefined	0.1%	0.01%	Undefined	Download and upload of video content.
5	Undefined	Undefined	Undefined	Undefined	Undefined	Download of data, e-mail and messaging exchange.
6	100 ms	50 ms	0.001%	0.0001%	0.0001%	Streaming of live TV content.
7	400 ms	50 ms	0.001%	0.0001%	0.0001%	Streaming of video content.

Y.1541, 2008) for high bit rate services, like IPTV. They were more recently revised in (ITU-T Recommendation Y.1541 Amendment 3, 2008).

However, even an IP network conforming to QoS Classes 6 or 7 is considered as not capable of providing sufficient loss and error ratios recommended for IPTV, as expressed in Table VIII.1 of (ITU-T Recommendation Y.1541 Amendment 3, 2008).

As a consequence of that, application layer error recovery mechanisms are required, running on edge equipment recovering from packet loss and errors.

Application/Transport layers reliability is emerging as an important issue in multimedia transmission services. Generally, reliability is achieved using three main mechanisms: Forward Error Correction (FEC), Retransmission based mechanisms or Hybrid solutions combining FEC and Retransmission. Retransmission solutions require a bidirectional channel and recover lost data by requesting retransmission of missing packets, from the source or, typically, from intermediate retransmission servers to deal with scalability issues. FEC approaches operate by compensating packet losses by adding redundant information to the source data.

The selection of the best solution depends on the delay tolerance of the type of service being delivered, on the bidirectional or unidirectional nature of the transmission mechanism and in the bandwidth or processing overhead that each of these mechanisms requires. In the following, we will analyze these solutions in more detail.

Retransmission Based Solutions

A retransmission-based erasure recovery process uses feedback packets sent by receivers to request the retransmission of lost packets, which are detected using packet headers fields, as supported by Real Time Protocol (RTP) sequence numbering. TCP could be optionally considered as a solution to implement a fully reliable transmission, however besides being restricted to unicast communications this imposes congestion control mechanisms that are the cause for bit rate and delay variations. As a consequence of that, RTP over UDP is more commonly used to transmit video streaming of live content over well managed networks, like IPTV and Mobile TV networks.

A logical way to implement retransmission would be to use the Real-Time Control Protocol (RTCP) (Schulzrinne et al., 2003) to send feedback messages requesting lost packets. However, the original RTCP specification included in the IETF RFC 3550 (Schulzrinne et al., 2003) did not consider the acknowledgement of specific

RTP packets. It was more recently extended in IETF RFC 4585 (Ott et al., 2004) a way to enable the transmission of negative acknowledgement packets (RTCP NACK) that specifically request retransmission of one or more lost packets. A new RTP payload format for retransmissions was also defined in IETF RFC 4588 (Rey et al., 2005).

Retransmissions require a return channel which may not always be available, as for instance happens in Digital Video Broadcasting applications. Additionally, a feedback implosion problem may arise in large scale point-to-multipoint transmissions due the potentially high number of negative acknowledgement packets. This can either be solved with a careful architectural design or using FEC solutions. In practical Mobile TV deployments the usage of re-transmission techniques, and most specifically the usage of TCP for live TV streaming has proved to be highly infective leading to a poor QoE.

Moreover, in terms of delay, the packet retransmission process introduces an additional delay equivalent to the sum between the mean Round Trip Time (RTT) and a jitter which results from network and end systems processing delays. A buffer must exist at the receiver to compensates for that delay enabling an in order insertion of these retransmitted packets.

Forward Error Correction (FEC) Based Solutions

Point-to-multipoint or broadcasted transmissions are commonly used to enable large scale distributions. In these distributions, however, if a proper reliability method is not considered, packet losses can significantly degrade the distribution efficiency. Forward Error Correction (FEC) is one of the candidate solutions to be considered that enable recovering from packet losses without the burdens caused by packet retransmissions, in these architectures.

At the transmitter side, a FEC encoder typically constructs a source block from which the repair packets are afterwards generated. This process introduces a fixed delay that depends on the source block size, the transmission bit rate and in the arrangement between FEC and data packets transmission. The relative code overhead introduced by FEC coding usually decreases when large source blocks are used, with the drawback of increasing the encoding delay. Accordingly, a trade-off between source block size and delay must be decided.

Application Layer – Forward Error Correction (AL-FEC) usually refers to packet erasure protection mechanism in which an additional amount of data is sent to account for a certain amount of packet losses at the IP layer.

Among the different AL-FEC coding solutions available, Raptor codes (Shokrollahi, 2006) are currently experiencing great popularity and widespread adoption. Raptor codes are a class of fountain codes which were designed for transmission of data over Binary Erasure Channels (BEC) (Elias, 1955). Their advantages include a wide range of source symbol values and code word lengths, small decoding complexity and high reception overhead efficiency. They are also more efficient in terms of processing than other FEC codes like Reed Solomon codes, not imposing any limitations in terms of the amount of protected data.

Due to their properties, Raptor codes were chosen by the Third Generation Partnership Project (3GPP) in the context of multimedia broadcast multicast services (MBMS) (3GPP TS 26.346, 2005). Raptor codes were also included in different AL-FEC specifications for streaming and download delivery within IPTV and mobile broadcasting services (Luby, 2008).

Besides these solutions, other AL-FEC specifications for streaming media exist. RFC 2733 (Rosenberg et al., 1999) specifies an RTP payload format for a generic FEC. It enables the generation of separate streams of FEC packets which result from the exclusive-or (parity check) of media packets. The payload format of FEC packets contains a 24 bits bitmask that specifies which media packets have been used to generate the FEC packet.

Different FEC solutions can also be combined as happens with the DVB AL-FEC specification (ETSI TS 102 034, 2007) developed for IPTV that considers a layered FEC approach. The base FEC layer consists of a packet-based interleaved parity code and enhancement FEC layers are based on the Raptor codes.

Generally, in terms of transmission rate, FEC coding consumes more bandwidth than retransmission, since repair packets are typically transmitted according to the maximum expected packet loss ratio, while retransmission packets are sent only by request. However, there are some hybrid solutions that might be used to merge the advantages of each of these solutions. An optimal equilibrium is dependent of specific binary error rates of the network being considered for the deployment of the solution.

Hybrid Solutions

Retransmission and Forward Error Correction (FEC) could be used in a way that they complement each other. When retransmission-based mechanisms are used together with FEC repair mechanisms, the same repair packet may be used by several receivers to recover from different lost packets. This enables a reduction in the average retransmission bit rate. Inversely, the introduction of retransmission mechanisms in FEC based mechanisms can be used to increase the loss recovery capability of some terminals that in some period of time could not recover from network losses due to receiving an insufficient amount of data and repair packets. The advantage of this solution stands in the fact that instead of sending an amount of FEC repair packets in accordance to highest losses experienced by some receivers, the server could adjust the number of FEC repair packets to meet the requirements of an average loss, filling the gaps of the other receivers using retransmissions.

Architectures and Solutions for QoE Assurance in Video Distribution

Currently, in IPTV, the video delivery is based in RTP/RTCP protocols. For large-scale environments, like IPTV, multicast is almost a necessity as routing mechanism. Regarding control mechanism RTCP is used as feedback mechanism. In particular, the receiver sends Receiver Reports (RR) with information about the quality of reception such as inter-arrival jitter, ratio of packet loss, round trip delay time, etc. These RR can also be extended to include some application dependant data (APP packet), where for instance, information about polling or usage data, can be added. Nevertheless, the communication has some shortcomings since it is based on unicast connections. Therefore, the feedback still puts some challenges that can be resolved by network engineering putting some feedback receivers on the edge of the network. Typically, the mechanisms described above are used for multicast only, since for unicast connections other approaches can be used, such as traffic shaping and QoS reservation mechanisms.

FUTURE RESEARCH DIRECTIONS

The definition of QoE metrics for video in transmission networks is still an open issue and several standardization entities are researching new models, to integrate as quality evaluation methods, in video distribution networks. However, the complex structure of compressed video combined with a great variety of parameters make this task more challenging when compared with other types of content, like voice for example. The varying nature of the transmission quality in wireless and mobile networks makes the understanding and the definition of normative frameworks for QoE even more imperative when compared with fixed and cable networks.

The heterogeneous nature of reception terminals and network conditions in wireless networks requires the development of appropriate video encoding solution. The H.264 Scalable Video Coding (ITU-T Recommendation H.264 and ISO/IEC 14496-10, 2007) has augmented the original H.264 encoder's functionality to generate several layers of quality. Enhancement layers may enhance the content represented by lower layers in terms of temporal resolution i.e. the frame rate, spatial resolution i.e. image size and the intrinsic quality of the video, specified as signal-to-noise ratio resolution.

By using H.264 SVC, different levels of quality could be transmitted efficiently over both wired and wireless networks (Song & Chen, 2007; Schierl et al., 2007), allowing seamless adaptation to available bandwidth and to the characteristics of the terminal. However, the transport of SVC presents many challenges (Monteiro et al., 2008) in terms of QoS, QoE assessment (Monteiro & Nunes, 2007) and resilience mechanisms which must be considered in order to take full advantage of its potential. Also, the decoding of SVC is very intensive computing task when compared to current video coding tools like MPEG-4 Part2 or even H.264/AVC MPEG-4 Part10, this means that the current generation of mobile devices cannot decode SVC in real time, and it is expectable that chip makers will be introducing optimized SVC decoders for the next generation of user terminals.

Meanwhile, as the both mobile TV and cabled IPTV services continue to mature, service providers are in need of an efficient way to configure, make diagnostics and manage the subscribers' end devices, whereas mobiles or set top boxes. Remote management (RM) frameworks are evolving to cover these requirements and are now moving from proprietary solutions to standards-based ones. Currently, the most widely used protocol standards are based on Broadband Forum's TR-069 (CPE WAN Management Protocol) (Broadband Forum TR-069, 2003) and TR-135 (Data Model for a TR-069-enabled STB) (Broadband Forum TR-135, 2003).

A remote end device management and diagnostics application is a strong requirement from service providers, as it brings higher operational efficiency and quality of service to its IPTV service. From a service provider point of view, it's an added value to be able to provide on-line assistance when a problem occurs on the subscriber's STB. On one hand, the problem is solved in a faster way providing a better quality of service to the subscriber and on the other hand the troubleshooting costs (operational costs) are lower to the Service Provider. Besides the benefits described above the usage of Remote Management Frameworks allows assessing the end devices' QoE and perform some configurations therefore some tuning could be done remotely to somehow overcome some conditions that can be degrading the quality. The range of parameters to be monitored through Remote Management Frameworks include information in the device level, information in the IP and transport layers and information in the service layer (e.g. IGMP, RTP, MPEG2 TS, packet lost, jitter). In regard to this, it should be noted that specific QoS and QoE related metrics for linear IPTV service are specified in (ATIS, 2007).

CONCLUSION

It was shown, that objective quality evaluation can be mainly divided into two groups, the FR metrics and the NR metrics. If the FR metrics are suitable whenever the reference content is present, that is not generally the case in IPTV deployments, except in the encoding step. Therefore, NR metrics play a crucial role by allowing not only defining the perceived video quality but also be incorporated in feedback mechanisms that allow for real time content adaptation that improves the global perceived video quality of the system and thus the overall QoE, helping to create more valuable IPTV and Mobile TV services. It is exactly in this context that FR and NR quality metrics are important, as they allow verifying the so called QoE, which is a more transversal parameter for video communication than QoS by itself. In spite of this, QoE is an inherently subjective component of a system and can not be easily measured or deterministically computed, it has to take into consideration highly subjective factors that for until very recently were not considered to be part of the telecommunications area of study.

Monitoring and QoE assessment when delivering multimedia contents, namely video services, is only part of the challenge to improve the service. When errors occur in the network, impairments are introduced in the delivered content and some approaches can be adopted such as the ones described before, i.e. the usage of FEC, retransmission or hybrid techniques. These allow mitigating or reducing the impacts of the impairments therefore actively contributing for improving the quality and until a certain extent also ensuring a constant or near-constant QoE level. This effort to improve or assure a certain level of QoE is fundamental for the success of any service, considering also those applications and services are becoming more quality-centric. On broader scope, this also allows exploring other paths, i.e. other that the pure technical challenges, moving forward with SLA for packet oriented networks. This is fundamental as the telecommunications players are migrating to all-IP networks and need to differentiate themselves from pure telecommunications providers where the performance of the network (QoS-based) was the main metric, towards service providers where the QoE of the service itself is the ultimate metric. This is, amongst other reasons, enforcing the usage of QoE as part of the engineering process when designing networks.

REFERENCES

ATIS. (2006). *ATIS-0800004, A Framework for Qos Metrics and Measurements Supporting IPTV Services*. Washington, DC: Alliance for Telecommunications Industry Solutions.

Bae, T. M., Thang, C. T., Kim, D. Y., Ro, Y. M., Kang, J. W., & Kim, J. G. (2006). Multiple Region-of-Interest Support in Scalable Video Coding. *Electronics and Telecommunications Research Institute Journal*, *28*(2), 239–242.

Broadband Forum TR-059. (2003). *DSL Evolution - Architecture Requirements for the Support of QoS-Enabled IP Services*. Retrieved from http://www.broadband-forum.org/technical/download/TR-059.pdf

Broadband Forum TR-069. (2003). *CPE WAN Management Protocol*. Retrieved from http://www.broadband-forum.org/technical/download/TR-069.pdf

Broadband Forum TR-126. (2006). *Triple-Play Services Quality of Experience (QoE) Requirements*. Retrieved from http://www.broadband-forum.org/technical/download/TR-126.pdf

Broadband Forum TR-135. (2003). *Data Model for TR-069 Enabled STB*. Retrieved from http://www.broadband-forum.org/technical/download/TR-135.pdf

Cranley, N., & Davis, M. (2007). An experimental investigation of IEEE 802.11e TXOP facility for real-time video streaming. In *Proceedings of Global Communications Conference* (pp. 2075-2080).

DVB Project DVB-HN A109. (2007). A109 *(TM3690r2), DVB-HN (Home Network) Reference Model Phase 1*. Retrieved from http://www.dvb.org/technology/standards/a109.tm3690r2.DVB-HN_ref_model.pdf

DVB Project DVB-HN CR. (2008). *CM639R2 DVB-HN Commercial Requirements Phase 1.* Retrieved from http://www.dvb.org/technology/standards/a132.Technical%20requirements%20for%20DVB-QoS.pdf

DVB Project HLTR. (2008, September). *High-level Technical Requirements for QoS for DVB Services in the Home Network DVB TM IPI.* Retrieved from http://www.dvb.org/technology/standards/a132.Technical%20requirements%20for%20DVB-QoS.pdf

Elias, P. (1955). Coding for two noisy channels. In *Proceedings of the 3rd London Symposium on Information Theory*, London, UK (pp. 61–76).

ETSI TS 102 034. (2007). *Transport of MPEG 2 Transport Stream (TS) Based DVB Services over IP Based Networks.* European Telecommunications Standardization Institute.

Furcht, B., et al. (2003). *The Handbook of Video Databases: Design and Applications*. Boca Raton, FL: CRC Press.

3GPP TS 26.346. (2005). *Multimedia Broadcast/Multicast Service (MBMS); Protocols and Codecs – Version 6.1.0.* Third Generation Partnership Project.

ITU-R. (2002). *Methodology for the Subjective Assessment of the Quality of Television Pictures* (Rec BT.500-11. Question ITU-R 211/11). Geneva, Switzerland.

ITU-R. (2004). Draft New Recommendation for Subjective Assessment of Streaming Multimedia Images by Nonexpert Viewers. (Document 6Q/57-E.) Source: EBU, Geneva, Switzerland.

ITU-T. (2004). *Objective Perceptual Video Quality Measurement Techniques for Digital Cable Television in the Presence of a Full Reference.* (Recommendation ITU-T J.144). Geneva, Switzerland.

ITU-T. (2005). *Advanced Video Coding for Generic Audiovisual Services*, v3, (Rec. H.264| ISO/IEC IS 14496-10 AVC). Geneva, Switzerland.

ITU-T Recommendation G.1070. (2007). *Opinion model for video-telephony applications.* International Telecommunication Union– Telecommunication.

ITU-T Recommendation G.107. (2008). *The E-Model, a computational model for use in transmission planning.* International Telecommunication Union– Telecommunication.

ITU-T Recommendation H.264 and ISO/IEC 14496-10. (2007). *Advanced Video Coding for Generic Audiovisual Services (MPEG-4 AVC).* International Telecommunication Union– Telecommunication and International Organization for Standardization/International Electrotechnical Commission, Joint Technical Committee 1. Version 7: April 2007.

ITU-T Recommendation J.148 (2003). *Requirements for an objective perceptual multimedia quality model.* International Telecommunication Union– Telecommunication.

ITU-T Recommendation J.144. (2004). *Objective perceptual video quality measurement techniques for digital cable television in the presence of a full reference.* International Telecommunication Union– Telecommunication.

ITU-T Recommendation P.862. (2001). *Perceptual evaluation of speech quality (PESQ): An objective method for end-to-end speech quality assessment of narrowband telephone networks and speech codecs.* International Telecommunication Union– Telecommunication.

ITU-T Recommendation P.564. (2007). *Conformance testing for voice over IP transmission quality assessment models.* International Telecommunication Union– Telecommunication.

ITU-T Recommendation Y.1541. (2006). *Network performance objectives for IP-based services*. International Telecommunication Union– Telecommunication.

ITU-T Recommendation Y.1541 Amendment 3. (2008). *Network performance objectives for IP-based services, Amendment 3: Revised Appendix VIII - Effects of IP network performance on digital television transmission QoS*. International Telecommunication Union– Telecommunication.

Kozamernik F., Sunna P., Wyckens E. and & Pettersen, D. I. (2005). Subjective quality of internet video codecs — phase II evaluations using SAMVIQ. *European Broadcasting Union (EBU) Technical Review*, 301.

Lee, K., Trong, S. T., Lee, B., & Kim, Y. (2008). QoS-Guaranteed IPTV Service Provisioning in IEEE 802.11e WLAN-based Home Network. In *Proceeding of Network Operations and Management Symposium Workshops*, Brazil (pp. 71-76).

Luby, M., Stockhammer, T., & Watson, M. (2008). Application Layer FEC In IPTV Services. *Institute of Electrical and Electronics Engineers Communications Magazine*, *46*(5), 94–101.

Monteiro, J. M., Calafate, C. T., & Nunes, M. S. (2008, September). Evaluation of the H.264 Scalable Video Coding in Error Prone IP Networks. *Institute of Electrical and Electronics Engineers Transactions on Broadcasting*, *54*(3), 652–659.

Monteiro, J. M., & Nunes, M. S. (2007, July). A Subjective Quality Estimation Tool for the Evaluation of Video Communication Systems. In *proceedings of IEEE Symposium on Computers and Communications (ISCC07), Mediawin Worshop*, Aveiro, Portugal (pp. 75-80).

Okamoto, J., et al. (2004). *An Objective Quality Assessment Method for Arbitrary Video Streams*. Picture Coding Symposium 2004, San Francisco, USA.

Ott, J., Wenger, S., Sato, N., Burmeister, C., & Rey, J. (2004). *Extended RTP Profile for Real-time Transport Control Protocol (RTCP)-Based Feedback* (RFC 4585). Internet Engineering Task Force.

Qiu, S., Rui, H., & Zhang, L. (2006). No-reference Perceptual Quality Assessment for Streaming Video Based on Simple End-to-end Network Measures. In *proceedings of the International Conference on Networking and Services*, California, USA (pp. 53–53).

Rey, J., Leon, D., Miyazaki, A., Varsa, V., & Hakenberg, R. (2005). *RTP Retransmission Payload Format* (RFC 4588). Internet Engineering Task Force.

Rosenberg, J., & Schulzrinne, H. (1999). *An RTP Payload Format for Generic Forward Error Correction* (RFC 2733). Internet Engineering Task Force.

Schierl, T., Stockhammer, T., & Wiegand, T. (2007, September). Mobile Video Transmission Using Scalable Video Coding. *Institute of Electrical and Electronics Engineers Transactions on Circuits and Systems for Video Technology*, *17*(9), 1204–1217.

Schulzrinne, H., Casner, S., Frederick, R., & Jacobson, V. (2003). *RTP: A transport protocol for real-time applications* (STD 0064, RFC 3550). Internet Engineering Task Force.

Shokrollahi, A. (2006). Raptor codes. *Institute of Electrical and Electronics Engineers Transactions on Information Theory*, *52*(6), 2551–2567.

Song, S., & Chen, C. W. (2007, September). Scalable H.264/AVC Video Transmission Over MIMO Wireless Systems with Adaptive Channel Selection Based on Partial Channel Information. *Institute of Electrical and Electronics Engineers Transactions on Circuits and Systems for Video Technology*, *17*(9), 1218–1226.

IEEE Std 802.11e. (2005). *IEEE Standard for Information Technology -Telecommunications and information exchange between systems –Local and metropolitan area networks – Specific requirements part 11: Wireless LAN Medium Access Control (MAC) and Physical Layer (PHY) specifications Amendment 8:medium access control(MAC) quality of service enhancements.*

Suzuki, T., Kutsuna, T., & Tasaka, T. (2008) QoE Estimation from MAC-Level QoS in Audio-Video Transmission with IEEE 8O2.lle EDCA. In *Proceedings of the Personal, Indoor and Mobile Radio Communications* (pp. 1-6).

Takahashi, A., Hands, D., & Barriac, V. (2008). Standardization Activities in the ITU for a QoE Assessment of IPTV. *Institute of Electrical and Electronics Engineers Communications Magazine*, *46*(2), 78–84.

VQEG. (2008). *The Video Quality Experts Group*. Retrieved on October 1, 2008, from http://www.vqeg.org

Wiegand, T. (2003). Rate- Constrained coder control and comparison of video coding standards. *IEEE Transactions on Circuits and Systems for Video Technology*, *13*(7).

Winkler, S. (2005). *Digital Video Quality: Vision Models and Metrics.* Hoboken, NJ: J. Wiley & Sons.

Winkler, S., & Mohandas, P. (2008). The Evolution of Video Quality Measurement: From PSNR to Hybrid Metrics. *Institute of Electrical and Electronics Engineers Transactions on Broadcasting*, *54*(3), 1–9.

Xiao, F. (2007). *DCT-based Video Quality Evaluation.* Final project for EE392J, Digital Video Processing, Stanford University.

Yamagishi, K., & Hayashi, T. (2008). Parametric Packet-Layer Model for Monitoring Video Quality of IPTV Services. In *IEEE International Conference on Communications 2008*, Beijing, China (pp. 110–114).

Chapter 17
Video Distortion Estimation and Content-Aware QoS Strategies for Video Streaming over Wireless Networks

Fulvio Babich
University of Trieste, Italy

Marco D'Orlando
University of Trieste, Italy

Francesca Vatta
University of Trieste, Italy

ABSTRACT

This chapter describes several advanced techniques for estimating the video distortion deriving from multiple video packet losses. It provides different usage scenarios, where the Peak Signal to Noise Ratio (PSNR) video metric may be used for improving the end user quality. The key idea of the presented applications is to effectively use the distortion information associated to each video packet. This allows one to perform optimal decisions in the selection of the more suitable packets to transmit. During the encoding process, the encoder estimates first the loss impact (for instance the amount of error propagation) of each packet. Afterwards, it generates side information as a "hint" for making video content aware transmission decisions. In this way, it is possible to define new scheduling schemes that give more priority to the packets with higher loss impact, and to assign fewer resources to the packets with lower loss impact. To this end, the usage of hint tracks, introduced in the MPEG-4 systems part, provides a syntactic means for storing scheduling information about media packets that significantly simplifies the operations of a streaming server. Moreover, the prioritization scheme may be used to minimize the overall error propagation under the delay constraint imposed by the video presentation deadline. The chapter also reviews recent research advances in the field of QoS mechanisms that adopt video specific metrics to improve the end user perceived quality.

DOI: 10.4018/978-1-61520-680-3.ch017

I. INTRODUCTION

The last decade has shown the pervasive expansion of several Information Technologies (ITs) including both a phenomenal growth in wireless communications and a revolution in multimedia technologies. Wireless networks have enabled a large variety of existing and emerging applications due to their low cost and flexible infrastructure. Several classes of different wireless technologies have been successfully deployed in different countries and different application scenarios. They are: Wide Area Networks (WANs), Local Area Networks (LANs), and Personal Area Networks (PANs). Cellular wireless networks belong to the class of WANs, Wireless LANs (WLANs), such as IEEE 802.11, and HIPERLAN belong to LANs, while Bluetooth, ZigBee, and Ultra Wide Bands (UWBs) belong to PANs. WANs offer greater mobility to the users, but lower data rates; LANs offer wider bandwidth, but limited coverage range; PAN technologies, instead, are usually deployed for cable substitution, and WLANs are generally used as the wireless replacement of the wired LANs. Wireless networks exhibit a large variation in channel conditions not only because of the different access technologies, but also due to channel impairments. These are mainly due to multipath fading, co-channel interference, noise, and so on, as well as competing traffic from other wireless users that share the same medium.

Besides, given the availability of wider bandwidth in wireless networks, the Internet multimedia applications are becoming more and more attractive to the mobile users. In fact, the Internet is becoming a truly multiservice network, in which infrastructure services requiring multimedia communications are emerging. These multimedia services range from voice over IP to video applications over IP. These include video conference, video surveillance, TV entertainment, and also interactive television (iTV), with services ranging from video-on-demand and interactive program guides to real-time shopping. Moreover, Internet-like data services, such as tele-medicine, tele-education, and tele-working services are becoming more and more popular.

Thus, as the possible use of these wireless networks spreads from simple data transfer to bandwidth intense, delay-sensitive, and loss-sensitive video applications, addressing Quality of Service (QoS) issues becomes crucial. To overcome channel impairments during the transmission, several different protection and adaptation strategies exist at different layers of the Open Systems Interconnection (OSI) protocol stack. Hence, to best understand how the user experience is influenced by these strategies, an evaluation of them is necessary.

From this point of view, this chapter discusses content aware QoS strategies for resource demanding services (including video applications). Here, the packet importance information is used as a criterion to improve the end user perceived quality. A detailed review of the recent advances in the research fields of content aware QoS strategies and scheduling techniques is presented. This allows a deeper comprehension of the aspects and trade-offs involved in the transmission of multimedia data over wireless networks. The motivation for this review is that most packet scheduling schemes, currently used in wireless networks, do not achieve the best possible quality for video transmission. This is due to the fact that they do not take the video content into consideration when making scheduling decisions. Although more complex, content-aware schemes could perform significantly better in terms of the end user perceived video distortion, and utilize more efficiently the available network resources. To make the best possible scheduling decisions, any such scheduling scheme should take into account the video encoding method, the channel conditions, as well as the decoding and error-resilience methods employed at the decoder.

II. CHAPTER OUTLINE

This chapter justifies the need of content aware scheduling techniques based on QoS strategies, and discusses some relevant issues. Moreover, it reviews several existing solutions proposed in the literature, and formalizes the main ideas for taking into account the packet importance to improve the end user perceived quality. A scheme for jointly estimating the video packet importance and selecting the more suitable packets to transmit is detailed. In particular, the prioritization of video packets, i.e., the determination of their importance, and, consequently, of their scheduling strategy, is based on the distortion of the decoded video stream with respect to the original one.

Content-aware scheduling techniques are important, since they not only lead to improved video performance over existing error prone networks, but they also provide valuable insights to design new algorithms and protocols for wireless multimedia systems. The approach may be twofold. On one hand, the prioritization scheme does not require the redesign or significant modifications of the existing transport and access protocols, and it can be directly applied with little modifications in the existing network architectures. This kind of prioritization schemes, based on rate-distortion optimized scheduling, belong to a variety of application-layer solutions that have been proposed to cope with the challenges raised by wireless networks. The raised challenges are mainly the available bandwidth and the link quality (due to multi-path fading, co-channel interference, and noise disturbances). The application-layer solutions include rate adaptation techniques, error resilience techniques, error concealment mechanisms, and joint source-channel coding. On the other hand, a different prioritization approach may be based on cross-layer design, which has been also widely investigated in the literature for wireless multimedia transmission. The results indicate that a significant gain in performance can be obtained, with respect to techniques considering a single layer optimization approach. It must be noted, however, that existing cross-layer solutions often overlook the important issue of packetization and its relationship to other protection strategies at various layers, as well as its impact on the rate distortion performance at the application-layer.

Finally, in a heterogeneous network including wireless links, packets may be lost due to channel impairments. Therefore, it is crucial to establish a relationship between packet losses and the distortion in the decoded video.

The chapter is organized as follows. After addressing, in Section III, the main issues related to the video transmission over wireless channels, in Section IV the multimedia characteristics and QoS requirements are presented, and in Section V the need for content-aware strategies is justified. Moreover, the QoS techniques, the packet scheduling algorithms, and the existing hinting mechanisms are reviewed in Sections VI, VII, and VIII, respectively. The complex problem of the loss distortion modeling is addressed in Section IX, in which the distortion estimation process and algorithms are described in detail. The performance measurements are addressed in Section X. Some practical application scenarios are proposed in Section XI, while, in Section XII, the implementation in a real test-bed, which uses the Distortion Estimation Algorithms (DEAs), is addressed. In Section XIII, another application scenario that may take advantage of DEAs, i.e., a multi-hop environment is described. Future research directions are discussed in Section XIV. Finally, Section XV summarizes and concludes the chapter.

III. ISSUES RELATED TO THE VIDEO TRANSMISSION OVER WIRELESS CHANNELS

Wireless networks are heterogeneous in bandwidth, reliability, and receiver device characteristics. In wireless channels, packets can be delayed

Figure 1. Lossy video transmission system

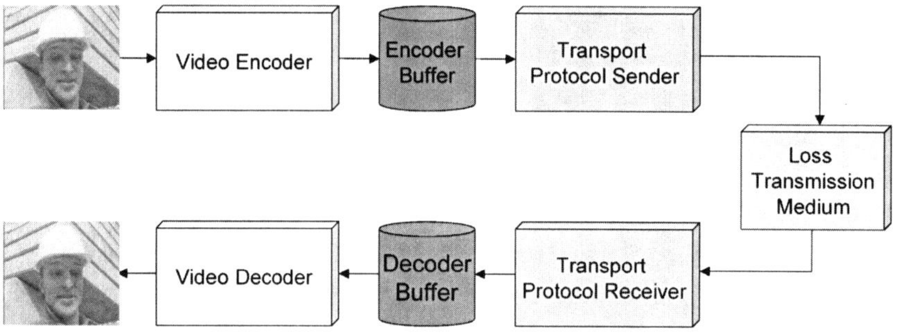

(due to queuing at different layers, for instance transport and Medium Access Control (MAC) layers, due to transmission or retransmission, and due to processing delays at the involved devices), lost, or also discarded, due to hardware limitations (for instance due to reduced display capabilities at the receiver end). For these reasons, the packet loss can reach the rate of 10% or more, and the delay and the resulting throughput for video bit streams can also vary in time significantly. This variability in the wireless resources has considerable consequences for video applications.

Video traffic may be classified into two main categories: interactive-video or videoconferencing (a "real-time" service in which alternate talking and listening is taking place) and streaming-video (both unicast and multicast).

The main components of a video communication system are shown in Figure 1. In the figure, it is possible to distinguish the encoder/decoder components, and a playout buffer, usually known as *de-jitter buffer*. This is introduced before the video decoder in order to compensate packet delays experienced during the network transmission. The playout buffer length allows one to trade between the rate of lost packets and the limiting delay, because, if the overall end-to-end delay increases, then the packet loss rate decreases, and vice versa.

A user may often experience an insufficient perceived quality, deriving from:

- The lack of some basic streaming requirements, such as wide bandwidth. Many Internet-like applications, for instance, High- Definition TV, require transmission bit rates of several Mbps. In this sense, video services are, in general, bandwidth demanding, both if encoded at a fixed or variable data rate. If the bandwidth required by the compressed video exceeds the channel capacity, packet losses will occur at the transport or access queues, causing distortions in the decoded data.
- Very stringent delay constraints: video packets must be available at the decoder before their deadline time to allow the reconstruction of an undistorted video. Delays less than 200 ms are needed for interactive like applications, such as videoconferencing, while video streaming applications tolerate delays until 5 s. Packets that reach the destination after their display time (deadline) are discarded at the receiver side or, if the receiver has significant computational resources, can be used only for supporting video concealment (Ghanbari, 1996).

Fortunately, video applications are elastic and may tolerate a certain amount of packet loss depending on several factors, such as the used sequence characteristics, the adopted compression scheme, and the error concealment strategy available at the receiver. For example, a streaming session through a channel with a typical packet loss rate of 5%, or less, may tolerate this loss without affecting the quality perceived by the user. This characteristic makes interactive like applications and streaming applications different with respect to traditional applications such as file transfer. Data networking, such as a file transfer, may be not time constrained, being the correct reception of the whole data set the mandatory requirement. E-mail, documents, pictures and all sorts of downloads are examples of data types that are transferred based on accuracy and not time.

For multimedia networking, instead, time is the most important aspect of the delivery together with the packet loss rate, which should be kept as low as possible.

As said above, delivery delay, called latency, must be kept below about 200 ms for interactive applications, such as videoconferencing. This implies that a challenge for the whole system design, including sender and receiver nodes, is to reduce the packet loss rate without increasing dramatically the end-to-end delay. For instance desirable strategies should rely on low delay mechanisms, such as Forward Error Correction (FEC). Conversely, high-delay mechanisms, such as end-to-end retransmissions, should be avoided.

For streaming video applications, instead, delays on the network can be in the less-than 5 s range, since streaming video is not perceived to be a real-time activity. This is due to the fact that memory buffering is used to collect packets and frames of streaming media to swamp out the latency effect. The delay caused by buffering is not perceived by the recipient because there is no reference of comparison available. However, the buffer is flushed out at regular intervals and latency must be less than the refresh interval, or video and audio will be disrupted. Since, for these applications, an end-to-end delay of a few seconds is acceptable, the effect of packet losses or packet errors may be mitigated using efficient Automatic Repeat reQuest (ARQ) techniques combined with FEC schemes (as described, e.g., in (Majumdar, Sachs, Kozintsev, Ramchandran, & Yeung, 2002) and (Yong, Guangwei, Peng, Hang, & Junyuan, 2008)).

As to summarize, multimedia networking requires sufficient available bandwidth, a very low bit error rate, low latency (with the help of buffering), and priority access to ensure quality of service. These requirements can be reduced to just two main elements: channel quality and priority access. Channel quality includes bandwidth, low bit error rate and low latency. Priority access stands alone as perhaps one of the most important QoS requirements and will be addressed in the next section.

IV. MULTIMEDIA CHARACTERISTICS AND QUALITY OF SERVICE REQUIREMENTS

As discussed in the previous section, compared to traditional data traffic, video traffic places different demands on QoS in a network, particularly in terms of delay, delay variation, and data loss. These are channel quality requirements. Another important QoS requirement, which stands alone, is the so called *priority access*.

In multimedia applications, the content of a packet is critical in determining the packet importance and, thus, its priority. In this sense, a content-aware utility function must be proposed, that accounts for the dependencies between video packets. Besides, it must also account for the effect that each video packet has on the final quality of the received video. This leads to a distortion aware scheduling scheme for packet based transmissions. In other words, the packets in the transmission buffer must be sorted on the basis of the contribution of each packet to the overall video quality.

To enhance the perceived user quality or to best use the network resources it is desirable that a video application interacts with the lower layers, which should take into account the stringent QoS requirements. To this end, the application may communicate some useful information, such as the packet importance, or timing issues related to the presentation deadline. This allows one to derive new strategies able to maximize the video performance in terms of end user satisfaction.

Thus, to achieve a high user quality several issues need to be investigated. In particular these issues can be classified as follows:

- **Easy adaptability** to wireless bandwidth fluctuations due to interference, multipath fading, mobility, competing traffic, etc.;
- **Robustness** to partial data losses caused by the fragmentation of video frames by lower layers and loss of some frames in the channel.

One possible way for addressing these issues is to adapt existing algorithms and protocols, at the lower layers of the protocol stack, to better support multimedia transmission. Conversely, application layer solutions may be modified to follow the wireless networks fluctuations. Another possibility is represented by the so-called cross-layer strategies that optimize the packetization, prioritization, and retransmission policies, considering jointly different levels of the protocol stack (for instance, deploying a joint application-layer packetization and a MAC-layer retransmission strategy). This can be done on the basis of the content characteristics, the channel conditions, and the specific features of the deployed video coder.

In video streaming applications, the non stringent requirements of the delay allow the transmitter to perform complex content-aware optimizations. For instance, advance rate-distortion optimized packet scheduling and retransmissions techniques, such as in (Majumdar, Sachs, Kozintsev, Ramchandran, & Yeung, 2002), are based on the importance of the video packets from the perspective of the end user. Recently, the advances in this research field have been shown to be able to optimize both network resources and end user quality. The base work on this subject is the one by Chou and Miao (Chou & Miao, 2006), considering the so-called Rate-Distortion Optimized (RaDiO) streaming. In (Chou & Miao, 2006), the authors consider the streaming as a stochastic process, with the goal of determining both which packets to send and when to send them. The constraint is that of minimizing the reconstructed distortion at the client for a given average transmission rate. The basic scenario considers a media streaming server that has stored a compressed video stream that has been packetized into data units. Each data unit has a size in bytes B_l and a deadline by which it must arrive at the client to be useful for decoding. The importance of each data unit is captured by its distortion reduction ΔD_l, a value representing the decrease in distortion that derives from the data unit decoding. The distortion is often expressed as Mean Squared Error (MSE). Observe that the distortion resulting from decoding a data unit may depend on the availability of other data units (ancestors).

The RaDiO framework can be used to choose the optimal set of data units to transmit at successive transmission opportunities. These transmission opportunities are assumed to occur at regular time intervals that depend on the available channel bandwidth. Because of decoding dependencies among data units, the importance of transmitting a packet at a given transmission opportunity often depends on which packets will be transmitted in the future. Therefore, the scheduler makes transmission decisions based on an optimized plan that may anticipate later transmissions. For simplicity, and in order to keep the system simple, only a finite number of data units participate in the optimization process (Chou & Miao, 2006).

A. Issues Related to QoS Evaluation

The evaluation of the quality as perceived by the user is usually performed to measure the performance of a video communication system. User satisfaction is, in fact, the right metric to evaluate a complete video service system. Given the different importance of the various elements in a scene involved in a video session, network based metrics such as throughput, packet loss rate, and delay cannot be considered suitable parameters to measure the perceived quality.

Subjective experiments are the best assessment technique to achieve reliable perceptual quality indications for a video communication system. However, the need for many different subjects, the time required to perform the experiments, and the need for reproducing the experiments make them difficult to employ. A number of procedures have been standardized for various media types (see (ITU-T, 1996), (ITU-T, 1999), and (ITU-T, 2000), for instance).

To overcome the costs and the complexity associated to the subjective tests, algorithms that compute objective quality measures have been proposed in the research community. For video, the MSE and the PSNR are the most used objective measures, given their limited computational complexity, despite their strong limitations. Actually, MSE and PSNR do not correlate well with the perceived quality, quantified through the Mean Opinion Score (MOS) values, especially when low-quality video signals are involved (Ong, Yang, Lin, Lu, & Yao, 2004). However, objective quality measures can be very useful to compare different solutions and algorithms. Moreover, these measures are used in rate-distortion optimized communication systems to select intra or inter coding modes or to decide which packet to transmit. In particular, rate-distortion optimized communication systems are based on the optimization of the expected distortion (which may use MSE or PSNR as distortion metrics even if any other mathematically tractable distortion metric will generally work). Given a packet loss probability of p_i, the expected distortion for packet i can be expressed as (Pahalawatta, & Katsaggelos A. 2007):

$$E\left\{D_i\right\} = \left(1 - p_i\right) \cdot E\left\{D_i^R\right\} + p_i \cdot E\left\{D_i^L\right\}, \quad (1)$$

where $E\{D_i^R\}$ is the expected distortion when the data packet i is received and $E\{D_i^L\}$ is the expected distortion when the packet i is lost.

Taking into account the packet interdependence, the contribution to the total distortion of a video packet may be non zero even if the packet itself has not been lost, but it depends on a packet that has been lost previously. Thus, the total distortion of a video packet sequence may be expressed by a function $f(\cdot)$ of all the single expected distortions (1) (Pahalawatta, & Katsaggelos A. 2007):

$$D_{\mathrm{TOT}} = f\left(E\left\{D_{i=[1,2,\cdots,N]}\right\}\right), \quad (2)$$

where N is the total number of video packets which are transmitted. It should be noticed that D_{TOT} takes into account the source distortion due to the video co-decoding algorithm as well as the channel distortion due to random channel errors.

V. NEED FOR CONTENT-AWARE STRATEGIES

In recent years, to address each of the aforementioned requirements, the research community has focused on adapting existing algorithms and protocols for multimedia compression and transmission over the scarce resources of wireless networks (Girod, Kalman, Liang, & Zhang, 2002). For instance, network adaptive video compression, bandwidth, and channel dependent bit rate adaptation, prioritization and layering mechanisms, error concealment strategies, rate–distortion modeling, streaming strategies, distortion and channel aware scheduling, link layer adaptation have been proposed and developed in real test beds. However, these solutions are unable to react to the limitations imposed by the wireless networks, when the interference is high, or when the stations are mobile. One reason may be that the resource management, adaptation, and protection strategies available in the lower layers of the OSI stack, usually at Physical (PHY) layer, MAC layer, and Network/Transport layers, are optimized without considering the specific characteristics of the video applications. Conversely, video applications, such as the compression mechanisms, the streaming algorithms, and so on, do not take into account the QoS mechanisms provided by the lower layers for error protection, scheduling, and resource management (Girod, Kalman, Liang, & Zhang, 2002). A system for a video application, that is designed without taking into account the multimedia transfer requirements, may lead to a simpler implementation, but usually may offer an unsatisfactory performance (in terms of objective and perceived quality) when packets are streamed over a limited network. Instead, significant improvements may be achieved in the same scenarios using low-complex techniques that consider the unique features of the video application. Video QoS, in fact, is not simply determined by the packet throughput at the receiver, but also by the video content, because the video compression algorithms are lossy, and spatial and temporal error concealment strategies of lost packets are implemented at the receiver. Consequently, in order to use efficiently the limited resources of the wireless networks for video transmission, a content-aware scheduling technique must be employed. A content-aware scheduling strategy aims at providing a method for assessing the importance (priority) of video packets. Using this method, the packets can be ordered on the basis of their contribution to the reduction of the expected distortion at the receiver. To this end, a lot of work has been dedicated in the literature to the accurate estimation of Equations (1) and (2).

For instance, in (Stockhammer, Wiegand, & Wegner, 2002) authors simulate multiple independent packet losses, compute the resulting distortions (1), and calculate the expectation (2). This method gains in accuracy at the cost of increasing the simulations number, the computation time, and the storage occupation. Another example can be found in (Wu, Hou, Li, Zhu, Zhang, & Chao, 2000), where a recursive algorithm to calculate the end-to-end distortion is presented. The distortion in (Wu, Hou, Li, Zhu, Zhang, & Chao, 2000) is calculated in terms of the expected mean absolute difference (MAD) of pixels.

An efficient algorithm that calculates the distortion in terms of MSE can be found initially in (Zhang, Regunathan, & Rose, 2000) and more recently in (Hua, 2007). It is called Recursive Optimal Per-pixel Estimate (ROPE) and it implements a recursive computation of the first and second order moments of the expected distortion. In this way, both the storage needs and the computational complexity are significantly reduced.

Some further improvements, with respect to the original ROPE algorithm, have been presented in (Leontaris, & Cosman, 2003), where, in addition to the first and second order moments of the expected distortion, the cross-correlation terms of the expected distortion are accurately estimated as well. Finally, further methods, which may be useful for expected distortion estimation, are presented in (Schmidt, & Rose, 2007) and (Schmidt, & Rose, 2008).

VI. REVIEW OF QOS TECHNIQUES

Several solutions have been proposed in the research community for enhancing the performance of multimedia over wireless networks. These solutions have been proposed at each layer of the protocol stack (see papers (Frossard, & Verscheure, 2001), (Li, Xu, Nahrstedt, & Liu, 1998) (Campbell, & Coulson, 1997) (Li, & van der Schaar, 2004) for a detailed review on the topics involved in each solution). For instance, the IEEE 802.11e standard (IEEE 802.11e/

D5.0 (2003)) has adopted an admission-control mechanism through which video applications can reserve time for transmitting their packets during each service interval. The reservation is performed statically. In practice, prior to the actual transmission, each packet is classified and the application assigns a category or a priority to the packet by declaring the multimedia traffic specification (TSPEC) parameter. This allocation strategy guarantees that the resources are divided among each of the wireless transmitters based on their selected parameters. Similarly, international telecommunications standardization committees (Telcordia), as well as existing overlay network infrastructures (Rejaie, Handley, & Estrin, 2000), (Cui, & Nahrstedt, 2003), enable application program interfaces to negotiate the needed QoS with the network using some reservation protocols such as, for instance, (Braden, Zhang, Berson, Herzog, & Jamin, 1997). However, it is important to note that these QoS negotiations are mainly designed for home entertainment and are performed only one time, prior to the stream transmission. For this reason, they do not take into account both the nature of the channel, which is unreliable, has time-varying characteristics, and the resource of which are strictly limited, and the nature of the video content, which is highly variable in bit rate, due to the fluctuations in video content and to the "intra" or "inter" coding modes.

To take into account both the nature of the channel and that of the video content, a cross-layer design may be useful. In this design, interdependencies between layers are implemented through the exchange of the appropriate information between layers, while building the appropriate amount of robustness into each layer. For example, routing protocols can avoid links experiencing deep fades, or the application layer can adapt its transmission rate based on the underlying network throughput and latency.

VII. REVIEW OF PACKET SCHEDULING TECHNIQUES

Packet-scheduling algorithms proposed in (Chou, & Miao, 2006), (Cheung, & Tan, 2006), (Chakareski, & Frossard 2005), (Miao, & Ortega, 2002) for video streaming applications were developed with the objective to optimize the rate-distortion (R-D) performance given the time-varying channel conditions and the video characteristics. In the context of these studies, packet scheduling is an optimization process in which packets are selected for first transmission or retransmission to minimize the end user distortion. A review and a comprehensive analysis of the R-D optimization process (also called RaDiO) via packet scheduling is accurately described in (Chou, & Miao, 2001). The very high complexity of this technique, which limits its applicability for real-time streaming, motivated successive studies (Miao, & Ortega, 2002), (Chakareski, & Frossard 2005), and (Cheung, & Tan, 2006), that propose low-complexity methods that may be applied in real-time streaming scenarios.

Other research activities propose network adaptation performed at the application layer through transcoding techniques or at the network layer by intelligent dropping strategies and packet marking techniques or QoS mapping. A distortion-based approach to packet marking is presented, e.g., in (De Martin, & Quaglia, 2001), where packets containing video data are individually examined and marked depending on the estimated distortion that their loss would introduce at the decoder and the desired level of perceptual QoS. To maximize perceptual QoS, the packets marked as "premium" are the most perceptually relevant. In (Zhai, Luna, Eisenberg, Pappas, Berry, & Katsaggelos, 2003) a similar approach is considered, with the aim of minimizing the end-to-end distortion while respecting cost constraints, or, alternatively, of minimizing the overall cost given the end-to-end distortion constraints.

Figure 2. MPEG-4 (.mp4) file format for a video-media stream

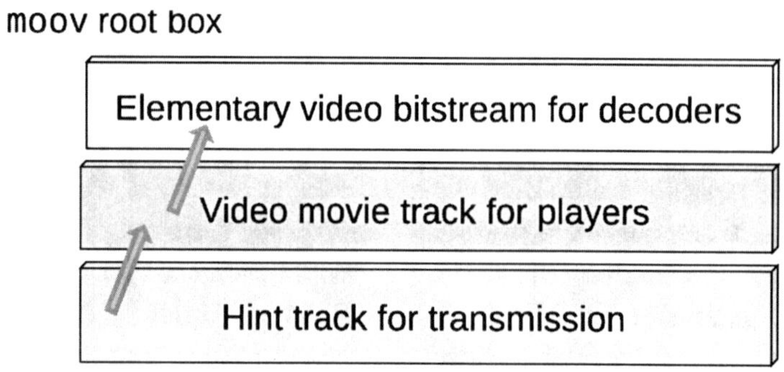

VIII. HINTING INFORMATION IN A PACKET

Video bitstreams can be created and stored for transmission using a file format such as specified in the standard MPEG-4 file format (Singer, Belknap, & Franceschini, 2001). Streaming may be supported by hint tracks, which are sets of structured metadata derived from the compressed bitstreams. A hint track contains information on packet payload offsets, sizes, protocol specific settings, and also packet deadline and, therefore, can significantly reduce the complexity involved in packetization and scheduling strategies during transmission. Hence, using hint track information, advanced packetization and scheduling algorithms can be deployed. However, existing hinting mechanisms do not specify the possibility to include distortion based information associated to each packet. This feature is necessary especially when the packetization and scheduling algorithms to deploy should take into account the packet importance. The hinting track concept is useful for wireless video transmission because it enables in particular:

- Real-time adaptation of the packet sizes at transmission time, after the encoding process;
- Real-time selection and prioritization of different portions of the bitstream (for instance different packets) based on the distortion impacts;
- Real-time selection of the packets based on delay constraints;
- Real-time optimized scheduling of video packets based on their deadline and the transmission of the previous packets.

A. MPEG-4 Hint Track and Proposed Integrations

An MPEG-4 format (.mp4) contains timed media information for multimedia presentation in streaming scenarios and storage applications. The format is deliberately designed with high flexibility and extensibility in order to allow management, editing, and presentation of the general media data.

The standard file format has a hierarchical structure and the basic building blocks adopted in the construction of the .mp4 files are called boxes. In particular, a box is a data structure that contains a certain type of media data, for instance a slice. Each box has a type name, reflecting the type of data contained. In addition, a box can contain other boxes related to the main one. The general structure of an .mp4 file format for streaming is shown in Figure 2. Normally, an .mp4 file starts with a root box called *moov*. The *moov* box contains other further boxes such as boxes for storing elementary bitstreams, boxes for storing synchronization information (also called "movie tracks"), and boxes for storing hints used by the streaming server to generate packets out of the elementary bitstreams (these boxes are called "hint tracks").

An .mp4 file can be viewed as a structure containing elementary bitstreams generated by encoders, movie tracks to guide the video player for local playback, and hint tracks for streaming the media over packet-based networks. The movie tracks contain information (timing and data pointers) that a player will use to extract the corresponding media data for presentation at the designated time. Hint tracks instead contain information such as timing and data for packet headers. Two later studies, (Chakareski, Apostolopoulos, Wee, Tan, & Girod, 2005) and (Van der Schaar, & Andreopoulos, 2005), have also proposed to use hint tracks for adaptive QoS streaming. In particular, in the seminal work proposed in (Chakareski, Apostolopoulos, Wee, Tan, & Girod, 2005), the use of Rate Distortion (R-D) hint track is recommended to store "pre-computed" characteristics of compressed media, so that the complexity of an optimization process at runtime, i.e., the so-called Rate-Distortion Optimized (RaDiO) streaming, can be significantly reduced. As anticipated in Section IV, the base works on this subject are the ones by Chou and Miao (Chou, & Miao, 2001), (Chou, & Miao, 2006) considering the so-called RaDiO streaming. The plan that controls the data unit transmissions is called transmission policy, π.

Assuming a time horizon of N transmission opportunities, π represents a set of length-N binary vectors π_l, with one vector for each data unit l. In this representation, the N binary elements of π_l indicate whether the data unit l will be transmitted at each of the next N transmission opportunities. The policy needs to take into account future acknowledgments that might arrive from the client to indicate that the packet has been received. Each transmission policy leads to its error probability, $\varepsilon(\pi_l)$, defined as the probability that data unit l arrives at the client side. Each policy is also associated to an expected number of times that the packet is transmitted under the policy π, called $\rho(\pi_l)$. The goal of the scheduler is to find a transmission policy π with the best tradeoff between expected transmission rate and expected distortion. The scheduler re-optimizes the entire policy π at each transmission opportunity to take into account new information since the previous transmission opportunity, and then executes the optimal π for the current time. An exhaustive search to find the optimal π is not generally tractable in terms of computational complexity. The search space grows exponentially with the number of considered data units, M, and the length of the policy vector, N, as explained in (Podolsky, McCanne, & Vetterli, 1998). Even though rates and distortion reductions are assumed to be additive (note that this assumption is valid only when losses are separated so that burst losses are not accounted), the exhaustive search would have to consider all 2^{MN} possible policies. Chou and Miao's RaDiO framework (Chou, & Miao, 2006) overcomes this problem by using an iterative algorithm. Their Iterative Sensitivity Adjustment (ISA) algorithm minimizes a Lagrangian cost function (Chou, & Miao, 2006) with respect to the policy π_l of one data unit while the transmission policies of other data units are fixed. Data unit policies are optimized one at a time until the Lagrangian cost function converges to a (local) minimum.

This basic algorithm has, as principal limitation, the computational complexity. To overcome this problem, several techniques have been proposed. Chou and Sehgal, for instance, have presented simplified methods to compute approximately optimized policies in (Chou, & Sehgal, 2002). An attractive alternative to ISA is a randomized algorithm, recently developed in (Setton, Noh, & Girod., 2006), (Setton, & Girod., 2008), in which heuristically and randomly generated candidate policies are compared at each transmission opportunity. The best policy from the previous transmission opportunity is one of the candidates, and thus past computations are efficiently reused. With a performance similar to ISA, the randomized algorithm usually requires much less computation.

Despite the enormous literature contributions in developing a solution for the RaDiO framework, the computational complexity associated to the algorithm that determines the optimal policy vector remains the main limiting factor, especially for a wireless router with limited CPU capabilities. To cope with this drawback, in (Babich, Comisso, D'Orlando, & Vatta, 2006), (Babich, D'Orlando, & Vatta, 2008), (Babich, D'Orlando, & Vatta, 2008b) the authors proposed three different video Distortion Estimation Algorithms (DEAs). These algorithms are able to estimate, at the encoder side, the end user distortion, taking into account the actual decoder behavior, the inter-frame error propagation and the loss pattern deriving from the transmission over the network. Compared to previous solutions, DEAs do not rely on the knowledge of future network behavior and require a low computational burden to perform the estimation. These features may make all these algorithms suitable for the implementation in real systems, urging the researchers to explore their behavior in a practical wireless scenario.

IX. LOSS DISTORTION MODELING

In order to accurately estimate video quality, it is important to find a relationship between the packet losses and the distortion in the decoded video. In this context, it is desirable to find a model able to estimate accurately the channel induced distortion. This is a challenging task because of the difficulties in mapping the loss probability into the received quality. First, it has to be assessed whether the expected distortion depends only on the average packet loss rate, or whether it also depends on the specific loss pattern. Many research activities implicitly assume that loss burst length does not matter, focusing on the average packet loss rate as the most important feature. For instance, in (Liang, Apostolopoulos, & Girod, 2008), the authors carefully analyze the distortion due to a single frame loss, accounting for error propagation, intra refresh and spatial filtering. Their model considers the effect of multiple losses as the superposition of multiple independent losses. With this linear or additive model, the expected distortion is proportional to the average packet loss rate. Using this approach it is possible to efficiently model the loss only when lost frames are sufficiently spaced. In many important applications, such as in low bit-rate video communication over a wireless link, each coded frame may fit within a single packet. In these cases, the losses may be bursty and may result in the loss of multiple frames. In (Liang, Apostolopoulos, & Girod, 2003), the authors show that, in general, longer bursts lead to larger distortions. However, there is still a noticeable difference between their model and the reconstructed quality. A full analytical model for the distortion is proposed in (Choi, Ivrlac, Steinbach, & Nossek, 2005), where it is assumed that in a Group Of Pictures (GOP) all frames after the lost frame are not available at the decoder. It follows that some subsequent frames that may be used for improving the video quality are not used by the decoder. To best understand the problem of distortion estimation, the following paragraph details with an example the model proposed in (Choi, Ivrlac, Steinbach, & Nossek, 2005).

A. Analytical Model for Estimating the Video Distortion

The *Foreman* video sequence is encoded using the H.264 video encoder with an encoding structure with one intra-frame (I-frame) followed by N_i-1 inter-frames (P-frames). This encoding structure is susceptible to error propagation, due to inter-frame dependencies, which is introduced by predictive coding. The concealment technique conceived by the authors consists in replacing each incorrect received frame and all subsequent ones in the GOP

Figure 3. Frame per frame distortion of the first GOP of Foreman sequence due to lost frames starting from position i

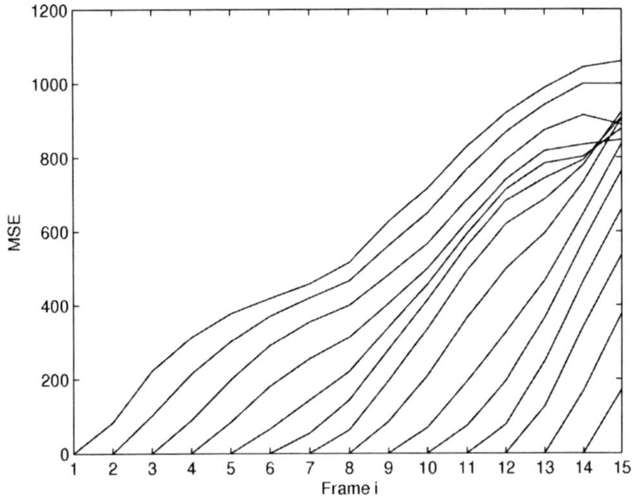

by the most recently correct received frame. This model is called previous frame error concealment, and is adopted to model the decoder behavior in presence of frame losses. Using this technique, if the *i*-th frame is the first frame lost in a GOP, then the *i* -th frame and all its dependents in the GOP are replaced by the (*i* -1)-th frame. Figure 3 shows the distortion D_i, measured using the MSE, that is obtained from the video test sequence *Foreman* encoded using H.26L/AVC source encoder (*N*=15 frames per GOP).

Each curve in the graph represents the picture quality corresponding to the loss of all frames starting from the frames indicated in abscissa. For example, consider the curve having MSE=0 up to *i*=7. All frames from *i*=1 to *i*=7 are received while all frames from *i*=8 to *i*=15 are lost (they are not used by the decoder). The MSE is zero for the first seven pictures and increases for *i*>7. This family of curves is used in the model to estimate the end distortion. It is worth noticing that all curves can be evaluated at the encoder side using only the frames stored in the encoder, without the need of the decoder. In fact each curve represents the Mean Square Difference (MSD) between the last correctly received frame and the current one. Observe that the decoders are able to improve picture quality also in presence of whole frame losses using the error mitigation techniques. For example, Figure 4 shows the performance of the decoder measured using MSE. The solid curve represents the behavior of the decoder when all frames, except the first, in the first GOP are not used by the decoder (the same curve is shown in Figure 3 for *i*=2). The dotted curves represent the video output when the decoder applies the Frame Copy (FC) concealment. For example, the first curve from the bottom represents the received quality when only the third frame is discarded. The curve immediately above is relative to the loss of exactly the third and the fourth frames. It is interesting to note that all curves present the same behavior: the information of the available frames is used to improve the output picture quality.

B. Encoder Internal Operations

Starting from the real behavior of the decoder described above, a new distortion model is presented in this section. An example clarifies how the estimation process works. The distortion is estimated considering the loss pattern effects

Figure 4. Decoder output performance due to burst losses in the first GOP. Solid curve: decoder output due to complete burst discarding. Dotted curves: decoder output due to partial burst discarding.

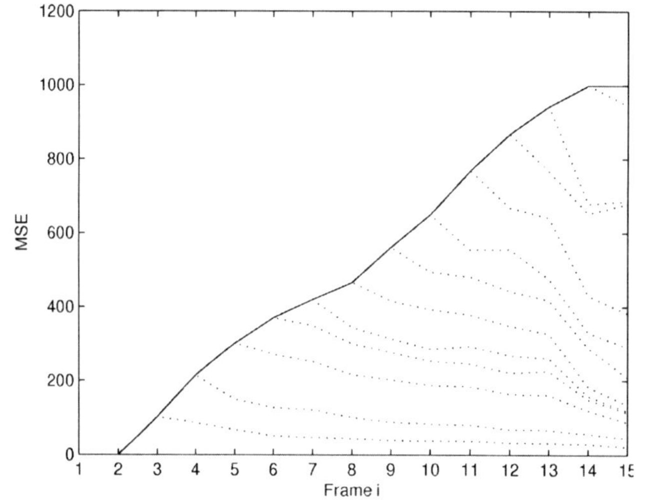

(including burst losses) and inter-frame error propagation. Compared to previous works, the presented model provides an accurate estimation of the channel induced distortion resulting from different loss events.

In order to explain better the modifications added to the encoder and the estimation algorithms internal operations, the following sections consider the sequence Quarter Common Intermediate Format (QCIF) *Foreman* as a reference for the described experiments. The sequence is encoded following the JVT H.264/AVC standard (and in particular using the reference software JM98, where the above mentioned modifications have been applied).

During the encoding procedure, two main outputs are provided by the encoder, as it is shown in Figure 5:

- The compressed bitstream
- A reference uncompressed sequence

Figure 5. Main outputs provided by the modified encoder

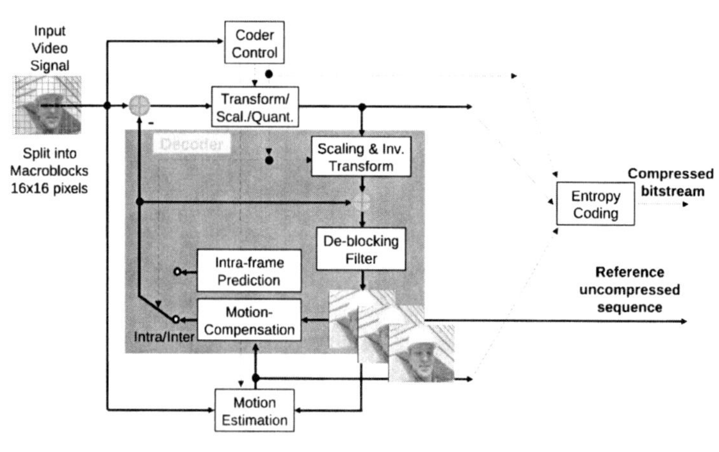

The compressed bitstream is packetized by the network transport layer, and then is transmitted over the channel. The reference sequence, instead, is used by the encoder to perform internal operations such as motion prediction, motion compensation as well as rate control. The uncompressed sequence, stored in the internal encoder buffer, is used extensively to perform the distortion estimation.

C. Distortion Estimation Process: Hypotheses

In the following it is assumed that the encoder is able to evaluate two amplitude arrays $A^1_{i,l}$ and $A^2_{i,l}$, with $i \geq l > 2$, defined by:

$$\begin{aligned} A^1_{t,l} &= MSD\left[f_t - f_{l-1}\right] \\ A^2_{t,l} &= MSD\left[f_t - f_{l-2}\right] \end{aligned}, \qquad (3)$$

where MSD[f_i - fl-1] represents the Mean Square Difference (MSD) between frames f_i and fl-1. The defined arrays give an estimate of the channel loss induced distortion, as explained in the following. More precisely, the MSD is used to estimate the actual MSE at the decoder. Assume that a single frame fits in a single packet. This simplifying hypothesis, which avoids one to distinguish between frames and packets, does not affect the generality of the presented estimation method. It follows that a single channel packet loss produces the loss of the whole frame. Observe that MSE takes into account the distortion introduced by the channel only, and the source compression distortion is neglected. In fact, the intent of the employed algorithms is to model the channel distortion. In the sequel it is shown that this assumption is acceptable in most of the channel conditions examined.

Let us examine the encoder operations. First, the bitstream is converted into prediction information and transform coefficients, allowing the reconstruction of the current frame for internal rate control operations. To do this, every uncompressed frame is forwarded also in a reference frame buffer, giving the chance to allow the prediction of the next frame inside the motion compensation process. The bitstream is then packetized, and the obtained packet is transmitted by the network transport layer. In case of successful reception, the packet is forwarded directly to the decoder for the decoding operations whereas, if the packet is loss, the simplest operation the decoder can perform is just to skip the decoding and not update the display buffer. In this case, the user will immediately recognize the loss, as the fluent motion and continuous display update is not maintained. However, this is not the only problem: not only the display buffer is not updated, but also the reference decoder buffer has a picture gap. Even in case of successful reception of the next packet, the corresponding decoded frame will differ from the reconstructed frame at the encoder side, given that the encoder and the decoder are referring to a different reference signal while decoding this packet. Therefore, the loss of a single packet has also effects on the quality of the subsequent frames.

D. Channel Distortion Estimation Algorithms

In order to allow the estimation algorithms to operate, assume that the application is able to determine the actual sequence loss pattern by exchanging some suitable signaling information with the lower layers, which adopt, for instance, an acknowledgment based transmission technique.

Assume that the following loss pattern occurs in a GOP: frames from index l_i to index i, and frames from index l_j to index j are lost, being $i < l_j - 1$. The considered scenario, based on two non overlapped bursts, may be extended to a generic number of bursts.

1. Step Distortion Algorithm

The first estimation algorithm is called Step Distortion Algorithm (SDA) and is able to estimate the distortion at the k-th frame in a GOP using the following method:

$$D(k) = \begin{cases} 0 & k < l_i \\ A^1_{k,l_i} & l_i \le k < i \\ A^1_{i,l_i} & i \le k < l_j \\ A^1_{i,l_j} & l_j \le k < j \\ A^1_{j,l_j} & k \ge j \end{cases}, \tag{4}$$

The SDA algorithm approximates the distortion envelope using a simple step function, and it is completely defined by the amplitudes at each time step k. In this way, the SDA assumes that the decoder is able at least to keep constant the distortion. Differently from the work in (Choi, Ivrlac, Steinbach, & Nossek, 2005), SDA assumes that the decoder has activated some error recovery techniques such as Frame Copy (FC) or Motion Copy (MC) error concealment.

2. Exponential Distortion Algorithm

The Exponential Distortion Algorithm (EDA) reproduces more accurately than SDA the distortion envelope caused by isolated losses.

Consider that, when a loss appears, the distortion ramps up in correspondence of the missed frame, since the decoder applies some recovery techniques to alleviate the visual effect. Due to error propagation, which is caused by predictive coding, the MSE associated with subsequent frames exhibits a nonzero value. More precisely, the distortion decreases as a consequence of the spatial filtering and the intra refresh until it eventually becomes zero at frames sufficiently apart from the lost one.

To take into account this amplitude decay effect, the EDA models the distortion at the k-th frame as follows:

$$D(k) = \begin{cases} 0 & k < l_i \\ A^1_{k,l_i} & l_i \le k < i \\ A^1_{i,l_i} e^{-b(k-i)} & i \le k < l_j \\ A^1_{i,l_j} & l_j \le k < j \\ A^1_{j,l_j} e^{-b(k-j)} & k \ge j \end{cases}, \tag{5}$$

where the parameter b is introduced to shape the error propagation effect. In particular, b can be split into two different parts, corresponding to the separate contributions due to the encoder and the decoder operations:

$$b = b_{enc} + b_{dec}. \tag{6}$$

From the encoder point of view, b_{enc} depends on the intra coded macroblock ratio, on the rate-control algorithm, on the number of reference frames stored in the encoder buffer to perform motion estimation and motion compensation, as well as on the intra refresh period. From the decoder point of view, instead, the parameter b_{dec} depends primarily on the employed mitigation scheme.

3. Advanced Distortion Algorithm

The SDA and the EDA provide an acceptable distortion approximation for isolated bursts of lost packets. When the distance between bursts get smaller (especially when the channel exhibits bad conditions), both the SDA and the EDA algorithms may lead to an optimistic evaluation of the channel induced distortion. A more precise estimation of the actual distortion is provided by the Advanced Distortion Algorithm (ADA). In this case, the distortion at the k-th frame in a GOP may be evaluated as:

Figure 6. SDA, EDA, and ADA algorithms in action

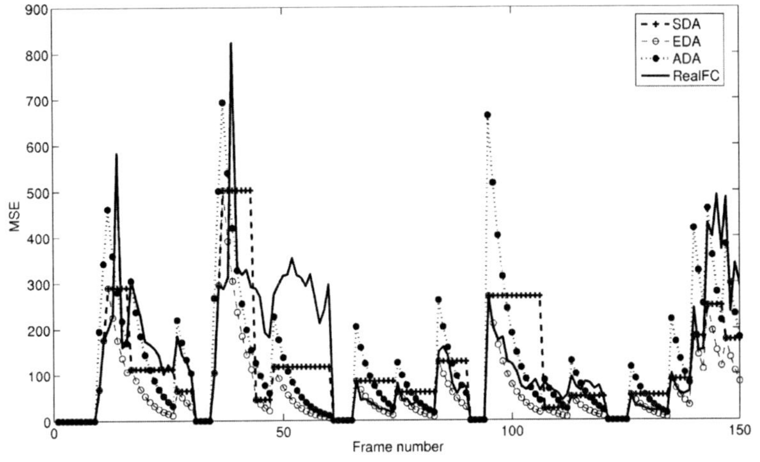

$$D(k) = \begin{cases} 0 & k < l_i \\ A_{k,l_i}^2 & l_i \le k < i \\ A_{i,l_i}^2 e^{-b(k-i)} & i \le k < l_j \\ A_{i,l_j}^2 & l_j \le k < j \\ A_{j,l_j}^2 e^{-b(k-j)} & k \ge j \end{cases} . \quad (7)$$

So the only modification, with respect to the EDA, is the selection of a different reference frame. In particular, ADA tries to provide a more accurate approximation of the distortion envelope using not the last received frame (as in the SDA and in the EDA), but the previous one. This simple modification, which implicitly introduces an additional distortion term in the estimation process, exploits the fact that successive frames in a scene contain little detail variations, and allows a better approximation of the distortion envelope.

As stated for EDA, a suitable choice of the parameter b may lead to a more accurate distortion estimation.

X. PERFORMANCE MEASUREMENTS

Several experiments have been conducted to assess the accuracy of the estimation algorithms detailed above. During the experiments, many encoding and decoding setting combinations are examined, to assess the generality of the proposed estimation techniques. An exhaustive explanation of the setup and of the validation tests performed is available in (Babich, D'Orlando, & Vatta, 2008). Figure 6 shows an example of how the distortion estimation algorithms SDA, EDA, and ADA work. The solid curve represents the real distortion obtained using the decoder, while the dotted marked curves represent the estimation using these three estimation algorithms.

In (Babich, D'Orlando, & Vatta, 2008) it is shown that DEAs are able to estimate the distortion with acceptable accuracy in most conditions, independently from the chosen sequence, and the adopted resolution. Both sequences with limited or significant motion are taken into consideration. It may be observed that large GOP sizes will result in severe visual error propagation during the sequence reproduction at the decoder side. However, DEAs are able to model the true distortion envelope also in this condition. The frame per frame accuracy, however, depends both on the chosen estimation algorithm and on the test conditions. For a large GOP size and large loss rates, only the ADA is able to reproduce the real distortion envelope accurately. Finally, estimation accuracy does not depend on frame types: the techniques introduced in (Babich, D'Orlando, & Vatta, 2008), in fact, may be used also to model the channel distortion in successive P-frames, even if there are B-frames in between.

XI. DISTORTION ESTIMATION IN PRACTICAL SCENARIOS

There are several application scenarios, where the algorithms introduced in (Babich, D'Orlando, & Vatta, 2008) may be used proactively by some agents on the network to enhance the user experience. The following paragraphs describe how to enhance the perceived quality by using the estimated distortion as a metric to influence the transmission schedule. The presented scenarios are either simulated or evaluated through experimental setups. In the developed test bed both the improvements obtained with respect to traditional delivery policies and the complexity with respect to optimal transmission techniques are discussed.

A. Bandwidth Adaptation Using Distortion Estimation

The distortion estimation algorithms may be used efficiently to fulfill the QoS requirements by bandwidth adaptation mechanisms. The scenario is commonly encountered in the Internet, and it occurs whenever the data rate on the incoming link at a network node exceeds the data rate on the outgoing link. During network congestion transient periods, router queues overflow, so that a bandwidth adaptation is required. The scenario is illustrated in Figure 7, where the incoming traffic at the node consists of multiple video streams multiplexed by the router on the single outgoing link. The main purpose is to maximize the quality at the end user side. The best performance is obtained adopting a scheduler able to choose at each transmission opportunity the packet with the higher distortion impact. A technique with very low additional complexity consists on adding some discarding capability in the router. Assume that each RTP/UDP (Real Time Protocol/User Data Protocol) video packet frame $P_{i,j}$, being i the packet number, and j the flow identifier (i.e., j=1,2,3 in Figure 7), reserves some bytes to store the associated distortion impact $A^1_{i,l}$ or $A^2_{i,l}$. The router, using the attached distortion value, may be able to satisfy the negotiated QoS by employing a suitable scheduling strategy on the outgoing link. More precisely, the transmission policy may grant a privilege to the packets with the higher distortion impact.

Figure 7. Bandwidth Adaptation trough packet dropping

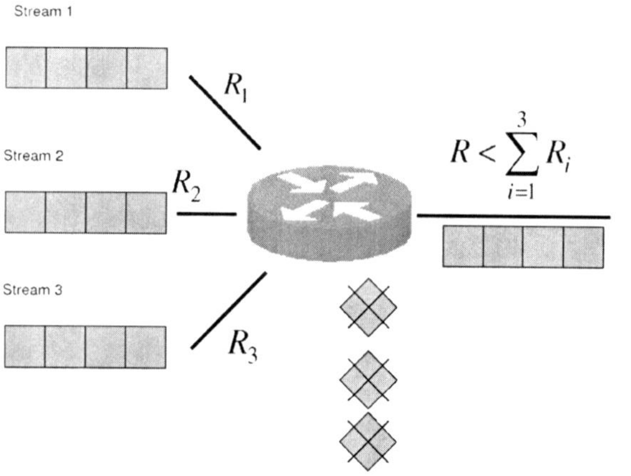

Therefore, the scheduling policy takes into account the user perceived quality, instead of relying on network parameters such as delay, jitter, throughput, or a combination of these. The network node implicitly performs a cross layer operation, by exploiting the application layer information that is included (hinted) in the payload of every packet. The resulting queue policy, based on application quality requirements, leads to smooth end user quality degradation.

B. Wireless Video Scheduling Using Distortion Estimation

Another application may take advantage of DEAs, providing high quality video delivery in wireless networks. Consider, for instance, a scenario in which multiple sources of video traffic communicate over a shared wireless medium. Taking into account the estimated distortion, each source can independently optimize the transmission schedule for its own packets, so that the quality of the streams sent over the shared channel is maximized. In this application, each source node encodes its own stream and runs a channel access scheme to contend resources to the other nodes. From the knowledge of the packet loss pattern, obtained through the DATA-ACK packet handshake adopted at the MAC layer, the estimated distortion may be updated using DEAs, and the scheduling policy may be reviewed.

XII. IMPLEMENTATION IN A REAL TEST BED

The adoption of a distortion estimation technique may depend on the required computational complexity. The computational burden deriving from most optimization algorithms, such as for instance RaDiO and its extensions, may be excessive for devices with consumer hardware and limited CPU resources. This section proposes a low complexity implementation aiming at optimizing the packet scheduling in a system with limited resources.

First, the developed tools and test bed setup are presented in detail. Then, a real campaign measurement is conducted to evaluate the overall system performance in terms of enhancements of the user experience during the video streaming sessions.

A. Distortion Estimation at the Encoder

The video encoder is modified to generate a resume file containing the description of the Network Abstraction Layer Units (NALUs), together with the associated distortion impact estimated using the EDA. The total distortion produced by the loss of each NALU is evaluated by integrating the estimated distortion (for instance using Eq. (5)) in the actual GOP. The packetization rule adopted in the system setup is obtained fitting a single NALU into the payload of a RTP packet, while the RTP header values are filled as defined in the RTP specification (Schulzrinne, Casner, Frederick, & Jacobson, 1996). The distortion impact produced by the loss of each packet is attached in the RTP payload, in order to share this information with the other network nodes that process the packet. The distortion impact associated to each packet is stored in a compressed manner using a single byte to minimize overhead in the payload of the packet. The quantized distortion D_q takes values from 0 to 255, where lower values indicate negligible distortion impact, while higher values indicate large distortion. If D is the distortion produced by the loss of a single packet, the quantized value is simply obtained using the following expression:

$$D_q = \left\lfloor \frac{D \cdot 255}{M} \right\rfloor, \tag{8}$$

where M represents the maximum distortion value obtained from an offline analysis of the encoded test sequences.

A simple transmission tool, that extracts both video slice packets and the associated distortion, is developed. In particular, the tool establishes an RTP connection with a specific IP address of one destination and then sends the H.264 encoded video packets to a specified UDP port. The transmitter sends each packet according to timing information by which the stream has been encoded. If the frame rate is 30 fps, all packets in one frame are sent in 33,3 ms. The distortion attached in each video packet, instead, measures the actual importance of each packet and may be used directly by other network nodes without performing additional operations such as decoding. In particular, an intermediate node may simply capture the packet, extract the first byte of the payload and perform the optimization.

To estimate the end user quality, the transmitter and the receiver dump all packets by means of standard capture tools such as *tcpdump* (Luis, 2008) and *wireshark* (Wireshark, 2008), by which it is possible to store packets allowing offline decoding and MSE evaluation. The packet trace files of both the transmitter and the receiver are fed into another tool that reconstructs the transmitted video as it is seen by the decoder, by comparing the traces and discarding the dropped or the excessively delayed packets from the original bitstream. The received application packets are then fitted into the decoder in order to produce the YUV output raw file. The decoder reads the packet stream and, for each packet contained in it, the standard decoding operations are performed, whereas, for each detected lost packet, the internal error recovery mechanisms are operated. More precisely, when a single packet loss, or a burst of losses occur, the decoder concealment mechanism tries to reconstruct the missing frames on the internal decoder buffer and on the user display buffer. Usually, for isolated packet losses, the error concealment mechanism is able to recover the entire frame, but, even in case of successful reception of the subsequent packets, the decoded frame differs from the reconstructed one, given the different reference signal taken into consideration by the encoder and by the decoder. However, the user perceived quality is satisfactory, because the display buffer shows a fluent video, without significant artifacts. The situation is more critical in presence of a burst of losses that covers two separate GOPs. In this case, both the decoder internal reference buffer and the display buffer present a picture gap due to the inability of the recovery mechanism to handle large bursts of losses. The user will immediately recognize the loss as a frozen frame output effect. The image on the screen stops and is kept constant (i.e., frozen) for the entire burst and until a complete frame refresh occurs.

During the display of the YUV video sequence each frame is stored in a file to allow offline evaluation of the objective video quality with the *psnr* tool. The tool takes as input the compressed YUV sequence and the one displayed after the decoding process and evaluates the objective video quality using the relation:

$$\overline{PSNR} = 20 \cdot \log 255 + 10 \cdot \log\left(N_f\right) - \log \sum_i D_i \tag{9}$$

where D_i represents the real picture MSE between the i-th frames of the compressed sequence and of the one displayed after decoding, and N_f is the number of frames considered in the video test sequence. It is important to underline that, for a right PSNR evaluation, the two sequences need to have the same number of frames. Moreover, it is important that encoder and decoder maintain a correct synchronization, to obtain a sensible result.

B. A Streaming Evaluation Test Bed

This section presents a detailed description of a streaming system implementation, including the developed software and the adopted hardware. In the system under test, the EDA algorithm is used to estimate the distortion at the encoder side, by applying the relations in Eq. (5) between frames available in the encoder buffer. A fixed value of the parameter b (6) is used, determined during an offline analysis of the transmitted sequence. The total distortion of each packet is evaluated, including both the single frame loss contribution and the propagation errors in the following frames.

The test scenario consists of several transmitters that send their own content to a single node through a common router. The data rate on the incoming link at the router node may exceed the data rate on the outgoing link, so that bandwidth adaptation is required. From the router point of view, all input streams will be multiplexed on the outgoing queue. An additional flow is added as a competing traffic using the *iperf* tool (Navlakha, & Ferguson, 2008). After multiplexing all incoming traffic, the router forwards it on the outgoing link to reach the destination. The outgoing link is a wireless ad-hoc network with a PHY link rate fixed at 1Mbps, that forces the router to discard some packets. Actually, the maximum User Datagram Protocol (UDP) throughput sustainable on the wireless link, measured with *iperf*, is around 850 kbps, while the multiplexed input rate exceeds this value.

Two different prioritization strategies are implemented in the router, as showed in Figure 8. In the first, all the incoming traffic is multiplexed on a single outgoing queue while, in the second, the RTP flows are forwarded on four different priority queues, on the basis of their distortion impact. The hardware and the software used include a desktop PC with Linux O.S.. Both the routing and scheduling capabilities are offered by the Click Modular Router (CMR) software application (Kohler, 2000). In the developed architecture, CMR captures the incoming packets from the Ethernet network interface, classifies them, and passes them to the corresponding queue. The developed *Extractor* module reads the distortion

Figure 8. Prioritization strategies at the router node. Single queue scheme (red). Multiple queues with distortion prioritization (blue)

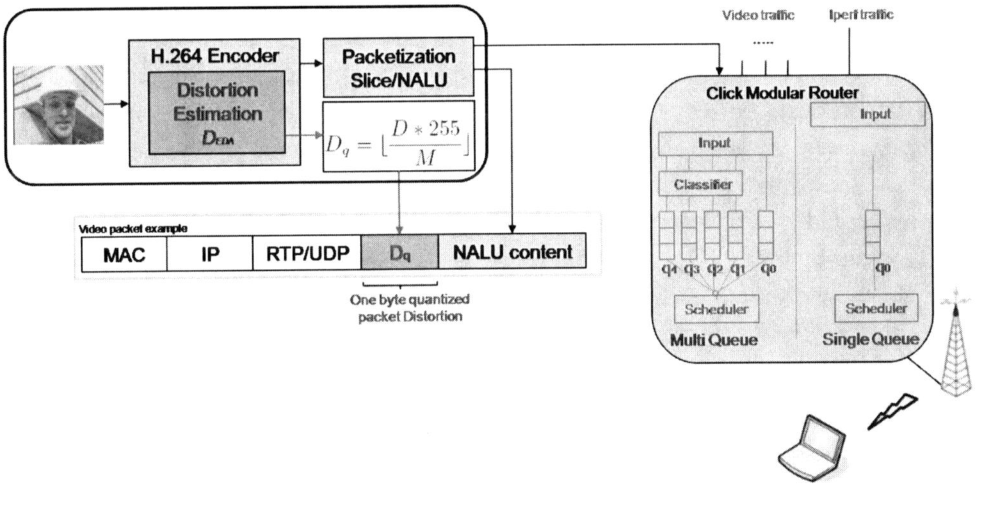

Table 1. Main characteristics of the four test video streams

Sequence	Length (s)	Avg. bit rate (kbps)	Y-PSNR (dB)
Foreman	300	157	36.35
Carphone	300	197	37.87
Miss America	300	65	35.70
Silent	300	75	36.28

Figure 9. Y-PSNR (dB) versus competing traffic rate of the two schemes

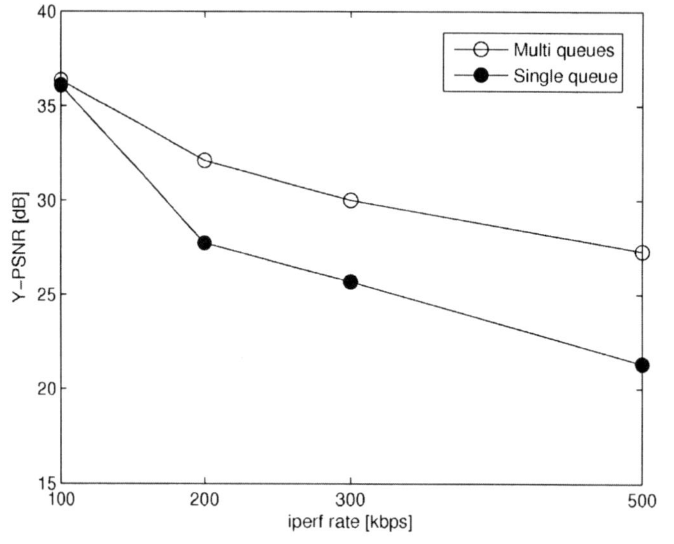

impact and differentiates the incoming packets. A *Classifier* module compares the distortion values of the video flows with predefined fixed thresholds evaluated during offline simulations. The status of each queue is evaluated, following the priority order, by the *Scheduler* at regular intervals. When a new packet is detected, the scheduler inserts the packet on the wireless card buffer.

The testing scenario consists on collecting the trace statistics for offline analysis during the transmission of the RTP video flows and the additional *iperf* competing traffic (Figure 8). Table 1 summarizes the main characteristics of the four video streams used during the experiment.

The sequences are encoded with a modified version of the H.264 video codec (JM98). Several simulations are run using different congestion levels by changing the rate of the competing *iperf* traffic.

C. Results

Figure 9 shows the average Y-PSNR (dB) of the four sequences resulting from the two prioritization schemes as a function of the competing traffic rate (kbps). It is possible to notice that the multiple queues with distortion prioritization scheme outperform the single queue scheme with a significant margin over the whole range of the competing traffic rate. This confirms that the distortion information may be exploited to significantly enhance the overall quality.

Observe that the performance improvement increases with the competing traffic rate. A 6 dB PSNR increment is obtained with a competing traffic rate of 500 kbps.

A comparison between the received quality as a function of the competing traffic rate for both the prioritized scheme and the single queue scheme is shown in Figure 10. From these results, it may be derived that the proposed distortion prioritization method makes the quality of the different video streams comparable, for the examined background traffic conditions. More precisely, from the figure it may be observed that, when the rate of the competing traffic increases, the prioritization scheme assigns more resources to *Foreman* and *Carphone*. These have a more significant impact on the overall quality because of the higher distortion values of their packets. As the competing

Figure 10. Y-PSNR (dB) vs. the competing traffic rate (kbps) for Foreman, Carphone, Miss America, and Silent sequences

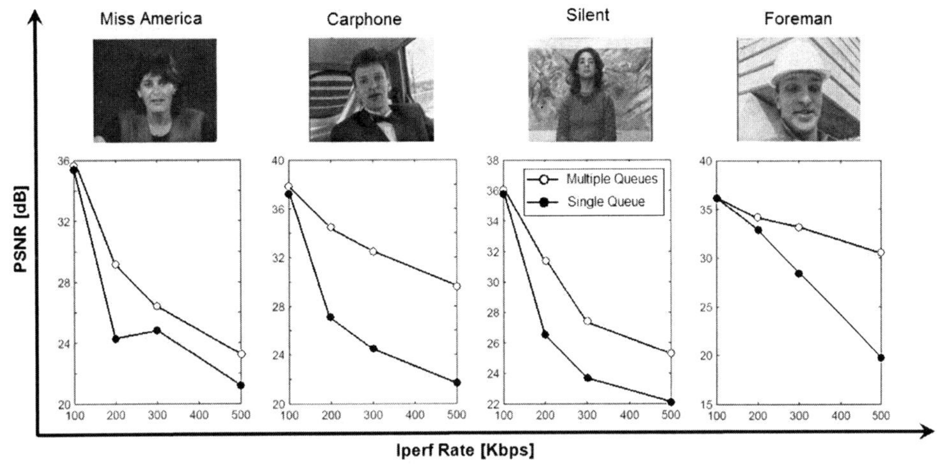

traffic rate decreases, the distortion prioritization scheme sends a larger percentage of *Miss America* and *Silent* packets, as shown in Figure 10. At lower competing traffic rates there is enough bandwidth for these two sequences, so that the scheduler can transmit also packets on lower priority queues. During the test a jitter analysis has been performed, confirming the improved efficiency of the priority, multi-queue scheme.

Observe that further improvements may be obtained by increasing the scheduler computational capabilities. For instance advanced rate distortion optimizations techniques may be used to select at each transmission opportunity the packet that:

- Minimizes the distortion with constraints on the output rate as described in (Chou, & Miao, 2006);
- Minimizes the distortion with constraints on the packet deadline as described in (Bucciol, Masala, Filippi, & De Martin, 2007);
- Minimizes the distortion with constraints on the average congestion level on the network as described in (Setton, & Girod, 2008).

Unfortunately the resulting computational complexity does not allow the implementation in a real system with consumer hardware, due to processing capability limitations. In fact, each of the above algorithms requires that the scheduler stores, to perform real time optimization, the instantaneous packet rate, and the inter-arrival time, or the single packet congestion level. This capability requires dedicated hardware. Really deployed solutions, instead, should require fewer resources and should save energy (Yuan, & Nahrstedt 2006).

Packet inspection, to examine the hinted distortion information, and packet selection, instead, may be performed in each commercial hardware that supports, for instance, the open source firmware DD-WRT (DD-WRT, 2009). A QoS scheme may be deployed by instructing the *iptables* service (iptables, 2009) to inspect packet header fields and to implement some basic operations such as rate limiting or traffic filtering. A more sophisticated QoS scheme may be obtained by using *iptables* in combination with the *tc* service (TrafficControl, 2009). The basic configuration uses *iptables* and its rules to select incoming packets and *tc* to manage the traffic.

Figure 11. Linear Multi-hop path with 4 hops and relative bandwidth

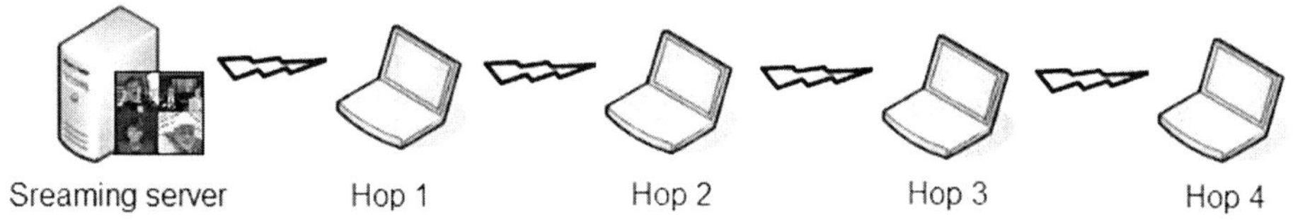

Iptables allows customizing packet inspection using headers at different OSI levels.

For instance it is possible to define rules that select incoming packets based on MAC level parameters (such as VLAN Ethernet tagging) or Network level header fields (such as the Type Of Service (TOS) header of the IP protocol) or Transport level header fields (such as the port of the UDP or the TCP protocol). Different packet selection strategies, required by EDAs, may be able to identify packets based on application layer data. A service with this capability is the application layer firewall Linux Level 7 packet classifier *L7-filter* (L7-filter, 2009) developed to manage P2P traffic inspection such as Kaza, Bittorrent and eDoney packets. This tool is intended to be used in conjunction with the *tc* service to implement an effective QoS scheme.

A custom *L7-filter* service may allow the RTP/UDP traffic inspection and the decoding of the distortion field attached in each NALU header. Therefore, the queue discipline, the traffic control mechanism and also the routing capabilities may be controlled directly, by configuring the *tc* service.

XIII. DEAS IN A MULTI-HOP ENVIRONMENT

DEAs may also be used to improve the received video quality in multi hop environments. In the next experiment a streaming server is used to send out a single stream on a four hops communication path. Every intermediate node receives the video content and retransmits it to the next hop. Being the single hops arranged in a decreasing available bandwidth order, the data receiving bandwidth is larger than the one used to send packets for every intermediate node. In order to avoid sending buffer overflow and the transmission of late video packets, intermediate nodes must drop some frames to shape the receiving bit-rate to the sending bandwidth. The frames received by each hop are selectively dropped according to their contribution to the decoding quality. Three different frame priority schemes are adopted in this experiment. In the first one, the frame priority is determined based on the frame type (Frame Drop Priority (FDP) scheme). Given the contribution to the decoding quality, type I packets have the top priority, followed by P and B packets. In the second and third scheme, the scheduling rule depends on distortion. In the second scheme each node simply extracts the distortion information contained in every frame. At each transmission opportunity a node scans all the received packets discarding excessive delayed ones. Then the packet with the highest distortion is selected and scheduled for the next transmission. This scheme is named Distortion Drop Priority (DDP). In the third prioritization scheme the sender has exact knowledge of the loss pattern of each hop and is able to communicate the new packet distortion impact to every hop using some signaling channel. Therefore, an intermediate node has the knowledge of the effective distortion produced by the loss of a new packet. This prioritization scheme is called Advanced Distortion Drop Priority (ADDP), and may be considered as a limiting

Figure 12. Y-PSNR in the nodes of the multi-hop path

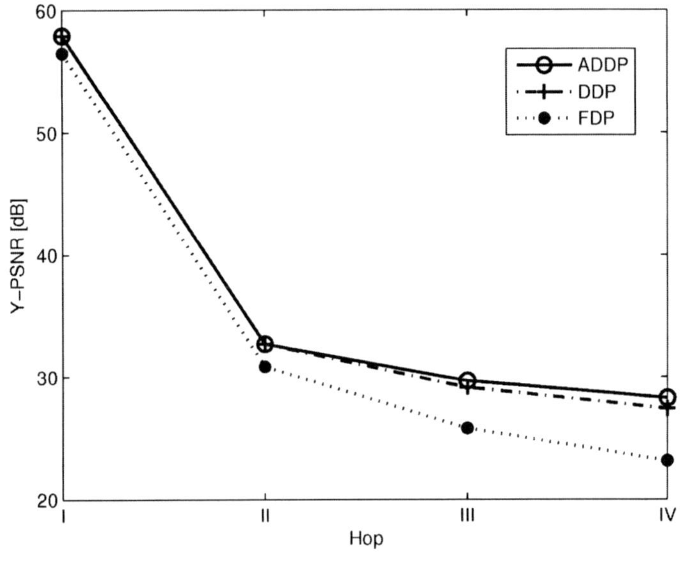

performance benchmark. In the experiment, the standard sequence *Foreman* with QCIF resolution is encoded using the H.264 encoder at 30 fps with average bit rate of 326 kbps.

The complete path is shown Figure 11 and the bandwidth of the links is fixed for convenience in a descending order at 300 kbps, 280 kbps, 260 kbps and 240 kbps.

Figure 12 shows the channel distortion, measured in terms of PSNR, and obtained decoding the received packets at each intermediate node using the three priority schemes. PSNR is obtained comparing the decoded sequence with the error free sequence so that the PSNR does not account the source compression distortion. The results put into evidence that the DDP scheme outperforms the FDP. In fact, FDP is not able to differentiate between two P frames while DDP captures the real distortion impact produced by the loss of every packet. Moreover, ADDP allows a minor improvement over DDP, given that the distortion values used by ADDP and DDP have a similar envelope (that differs only for a constant offset). Only in the third and fourth hop a little gain may be obtained. This experiment confirms that a content aware scheduling scheme with reduced computational complexity may provide significant gains in terms of perceived quality at the end user side.

XIV. FUTURE RESEARCH DIRECTIONS

Many research activities have been conducted in the field of network adaptive media transport but the open challenge in developing practical content-aware scheduling algorithms is the required computational complexity. Expected future schedulers should respond dynamically to the rapidly changing channel conditions, transmitting at each transmission opportunity the more suitable packets. Today the computational complexity of the available network adaptive media transport mechanisms, such as RaDiO and its derivations, requires dedicated hardware and maybe also complex software.

The H.264 Scalable Video Codec (SVC) codecoding techniques, and especially fine granularity scalability, could be easily adapted to channel dependent content-aware scheduling schemes, since they provide a natural packet prioritization strategy for the scheduler.

Therefore, developing resource distortion optimized scheduling schemes for scalable video coding techniques is an important area of today research. Packet selection strategies, based on the distortion importance of each sub stream, may guarantee a proper level of end user satisfaction, by keeping the QoS above a suitable threshold value. Another important research field is related to obtaining objective measures for video quality assessment that are perceptually relevant. The particular types of errors that can occur due to video packet losses are specific to the block-based motion compensation technique adopted in every modern video encoder as well as to the spatial and temporal error-concealment methods used at the decoder. The widely adopted video quality measures MSE or PSNR are not suitable to measure the perceptual distortions caused by such errors. So it is desirable to find additional objective metrics able for instance to take into account not only the frame content but also the temporal evolution of the video sequence.

XV. CONCLUSION

This chapter has analyzed several issues regarding QoS in video applications. The QoS treated in the present chapter is also called application-level QoS due to the fact that the objective is to measure the quality perceived by the end user. After a preliminary review of the main issues involved in delivering multimedia content over packet networks, three different distortion estimation algorithms have been proposed providing technical details and validation results. The chapter has presented a technique for delivering the distortion information to the network nodes. The information stored in the compressed video packets can be easily parsed and decoded by each network node. This information, called "hint" (or distortion impact), allows streaming servers to simply read the distortion information from packets instead of estimating them on a real-time basis. Wireless video may become the next killer application, and QoS is certainly the most important enabler in this picture.

REFERENCES

L7-filter. (2009). *Application Layer Packet Classifier for Linux*. Retrieved January 2009, from http://l7-filter.sourceforge.net/

Babich, F., Comisso, M., D'Orlando, M., & Vatta, F. (2006). Distortion Estimation Algorithms (DEAs) for wireless video streaming. In *Proceedings of the IEEE Global Telecommunications Conference GLOBECOM*, San Francisco, USA (pp. 1- 5).

Babich, F., D'Orlando, M., & Vatta, F. (2008). Video Quality Estimation in Wireless IP Networks: Algorithms and Applications. *ACM Transactions on Multimedia Computing, Communications, and Applications (TOMCCAP)*, 4(1).

Babich, F., D'Orlando, M., & Vatta, F. (2008b). Applications of Video Distortion Estimation Algorithms for Efficient Video Streaming. In *Proceedings of the IEEE Global Telecommunications Conference, GLOBECOM*, New Orleans, USA.

Braden, R., Zhang, L., Berson, S., Herzog, S., & Jamin, S. (1997). *Resource ReSerVation Protocol (RSVP) - Version 1 Functional Specification*. IETF RFC 2205.

Bucciol, P., Masala, E., Filipi, E., & De Martin, J.C. (2007). Cross Layer Perceptual ARQ for Video Communication over 802.11e Wireless Networks. *Advances in Multimedia, 2007*(1), 1-12.

Campbell, A., & Coulson, G. (1997). QoS adaptive transports: Delivering scalable media to the desktop. *IEEE Network*, *11*(2), 18–27. doi:10.1109/65.580910

Chakareski, J., Apostolopoulos, J. G., Wee, S., Tan, W., & Girod, B. (2005). Rate distortion hint tracks for adaptive video streaming. *IEEE Transactions on Circuits and Systems for Video Technology*, *15*(10), 1257–1269. doi:10.1109/TCSVT.2005.854227

Chakareski, J., & Frossard, P. (2005). Distributed packet scheduling of multiple video streams over shared communication resources. In *Proceedings of Multimedia Signal Processing* (pp. 1- 4).

Cheung, G., & Tan, W. (2006). Packet scheduling of streaming video with flexible reference frame using dynamic programming and integer rounding. In *Proceedings of the IEEE International Conference of Multimedia Expo* (pp. 1969-1972).

Choi, L., Ivrlac, M., Steinbach, E., & Nossek, J. (2005). Analysis of distortion due to packet loss in streaming video transmission over wireless communication links. In *IEEE International Conference on Image Processing (ICIP)* (Vol. 1, pp. 189-192).

Chou, P. A., & Miao, Z. (2006). Rate-distortion optimized streaming of packetized media. *IEEE Transactions on Multimedia*, *8*(2), 390–404. doi:10.1109/TMM.2005.864313

Chou, P. A., & Sehgal, A. (2002). Rate-distortion optimized receiver-driven streaming over best-effort networks. In *Proceedings of the International Packet Video Workshop*, Pittsburgh, PA, USA.

Control, T. (2009). *Linux Advanced Routing & Traffic Control*. Retrieved January 2009, from http://lartc.org/

Cui, Y., & Nahrstedt, K. (2003). Layered peer-to-peer streaming. In *Proceedings of ACM NOSSDAV* (pp. 162–171).

DD-WRT. (2009). *DD-WRT*. Retrieved January 2009, from http://www.dd-wrt.com/dd-wrtv3/

De Martin, J. C., & Quaglia, D. (2001). Distortion-based packet marking for MPEG video transmission over DiffServ networks. In *Proceedings of the IEEE International Conference on Multimedia and Expo* (pp. 399- 402).

Frossard, P., & Verscheure, O. (2001). Joint Source/FEC Rate Selection for Quality-Optimal MPEG-2 Video Delivery. *IEEE Transactions on Image Processing*, *10*(12), 1815–1825. doi:10.1109/83.974566

Ghanbari, M. (1996). Postprocessing of late cells for packet video. *IEEE Transactions on Circuits and Systems for Video Technology*, *6*(6), 669–678. doi:10.1109/76.544738

Girod, B., Kalman, M., Liang, Y., & Zhang, R. (2002). Advances in Channel-Adaptive Video Streaming [Invited paper]. *Wireless Communications and Mobile Computing*, *2*(6), 549–552. doi:10.1002/wcm.87

Hua, Y. (2007). Advances in Recursive Per-Pixel End-to-End Distortion Estimation for Robust Video Coding in H.264/AVC. *IEEE Transactions on Circuits and Systems for Video Technology*, *17*(7), 845–856. doi:10.1109/TCSVT.2007.897116

IEEE802.11e/D5.0. (2003). Draft Supplement to Part 11: Wireless Medium Access Control (MAC) and physical layer (PHY) specifications: Medium Access Control (MAC) Enhancements for Quality of Service (QoS).

Iptables. (2009). *Iptables network firewall*. Retrieved from January 2009, From http://www.netfilter.org/

ITU-R. (2000). *Methodology for the subjective assessment of the quality of television pictures*. ITU-R Rec. BT.500–10.

ITU-T. (1996). *Methods for subjective determination of transmission quality*. ITU-T Rec. P.800.

ITU-T. (1999). *Subjective video quality assessment methods for multimedia applications*. ITU-T Rec. P.910.

Kohler, E. (2000). *The Click Modular Router*. Ph.D. dissertation delivered at MIT.

Leontaris, A., & Cosman, P. C. (2003). Video compression with intra/inter mode switching and a dual frame buffer. In *Proceedings of the IEEE Data Compression Conference* (pp. 63–72).

Li, B., Xu, D., Nahrstedt, K., & Liu, J. W. S. (1998). End-to-end QoS support for adaptive applications over the Internet. In *Proceedings of SPIE (Symp. on Voice, Video Data Communications)*, Boston, MA (pp. 166–176).

Li, Q., & van der Schaar, M. (2004). Providing Adaptive QoS to Layered Video over Wireless Local Area Networks through Real-Time Retry Limit Adaptation. *IEEE Transactions on Multimedia*, *6*(2), 278–290. doi:10.1109/TMM.2003.822792

Liang, Y., Apostolopoulos, J. G., & Girod, B. (2003). Analysis of Packet Loss for Compressed Video: Does Burst Loss Matter? In *Proceedings of the IEEE International Conference on Acoustics, Speech, and Signal Processing (ICASSP)*, Hong Kong, China.

Liang, Y. J., Apostolopoulos, J. G., & Girod, B. (2008). Analysis of Packet Loss for Compressed Video: Effect of Burst Losses and Correlation Between Error Frames. *IEEE Transactions on Circuits and Systems for Video Technology*, *18*(7), 861–874. doi:10.1109/TCSVT.2008.923139

Luis, M. G. (2008). *TcpDump Libpcap*. Retrieved October 13, 2008, http://www.tcpdump.org/tcpdumpman.html.

Majumdar, A., Sachs, D. G., Kozintsev, I. V., Ramchandran, K., & Yeung, M. M. (2002). Multicast and unicast real-time video streaming over wireless LANs. *IEEE Journal on Selected Areas in Communications*, *12*, 524–534.

Miao, Z., & Ortega, A. (2002). Expected runtime distortion based scheduling for delivery of scalable media. In *Proceedings of International Packet Video Workshop*, Pittsburgh, PA.

Navlakha G., & Ferguson J. (2008). *IPerf Tool*.

Ong, E., Yang, X., Lin, W., Lu, Z., & Yao, S. (2004). Video quality metric for low bitrate compressed videos. In *Proceedings of the IEEE International Conference on Image Processing*, Genova, Italy (Vol. 5, pp. 3531–3534).

Pahalawatta, P. V., & Katsaggelos, A. K. (2007). Review of content-aware resource allocation schemes for video streaming over wireless networks. *Wireless Communications and Mobile Computing*, *7*, 131–142. doi:10.1002/wcm.469

Podolsky, M., McCanne, S., & Vetterli, M. (1998). *Soft ARQ for layered streaming media, Technical Report UCB/CSD-98-1024*. University of California, Computer Science Department, Berkeley, CA.

Rejaie, R., Handley, M., & Estrin, D. (2000). Layered quality adaptation for Internet video streaming. In *IEEE Journal on Selected Areas in Communications - Special Issue Internet QoS*, *18*(12), 2530–2543.

Schmidt, J. C., & Rose, K. (2007). First-order distortion estimation for efficient video streaming at moderate to high packet loss rates. In *Proceedings of Packet Video Workshop* (pp. 318-325).

Schmidt, J. C., & Rose, K. (2008). Macroblock-based retransmission for error-resilient video streaming. In *Proceedings of the IEEE International Conference on Image Processing* (pp. 2308-2311).

Schulzrinne, H., Casner, S., Frederick, R., & Jacobson, V. (1996). *A Transport Protocol for Real-Time Applications*. Request for Comments (standard) RFC 1889, Internet Engineering Task Force (IETF).

Setton, E., & Girod, B. (2008). *Peer-to-Peer Video Streaming*. Berlin, Germany: Springer.

Setton, E., Noh, J., & Girod, B. (2006). Congestion-distortion optimized peer-to-peer video streaming. In *Proceedings of the IEEE International Conference on Image Processing (ICIP)*, Atlanta, GA.

Singer, D., Belknap, W., & Franceschini, G. (2001). *ISO Media File Format Specification: MP4 Technology Under Consideration for ISO/IEC*.

Stockhammer, T., Wiegand, T., & Wegner, S. (2002). Optimized transmission of H.26L/JVT coded video over packet-switched networks. In *Proceedings of the IEEE International Conference on Image Processing* (pp. 173–176).

Technologies, T. (n.d.). *Next Generation Network (NGN) Services* [white paper]. Retrieved from http://www.mobilein.com/NGN_Svcs_WP.pdf

Van der Schaar, M., & Andreopoulos, Y. (2005). Rate-distortion-complexity modeling for network and receiver aware. *IEEE Transactions on Multimedia*, *7*(3), 471–479. doi:10.1109/TMM.2005.846790

Wireshark. (2008). *Wireshark: Network Protocol Analyzer*. Retrieved October 2008, from http://www.wireshark.org/

Wu, D., Hou, Y. T., Li, B., Zhu, W., Zhang, Y.-Q., & Chao, H. J. (2000). An end-to-end approach for optimal mode selection in Internet video communication. *IEEE Journal on Selected Areas in Communications*, *18*(6), 977–995. doi:10.1109/49.848251

Yong, J., Guangwei, B., Peng, Z., Hang, S., & Junyuan, W. (2008). Performance Evaluation of a Hybrid FEC/ARQ for Wireless Media Streaming. In *Proceedings of the IEEE International Conference on Circuits and Systems for Communications (ICCSC)*, (pp. 90–94).

Yuan, W., & Nahrstedt, K. (2006). Energy-Efficient CPU Scheduling for Multimedia Applications. *ACM Transactions on Computer Systems*, *24*(3), 1–40. doi:10.1145/1151690.1151693

Zhai, F., Luna, C. E., Eisenberg, Y., Pappas, T. N., Berry, R., & Katsaggelos, A. K. (2003). A Novel Cost–Distortion Optimization Framework for Video Streaming over Differentiated Services Networks. In *Proceedings of the IEEE International Conference on Image Processing*, Barcelona, Spain (pp. 14-17).

Zhang, R., Regunathan, S. L., & Rose, K. (2000). Video coding with optimal inter/intra-mode switching for packet loss resilience. *IEEE Journal on Selected Areas in Communications*, *18*(6), 966–976. doi:10.1109/49.848250

KEY TERMS AND DEFINITIONS

Distortion Estimation: Distortion estimation is an important aspect to consider when delivering a video content over an IP network. Usually distortion estimation refers to the evaluation of annoying artifacts in the received stream due to compression and channel failures.

Error Resilience: Error Resilience is an amount of techniques that may be included in each modern video encoder to alleviate the effect introduced during the transmission over an error prone network.

H.264/AVC: H.264/AVC is the last standard for video compression and is the latest block-oriented motion-compensation-based standard developed by the ITU-T Video Coding Experts Group (VCEG) together with the ISO/IEC Moving Picture Experts Group (MPEG). The final drafting work on the first version of the standard was completed in May 2003.

Quality of Service (QoS): Quality of service is the ability to provide different priority to different applications, users, or data flows, or to guarantee a certain level of performance to a data flow.

Video Quality: Video quality is a characteristic of a video and represents a formal or informal measure of the perceived video degradation (obtained comparing the received video with the original one). Each video processing system such as compression may introduce some amounts of distortion or artifacts in the video signal, so video quality evaluation is an important problem.

Video Streaming: Video streaming is a video that is constantly received by an end-user while it is being delivered by a streaming server.

Wireless Networks: Wireless network refers to any type of computer network that is wireless, and is associated with a IP based network whose interconnections between nodes do not use wires.

Chapter 18
Perceptual Quality Assessment of Packet-Based Vocal Conversations over Wireless Networks:
Methodologies and Applications

Sofiene Jelassi
University of Sousse, Tunisia & University of Pierre et Marie Curie, France

Habib Youssef
University of Sousse, Tunisia

Guy Pujolle
University of Pierre et Marie Curie, France

ABSTRACT

In this chapter, the authors describe the intrinsic needs to effectively integrate interactive vocal conversations over heterogeneous networks including packet- and circuit- based networks. The requirement to harmonize transport networks is discussed and a foreseen architecture multi -operators and -services is presented. Moreover, envisaged remedies to the ever increasing network complexity are also summarized. Subjective and objective methodologies to evaluate voice quality under listening and conversational conditions are thoroughly described. In addition, software- and emulation- based frameworks developed in order to evaluate and improve voice quality are rigorously described. This chapter stresses parametric model-based assessment algorithms due to their ability to be useful for on-line network management. In particular, the authors describe parametric assessment algorithms over last-hop wireless Telecom networks and packet-based networks. The last part of this chapter describes several management applications which consider users' preferences and providers' needs.

DOI: 10.4018/978-1-61520-680-3.ch018

1. INTRODUCTION

Next-generation network infrastructure should cater simultaneously to a multitude of services having different *quality of service* needs. In fact, next-generation networks should be well-engineering to deliver services as a *triple-play* package which includes voice, video, and data or *quadruple-play* when wireless access facility is included. Services over next-generation networks could be delivered over a heterogeneous infrastructure using a wide variety of wired and wireless mobile access devices. New generation of services should offer, on the one hand, for provider additional revenue and more management flexibility, and on the other hand, for consumer personalized, ubiquitous, reliable and secure services. From consumer perspective, new services should ultimately provide, at a reduced price, a good quality of experience.

There are several *pitfalls* which should be properly addressed in order to successfully achieve intended goals. In fact, the high service flexibility entails enormous complications at network design, management, and evaluation stages. To cope with the ever increasing network complexity, several ongoing projects have been launched within standardization bodies, academic institutions, as well as industrial enterprises in order to define and standardize new architectures and management policies dedicated for next-generation networks. The ultimate goal of new proposals is to offer a good *Quality of Experience* (QoE) for subscribers while optimizing network resource utilizations. The estimation of QoE is of keen economical importance since it could be used to quantify the suitability of new proposals and technologies which will be adopted for next-generation networks. Moreover, QoE could be used by new management policies for quality monitoring, tuning, planning, and enhancement in a user-friendly way.

The remainder of this chapter is organized as follows. Section 2 presents a number of networking multimedia services which could be catered to consumers over next generation networks. Section 3 discusses some convergence scenarios and gives a brief survey about foreseen mobile wireless network architecture. Section 4 goes over the QoS provision methodologies for delay-sensitive services. Section 5 provides an in-depth description of the assessment methodologies used to evaluate vocal services. A comprehensive description of assessment frameworks of voice conversations is given in Section 6. A thorough description of parametric assessment algorithms over mobile Telecom networks and packet-based, best-effort, networks is given in Section 7. Several management applications over wireless networks using QoE are described in Section 8. We conclude in Section 9.

2. NETWORKING MULTIMEDIA SERVICES

The actual trend of network evolution is characterized by the *convergence* of Internet and Telecom services. This convergence is driven by standardization bodies as well as industry due in part to the expected value-added (Chauveau, 2005). Basically, this is performed by integrating/adding Telecom services over IP infrastructure. Technically, this integration is merely done by dividing original digitized stream into media units which constitute the payload part of carried IP packets. Moreover, the flexibility of packet-based networks enables providing other services such as radio over IP, IPTV, and video/music streaming.

Telecom services such as conversational services (vocal/video) and instantaneous vocal/video messaging are characterized by their sensitivity to delay. However, the *unmanaged* packet-based networks such as the Internet are suited to deliver delay-insensitive services such as E-Mail, FTP, and WWW. This service is sometimes called *elastic* media since delay and delay variation do not greatly affect the quality of service. Indeed,

Figure 1. Services classification with respect to packet loss tolerance and delay (ITU-T Recommendation G.1010, 2001)

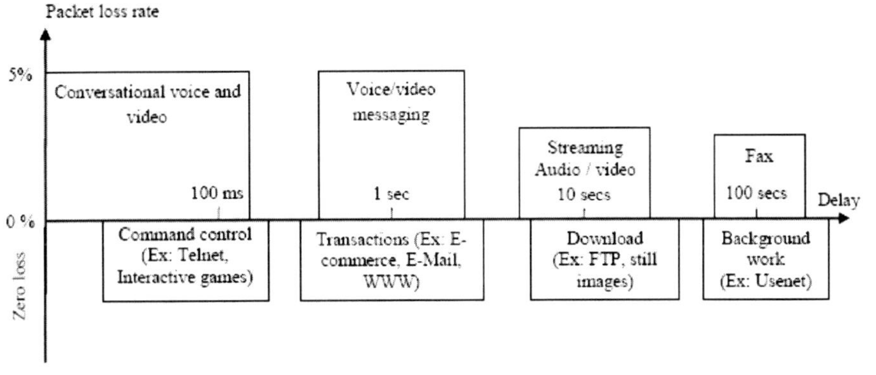

IP data networks are designed to reliably deliver bulk media with little consideration to transit delay. Thus, packets are usually subjected to variable and unbounded delays over IP networks. These properties are unsuitable for *delay-sensitive* services which require the reception of each media unit before its deadline, but may tolerate some packet losses. Actually, media units of delay-sensitive services are carried using the unreliable UDP transport protocol. UDP protocol does not provide congestion and flow control mechanisms which could certainly lead, in a large scale environment, to unfairness problems and threat network stability.

Basically, networking services are classified in terms of their sensitivity to both *packet loss* and *delay*. Figure 1 illustrates the classification made by the (ITU ITU-T Recommendation G.1010, 2001). A rule-of-thumb, for E-commerce applications and E-mail, users will tolerate service-time delays up to two seconds. A similar level of acceptable delay applies to messaging services, though a certain degree of packet loss can be tolerated.

The ITU-T recommendation G.1010 highlights the discrepancy between *conversational* and *streaming* services. In fact, conversational and streaming applications share several features regarding their sensitivity to delay and losses and the type of processed media. Moreover, both services have an obvious similarity in the play-out process. Indeed, media units under conversational or streaming modes are played at run-time while the remainder part of delivered media is either inside the network or at the sender side. To do that, both applications delay temporary received media units in a play-out buffer before playing them at a fixed or adaptive rate. Despite these apparent similarities, conversational and streaming applications have several intrinsic differences. First, data exchange is bilateral under conversational service class and unilateral under streaming service class. In fact, conversational services are directly set up between end-users. However, streaming services are set up between a server and a set of authorized and authenticated clients. The streaming servers are responsible for multimedia content storage and distribution. Second, conversational data are acquired at run-time which reduces considerably the ability to schedule efficiently packet transmissions. In contrast, streamed data are prepared in advance and optimally preserved on a storage device. This will offer more flexibility to efficiently schedule packet transmissions according to network and buffer receiver states. Further, streaming service users are allowed to perform VCR type operations such as Fast Forward and Backward, Jump Forward and Backward, and Pause, which are obviously inapplicable under conversational service. A major discrepancy between conversational and streaming service class stems from users tolerance of delay and loss. As illustrated in Figure 1, for streaming services, consumers can tolerate an overall delay of as long as 10 sec before the start of the play-out process. However, for conversational services, consumers impose a stringent overall delay due to interactivity. In fact, clients expect to experience a *mouth-to-ear* (M2E) delay in the order of 150 msec, as in circuit-based conversations. Moreover, streaming applications use a buffering delay in the order of 5-10 sec, whereas, conversational applications use a buffering delay in the order of 20 – 50 msec.

Undoubtedly, interactive voice conversations are the most popular service offered by Telecom providers which will be greatly affected by its integration over IP networks. To successfully provide interactive vocal conversations over data networks, users should experience a perceptual quality similar to the perceptual quality offered by Telecom providers.

A potential source of quality improvement consists of using a wide-band instead of a narrow-band voice CODEC which is commonly used over telephone networks (Linden, n.d.). In fact, in a conventional narrow-band telephone system, only the spectral bandwidth limited between 300 Hz and 3400 Hz are processed and transmitted. This bandwidth limitation explains why telephone speech signal is deemed weak, unnatural and lack crispness. In fact, even without increasing the sampling frequency, the voice quality may be enhanced by expanding the lower band down to 50 Hz. This improves the bass of speech and has a major impact on the naturalness, presence, and comfort of conversation. Expanding the upper bound to 4000 Hz improves the naturalness and crispness of sound. All in all, more natural voice and higher intelligibility can be achieved just by extending the bandwidth within the limitations of the narrowband speech. This constitutes the first step toward "face-to-face" communication quality offered by wideband speech (Linden, n.d.).

There are several potential pitfalls when deploying voice over IP (VoIP) networks which should be properly considered. First, M2E delay over IP networks may be quite large which harms users' interactivity. In fact, in contrast to Telecom conversation where total delay is related to geographic location of users, M2E delay over wide area IP networks depends, in addition to users' location, on several other factors such as access network, management policy, link properties, conversation time, and the features of intermediate nodes. Indeed, even if communicating nodes are located in the same office branch, M2E delay could be pretty large due to network congestion. Figure 2 sketches potential sources of delay sustained by packet-based voice conversations over a typical IP network. This Figure shows that delay at a terminal node constitutes an important amount of incurred total delay. This source of disturbance should be properly attenuated in order to improve the users' quality of experience. This could be performed for example by defining adequate management policies at access and core networks which consider properly the features of delay-sensitive services. Moreover, delay sustained at terminal nodes should be smartly reduced without impairing the intelligibility of voice signals.

Apart from the M2E delay, voice packets over wide area IP networks sustain a variable network delay, known as *delay jitter*. This source of disturbance entails the occurrence of gaps due to late arrivals (Melvin, 2004). To prevent a high late arrival ratio, the receiver node uses a de-jittering buffer where early packets are temporary delayed (Melvin, 2004). Buffering delay is subject to a critical trade-off between delay and late loss ratio. Indeed, reducing buffering delay results in a reduction of total delay at the expense of a potential increase of late arrivals, and vice versa. The de-jittering mechanism could be made on a per-hop basis, which requires upgrading intermediate nodes.

Another potential source of disturbance observed over wide area IP networks is *packet*

Figure 2. Delay and loss over best-effort packet-based configuration

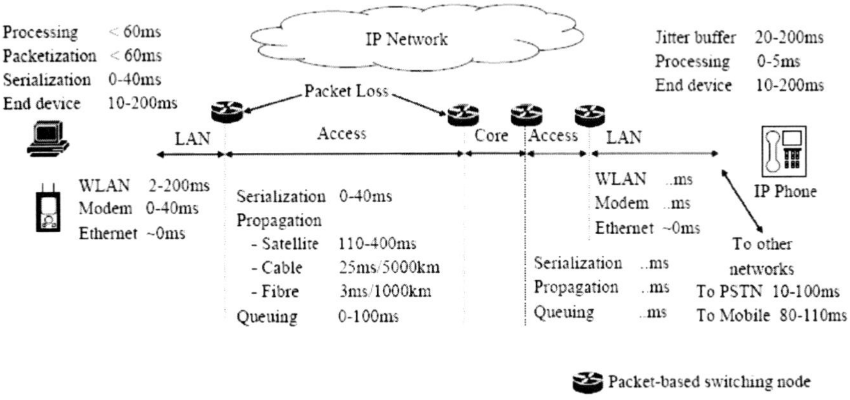

losses (Roychoudhuri, 2003). In fact, in contrast to Telecom networks where losses occur essentially due to hand-over or signal interference, packet losses over IP networks occur often at waiting queues of intermediate nodes. Moreover, packet losses could arise in a random way when data are carried over a wireless link. Further, packets reaching the receiver side after their play-out instant are useless and assumed as lost. This source of disturbance impairs significantly the intelligibility of heard voice signals. In addition, several empirical studies which have been performed to characterize packet loss behavior over IP networks show that packet losses are bursty (Roychoudhuri, 2003; Bolot, 1993). This means that packet losses arise in sequence, i.e., several consecutive packets are lost. This specific feature increases service degradation at perceptual level due to the experience of pretty large gaps. There are several remedies to either cancel or reduce the effect of packet losses on quality of experience. Recovering techniques can be classified as sender-based and receiver-based reparation schemes (Liao et al., 2001). Sender-based recovering techniques, such as retransmission, forward error correction, and inter-leaving require active cooperation of sender side in order to recover lost packets. In contrast, received-based recovering techniques such as repetition, noise insertion, or interpolation repair passively missing fragments without cooperation of sender side. Practically, sender-based recovering schemes are used to repair burst losses, whereas receiver-based recovering schemes are used to repair individual losses (Sat & Wah, 2006).

Next, we describe foreseen architectures of Next Generation Networks (NGNs) to accommodate delay-sensitive services such as VoIP or IPTV traffics.

3. ARCHITECTURE OF NEXT-GENERATION NETWORKS

The world-wide integration and provision of multimedia services over packet-based networks including last- and multi- hop wireless networks requires the design of new architectures, signaling protocols and QoS-aware management protocols. In order to justify this requirement, we present in Figure 3 three simple possible scenarios of vocal conversations carried over at least one packet-based network. In Scenario 0, a vocal conversation is established between two IP terminals over two IP clouds. Each IP terminal is equipped with a sound card and the adequate software to play received voice packets such as Skype or GoogleTalk (Sat & Wah, 2006). Moreover, each IP terminal is equipped with the adequate network card interface which allows acceding to the IP network. The

Figure 3. Basic convergence scenarios

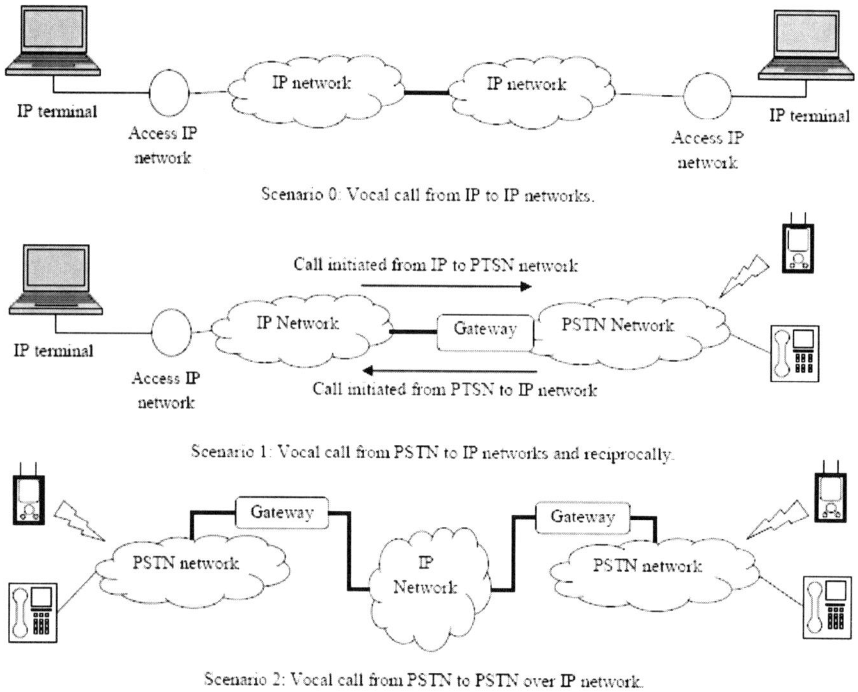

consumers could access to the network through a wire or wireless link. Vocal streams could be multiplexed with other streams through a *backbone* link which connects geographically distant sites. Backbone links constitute the core part of delivering networks and provide generally a high speed connection. Scenario 1 illustrates the establishment of a vocal conversation between a conventional vocal terminal and an IP terminal (see Figure 3). The vocal terminal can be a classical telephone handset or a Telecom hands-free terminal. In this scenario, it is required to deploy a dedicated *gateway* in order to interconnect packet- and circuit- based networks. Finally, scenario 3 shows the establishment of a vocal conversation between two circuit-based terminals. In this case, produced data are transmitted through a packet-based network located between two circuit-based networks. As a consequence, an additional number of gateways are required. The gateway functionalities and selection when several alternatives may be used are closely related to the manufacturer such as Cisco, Micom, 3COM, etc. (ETSI Technical report, 2000).

The described simple scenarios show the need to harmonize packet- and circuit- based networks in order to allow a successful convergence. *Harmonization* lies from signaling, data conversion, data delivery, telephone routing, synchronization, mobility management, to service billing. Significant efforts have been made within standardization organization such as ITU, ETSI, and IETF to *normalize* the convergence. This will result in a unified network infrastructure which could be used to achieve providers and consumers needs.

The actual trend consists of keeping the legacy circuit switched networks (fixed and mobile) and to create a high speed packet-based IP network which is used to inter-connect heterogeneous networks. Figure 4 illustrates the foreseen architecture which includes at the core an IMS (*IP Multimedia Subsystem*) system (ETSI Working Group, n.d.; Bertrand, 2007). Services provided by broadband wireless and wired as well as fixed and mobile Telecom access should be seamlessly delivered over heterogeneous access networks. This should consider the specific requirements of each service. This enables optimizing network resource utilization, assuring service ubiquity, and improving the quality of experience. In addition, new business services and applications are rapidly deployed and managed in a flexible way.

Jain (2006) outlines the need to re-architect IP networks in order to cope with the ever increasing complexity of next-generation networks. Indeed, today networks represent a highly complex and dynamic environment due to their size, heterogeneity, traffic diversity, etc. As a consequence, the initial design of packet-based networks is unsuitable to support actual and new generation services. In fact, IP networks are designed by moving complexity toward the edge node and keep core nodes as simple as possible which leads to a static and cumbersome environment. There are several approaches to prevent the Internet ossification observed in the last few years (Peterson et al., 2005; Al-Agha, 2008). The actual tendency consists of using intelligent intermediate nodes to build an autonomic communication environment with separation between the control and data planes (Al-Agha, 2008). Such system should exhibit a high degree of flexibly to adapt its behavior according to the environment dynamics (changes in topology, technologies, service demands, application context, etc). The idea consists of designing networks that are self-piloting, self-healing, self-configuring, self-optimizing, and self-protecting (Al-Agha, 2008).

In the following section, we briefly describe the general approaches proposed by the research community for the adequate provision of quality of service over existing IP networks.

4. QOS PROVISION APPROACHES

QoS provision is mainly required for delay-sensitive services. There are two schools to improve quality of delay-sensitive services which could be likely combined: reactive and predictive strategies.

4.1 Reactive QoS Provision Strategy

The improvement of QoE could be performed by adequately engineering quality control algorithms at application layer of the sender and receiver sides. The installed quality control algorithms attempt to properly react to hide the impairments introduced during the delivery process without explicit network assistance. To do that, the sender could for example adjust its transmission rate according to the measured available bandwidth (Hoene, 2005). This allows to efficiently use the shared bandwidth and reduces both delay and loss sustained by the receiver. Moreover, the sender may dynamically adapt its recovering mechanism and coding algorithm according to network channel conditions. This is especially useful over lossy wireless channels where losses arise in random way due to the high sensitivity of unguided media to the surrounding environment. Control information about channel and network conditions can be communicated to the sender through a feedback sent by the receiver periodically or when a triggering condition arises (Sat & Wah, 2006). On the other hand, the receiver could remove the introduced delay jitter through an intelligent monitoring of the de-jittering buffer. Moreover, the receiver could hide at perceptual level individual missing segments using an appropriate Packet Loss Concealment (PLC) algorithm. More sophisticated recovering algorithms repair missing frames through signal processing (Liao et al., 2001).

The advantage of the reactive strategy stems from the fact that quality improvement is done by only upgrading edge nodes. This is suitable for large scale environments such as the Internet. Moreover, the network impairment concealment is performed in a user-friendly way according to the specific features of each service. The short-

Figure 4. Future network architecture (Bertrand, 2007)

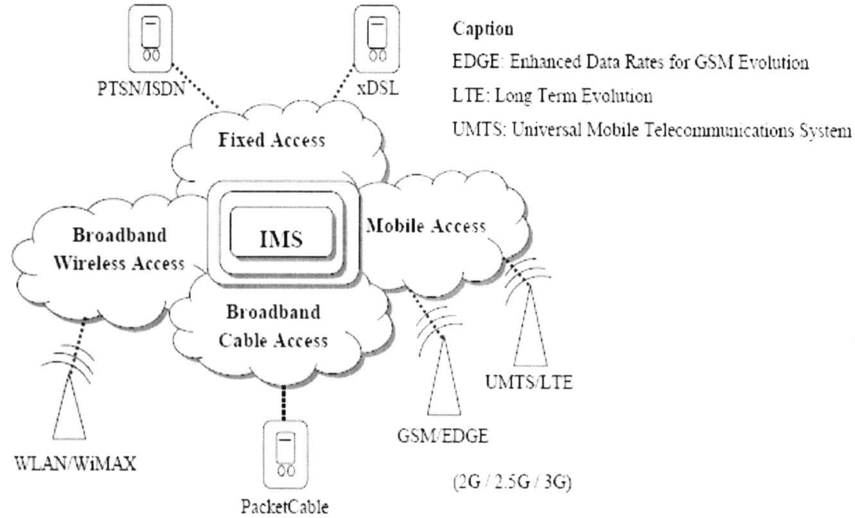

coming of the reactive strategy stems from the fact, that sometimes quality control algorithms at application layer are unable to hide network impairments under poor conditions. In such a case, it is needful to introduce quality control algorithms inside the network to improve the perceptual quality of services.

4.2 Predictive QoS Provision Strategy

The predictive strategy of quality provision paves the ground to satisfy the requirements of each delivered service in terms of reliability, delay, and delay jitter. Specifically, the predictive strategy installs adequate quality control algorithms at intermediate nodes. Moreover, it could define new components, protocols, and management policies to adequately handle each individual stream according to its requirements. Basically, there are two schools to achieve quality of service: resource reservation and service differentiation. In this sense, the IETF has defined two architectures to provide QoS over the Internet at packet layer: IntServ and DiffServ. IntServ QoS management protocol offers the required quality of service by reserving resources throughout the forwarding path (Blake et al., 1998). The reservation of resources at intermediate nodes is performed using the companion protocol RSVP which executes an admission control procedure at each intermediate node. Routers supporting IntServ should maintain in their control tables all information needed for the identification of each served flow. IntServ exhibits several shortcomings regarding scalability and efficiency of resource utilization. To avoid IntServ drawbacks, DiffServ was proposed and standardized as a scalable and resource-effective means for quality of service provision (Bradenet al., 1994). DiffServ QoS management protocol specifies two categories of nodes: edge and core nodes. The consumers are connected to the network through an edge node which performs admission and monitoring procedures. DiffServ uses SLA (Service Level Agreement) to properly characterize each stream. The edge node associates for each stream the adequate priority level which will be set into the header of each forwarded packet. This information will be later used by core nodes which deploy priority queuing discipline. Both IntServ and DiffServ have been adapted in order to achieve quality of service over last- and multi- hop wireless networks (Xu et al., 2003; Mirhakkak et al., 2000). Basically, these adaptations are performed to consider the network dynamics sustained by services over mobile networks. In reality, in the context of wireless networks, mobility induces frequent path switching, especially over a MANET where all nodes are continuously moving. This entails service interruption and establishment of routes with different quality of service depending on node location and load. Providing a constant quality of service in such environment is a serious challenge.

Apart from quality provision at network layer, there are significant efforts done to improve quality of service at link layer. This is mainly done to enhance the quality of service over shared wireless channels. In this sense, the IEEE has defined the norm 802.11e to improve the quality of service over WLAN (Wireless Local Area Networks) Gu & Zhang, 2003]. This is done through the deployment of priority queuing discipline on each wireless station at MAC layer to properly schedule frame transmissions. The parameters of MAC protocol are adequately adapted according to the service needs Gu & Zhang, 2003]. Basically, the adjustment of parameters gives high priority to delay-sensitive frames to accede in a distributed way to the shared medium. WiMAX is another QoS-aware protocol that has been standardized by the IEEE under the norm 802.16 (Cicconetti et al., 2006). This protocol enables long range communication and more sophisticated QoS support at MAC layer. Several different types of services can be used in WiMAX networks. The standard defines two basic operational modes: point-to-multi-point (PMP) and mesh. In the PMP mode, the subscriber stations (SS) are only allowed to communicate through the base stations (BS). It is expected that network providers will use PMP mode to connect customers to the Internet. In the mesh mode, subscriber stations can communicate with each other and with the base stations. The basic approach for providing the QoS guarantees in a WiMAX network is that the BS performs cleverly the scheduling for both the uplink and the downlink (Cicconetti et al., 2006).

A pre-requisite to the successful integration of voice services over IP networks is the reliable assessment of the user quality of experience. In the following section, we provide a thorough discussion of methodologies recommended by standardization bodies and their extensions.

5. METHODOLOGIES FOR VOCAL SERVICE QUALITY EVALUATION

It is extremely important for user and provider perspectives to determine the service quality of experience. This could be used for planning, maintenance, diagnosis, QoS management as well as billing and complaining. Basically, quality of experience could be evaluated subjectively or objectively. Subjective approaches need the involvement of human subjects. In contrast, objective approaches evaluate *automatically* the perceptual quality using models and algorithms running over a calculator. In this section, we describe both methodologies and specific requirements to evaluate interactive vocal conversation service.

5.1 Subjective-Driven Vocal Service Quality Evaluation

The subjective calculation of speech quality is performed using subjective trials. The evaluation of perceptual quality can be performed to quantify either the Listening Quality (LQ) or the Conversational Quality (CQ). LQ includes only impairments affecting the intelligibility of a voice sequence such as noises, coding, and losses. However, CQ includes in addition to the impairments captured by LQ those which affect the interactivity between communicating parties such as echoes and one-way delay. The ITU-T recommendation P.800 gives an in-depth description of how these trials should be conducted (ITU-T Recommendation P.800, 1996).

5.1.1 Listening Tests for Telephony

A listening subjective essay includes a set of human subjects which are asked to rate heard voice quality under specific degradation conditions. Typically, each original high-quality voice recording consists of speech sentence pairs of around 5 to 8 sec duration from a single talker where two male and two female talker recordings are used to evaluate each condition under test. Noise may be added to simulate a noisy environment such as car or air-conditioner noises. The original signals are processed through a filter modeling the handset sent path. Then, they are encoded using a speech encoder, under specific configuration, and properly distorted in order to mimic impairments occurring in the network. Finally, received signals are decoded. The degraded signals are presented to 24 to 32 subjects using a standardized telephone receiver, and subjects vote on the quality of each voice sequence. A widely used scale of five-point absolute category rating (ACR) listening quality (LQ) is used during voting sessions (see Table 1). The produced scores are statistically analyzed in order to obtain the mean opinion score of listening quality using subjective tests, denoted as MOS-LQS, of examined speech sequences. During the subjective experience, several questions could be asked to subjects in order to evaluate different dimensions of vocal quality such as the overall perceived quality, the naturalness of heard voice sequences, and degradation due to noises.

5.1.2 Conversational Quality Tests

The listening trails involve a human subject passively. Thus, such trials are unsuitable to measure impairments that occur or emerge in inter-personal communications. In general, a user's view of the quality of a conversation over a telephone connection is made using three distinct attributes:

- **Listening quality:** How does the subject perceive the voice from the other side of the link (noise, distortion)?

Table 1. Absolute category rating opinion scale

Listening quality	score
Excellent	5
Good	4
Fair	3
Poor	2
Bad	1

- **Talking quality:** How does the subject perceive his/her own voice (echo, side-tone, background noise switching)?
- **Interaction quality:** How well can both parties interact with each other (delay, double-talk distortions).

In general, conversational trials use pairs of human subjects, talking over a test network while performing some kind of interactive task, before voting (independently), normally using the quality scale. This allows tests to consider all the properties of the network from talker's mouth to ear, which include side-tone and handset acoustics, echo, delay jitter, and delay. However, conversational tests are relatively rare because they are slower, more expensive, and complex compared to listening tests.

It is well-recognized that subjective trials are expensive, unbiased, cumbersome, and time-consuming Rix et al., 2006]. Obviously, subjective approaches are useless for quality monitoring and management at run-time. To avoid subjective trials, significant research work has been performed to design instrumental algorithms which estimate or predict automatically the perceptual quality Rix et al., 2006].

5.2 Automatic-Driven Vocal Service Quality Evaluation

There are several algorithms which are designed to evaluate *automatically* speech quality Rix et al., 2006]. The performance of objective assessment algorithms is evaluated in terms of correlation between known subjective and objective scores. Moreover, the deviation between known subjective and objective scores, which is basically calculated using the mean square error (MSE), could be used as indicator to evaluate an objective assessment strategy.

There are several ways to classify existing objective assessment algorithms according to their inputs, processing, and system characterization. We select the classification made by Rix, A. et al in Rix et al., 2006] which defines two categories: *black box* signal approach and *glass box* system approach (see Figure 5).

5.2.1 Black Box Signal Approach

Black box signal approach estimates the perceptual quality by only processing voice signals without characterizing the underlying system. This assessment strategy is intended initially to evaluate speech quality over Telecom networks then extended to evaluate voice quality over packet-based networks (Hoene, 2005). Black box signal approach, called sometimes as *end-to-end assessment approach*, could be *intrusive* or *non-intrusive* (see Figure 6).

Intrusive models compare an original test signal with a degraded version that has been processed by a system. These methods are also called *comparison-based* or *full reference* models. Intrusive models transform original and degraded voice signals using perceptual models in order to consider key properties of human hearing. Next, they compute the difference in the transformed space which will be used to estimate the MOS score. The standardized and widely-used ITU-T

Figure 5. Overview of black box signal approach and glass box system parameter approach Rix et al., 2006

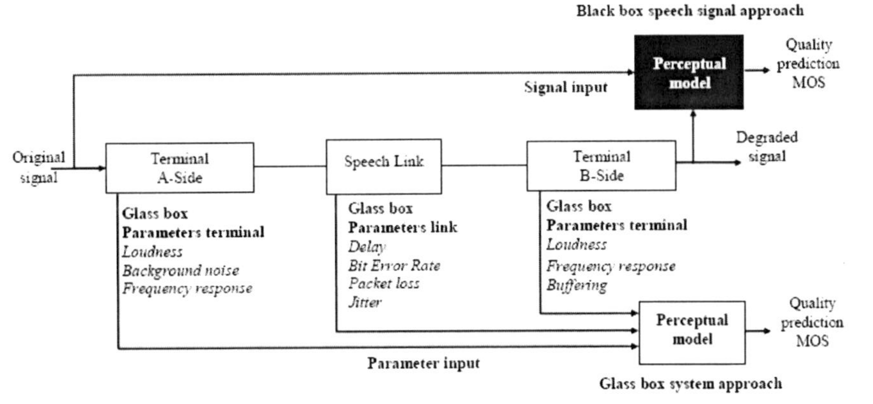

Figure 6. Intrusive and non-intrusive tests (a) Illustration of an intrusive test (b) Illustration of a non-intrusive test

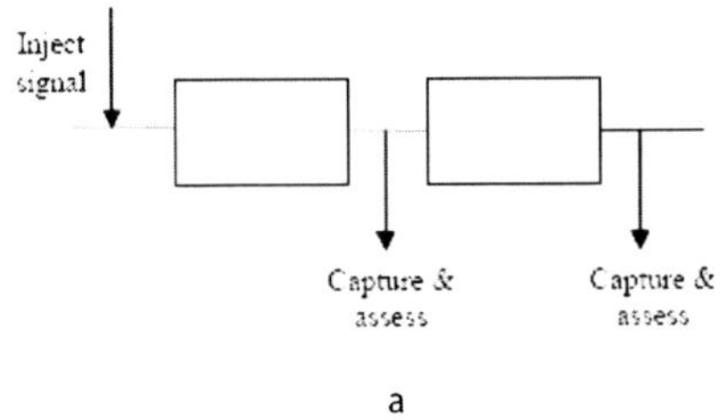

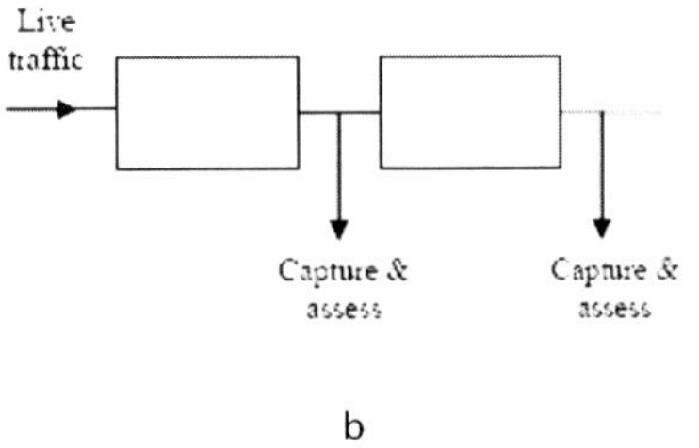

PESQ (Perceptual Evaluation of Speech Quality) assessment algorithm, which has been defined in the recommendation P.862, follows intrusive strategy (ITU-T Recommendation P.862, 2001). There are several amendments which are introduced by researchers in order to improve the accuracy over a wide range of networks, impairment situations, languages, and genders. Empirical studies prove that PESQ algorithm scores are highly correlated with subjective score (0.93) (ITU-T Recommendation P.862, 2001).

Technically, the intrusive approach injects a test signal into a system which is captured and assessed at several points (see Figure 6a). This requires taking out of service the system under tests. The main drawback of intrusive algorithms is the requirement to accede to the original speech sequences which are frequently unavailable. Hence, an intrusive algorithm is unable to evaluate live speech quality at run-time. Moreover, injected signals will surely result in network load increasing.

Non-intrusive signal-based models (also known as *no-reference* or *single-ended* models), which are in their infancy compared to intrusive models, estimate MOS score by only processing the degraded output speech signal of a live voice conversation. Basically, non-intrusive signal-based approach extracts a set of key parameters for the analysis of artifacts observed in an examined sequence such as interruption, mute, time clipping, and strong additive noise Rix et al., 2006]. Next, the subjective quality is estimated using a cognitive linear combination of all extracted signal features. Non-intrusive signal-based models are relatively less accurate than intrusive signal based models, but their single-ended property is useful under several usage scenarios where intrusive approaches are inapplicable such as service monitoring at run-time (see Figure 6b).

5.2.2 Glass Box System Parameter Approach

Glass box system parameter approach, which is also known as *non-intrusive parametric* model, does not require processing original or degraded voice signals, but estimates subjective quality from measured properties of the underlying transport network and/or terminal such as echo, delay, speech levels, noises, VoIP network characteristics, or cellular radio measure (see Figure 5). Parametric models are widely used for planning purposes to construct MOS estimates based on tabulated values such as CODEC type, bit-rate, delay, packet loss statistics, etc Rix et al., 2006].

E-Model is undoubtedly the most known parametric model-based approach. It was developed and standardized in 1998 by the ITU-T in Recommendation G.107 (ITU-T Recommendation G.107, 2003). Since its first release, several revisions have been made in order to increase its accuracy over a wide range of networks. E-Model is an end-to-end parametric assessment tool widely used for planning purposes to predict the conversational quality of vocal services over planned *telephone* networks (Sat & Wah, 2006). E-Model rates the transmission quality by combining sources of impairment experienced through the mouth to ear path, and providing as output a rating factor, denoted R. The rating factor is a scalar ranging from 0 to 100 corresponding respectively to the worst and the best transmission quality. A planned configuration resulting in a rating factor value below 60 is not recommended (ITU-T Recommendation G.107, 2003). Obviously, the rating factor R is strongly related to the MOS score. Figure 7 illustrates graphically the relationship between R and MOS. There are mathematical functions which enable to convert R to MOS, and conversely (Hoene, 2005).

Actually, the calculation of the R factor is based on twenty-one input parameters and includes complex mathematical formulas which are defined and obtained based on subjective

Figure 7. Relationship between R and MOS (ITU-T Recommendation G.107, 2003)

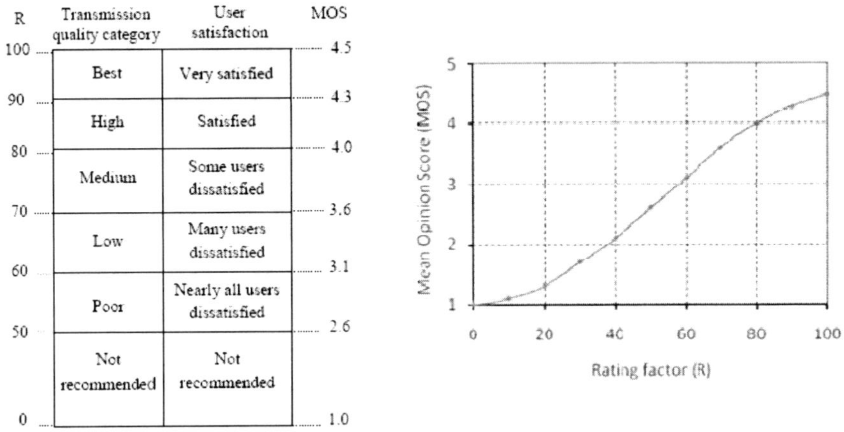

experiences (ITU-T Recommendation G.107, 2003). To simplify the calculation, the ITU-T has recommended fourteen default parameters that are independent of the transmission quality of the transport network. For the sake of simplicity, E-Model assumes that sources of impairment are additive on psychological scale. Hence, the rating factor is computed as follows:

$$R = R_0 - I_s - I_d - I_e + A \quad (1)$$

where, R_0 represents the transmission rating computed based on the basic signal-to-noise ratio which accounts for noise at the sender and receiver sides, I_s captures the sum of all impairments which may occur more or less simultaneously with the voice signal such as quantization and loud sounds, I_d captures impairments affecting interactivity such as delay and echoes, I_e captures impairments affecting the intelligibility of vocal stream such as CODEC used and the packet loss ratio, and A represents an advantage factor that accounts for user willingness to accept some quality degradation in return for ease of access. The value of A varies between 0 and 20 corresponding respectively to a wired network and two satellite hops. The reduced formula when default values are assigned is given by:

$$R = 93.2 - I_d - I_e + A \quad (2)$$

The advantage of E-Model stems from its high computational efficiency. However, the accuracy of E-Model is questionable (Takahashi et al., 2004). This is in part due to additive property adopted by E-Model which entails the production of inaccurate scores under several circumstances. Indeed, this assumption is done for the sake of simplicity and to make the model formally tractable. That is why, ITU-T recommends E-Model only for planning purposes in order to provide an initial feedback about conversational voice quality.

In order to improve the accuracy of parametric model-based assessment algorithms, it is possible to use a mixed approach where parameters could be estimated from voice signals. Thus, the resulting rating factor will reflect the effect of degradation according to specific content.

6. FRAMEWORKS FOR VOCAL QUALITY ASSESSMENT

The evaluation of *vocal quality* over next-generation networks requires sophisticated assessment frameworks in order to evaluate the suitability of designed architectures, management policies, QoS control algorithms, etc. In this section, we describe two frameworks which could be used to evaluate new proposals at perceptual level.

6.1 Software-Based Vocal Quality Assessment Frameworks

Software-based assessment frameworks can be used in order to evaluate speech quality over planned networks. The simulation software could be *application-centric* or *network-centric*. Application-centric frameworks are developed in order to evaluate the performance of components deployed at the ends of a connection such as CODEC used, play-out algorithm, recovering strategy, acoustic echo canceller, etc. The impairments introduced by the transport network are properly modeled to mimic real behavior of disturbing sources such as burst packet losses observed over the Internet and random bit errors observed over the wireless Telecom network.

Figure 8 outlines the basic components of an *application-centric* software-based assessment framework (Roychoudhuri et al., 2006), 28]. As we can see, the assessment framework includes a database of high-quality original voice sequences which are optimally stored to facilitate accessibility and analyses. Raw voice sequences are encoded using a dedicated CODEC under specific configuration parameters such as rate and voice activity detector (VAD) threshold. The produced bitstream is impaired by introducing sources of distortion such as noise, delay, and loss. The framework illustrated in Figure 8 introduces distortions due to losses solely. Specifically, the plotted framework enables introducing packet losses observed over packet-based networks such as Internet and bit errors observed over wireless circuit-based networks such as GSM. Loss simulators should be adequately parameterized in order to accurately identify the effect of each input factor on the response variable, i.e., the perceived quality.

Impaired received stream is decoded using the adequate CODEC and the final degraded speech sequence is generated (see Figure 8). Distorted voice sequences could be assessed subjectively using human subjects or objectively using an automatic assessment algorithm. The assessment algorithm inputs depend on the strategy used by the assessment tool to evaluate a speech sequence. Generally, the standardized intrusive end-to-end

Figure 8. Application-centric framework for vocal quality assessment

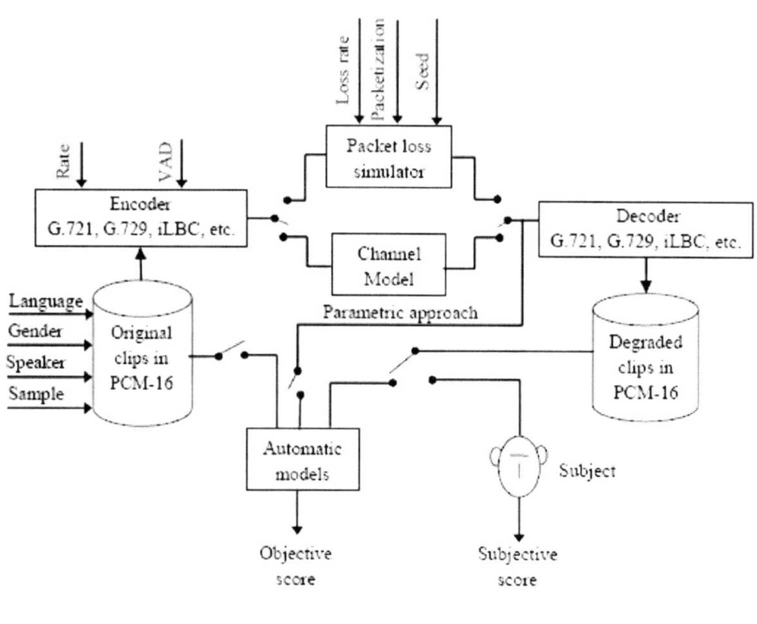

Figure 9. Network-centric assessment framework of vocal conversation

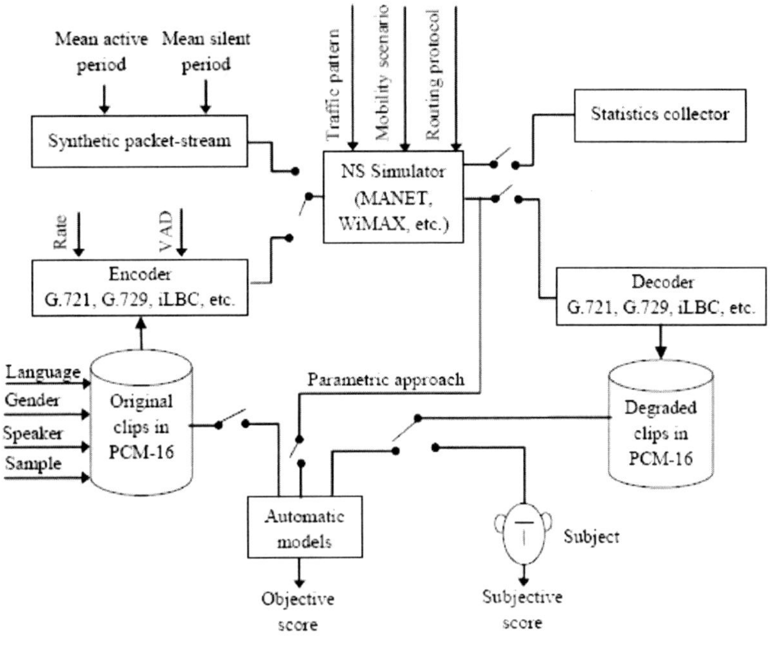

assessment algorithm PESQ is used to quantify the effect of introduced distortions on perceived quality (Roychoudhuri et al., 2006), 28].

As we can see, application-centric software is only suitable to evaluate vocal quality under common conditions. In order to consider the specificity of each network and configuration, Wanstedt et al. (2002) have proposed a *network-centric* assessment framework which enables introducing distortions according to a specific network configuration, technology, and load. Figure 9 outlines the main components of the proposed assessment framework which is basically designed to evaluate voice conversations over mobile ad-hoc networks, known as MANET. However, it can also be useful to evaluate other heterogeneous environments including WiMAX, WLAN, Bluetooth, and UMTS. The network part of the system is simulated using the open source, event-driven, Network Simulator, commonly known as NS2, which is widely used to study various network architectures, including MANETs ("NS Simulator Homepage," n.d.). As shown in Figure 9, users could define the traffic pattern and movement scenario. Moreover, the network parameters such as network size, number of nodes, routing protocol, signal propagation model, and specific node parameters such as data rate, transmission range and MAC protocol could be easily defined.

The developed framework enables the transmission of real or synthetic vocal sequences. Real vocal sequences are encoded, packetized, and sent through the NS2 simulator. However, synthetic vocal sequences are artificially generated according to an ON / OFF source where mean active and silent periods are specified with respect to the ITU-T recommendation P.59. This model is actually obsolete since modern CODECs send periodically voice packets even during silent periods, which are called comfort noise packets. The ON/OFF model could be easily adapted to mimic modern CODECs by periodically generating voice packets during OFF periods (Yih-Guang, 2007). In order to accurately evaluate vocal conversations, the simulator NS2 has been extended to adequately packetize and de-packetize real voice sequences at sender and receiver sides, respectively. At receiver side, several play-out algorithms have been implemented in order to study their suitability over a MANET ("NS Simulator Homepage," n.d.). Using this assessment framework, new management policies, handover controls, play-out algorithms, and QoS engineering algorithms, designed to improve perceived vocal quality over a MANET, could be effectively evaluated under realistic conditions (Wanstedt et al., 2002).

The assessment framework described in Wanstedt et al. (2002) could easily be extended to evaluate voice service quality as well as other delay-sensitive services over next-generation networks (NGN). Indeed, several features and technologies of NGN are introduced within NS2 distribution which will greatly assist network designers and architects.

6.2 Emulation-Based Vocal Quality Assessment Frameworks

The assessment of vocal quality could be done based on *emulation*. This approach needs the deployment of adequate experimental test-beds which mimic in a controlled way real infrastructures. Experimental test-beds allow assessing conversational voice quality in a more realistic way than software-based approaches. However, typical experimental test-beds are costly in time, space, and price. Moreover, running experiments using test-beds require an important engineering effort. That is why, sophisticated test-beds are solely built and used by standardization bodies and corporations specialized in quality (ETSI report, 2002; Takahashi et al., 2006).

The ETSI used extensively emulation-based strategy to evaluate voice conversations carried over hybrid packet- and circuit- based networks (ETSI report, 2002). Especially, a multitude of experimental test-beds has been set-up during the 2nd ETSI TIPHON VoIP Speech Quality Test Event. An example of deployed test-beds during this event is illustrated in Figure 10. The shown configuration is set-up to evaluate voice conversations delivered over an IP network using ISDN as access technology. A PBX is properly configured in order to set-up 2-way vocal calls using two handset phones. Vocal data are transmitted through two gateways which achieve a seamless conversion between circuit- and packet- based parts. A packet network emulator has been used to introduce distortions observed over IP networks such packet losses, delay, and delay jitter. A computer-based monitor agent is used to extract features of impaired packetized vocal stream. An acoustic system is used to properly acquire and record original and degraded voice sequences. The injected vocal traffic could be a stored voice sequence or a real voice generated by a subject at run-time.

Emulation-based strategy could be used to evaluate the effect of handover on quality of delivered services (Malden Electronics, 2008). Handover represents an intrinsic feature of wireless systems which could be performed either between radio cells belonging to the same network or between heterogeneous overlapping networks. It is well recognized that handover influences significantly the users' experience. The handover can be examined

Figure 10. Test-bed for speech quality evaluation (ETSI report, 2002)

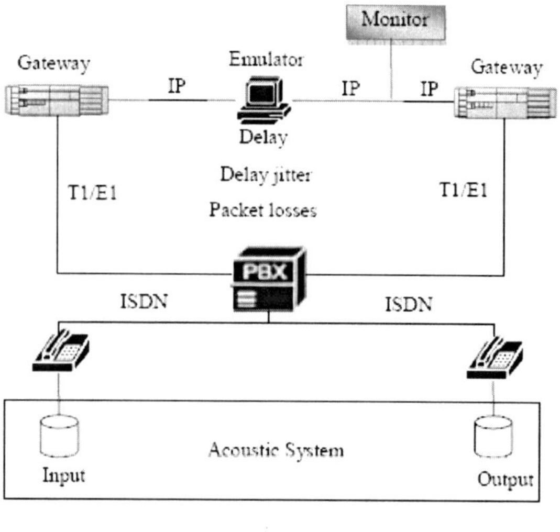

from either network perspective or terminal/user perspective. The measurements of the handover from the users' point of view offer two principal benefits (Malden Electronics, 2008):

- The *handover effect* is measured in terms of the impact on speech quality; signal interruption time and changes in quality, speech level, noise level, and delay can all be clearly seen.
- The method works with all combinations of access network technologies. Intrusive measurements can be used to assess transparently handover in GSM/GSM (cell handover), GSM/WLAN, WCDMA/WLAN, GSM/WCDMA and other technologies.

Figure 11 outlines an example of drive-test configuration done in lab or urban environments which mimics vocal calls between mobile and fixed users (Malden Electronics, 2008). The mobile user accedes to the network infrastructure via a wireless interface, whereas, the fixed user accedes to the network via PSTN network. Two wireless access stations are adequately configured (covering range, data rate, recovering strategy, handover approach, etc.) to imitate target scenarios. Initially, mobile equipment is connected to fixed infrastructure through the wireless access station A. Throughout ongoing vocal calls, the mobile node moves toward wireless access station B with a specific velocity. This is done to mimic pedestrian and vehicular mobility under specific environments. Once the mobile node is getting out of the coverage range of wireless access station A, a handover is performed. After a handover occurrence, the mobile node becomes served by wireless access station B. Dedicated monitoring equipment collects several measurements about measured speech quality. At the end of vocal call, human subjects could be asked to rate their experience. Alternatively, the quality could be obtained using objective assessment algorithms which should process collected measurements of interest before, during, and after the handover which include, among others, speech quality, delay / delay variation, speech level.

Next, we look into the use of parametric model-based assessment algorithms of users' quality of experience, in the context of cellular Telecom systems as well as packet-based networks.

7. PARAMETRIC MODEL-BASED VOCAL QUALITY PREDICTION

Parametric model-based assessment algorithms are highly incited by industrials. In fact, several features of parametric models are attractive for a multitude of applications. For instance, parametric model-based algorithms could be easily used for network diagnosis, maintenance, and

Figure 11. Drive-test evaluation in lab/urban environment of speech quality during a handover

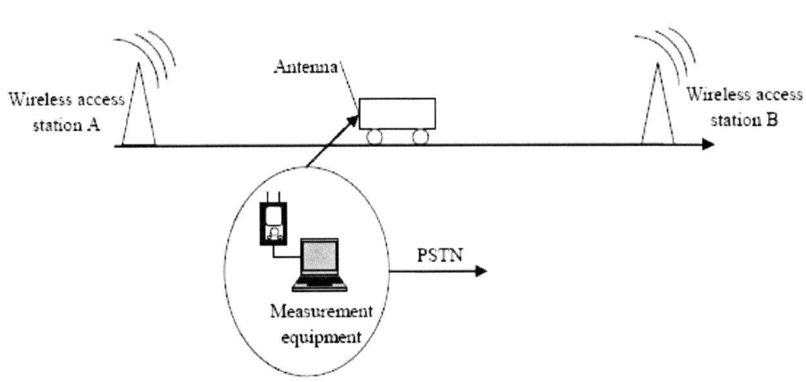

optimization, as well as quality monitoring at run-time. This is performed without disturbing the transport network or acceding to signal layer which is preferred for security reasons. Moreover, parametric models are characterized by a reduced complexity. In this section, we describe how parametric model-based assessment algorithms are developed and used in the context of cellular Telecom systems and packet-based networks. This description is meant as a guideline for further investigations of parametric models, specifically for next-generation networks.

7.1 Vocal Quality Assessment over Cellular Telecom Networks

Initially, cellular Telecom operators measure and benchmark their QoS network performance mostly using the dropped call and bit error rates, denoted respectively as DCR and BER (Barile et al., 2006). DCR measures the rate of lost connections and BER is used to estimate speech quality. However, in a stable network, DCR is almost always near 0%. Hence, DCR parameter is inadequate to benchmark the QoS provided to users from different providers. Moreover, using BER only does not allow full characterization of perceptual quality.

Traditionally, speech quality over GSM systems was measured using RxQual metric (3rd Generation Partnership Project, 1999). The value

Table 2. RxQual threshold levels

RxQual	% of error bits
0	< 0.2%
1	< 0.4%
2	< 0.8%
3	< 1.6%
4	< 3.2%
5	< 6.4%
6	< 12.8%
7	> 12.8%

of RxQual is calculated through a logarithmic mapping of the BER into a scale varying from 0 (excellent quality) to 7 (worst quality). The relation of RxQual to BER is given in Table 2 (3rd Generation Partnership Project, 1999). According to GSM specifications, RxQual measure is available in the operation and maintenance center, denoted as OMC, for the uplink direction and is usually also part of the standard measurement reports sent from the mobile station (3rd Generation Partnership Project, 1999). The availability of RxQual measures makes it useful to supervise all ongoing calls in the radio network.

RxQual is a very basic measure. In fact, it simply reflects the average BER over a certain period of time (≈0.5 sec). However, speech quality assessment is a complex process which is influenced by many factors. In particular, RxQual fails to consider the following factors (Ericsson, 2008):

- **The distribution of bit errors over time:** For a given BER, if the BER fluctuates very much, the perceived quality is lower than if the BER remains rather constant most of the time. Different channel conditions entail different BER distributions. However, RxQual measure is unable to capture the effect of loss distribution since it merely represents the average BER.
- **Frame erasures:** The perceived quality is sensibly affected in a negative way when entire speech frames are lost.
- **Handovers:** Handovers entail the loss of some frames which generally produces audible disturbances. This does not show at all in RxQual since according to GSM specification BER measurements are suppressed during handovers.
- **The choice of speech CODEC:** The general quality level and the highest quality vary widely between speech CODECs. Moreover, each codec has its own strengths and weaknesses with respect to the types of input and channel conditions.

In short, RxQual is unable to precisely capture several factors that influence sensibly speech quality judged by mobile users. To accurately reflect speech quality rather than radio channel conditions, SQI (Speech Quality Index) metric has been defined (Karlsson et al., 1999). SQI is conceived specifically to evaluate speech quality over the last wireless hop of a vocal connection (see Figure 12). In fact, intrusive end-to-end assessment algorithms used classically over PSTN estimate speech quality of combined sources of distortion which include radio network, switches, and user equipments. Therefore, end-to-end speech quality measurements do not offer direct relation with the radio channel, which is unsuitable for cellular network optimization and diagnosis.

SQI assessment algorithm, which falls in parametric model-based assessment category, was designed to consider all features of current radio channel conditions (Ericsson, 2008). The following input factors have been proposed in the literature in order to compute SQI metric:

- **RxQual:** This measure, which is used classically for vocal call assessment at run-time, is calculated through a logarithmic mapping of channel bit error rate averaged over a period of 480 ms. In the GSM system, values below four are preferred since at BER less than 1.6% nearly all bit errors can be recovered by the channel decoder (see Table 2).
- **RxLev:** The received power level at the mobile station is measured in dBm (relative to 1 mW) and mapped linearly to a RxLev index ranging from 0 to 63 (Werner et al., 2003). Received power measure, which reflects the radio channel in terms of path loss and slow fading, are reported to the serving base station periodically every 480 ms (Werner et al., 2003).
- **FER:** This measure corresponds to frame erasure rate observed during a monitoring period.
- **LFER:** This metric represents the length of erased frames computed as the mean sequence length of consecutively erased speech frames sustained over the last monitoring period.
- **MxLFER:** This measure represents the maximum length of erased frames during the last 2.5 sec.
- **MnMxLFER:** This metric corresponds to the mean of maximum length of erased frames which is a combination of local maximum sequence lengths of erased speech frames during four intervals of equal length. A large value over short periods is considered as a potential indicator of severe quality degradation at user level.

Figure 13 illustrates graphically the methodology adopted by SQI assessment speech quality algo-

Figure 12. Range of end-to-end vs. air interface speech quality measurement

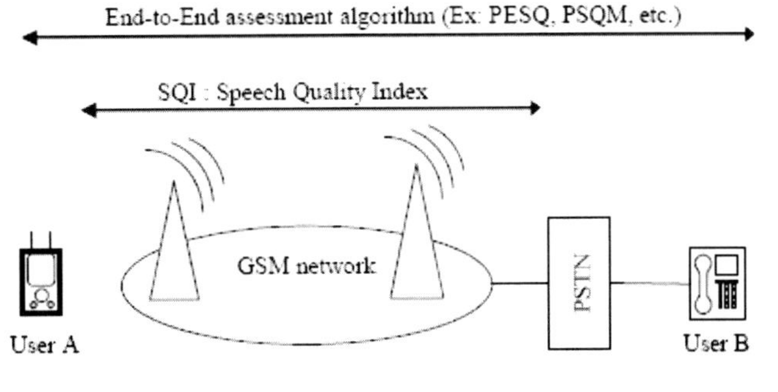

Figure 13. Principal of SQI, input parameters and basic processing steps

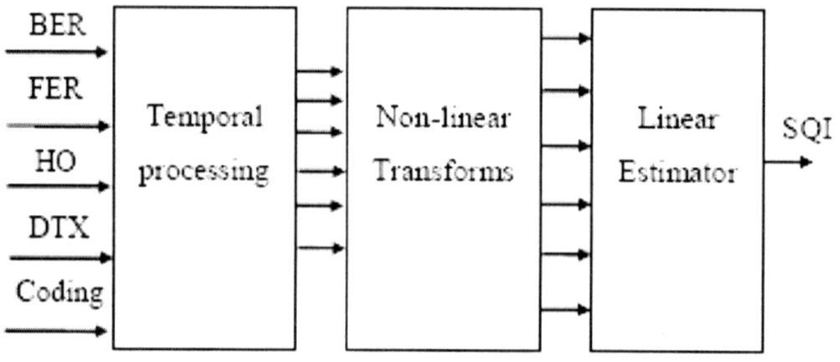

rithm to quantify sustained speech quality. First, raw radio channel measurements such as BER, FER, and HO (Handover) are processed in temporal domain. Then, collected set of measurements is non-linearly transformed in order to emphasize/deemphasize the effect of each measured factor with respect to the features of auditory human sensor system. Finally, the speech quality score is calculated through a linear combination of transformed measurements. In order to accurately estimate perceptual quality, non-linear transformation functions and final linear combination rule should be adequately established and calibrated for each speech CODEC. The SQI value is reported to the base access station every 480ms.

To determine adequate non-linear transformation functions and final linear combination, a modeling sequence should be performed *off-line* using existing subjective listening results of speech quality collected from the studied network. Figure 14 outlines the steps followed to build appropriate models to derive SQI value at run-time. The radio quality data is analyzed to determine what parameters are important, their distributions as well as their inter-relations. Sometimes, input parameters should be transformed to increase their correlation with subjective speech quality results. For instance, BER is linearly related to speech quality and can be used without transformation. However, FER behaves non-linearly and requires a transformation. For example, the square root of FER correlates well with speech quality. Moreover, it is observed empirically that MxLFER correlates linearly with subjective results (Wanstedt et al., 2002). After single factor analyses, a multivariate model of speech quality is developed using multiple linear regression. Further transformations of variables and consideration of outliers may be necessary to produce a robust model. A likely form of SQI model for a given CODEC could be given by:

$$SQI = a \times BER + b \times FER^{x} + c \times MxLFER + d \quad (3)$$

Figure 14. Schematic of receiver side of modeling sequence

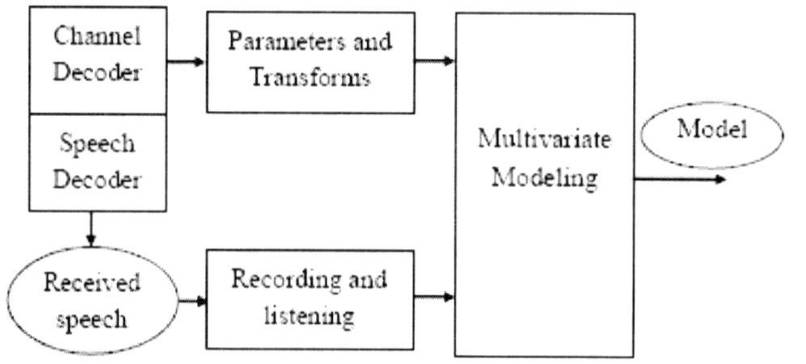

where a, b, c, and d are four real constants obtained using a multiple regression, x is a scalar exponent used to emphasize correlation between FER and subjective results. Generally, a representative selection of sample should be performed to ensure that all parts of the data space are equally represented in order to avoid production of biased models. For example, the original sample distribution could be flattened by cropping the original histogram above a certain level (Wanstedt et al., 2002). The obtained models should be properly validated *off-line* using a different set of subjective results, which is not used during the models development step.

In Karlsson et al. (1999), the authors propose the following model to estimate SQI using GSM Full-Rate CODEC when handover and DTX events are not accounted for:

$$\mathrm{SQI} = 20.67 - 57.2 \times \mathrm{BER} - 29.3 \times \sqrt{\mathrm{FER}} - 0.11 \times \mathrm{MxLFER} \tag{4}$$

where, the average BER value corresponds to the average bit error rate occurring during the last 2.5 sec, the square root of FER is limited to the maximal value 0.66. According to the conducted empirical study, the developed parametric model provides better performance than end-to-end intrusive algorithm PSQM. In fact, PSQM assessment algorithm was designed to assess speech quality over fixed Telecom networks. Therefore, it will be surely unsuitable to measure speech quality over radio channel.

In Werner et al. (2003), authors propose another model to estimate precisely speech quality over GSM systems. The model is obtained using a multiple linear regression applied on a set of key parameters described previously after linearization (Werner et al., 2003). Specifically, the following model has been proposed to estimate speech quality over GSM system at run-time:

$$\mathrm{SQM} = \mathrm{T}_1 \times \mathrm{f}_1\left(\mathrm{L}_6\left(\mathrm{RxQual}\right)\right) + \mathrm{T}_2 \times \mathrm{f}_2\left(\sqrt{\mathrm{FER}}\right) + \mathrm{T}_3 \times \mathrm{f}_3\left(\mathrm{L}_1\left(\mathrm{MnMxLFER}\right)\right) + \mathrm{B} \tag{5}$$

where, SQM stands for Speech Quality Measure which depends on subjective measurement approach, T_1, T_2 T_3, and B are the weighing factors which are optimized over available data set. The function f_n corresponds to polynomial of degree $m \in \{2..6\}$. L_p corresponds to the Euclidian norm of order p which is calculated as follows:

$$\mathrm{L}_p(\mu) = \left[\frac{1}{N}\sum_{k=1}^{N} \mu(k)^p\right]^{1/p} \tag{6}$$

where, ε corresponds to the input parameter, N the number of measures done during an assessment period.

The perceptual impact of inter-cell handover on speech quality, expressed in term of MOS score, has been studied by Barile et al. (2006). Specifically, authors introduce a new Boolean variable, denoted as HO, into the developed speech quality model to account for a handover instance on user experience. The value of HO is set to 1 when a handover occurs during the assessment interval, otherwise it is set to 0. A large set of voice sequences has been collected from an existing GSM infrastructure covering a wide range of channel conditions. The assessment of voice sequences has been done *automatically* using the intrusive PESQ assessment algorithm. As in Werner et al. (2003), a multiple linear regression after linearization has been performed to derive suitable speech quality models. Specifically, M. Barile et al propose the following model for GSM EFR (Enhanced Full Rate) and HR (Half Rate) speech CODECs, working respectively at a coding bit-rate equal to 12.2 kbps and 6.5 kbps:

$$\mathrm{MOS} = \mathrm{A}_0 + \mathrm{A}_1\mathrm{L}_{p_1}\left(\mathrm{RxQual}\right) + \mathrm{A}_2\mathrm{L}_{p_2}\left(\mathrm{FER}\right) + \mathrm{A}_3\mathrm{HO} \tag{7}$$

Table 3. p-values that maximize the correlation coefficient

COEFFICIENT	EFR		HR	
	UL[(1)]	DL[(2)]	UL	DL
p_1	1/2	1	2	2
p_2	1	1	-	-

[(1)] UL: Upper link [(2)] DL: Down link

where, A_0, A_1, A_2, A_3 correspond to weighting coefficients which vary according to CODEC used. The constant A_2 is equal to 0 for the HR CODEC. L_p corresponds to Euclidian norm of order p calculated using Equation (6). The suitable values for p_1 and p_2 are reported in Table 3. The weighting coefficients are obtained using a multiple linear regression which minimizes the root square error over all data sets. The obtained results are summarized in Table 4.

Note finally, that the described approach to model speech quality could be used to develop adequate models over a new generation cellular Telecom system such as UMTS, EDGE, and LTE.

7.2 Vocal Quality Assessment over Packet-Based Networks

The rapid growth of packet-based vocal applications (VoIP) has incited industrials and academics to study quality of service over packetized networks. Indeed, conventional vocal assessment algorithms were intended to evaluate conversational voice service over Telecom networks. However, features of packet-based networks differ radically from circuit-switched networks. Therefore, original assessment algorithms are obviously unsuitable to evaluate packetized vocal conversations. There are several tentative in the literature to adapt original assessment algorithms in order to properly evaluate packet-based vocal conversations (Hoene, 2005).

Guided by industrial needs, most efforts aimed at designing assessment tools, which rely only on *packet-layer information* and specifically on the header content of each delivered vocal packet. The assessment tools could be able to evaluate at run-time a live vocal conversation in a non-intrusive way. Moreover, it could be deployed anywhere in the path separating end communicating nodes. The accuracy of produced objective scores should be pretty correlated with subjective results. Almost all proposed packet-based vocal assessment algorithms attempted either adapting original ITU-T E-Model, or using its paradigm, claiming that the effect of several sources of distortion is additive in psychological scale (Hoene, 2005; Cole & Rosenbluth, 2001).

To the best of our knowledge, Cole, R. G. et al made the first revisions of E-Model in order to make it suitable for quality monitoring of packetized voice streams (Cole & Rosenbluth, 2001). Authors argue that delay and equipment impairment factors are more relevant to evaluate vocal conversations over IP networks. They suggest using default values for all factors that are not related to the quality of the transport network such as room and circuit noises. Authors

Table 4. Coefficients of Barile, M. et al models

COEFFICIENT	EFR		HR	
	UL	DL	UL	DL
A_0	3.772	3.683	3.271	3.257
A_1	-0.009	-0.044	-0.003	-0.014
A_2	-0.023	-0.041	-	-
A_3	-0.115	-0.071	-0.156	-0.241

propose the following formula to compute the rating factor:

$$R = 93.2 - I_d\left(T_a\right) - I_e\left(CODEC,plr\right) + A \quad (8)$$

where, T_a corresponds to the mean end-to-end delay observed during an assessment period (around 10 sec), plr and CODEC represent the end-to-end packet loss rate and CODEC used. In order to estimate the rating factor at run-time using Equation (8), adequate models of delay and equipment impairment factors should be developed. These models are built and calibrated according to a set of available subjective results (ITU-T Recommendation G.107, 2003). In Cole and Rosenbluth (2001), the authors propose the following delay distortion model, obtained based on linear regression:

$$I_d(T_a) = 0.024 \times T_a + 0.11 \times (T_a - 177.3) \times H(T_a - 177.3) \quad (9)$$

$$\text{where} \begin{cases} H(x) = 1 \text{ if } x < 0 \\ H(x) = 0 \text{ if } x \geq 0 \end{cases}$$

Lingfen, S., et al criticized the inaccuracy of this model to quantify the effect of delay beyond 400 ms. Therefore, they propose the following model, obtained using polynomial regression, which is able to accurately quantify the effect of one-way delays reaching 600 ms (Sun & Ifeachor, 2006):

$$I_d\left(T_a\right) = -2.468 \times 10^{-14} \times T_a^6 + 5.062 \times 10^{-11} \times T_a^5 - 3.903 \times 10^{-8} \times T_a^4 + 1.344 \times 10^{-5} \times T_a^3 - 0.001802 \times T_a^2 + 0.103 \times T_a - 0.1698 \quad (10)$$

The equipment impairment factor I_e is related to coding technology, packet loss behavior, and de-jitter buffer and packet loss concealment algorithms. There are several mathematical models which are proposed in the literature in order to quantify the effect of equipment factor on users' perceived quality (Cole & Rosenbluth, 2001; Sun & Ifeachor, 2006; Broom, 2006). In order to avoid extensive subjective experiment required to develop I_e models, Lingfen, S., et al proposed an instrumental (objective) approach to derive equipment model under any configuration (Sun & Ifeachor, 2006). A similar approach has been standardized by ITU-T under Recommendation P.834. Essentially, the general form of I_e is given by:

$$I_e\left(CODEC,plr\right) = a + b \times \ln\left(1 + c \times plr\right) \quad (11)$$

where, a, b, and c are real fitting coefficients, which are obtained using regression applied on a known set of subjective scores. For instance, the recommended coefficients for the G.711 voice CODEC equipped with a packet loss concealment (PLC) algorithm, under a random loss condition, are a= 0, b= 30, and c= 15 (Cole & Rosenbluth, 2001).

Cole, R. G. et al adaptation does not account for several features of packet-based, best-effort, networks and especially the effect of loss behavior and temporal impairment location on users' experience. In order to properly evaluate users' experience, a set of new key concepts have been defined (Clark, 2001):

- **Instantaneous quality:** This is the measured voice quality due to packet loss, delay, CODEC used, and other impairments at some moment during the call
- **Perceptual quality:** This corresponds to the quality that would be reported by users at some time during the call
- **Time varying loss behavior:** IP packet losses are bursty in nature and, according to Bolot (1993), they oscillate between a burst and gap state. Burst is defined as a period of time bounded by lost and/or discarded packet with a high rate of losses. Gap is defined as a period of time between two bursts.
- **Recency effect:** In MOS experiments carried by Telecom operator, it was noticed

Figure 15. Perceptual effect due to time varaying loss behavior (Clark, 2001)

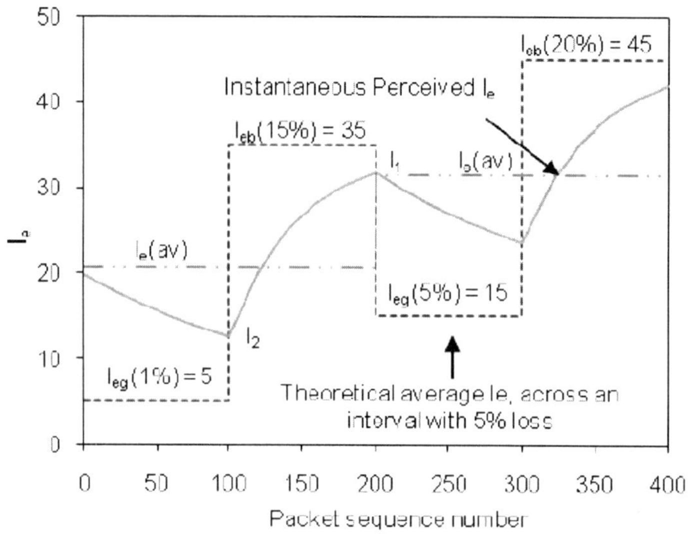

that the perceived quality varies according to loss location during the conversation time. Hence, impairments that occur at the end of a call are more annoying than impairments that occur at the beginning of a call.

A recommendation called provisionally P.VTQ by the ITU-T calls for a single-ended algorithm to evaluate packetized vocal conversations (Broom, 2006). P.VTQ gives the target performance and properties of such an assessment tool. There are two competing commercial proposals for normalization: VQmon developed by Telechmy and PsyVoIP developed PsyTechnics (Clark, 2001; Barriac, 2003).

7.2.1 VQmon of Telechmy

The single-ended vocal monitoring tool, denoted VQmon (Voice Quality monitoring), is specifically intended to evaluate packetized vocal conversations. It modifies the way used to compute I_e factor of E-Model to account for distortions and users' behavior over IP networks. The calculation of I_e considers users' experience at the transition between high and low loss periods, called respectively *burst* and *gap* periods. In fact, it is observed by Telecom operators, that when a transition occurs from *good* to *bad* network state at some moment during a conversation, then clients will not be immediately affected by network quality degradation. However, after a certain period, the listener would become annoyed with the voice quality degradation. The same psychological process is observed when a transition occurs from bad to good network state. Figure 15 illustrates the evolution of perceptual disturbances due to I_e reported by users over periods characterized by low and high loss rates (Clark, 2001).

These perceptual effects are modeled by VQmon using exponential decay/rise, calibrated based on subjective experiences done by France Telecom. Precisely, a time constant of 5 sec (resp. 15 sec) is used to model the transition from good (resp. bad) to bad (resp. good) states. At the end of an assessment period, VQmon calculates the average equipment distortions as follows (Clark, 2001):

$$I_e(av) = \left(bI_{eb} + gI_{eg} - t_1\left(I_{eb} - I_2\right)\left(1 - e^{-b/t_1}\right) + t_2\left(I_1 - I_{eg}\right)\left(1 - e^{-g/t_2}\right)\right) \Big/ \left(b + g\right) \quad (12)$$

where, I_{eb} and I_{eg} correspond to the instantaneous values of I_e during burst and gap periods which are calculated with respect to the mean packet loss rate (see Figure 15), t_1 and t_2 are two time constants used to calibrate exponential curves, b and g represent respectively the mean burst and gap durations. I_1 and I_2 correspond respectively to the values of equipment impairment factor at the transition from bad to good and conversely (see Figure 15). Based on empirical curves plotted in Figure 15, the values of I_1 and I_2 are computed as follows:

$$I_1 = I_{eb} - \left(I_{eb} - I_2\right)e^{-b/t1} \tag{13}$$

$$I_2 = I_{eg} + \left(I_1 - I_{eg}\right)e^{-g/t2} \tag{14}$$

In order to determine the mean gap and burst durations, g and b, as well as burst and gap loss rates, a 4-state Markov chain is used by VQmon. The chain enables capturing the alternating behavior of packet losses during a monitoring period. The transition probabilities are computed at run-time using a set of counters updated using an efficient packet-loss driven algorithm. At the end of a monitoring period, VQmon computes required measures using the calibrated Markov chain (Clark, 2001). The *recency* effect described previously is incorporated by VQmon in the calculation of I_e factor at the end of vocal call as follows:

$$I_e\left(\text{end of call}\right) = I_{e\text{-}w}\left(\text{av}\right) + \left(k \times \left(I_1 - I_e\left(\text{av}\right)\right)\right) \times e^{-y/t_3} \tag{15}$$

where, $I_{e\text{-}w}$ corresponds to the weighted average of $I_e(av)$ observed during a vocal call, where the weights correspond to the period duration, y represents the duration since the last burst loss period, t_3 is used to calibrate the exponential decay from the recent significant distortion which varies between 30 and 60 sec, and k is a nominal constant which is set to 0.7. Therefore, the rating factor at the end of a packet-based vocal conversation is calculated as follows:

$$R = 93.2 - I_e\left(\text{end of call}\right) - I_d \tag{16}$$

where I_d represents the weighted instantaneous delay impairment factor where weights correspond to period duration. Empirical study proves that VQmon correlates well with subjective scores under a wide range of conditions. However, the accuracy of VQmon remains invalid under several circumstances.

7.2.2 PsyVoIP of Psytechnics

PsyVoIP has been proposed by Psytechnics Corporation as a single-ended assessment tool of packetized vocal conversations (Barriac, 2003). The major contribution of PsyVoIP consists of considering precisely the role played by edge-device to reduce network impairments, namely delay, delay jitter, and losses. Indeed, PsyVoIP designers argue that VQmon is unable to properly account for de-jitter buffer and packet loss algorithms used by an edge-device which are often guarded secret by the manufactures. Broom, S. R. showed empirically that edge-device produce different perceptual quality under the same level of network distortions namely mean packet loss rate and delay jitter (Barriac, 2003). To account for different edge-device, Broom, S. R. proposes to calibrate PsyVoIP off-line according to the specific feature of each edge-device. The monitoring VoIP architecture of PsyVoIP is illustrated in Figure 16. The role of each module is described is the following sections.

1. **IP call handler:** This module is used to identify vocal packets belonging to the monitoring call. This is done based on packet header information. Non-VoIP packets are simply discarded.

Figure 16. VoIP monitor architecture (Barriac, 2003)

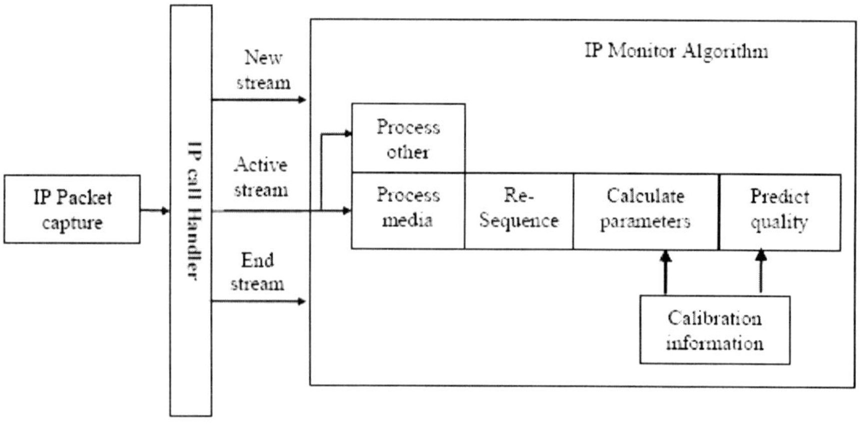

2. **Process media:** This module extracts relevant information from the header, and optionally from the payload if available/ allowed.
3. **Process others:** This module handles other packets which do not contain payload data, but are used for control and management of vocal stream such RTCP packets.
4. **Re-sequence module:** This module re-sequences incoming packets to enable detecting lost and out-of-order packets. The output of the re-sequence buffer is a container for each packet in the stream. A packet present/lost indicator is used to identify lost packets.
5. **Calculate parameters:** This module is of keen importance for a flexible monitoring architecture. It calculates relevant parameters on a stream-by-stream basis using packet information from the output of the re-sequence buffer. Each parameter is calculated on a packet-by-packet basis from a series of *base parameters* over a window of N packets (Barriac, 2003).
6. **Predict quality:** The final step consists of producing a quality prediction using internal parameter values over the last window as follows:

$$Q_n = Q^{MAX}\text{-}h\left(\sum_{m=1}^{P} k_m\left(p_{m,n}\right)\right) \quad (17)$$

7. Where, $p_{m,n}$ represents the value of mth internal parameters at n^{th} packet, k_m is a non-linear function developed specifically for each edge-device, P corresponds to the number of parameters, h is a monotonic and nonlinear mapping function which is applied to produce a quality score on a MOS scale. Since each parameter is based on quality degradations, the final quality score is produced as a maximum achievable MOS, minus the degradation. This maximum value is configurable, and enables to easily account for the performance of different CODECs.
8. **Calibration information:** This module is the key to being able to account for different VoIP devices. This information describes the parameters, functions, and weightings required to be able to predict the speech quality from a stream of VoIP packets for the specified edge-device. The monitor can be configured so that each stream can be assigned to a different set of calibration information.

Empirical study conducted by Broom, S. R. shows that PsyVoIP achieves better correlation with subjective results then a generic assessment

Figure 17. QoS loop control management for VoIP using subjective metric (Meddahi et al., 2006)

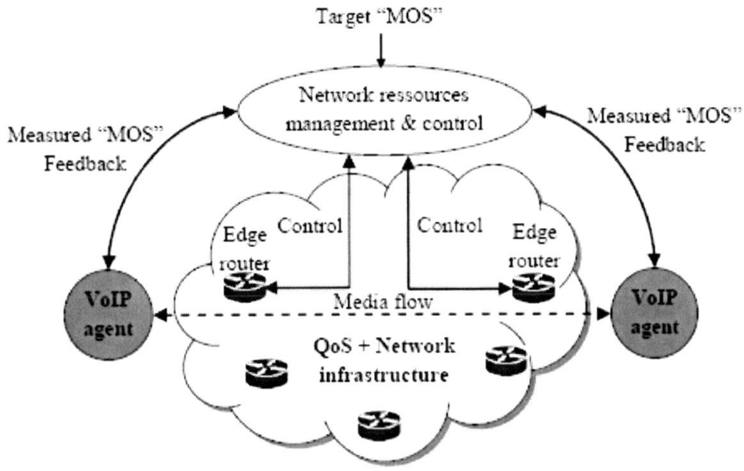

approach, which is not calibrated for a specific edge-device. In our opinion, the benefits of VQ-mon and PsyVoIP should be properly combined in order to improve the accuracy and flexibility of assessment methodology for VoIP applications.

8. VOCAL QUALITY MEASUREMENT APPLICATIONS

Apart from the network vocal quality evaluation, the measured speech quality could be useful for several purposes such as network and edge-device management and diagnosis. In this section, we give several usage scenarios where speech quality measurement could be used to deliver effectively vocal services over next-generation networks.

8.1 Quality-Based QoS Management of VoIP calls

Meddahi, A., et al argue that generic QoS management mechanisms at network layer such as DiffServ and IntServ are unsuitable to provide specific-service QoS needs (Meddahi et al., 2006). In fact, a good network state expressed in terms of objective measures such as delay, delay jitter, or packet loss does not allow inducing consistently that consumers incur a good perceptual quality. Moreover, conventional QoS approaches lack flexibility to handle network dynamics, providers' needs, and consumers' preference.

To efficiently manage QoS over VoIP networks, Meddahi, A., et al propose a new QoS architecture that integrates a subjective score into the QoS control loop (see Figure 17). The architecture goal is to maintain a constant service quality in terms of MOS score during the entire voice conversation. Authors use P-E-Model[1] in order to quantify at run time VoIP subjective quality. The estimated scores are used by the signaling protocol in order to dynamically control and optimize shared network resources such as queuing allocation and congestion thresholds (see Figure 17).

As sketched in Figure 17, the QoS management is based on automatic loop control that involves communication between VoIP agents and the QoS infrastructure (QoS controller, edge routers, etc.). VoIP agents integrate the P-E-Model and evaluate periodically the current MOS from received RTP/RTCP media packets. The MOS score reflects *instantaneous* impairment factors such as packet loss and delay. A QoS report including instantaneous MOS is transmitted by VoIP agent to QoS infrastructure, which reacts according to the observed MOS variation (degradation or improvement based on target MOS) by adjusting edge network resources, e.g. queuing allocation for the *voice call*. The management and control goal is to keep a stable MOS value during a VoIP call, under well-recognized network IP dynamics. If the observed MOS for a specific voice call is degraded or improved, the QoS controller allocates dynamically more or less network resources.

8.2 Quality-Based Network Selection

Mobile consumers over next generation networks could be served at one moment by several overlapping heterogeneous wireless networks. In such a case, mobile users should choose the access network that will likely achieve good quality. The network selection/switching procedure can be performed either at the start or during the service. An inter-network hard handover occurs when users switch from one network to another due to specific reasons related to both consumers and providers. A handover could be managed in network- or host- centric way. In a traditional network-centric approach, the infrastructure monitored by providers decides when a handover is required through a set of control algorithms. However, in a host-centric approach, end-nodes have the ability to perform a handover when quality of service becomes unsteady and unsatisfactory.

Murphy, L. et al argue that a host-centric network selection approach is more suitable to support delay sensitive services and especially packetized vocal conversations (Murphy et al., 2007). In fact, in such a case, inter-handover will be performed according to specific needs of each service. In particular, for delay-sensitive services such as vocal call, handover should be performed seamlessly by reducing/canceling service interruption. Obviously, an intelligent handover monitoring entails a significant improvement of call quality. Indeed, simple policy such as “wireless LAN if available, otherwise 3G” will likely dissatisfy both consumers and providers.

To properly react according to packetized vocal call service needs, Murphy, L. et al developed an end-point controlled network selection management policy. In their scheme, handover decisions are delegated to terminal equipments according to their specific needs. To assure a seamless handover, authors use the message-based, multi-streamed, multi-homed, and reliable SCTP transport protocol. In contrast to standard TCP transport protocol, SCTP allows delivering out-of-order packets to applications which is more suitable for multimedia applications.

Murphy, L. et al exploit the multi-homing feature of SCTP to transparently manage handover over several heterogeneous overlapping networks. Authors introduce several revisions to the initial specification of SCTP in order to accommodate delay-sensitive application needs, especially, in the path selection strategy. Precisely, revised SCTP creates a *primary path* with specified destination and source addresses. Secondary destination addresses are associated with specified source addresses to create secondary paths. In addition, SCTP monitors at run-time delay and jitter on all active paths and makes this information available to the application.

A quality-based network selection controller has been developed in order to decide the necessity to perform a switching from one path to another. To do that, Murphy, L. et al conceived adequate perceptual models which map objective measurements such as delay and delay jitter into a rating score which quantifies precisely the incurred quality of service. In the context of VoIP applications, authors calculate appropriate measurements as follows. Sampled network delays of each active path are smoothed-out based on an exponentially weighted moving average as follows:

$$\hat{T}^{i}_{net} = \pm \times \hat{T}^{i-1}_{net} + \left(1 - \pm\right) \times T^{i}_{net} \tag{18}$$

where, $\hat{T}^{i}_{net}$ and T^{i}_{net} represent respectively the smoothed network delay and measured network delay upon the reception of ith packet, and α is a weighting factor[2]. Jitter metric is calculated based on the definition specified in the RFC 1889 which calculates jitter as follows:

$$\hat{J}_i = \hat{J}_{i-1} + \left(\left| T_{net}^{i} - T_{net}^{i-1} \right| - \hat{J}_{i-1} \right) / 16 \quad (19)$$

where $\hat{J}_i$ corresponds to mean jitter delay. The application is able to accede to both metrics, and can decide which path is most suitable to its needs.

A utility/rating function has been defined rigorously by authors to convert collected objective measures into *utility/quality* scale. The utility is calculated additively according to threshold values specified in Table 5. Moreover, Murphy, L. et al include network cost in the calculation of utility score. To do that, consumers consider the cost of receiving 30 packets using the active path, which is adequately defined by authors. This value is subtracted from the total utility score. Finally, the client performs its decision that maximizes the difference between utility and cost as follows:

$$\underset{j \in paths}{\mathrm{Max}} \left(U_j - C_j \right) \quad (20)$$

where, U_j and C_j represent respectively the utility and cost of jth path. Technically, the calculation of utility score is based on a transport protocol control message sent to application layer every 300 ms. Heartbeat messages are sent over secondary paths to collect required measurements. Every 9 seconds, the performance of two paths is compared and the client decides according to Equation (20) if a handover is necessary or not.

Table 5. Utility threshold levels

Objective measure	Utility
Primary Path	+1
Delay < 55 ms	+3
Delay < 70 ms	+1.5
Jitter < 5 ms	+8
Jitter < 10 ms	+4
Jitter < 20 ms	+1

Figure 18 illustrates graphically a likely envisaged scenario which has been implemented by authors during their experimental study. In this scenario, the client could be served either by WiMAX or Wi-Fi systems. Appropriate equipments have been deployed and configured such as outdoor and indoor units, server, router, and Wi-Fi and WiMAX access points in order to evaluate network selection strategy. Throughout a vocal call, the client is allowed to switch from WiMAX system to Wi-Fi system and conversely. A set of background stations has been used to vary the network load of Wi-Fi system. A 2-way voice call has been set-up between the mobile client and a fixed user through a dedicated server. The data streams were initially sent using Wi-Fi access link. As the Wi-Fi link quality degraded due to mobility or congestion, the call is seamlessly transferred to a WiMAX link in the sense that no packets were lost and there was no interruption to the packet stream.

Experimental study show that developed quality-aware handover maintains a high call quality by selecting properly access network. Authors prove that handover are seamlessly performed without disturbing ongoing service. This scheme clearly outperforms conventional approach which bases its decision only on signal strength. Indeed, the client could sustain a very bad quality even when signal strength is high due to the shared nature of wireless data networks.

8.3 Quality-Based Handover Management

In order to offer service ubiquity for users and resource management efficiency for providers, Marsh, I. et al conceived a perceptual-aware, network-centric, handover controller to switch a vocal session between two overlapping networks (Marsh et al., 2006). Precisely, a handover is performed between overlapping WLAN and GSM networks. This allows, on one hand, exploiting relatively the high capacity of a WLAN, and on

Figure 18. Network selection between Wi-Fi and WiMAX based on client and link quality (Murphy et al., 2007)

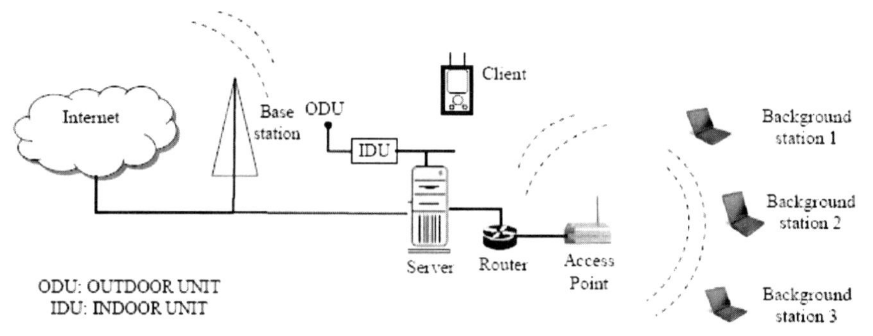

Figure 19. Handover scenario between WLAN and GSM networks (Marsh et al., 2006)

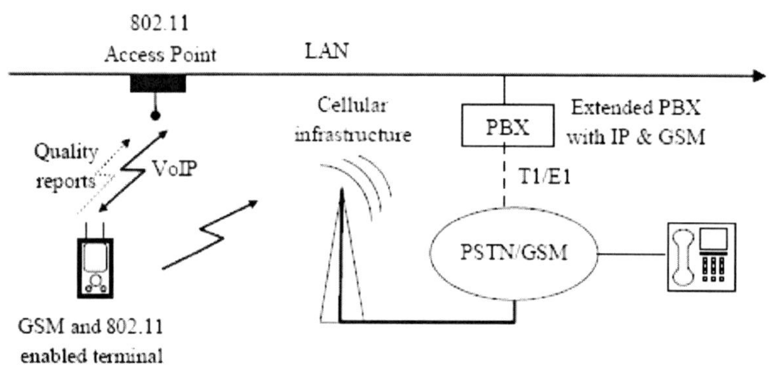

the other hand, reducing the GSM network load and cost.

Figure 19 outlines the scenario examined by authors to enable inter-network handover. In their envisaged scenario, a mobile subscriber initiates a voice session to a land PSTN subscriber using a WLAN as last-wireless hop. In fact, at the start of the voice session, authors assume that the mobile subscriber belongs to the coverage area served by the WLAN access point. Next, when the quality of vocal service becomes under a certain critical threshold due to mobility or congestion, then a handover is performed. In such a case, the mobile subscriber is related to the land subscriber using GSM infrastructure. The hands-free terminal is equipped with two wireless card interfaces in order to allow connection to WLAN and GSM networks. As illustrated in Figure 19, the mobile terminal sends adequate quality reports to an extended PBX that analyzes received feedbacks. Once an unsatisfied score is detected, the PBX indicates to the mobile terminal the requirement to perform a handover. To do that seamlessly, a vocal channel is opened using GSM infrastructure between the mobile terminal and PBX, which is responsible to relay received vocal information toward the fixed subscriber.

In order to estimate service quality, Marsh, I. et al argue that a single objective metric such as packet loss, signal strength, or delay jitter does not offer sufficient reliability to decide the need to initiate a handover. To accurately estimate service quality, authors develop a linear combination which maps all available link layer metrics to a single score called *handover score*. Primary factors used to calculate handover scores are called *handover contributors*. In Marsh et al. (2006), authors indicate the following contributor factors to estimate service quality:

- **Received signal strength indicator:** The signal-to-noise ratio is a good indicator about quality of service, especially over wireless Telecom networks. This metric may entail inaccurate estimates about quality over wireless data networks. In fact, over wireless Telecom networks, high signal strength indicates that users sustain potentially a good perceptual quality. This rule is questionable over wireless data networks where perceived quality could be poor in spite of high measured signal strength due to, for example, congestion-induced packet losses. In this quality-based handover scheme, the mobile terminal records received signal strength periodically. The obtained value is scaled according to handover score defined by authors. Specifically, the measured received signal strengths are mapped to values varying from 0 to +90.
- **Delay jitter:** An increasing delay jitter is a good indicator of poor quality. According to a preliminary empirical study, authors assign a score of +10 and 0 to good and negligible jitter conditions, respectively. Moreover, a score of -10 and -20 is assigned to poor and very poor jitter conditions, respectively.
- **Packet loss:** High packet loss rate indicates that users sustain undoubtedly a very poor quality. According to a preliminary empirical study, a decreasing score step of -10 is assigned to sustained packet loss rate with an increasing step of 8%. A long bad period is accounted for by increasing properly the contribution of packet loss.
- **RTCP losses:** The mobile terminal will likely sustain reception problems when the monitoring node does not receive RTCP quality reports. Authors indicate that three or more consecutive losses of RTCP feedback are quite significant to reduce aggressively the overall handover score. Precisely, a decreasing score step of -10 is assigned to each consecutively lost RTCP report.
- **Transmission rates:** Actual wireless data interfaces are able to reduce their data rate according to network dynamics. This factor could be considered in the calculation of handover score. Particularly, selecting lower rates such as 2 and 1 Mbps indicates an increase of connection loss probability. This factor should be adequately considered in the computation of handover scores.

At the end of a monitoring period, the PBX computes the handover score by linearly combining handover contributor as follows:

$$\text{Handover score} = \text{Signal} + \text{Loss} + \text{Jitter} + \text{Report Loss} \quad (21)$$

where the handover scores vary from -100 to 100. A large positive score indicates a good perceived quality. The mobile users are allowed to specify the lower acceptable threshold score. As a consequence, a handover is only performed when the calculated handover score falls below the defined threshold. In Marsh et al. (2006), authors use a threshold score equal to +30 as a default value. An increasing threshold results in the improvement of average quality at the expense of a higher communication cost, since the system will switch the vocal session to GSM system earlier. Conversely, a decreasing threshold results in the reduction of communication cost at the expense of longer periods of degraded quality.

Over a series of 100 subjective experiences, 68 cases indicate that the developed handover scheme performs handover operation at the desired time. In 10 cases, human subjects indicate that the handover is performed with delay, i.e., the subject perceived poor quality for a brief period while waiting the occurrence of handover. In 7 cases, the trigger suggested as unnecessary.

The remaining 15 cases do not trigger handover operation which is optimal. Therefore, in 83% of cases the algorithm made the ideal decision which is pretty acceptable in practice.

8.4 Quality-Based De-Jittering Management

Packet-based networks disturb original vocal packet stream by introducing delay jitter, which is removed at receiver side using a de-jitter buffer. In the context of VoIP, the goal of a de-jitter buffer algorithm is to optimize the trade-off delay / late arrivals. Basically, de-jitter buffer algorithms could be fixed or adaptive. Fixed de-jitter algorithms maintain a constant play-out delay during a voice session which could be set in advance or dynamically computed at the start of a voice session (Melvin, 2004). In contrast, adaptive de-jitter algorithms adjust the play-out delay according to the observed network delay and delay variation trend. In order to seamlessly adapt the play-out delay, adjustments are performed during silent periods only (Melvin, 2004).

Classically, delay adjustments are made using a set of objective measures such as delay variation and packet late ratio. Actually, it is well-accepted that such algorithms are unsuitable since they aim at optimizing objective measures, which could often lead to a poor perceptual quality. Thereby, it is more appropriate to adjust the play-out delay in a user-friendly way. Following this observation, several emergent de-jitter buffer management algorithms are proposed in the literature which aim at maximizing the perceptual quality (Broom, 2006; Barriac, 2003).

In this sense, Fujimoto, K. et al developed a de-jitter buffer algorithm which selects the play-out delay that maximizes the perceptual quality (Broom, 2006). To this end, authors build a parametric model which maps objective measures, namely the play-out delay and packet loss rate into a MOS score. The following model has been developed by authors in order to quantify subjective quality of G.711 voice CODEC:

$$MOS\ (e_{e2e}, d_{e2e}) = 4.10\text{-}0.195 \times e_{e2e} + 2.64 \times 10^{-3}\ d_{e2e}\ \text{-}1.86 \times 10^{-5} d^{2}_{e2e} + 1.22 \times 10^{-8} d^{3}_{e2e} \quad (22)$$

where e_{e2e} and d_{e2e} correspond to end-to-end packet loss ratio and play-out delay. The developed parametric model needs to properly measure the overall packet loss ratio. Formally, the overall packet loss ratio is computed as follows:

$$e_{e2e} = e_{net} + e_{de\text{-}jitter} \quad (23)$$

where e_{net} corresponds to the ratio of lost packets over the delivering network and $e_{de\text{-}jitter}$ corresponds to the ratio of ignored packets at the de-jitter buffer. The ratio of ignored packets is related to the play-out delay. This relation could be expressed using delay Cumulative Distribution Function (CDF), which is defined as $F(x) = P(X \leq x)$. Therefore, for a play-out delay d, the ratio of ignored packets could be calculated as follows:

$$e_{de-jitter} = P\left(X > d\right) = 1 - F\left(d\right) \quad (24)$$

Fujimoto, K. et al argue that network delay distribution over an IP network could be fairly modeled using a Pareto distribution. Therefore, late packet ratio is computed as follows:

$$e_{de\text{-}jitter} = \left(\frac{k}{d}\right)^{\pm} \qquad d \geq k \quad (25)$$

Where, d represents the play-out delay used during the last talk-spurt, k and α are the distribution parameters which are given by:

$$k = \min\left(d_1, d_2, \ldots, d_n\right) \text{ and}$$
$$\pm = n\left[\sum_{i=1}^{n} \log\left(\frac{d_i}{k}\right)\right]^{-1} \quad (26)$$

where d_i represents the ith one-way network delay stored in the history, which is updated once a new measure is available, and n represents the history length. The quality model could be expressed using one input parameter as follows:

$$Q\ (d_{e2e}) = 4.10\text{-}0.195 \times e_{net}\ \text{-}19.5 \times \left(\frac{k}{d_{e2e}}\right)^{\alpha} + 2.64 \times 10^{-3}\ d_{e2e}\ \text{-}1.86 \times 10^{-5} d_{e2e}^{2} + 1.22 \times 10^{-8} d_{e2e}^{3} \quad (27)$$

Fujimoto, K. et al compute the d_{e2e} that maximizes the quality model (see Equation 27) during the next talk-spurt. An empirical study over existing wide area IP networks proves that the algorithm developed by Fujimoto, K. et al outperforms traditional de-jitter algorithms in terms of MOS score.

9. CONCLUSION AND OPEN ISSUES

Services over next-generation networks are characterized by intrinsically different QoS needs. They could be offered by several operators using heterogeneous networks. This leads to network convergence which allows offering a multitude of services to fixed and mobile users using a wide range of access devices. In order to successfully reach intended goals such as ubiquity and flexibility of services, harmonization between technologies and networks should be made. Such a complex networking environment requires sophisticated management policies in order to optimize both, the perceived quality and resource utilization.

In this chapter, we studied interactive vocal conversations over heterogeneous networks. The inherent needs to successfully offer vocal conversations over next-generation networks have been discussed. The requirement to harmonize different networks and technologies has been illustrated, and foreseen architecture has been discussed. We gave several envisaged ways to accommodate vocal conversation service. Subjective and objective methodologies for voice quality assessment have been thoroughly described. Software- and Emulation- based frameworks developed for voice quality evaluation and improvement has been reported. For the sake of voice service management, parametric model-based assessment algorithms are more relevant. That is why parametric assessment algorithms have been thoroughly described, especially over last-hop wireless Telecom networks and packet-based networks. The final part of this chapter described several ways to manage voice service under a multitude of envisaged scenarios.

Network management policies based on perceptual quality are in their infancy. Significant challenges are still unsolved in order to allow a worldwide spread of perceptual-based quality management. First of all, the accuracy of objective models remains questionable under several circumstances. Therefore, sophisticated and dedicated perceptual models over next generation networks, which map objective measures to a subjective score should be developed and properly calibrated. Moreover, estimated perceived quality should be properly considered by policy enforcer node, which should perform adequate physical operations in the network to maintain an acceptable perceived quality. In addition, management of multimedia sessions over heterogeneous networks is an important challenge. For instance, each stream (video / audio) could be served by different providers then multiplexed at terminal nodes. This requires quality-aware sophisticated applications running over each terminal. Last and not least, in our opinion, perceptual quality-based management should be used over mobile ad-hoc networks to assist transport and routing protocols to maintain a stable perceptual quality and to prevent annoying service interruptions.

REFERENCES

Barile, M., Camarda, P., Dell'Aquila, R., & Vitti, N. (2006, September). Parametric Models for Speech Quality Estimation in GSM Networks. In *Proceedings of software in telecommunications and computer networks, SoftCOM 2006* (pp. 204 - 208).

Barriac, V. (2003, October). Non-intrusive evaluation of voice quality in IP environments: On-going work in standardization. In *ITU-T Workshop*, Oct. 2003. Retrieved from http://www.itu.int/ITU-T/worksem/qos/index.html

Bertrand, G. (2007 May). *The IP multimedia subsystem in next generation networks*. Retrieved from http://www.irisa.fr/armor/lesmembres/Alia/IMS_principes_et_architecture/

Blake, S., Black, D., Carlson, M., Davies, E., Wang, Z. & Weiss, W. (1998, December). *An architecture for differentiated services*. IETF RFC 2475.

Bolot, J. C. (1993, September). End-to-end packet delay and loss behaviour in the Internet. In *Proceedings of SIGCOMM 1993*, New York, USA.

Braden, R., Clark, D. & Shenker S. (1994, June). *Integrated services in the Internet architecture: an overview*. IETF RFC 1633.

Broom, S. R. (2006, November). VoIP Quality Assessment: Taking Account of the Edge-Device. *IEEE Transactions on Audio, Speech, And Language Processing, 14*(6), 1977–1983. doi:10.1109/TASL.2006.883233

Chauveau, D. (2005, June). *NGN, strategies for shaping future regulation: the value added of standardisation*. NGN Workshop.

Cicconetti, C., Lenzini, L., Mingozzi & Eklund, C. (2006, March). *Quality of service support in IEEE 802.16 networks*. IEEE Network Magazine, pp. 50-55.

Clark, A. D. (2001). Modeling the Effects of Burst Packet Loss and Recency on Subjective Voice Quality. In *Proceedings of IP Telephony Workshop*, Columbia, USA.

Cole, R. G., & Rosenbluth, J. H. (2001, April). Voice over IP Performance Monitoring. *Computer Communication Review, ACM Sigcomm Computer . Communication Review, 31*(2).

Ericsson. (2008, December). *Measuring Speech Quality* [White paper]. Retrieved from http://www.ericsson.com/tems

ETSI report. (2002, April). *2nd ETSI TIPHON VoIP Speech Quality Test Event*. Sophia Antipolis, France.

ETSI Technical report. (July, 2000). *Telecommunications and Internet Protocol Harmonization Over Network (TIPHON); End to end Quality of Service in TIPHON Systems*. Part 1 of general aspects of Quality of Service (QoS), France, TR 101 329-1, V 3.1.1.

ETSI Working Group. (n.d.). *Telecommunications and Internet converged Services and Protocols for Advanced Networking (TISPAN)*. Retrieved from http://www.etsi.org/tispan/

Fujimoto, K., Ata, S., & Murata, M. (2002, November). Adaptive playout buffer algorithm for enhancing perceived quality of streaming applications. In *Proceedings of IEEE Globecom2002*.

Gu, D., & Zhang, J. (2003, June). QoS enhancement in IEEE802.11 wireless local area networks. *IEEE Communications Magazine, 41*(6), 120–124. doi:10.1109/MCOM.2003.1204758

Hoene, C. (2005, December). *Internet telephony over wireless links*. PhD thesis, Technical University of Berlin, Germany.

ITU-T Recommendation G.1010. (2001, November). *Quality of service and performance: End-user multimedia QoS categories*.

ITU-T Recommendation G.107. (2003, March). *The E-Model a Computational Model for Use in Transmission Planning.*

ITU-T Recommendation P.800. (1996, August). *Methods for subjective determination of transmission quality.*

ITU-T Recommendation P.862. (2001, February). *Perceptual Evaluation of Speech Quality (PESQ), an objective method for end-to-end speech quality assessment of narrowband telephone networks and speech CODECs.*

Jain, R. (2006, October). Internet 3.0: Ten problems with current Internet architecture and solutions for the Next Generation. In *Proceedings of IEEE Military Communications Conference (Milcom 2006)*, Washington, DC, USA.

Jelassi, S., & Youssef, H. (2007, October). EVOM: A software based platform for voice transmission and quality assessment over wireless ad-hoc networks. In *the proceeding of 10th IEEE/ACM Annual International Symposium on Modeling, Analysis and Simulation of Wireless and Mobile Systems (MSWiM 2007)*, Chania, Crete Island, Greece.

Karlsson, A., Heikkilä, G., Minde, T. B., Nordlund, M., & Timus, B. (1999, June). Radio link parameter based speech quality index – SQI. In *Proceedings of IEEE Workshop on Speech Coding.*

Liao, W.-T., Chen, J.-C., & Chen, M.-S. (2001). Adaptive recovery techniques for real-time audio streams. In *INFOCOM 2001.*

Linden, J. (n.d.). *VoIP: Better than PSTN?* Retrieved from http://www.analogzone.com/nett0307.pdf

Malden Electronics. (2008, December). *Multi-DSLA Application – Wireless handover* [white paper]. Retrieved from http://www.malden.co.uk/documents/MultiDSLAWirelessHandover.pdf

Marsh, I., Grönvall, B., & Hammer, F. (2006, May). The design and implementation of a quality-based handover trigger. In *5th International IFIP-TC6 Networking Conference (NETWORKING 2006)*, Coimbra, Portugal.

Meddahi, A., & Afifi, H. (2006, November). Packet-E-Model: E-Model for VoIP quality evaluation. *Elsevier Computer Networks Journal, 50*, 2659–2675. doi:10.1016/j.comnet.2005.10.008

Meddahi, A., Afifi, H., & Vanwormhoudt, G. (2006, June). "MOSQoS": Subjective VoIP Quality Feedback Control and Dynamic QoS Adaptation. In *Proceedings of IEEE ICC*, Istanbul, TURKEY (pp. 2034-2039).

Melvin, H. (2004, October). *The use of synchronized time in voice over Internet Protocol (VoIP) applications.* Ph.D. Thesis, University College Dublin, Ireland.

Mirhakkak, M., Schult, N., & Thomson, D. (2000). Dynamic Quality-of-Service for mobile ad-hoc networks. In *Proceedings of the 1st ACM International Symposium on Mobile Ad Hoc Networking and Computing*, Boston, Massachusetts (pp. 137-148).

Murphy, L., Noonan, J., Perry, P., & Murphy, J. (September, 2007). An application-quality-based mobility management scheme. In *Proceedings of 9th IFIP / IEEE International Conference on Mobile and Wireless Communications Networks (MWCN 2007)*, Cork, Ireland.

Peterson, L., Shenker, S., & Turner, J. (April, 2005). Overcoming the Internet impasse through virtualization. *IEEE Computer Magazine, 38*(4).

3rd Generation Partnership Project. (1999). *Technical Specification Group GSM/EDGE Radio Access Network; Radio subsystem link control (Release 1999).*

Al-Agha, K. (2008, November). *Internet new generation and Post-IP architecture.* Tutorial presented at the Second IEEE / IFIP International Conference on New Technologies, Mobility, and Security IEEE/IFIP NTMS 2008, Tangier, Morocco.

Rix, A., Beerends, J., Kim, D., Kroon, P., & Ghitza, O. (2006, November). Objective assessment of speech and audio quality: Technology and Applications. *IEEE Transactions on Audio, Speech, and Language Processing, 14*(6), 1890–1901. doi:10.1109/TASL.2006.883260

Roychoudhuri, L., Al-Shaer, E., Hamed, H., & Brewster, G. (2003, May). Audio transmissions over the Internet: experiments and observations. In *IEEE Proceedings of International Conference on Communications (ICC 2003)*, Anchorage, Alaska, USA.

Roychoudhuri, L., Al-Shaer, E., & Settimi, R. (2006, March). Statistical measurement approach for on-line audio quality assessment. In *Proceedings of Passive and Active Measurement (PAM'06)*.

Sat, B., & Wah, B. W. (2006). Analysis and evaluation of the Skype and Google-Talk VoIP system. In *Proceedings of IEEE international conference on Multimedia and Exposition*.

Simulator Homepage, N. S. (n.d.). Retrieved August 23, 2008, from http://www.isi.edu/nsnam/ns

Sun, L., & Ifeachor, E. (2004, June). New models for perceived voice quality prediction and their applications in play-out buffer optimization for VOIP networks. In *Proceedings of IEEE International Conference on Communication (ICC'04)*.

Sun, L., & Ifeachor, E. C. (2006, August). Voice Quality Prediction Models and Their Application in VoIP Networks. *IEEE Transactions on Multimedia, 8*(4).

Takahashi, A., Kurashima, A., & Yoshino, H. (2006, November). Objective assessment methodology for estimating conversational quality in VoIP. *IEEE Transactions on Audio, Speech, and Language Processing, 14*(6).

Takahashi, A., Yoshino, H., & Kitawaki, N. (2004, July). Perceptual QoS assessment technologies for VoIP. *IEEE Communications Magazine, 42*(7), 28–34. doi:10.1109/MCOM.2004.1316526

Wanstedt, S., Pettersson, J., Xianchun, T., & Heikkila, G. (2002). Development of an objective speech quality measurement model for the AMR CODEC. In *Proceedings of MESAQIN 02*, Prague, Czech Republic.

Werner, M., Kamps, K., Tuisel, U., Beerends, J. C., & Vary, P. (2003, September). *Parameter-based speech quality measures for GSM.* In Proceedings of 14-th IEEE PIMRC 2003.

Xu, K., Tang, K., Bagrodia, R., Gerla, M., & Bereschinsky, M. (2003, October). Adaptive bandwidth management and QoS provisioning in large scale ad-hoc networks. In *Proceedings of IEEE MILCOM 2003*, Boston, Massachusetts.

Yih-Guang, J., Yang-Han, L., Ming-Hsueh C., Hsien-Wei, T., Jheng-Yao, L., & Chih-Wei, S. (2007, July). *Mean Traffic Bit Rate with ON-SID Modeling of VoIP Traffic*, IEEE 802.16 Broadband Wireless Access Working Group.

ENDNOTES

1. P-E-Model is an adaptation of original E-Model to evaluate VoIP conversations [46]
2. α is set to 0.94 according to a preliminary experiments

Chapter 19
Quality of Service Provisioning in the IP Multimedia Subsystem

Richard Good
University of Cape Town, South Africa

David Waiting
Telkom South Africa Ltd, South Africa

Neco Ventura
University of Cape Town, South Africa

ABSTRACT

The 3GPP IMS defines a network architecture that allows rapid provisioning of rich multimedia services. While standardization of the IMS core architecture is largely complete, there are several areas that are still to be addressed before effective deployment can be realized. In particular a QoS framework is required that efficiently manages scarce network resources, ensures reliability and differentiates IMS services from web-based services. This chapter reviews the most promising candidate resource management frameworks, performs architectural alignment and defines a set of generic terms and elements to provide a convenient point of departure for future research. This harmonization of standardized architectures is critical to avoid interoperability concerns that could cripple deployment. Further challenges are discussed, in particular the vertical and horizontal co-ordination of resources, and current research works that address these challenges are presented.

1. INTRODUCTION

The IP Multimedia Subsystem (IMS) defines a network architecture that promises to revolutionize inter-personal communication and enable convergence. With the aid of IMS, innovative rich services can be delivered to customers over a variety of access technologies and handsets. New applications that harness the power of voice, video and text messaging will enable customers to interact in ways never before imagined and provide new revenue streams for network operators who are currently experiencing dwindling voice revenues. IMS is seen as the silver bullet to resurrect the fortunes of the once-mighty telecommunications operators by reducing costs and luring customers away from increasingly popular Internet services. Despite these grand promises there are several hurdles to

DOI: 10.4018/978-1-61520-680-3.ch019

overcome before circuit-switched technologies can be moth-balled once and for all.

Since the inception of voice telephony networks there have been several technological breakthroughs that have provided a richer user experience and wider availability. Digital switching technologies have steadily been introduced to replace legacy telephone exchanges and voice is now carried by means of digital pulse-coded streams. The advent of the intelligent network improved the operators' ability to provide enhanced voice services, and the introduction of ISDN provided voice and data over a single channel. Mobile technologies including GSM and UMTS have brought voice directly to the customer, and offer other services such as SMS and MMS to increase communication options. However, one of the biggest areas of growth for network operators has not been in voice but in data communications, to meet the rapidly expanding requirements of customers wishing to make use of Internet applications.

From its humble beginnings as a research tool for the US department of defense, the Internet has blossomed into an indispensable tool for both work and leisure. The web has seen huge gains in popularity due to the surge in sites offering user-generated content. Music and video can be streamed across the Internet directly into the consumer's home and interactive games can be played with opponents across the world. This, together with rampant peer-to-peer file sharing, has necessitated faster access and core network technologies to keep up with user demand. Dial-up home connections have migrated to ADSL and even fiber links, and GSM/EDGE wireless links have been replaced with HSPA and WiMAX networks with speeds of several megabits per second. However, the packet-switched portions of these networks are designed for data, not real-time communications such as Voice over IP (VoIP). With VoIP, voice samples are encoded, packetized and transmitted over an IP network. That is not to say that modern networks cannot handle such traffic; the popularity of Internet-based VoIP applications has shown that for the most part they can. The problem lies in the fact that the quality is not guaranteed and is therefore not a replacement for legacy voice networks that offer predictable voice quality. This is a problem for the potential IMS operator who must ensure that any replacement of existing technologies offers an equal, if not better, experience to their customers.

But the attraction of IMS is not only its ability to replicate existing voice services over an IP network. It is envisaged that it will enable a host of rich multimedia services such as high definition video broadcasts, interactive gaming, file sharing and music streaming, all accessed from increasingly sophisticated devices that include cameras, motion sensors, global positioning systems and touch screens. These services all have unique QoS requirements that must be met to ensure a pleasurable and predictable experience. The challenge is to make best-effort IP networks into networks that can meet strict delay, packet-loss and jitter bounds, without sacrificing the flexibility and benefits of packet-switching. This will mean that operators need only maintain a single network for both real-time and non real-time traffic, thereby reducing capital and operating costs significantly. Wireless operators can make better use of scarce and expensive frequency spectrum by leveraging the efficiencies of VoIP. This is obviously an attractive proposition for operators who are under increasing financial pressure due to new market entrants and worldwide deregulation of the telecommunications industry. Therefore, it is clear that operators must adopt a sound QoS framework to ensure reliability of services and maximum returns on their expenditure.

Despite the fact that IMS specifications are largely complete, end-to-end QoS provisioning mechanisms are sorely lacking. The standardization of IMS/NGN resource management frameworks has been fragmented involving numerous standardization bodies with overlapping scope. This has resulted in weak functional and interface specifications that have led to interoperability concerns. Harmonization of the IMS/NGN resource management frameworks will be critical to provide a platform to flexibly control resources and provide a sound business case for deploying IMS services. Furthermore the standardized architectures have notable shortcomings regarding QoS provisioning for advanced multimedia services in multiple domain scenarios. These challenges need to be considered and addressed by operators when formulating their NGN strategies.

The remainder of this chapter is organized as follows; first, a brief overview of the IMS standardization, protocols and architecture. Thereafter the authors describe the three primary candidate frameworks for provisioning QoS in the IMS, that is, the 3GPP Policy Control and Charging (PCC) framework, the TISPAN Resource and Admission Control Subsystem (RACS), and the ITU-T Resource and Admission Control Functions (RACF). The authors propose a federated QoS framework, the Common PCC that encompasses the most important elements from each of the standardization bodies. Subsequently the chapter covers the deployment challenges for QoS frameworks, specifically the vertical and horizontal coordination of resources, after which the chapter is concluded.

2. THE IP MULTIMEDIA SUBSYSTEM

IMS was first standardized in Release 5 of the 3rd Generation Partnership Project (3GPP) specifications with the priority firmly focused on mobile networks. 3GPP2 and other technical bodies standardize their own versions of IMS that usually exhibit only minor variations to the 3GPP version. In 2004 ETSI formed the Telecoms and Internet converged Services and Protocols for Advanced Networks (TISPAN) technical committee with the aim of standardizing IMS for fixed broadband networks; this work was then fed back back to the 3GPP. TISPAN's work was instrumental in providing a road-map for PSTN to VoIP migration. However, in a bid to prevent fragmentation of the IMS standards, the 3GPP and TISPAN have since decided to pool their efforts to create Common IMS, which forms part of the 3GPP Release 8 specifications. For simplicity the focus of this chapter is primarily on the 3GPP Common IMS.

The 3GPP looks to the Internet Engineering Task Force (IETF) for the protocols required in the IMS. Therefore, many of the protocols traditionally found in the Internet environment, such as Session Initiation Protocol (SIP), Diameter and the Real-Time Protocol (RTP), have been incorporated into the IMS specifications. The preferred approach by the 3GPP is to use Internet standards unchanged; however, when no suitable protocol exists or modifications to existing protocols are required, the 3GPP collaborate with the IETF to publish additional RFCs that fill these gaps (Rosenbrock et. al., 2001). Many of these new specifications do not even mention IMS as they are developed to be applicable to general networking environments.

IMS aims to address many problems associated with the next-generation network (NGN): interoperability, access-awareness, security, policy support, interworking and QoS. In order to achieve these goals the IMS specifies many logical elements and interfaces. The implementation of these elements is left to the individual equipment vendor, but the functionality must be consistent throughout. As is typical with any NGN architecture, the control and user planes are handled separately and utilize different protocols. This separation leads to a more efficient and scalable architecture, but it requires that there be suitable communication between the planes to ensure that there are bearers available for the sessions negotiated at the signaling level.

A protocol used extensively for signaling in the IMS is SIP; it allows for multimedia sessions of any type to be initiated, modified and terminated. The type of media session is described by the Session Description Protocol (SDP), which is carried in the respective bodies of the actual SIP messages. Three Call Session Control Functions (CSCFs) are defined in the IMS: the Proxy-CSCF (P-CSCF), the Interrogating-CSCF (I-CSCF) and the Serving-CSCF (S-CSCF). The CSCFs are primarily SIP servers as they act as SIP routers and registrars, but may also contain interfaces to other protocols, such as Diameter, the protocol that fulfills authentication, authorization and accounting needs in the IMS.

The P-CSCF is the first point of contact for the IMS user equipment (UE) and as such all incoming and outgoing SIP traffic to and from the UE traverses this proxy. Its main tasks are to ensure integrity of the signaling and to provide security by setting up encrypted channels to the UE. The P-CSCF is a trusted network element and asserts the identity of the UE so that other elements do not need to. Another task of the P-CSCF is to facilitate signaling compression. SIP is a verbose text-based protocol, which makes it easy to debug, but unfortunately adds a great deal of signaling overhead during session set up. Signaling compression reduces the amount of traffic flowing between the UE and the network equipment and provides the benefit of conserving bandwidth and reducing call setup times if the terminal is connected wirelessly.

The I-CSCF serves two important tasks. First, it lies at the edge of an administrative domain and handles all incoming session requests from other domains, routing these requests to the correct next hop. A DNS query for a domain always returns the address of the I-CSCF. The I-CSCF may also perform a topology hiding function to remove sensitive information from outgoing SIP requests to other domains. The second function of the I-CSCF is to assign an appropriate S-CSCF to the UE when it registers. The I-CSCF queries the Home Subscriber Server (HSS) for the list of available S-CSCFs and their capabilities and makes an assignment decision based on the needs of the UE.

The S-CSCF handles the complex routing of SIP requests. It may route a request either within a particular domain or to an external domain. Alternatively, depending on the user's subscription policy, the request may be routed to one or more application servers for further processing, or if appropriate, the request may be routed to a SIP proxy that allows it to break out to the PSTN. The S-CSCF acts as a SIP registrar maintaining a binding between public user identities and IP addresses.

In the IMS architecture signaling and media are decoupled. IMS terminals communicate real-time media over the packet-switched network typically over a very different path that the SIP signaling traverses. In the case of voice, the sound samples are encoded and transported to the corresponding party by means of RTP over IP. Unlike traditional Time-Division Multiplexed (TDM) telephony, in IMS there are no dedicated circuits assigned to each voice call. Real-time multimedia packets must compete with other IP traffic, such as web browsing, gaming and peer-to-peer. It is important that each of these services be assigned to a different class of traffic so that services with strict QoS bounds may be given priority at the bearer level.

3. IMS/NGN RESOURCE MANAGEMENT FRAMEWORKS

An important motivation for QoS management in the IMS is the widespread proliferation of Web 2.0 services. The Internet revolution has led the transition of the World Wide Web from a collection of websites to a complete computing platform serving Internet applications. This poses a threat to IMS service deployment; operators will need to justify charging for services that are typically available free of charge through service differentiation. Increased reliability through efficient management of resources is the main driver for this differentiation.

The IETF have defined a Policy Based Network Management (PBNM) architecture for all areas of network management. This architecture can be applied to any scenario where access to a resource needs to be restricted and distributed, and automated management of this access is desirable. The framework has been adopted by IMS and NGN standardization bodies, including 3GPP, TISPAN and ITU-T, to form the basis of their resource and admission control frameworks.

Architectural alignment and harmonization between the various standardized frameworks will be critical to avoid interoperability concerns that could cripple deployment. In order to address the issue of harmonization between architecture specifications, a comprehensive snapshot of the state of the art regarding mediation between QoS control elements and transport layer resources in the IMS/NGN framework, is necessary.

3.1 3GPP Policy Control and Charging Framework

Along with the introduction of IMS technology as part of their Release 5 specification, 3GPP exposed resource management functions to applications through the Service Based Local Policy (SBLP) architecture. This architecture was further developed in Release 6 and 7, the SBLP architecture was combined with the Flow Based Charging architecture to create the PCC architecture. Release 8 extends the scope of this framework to mediate with a number of IP service elements, to support a greater number of access technologies and QoS models, and to support inter-domain communication.

3.1.1 Functional Elements

The Release 7 PCC architecture has three critical components: an Application Function (AF), a Policy and Charging Rules Function (PCRF) and a Policy and Charging Enforcement Function (PCEF) (3GPP, 2008a). This architecture is depicted in Figure 1. The AF logically resides in the service control layer and represents any element that might request resources. These elements lie on the signaling path and, based on extracted service information, they create authorization requests that are passed to the PCRF in the resource control layer.

Figure 1. The 3GPP PCC functional architecture

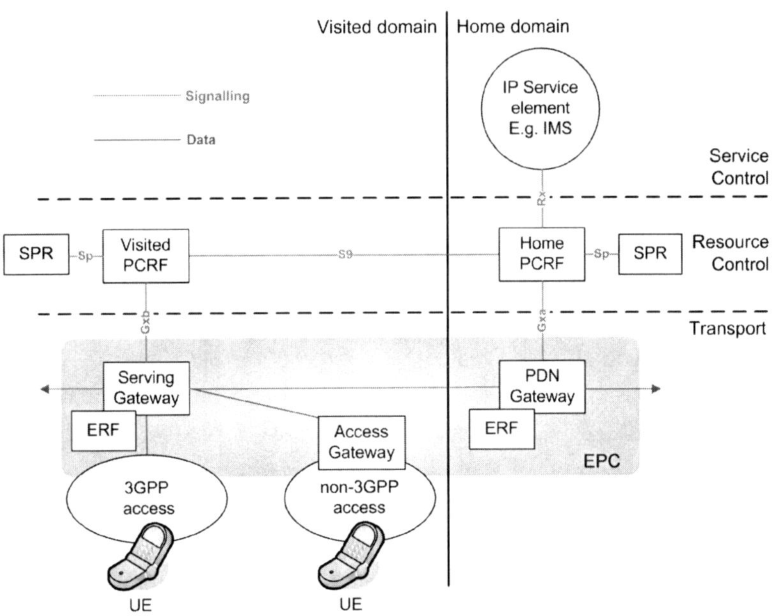

The PCRF instantiates the Policy Decision Point (PDP) specified in the IETF PBNM model and performs policy control consisting of authorization, binding, establishment of transport layer paths and QoS control. The PCRF receives authorization requests and, upon extracting the service information, performs authorization based on policies stored in a policy repository. The format, content, provisioning, storage and retrieval of these policies is regarded as network operator specific and therefore not standardized, though the definition of a Subscription Profile Repository (SPR) implies that subscription profile related policies should be present. Upon authorization the PCRF defines a PCC Rule that contains service data flow filters to identify packet flows that constitute a service data flow.

The PCC Rule contains parameters that describe how the service data flow should be treated in the transport layer. Session binding is performed using user-identity based identification; this scheme uses the UE IP address, or any kind of UE identity to identify the home domain and hence the QoS elements involved. The manner in which the PCC Rule is enforced depends on the QoS reservation procedure in use; there are two models for requesting QoS enabled paths; end-point initiated establishment (pull mode) and network-initiated establishment (push mode). In pull mode, intelligent UEs make resource requests from the transport plane. In push mode, the entities involved with session negotiation make the requests for resources; the PCRF installs or pushes the PCC rules to the PCEFs, which logically reside on the transport layer devices. The installed PCC rules identify service data flows based on the service data flow filters and the associated flows are treated accordingly, this is referred to as QoS control.

The Evolved Packet Core (EPC) is central to the Services Architecture Evolution (SAE) work item currently under standardization by the 3GPP, where the SAE forms the All-IP based core network for the Long Term Evolution (LTE) architecture. The EPC supports mobility between heterogeneous access networks, and incorporates an evolved QoS concept that is aligned with the PCC framework. In this evolved architecture the IMS is seen as one of several IP service elements, hence the AF is no longer limited to IMS specific elements. The PCRF has its functionality split into home domain and visited domain functions to offer breakthrough for data in both domains. This new system introduces service level QoS parameters that are conveyed in the PCC rules; in particular a QoS Class Identifier (QCI), an Allocation and Retention Priority (ARP) and authorized guaranteed and maximum bit rate values for uplink and downlink (3GPP, 2008b). The QCI is a scalar that represents the QoS characteristics that the EPC is expected to provide for each service data flow.

The interaction between the PCRF and the transport layer has been extended. The PCEF, as the element residing in the transport layer, is separated into the Serving Gateway, the Packet Data Network (PDN) Gateway and the evolved Packet Data Gateway (ePDG). The Serving Gateway is a router that resides on the local network to which the end-user is attached, it performs connectivity provisioning including access control and resource provisioning. The PDN Gateway has similar functionality but is located in the home network of the end-user. The ePDG authenticates end-users connecting via untrusted network access and monitors traffic.

3.1.2 Reference Point Definitions

3GPP define reference points that describe functional requirements and in depth protocol specifications for the associated interface, to provide interaction between the aforementioned logical elements. The Rx reference point is defined as method of interaction between the AF and the PCRF; the interface extends the Diameter base protocol and defines new commands and Attribute Value Pairs (AVP). To provision PCC rules and install them on the logical elements in the transport layer, 3GPP define the Gx reference point, also an extension of the Diameter base protocol. Additional Diameter commands and AVPs allow a PCRF to provision and install PCC rules on transport layer devices through Diameter resource requests.

With the PCRF split into home and visited functionality as of Release 8, the S9 interface has been introduced to support inter-domain communication between PCRFs in neighboring domains. This reference point is Diameter based and in the early stages of development, it supports basic roaming scenarios and allows a PCRF to request resources in a neighboring domain. The Sp reference point lies between the SPR and the PCRF; it allows a PCRF to request subscription information related to an authorization request and its definition is deemed as operator specific (3GPP, 2008a).

3.2 TISPAN Resource and Admission Control Subsystem

The RACS was included as part of the TISPAN Release 1 specification finalized in 2005, largely based on the early 3GPP PCC framework. Release 2 was finalized in 2008 and included in depth control scenarios and protocol specifications; in particular the scope of the RACS was extended to access and core networks, as well as to points of interconnection between networks, in order to support end-to-end QoS provisioning.

3.2.1 Functional Elements

The Release 2 specified RACS consists of two critical components: the Service-Based Policy Decision Function (SPDF) and the generic Resource and Admission Control Function (x-RACF) (TISPAN, 2008b). These elements interact with an AF in the service control layer and the transport processing functions in the transport layer, and support session and non-session based authorization. The RACS architecture is shown in Figure 2.

The role of the IETF PBNM defined PDP is split in the RACS architecture. The SPDF acts as a single point of contact and final policy decision point for the administrative domain, while the x-RACF acts as a local policy decision point regarding subscriber access admission control and resource handling control. Upon receiving authorization requests the SPDF applies operator specific policies to perform admission control. If the request is authorized the SPDF checks resource availability by querying the x-RACF. As of Release 2, the x-RACF has two functional specializations, the Access-RACF (A-RACF) and the Core-RACF (C-RACF).

The A-RACF retrieves the authenticating user's QoS profile from the Network Attachment Subsystem (NASS) and authorizes the request; this element is deployed in the access network domain where network resources may be provisioned on a per subscriber basis. The C-RACF is deployed in the core network and allocates network resources, but not on a per-subscriber basis.

The transport processing functions are divided into the Resource Control Enforcement Function (RCEF), the Border Gateway Function (BGF) and the Basic Transport Function (BTF). The BTF consists of elementary forwarding functions and elementary control functions. The BGF is a gateway between different IP transport domains and is under the control of the SPDF. The RCEF exists in the access network domain or IP edge nodes and is under the control of the x-RACF. The RACS framework supports both push and pull mode resource reservation mechanisms.

An important functional requirement for the RACS is an architecture for resource monitoring and QoS reporting. QoS reporting is the ability of a network element to gather QoS metrics related to a single service instance, while resource monitoring is the ability to monitor topologies and transport segments under RACS control. A Topology and Resource Information Specification (TRIS) should be maintained to hold information related to physical topology, logical topology and

Figure 2. The logical elements of the TISPAN RACS framework

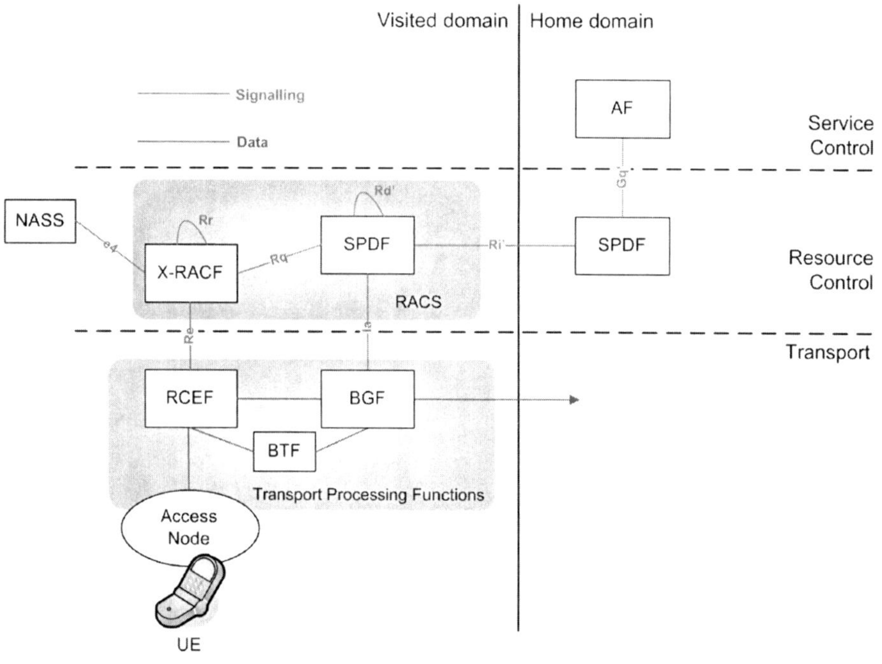

routing information (TISPAN TR 182 022, 2007). QoS reporting should be supported by all transport processing functions, information should be collected for each service instance and interfaces between QoS reporting collectors and QoS reporting users (e.g. RACS) should be implemented. Detailed logical architectures for these functional requirements are not specified.

3.2.2 Reference Point Definitions

Release 2 defines the Gq' reference point between the AF and SPDF to exchanges session based policy information. The Gq' interface is based on the 3GPP Release 6 Gq Diameter application; a specific Diameter application is defined that instantiates new commands and AVPs. The SPDF queries the x-RACF via the Diameter based Rq reference point and issues resource requests in the core and access networks to the x-RACF, indicating IP QoS characteristics.

The Diameter based e4 reference point is instantiated to allow the x-RACF to query the Network Attachment Subsystem (NASS) for user QoS profiles. The SPDF installs policy decisions and configures the BGF in the transport layer via the Ia reference point; this interface defines a profile of the Gateway Control Protocol (H.248) to control the various capabilities of the BGF (TISPAN, 2008a). The Re reference point lies between the x-RACF and the RCEF; this Diameter based interface allows policy rules to be requested by, or pushed to, the RCEF.

The RACS end-to-end QoS support allows for basic roaming scenarios and the Ri' reference point is implemented for inter-domain communication between SPDFs. TISPAN have released a draft version of the Resource Connection Initiation Protocol (RCIP) to facilitate inter-domain communication via the Ri' interface, but this is an ongoing standardization work (Callejo-Rodriguez & Enriquez-Gabeiras, 2008).

3.3 ITU-T Resource and Admission Control Functions

The ITU-T, as an inter-government, public-private partnership, promotes global convergence of, and consensus on, technologies and services. In 2004 it began developing the Resource and Admission Control Function (RACF) based largely on the early work of 3GPP and TISPAN. The ITU-T QoS control architecture defines a high level reference framework and covers the broad aspect of extending the region of control to core and access networks, and defines additional control scenarios.

3.3.1 Functional Elements

The Service Control Function (SCF) is responsible for the application signaling for service setup and logically resides in the service stratum. The SCF derives the QoS needs of the requested service and sends them to the RACF in the transport stratum for authorization. Service information is extracted from application signaling for session and non-session based services, and is used to create authorization requests. The RACF has two functional entities: The Policy Decision-Functional Entity (PD-FE) and the Transport Resource Control-Functional Entity (TRC-FE) (ITU-T Rec. Y. 2111, 2006). The PD-FE is the single contact point between any SCF and the transport stratum. This element performs authorization, reservation and commitment of network resources. The TRC-FE monitors the network topology and the resource state of the network. It performs technology-dependent admission control on behalf of the PD-FE. The RACF framework is illustrated in Figure 3.

Authorization decisions taken at the PD-FE are subject to operator specific policies and are based on service information, transport subscription information and transport network information. The transport functions are divided into the Transport Resource Enforcement-Functional Entity (TRE-FE) and the Policy Enforcement-Functional Entity (PE-FE). The TRE-FE is dynamically controlled by the TRC-FE to perform polling of network usage, bandwidth reservation and allocation, and traffic shaping. In this way the TRC-FE carries out QoS reporting and resource monitoring and can provide transport network information to the PD-FE. The TRC-FE maintains a Network Topology and Resource Database (NTRD) based on information provided by the TRE-FE. The TRE-FE also enforces policy rules received

Figure 3. The functional architecture of the ITU-T RACF

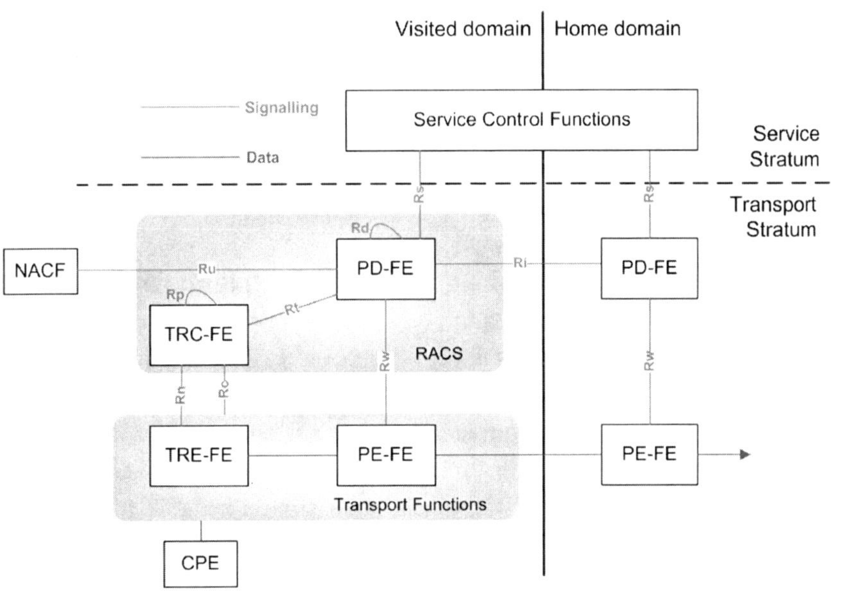

from the TRC-FE at the technology-dependent aggregation level.

Once the request is authorized, the PD-FE pushes service definitions in the form of policy rules to the PE-FE, located at the edge or border of an administrative domain in the core or access network. The RACF architecture supports both push and pull mode operation. The RACF specifies two end-to-end QoS control scenarios. In the first scenario QoS requirements for a given service can be passed over the end-to-end path through the application signaling or via an interdomain reference point. In the second scenario the requirements traverse the end-to-end path through path coupled QoS signaling. Detailed logical architectures for these high level functional requirements are not specified.

3.3.2 Reference Point Definitions

The Diameter based Rs reference point is instantiated to allow QoS resource request information to be exchanged between the SCF and the PD-FE in the same or different domains. The Ru reference point allows the PD-FE to query the NACF for subscription information. This Diameter based interface allows the PD-FE to retrieve access network specific profile information and user subscription information to incorporate into the authorization decision. The PD-FE interacts with the TRC-FE via the Rt reference point. This interface allows the PD-FE to determine via the TRC-FE whether or not the requested resources are available for a given media flow, and to request relevant TRC-FEs to detect and monitor the usage of a particular media flow. The interface is based on Diameter and the definition is very similar to that of the Rs interface between the SCF and the PD-FE.

To collect the network topology and resource status information that populates the NTRD, the Rc reference point is instantiated between the TRC-FE and the transport functions. The Rn reference point, between the TRC-FE and TRE-FE carries policy decisions that are enforced at the TRE-FE at the technology-dependent aggregation level. The PE-FE is the key injection node to enforce dynamic QoS rules in the transport layer, the Rw reference point allows the PD-FE to install the final decisions, either by push or pull mode, to the PE-FE. The functional definitions of these reference points are yet specified (Rothenberg & Roos, 2008).

The Ri reference point facilitates inter-domain communication between PD-FEs, and are used when an SCF is not capable of interacting with the PD-FEs in each domain traversed by the media flow. The functional definition of this reference point is an ongoing standardization work.

3.4 Generic QoS Management Framework

It is clear that common attributes exist between the developed frameworks. It is important that the same harmonization that resulted in the single set of Common IMS specifications takes place in the resource management sphere. The harmonization of the RACS Gq' reference point and the PCC Rx reference point is an ongoing joint initiative between TISPAN and 3GPP (3GPP, 2008c), and is a proposed work item under 3GPP Release 9 specifications. However, the overall harmonization of resource management frameworks is not investigated. This section examines common functional elements and reference points and defines a generic architecture, dubbed by the authors as the Common PCC framework.

3.4.1 Architectural Alignment

As can be observed in Figure 4, the separation of application, resource and transport control layers is consistent in all NGN resource management frameworks. An element in the service control layer should be able to intercept signaling and request resource authorization from the resource control layer.

Figure 4. Common attributes exist between the PCC, RACS and RACF frameworks

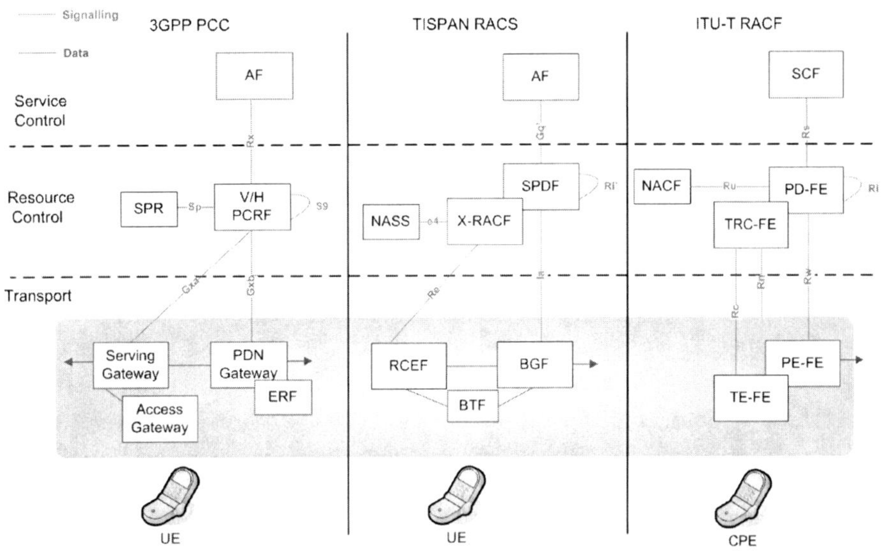

The interaction between the service control and resource control layers for session based services, by the Rx, Gq' and Rs reference points in the 3GPP, TISPAN and ITU-T architectures respectively, is based on the Diameter protocol for all examined architectures. It is likely that a harmonized interface will be similar to the 3GPP Rx interface as this is the most extensively defined.

When considering multiple transport technology deployments and the interconnection of administrative domains, the division of the policy decision element into two functions, as done by the RACS and RACF, is justified. This supports network scalability while also providing a further layer of abstraction. The reference to operator specific policies is common in all architectures, and in the broad sense the 3GPP SPR, TISPAN NASS and ITU-T NACF have similar functions. It is clear that a policy repository is necessary that includes, among other operator specific policies, subscription and QoS specific profile information. Policies will control other aspects of the network, and it is expected that access to the policy repository will be facilitated by more than one protocol to cater for the wide range of application scenarios.

The QoS resource control performed at the PDP element, though described using different nomenclature, is essentially identical for each of the architectures; it consists of service based admission control (authorization), resource based admission control (reservation), and enforcement of reserved resources (commitment). Additionally, the information available to carry out QoS resource control is similar for all three architectures and includes service specific information, subscription specific information, and transport network specific information.

The transport layer elements are diverse in the architectures. A border gateway element that provides connectivity capabilities is essential in all architectures, and by housing a gateway in the home and visited domain, data breakthrough can be supported in both domains. A specialized transport layer element in the access or IP edge nodes is also desirable; these elements enforce policy decisions at the technology dependent aggregation level. The interaction between the resource control and transport layer elements, by the Gx, Re/Ia and Rw/Rc reference points, encompasses more than one protocol. The RACS utilizes H.248 for controlling the BGF and Diameter for controlling the RCEF. 3GPP specify Diameter for the Gx reference point; this Diameter application has been extensively defined and much work has gone into making it applicable to a wide range of transport plane technologies. There are several candidate protocols to collect transport layer topology and resource status information, although none are specified by any of the standardization bodies. Hence this interaction is unspecified in the Common PCC framework.

End-to-end QoS support is elementary in all of the architectures. In the RACF architecture the inter-domain reference point is not specified, the RACS is currently standardizing RCIP for this interface, while the 3GPP S9 reference point defines a Diameter application for inter-domain resource authorization.

3.4.2 Common PCC Framework

The Common PCC framework includes an AF in the service control layer that extracts QoS information from application signaling to create authorization requests. A PDF in the resource control layer acts as a single point of entry for authorization requests from AFs and neighboring PDFs, and is split into home and visited functions. An x-RACF element performs technology dependent admission control and monitors transport functions to maintain a TRIS. A policy repository is defined that contains operator specific policies including subscription and QoS profile policies.

The transport layer comprises a BGF spit into home and visited functions to provide connectivity capabilities. RCEF functionality is incorporated at the access or IP edge node to enforce policies at the technology dependent aggregation level.

The 3GPP Diameter based Rx interface is used for interaction between the AF and the PDF; AVPs and messages defined by the Gq' and Rs Diameter applications are incorporated. The TISPAN Diameter based Rq interface facilitates interaction between the PDF and x-RACF, while the 3GPP Gxa and Gxb reference points allow communication between the PDF and BGF, and x-RACF and RCEF, respectively.

The Rc interface allows an x-RACF element to query the transport layer functions and populate a TRIS with transport layer topology and resource information. The protocol for this interface is not specified. The Sp interface allows interaction between the PDF and the policy repository. The protocol for this interaction is deemed operator specific and therefore not specified. The 3GPP Diameter based S9 interface is utilized for inter-domain interactions between neighboring PDFs.

The Common PCC framework supports push and pull mode operation. Session binding is based on user-identity based identification. QoS resource control, performed at the PDF, includes resource authorization, reservation and commitment. Service level QoS parameters are conveyed in PCC rules and include a QCI, an ARP and authorized guaranteed and maximum bit rate values for uplink and downlink. Figure 5 shows the Common PCC logical architecture; the defined terms and functional elements allow for more coherent and focused future research.

4. DEPLOYMENT CHALLENGES

The high level requirements of the IMS resource management framework include: minimal effect on session setup delay, backwards compatibility, convergence towards agnostic access, and rapid time to market of new services (Ludwig et. al., 2006). Apart from interoperability concerns and the need for harmonization between the resource management frameworks addressed earlier, there are a number of shortcomings in the standardization work. These shortcomings can be broadly separated into two areas: vertical coordination and horizontal coordination of resources. Vertical coordination refers to the interaction between the applications requesting resources and the transport layer resources that will carry the application traffic, while horizontal coordination

Figure 5. The Common PCC Framework encompasses work done by all standardization bodies and defines a generic set of terms and functional elements

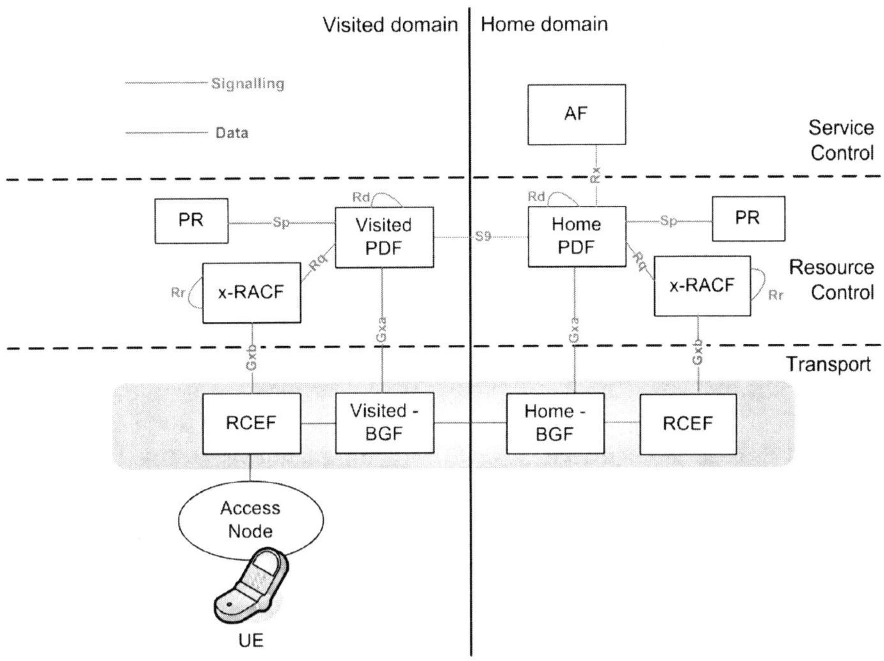

refers to the ability to provide seamless end-to-end QoS connectivity across administrative domains (Rothenberg & Roos, 2008). These areas cover the major deployment challenges faced when facilitating inter- and intra-domain policy controlled resource management across heterogeneous transport technologies.

4.1 Vertical Coordination of Resources

To ensure the rapid development and deployment of new services it is critical that no new QoS standardization is required when deploying advanced multimedia rich services with new requirements. Policy refinement refers to the translation of policies at different levels of the management hierarchy, and, in the context of general policy based management, has been under study for some time. While it is clear that technology independent policies should fully characterize a network path, and technology specific policies should include transport specific classifiers and/or link layer QoS information, there is no standardized method to perform this policy refinement and effectively map QoS descriptors across different layers of the policy life cycle. There are numerous proposals for policy information representation, but the IMS and resource management specifications do not specify any particular model. These are all open issues within the Common PCC framework that need to be addressed before wide scale deployment becomes realistic.

4.1.1 A Framework for SIP Session Policies

In the Internet Draft "A Framework for Session Initiation Protocol (SIP) Session Policies", Hilt, Camarillo & Rosenberg (2008) present standardized mechanisms by which SIP proxy servers can define or influence policies on sessions. The authors suggest that the adopted IMS policy framework, whereby sessions are described using the Session Description Protocol (SDP), prohibits the innovative creation of new services. If any service requires a new SDP extension to describe itself, or the use of a separate description format, it would be necessary to upgrade all SIP proxy and policy control elements in the network, thus breaking a major SIP design principle. Additionally the policy mechanisms assume that SIP proxy servers have access to the SDP bodies of the SIP messages. This means that end-to-end encryption mechanisms like Secure / Multipurpose Internet Mail Extensions (S/MIME) are not supported, and end-users must use SDP as their session description format.

This Internet Draft describes extensions that allow proxy servers to communicate different policies to the end-users without accessing end-to-end bodies such as session descriptions. A policy server is defined that delivers session policies to the end-user; these policies can accept the session, reject the session or propose changes to the session parameters that would deem the request acceptable. XML document formats and event packages to represent and exchange session policies have been defined (Hilt & Camarillo, 2008).

The signaling between endpoints, and the policy exchange between an endpoint and a policy server are decoupled because the end-user and the policy server communicate directly over a dedicated policy channel. This decoupling means that separate encryption mechanisms can be used on the signaling path and the policy channel, and proxy servers need not access end-to-end bodies nor be upgraded to deploy services with new requirements.

Figure 6 demonstrates typical signaling in the session policies framework for session initiation with session-specific policies. The originating end-user retrieves the originating policy server's Uniform Resource Identifier (URI) from its proxy server (1-3), and contacts the policy server over the dedicated policy channel (4, 5). The session is described in the XML document included in the body of the Subscribe request. Once the session

Figure 6. Session initiation with session-specific policies

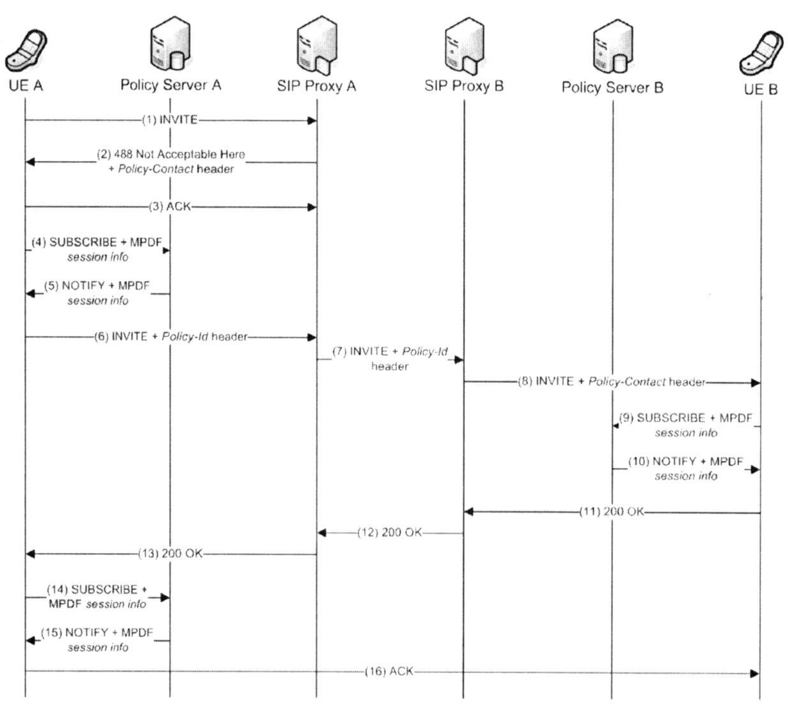

is accepted, a new Invite request is created and forwarded to the terminating proxy server, which informs the terminating end-user of the terminating policy server's URI (7). The terminating end-user confers with the policy server in the terminating domain (9, 10), and once authorized sends a 200 Ok response (11-13). The originating end-user again confers with the originating policy server (14, 15), this time including the session description from the terminating end-user, and eventually the session is initiated. The 200 Ok messages sent in response to the Subscribe and Notify requests are omitted from the diagram for simplicity. The main disadvantages of the session policies framework are the additional signaling required to initiate sessions because of introduced round-trip times between the end-user and policy server, and the added complexity of storing, creating and interpreting policy documents at the end-user.

4.1.2 Advanced QoS Negotiation

Future IMS services are expected to be exceedingly multimedia rich and customized to meet end-user preferences and capabilities. The regular deployment of new services will result in a constantly changing network dynamic and the Common PCC framework will have to cater for complex and unpredictable QoS requirements.

Skorin-Kapov et. al. (2007) describe enhancements to the standardized IMS QoS negotiation procedure necessary to address these dynamically changing QoS requirements. The authors point out the need to incorporate end-user preferences, network constraints and service requirements into the QoS negotiation procedure. A networked virtual reality service is used to illustrate these requirements; end-user preferences would allow a user to set the relevance of different events such as timing constraints for displaying and interacting with the virtual service, audio and text chat. Network constraints require that the service adapt to dynamic changes in the network occurring during service execution. In terms of the virtual reality service if the available bandwidth is unexpectedly reduced (e.g. due to a wireless link) the desired action might be to drop audio chat or switch to a text-based chat to maintain the maximum quality of experience.

Service requirements might require that the application be available in several customized versions, the virtual reality service could be offered as a low-cost version suitable for dial-up access, and a default version with attractive graphics.

The authors argue that current standards lack techniques to address these issues in a comprehensive manner. A model is proposed for dynamic negotiation and adaptation of QoS requirements, which uses generic client and service profiles as a basis. A client profile specifies end-user terminal and access network constraints, and application related preferences; such preferences are set by the end-user, though a generic client profile is defined. A service profile specifies different supported configurations of a service to address diverse end-user capabilities; such preferences are set by the application developer, though a generic service profile is defined.

To incorporate these concepts into the IMS architecture, the authors propose the addition of a QoS parameter matching and optimization (QMO) Application Server (AS). The QMO AS examines the service and client profiles to determine feasible service parameters and suitable service versions, and to perform optimization. As with every other AS, the decision whether or not to involve the QMO AS for a particular service is taken by the S-CSCF. When an end-user initiates a session the client profile specific to the requested service is encapsulated in the Invite request; this request is forwarded to the terminating AS via the QMO AS. The terminating AS defines the service profile and encapsulates the information in the 183 Session Progress response that is conveyed to the end-user via the QMO AS. The service and client profiles are represented using XML-based SDP-next-generation (SDPng). Typical IMS session negotiation takes place, but additional interactions between the QMO AS and the terminating AS facilitate optimization and the selection of feasible service versions, based on the provided client and service profiles.

This configuration has the advantage of customized service delivery, optimized for the end-user preferences and terminal capabilities. The definition of generic client and service profiles allows new services to inherit advanced QoS negotiation and optimization support. The matching and optimization procedures carried out during session initiation and renegotiation increases signaling traffic, and the model also suffers from poor scalability; for a large number of users, running QMO procedures separately for each session will be time-consuming and costly.

4.1.3 Policy Refinement

The creation of the PCC rule based on service information, subscription information and transport network information is critical, as it is this rule that exhaustively defines how service data flows related to an application should be treated in the transport layer. Albaladejo et. al. (2008) point out that this process is not specified in the 3GPP standards and is left for operator configuration. In particular the authors argue that there is no universal interpretation of how to map the content received in the Media-Component-Description Attribute Value Pair (AVP) in the authorization request, into PCC rules.

They propose two solutions for creating PCC rules for a session. The first creates a PCC rule for each Media-Component-Description AVP, while the second approach creates a PCC rule for each associated Media-Sub-Component AVP. The first approach complicates the structure and its operation; as there are essentially an unlimited number of Media-Sub-Component AVPs, the number of Flow-Description AVPs within each service data flow is also unlimited. The second approach limits the number of Flow-Description AVPs in each PCC rule to two, but splits media components into separate PCC rules. Through experimental analysis, the authors found that despite complicating the operation, the first approach, where the PCC rule is based on the Media-Component-Description AVP, was better suited to the IMS environment, as the second approach suffered serious scalability problems when more than one Media-Component-Description AVP was included in the authorization request.

4.2 Horizontal Coordination of Resources

End-to-end QoS coordination can be facilitated at any of the levels within the IMS model. NGN architectures supporting IMS service control will likely implement some form of resource control in most domains and network segments; therefore it makes sense to exploit these already implemented mechanisms at the resource control layer. However proprietary interfaces in network equipment, resulting in highly vendor specific solutions, may hamper deployment. When combined with the lack of a general interface specification between service control functions and resource management functions, this could lead to general interoperability issues.

Essentially two end-to-end QoS control scenarios exist. In the first scenario QoS requirements for a given service are passed over the end-to-end path through the application signaling via the inter-domain reference points. In the second scenario QoS requirements are passed over the end-to-end path through path-coupled QoS signaling. Path-coupled signaling requires modification to all routing devices in the transport layer, limiting the applicability in the short to medium term. However the first approach, using Common PCC elements and interfaces, does not facilitate end-to-end resource reservation across multiple domains, nor does it link the service control inter-domain routes with the routes followed by the media in the transport layer.

4.2.1 IETF NSIS/NSLP

The IETF is working on an end-to-end signaling protocol suite, the Next Steps In Signaling (NSIS), with QoS as its first use case. The QoS NSIS Signaling Layer Protocol (NSLP) extends the Resource Reservation Protocol (RSVP), addressing many shortcomings including scalability. The Internet Draft "NSLP For Quality of Service Signaling" (Manner, Karagiannis & McDonal, 2008), defines a protocol that establishes and maintains state at nodes along the path of a data flow for the purpose of providing some forwarding resources for that flow.

The QoS NSLP extends the set of reservation mechanisms to meet the requirements stipulated in RFC 3726, "Requirements For Signaling Protocols" (Brunner, 2004). In particular, support for sender and receiver-initiated reservation is incorporated. The Internet Draft defines three nodes: a QoS NSLP NSIS Entity (QNE), any element that supports QoS NSLP; a QoS NSLP Initiator (QNI), the first node in a sequence of QNEs that issues a reservation request for a session; and a QoS NSLP Responder (QNR), the last node in a sequence of QNEs that receives a reservation request for a session. A QoS NSLP signaling session consists of a single QNI, any number of QNEs and a single QNR and can span the end-to-end data path, or a segmented portion of the path.

It is important to note that a distinction is made between the operation of the signaling protocol, and resource allocation and management techniques. A Resource Management Function (RMF) is defined that is responsible for all resource provisioning, monitoring and assurance functions in the network and is particular to a specific QoS Model (QOSM). This means that the QoS NSLP is independent of a specific QOSM, and all information related to RMF functions is carried in a QoS Specification (QSPEC) object that is encapsulated in the NSLP messages. The QSPEC object is conveyed in QoS NSLP messages but is opaque to the NSLP signaling as it is only interpreted by the RMF, where the information is used to provision resources.

QoS NSLP is a candidate for the resource reservation protocol used by the Common PCC Framework for the pull mode approach to requesting QoS-enabled paths (Alfano, McCann & Towle, 2006). Unfortunately practical issues limit the applicability of this approach in real world scenarios. The introduction of the QoS NSLP requires modification to all routing devices in the transport layer. Network operators heavily invested in legacy networks will be hesitant to commit the necessary capital expenditure for such an overhaul, and link layer QoS signaling, like PDP context activation in UMTS, goes against the principle of separating core procedures from the subtleties of the access network.

4.2.2 Future Internet: EuQoS Project

There are a number of research initiatives examining the evolution of the Internet with regard to its structure and management in the future, the Future Internet. The provisioning of advanced QoS connectivity services has been identified as a key driver for the operator's business role in the Future Internet (Callejo-Rodriguez & Enriquez-Gabeiras, 2008).

The End-to-End Quality of Service Support over Heterogeneous Networks (EuQoS) project is an ongoing European research initiative aimed at building an entire QoS framework, addressing all relevant network layers, protocols and technologies (Masip-Bruin et. al, 2007). The framework has been prototyped and tested in a multi-domain environment distributed across Europe. The project has a broad scope but deals with QoS routing or finding a feasible path between a source and destination node satisfying one or more QoS constraints. Although the project does not specifically apply to the IMS architecture, IMS is based largely on Internet protocols and these solutions can be adapted.

The EuQoS framework defines a virtual network layer to decouple network decisions from network technologies. This layer is split into technology independent and technology specific layers. The technology independent layer houses Resource Managers (RM) that manage QoS reservation and authorization for each domain, while the technology specific layer houses Resource Allocators (RA) that enforce specific decisions on transport layer devices. Essentially the RM acts as a PDF, and RAs act as distributed PEPs.

In order to check the availability of resources, EuQoS uses path-coupled signaling based on NSIS extensions defined in the Internet Draft "GIST Extensions for Hybrid On-path Off-path Signaling (HyPath)". The authors argue that the path coupled signaling must reach all RMs along the path to ensure end-to-end resource availability, even though these RMs may not lie on the data path. The extension, known as EQ-NSIS, allows some routers to re-direct the end-to-end signaling to RMs that are not necessarily on the data path. This approach is similar to the pull mode operation used by the Common PCC framework in coordination with NSIS, however it also provides a means for RMs or PDFs to discover ingress and egress points through which the data-path will pass in its domain, and supports non-NSIS domains.

Extensions to the Border Gateway Protocol-4 (BGP-4) were defined to take into account intra- and inter-domain QoS information. The extensions, known as EQ-BGP, create a road map of available QoS paths between source and destination that are advertised to neighboring domains. The protocol includes an optional path attribute that conveys information about the QoS capabilities of a path, and a QoS assembling function for computing aggregated values of QoS parameters for end-to-end routing paths. The EQ-NSIS and EQ-BGP extensions facilitate the discovery and advertisement of QoS routes, though as with any path-coupled approach to end-to-end QoS routing, significant modification to the legacy transport layer is necessary. Additionally the EuQoS framework requires the sharing of potentially sensitive QoS and topology information with neighboring domains.

Tailoring the Common PCC framework to ensure rapid and innovative service creation, and to provision end-to-end QoS-enabled paths, is an active area of research. The policy of the 3GPP to adopt IETF standards wherever possible means that the IETF work on SIP session policies and NSIS/NSLP is of particular importance. Mechanisms to facilitate deployment of advanced QoS services and the development of the inter-domain reference points to enabled QoS connectivity across multiple domains are also necessary.

5. CONCLUSION

The IMS, as the candidate technology to facilitate the move to an all-IP infrastructure, provides ubiquitous access through wireless technologies and allows for innovative and rapid service creation through Internet ideologies. An IMS QoS framework is necessary to support the rapidly expanding requirements of advanced multimedia services, and to incorporate strict delay, packet-loss and jitter bounds into the typically best-effort IP network.

The centralization of IMS standardization has helped alleviate interoperability concerns and there is widespread belief that the number of commercial IMS deployments will grow rapidly in the near future. This will be fueled in part by the adoption of LTE and EPC technologies, of which IMS is a central IP service element. However, IMS as an emerging technology still faces challenges. The most critical is that posed by the widespread proliferation of Web 2.0 services. IMS operators will need to justify charging for services that are typically available free of charge in the Internet space; reliability and guaranteed transport of multimedia services through efficient management of resources will be critical to differentiate IMS services.

The standardization of the IMS/NGN resource management framework has been fragmented resulting in weak functional and interface specifications. The standardized architectures have notable shortcomings regarding support for advanced multimedia services and end-to-end QoS mechanisms. This chapter has highlighted the work of several experts in the field who are currently addressing these challenges. While it is important to leave room for flexibility and vendor innovation, it is critical that these proposals be further developed and integrated into the Common PCC specification once they are mature, to prevent general interoperability issues that could render end-to-end QoS provisioning for advanced multimedia services almost impossible.

REFERENCES

Albaladejo, A., Gouveia, F., Corcici, M., & Magedanz, T. (2008, November). The PCC Rule in the 3GPP IMS Policy and Charging Control Architecture. In *Proceedings of 2008 IEEE Global Communications Conference* (pp. 1-5).

Alfano, F., McCann, P., & Towle, T. (2006). IMS Service-Based Bearer Control. *Bell Labs Technical Journal*, *10*(4), 151–166. doi:10.1002/bltj.20131

Brunner, M. (2004, April). *Requirements for Signaling Protocols* (IETF Request For Comments 3726).

Callejo-Rodrigues, M., & Enriquez-Gabeiras, J. (2008, October). Bridging the Standardization Gap to Provide QoS in Current NGN Architectures. *IEEE Communications Magazine*, *46*(10), 132–137. doi:10.1109/MCOM.2008.4644130

ETSI TISPAN. (2007, December). *TR 182 022 Architectures for QoS handling* (Technical Report).

ETSI TISPAN. (2008a, June). *ES 28 018 H.248 Profile for controlling Border Gateway Functions (BGF) in the Resource and Admission Control Subsystem (RACS)* (ETSI Standard).

ETSI TISPAN. (2008b, July). *ES 282 003 Resource and Admission Control Subsystem (RACS) Functional Architecture* (ETSI Standard).

3GPP. (2008a, December). *TS 23.203 Policy and Charging Control Architecture v8.4.0* (3GPP Technical Specification).

3GPP. (2008b, December). *TS 23.401 General Packet Radio Service enhancements for Evolved Universal Terrestrial Radio Access Network V8.4.1* (3GPP Technical Specification).

3GPP. (2008c, February). *TR 23.822 Framework for Gq'/Rx Harmonization V0.2.0* (3GPP Technical Recommendation).

Hilt, V., & Camarillo, G. (2008, July). *A Session Initiation Protocol (SIP) Event Package for Session-Specific Session Policies - draft-ietf-sipping-policy-package-05* (IETF Internet).

Hilt, V., Camarillo, G., & Rosenberg, J. (2008, November). *A Framework for Session Initiation Protocol (SIP) Session Policies - draft-ietf-sip-session-policy-framework-05* (IETF Internet Draft).

ITU-T. (2001). *Rec. Y. 2111 Resource and Admission Control Functions in Next Generation Networks.*

Ludwig, R., Ekstrom, H., Willars, P., & Lundin, N. (2006, September). An Evolved 3GPP QoS Concept. In *Proceedings of 2006 IEEE Vehicular Technology Conference* (pp. 388–392).

Manner, J., Karagiannis, G., & McDonal, A. (2008, February). *NSLP for Quality-of-Service Signaling - draft-ietf-nsis-qos-nslp-16* (IETF Internet Draft).

Masip-Bruin, X., Yannuzzi, M., Serral-Gracia, R., Domingo-Pascual, J., Enriquez-Gabreiras, J., & Callejo, M. (2007, February). The EuQoS System: A Solution for QoS Routing in Heterogeneous Networks. *IEEE Communications Magazine*, *45*(2), 96–103. doi:10.1109/MCOM.2007.313402

Rosenbrock, K., Sanmugam, R., Bradner, S. & Klensin, J. (2001, June). *3GPP-IETF Standardization Collaboration* (IETF Request For Comments 3113).

Rothenberg, C., & Roos, A. (2008, March). A Review of Policy-Based Resource and Admission Control Functions in Evolving Access and Next Generation Networks. *Journal of Network and Systems Management*, *16*(1), 14–45. doi:10.1007/s10922-007-9096-3

Skorin-Kapov, L., Mosmondor, M., Dobrijevic, O., & Matijasevic, M. (2007, July). Application-Level QoS Negotiation and Signaling for Advanced Multimedia Services in the IMS. *IEEE Communications Magazine*, *45*(7), 108–116. doi:10.1109/MCOM.2007.382669

Section 5
Ad-Hoc/Mesh

Chapter 20
Quality of Service (QoS) Routing in Mobile Ad Hoc Networks

R. Asokan
Kongu Engineering College, India

A. M. Natarajan
Bannari Amman Institute of Technology, India

ABSTRACT

A Mobile Ad hoc NETwork (MANET) consists of a collection of mobile nodes. They communicate in a multi-hop way without a formal infrastructure. Owing to the uniqueness such as easy deployment and self-organizing ability, MANET has shown great potential in several civil and military applications. As MANETs are gaining popularity day-by-day, new developments in the area of real time and multimedia applications are increasing as well. Such applications require Quality of Service (QoS) evolving with respect to bandwidth, end-to-end delay, jitter, energy etc. Consequently, it becomes necessary for MANETs to have an efficient routing and a QoS mechanism to support new applications. QoS provisioning for MANET can be achieved over different layers, starting from the physical layer up to the application layer. This chapter mainly concentrates on the problem of QoS provisioning in the perception of network layer. QoS routing aims at finding a feasible path, which satisfies QoS considering bandwidth, end-to-end delay, jitter, energy etc. This chapter provides a detailed survey of major contributions in QoS routing in MANETs. A few proposals on the QoS routing using optimization techniques and inter-layer approaches have also been addressed. Finally, it concludes with a discussion on the future directions and challenges in QoS routing support in MANETs.

1. INTRODUCTION

The recent developments in the information super highway have made connectivity possible between users at anytime and anywhere in the world. From the Advanced Research Project Agency NETwork (ARPANET) to the present day 4G networks, Communication Networks have greatly influenced every facet of human life like commerce, industry, defence, government, home, recreation. Networking solutions have become an integral part of modern living. Mobile networks are required to support the

DOI: 10.4018/978-1-61520-680-3.ch020

seamless delivery of data, high quality voice and video. The mobile communication is generally and widely supported by wired fixed infrastructure. The mobile devices use single-hop wireless radio communication to access a base station that connects it to the wired infrastructure. In contrast, MANET does not use any fixed infrastructure. Mobile ad hoc networks are formed by autonomous system of mobile hosts connected by wireless links with no supporting fixed infrastructure or central administration.

The nodes of MANET intercommunicate through single-hop and multi-hop paths in a peer-to-peer fashion. Communication between these nodes are either direct or through intermediate nodes acting as routers. Thus, the nodes operate as both hosts as well as routers. Due to the limited range of transmission and several nodes may be needed to route a packet to its destination. Since the nodes are mobile, the creation of routing paths is altered by the addition and deletion of nodes. The topology of the network changes rapidly and unexpectedly. The concept of MANET is used in many application environments without requiring any infrastructure support.

1.1 Quality of Service (QoS)

QoS is defined as a set of service requirements that needs to be met by the network while transporting a packet stream from a source to its destination. The network is expected to assurance a set of measurable specified service attributes to the user in terms of end-to-end delay, bandwidth, probability of packet loss, delay variance (jitter), etc.

The QoS metrics can be classified as additive metrics, concave metrics and multiplicative metrics. Let m(u,v) be the performance metric for the link (u,v) connecting node u to node v and path $(u,u_1,u_2 \dots u_k,v)$ a sequence of links for the path from u to v. A constraint is additive if $m(u,v) = m(u,u_1) + m(u_1,u_2) + \dots + m(u_k,v)$. The end-to-end delay is an additive constraint because it is the accumulation of all delays of the links along the path.

A constraint is concave if $m(u,v) = \min\{m(u,u_1), m(u_1,u_2), \dots, m(u_k,v)\}$. The bandwidth bw(u,v) requirement for a path between node u and v is concave. To find a QoS feasible path for a concave metric, the available resource on each link should be at least equal to the required value of the metric.

A constraint is multiplicative if $m(u,v) = m(u,u_1) \times m(u_1,u_2) \times \dots \times m(u_k,v)$. The probability of a packet prob(u,v), sent from a node u to reach a node v, is multiplicative, because it is the product of individual probabilities along the path. Bandwidth and energy are concave metric, while cost, delay, and jitter are additive metrics. Bandwidth and energy are concave in the sense that end-to-end bandwidth and energy are the minimum of all the links along the path. The reliability or availability of a link is a multiplicative metric (Baoxian & Hussein, 2005).

To support QoS, the link state information such as delay, bandwidth, jitter, cost, loss ratio and error ratio in the network should be available and manageable. However, receiving and managing the link state information in a MANET is difficult, because the quality of a wireless link changes with the surrounding circumstance. In addition, the resource limitations and the mobility also add to the complexity. These networks have certain unique characteristics that pose several difficulties in provisioning QoS. Some of the characteristics are

Dynamic Topology

Nodes can be extremely mobile as a result the topology of the network changes frequently and dynamically. Topology information has a limited lifetime and must be updated frequently to allow data packets to be routed to their destinations. Because the nodes have do not have any restriction on mobility, the network changes dynamically. Hence the admitted QoS sessions may suffer due to frequent path breaks, thereby requiring such sessions to be re-established over new paths. The delay incurred in reestablishing a QoS session may cause some of the packets belonging to the session to miss their delay targets which is not acceptable for applications that have stringent QoS requirements.

Imprecise State Information

The nodes maintain both the link-specific state information and flow-specific state information. The link specific information includes bandwidth, delay, delay jitter, loss rate, error rate, stability and cost for each link. The flow-specific information includes session ID, source address, destination address and QoS requirements of the flow. The state information is inherently imprecise due to dynamic changes in network topology and channel characteristics. Hence routing decisions may not be accurate.

Lack of Central Co-ordination

Unlike WLANS and cellular networks mobile ad hoc networks do not have central controllers to co-ordinate the activity of nodes. The main advantage of MANET is that it may be set up suddenly, without planning and nodes can change their position dynamically. This makes difficult to provide any form of centralized control and operate in a completely distributed manner. This further complicates QoS provisioning in MANETs.

Hidden Terminal Problem

The hidden terminal problem is inherent in ad hoc wireless networks. This problem occurs when packets originating from two or more sender nodes, which are not within the direct transmission range of each other, collide at a common receiver node. It necessities the retransmission of the packets, which may not be acceptable for flows that have stringent QoS requirements.

Limited Resource Availability

Resources such as computational power, bandwidth, battery life, storage space and processing capability are limited in MANET nodes compared to devices used in wired networks. This will affect the performance of the QoS providing mechanism. Since low memory capacity restricts the amount of QoS information that can be stored, requiring more frequent updates, which in turn increase the overhead. Therefore efficient resource management methods are required for optimal utilization of these limited resources.

Error-Prone Shared Radio Channel

The radio channel is a broadcast medium by nature. During propagation through the wireless medium, the radio waves suffer from several impairments such as attenuation, multipath propagation and interference.

Insecure Medium

Due to the broadcast nature of the wireless medium, communication through a wireless channel is extremely insecure. Therefore, security is a significant issue in ad hoc wireless networks, particularly for military and tactical applications. These networks are susceptible to attacks such as eavesdropping, spoofing, denial of service, message distortion and impersonation. Without sophisticated security mechanisms, it is very difficult to provide secure communication.

MANETs are expected to become an essential part of the computing environment in the near future. Numerous challenges must be overcome to realize the practical benefits of ad hoc networking. These include effective routing, medium access, mobility management and security and QoS issues. MANETs are expected to further support a wide range of real time applications. Many routing protocols have been developed for MANET to establish and maintain multi-hop routes between nodes. Most of the protocols such as Ad hoc On-demand Distance Vector protocol (AODV), Dynamic Source Routing protocol (DSR), Temporally Ordered Routing Protocol (TORA) establish and maintain routes on a best-effort basis.

The support for QoS services will be an important and desirable component of MANETs. Therefore the support for QoS will be an important and essential requirement of MANETs. QoS routing plays an important role in providing QoS in MANETs. Although quite a lot of research has been undertaken to support QoS in the Internet, traditional wireless networks are not suitable for MANET. The exclusive characteristics of MANET like dynamically varying network topology, lack of precise information, lack of central control, limited resource availability and hidden terminal problems create several difficulties in provisioning the QoS. Although some work has been carried out to address this critical issue, research in this area is far from exhaustive. Hence the QoS routing support for MANETs remains an open problem.

1.2 Quality of Service (QoS) in MANETs

QoS provisioning can be done over different layers in the protocol stack starting from the physical layer up to the application layer. Each layer take cares of different measurements in QoS provisioning. For example, the physical layer take cares of transmission quality. The link layer handles the variable bit error rate. The network layer deals with the change in bandwidth and delay. The transport layer focuses on the delay and packet loss due to transmission errors. The application layer aimed at the frequent disconnections and reconnections (Prasant, Jian, & Chao, 2003).

QoS Support in Physical Layer

The signal-to-noise ratio of wireless medium in mobile ad hoc networks fluctuates with respect to time. Therefore, adaptive modulation is required to alter many parameters based on the current channel state to obtain better performance from wireless medium. As a result, one of the major challenges in supporting QoS over wireless medium is channel estimation. It includes perfect channel estimation in the receiver and reliable feedback of the estimation to the transmitter. Therefore the transmitter and receiver can be properly synchronized. Because of the time-varying fading channel, coding schemes developed for a fixed channel is not suitable for MANETs. The channel coding required to solve the problems introduced by channel, multipath fading and mobility.

QoS Provisioning at the MAC Layer

The Medium Access Control (MAC) protocol determines which node should transmit next on the broadcast channel when several nodes are competing for transmission on that channel. The existing MAC protocols for ad hoc wireless networks use channel sensing and random back-off schemes, making them suitable for best-effort data traffic. MAC layer aimed at providing QoS guarantee for real-time traffic support. One of the main problems which occurred in MANETs was the hidden and exposed terminal problems. This can be handled by a fully distributed scheme. Karn (1990) proposed Multi-hop Access Collision Avoidance (MACA) to solve the problem by using request-to-send and clear-to-send (RTS/CTS) dialogs. This scheme does not completely eliminate the hidden terminal problem. So an extended approach, namely, MACA for Wireless (MACAW) was proposed to provide faster recovery from hidden terminal collisions (Bharghavan, Demers, Shenker & Zhang, 1994).

Some synchronous methods are proposed for multihop wireless network, they are cluster TDMA the virtual network and SWAN (Gerla &Tsai 1995; Alwan et al., 1996). These protocols support real-time traffic since slots can be reserved and QoS routing is used to find the route with sufficient bandwidth. The downside is that strict time framing and global synchronization introduce much implementation complexity and cost. Multi-hop Access Collision Avoidance with Piggyback Reservation (MACA/PR) provides guaranteed bandwidth support (via reservation) to real-time traffic. It permits to establish real time connections over a single hop only. However, it should work with QoS routing algorithm and a fast reservation setup mechanism (Lin & Gerla, 1997).

QoS Provisioning at the Network Layer

QoS implemented in the network layer aims to find a route which provides the required quality. The metrics used to select the route is not only the number of hops along the route but also other metrics like delay, bandwidth, network life time and data rate. QoS routing is a scheme that takes into consideration the appropriate information about each link and based on that information select paths that satisfies the QoS requirements of a flow. QoS routing protocols have a key part in a QoS mechanism, because it is their function to find nodes that can serve the application's requirements. Many routing protocols have been developed to discover and retain routes between source and destination nodes. The main objective of the QoS routing protocols is to establish a path from a source to the destination that satisfies the needs of the desired QoS. These protocols work with the resource management methods to establish paths through the network that meet end-to-end QoS requirements.

Transport Layer Issues for QoS Support

TCP designed for the Internet performs well based on the assumption that most packet losses are due to network congestion. This assumption is not true in the context of wireless networks, where packet losses are mostly due to wireless channel noise and route changes. Whenever a TCP sender detects any packet loss, it will activate its congestion control and avoidance algorithms, which makes TCP performs poorly in term of end-to-end throughput. A lot of work has been done to improve TCP performance in mobile wireless networks (Balakrishnan, Padmanabhan, Seshan & Katz, 1996; Chen et al, 2001). These protocols are not suitable for use in infrastructure-less environments such MANETs where no base stations exist. Some of these protocols are dependent on explicit feedback mechanisms to distinguish error losses from congestion losses such that appropriate actions can be taken when packet losses occur, while other protocols are based on implication and estimation from observation (Chandran, Raghunathan, Venkatesan & Prakash, 2001;Liu & Singh, 2001).

Application Layer Issues for QoS Support

Much work has been done in using application layer techniques for adaptive real-time audio/video streaming over the Internet. These application techniques include methods based on compression algorithm features, layered encoding, rate shaping, adaptive error control, and bandwidth smoothing. Most of these techniques were investigated in the context of Internet. Considering the unique characteristic of MANETs, it is conceivable that some modification and improvement must be made to these techniques for use in MANETs. Other techniques are also under investigation, such as joint source-channel coding and joint source-network coding. These joint coding approaches attempt to consider both source characteristics and current channel/network states to achieve better overall performance in transmitting image and real-time audio/video over MANETs.

2. QOS ROUTING PROTOCOLS

Ad hoc routing protocols can be classified into three major categories based on the routing information update mechanism. In proactive or table driven protocols, every node maintains the network topology information in the form of routing tables by periodically exchanging routing information. Whenever a node requires a path to a destination, it runs an appropriate path finding algorithm on the topology information it maintains. The Destination Sequenced Distance Vector routing protocol (DSDV), Wireless Routing Protocol (WRP), Source Tree Adaptive Routing protocol (STAR) and Cluster-head Gateway Switch Routing protocol (CGSR) are some examples for the protocols that belong to this category.

Protocols fall under reactive or on demand protocols category do not maintain the network topology information. They obtain the necessary path when it is required, by using a connection establishment process. The Dynamic Source Routing protocol (DSR), Ad hoc On-demand Distance Vector routing protocol (AODV), Temporally Ordered Routing Algorithm (TORA) and Associativity Based Routing (ABR) are some of the protocols that belong to this category. Protocols belonging to hybrid routing protocols category combine the best features of the above two categories (Royer & Toh, 1999).

2.1 Single Metric

2.1.1 Bandwidth

Core-Extraction Distributed Ad Hoc Routing (CEDAR)

Sivakumar, Sinha, and Bhargavan (1999) have proposed the Core-Extraction Distributed Ad Hoc Routing (CEDAR) algorithm. It dynamically establishes the core of the network and then incrementally propagates the link states of stable high bandwidth links to the nodes of the core. The route computation is on-demand basis and is performed by the core nodes using only local state. CEDAR has three key components:

- **Core extraction:** A set of nodes is elected to form the core that maintains the local topology of the nodes in its domain, and also perform route computations. The core nodes are elected by approximating a minimum dominating set of the ad hoc network.
- **Link state propagation:** QoS routing in CEDAR is achieved by propagating the bandwidth availability information of stable links in the core. The basic idea is that the information about stable high-bandwidth links can be made known to nodes far away in the network, while information about the dynamic or low bandwidth links remains within the local area.
- **Route computation:** Route computation first establishes a core path from the domain of the source to the domain of the destination. Using the directional information provided by the core path, CEDAR iteratively tries to find a partial route from the source to the domain of the furthest possible node in the core path satisfying the requested bandwidth. This node then becomes the source of the next iteration.

In the CEDAR approach, the core provides an efficient and low-overhead infrastructure to perform routing, while the state propagation mechanism ensures the availability of link-state information at the core nodes without incurring high overheads. The bandwidth is used as the only QoS parameter for routing.

QoS Support Using Bandwidth Calculations

Lin and Liu (1999) have proposed an available bandwidth calculation algorithm for ad hoc networks with Time Division Multiple Access (TDMA) for communications. This algorithm involves end-to-end bandwidth calculation and bandwidth allocation. Here, only bandwidth is considered to be the QoS parameter. Bandwidth is measured in terms of the number of free slots available at node. The purpose of this algorithm is to find a shortest path satisfying the bandwidth requirement. The transmission is organized into frames each containing a fixed number of time-slots. The network is synchronized on a frame and slot basis. Using this algorithm, the source node can determine the resource availability for supporting the required QoS to any destination in the ad hoc networks. This approach is particularly useful in call admission control. In TDMA systems, time is divided into slots, which in turn are grouped into frames. Each frame contains two phases: control phase and data phase. During the control phase, each node takes turns to broadcast its information to all of its neighbors in a predefined slot. So at the end of control phase, each node has learned the free slots between itself and its neighbors. This information helps nodes to schedule free slots, verify the failure of reserved slots and drop expired real-time packets Based on this information, bandwidth calculation and assignment can be performed distributive. The data phase is used for data transmission and reception of data packets. This protocol gives an efficient allocation scheme. The standby routing mechanism can reduce the packet loss during path breaks.

On-Demand QoS Routing (OQR) Protocol

Lin (2001) proposed an admission control scheme over an On-demand QoS Routing (OQR) protocol to guarantee bandwidth for real-time applications. Since routing is on-demand in nature, there is no need to exchange control information periodically and maintain routing tables at each node. Similar to the Bandwidth Routing (BR) protocol, the network is time-slotted and bandwidth is the key QoS parameter. The path bandwidth calculation algorithm proposed in BR is used to measure the available end-to-end bandwidth. QoS routing protocol produces lower control overhead.

Multi-Path QoS Routing

Liao, Tseng, Wang, and Sheu (2001) have suggested a multipath QoS routing protocol. This protocol attempts to discover multiple paths that jointly satisfy the bandwidth requirements. The original bandwidth request is essentially split into several sub-bandwidth requirements. Each sub-path is then accountable for one sub-bandwidth requirement. This protocol is on-demand and it uses the local bandwidth information available at each node for discovering routes. A ticket-based approach is used to search for multiple paths. In this method, a number of probes are sent out from the source, each carrying a ticket. Each probe is responsible for searching one path.

The number of tickets sent controls the amount of flooding that is done. Each probe travels along a path that contains the necessary bandwidth. The source initially sends a certain number of tickets each containing the total bandwidth requirement. The tickets are sent along links that contain sufficient bandwidth to meet the requirement. When an intermediate node receives a ticket, it checks to see which links have enough bandwidth to meet the requirement. If it finds some, it then chooses a link, reserves the bandwidth and forwards the ticket on the link. If the links do not have the required bandwidth, the node reserves bandwidth along multiple links in such a way that the sum of the reserved bandwidths equals the original requirement.

In this way, the bandwidth requirement is split into sub-bandwidth requirements, equaling the bandwidths reserved along each of the links. The original ticket is split into sub-tickets, with each sub-ticket being forwarded along one of the links. Each sub-ticket is then responsible for finding a multi-path satisfying the sub-bandwidth requirement. If links cannot be found to satisfy the bandwidth requirements, the intermediate node drops the ticket. This means that links with more available bandwidth are preferred. The multipath QoS routing algorithm is suitable for ad hoc networks with very limited bandwidth where a single path satisfying the QoS requirements is unlikely to exist.

INORA

INORA(INSIGNIA + TORA) is a network layer QoS support mechanism has been proposed (Dharmaraju, Chowdhury, Hovareshti & Baras, 2002).

It makes use of the INSIGNIA in-band signalling mechanism and the TORA routing protocol for MANETs. In INORA, QoS signalling is used to reserve and release resources and set up, remove and renegotiate flows in the network. These reservations can be either hard state or soft state. The latter is more desirable in MANETs due to their dynamic nature. The INORA protocol operates the signalling mechanism separately from the TORA routing protocol. This provides decoupling of the two mechanisms and there is no interaction between. TORA provides the route between the source and the destination of a flow.

Then the signalling mechanism (INSIGNIA) establishes resources for the route provided by TORA. INORA tries to find paths in the network that can satisfy the desired QoS requirements. In INORA, INSIGNIA asks TORA for alternative routes when the current route is not able to meet the QoS requirements. The INORA scheme provides load-balancing in the network which aids in the performance of non-QoS flows. Future work will try to alleviate congestion in the wireless network by establishing QoS flows which avoid congested neighborhoods. The decoupling between the signalling and routing protocols allows for more flexibility in the design to incorporate load-balancing, congestion control, class-based admission control and so on. This added flexibility comes at the price of more overhead.

QoS-TORA

Gerasimov and Simon (2002) proposed the protocol named QoS-TORA. It is designed to work in a TDMA network where the bandwidth of a link is measured in terms of slot reservations in the data phase of the TDMA frame. The simulation result shows considerable improvements in the probability of being able to find an end-to-end QoS path. The simulation also shows that QoS-TORA provides higher throughput under higher mobility circumstances.

Chen and Heinzelman (2005) proposed a QoS aware routing protocol. The authors introduce the bandwidth estimation by disseminating bandwidth information through hello messages. The authors compare hello bandwidth estimation and listen bandwidth estimation methods of estimating bandwidth. These methods work equally well in static topologies by using large weight factors to reduce the congestion and minimize the chance of lost hello messages incorrectly signaling a broken route. While hello performs better in terms of end-to-end throughput, listen performs better in terms of packet delivery ratio.

2.1.2 Power

Power-Aware Multiple Access Protocol (PA-MAS)

A Power-Aware Multiple Access Protocol (PA-MAS) has been proposed (Singh & Raghavendra, 1998). Here, a node turns off its radio interface for a specific duration of time, when it knows that it will not be able to send and receive packets during that time because of the possibility of multiple access interference. The sleep time is of the order of packet duration, which could be very small. This approach would be quite viable for low bandwidth mobile networks, where small packets can be combined to form large packets or in radios with fast settling periods.

Conditional Max-Min Battery Capacity Routing (CMMBCR)

Conditional Max-Min Battery Capacity Routing (CMMBCR) algorithm proposed (Toh, 2001). This algorithm chooses the route with minimal total transmission power if all nodes in the route have remaining battery capacities higher than a threshold, otherwise routes including nodes with the lowest remaining battery capacities are avoided. This method considers both the total transmission energy utilization of routes and the residual power of nodes. When all nodes in some probable routes have enough residual battery capacity, a route with minimum total transmission power among these routes is selected. Since less total power is required to forward packets for each connection, the relaying load for most nodes must be reduced and their lifetime will be extended. However, if all routes have nodes with low battery capacity, a route including nodes with the lowest battery capacity must be avoided to extend the lifetime of these nodes with MMBCR applied.

Maximum Residual Packet Capacity (MRPC) Protocol

The Maximum Residual Packet Capacity (MRPC) protocol is proposed (Misra & Banerjee, 2002). It considers battery charge as well as link reliability during route selection. This protocol identifies the capacity of a node not just by its residual battery energy, but also by the expected energy spent in reliably forwarding a packet over a specific link. Such a formulation better captures scenarios where link transmission costs also depend on physical distances between nodes and the link error rates. As routes are discovered, the lifetime of the path is accumulated by calculating the lifetime of each link. Using a max-min formulation, MRPC selects the path that has the largest packet capacity at the critical node. This protocol results not only in load balancing, increasing the life of the network and avoiding congestion, but also yields closer-to-optimal energy consumption per packet, as well as lower packet delay and packet loss probability, due to the preference for more reliable links.

Localized Energy Aware Routing (LEAR) Protocol

The Localized Energy Aware Routing (LEAR) routing protocol is based on DSR, but modifies the route discovery procedure for balanced energy consumption (Woo, Yu, Youn & Lee, 2001). In DSR, when a node receives a route-request message, it appends its identity in the message's header and forwards it toward the destination. Thus, an intermediate node always relay messages if the corresponding route is selected. However, in LEAR, a node determines whether to forward the route-request message or not depending on its residual battery power. The destination node will receive a route-request message only when all intermediate nodes along a route have good battery levels.

2.1.3 Other Metrics

Delay Sensitive Adaptive Routing Protocol (DSARP)

A novel routing protocol for ad-hoc networks, Delay Sensitive Adaptive Routing Protocol (DSARP), is presented, which not only provides a reliable route for delay-sensitive traffic, but also can select the route based on the constrained condition: the shortest route and the lowest average delay (Sheng, Li & Shi,2003). Its basic operation is very similar to DSR. But it provides delay guarantees for time-sensitive applications. Simulation results show that DSARP outperforms the dynamic source routing protocol used in ad-hoc wireless networks.

Adaptive QoS Routing Algorithm (ADQR)

Hwang, Lee and Varshney (2003) proposed an Adaptive QoS Routing algorithm (ADQR). ADQR differs from other QoS routing protocols by using signal strength to predict the route breaks and initiate a fast reroute of data. Three levels of signal strength, Th_1, Th_2 and S_r ($Th_1 > Th_2 > S_r$), are defined. S_r is the minimal signal strength to receive a data packet. Three different classes are also defined for nodes, links and routes. If the received signal strength from a neighbor node is higher than Th_1, that neighbor node is in the first node class. If the received signal strength from the neighbor is between Th1 and Th_2, that neighbor node is in the second node class. If the signal strength is between Th_2 and S_r, that neighbor node is in the third node class. Links between the first node class nodes are in the first link class; links between the second node class nodes are in the second link class; and links between the third node class nodes are in the third link class. Also, three route classes are defined, where the bottleneck link determines the path class. Each node keeps a neighbor table, which records the node's neighbors and their corresponding cumulative signal strength. ADQR uses a fast route maintenance scheme, called two-phase monitored rerouting, which is composed of pre rerouting and rerouting. The pre rerouting phase occurs when the route changes from first route class to second route class, and the rerouting phase is invoked when the route changes from second route class to third route class. In pre rerouting, the source node finds alternate paths in advance, before the current path becomes unavailable, and in rerouting, the source node. Switches to one of these alternate paths in advance of the current path becoming unavailable.

2.2 Multiple Metrics

2.2.1 Bandwidth and Delay

Distributed Quality of Service Routing

A distributed routing framework to study Delay-Constrained Least-Cost (DCLC) and Bandwidth-Constrained Least-Cost (BCLC) path problems based on selective probing is presented (Chen & Nahrstedt, 1999). While determining a QoS-aware routing path, this algorithm tries to limit the amount of flooding (routing) messages by issuing a certain amount of logical tickets.

Each node maintains such end-to-end state information as delay, bandwidth and cost for every possible destination through the use of an underlying distance-vector protocol. In this framework, when a connection request arrives, probes are flooded selectively throughout the network along the paths that satisfy the QoS and optimization requirements. Each probe arriving at a destination detects a feasible path. The path establishment process, restoration process in case of link-failures, and the need to maintain state information with a use of distance-vector protocol lead to very high signalling cost and hence will affect routing performance.

The need to maintain redundant paths for the same flow affects badly the scalability of this framework. In addition, this work considers only the type of ad hoc networks whose topologies are not changing drastically and unpredictably and hence the proposed mechanism is mostly applicable to semi-stationary ad hoc networks. In this method the delay and bandwidth are used for QoS routing but not together. They are implemented as different algorithms.

This work is later extended by adopting fuzzy logic to model the imprecise state information (Raju, Hernandes & Zou, 2000). Accordingly, a rule-based fuzzy logic control model is employed in order to determine the maximum number of probes that can be used in the feasible path discovery process between a given source-destination pair.

QoS-Enabled Ad Hoc On-Demand Distance Vector Routing Protocol

Perkins et al. have extended the basic Ad hoc On-demand Distance Vector (AODV) routing protocol to provide QoS support in ad hoc wireless networks (Perkins, Royer & Das, 2000). To provide QoS, packet formats have been modified in order to specify the service requirements, which must be met by the nodes forwarding a Route Request or a Route Reply. Several modifications have been carries out for the routing table structure and Route request and Route reply messages in order to support QoS routing. Each routing table entry corresponds to a different destination node. The following fields are appended to each routing table entry:

- Maximum delay
- Minimum available bandwidth
- List of sources requesting delay guarantees
- List of sources requesting bandwidth guarantees

Maximum delay extension field: The maximum delay extension field is interpreted differently for Route Request and RouteReply messages. In a RouteRequest message, it indicates the maximum time allowed for a transmission from the current node to the destination node. In a RouteReply message, it indicates the current estimate of cumulative delay from the current intermediate node forwarding the RouteReply, to the destination.

Minimum bandwidth extension field: In a Route Request message, this field indicates the minimum bandwidth that must be available along an acceptable path from the source to the destination. In a RouteReply message, it indicates the minimum bandwidth available on the route between the node forwarding the RouteReply and the destination node. Using this field, the source node finds a path to the destination node satisfying the minimum bandwidth constraint.

List of sources requesting qos guarantees: A QoSLost message is generated when an intermediate node experiences an increase in node traversal time or a decrease in the link capacity. The QoSLost message is forwarded to all sources potentially affected by the change in the QoS parameter.

The advantage of this protocol is the simplicity of extension of the AODV protocol that can potentially enable QoS provisioning. However, as no resources are reserved along the path from the source to the destination, this protocol is not suitable for applications that require hard QoS guarantees. Further, node traversal time is only the processing time for the packet, so the major part of the delay at a node is contributed by packet queuing and contention at the MAC layer. Hence, a packet may experience much more delay than this when the traffic load is high in the network.

Quality of Service Optimized Link State Routing (QOLSR)

Optimized Link State Routing (OLSR) is a proactive routing protocol that employs an efficient link state packet mechanism called Multipoint Relaying (MPR) proposed (Jacquet et al, 2001). This protocol optimizes the pure link state protocol. When a node wants to send topology updates, it selects a group of neighbouring nodes to retransmit the routing packets called the multipoint relays of the source node. If a node receives a topology update packet from a node for which it is not a multipoint relay, it will update its topology with the information in the packet but will not rebroadcast the packet. Each node determines a route that is optimal in terms of hop-count to every known destination in the network and greatly reduces network routing overhead since not all nodes forward routing messages. OLSR also reduces the size of routing packets by letting a node send only routing updates for nodes that selected the node as a multipoint relay. This means that a node can only be reached through its multipoint relay nodes. When a packet has to be sent to a destination node, OLSR calculates the shortest path to the node using the topology information in its routing tables.

Delay and bandwidth metrics are taking into account as QoS constraints for the QOLSR protocol (Badis & Agha, 2005). Such metrics are included on each routing table entry corresponding to each destination. These values are estimated using the periodic HELLO messages. The total expected MAC delay of a packet is a product of the average estimated delay of one packet and the total number of packets awaiting transmission. The idle time and window duration are calculated to produce the link utilization factor and the permissible bandwidth measurements.

Ad Hoc QoS On-DEMAND ROUTING (AQOR)

Ad hoc QoS On-demand Routing (AQOR) is a resource reservation and signaling algorithm proposed (Xue & Ganz, 2003). AQOR provides end-to-end QoS support in terms of bandwidth and end-to-end delay in MANETs. They introduce detailed computation algorithms for available bandwidth calculation and end-to-end delay in an unsynchronized wireless environment. The wireless channel is a shared medium and can only be used one at a time by the nodes within transmission range. The bandwidth calculation is based on the aggregate traffic of the neighborhood and is performed on the MAC Layer. AQOR proposes to use HELLO-packets to keep an updated view of the neighborhood. It reserves bandwidth on each node along a path that is being used by the source. The reservation has been done in the route discovery phase but doesn't actually take place until the first packet has been forwarded at a node. AQOR proposes an adaptive route recovery model when a QoS violation has been detected. This model makes the destination do a reverse route exploration. The bandwidth calculation and resource reservation model in AQOR showed promising results.

2.2.2 Path and Power

A framework of Self-Healing and Optimizing Routing Techniques (SHORT) for MANETs has been proposed (Gui & Mohapatra, 2003). SHORT techniques for both AODV and DSR algorithm have been analyzed and evaluated in the literature. Simulation results show that higher delivery rate and longer network lifetime can be achieved by adopting SHORT. SHORT is a technique that optimizes the route length results in significant performance gain over the underlying routing protocols. The proposed schemes monitor the routing path and try to shorten the path length as and when it is feasible. SHORT improves the performance by monitoring routing paths continuously and redirecting the path whenever a short-cut path is available.

3. ENERGY AND DELAY AWARE PROTOCOLS

On-demand routing protocols normally pick the shortest path route during the route detection process and then stick to the route till they break. Continuous use of the route may drain the battery power. This is particularly true if one or more nodes are on other routes as well. Each message transmission and reception drains battery power. If a node runs out of battery power and fails to forward any messages, it naturally falls out of the network. As each node in a MANET performs the routing function for establishing communication link, the death of even a few of the nodes due to energy exhaustion might cause disturbance of service in the entire network. In such cases, the route breaks and the protocol finds an alternate route via another route discovery. This will affect the operational life time of ad hoc network. In a conventional routing algorithm, which is unaware of energy budget, connections between two nodes are established through the shortest path routes. In this situation, routing protocol has to take into consideration of the residual energy. The delay is the total latency experienced by a packet to traverse the network from the source to the destination. At the network layer, the end-to-end packet latency is the sum of processing delay, packetization delay, transmission delay, queuing delay and propagation delay. Asokan and Natarajan (2008) proposed a new energy and delay aware protocols called Energy and Delay aware Adhoc On demand Distance Vector Routing (EDAODV) and Energy and Delay aware Dynamic Source Routing (EDDSR) based on extension of AODV and DSR.

3.1 Ad Hoc On-Demand Distance Vector (AODV) Routing

The Ad hoc On-demand Distance Vector (AODV) routing protocol builds on the DSDV algorithm (Perkins & Royer, 1999). AODV is an improvement on DSDV because it typically minimizes the number of required broadcasts by creating routes on a demand basis, as opposed to maintaining a complete list of routes as in the DSDV algorithm. AODV is a pure on-demand route acquisition system, since nodes that are not on a selected path do not maintain routing informa-

Figure 1. Propagation of RREQ and RREP in AODV

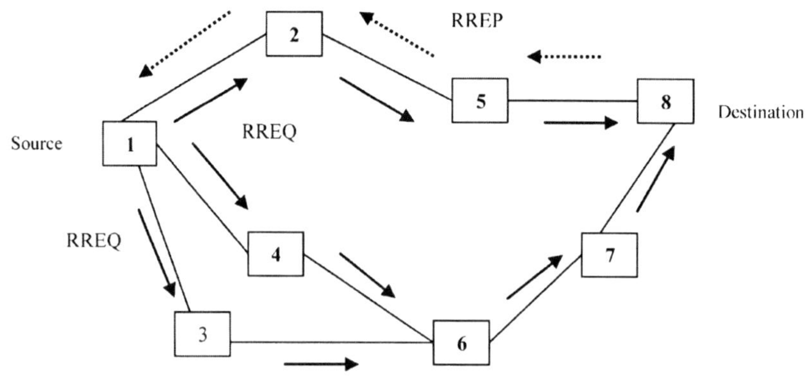

tion or participate in routing table exchanges. When a source node desires to send a message to some destination node and does not already have a valid route to that destination, it initiates a path discovery process to locate the other node. It broadcasts a route request (RREQ) packet to its neighbors, which then forwards the request to their neighbors and so on, until either the destination or an intermediate node with a fresh enough route to the destination is located. AODV utilizes destination sequence numbers to ensure that all routes are loop-free and contain the most recent route information. Intermediate nodes can reply to the RREQ only if they have a route to the destination whose corresponding destination sequence number is greater than or equal to that contained in the RREQ.

During the process of forwarding the RREQ, intermediate nodes record in their route tables the address of the neighbor from which the first copy of the broadcast packet is received, thereby establishing a reverse path. If additional copies of the same RREQ are later received, these packets are discarded. Once the RREQ reaches the destination or an intermediate node with a fresh enough route, the destination/intermediate node responds by uni-casting a route reply (RREP) packet back to the neighbor from which it first received the RREQ as shown in Figure 1.

As the RREP is routed back along the reverse path, nodes along this path set up forward route entries in their route tables, which point to the node from which the RREP came. These forward route entries indicate the active forward route. Associated with each route entry is a route timer, which will cause the deletion of the entry if it is not used within the specified lifetime. Because the RREP is forwarded along the path established by the RREQ, AODV only supports the use of symmetric links. The main advantage of AODV is that routes are obtained on demand and destination sequence numbers are used to find the latest route to the destination. One of the disadvantages of AODV is that intermediate nodes can lead to inconsistent routes if the source sequence number is very old and the intermediate nodes have a higher but not the latest destination sequence number, thereby causing stale entries.

3.2 Dynamic Source Routing (DSR)

The Dynamic Source Routing (DSR) protocol is an on-demand routing protocol that is based on the concept of source routing (Johnson & Maltz, 1996). The protocol consists of two major phases: route discovery and route maintenance. When a mobile node has a packet to send to some destination, it first consults its route cache to determine

Figure 2. Propagation of route request in DSR

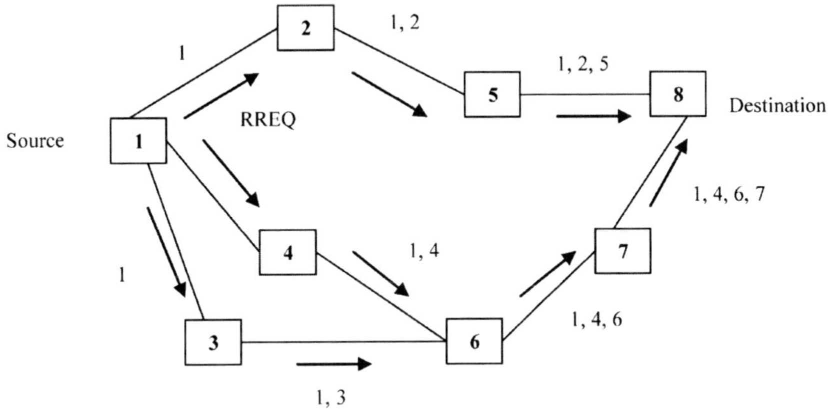

whether it already has a route to the destination. If it has an unexpired route to the destination, it will use this route to send the packet.

On the other hand, if the node does not have such a route, it initiates route discovery by broadcasting a route request (RREQ) packet as shown in Figure 2. This route request contains the address of the destination, along with the source node's address and a unique identification number. Each node receiving the packet checks whether it knows of a route to the destination. If it does not, it adds its own address to the route record of the packet and then forwards the packet along its outgoing links. To limit the number of route requests propagated on the outgoing links of a node, a mobile only forwards the route request if the request has not yet been seen by the mobile and if the mobile's address does not already appear in the route record.

A route reply (RREP) is generated when the route request reaches either the destination itself, or an intermediate node which contains in its route cache an unexpired route to the destination as shown in Figure 3. By the time the packet reaches either the destination or such an intermediate node, it contains a route record yielding the sequence of hops taken. If the node generating the route reply is the destination, it places the route record contained in the route request into the route reply. If the responding node is an intermediate node, it will append its cached route to the route record and then generate the route reply.

Figure 3. Propagation of route reply in DSR

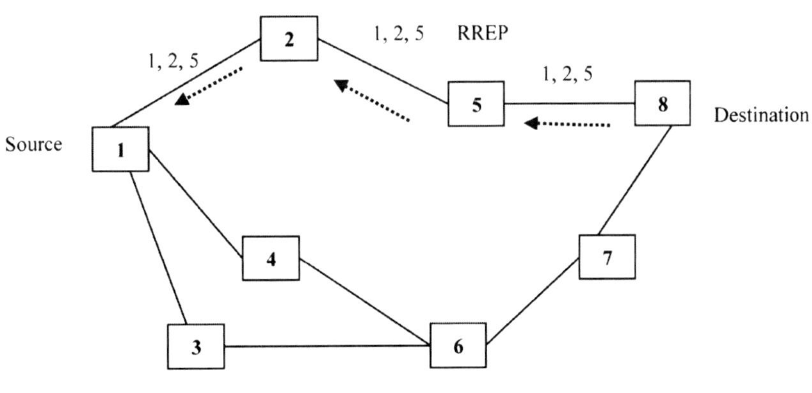

This protocol uses a reactive approach which eliminates the need to periodically flood the network with table update messages which are required in a table-driven approach. The disadvantage of this protocol is that the route maintenance mechanism does not locally repair a broken link. Also, considerable routing overhead is involved due to the source-route mechanism employed in DSR. This routing overhead is directly proportional to the path length.

3.3 Energy and Delay Extension in AODV and DSR

The minimum energy and maximum delay fields are added with the RREQ for each destination. A source transmits a RREQ packet with QoS energy and delay extension as shown in Figure 4. The energy extension indicates the minimum energy required to be available on the entire path between the source and its destination.

The minimum energy is selected by the node, which initially requests the route. The application and duration of transmission are the factors that determine the minimum energy. The percentage of the initial energy is taken as the energy metric in the QoS specification. The extension of delay gives the maximum delay allowed between the source and its destination. As shown in Figure 4, the QoS energy extension is 30% (0.3) of the node's initial energy and the maximum delay is 100 milliseconds (ms). Both minimum energy and maximum delay verifications of RREQ have been done in each node. RREQ packets are discarded if one of the constraints cannot be satisfied.

Before forwarding the RREQ packet, an intermediate node compares its available energy to the energy field indicated in the QoS extension. If the required energy is not available, the packet is discarded and the process is stopped. If the energy requirement is satisfied, then the delay is estimated. If it exceeds the QoS delay, the packet is discarded. Otherwise, the node subtracts its Node Traverse Time (NTT) from the delay bound provided in the extension. The delay value in RREQ packet indicates the delay allowed for a transmission between the source and its destination. The RREQ is forwarded with updated QoS delay extension.

In response to the RREQ, the destination sends an RREP packet with its measured available energy and initial delay corresponding to its NTT. The delay in RREP packet indicates the estimate of the cumulative delay allowed for a transmission between the intermediate node, which forwards the RREP and destination. Each intermediate node forwarding the RREP compares the energy field of the extension with its own available energy on the selected route. It keeps the minimum between these two values to propagate the RREP as shown in Figure 5. This value is recorded in the routing table for the destination. In case of delay, the intermediate node adds its own NTT to the delay field and records this value in the routing table for the concerned destination before forwarding the RREP. This entry update allows an intermediate

Figure 4. QoS route request for energy and delay

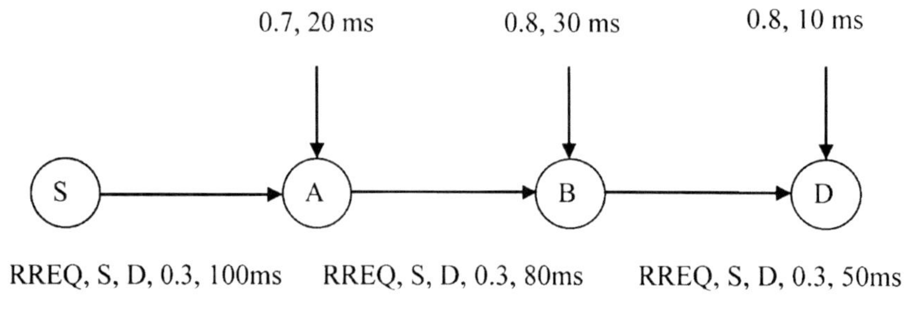

Figure 5. QoS route reply for energy and delay

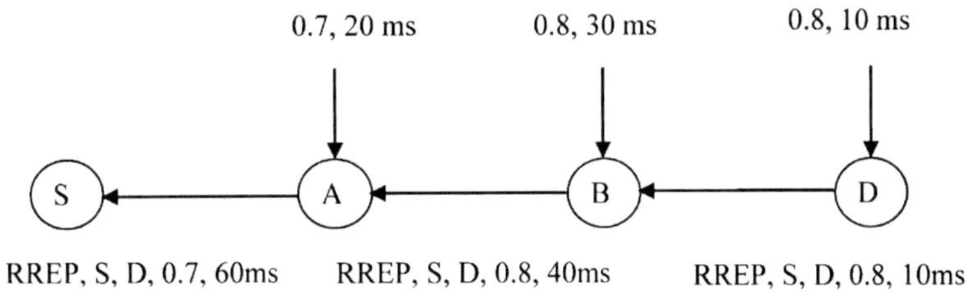

node to answer the next RREQ by comparing the maximum delay field in the table.

The flow chart as shown in Figure 6 describes the sequence of operation. Energy and delay metrics are used in AODV route discovery. Each RREQ packet flooded in the network builds up the cost for the path traversed so far by the packet. Each routing table entry also maintains energy and delay for that route. In regular AODV, the node acts only on the very first RREQ received per route discovery. Duplicates of the RREQ received via alternate routes are ignored. However, use of these new cost metrics requires that AODV acts on all such duplicates if they carry a lower cost metric. If a RREQ arrives with a lower cost metric, it is forwarded when the node is not the destination and does not have a route to the destination, otherwise it replies.

Energy and delay metrics are used as QoS extension in DSR route discovery. Every node receiving the route request searches through its route cache for a route to the requested destination with the minimum energy and maximum delay. Each node's route cache will have the energy and delay values. If both energy and delay constraints are satisfied then intermediate node forwards the RREQ to the next node. Otherwise, it is discarded.

The performance of the proposed protocols is evaluated using the Network simulator (Ns-2). Table 1 lists the simulation parameters and environments used.

End-to-End Delay

Figure 7 depicts the effect of mobility on end-to-end delay for two QoS requirements, 250 and 350 milliseconds (ms). The end-to-end delay increases as the node speed increases. In AODV and EDAODV a steep rise when the translation of mobility occurs 60 to 70 Km/hr. This is due more broken inks and frequent re-routing and thus causes more packet loss and larger end-to-

Table 1. Simulation parameters

Simulation area	**670m x 670m**
Transmission range	250 m
Mobility model	Random way point
Speed	0-20 meter/second
Routing protocols	AODV, DSR, EDAODV and EDDSR
MAC	IEEE 802.11
Traffic source model	Constant bit rate
Channel data rate	2 Mbps
Initial energy	20 Joules

Figure 6. Flow chart for EDAODV and EDDSR

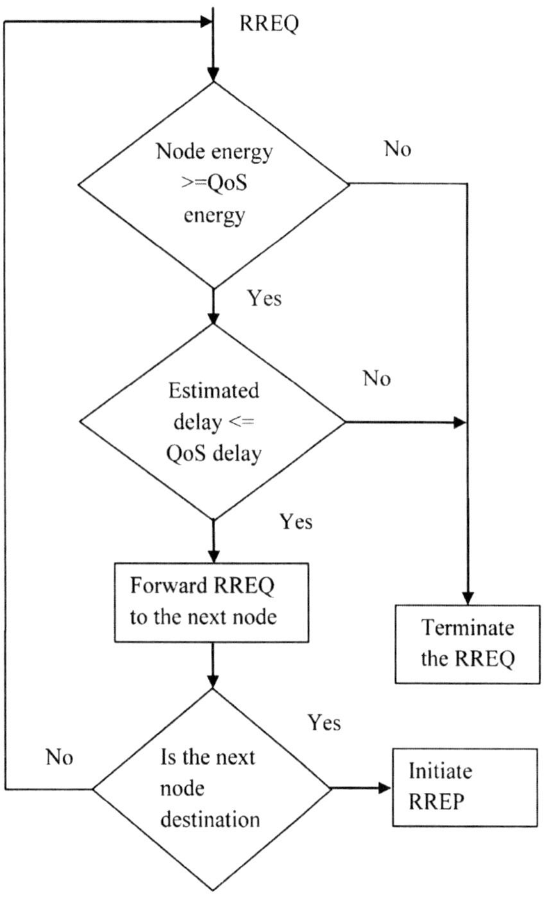

Figure 7. Effect of mobility on end-to-end delay (a) 250 ms QoS delay (b) 350 ms QoS delay

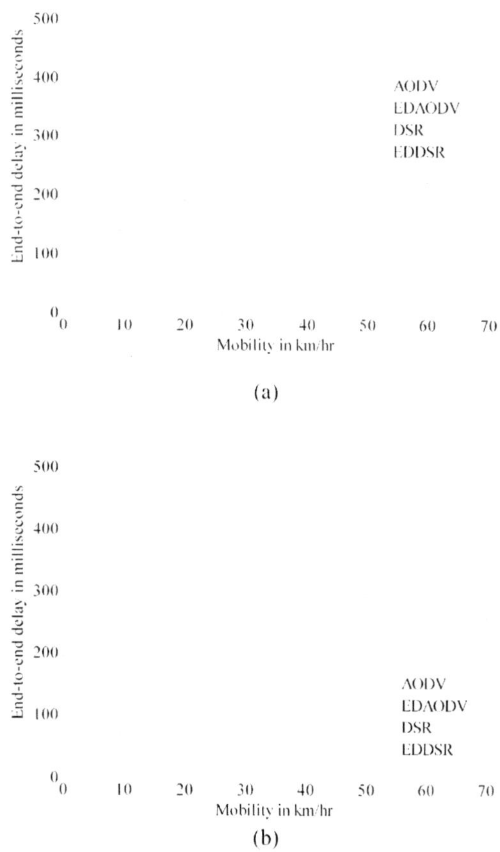

end delay. For EDAODV the end-to-end delay is about 75 ms less in 250 ms QoS delay and 80 ms less in 350 ms QoS delay. EDDSR satisfies the requirement in 250 ms QoS delay and 30 ms less delay in 350 ms QoS delay. It is observed that two QoS requirements 250 ms and 350 ms are satisfied in EDDSR and EDAODV. But, in AODV and DSR, the QoS requirements are not satisfied. This is because end-to-end delay verification of RREQ has been done and RREQ packets are discarded if delay constraint cannot be satisfied in the particular path. There is a sudden dip in the graph when the pause time takes place from 100 to 200 seconds. This is due to sudden increase in the number of link breakages.

Remaining Energy

Figure 8 expounds the effect of pause time on remaining energy under four protocols. The remaining energy at the end of simulation is much higher for EDAODV and EDDSR than for AODV and DSR. In EDAODV, the improvement is about 8 times for low pause time and up to 5 times for high pause time. In case of EDDSR, the improvement is about 60 times at low pause time and 6 times at high pause time. This is because the minimum energy verification of RREQ has been done in each node. Before forwarding the RREQ packet, an intermediate node compares its available energy to the energy field indicated in the QoS extension. If the required energy is not available, the packet is discarded and the process is stopped. However,

Figure 8. Effect of pause time on remaining energy

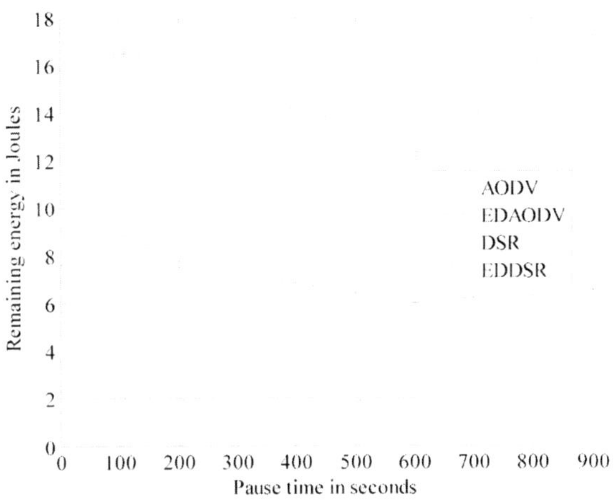

Figure 9. Illustration of Route in TORA

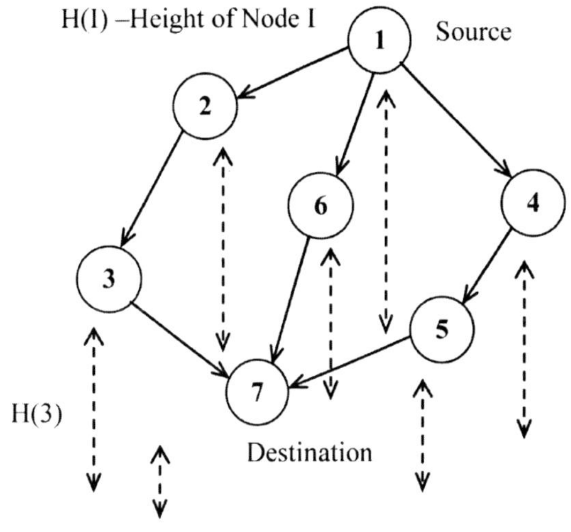

these improvements strongly depend on the initial energy and the simulation time.

3.4 Energy and Delay Aware TORA (ED-TORA)

Energy and delay aware protocol called Energy and Delay aware TORA (EDTORA) based on extension of TORA is proposed (Asokan & Natarajan, 2007). TORA is a source-initiated on-demand routing protocol, which uses a link reversal algorithm and provides loop-free multi-path routes to a destination node (Park & Corson, 1997). Each node maintains its one-hop local topology information and also has the capability to detect partitions. TORA has the unique property of limiting the control packets to a small region during the reconfiguration process initiated by a path break. TORA has three main functions: establishing, maintaining and erasing routes. The route establishment function is performed only when a node requires a path to a destination but does not have any directed link. This process establishes a destination-oriented, Directed Acyclic Graph (DAG) using a query/update mechanism.

Let us consider the network topology shown in Figure 9. When node 1 has data packets to be sent to the destination node 7, a query packet is originated by node 1 with the destination address included in it. This query packet is forwarded by intermediate nodes 2,3,4,5,6 and reaches the destination node 7. The destination node 7 originates an update packet. Each node that receives the update packet sets its distance to a value higher than the distance of the sender of the update packet. Once a path to the destination is obtained, it is considered to exist as long as the path is available, irrespective of the path length changes due to the reconfigurations that may take place during the course of the data transfer. When an intermediate node discovers that the route to the destination node is invalid, it changes its distance value to a higher value than its neighbors and originates an update packet. Minimum energy and maximum delay fields are also added with the query packet.

A source requiring minimum energy and maximum delay transmits a query packet with QoS energy and delay extension. Both minimum energy and maximum delay verifications of a query have been done in each node. Query packets are discarded if one of the constraints cannot be satisfied. Before forwarding the query packet, an intermediate node compares its available energy to the energy field indicated in the QoS extension.

Figure 10. Flow chart for EDTORA

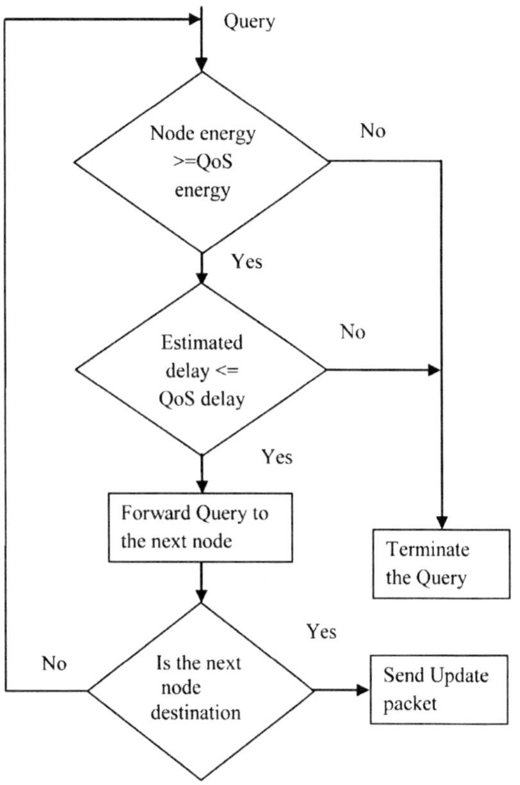

If the required energy is not available, the packet is discarded and the process stops. If the energy constraint is satisfied then the delay is estimated and if it exceeds the QoS delay, the packet is discarded. Otherwise, the node subtracts its NTT from the delay bound, provided in the extension. The delay value in query packet indicates the delay allowed for a transmission between the source and destination. The query packet is forwarded with updated QoS delay extension. The flow chart as shown in Figure 10 describes the sequence of operation. The performance of TORA and EDTORA is evaluated using the Network simulator (Ns-2). Table 2 lists the simulation parameters and environments used.

End-to-End Delay

Figure 11 exhibits the effect of mobility on end-to-end delay for QoS requirement at 250 ms for pause time of 5 seconds. The end-to-end delay increases as the node speed increases. Higher mobility causes more broken links and frequent re-routing and thus causes more packet loss and larger end-to-end delay. It is observed that QoS requirement is satisfied in EDTORA. The end-to-end delay is about 110 ms less than QoS requirement (250 ms) for EDTORA. But TORA exceeds the QoS requirement.

Figure 11. Effect of mobility on end-to-end delay for pause time of 5 seconds

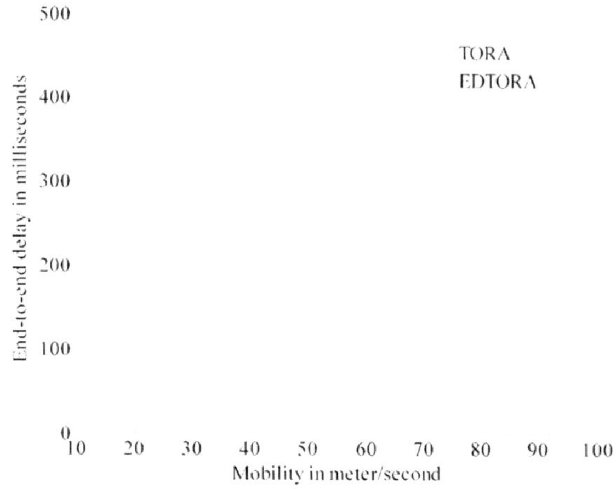

Network Lifetime

Figure 12 points out the time at which certain number of nodes die, when simulating two protocols. It can be seen from the graph that TORA nodes die earlier than EDTORA nodes. The first node dies at 35 seconds in TORA and at 65 seconds in EDTORA. At 100 seconds simulation time, 41 nodes die in TORA while only 6 nodes die in EDTORA. There is a sudden increase in the number of dead nodes between 30 to 42 seconds in TORA and from 42 to 60 seconds in EDTORA. Because, when the time reaches to 30 seconds in TORA the nodes with low in energy level dies. In EDTORA, more number of nodes die when the time reaches 42 seconds. This is due to the minimum energy verification of query packet has been done in each node. Before forwarding the

Table 2. Simulation parameters

Simulation area	**670m x 670m**
Transmission range	250 m
Mobility model	Random way point
Speed	0-20 meter/second
Routing protocols	TORA and EDTORA
MAC	IEEE 802.11
Traffic source model	Constant bit rate
Channel data rate	2 Mbps
Initial energy	20 Joules

Figure 12. Number of nodes dead vs time

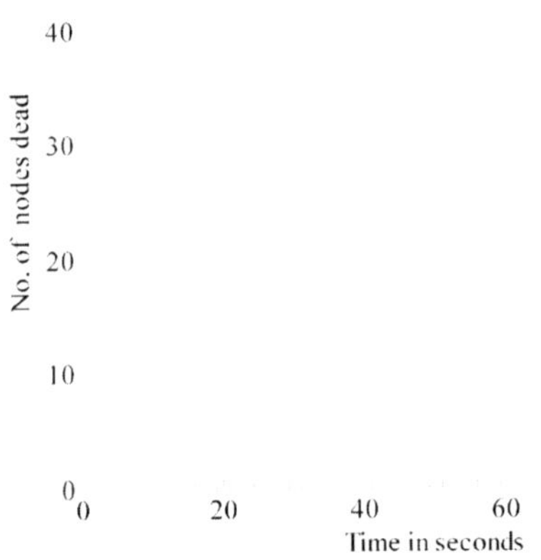

query packet, an intermediate node compares its available energy to the energy field indicated in the QoS extension. If the required energy is not available, the packet is discarded and the process is stopped. However, these improvements strongly depend on the initial energy and the simulation time.

The simulation results show that these protocols satisfy the energy and delay QoS requirements. It has been found that these protocols give better performance than existing protocols in terms of end-to-end delay, energy, packet delivery ratio and packet loss.

4. QOS ROUTING PROTOCOLS USING OPTIMIZATION TECHNIQUES

Genetic Algorithm (GA) Based Routing Method

A Genetic Algorithm (GA) based routing method for Mobile Ad hoc Networks (GAMAN) is proposed (Barolli, Koyama, Suganuma & Shiratori, 2003). It is a source-based routing algorithm. Few nodes are involved in route computation by using small population size. The nodes in sub-population care only about the routes. The broadcast is avoided because the information is transmitted only for the nodes in a population. The GA explores different routes and they are ranked by sorting. Therefore, the first route is the best one, but other routes ranked are used as backup routes. By using a tree based GA method, the loops are avoided. This algorithm uses the delay and transmission success rate as QoS parameters.

This algorithm is enhanced by adding an effective topology extraction to reduce the search space of GAMAN and it is called as E-GAMAN algorithm. Robustness rather than optimality is the primary concern of E-GAMAN. In the case of MANETs, it is better to find a route very fast in order to have a good response time to the speed of topology change, than to search for the optimal route but without meaning, because the

Figure 13. Ants attempt to take the shortest path after an initial searching time

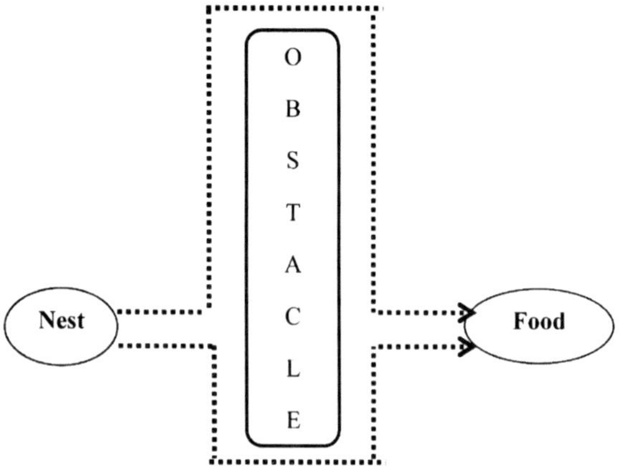

network topology is changed and this route does not exist anymore.

A heuristic and distributed route discovery method named RLGAMAN that supports QoS requirement for MANETs proposed (Peng & Deyun, 2006). This method integrates a distributed route discovery scheme with a reinforcement learning (RL) method that only utilizes the local information for the dynamic network environment and the route expand scheme based on genetic algorithms (GA) method to find more new feasible paths and avoid the problem of local optimize. Simulations under various load and packet loss conditions are reported and this approach can provide improvements to network QoS.

Ant Colony Optimization (ACO)

Ants have always been a fascinating subject for human beings. Individually, they are simple creatures with limited memory and behaviour that sometimes seems to have a random component. However, collectively, ants consistently achieve remarkable feats of cooperation, coordination and construction.

ACO is a subset of Swarm Intelligence. The basic idea of ACO is taken from the food searching behavior of real ants (Liu, Kwiatkowska & Constantinou, 2004). When ants are on the way in search for food, they start from their nest and walk toward the food. When an ant reaches an intersection, it has to decide which branch to take next. While walking, ants deposit a pheromone, which ants are able to smell, which marks the route taken. The concentration of pheromone on a certain path is an indication of its usage. With time, the concentration of pheromone decreases due to diffusion effects. Figure 13 shows a scenario with two routes from the nest to the food source. Since the lower route is shorter than the upper one, the ants, which take this path, will reach the food source first. On their way back to the nest, the ants again have to select a path. After a short time, the pheromone concentration on the shorter path will be higher than on the longer path, because the ants using the shorter path will increase the pheromone concentration faster. The shortest path will thus be identified and eventually all ants will only use this path. This behavior of the ants can be used to find the shortest path in networks.

The basic idea behind ACO algorithms for routing is the acquisition of routing information through the sampling of paths using small control packets, which are called ants. The ants are generated concurrently and independently at the nodes, with the task to test a path from a source node to an assigned destination node (Streltsov & Vakiki, 1996). The routing tables contain for each destination a vector of real valued entries, one for each known neighbor node. These entries are a measure of the goodness of going over that neighbor on the way to the destination. They are termed pheromone variables and are continually updated according to path quality values calculated by the ants. The repeated and concurrent generations of path sampling ants result in the availability at each node of a bundle of paths, each with an estimated measure of quality. In turn, the ants use the routing tables to define which path to their destination they sample: at each node they stochastically choose a next hop, giving higher probability to links with higher pheromone values. Routing tables are also called pheromone tables.

Ant Based Control (ABC) is another stigmergy based ant algorithm designed for telephone networks (Schoonderwoerd, Holland, Bruten & Rothkrantz, 1996). The basic principle relies on mobile routing agents, which randomly explore the network and update the routing tables according to the current network state. The routing table stores probabilities instead of pheromone concentrations.

AntNet proposed by Di Caro and Dorigo (1998) is a routing algorithm proposed for wired datagram networks based on the principle of ACO. In AntNet, each node maintains a routing table and an additional table containing statistics about the traffic distribution over the network. The routing table maintains for each destination and for each next hop a measure of the goodness of using the next hop to forward data packets to the destination. AntNet uses two sets of homogeneous mobile agents called forward ants and backward ants to update the routing tables. The forward ants use heuristics based on the routing table to move between a given pair of nodes then they collect information about the traffic distribution over the network. The backward ants retrace the paths of forward ants in the opposite direction. At each node, the backward ants update the routing table and the additional table containing the statistics about the traffic distribution over the network.

The Ant colony based Routing Algorithm (ARA) suitable for MANETs, is based on both swarm intelligence and ant-colony meta-heuristics. ARA consists of three phases: route discovery, route maintenance, and route failure handling. In the route discovery phase, new routes between nodes are discovered with the use of forward and backward ants, similar to AntNet. Routes are maintained by subsequent data packets, i.e., as the data traverse the network, node pheromone values are modified so that their paths are reinforced (Gunes, Sorges & Bouazizi, 2002).

Probabilistic Emergent Routing Algorithm (PERA) works in an on-demand way, with ants being broadcast towards the destination at the start of a data session (Baras & Mehta, 2003). Multiple paths are set up, but only the one with the highest pheromone value is used by data and the other paths are available for backup.

Termite is another ant-based routing algorithm that is similar to ARA proposed (Roth & Wicker, 2003). However, unlike the ARA, pheromone is not considered in the route discovery phase. Instead of the forward and backward ants, RREQ and RREP control packets are used to discover the routes. The RREQ packet randomly walks, not floods, through the network to discover a route to the destination. Pheromone levels are used for routing data packets and proactive seed packets are introduced for route maintenance.

In ACO routing algorithms, routing information is gathered through a stigmergic learning process using ant agents (Liu, Kwiatkowska & Constantinou, 2004). These are lightweight agents, which are generated concurrently and independently by the nodes, with the task to sample path to an assigned destination. An ant, going from its source ‘s’ to a destination ‘d’ collects information about the quality of the path it follows (e.g. end-to-end delay) and by retracing its way back from ‘d’ to ‘s’, it uses this to update the routing information at intermediate nodes. Routing information is expressed in the form of tables kept locally at each node.

Paul and Gaorge (2005) proposed an ant-based multipath routing protocol that considers both energy and latency. They consider MANET with dual-priority traffic namely: latency-critical and not latency-critical. For latency-critical traffic, energy pheromone and delay pheromone metrics are combined after being normalized. For the latter, not latency-critical traffic, only energy pheromone metric is used.

Ad hoc Networking with Swarm Intelligence (ANSI) is a congestion-aware routing protocol, which, owing to the self-configuring mechanisms

Figure 14. ADSR route request packet format

IP Header	DSR Fixed Header	Source Address	Sequence Number	Destination Number	Forward Ant	Route Record	Hop Count	Route Address Add1, Add2 -- Addn

of Swarm Intelligence (Rajagopalan & Shen, 2005). It is able to collect more information about the local network and make more effective routing decisions than traditional MANET protocols. ANSI is thus more responsive to topological fluctuations.

4.1 Ant Dynamic Source Routing and Ant Temporally Ordered Routing Algorithm

ACO based routing protocols called Ant Dynamic Source Routing (ADSR) and Ant Temporally Ordered Routing Algorithm (AntTORA) are developed to support multiple QoS routing metrics like delay, jitter and energy.

4.1.1 Ant Dynamic Source Routing (ADSR)

DSR is an on-demand routing protocol that is based on the idea of source routing (Johnson & Maltz, 1996). Mobile nodes are required to maintain route caches that contain the source routes of which the mobile is aware. Entries in the route cache are continually updated as new routes are learnt. The protocol consists of two major phases: route discovery and route maintenance. ADSR protocol is described and the performance of DSR and ADSR is analyzed (Asokan, Natarajan & Venkatesh, 2008).

Route Discovery

When a node has a packet to send to the destination, it first consults its route cache to determine whether it previously has a route to the destination. If it has an unexpired route to the destination, it will use this route to send the packet. On the other hand, if the node does not have such a route, it initiates route discovery by broadcasting a route request packet. This route request contains the address of the destination, along with the source node's address and a unique identification number. Each node receiving the packet checks whether it knows a route to the destination. If it does not, it adds its own address to the route record of the packet and then forwards the packet along its outgoing links. To limit the number of route requests propagated on the outgoing links of a node, a mobile only forwards the route request if the mobile has not yet seen the request and if the mobile's address does not already appear in the route record. In ADSR, FANT packets are added in the route request as shown in Figure 14.

FANT packets containing energy, delay and jitter information, a separate pheromone level will be maintained for each metric. Thus, energy, delay and jitter pheromone levels are added in route request. A route reply is generated when the route request either reaches the destination itself, or reaches an intermediate node, which contains in its route cache an unexpired route to the destination. By the time the packet reaches either the destination or such an intermediate node, it contains a route record yielding the sequence of hops taken. If the node generating the route reply is the destination, it places the route record contained in the route request into the route reply.

If the responding node is an intermediate node, it will append its cached route to the route record

Figure 15. ADSR route reply packet format

IP Header	DSR Fixed Header	Source Address	Destination Address	Backward Ant	Dynamic Route Record	Sequence Number	Reply Address Address source Add1, Add2 ---- Addr dest.

and then generate the route reply. To return the route reply, the responding node must have a route to the initiator. If it has a route to the initiator in its route cache, it may use that route. Otherwise, if symmetric links are supported, the node may reverse the route in the route record. If symmetric links are not supported, the node may initiate its own route discovery and piggyback the route reply on the new route request. In ADSR, BANT packets are added in the route reply as shown in Figure 15. BANT packet headers have fields to track the residual energy, cumulative delay and jitter based on backlog information of queued packets destined to the packet's source.

Route Maintenance

Route maintenance is accomplished using route error packets and acknowledgments. Route error packets are generated at a node when the data link layer encounters a transmission problem. When a route error packet is received, the hop in error is removed from the node's route cache and all routes containing the hop are truncated at that point. In addition to route error messages, acknowledgments are used to verify the correct operation of the route links. The performance of DSR and ADSR protocol is evaluated using the Network simulator (Ns-2). Table 3 lists the simulation parameters and environments used.

End-to-End Delay

Figure 16 elucidates the effect of mobility on end-to-end delay. The end-to-end delay increases as the mobility increases. This higher mobility causes more link breaks and frequent re-routing, thus causing larger end-to-end delay. ADSR shows better performance in all the mobility conditions and its maximum improvement over DSR is around 44%.

Jitter

The variations in jitter under different mobility conditions are shown in Figure 17. The jitter is increased at higher mobility due to breaking of more links and frequent re-routing. ADSR gives a better performance than DSR in all the mobility

Table 3. Simulation parameters

Simulation area	**500m x 500m**
Number of nodes	100
Node communication range	50m
Mobility model	Random waypoint
Speed	0-100 meter/second
Routing protocols	DSR, ADSR
MAC	IEEE 802.11
Data rate	2 Mbps
Traffic source model	Constant bit rate

Figure 16. Effect of mobility on end-to-end delay

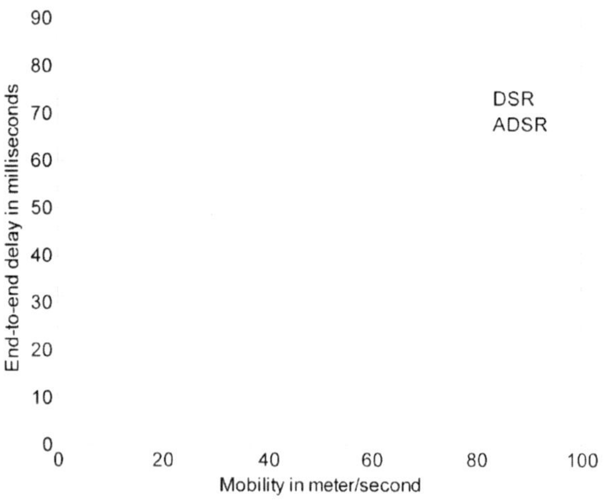

Figure 17. Effect of mobility on jitter

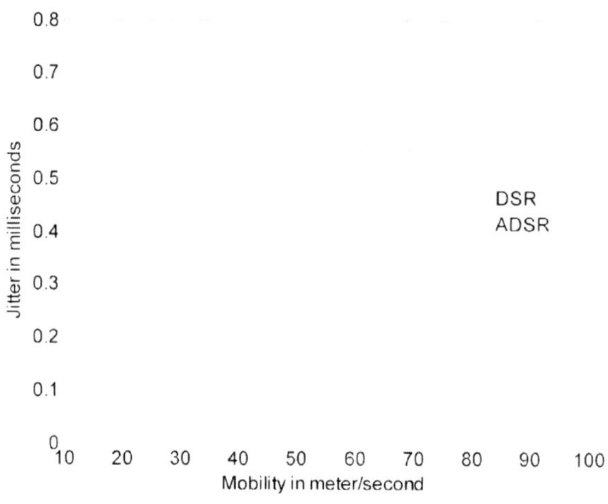

conditions. The reduction in jitter varies from 62% to 21%. This is due to addition of jitter pheromone in the route request and route reply.

4.1.2 Ant Temporally Ordered Routing Algorithm (AntTORA)

TORA is a source-initiated on-demand routing protocol, which uses a link reversal algorithm and provides loop-free multipath routes to a destination node proposed (Park & Corson, 1997). The exchange of routing information is restricted to a region within one hop distance of the node where the topological change occurred. Each node maintains its one-hop local topology information and has the capability to detect partitions. TORA has the unique property of limiting the control packets to a small region during the reconfiguration process initiated by a path break. TORA provides multiple routes for any desired source/destination pair.

Route Creation

The creation of routes basically assign directions to links in an undirected network or portion of the network, building a DAG routed at destination. TORA associates a height with each node in the network. All messages in the network flow downstream, from a node with higher height to a node with lower height. Routes are discovered using QRY and UPD packets. When a node with no downstream links needs a route to a destination, it will broadcast a QRY packet. In AntTORA, the FANT packets are added in the QRY packet (Asokan & Natarajan, 2008). Separate pheromone level will be maintained for FANT packets containing energy, delay and jitter. Thus, energy, delay and jitter pheromone levels are added in QRY packet. The QRY packet format of TORA and AntTORA are given in Figure 18 and 19.

This QRY packet will propagate through the network until it reaches a node that has a route or the destination itself. Such a node will then broadcast a UPD packet that contains the node

Figure 18. TORA QRY packet format

Version #	Type	Reserved	Destination IP Address

Figure 19. AntTORA QRY packet format

Version #	Type	Reserved	Destination IP Address	Forward ant

Figure 20. TORA UPD packet format

Version#	Type	Reserved	Destination IP Address	Mode #	H.tau, H.oid, H.r, H.delta, H.id

Figure 21. AntTORA UPD packet format

Version#	Type	Reserved	Destination IP Address	Mode #	Backward ant H.tau, H.oid, H.r, H.delta, H.id

height. In AntTORA, BANT packets are added in the UPD packet. BANT packet headers have fields to track the residual energy, cumulative delay and jitter based on backlog information of queued packets destined to the packet's source. The UPD packet format of TORA and AntTORA are shown in Figure 20 and 21.

TORA associates for each destination a metric to each node. This metric can be interpreted as height H (i) of the node 'i'. The height is composed of five different parameters. The first three parameters define the reference level and the other two the offset level. The parameters in the packet formats have the following meaning:

- **Version:** The TORA version number.
- **Type:** TORA packet type. For QRY packet this field is set to 1 and UPD packet this field is set to 2.
- **Reserved:** Field reserved for future use.
- **Destination IP Address:** The IP address for which a route is being requested.
- **Mode:** The mode of operation associated with the destination IP address.
- **H.tau:** Time of the last reference level update.
- **H.oid:** Identification (ID) of the node which defined the last reference level.
- **H.r:** Flag if the reference level was reflected.
- **H.delta:** To separate nodes with equal reference levels.
- **H.id:** Unique node ID.

Every node receiving this UPD packet will set its own height to a larger height than specified in the UPD message. The node will then broadcast its own UPD packet. This will result in a number of directed links from the originator of the QRY packet to the destination. This process can result in multiple routes.

Route Maintenance

When a node discovers link failure, it sets its own height, higher than that of its neighbors and issues an update to that effect reversing the direction of the link between them. If it finds that it has no downstream neighbors, the destination is presumed lost and it issues a clear packet to remove the invalid links from the rest of the network. ADSR and AntTORA protocols produce better results than the existing protocols-DSR and TORA in terms of end-to-end delay, energy, jitter and throughput. The routing overhead of these protocols marginally increases due to the addition of FANT and BANT packets. The performance of TORA and AntTORA is evaluated using Network simulator (Ns-2). Table 4 lists the simulation parameters and environments used.

End-to-End Delay

Table 5 shows the effect of mobility on end-to-end delay of TORA and AntTORA. The end-to-end delay increases as mobility increases from 10 m/s to 100 m/s. A higher mobility causes more link breaks and frequent re-routing, thus causing larger end-to-end delay. Broken links may cause additional route recovery process and route discovery process. AntTORA shows better performance in all the mobility conditions. The improvement over TORA varies from 47% to 89%.

Table 4 Simulation parameters

Simulation area	500m x 500m
Number of nodes	100
Node communication range	50m
Mobility model	Random waypoint
Speed	0-100 meter/second
Routing protocols	TORA and AntTORA
MAC	IEEE 802.11
Data rate	2 Mbps
Traffic source model	Constant bit rate

Table 5. Effect of node mobility on end-to-end delay

Mobility (meter /second)	End-to-end delay (milliseconds)	
	TORA	AntTORA
10	67.86	7.45
20	71.89	9.34
30	72.47	28.10
40	74.60	19.50
50	76.91	21.60
60	76.87	28.10
70	78.68	32.34
80	79.40	35.14
90	80.31	37.62
100	81.43	42.84

Jitter

Figure 22 shows the variations in jitter under various node speeds. The jitter is increased at higher mobility due to breaking of more links and frequent re-routing. The jitter is reduced in AntTORA by 87% to 95%. The average jitter rises when the network density increases. The reduction in jitter over TORA is maximum when the number of nodes reaches 100. This is due to jitter pheromone included in the QRY and UPD packets of AntTORA.

Energy

Node energy is the average residual node energy across the path. Table 6 expresses the effect of mobility on residual node energy. Residual node energy decreases with the increase in mobile speed, due to more link failure. The improvement over TORA is varied in the range of 2% to 12%.

Figure 22. Node mobility vs. jitter

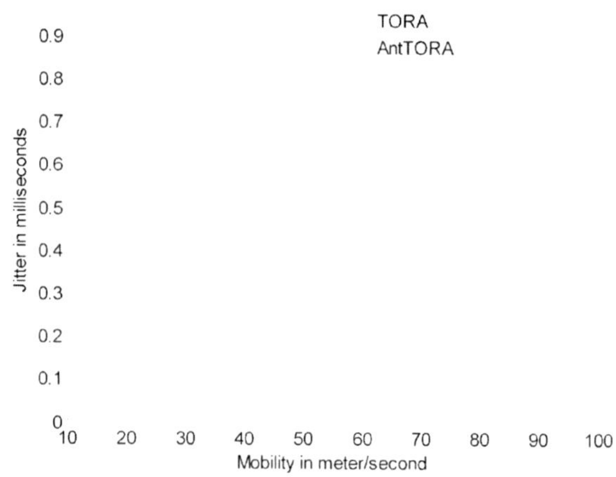

Table 6. Effect of mobility on energy

Mobility (meter /second)	Energy (Joules)	
	TORA	AntTORA
10	9.00	9.45
20	8.84	9.13
30	8.94	9.12
40	8.74	9.05
50	8.58	8.83
60	8.00	8.74
70	7.84	8.51
80	7.64	8.40
90	7.46	8.31
100	7.44	8.02

5. CROSS LAYER

The first major work on MANET QoS was the INSIGNIA framework, where resources are reserved in an end-to-end manner through a Resource Reservation Protocol (RSVP)(Zhang, Lee Gahng-Seop & Campbell, 2000). This QoS framework is designed to support adaptive services as a primary goal in ad hoc networks. It allows packets of audio, video and real-time data applications to specify their maximum and minimum bandwidth needs and plays a central role in resource allocation, restoration control and session adaptation between communicating mobile hosts. Based on availability of end-to-end bandwidth, QoS mechanisms attempt to provide assurances in support of adaptive services.

To support an adaptive service, the INSIGNIA framework establishes and maintains reservations for continuous media flows and micro-flows. To support these communication services, the INSIGNIA QoS framework comprises a number of architectural components, namely in-band signalling, admission control, packet forwarding, routing protocol, packet scheduling and Medium Access Control (MAC). A key component of this QoS framework is the INSIGNIA signalling system–an RSVP like signalling system that supports fast reservation, restoration and adaptation algorithms that are specifically designed to deliver adaptive service. The admission control module is responsible for allocating bandwidth to flows based on the maximum and minimum bandwidth requested. Once resources have been allocated, they are periodically refreshed by a mobile soft-state mechanism through the reception of data packets. The packet-forwarding module classifies incoming packets and forwards them to the appropriate module.

EARA-QoS is an on-demand multipath routing algorithm for MANETs, inspired by the ant foraging intelligence(Liu, Kwiatkowska & Constantinou, 2005). This algorithm incorporates positive feedback, negative feedback and randomness into the routing computation. Positive feedback originates from destination nodes to reinforce the existing pheromone on good paths. Ant-like packets, analogous to the ant foragers, are used to locally find new paths. Artificial pheromone is laid on the communication links between nodes and data packets are biased towards strong pheromone, but the next hop is chosen probabilistically. To prevent old routing solutions from remaining in the current network status, exponential pheromone decay is adopted as the negative feedback. By adopting the cross-layer optimization concept, both the network layer and the MAC layer information are used to compute routes that avoid the congested areas. The core of this QoS provisioning technique is the service class differentiation based queuing scheme. The results of simulation experiments show that this algorithm performs fairly well under situations of various nodal mobility, network density and data loads.

The integrated Mobile Ad-hoc QoS framework (iMAQ) is a cross-layer architecture to support the transmission of multimedia data over a MANET (Chen, Shah, & Nahrstedt, 2002) The framework involves an ad hoc routing layer and a middleware service layer. At each mobile node, these two layers share information and collaborate to provide QoS assurances to multimedia traffic. The network layer is facilitated with a predictive location-based QoS routing protocol. The middleware layer communicates with the network layer and applications to provide QoS support and maximize overall system QoS satisfaction. The middleware layer also uses location information from the lower network layer and tries to predict network partitioning. In order to provide better data accessibility, it replicates data between different network groups before partitioning occurs.

It proposes a general QoS framework for MANETs. This framework is hybrid in nature such that it combines the advantages of per-flow provisioning schemes as in IntServ and per-class provisioning schemes as in DiffServ. Accordingly, with FQMM, every source node plays the role of an ingress node for the flows it originates. It is hence responsible for such processes as classification, metering and marking of its own traffic. The other intermediate nodes perform traffic shaping according to those marks. Like DiffServ, FQMM has service differentiation. However, the FQMM model tries to improve the per-class granularity of DiffServ to per-flow granularity for certain classes of traffic. Accordingly, high-priority traffic is given per-flow provisioning, while other lower priority traffic is given per class provisioning. However, per-flow granularity is preserved for a small portion of traffic. (Xiao, Seah & Chua, 2000).

The FQMM model proposes a relative and adaptive differentiation traffic profile. The goal of such traffic profile is to keep consistent differentiation among sessions, which could be per-flow, or per aggregate of flows. Since it is deemed that an absolute traffic profile is not possible due to the inadequate bandwidth availability, FQMM favours a traffic profile being defined as the relative percentage of the effective link capacity in order to keep the differentiation among sessions predictable and consistent. A token bucket is used as the traffic profiler, and hence the use of a token bucket-metering algorithm allows packets to be marked as in-profile and out-of-profile. In case of network congestion, out-of-profile packets are discarded with a higher probability than in-profile packets.

6. CONCLUSION

A mobile ad hoc network is characterized by mobile nodes capable of communicating over wireless medium and establishing a network without a pre-existing infrastructure. In MANETs, the resources like bandwidth, computation power, memory and battery are to be used to achieve better performance. Challenges faced by MANETs include routing, QoS provisioning, energy efficiency, security and multicasting. This chapter focuses on QoS provisioning in network layer. The most popular on demand routing protocols in MANETs, such as DSR, AODV and TORA and the different QoS parameters were presented. QoS routing protocols are classified based on the metrics used. The concept, strengths and drawbacks of these protocols are also discussed. QoS routing protocols using optimization techniques are also described. The majority of the work reported in this chapter focuses on the design and performance evolution in terms of traditional metrics such as bandwidth, delay, energy (or) bandwidth and delay. There are few that attempt to optimize multi-constraint routing such as ACO and Genetic algorithm. These methods have limited applicability due to the overhead and energy cost of collecting enough state information. More research is required to establish QoS with various networking environments and topologies. Research in this field provides considerable challenge and potential to enhance the growth of mobile ad hoc networks. The challenges include resource availability, location management, cross layer QoS, support for heterogeneous nodes and security.

REFERENCES

Alwan, A., Bagrodia, R., Bambos, N., Gerla, M., Kleinrock, L., Short, J., &Villasenor, J.,(1996, April). Adaptive Mobile Multimedia Networks. *IEEE Personal Communications, 34-5.1*

Asokan, R., & Natarajan, A. M. (2007, July). An approach for reducing the end-to-end delay and increasing network lifetime in mobile ad hoc networks. *International Journal of Information Technology*, *4*(2), 121–127.

Asokan, R., & Natarajan, A. M. (2008, March). Performance evaluation of energy and delay aware quality of service (QoS) routing protocols in mobile ad hoc networks. *Int. Journal of Business Data Communications and Networking*, *4*(1), 52–63.

Asokan, R., & Natarajan, A. M. (2008, October). Performance Analysis of Quality of Service Enabled Temporally Ordered Routing Algorithm using Ant Colony Optimization in Mobile Ad hoc Networks. *Journal of Advances in Engineering Science*, *2*(2), 11–18.

Asokan, R., Natarajan, A. M., & Venkatesh, C. (2008, July). Ant Based Dynamic Source Routing Protocol to Support Multiple Quality of Service (QoS) Metrics in Mobile Ad Hoc Networks. *International Journal of Computer Science and Security*, *2*(3), 48–56.

Badis, H., & Agha, K. A. (2005). QOLSR, QoS Routing for Ad Hoc Wireless Networks using OLSR. *Wiley European Trans. Telecommunications*, *15*(4), 427–442. doi:10.1002/ett.1067

Balakrishnan, H., Padmanabhan, V., Seshan, S., & Katz, R. (1996). A comparison of mechanisms for improving TCP performance over wireless links. In *Proc. of ACM SIGCOMM'96,* Stanford, CA.

Baoxian, Z., & Hussein, T. M. (2005, October). QoS routing for wireless ad hoc networks: problems, algorithms, and protocols. *IEEE Communications Magazine*, *42*(10), 110–117. doi:10.1109/MCOM.2005.1522133

Baras, J. S., & Mehta, H. (2003, March). PERA: A probabilistic emergent routing algorithm for mobile Adhoc Networks. In *Proceedings of WiOpt 03: Modeling and optimization in mobile adhoc and wireless networks.*

Barolli, L., Koyama, A., Suganuma, T., & Shiratori, N. (2003, June). GAMAN: A GA based QoS routing method for mobile ad hoc networks. *Journal of Interconnection Networks*, *4*(3), 251–270. doi:10.1142/S0219265903000866

Bharghavan, V., Demers, A., Shenker, S., & Zhang, L. (1994). MACAW: A Media Access Protocol for Wireless LANs. In *Proc. ACM SIGCOMM'94* (pp. 212–225).

Chandran, R., Raghunathan, K. S., Venkatesan, S., & Prakash, R. (2001, February). A feedback-based scheme for improving TCP performance in ad hoc wireless networks. *IEEE Personal Communications Magazine*, *8*(1), 34–39. doi:10.1109/98.904897

Chen, K., Shah, S. H., & Nahrstedt, K. (2002). Cross Layer Design for Data Accessibility in Mobile Ad Hoc Networks. *Journal of Wireless Communications*, *21*, 49–75. doi:10.1023/A:1015509521662

Chen, L., & Heinzelman, W. B. (2005, March). QoS-Aware Routing Based on Bandwidth Estimation for Mobile Ad Hoc Networks. *IEEE Journal on Selected Areas in Communications*, *23*(3), 561–572. doi:10.1109/JSAC.2004.842560

Chen, S., & Nahrstedt, K. (1999, August). Distributed Quality-of-Service Routing in Ad Hoc Networks. *IEEE Journal on Selected Areas in Communications*, *17*(8), 1488–1505. doi:10.1109/49.780354

Chen, W.P., Hsiao, Y.C., Hou, J.C., Ge, Y., & Fitz, M.P. (2001, April). Syndrome: A light-weight approach to improving TCP performance in mobile wireless networks. *The journal of Wireless communications and Mobile computing.*

Dharmaraju, D., Chowdhury, A. R., Hovareshti, P., & Baras, J. S. (2002, August). INORA-a unified signaling and routing mechanism for QoS support in mobile ad hoc networks. In *Proceedings of 31st International Conference on Parallel Processing Workshops* (pp. 86–93).

Di Caro, G., & Dorigo, M. (1998, December). AntNet: Distributed stigmergetic control for communications networks. *Journal of Artificial Intelligence Research, 9*, 317–365.

Gerasimov, I., & Simon, R. (2002, April). A bandwidth reservation mechanism for on-demand ad hoc path finding. In *Proceedings of the IEEE/SCS 35th Annual Simulation Symposium,* CA (pp. 27-34).

Gerla, M., & Tsai, J. T. (1995). Multicluster, Mobile, Multimedia Radio Network. *ACM Baltzer Journal of Wireless Networks, 1*(3), 255–265. doi:10.1007/BF01200845

Gui, C., & Mohapatra, P. (2003, June). SHORT: self-healing and optimizing routing techniques for mobile ad hoc networks. In *Proceedings of the 4th ACM International Symposium on Mobile Ad Hoc Networking & Computing* (pp. 279–290).

Gunes, M., Sorges, U., & Bouazizi, I. (2002, August). ARA - The Ant-colony based Routing Algorithm for MANETs. In *Proceedings of International Conference on Parallel Processing Workshops* (pp. 79-85).

Hwang, Y., Lee, H., & Varshney, P. (2003). An Adaptive QoS Routing Protocol with Dispersity for Ad hoc Networks. In *Proceedings of the 36th Hawaii International Conference on System Sciences.*

Jacquet, P., Muhlethaler, P., Clausen, T., Laouiti, A., Qayyum, A., & Viennot, L. (2001, December). Optimized Link State Routing Protocol for Ad hoc Networks. In *Proceedings of the IEEE International Multi Topic Conference* (pp. 62-68).

Johnson, D. B., & Maltz, D. A. (1996). Dynamic source routing in ad hoc wireless networks. *Mobile Computing, 353*(5), 153–181. doi:10.1007/978-0-585-29603-6_5

Karn, P. (1990, April). MACA -a New Channel Access Method for Packet Radio. In *Proceedings of ARRL/CRRL Amateur Radio 9th Computer Networking Conference* (pp. 134–140).

Liao, W. H., & Tseng, Y. C. Wang, S.L., & Sheu, J.P. (2001). A multi-path QoS routing protocol in a wireless mobile ad hoc network. In *Proceedings of IEEE Intl. Conf. on Networking* (pp. 158–167).

Lin, C. R. (2001, April).On-Demand QoS Routing in Multihop Mobile Networks. *Proceedings of. IEEE International Conference on Computer Communications,*3, 1735–1744.

Lin, C. R., & Liu, J. S. (1999, August). QoS routing in ad hoc wireless networks. *IEEE Journal on Selected Areas in Communications, 17*(8), 1426–1438. doi:10.1109/49.779924

Lin, R., & Gerla, M.,(1997).Asynchronous Multimedia Multihop Wireless Networks. *IEEE INFOCOM 1997.*

Liu, J., & Singh, S. (2001, July). ATCP: TCP for mobile ad hoc networks. *IEEE JSAC, 19*(7), 1300–1315.

Liu, Z., Kwiatkowska, M. Z., & Constantinou, C. (2004). A Swarm Intelligence routing algorithm for MANETs. *Proc. of the IASTED Intl. Conference*, 484–489.

Liu, Z., Kwiatkowska, M. Z., & Constantinou, C. (2005). A biologically inspired QoS routing algorithm for adhoc networks, *International Conference on Advanced Information Networking and Application* 1, 426-431.

Misra, A., & Banerjee, S. (2002).MRPC: Maximizing Network Lifetime for Reliable Routing in Wireless Environments. *Proc. IEEE Wireless Communication and Networking Conf.*, Orlando.

Park, V. D., & Corson, M. S. (1997, April). A highly adaptive distributed routing algorithm for mobile wireless networks. *Proceedings of IEEE INFOCOM '97*,3, 405– 1413.

Paul, J., & Gaorge, K. (2005). Pheromone Aided Robust Multipath and Multi-priority Routing in Wireless MANETs. In *Proceedings of the 2nd ACM international workshop on Performance Evaluation of Wireless Ad hoc, Sensor and Ubiquitous Networks*, 106-113.

Peng, F., & Deyun, Z. (2006, February). Hybrid Optimize Strategy based QoS Route Algorithm for Mobile Ad hoc Networks. *Journal of Computer Science*, *2*(2), 160–165. doi:10.3844/jcssp.2006.160.165

Perkins, C. E., & Royer, E. M. (1999, February). Ad-hoc On Demand Distance Vector Routing. In *Proceedings of the 2nd IEEE Workshop on Mobile Computing Systems and Applications* (Vol. 3, pp. 90-100).

Perkins, C.E., Royer, E.M., & Das, S.R. (2000, July). Quality of Service for Ad hoc On Demand Distance Vector Routing. *IETF internet draft, draft-ietf-manet-aodvqos-00.txt*.

Prasant, M., Jian, L., & Chao, G. (2003, March). QoS in ad hoc wireless network. *IEEE Wireless Communications Magazine*, 44-52.

Rajagopalan, S., & Shen, C. C. (2005). ANSI: A Unicast Routing Protocol for Mobile Networks using Swarm Intelligence. In *Proceedings of the International Conference on Artificial Intelligence* (pp. 24-27).

Raju, G. V. S., Hernandes, G., & Zou, Q. (2000). Quality of Service Routing in Ad Hoc Networks. In . *Proceedings of the IEEE Wireless Communications and Networking Conference*, *1*, 263–265.

Roth, M., & Wicker, S. (2003, June). Termite: Ad hoc Networking with Stigmergy. In . *Proceedings of the IEEE Global Communications Conference*, *5*, 2937–2941.

Royer, E. M., & Toh, C. H. (1999, April). A review of current routing protocols for ad hoc mobile wireless networks. *IEEE Personal Communications*, *6*(2), 46–55. doi:10.1109/98.760423

Schoonderwoerd, R., Holland, O. E., Bruten, J. L., & Rothkrantz, L. J. (1996). Ant-based load balancing in telecommunications networks. *Adaptive Behavior*, *5*, 169–207. doi:10.1177/105971239700500203

Sheng, M., Li, J., & Shi, Y. (2003, January). Routing protocol with QoS guarantees for ad-hoc network. *Electronics Letters*, *39*, 143–145. doi:10.1049/el:20030017

Singh, S., & Raghavendra, C.S. (1998, July). PAMAS - power aware multi-access protocol with signaling for ad hoc networks. *ACM Computer Communication Review*.

Sivakumar, R., Sinha, P., & Bharghavan, V. (1999, August). CEDAR: a core extraction distributed ad hoc routing algorithm. *IEEE Journal on Selected Areas in Communications*, *17*(8), 1454–1465. doi:10.1109/49.779926

Streltsov, S., & Vakiki, P. (1996, March). Variance reduction algorithms for parallel replicated simulation of uniformized Markov chains. *Discrete Event Dynamic Systems: Theory and Applications*, *6*, 159–180. doi:10.1007/BF01797237

Toh, C. K. (2001, June). Maximum battery life routing to support ubiquitous mobile computing in wireless ad hoc networks. *IEEE Communications Magazine*, *39*(6), 138–147. doi:10.1109/35.925682

Woo, K., Yu, C., Youn, H. Y., & Lee, B. (2001). Non-Blocking, Localized Routing Algorithm for Balanced Energy Consumption in Mobile Ad Hoc Networks. In *Proceedings of Int'l Symp. on Modeling, Analysis and Simulation of Computer and Telecommunication Systems* (pp. 117-124).

Xiao, H., Seah, W. K. G., & Chua, K. C. (2000, May). A flexible quality of service model for mobile ad hoc networks. In *Proc. of IEEE Vehicular Technology Conference (VTC)*, 1, 445–449.

Xue, Q., & Ganz, A. (2003, February). Ad hoc QoS on-demand routing (AQOR) in mobile ad hoc networks. *Journal of Parallel and Distributed Computing*, *63*(2), 154–165. doi:10.1016/S0743-7315(02)00061-8

Zhang, X., Lee, S. B., Gahng-Seop, A., & Campbell, A. T. (2000, April). INSIGNIA: an IP based quality of service framework for mobile ad hoc networks. *Journal of Parallel and Distributed Computing*, *60*(4), 374–406. doi:10.1006/jpdc.1999.1611

Chapter 21
QoS and Energy-Aware Routing for Wireless Sensor Networks

Shanghong Peng
University of Guelph, Canada

Simon X. Yang
University of Guelph, Canada

Stefano Gregori
University of Guelph, Canada

ABSTRACT

Quality of service (QoS) and energy awareness are key requirements for wireless sensor networks (WSNs), which entail considerable challenges due to constraints in network resources, such as energy, memory capacity, computation capability, and maximum data rate. Guaranteeing QoS becomes more and more challenging as the complexity of WSNs increases. This chapter firstly discusses challenges and existing solutions for providing QoS and energy awareness in WSNs. Then, a novel bio-inspired QoS and energy-aware routing algorithm is presented. Based on an ant colony optimization idea, it meets QoS requirements in an energy-aware fashion and, at the same time, balances the node energy utilization to maximize the network lifetime. Extensive simulation results under a variety of scenarios demonstrate the superior performance of the presented algorithm in terms of packet delivery rate, overhead, load balance, and delay, in comparison to a conventional directed diffusion routing algorithm.

INTRODUCTION

Wireless sensor networks represent a new paradigm in wireless technology, drawing significant attention and research from diverse fields of engineering. Many new applications are emerging and the rapid deployment of such networks is underway with busy researchers and engineers creating and optimizing WSN technology all around the world (Culler, 2003). The vision of many researchers is to create sensor-rich ubiquitous computing and smart environments through planned or ad-hoc deployment of thousands of sensor nodes, each with a short-range wireless communications channel, and capable of detecting ambient conditions, such as temperature, movement, sound, light, or the presence of certain objects.

DOI: 10.4018/978-1-61520-680-3.ch021

The continuous decrease in the size, power dissipation, and cost of sensors has motivated intensive research in the past few years addressing collaboration among sensors in data gathering and processing. Networks of autonomous sensor nodes are having a significant impact on the efficiency of many surveillance and security applications, including environmental monitoring and natural disaster prevention.

However, energy supply is the primary constrained factor in many WSNs powered through batteries or environmental energy sources. To maximize the lifetime of WSNs, it is crucial to develop energy-efficient algorithms that optimize the overall energy consumption. Recently, significant research efforts have been devoted to the power optimization design for WSNs (Heinzelman et al., 2001; Chen, 2008). Since wireless transmission is the dominant power-dissipating operation in WSNs, special cares must be taken to design energy-aware wireless transmission systems.

On the other hand, while much of the existing research in WSNs has been focusing on energy minimization and lifetime maximization, such as in the Low Energy Adaptive Clustering Hierarchy (Heinzelman et al., 2000) and the Power-Efficient Gathering in Sensor Information Systems (Lindsey & Raghavendra, 2002), less efforts were devoted to optimize the quality of service of wireless communication systems. However, in many WSN applications, especially those in surveillance intelligence, data gathering is often required to be timely and reliable (Yu et al., 2004; Tavares et al., 2008; Sohraby et al., 2007). In addition, different applications have different transmission quality requirements on the end-to-end latency, jitter, and packet loss ratio (Aghdasi et al., 2008; Brun et al., 2006; Kumar & Rajesh, 2008; Seo et al., 2007). Some applications may also have dynamic QoS requirements. For example, in object tracking applications, the end-to-end latency requirement is dynamic, since the moving speed of the object is time-varying. Therefore, QoS provisioning is an important issue for WSNs.

Traditionally, the problems of QoS provisioning and power optimization are considered separately at different layers of the OSI (Open Systems Interconnection) reference model (i.e., protocol stack), which is often not efficient in energy utilization. Therefore, a novel adaptive routing algorithm for WSNs is presented, where the QoS requirements and node energy level are jointly designed.

This chapter presents one of the first biologically inspired methods on the joint design of the QoS requirements and energy awareness for WSNs. After providing some background information, a new QoS and Energy-Aware Routing algorithm (QEAR) is described. Next, some simulation results are analyzed in various application scenarios compared with the existing state-of-the-art Directed Diffusion (DD) routing algorithm. Finally, we conclude the chapter with a brief discussion.

BACKGROUND

In a wireless sensor network, groups of sensor nodes need to collaborate together and form a network, which can offer some specific services, such as data collection, environmental surveillance, and target tracking. Consequently, the primary goal for WSNs is to establish one or more routes between two nodes so that they can communicate reliably and efficiently. Such a network is characterized by the following challenges:

- The network topology can change dynamically due to the failure and random movement of nodes;
- Any node may "leave or join" the network (i.e., sleep or active mode) and the protocol must be adaptable accordingly;
- Although no guarantee of service can be provided, the protocol must be able to maximize the reliability of packet in the network for the given conditions.

With these factors in mind, the key design parameters of a routing algorithm for WSNs (Chen & Varshney, 2004; Akkaya & Younis, 2005; Medhi & Ramasamy, 2007) are:

- **Effective routing:** This is the foremost requirement of the protocol – to successfully discover a route and deliver the packet from the source to the destination. Some of the metrics for effective routing include packet delivery ratio, percentage of optimal routes taken, and average end-to-end delay.
- **Routing overhead:** Because sensor nodes typically have low computational capability and memory, wireless sensor networks could not support diffusion communication which is widely deployed in the wired networks. For example, distance-vector routing protocol uses the Bellman-Ford algorithm and link-state protocol use the Dijkstra's algorithm (Dijkstra, 1959) to calculate shortest (i.e., lowest cost) paths. So not only the network bandwidth used by the routing messages must be considered, but also how much processing power and memory is required in the nodes.
- **Congestion avoidance:** Strongly related to the previous parameter, this is to ensure that the routing algorithm does not congest a particular route or node thereby leading to packet drops or even failure of the nodes.
- **Energy consumption:** In many cases, nodes are battery-powered and batteries are recharged using energy harvested from the environment (e.g. solar or vibration). Furthermore, compared with the data acquisition, storage, and processing modules in a node, the RF transceiver is most power-hungry module. Hence, effective and efficient data transmission is very important for maximizing the network lifetime. For the routing algorithm, it must be designed to discover the optimal route based on multiple constraints, such as link cost, available bandwidth, and the available energy at each node.
- **Load balancing:** Some of the nodes may be strategically located resulting in being present in most of the optimal routes of communication. Such nodes may get overloaded leading to network congestion. Hence there is a need for balancing loads among the nodes through using suboptimal routes, which will lead to a more even load distribution.

Based on the above requirements, it is crucial to develop an adaptive optimized routing algorithm, which is responsible for not only packet routing, but also the overall network management. In this attempt, this chapter looks at the characteristics of emergent behaviors found in natural environments by studying the foraging patterns of the ants and proposes a solution to the network service of quality and lifetime problems through routing.

The basic idea of the ant colony optimization (ACO) metaheuristic (Dorigo et al., 1996; Dorigo & Stützle, 2004) is taken from the food searching behavior of real ants. When ants are on their way searching for food, they start from their nest and proceed toward the food. When an ant reaches an intersection, it has to decide which way to take next. While walking, ants deposit pheromone, which marks the route taken. Subsequently, more ants are attracted by these pheromone trails and in turn reinforce them even more. As a result of this autocatalytic effect, the optimal solution emerges rapidly. This approach has been successfully applied to many combinatorial optimization problems, such as symmetric and asymmetric traveling salesman problems (Dorigo & Gambardella, 1997; Gambardella & Dorigo, 1996), and vehicle routing problems (Gambardella et al., 1999). Besides, it has been applied to solve routing problems for communication networks, such as the AntNet algorithm (Di Caro & Dorigo 1998), the AntHocNet algorithm (Di Caro et al., 2005), the

Ant-colony based Routing Algorithm (Günes et al., 2002), the Global Positioning System Ant-Like routing algorithm (Câmara & Loureiro, 2001), and the Emergent Ad hoc Routing Algorithm enhanced with QoS (Liu et al., 2005). Similarly, its optimization principle can be applied easily to develop the energy-aware routing algorithm with multiple QoS constraints for WSNs. In fact, this is a positive feedback control process to establish and maintain the route dynamically.

IMPLEMENTATION OF A BIO-INSPIRED QOS AND ENERGY-AWARE ROUTING ALGORITHM FOR WSNS

In wireless sensor networks, routing is one of the key issues for researchers and scientists due to their highly dynamic and distributed nature. In particular, energy efficient routing may be the most important design criteria for WSNs, since the energy available to each sensor node is limited. Power failure of a node not only affects the node itself, but also its ability to forward packets and, eventually, the overall network lifetime. For this reason, many research efforts have been devoted to developing energy-aware routing protocols.

However, researchers did not focus on QoS requirements, which are critical for real-time high-bandwidth applications, such as multimedia streams and voice. In order to solve the above two main problems, a routing algorithm for WSNs using the ant colony optimization metaheuristic is described in this section.

In order to develop the routing algorithm, it is important to reduce the memory used in sensor nodes, balance the whole network load, and consider the energy level of the paths found by ants. Besides, some QoS requirements (e.g., for video streams and target tracking) should be taken into account such as packet delivery ratio, delay and delay jitter of the end-to-end. At the same time, routing overhead needs to be controlled effectively. For a mobile ad-hoc network, the overhead is usually relatively high, sometimes accounting for more than 80% of the network traffic. Therefore, it is crucial to consider the energy status of the node such as the average energy and minimum remaining energy on a link, quality of service of the network such as available maximum bandwidth, packet loss rate, and delay of packets.

QoS Guarantees

It has been proved that in a sensor node the tasks related to communications (i.e., transmitting and receiving data) spend much more energy than those related with data processing and memory access. Since one of the main concerns in WSNs is to maximize the lifetime of the network, which means saving as much energy as possible, it is preferable that the routing algorithm performs as much processing as possible in the network nodes, rather than transmitting the rough data through the network to the sink node to be processed there. Therefore, our objective is to satisfy certain QoS constraints, balance the whole network load, and consider the energy level of the paths. The QEAR algorithm is composed of two main parts: route discovery and route maintenance.

Routing Discovery

For wireless sensor networks, which are often deployed in an ad-hoc fashion, routing typically begins with neighbor discovery. Nodes send rounds of messages (packets) and build local neighbor tables. These tables include the minimum information of each neighbor's ID and location. Usually, this means that nodes must know their location prior to neighbor discovery. Other typical information in these tables includes nodes' remaining energy, delay via that node, and an estimation of link quality. Once the tables exist, messages are directed from a source location to a destination address. In the QEAR algorithm, the route discovery process is performed in the following six stages:

Figure 1. Route discovery phase I (i.e., propagation of forward ants from source to destination)

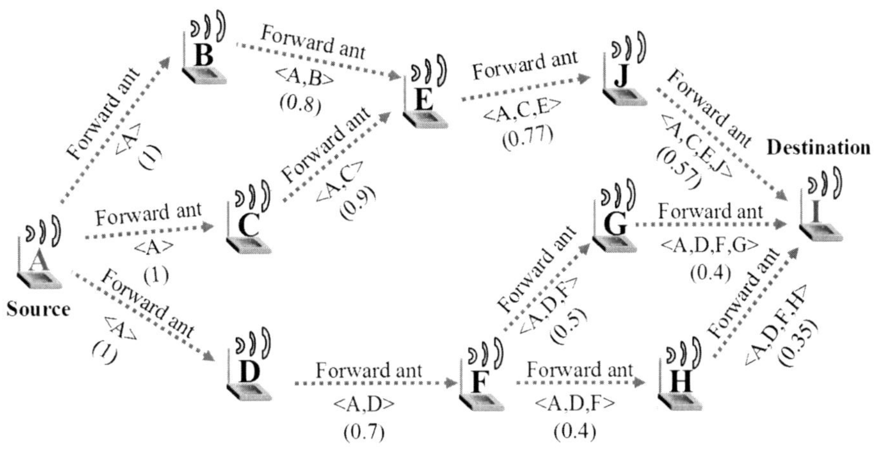

Stage I: When the sink node detects an event of interest that has to be transmitted, it first checks the cache for existing routes. When no routes are known, it broadcasts forward request ants as shown in Figure 1. This process can be compared to ants initially spreading out from the nest in all directions in search of a food source.

Stage II: Each forward ant searches for the destination by selecting the next hop node according to the link probability distribution function given by

$$p_k(r,s) = \begin{cases} \dfrac{[T(r,s)]^{\alpha}\,[E(s)]^{\beta}}{\sum_{u \notin M_k} [T(r,u)]^{\alpha}\,[E(u)]^{\beta}}, & \text{if } s \notin M_k, \\ 0, & \text{otherwise}, \end{cases} \qquad (1)$$

where $p_k(r,s)$ is the probability with which a forward ant k chooses to move from node r to node s; T is the routing table at each node that stores the amount of pheromone trail on connection (r,s); $E(s)$ is the visibility function which equals the remaining energy level of node s; and α and β are parameters that control the relative importance of pheromone concentration versus visibility. Initially all the links have equal probability. The identifier of every visited node is saved onto a memory M_k and carried by the ant.

The selection probability is a trade-off between visibility (i.e., nodes with more energy should be chosen with higher probability) and actual pheromone concentration (i.e., if on link (r,s) there has been a lot of ant traffic, the link should be used with higher probability).

Stage III: While moving forward, each forward ant records a list of nodes it has visited and tries to avoid traversing the same node. This helps forming loop-free routes.

Stage IV: Once a forward ant reaches the destination node, it becomes a backward ant that returns through the links that the forward ant had traversed, as shown in Figure 2. Thus, when different forward ants reach the destination through different routes, the destination sends a backward ant for each of them. This is to ensure that multiple paths exist between the source and the destination. Similarly, in the case of ants, initially multiple paths exist between the nest and the food source. Gradually, the best path (which for ants is the shortest path) gets strengthened through increased pheromone.

Stage V: During the backward travel, the pheromone is distributed to each node in the path as follows,

$$\Delta T_k = \frac{e_M E_{Min_k} + e_A E_{Avg_k}}{d_F D_{F_k} + d_B D_{B_k}} \cdot \frac{af_b}{bf_d + cf_j + df_{pl}}, \qquad (2)$$

Figure 2. Route discovery phase II (i.e., propagation of backward ants from destination to source)

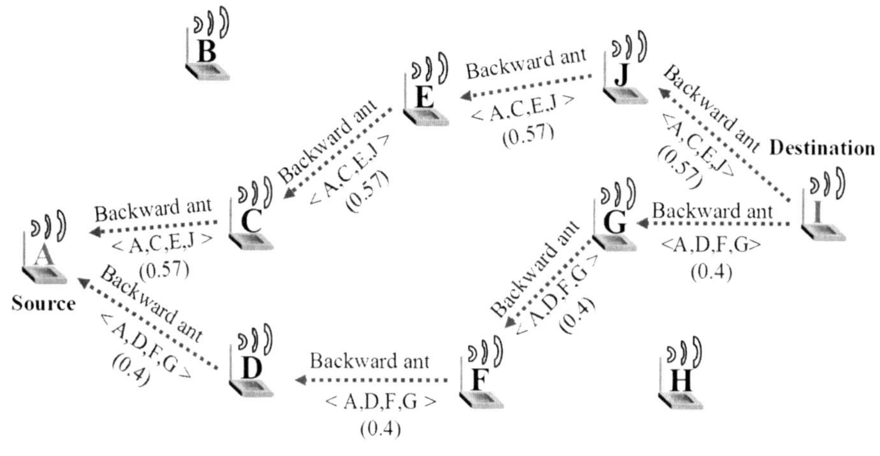

where ΔT_k is the amount of the pheromone dropped by ant k, which is computed at the destination node during its journey; e_M (with $0 < e_M \leq 1$) and e_A (with $0 < e_A \leq 1$) are the weights of the minimum energy of the nodes in the ant k's path E_{Mink} and the average energy of the nodes in its path E_{Avgk}, respectively; d_F (with $0 < d_F \leq 1$) and d_B (with $0 < d_B \leq 1$) are the weights of D_{Fk} and D_{Bk}, respectively; D_{Fk} is the distance travelled by the forward ant k (i.e., the number of nodes stored in its memory); and D_{Bk} is the travelled distance (i.e., the number of visited nodes), by backward ant k until node r. This parameter will force the ant to lose part of the pheromone strength during its way to the source node. The idea behind this behavior is to build a better pheromone distribution (i.e., nodes near the sink node will have more pheromone levels) and will force remote nodes to find better paths. Such behavior is extremely important when the sink node is able to move, since the pheromone adaptation will be much faster.

Calculating ΔT_k only as a function of the energy levels of the path, does not lead to optimized routes, since a path with eight nodes can have the same energy average as a path with only four nodes. Therefore, ΔT_k must be calculated as a function of both parameters: the energy levels and the length of the path. This can be achieved by introducing the parameter D_{Fk} and D_{Bk}.

Parameters a, b, c, and d (with $0 < a \leq 1$, $0 < b \leq 1$, $0 < c \leq 1$, $0 < d \leq 1$) are the weights of reward function f_b and penalty functions f_d, f_j, f_{pl}, respectively, which denote the relative importance of available bandwidth, delay, delay jitter, and packet loss in the objective function.

$$f_b = \begin{cases} 1, & \text{if } bandwidth(r,s) - B_{\min} \geq 0, \\ r_b, & \text{otherwise,} \end{cases} \tag{3}$$

$$f_d = \begin{cases} 1, & \text{if } delay(r,s) - D_{\max} \geq 0, \\ r_d, & \text{otherwise,} \end{cases} \tag{4}$$

$$f_j = \begin{cases} 1, & \text{if } delay\ jitter(r,s) - J_{\max} \geq 0, \\ r_j, & \text{otherwise,} \end{cases} \tag{5}$$

$$f_{pl} = \begin{cases} 1, & \text{if } packet\ loss(r,s) - P_{\max} \geq 0, \\ r_{pl}, & \text{otherwise,} \end{cases} \tag{6}$$

where f_b is the reward function of bandwidth metric. If the individual can satisfy the bandwidth constraint B, then the value of f_b is 1, otherwise the value is r_b (with $0 < r_b < 1$). f_d is the penalty function of delay metric. If the individual can satisfy the delay constraint D, then the value of f_d is 1, otherwise the value is r_d (with$0 < r_d < 1$).

Similarly, f_j and f_{pl} are the penalty functions of delay jitter and packet loss metric, respectively. If the individual can satisfy the delay jitter constraint J, then the value of f_j is 1, otherwise the value is r_j (with $0 < r_j < 1$). If the individual can satisfy the packet loss constraint P, then the value of f_{pl} is 1, otherwise the value is r_{pl} (with $0 < r_{pl} < 1$). The values of r_b, r_d, r_j, and r_{pl} decide the degrees of reward and punishment.

There are four constraints including available minimum bandwidth B_{min}, maximum delay D_{max}, maximum delay jitter J_{max}, and maximum packet loss P_{max}.

The pheromone in this algorithm is also used as a way to record the traffic load in each path on global behavior by available bandwidth w. The pheromone effects make the forward ant avoid choosing the path with a heavy traffic load, and balance the energy consumption across the whole network. Besides, it takes into account three crucial metrics of QoS provisioning: end-to-end delay t, delay jitter J and packet loss rate l. These parameters are calculated as follows,

$$w = \min\left\{w_i, i=1,2,\ldots,j\right\}, \tag{7}$$

$$t = \sum_{i=1}^{j} t_i, \tag{8}$$

$$J = E\left[\left\| \sum_{k=1}^{n} \Delta_k \right\|\right], \tag{9}$$

$$l = 1 - \prod_{i=1}^{j} \left(1 - l_i\right), \tag{10}$$

where i (with $1 \le i \le j$) is a link that the ant visited; w_i, t_i, and l_i are the bandwidth, delay, and packet loss rate of the ith link on the path, respectively; J is the end-to-end delay jitter; and Δ_k is the variation of the inter-packet delay at node k.

Stage VI: In the next route exploration round, the link probability distribution of each intermediate node will be updated according to the pheromone concentration. By performing this algorithm several iterations, each node will be able to know which neighbors are the best in terms of the optimal function, as shown in Equation (11). Once a node r receives a backward ant coming from a neighboring node s, it updates its routing table in the following manner:

$$T_k(r,s) = (1-\rho)\cdot T_k(r,s) + \Delta T_k, \tag{11}$$

where ρ (with $0 \le \rho \le 1$) is a coefficient that represents the evaporation of pheromone concentration since the last time $T_k(r,s)$ was updated.

Routing Maintenance

Each sensor node maintains a neighbor table and an event table. The event table contains a list of events that the node has observed. The neighbor table can be maintained by actively initiating a Hello message or passively eavesdropping on network broadcasts. In the proposed algorithm, each node will periodically exchange a Hello message in order to maintain the route table and timely response to the topology change. It usually includes the geographic location, the remaining energy, available bandwidth, buffer size, and the pheromone concentration with its neighboring nodes.

If a node receives a Hello message from a new node n, it will add n as a destination in its routing table. After that, it expects to receive a Hello message from the node n at every Hello period. After missing some Hello messages continuously, the node n will be removed. Using these messages, nodes know about their immediate neighbors and have pheromone information about them in their routing tables. Additionally, they can detect broken links rapidly and clean up old pheromone entries from their routing tables. Therefore, the QEAR algorithm can implement robust routing and reliable packet delivery.

Regarding route failure handling, the QEAR algorithm assumes that the medium access control (MAC) layer is compliant to the IEEE 802.15.4 standard. Thus, it allows the sensor node to detect a link failure by the missing acknowledgment on the MAC layer, and to deactivate that link by resetting the pheromone concentration to zero. Then, the routing table is checked for different links to the destination and the packet is forwarded accordingly. If no other route exists, the current node notifies the sender about the route failure, which thereby initiates a new route discovery process.

In summary, the presented routing algorithm improves the use of the precious energy resources on sensor nodes. At the same time, it takes into account QoS metrics, such as delay, jitter, and packet loss. Since the available bandwidth is considered, this algorithm is especially well suited for real-time high bandwidth traffic requirement, such as voice and video transmissions.

Energy Threshold Management

In wireless sensor networks, the energy management problem has been studied intensively. Various approaches for reducing the energy expenditure have been presented in literature; several papers minimize the transmitter power (a significant energy drain for WSN nodes) while maintaining connectivity. Several routing protocols showed significant improvements in the network lifetime for mobile ad-hoc networks by choosing routes that avoid nodes with low battery and balancing the traffic load. Approaches of the MAC layer are geared towards reducing idle listening power and decreasing the number of collisions. Application layer approaches show dramatic energy savings for several classes of applications. Other papers show that cross-layer approaches may also be very effective at conserving energy. In this chapter we focus on routing strategies that maximize the lifetime of the WSN.

As shown in Equation (2), the QEAR algorithm takes into account not only the remaining energy of all nodes in the paths from the source to destination but also the average energy of these nodes. In addition, an energy threshold level is set up. If the node has sufficient residual energy, it will go on participating in forwarding data in the current route as an intermediate node. When a node's residual energy has fallen below a threshold (e.g., 10%), it will construct an Energy Low packet. At the same time, a delay timer is enabled. After it expires, the node will send it to the source node. This mechanism attempts to prolong network's lifetime by preventing hot spot nodes from consuming more energy for forwarding data packet. While the source node receives the Energy Low message, it will choose any other route it happens to know in its route cache. If there is no route stored in the cache, it will start route discovery process to discover a new route. Thus, the selected routes have good energy levels. It effectively avoids the packet loss at the intermediate nodes and saves the overhead of packet retransmission.

PERFORMANCE ANALYSIS OF A BIO-INSPIRED QOS AND ENERGY-AWARE ROUTING ALGORITHM FOR WSNS

In this section, the performance of the bio-inspired QEAR algorithm is evaluated through a number of simulations in comparison with the existing state-of-the-art directed diffusion routing algorithm (Intanagonwiwat et al., 2000; Heidemann et al., 2001; Intanagonwiwat et al., 2003).

Assumptions

To model the lifetime of the general sensor networks considering the real-time and mobile applications such as motion detection and target tracking, the following assumptions are made:

- A static network of homogeneous sensor nodes and a sink node distributed over a given region with uniform density.
- The sensor network works with a query-driven model. In the interested area the

sensor nodes could detect the events and send their readings to the sink in a multi-hop fashion. Each node generates one data packet per time unit called a round.

- The delay per hop is the same along a path that packets take through the network.
- Each sensor node has a battery with finite energy, whereas the sink has an unlimited amount of energy available to it.

All nodes transmit at the same constant power. Hence, all nodes have the same radio transmission range, the same energy consumption for receiving one packet, and the same energy consumption for transmitting one packet.

Simulation Parameter Setup

In order to demonstrate the performance of the QEAR algorithm, it is compared to a conventional routing algorithm: Direct Diffusion. Both of them are reactive routing protocol based on queries. In fact, in huge sensor networks where the number of nodes can easily reach more than 1000 units, the memory of ants would be so big that it would be unfeasible to send ants through the network. Besides, the multiple constrained routing optimization is a NP-hard problem. Therefore, the small- and mid-scale networks are chosen for the simulation experiments.

The simulation program is executed on a standard 1.6 GHz PC using OMNeT++ (version 3.2). It needs a few seconds of CPU time for a single simulation run. The radio communication channel was modeled with a duplex transceiver. The network stack of each node consists of the IEEE 802.15.4 MAC layer with 30-meter transmission range and a network interface.

In this simulation, we assume $\rho = 0.7$, $\alpha = \beta = a = b = c = d = 1$, $e_M = e_A = d_F = d_B = 0.5$. CBR (Constant Bit Rate) traffic with different deadlines is used. The payload size is a constant 64 bytes. Moreover, data aggregation is not considered in this experiment. This means the sensed data are transmitted unchanged to sink. The main parameters used for simulation are given in Table 1.

Table 1. Simulation parameters

Components	Setting
Simulator	OMNeT++
Area	(80m×80m), (150m×150m), (200m×200m)
Number of nodes	10, 50, 200
Node placement	Uniform
Payload size	64 Bytes
Application	Many-to-one CBR streams
Routing protocol	DD, QEAR
MAC protocol	IEEE 802.15.4 MAC
Radio model	Two-ray ground
Radio range	30 m
Bandwidth	250 kb/s
Run time	200 s
Confidence interval	95%

The key performance metrics evaluated in the simulations are:

- Packet delivery ratio (i.e., ratio of number of packets received to number of packets sent);
- Average hop count per packet (i.e., ratio of number of packets related to messages which are forwarded by intermediate nodes including sink node to number of messages sent);
- Routing overhead (i.e., ratio of number of control packets to number of messages sent);
- Energy consumption or load distribution (i.e., comparison of the number of packets forwarded or sent by the individual nodes);
- Average end-to-end delay (i.e., average time of the messages which take to travel from the source and sink node).

Table 2. Traffic statistics for DD and QEAR under different scenarios

Metrics	Directed Diffusion			QEAR		
	10 nodes, 2 sources	50 nodes, 5 sources	200 nodes, 20 sources	10 nodes, 2 sources	50 nodes, 5 sources	200 nodes, 20 sources
Packet delivery ratio	99.97%	96.25%	92.73%	99.98%	99.41%	99.07%
Average hop count/packet	2.31	3.25	7.42	2.42	3.58	7.76
Control packet count	2476	9176	175261	2341	5656	112554
Message packet count	62096	149258	1480327	62278	144609	1475011
Routing overhead	4.0%	6.1%	11.4%	3.7%	4.0%	7.6%
Load unbalance	7,379,569	5,336,136	46,995,907	2,221,663	918,958	22,319,072

QoS Metric and Load Balancing Analyses

The first set of simulations is performed on a 10-node network in an 80 m by 80 m area. The results are the average of the performance metrics over three simulations with different 10-node scenarios with the same traffic. These sensor nodes are uniformly deployed across the area. The traffic load is 2 sources sending 64-byte data at an interval of 0.1 seconds.

Table 2 shows traffic statistics for the directed diffusion routing algorithm and the QEAR algorithm under different scenarios. It compares the performance of the QEAR algorithm with the DD algorithm. At the left column it indicates the results of the first set of simulations. Both algorithms have the similar performance in terms of packet delivery ratio and average hops; however, the directed diffusion algorithm has higher overhead due to the continuous exchange of Hello messages in the route maintenance phase for updating the quality of a particular route.

The greatest difference is in load balancing and energy management. Figure 3 shows the number of packets sent or forwarded by each node in the network (except the sink node). Simulation results of DD show an uneven distribution of packets among nodes. Some of the nodes (e.g., nodes 8 and 9) forward an extremely high number of packets that result in high-energy utilization, while others are relatively idle.

In order to evaluate the network load balance performance, the average load unbalance is calculated by

$$U_b = \frac{1}{j}\sum_{i=1}^{j}\left(N_i - N_A\right)^2, \quad (12)$$

where U_b is the average load unbalance in the network; i is the index of a node; j is the total number of nodes in the network; N_i is the number of forwarded packets at the node I; and N_A is the average number of forwarded packets at the node in the network.

Combining Figure 3 with Table 2, in the DD algorithm, the average load unbalance is up to 7,379,569. In the QEAR algorithm, however, the network load is evenly distributed among most of the nodes because multiple routes are used to send packets. The average load unbalance of the QEAR algorithm is only 2,221,663. Its load balance ability is much better than DD (more than triple). It has to be noticed that some nodes have very low packets sent or forwarded, since they do not lie in the route from source to destination.

The second set of simulations is performed on a 50-node network in a 150 m by 150 m area. In addition, 5 CBR traffic sources are used which send 64-byte packets at an interval of 0.1 seconds.

At the middle column of Table 2 depicts traffic statistics for DD and QEAR in 50-node scenario. It means that the QEAR algorithm has not only a lower routing overhead than DD (about 2% less),

Figure 3. Network load distribution for DD and QEAR (10-node scenario)

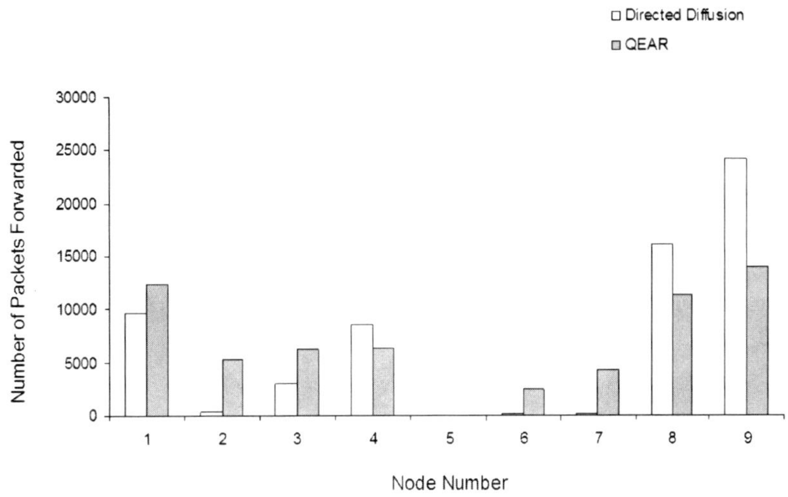

Figure 4. Network load distribution for DD and QEAR (50-node scenario)

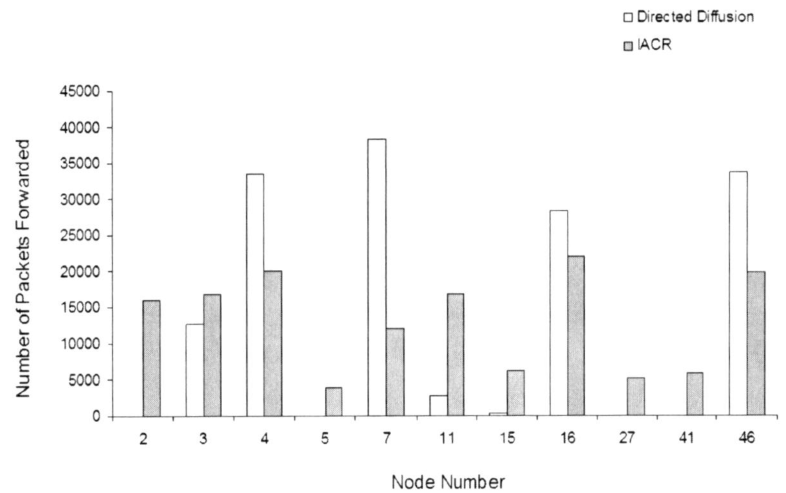

but also a better performance in terms of packet delivery ratio (up to 99.4%). What's more, the increase in average hop count per packet is only around 0.3 hops.

In Figure 4, it can also be observed that the load distribution is uneven in the DD algorithm with only 16% of nodes involved in active traffic (only nodes with more than 1,000 packets forwarded are plotted in the figure). Some nodes handle heavy traffic forwarding (more than 35,000 packets); however, in the QEAR algorithm, more nodes (about 24%) are involved in active traffic forwarding, and the traffic is more distributed among these nodes.

Besides, combining Figure 4 with Table 2, it is shown that the average load unbalance of the QEAR algorithm is only 918,958, compared with 5,336,136 for the DD algorithm. Thus, the QEAR algorithm provides better load and energy balancing than the DD algorithm.

The third set of simulations is performed on a 200-node network in a 200 m by 200 m area. In addition, 20 CBR traffic sources are used which send 64-byte packets at an interval of 0.1 seconds.

The right column of Table 2 indicates that the QEAR algorithm continues holding a very high packet delivery ratio, up to 99.07%. It benefits from a multi-path routing mechanism (probability forwarding). Packet delivery ratio of the DD algorithm, however, decreases sharply, only 92.7%. This metric is the most important in communications systems. Concerning routing overhead, the QEAR algorithm keeps getting lower (3.8% less). As to the number of average hop per packet, the QEAR algorithm is only around 0.3 hops more than the DD algorithm. Combined with these metrics, the QEAR algorithm performs much better than the DD algorithm.

In Figure 5, it is obvious that the load distribution is uneven in the DD algorithm with only 5.5% of nodes involved in active traffic (only nodes with more than 20,000 packets forwarded are plotted in the figure). Some nodes handle heavy traffic forwarding (more than 250,000 packets); however, in the QEAR algorithm, more nodes (about 9%) are involved in active traffic forwarding, and the traffic is distributed more evenly among these nodes.

Besides, as shown in Table 2 and Figure 5, the average load unbalance of the QEAR algorithm is only 22,319,072, compared with 46,995,907 for the DD algorithm. In one word, the QEAR algorithm outperforms the QEAR algorithm and can provide good energy balancing in a large-scale wireless sensor network.

At last, the average end-to-end delay from source to sink is simulated, as shown in Figure 6. It is obvious that the DD algorithm has higher latency, up to 0.2 seconds in a 50-node, 5 traffic sources scenario; however, in the QEAR algorithm, the packets are forwarded to sink in 0.1 seconds on average. Once more, it verifies that the new transmission model has better multi-constraint based communication performance.

Parameter Sensitivity Analyses

Various previous works have shown that the use of ACO to make next-hop decisions offer improved performance compared to next-hop decisions based purely on hop count. The aim of this subsection is to examine the influence of QEAR parameters on a network's performance. The experiments performed will start off with the QEAR algorithm where the probability of choosing a next-hop node is calculated by using Equation (1). Once

Figure 5. Network load distribution for DD and QEAR (200-node scenario)

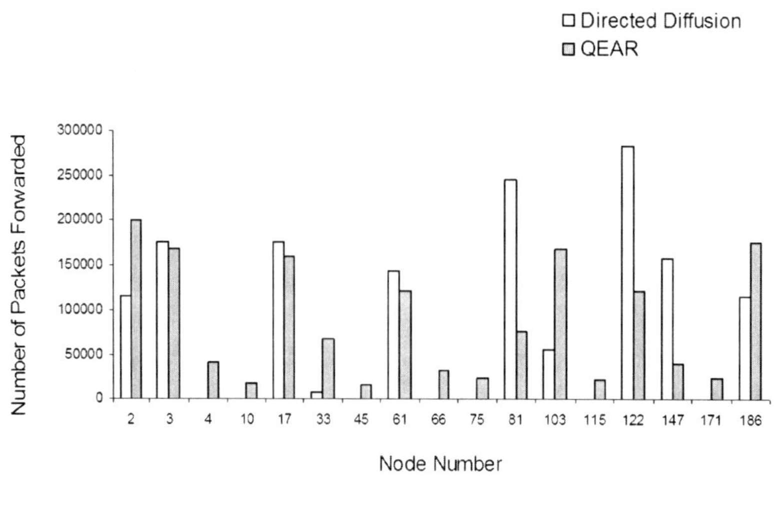

the probabilities are calculated, the next hop node is selected using roulette-wheel selection. These tunable parameters are discussed as follows.

Initialization of the Random Variants

The QEAR algorithm uses the probabilistic routing idea; however, the routing tables store pheromone concentrations, which are transformed into probabilities, as shown in Equation (1). The presented approach [refer to Equation (2) and Equation (11)] proposes an adaptive and dynamic pheromone increase and a decrease for the packet source, both within boundaries to avoid extreme pheromone differences.

It starts by setting all pheromone levels to an initial value. The system is then allowed to run for a fixed period before nodes start generating data packets, in order for the nodes to establish routes to other nodes. They implement a relative pheromone updating scheme by making the amount of pheromone updated by each ant inversely proportional to the "age" of the ant. The rational behind this is that ants with shorter paths will have more influence on the routing tables, and vice versa. They also implement a mechanism for relieving congestion by delaying ants' route to a congested node. This gives the congested node time to decongest, and also "ages" the ants so that the pheromone deposited by them will be less. Therefore, it decreases the probability of future visits to the congested node.

In this simulation, the following values have been selected: initial pheromone = 1, maximum pheromone = 1,000, minimum pheromone = 0.1, initial decay period = 1 second. Later, the decay period will be longer as follows:

$$T_{new} = e^{r_d - 0.5} T_{old}, \tag{13}$$

where T_{new} is the new decay period; T_{old} is the old decay period; r_d is the random value between zero and one.

If a packet arrives from an unknown source, a row is created in the routing table for the new destination. If this newly discovered node happens to be neighbor, in addition to the row also an additional column is appended. The specified initial pheromone value is assigned to the new entries. The concentration must not fall below the minimum pheromone, even if no traffic is sent by the source for some time. This procedure helps to detect idle nodes easily.

Pheromone Evaporation Coefficient

Pheromone evaporation allows the QEAR algorithm to "forget" old solutions gradually over time. It plays an important role to allow routes to become less attractive over time so that stale routes are less likely to be used.

For small values of the evaporation coefficient of pheromone concentration ρ, pheromone evaporates slowly. Nodes will therefore accumulate more routes in their routing tables, but the routes may not be valid anymore. For large values of ρ, the routes in a nodes routing table are more likely to be valid, but the node may delete valid routes before they can be exploited.

The experiment is performed on a 50-node network in a 150 m by 150 m area. In addition, 5 CBR traffic sources are used which send 64-byte packets at an interval of 0.1 seconds. The parameter ρ is stepped from 0 to 1 in 0.1 interval. Table 3 depicts the measured packet delivery ratios with 95% confidence intervals. In a static network, the algorithm manages to deliver up to 99% of the packets with ρ = 0.6 and 0.7. With different pheromone evaporation coefficients, packet delivery ratios are not changed sharply. It means that the pheromone evaporation coefficient has little effect on the performance of a static network. Since link failures occur infrequently, the routes accumulated by nodes stay valid much longer, so the need to remove old routes with the evaporation mechanism is small. However, in a mobile network, invalid routes accumulated at nodes cause

Figure 6. End-to-end delay for DD and QEAR (50-node scenario)

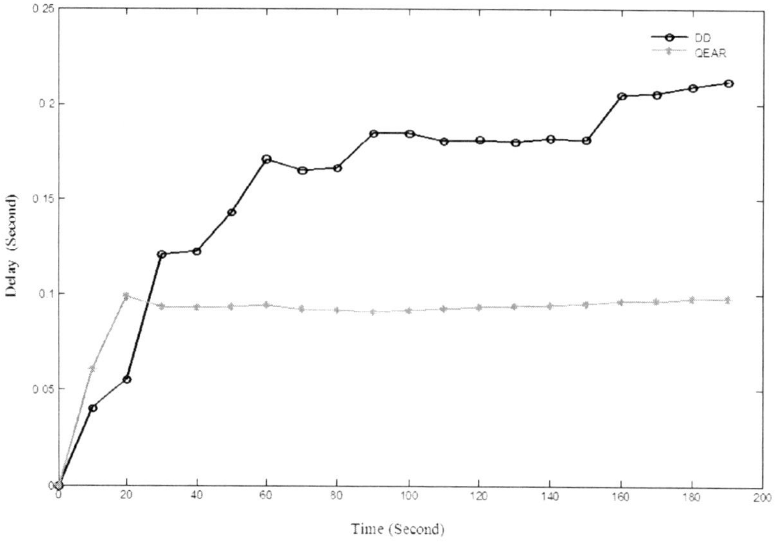

packets to be dropped much more frequently. It is better to increase ρ. Thus, the configured ρ in this simulation is suitable.

Coefficients Related to Heuristic Information

As shown in Equation (1), α and β are coefficients related to heuristic information. A method of controlling exploration and exploitation is suggested by Schoonderwoerd et al. (1996), based on a pseudorandom-proportional action rule. This concept was first introduced by Gambardella and Dorigo (1996) in an ant colony system to explicitly control an algorithm's exploration and exploitation characteristic. Another method in the sensitivity analysis of the model parameters is to use automatic procedures for parameter tuning (Birattari et al., 2002). Therefore, based on these empirical values, α and β are assumed to be 1 in the simulations.

Table 3. Impact of ρ on packet delivery ratio

ρ	Packet delivery ratio
0	0.97
0.1	0.97
0.2	0.98
0.3	0.98
0.4	0.98
0.5	0.98
0.6	0.99
0.7	0.99
0.8	0.98
0.9	0.98
1	0.98

Coefficients Related to Energy and QoS Information

Based on above discussed methods, parameters related to energy and QoS information are defined according to the empirical theory. Thus, they are assumed as $a = b = c = d = 1$, $e_M = e_A = d_F = d_B = 0.5$.

FUTURE RESEARCH DIRECTIONS

In contrast to a conventional communication and data network, a wireless sensor network generally consist of a large number of sensor nodes (e.g., up to 1,000 nodes), which are deployed to realize a common goal (e.g., sensing an event of interest or measuring data correlated to physical phenomena). The sensor nodes are cooperative intrinsically and should work together to meet their application needs. In fact, two characteristics of wireless sensor networks can be exploited to improve communication efficiency: the cooperation among the sensor nodes and application-specific performance metrics. For WSNs, efficient resource usage not only means efficient bandwidth utilization, but also a minimal usage of energy. Thus, QoS and energy management in WSNs should also include various control mechanisms besides QoS guarantee and energy threshold mechanisms.

Future research will focus on investigating how to offer longer network lifetime, adapt the QoS required, and ensure efficient routing of information in a wireless sensor network with various traffic types, varying load, and mobile nodes. In addition, use of in-network processing could help to shape the behavior of data transmission. The use of aggregation and correlation of data can give further advantages to enhance QoS performance and prolong the lifetime of wireless sensor networks if methods to minimize the delay are developed.

CONCLUSION

In this chapter, a new QoS and energy-aware routing algorithm is introduced. It solves the multi-constrained routing problem in wireless sensor networks based on a biological evolution algorithm, such as ant colony optimization. The QEAR algorithm is implemented, which considers the features of WSNs (i.e., limited energy levels, low processing and memory capabilities). The lightweight ants are used to find routing paths between the sensor nodes and the sink node, which are optimized in terms of distance, delay, delay jitter, packet loss, bandwidth, and energy levels. These special ants balance communication loads and maximize energy savings, contributing to the extended lifetime of the wireless sensor networks. The simulation results show that the QEAR algorithm outperform the directed diffusion algorithm in different WSN scenarios, without causing much more overhead. The QEAR algorithm also seems more scalable than the directed diffusion algorithm: increasing the number of nodes, its performance advantage increases and its overhead grows slower than DD's. Other important considerations and limits of routing design for WSNs are articulated in the context. Future trends of the routing algorithm are also discussed.

REFERENCES

Aghdasi, H. S., Abbaspour, M., Moghadam, M. E., & Samei, Y. (2008). An energy efficient and high-quality video transmission architecture in wireless video-based sensor networks. *Sensors*, *8*, 4529–4559.

Akkaya, K., & Younis, M. (2005). A survey of routing protocols in wireless sensor networks. *Elsevier Ad Hoc Network Journal*, *3*(3), 325–349.

Birattari, M., Stützle, T., Paquete, L., & Varrentrapp, K. (2002). A racing algorithm for configuring metaheuristics. In *Proceedings of the Genetic and Evolutionary Computation Conference (GECCO 2002)*, New York, USA (pp. 11-18).

Brun, O., Bockstal, C., & Garcia, J. M. (2006). A simple formula for end-to-end jitter estimation in packet-switching networks. In *Proceedings of the International Conference on Networking, International Conference on Systems and International Conference on Mobile Communications and Learning Technologies (ICNICONSMCL'06)*, Morne, Mauritius (pp. 14).

Câmara, D., & Loureiro, A. A. F. (2001). GPS/ant-like routing in ad hoc networks. *Telecommunication Systems*, *18*(1-3), 85–100. doi:10.1023/A:1016739402641

Chen, D., & Varshney, P. K. (2004). QoS Support in Wireless Sensor Networks: A Survey. In *Proceedings of the International Conference on Wireless Networks (ICWN '04),* Las Vegas, USA (pp. 227-233).

Chen, M., Kwon, T., Mao, S., Yuan, Y., & Leung, V. C. M. (2008). Reliable and energy-efficient routing protocol in dense wireless sensor networks. *International Journal of Sensor Networks, 4*(1-2), 104–117. doi:10.1504/IJSNET.2008.019256

Culler, D. (2003). 10 emerging technologies that will change the world. *Technology Review, 106*(1), 33–49.

Di Caro, G., & Dorigo, M. (1998). AntNet: Distributed stigmergetic control for communications networks. [JAIR]. *Journal of Artificial Intelligence Research, 9,* 317–365.

Di Caro, G., Ducatelle, F., & Gambardella, L. (2005). AntHocNet: An adaptive nature-inspired algorithm for routing in mobile ad hoc networks. *European Transactions on Telecommunications . Special Issue on Self-Organization in Mobile Networking, 16*(5), 443–455.

Dijkstra, E. W. (1959). A note on two problems in connection with graph. *Numerische Mathe, 1,* 269–271. doi:10.1007/BF01386390

Dorigo, M., & Gambardella, L. M. (1997). Ant colony system: A cooperative learning approach to the traveling salesman problem. *IEEE Transactions on Evolutionary Computation, 1*(1), 53–66. doi:10.1109/4235.585892

Dorigo, M., Maniezzo, V., & Colorni, A. (1996). The ant system: Optimization by a colony of cooperating agents. *IEEE Transactions on Systems, Man, and Cybernetics-Part B, 26*(1), 29–41. doi:10.1109/3477.484436

Dorigo, M., & Stützle, T. (2004). *Ant Colony Optimization*. Cambridge, MA: MIT Press.

Gambardella, L. M., & Dorigo, M. (1996). Solving symmetric and asymmetric TSPs by ant colonies. In *Proceedings of IEEE Conference on Evolutionary Computation (ICEC'96),* Nagoya, Japan (pp. 622-627).

Gambardella, L. M., Taillard, E., & Agazzi, G. (1999). MACS-VRPTW: A multiple ant colony system for vehicle routing problems with time windows. In D. Corne, M. Dorigo, & F. Glover (Eds.), *New Ideas in Optimization* (pp. 63-76). London: McGraw-Hill.

Günes, M., Sorges, U., & Bouazizi, I. (2002). ARA – the ant-colony based routing algorithm for MANETs. In *Proceedings of the ICPP International Workshop on Ad Hoc Networks (IWAHN),* Vancouver, Canada (pp. 79–85).

Heidemann, J., Silva, F., Intanagonwiwat, C., Govindan, R., Estrin, D., & Ganesan, D. (2001). Building efficient wireless sensor networks with low-level naming. In *Proceedings of the eighteenth ACM Symposium on Operating Systems Principles,* Banff, Alberta, Canada (pp. 146-159).

Heinzelman, W., Chandrakasan, A., & Balakrishnan, H. (2000). Energy-efficient communication protocol for wireless microsensor networks. In *Proceedings of the Hawaii International Conference System Sciences (HICSS'00),* Hawaii, USA (pp. 4-7).

Intanagonwiwat, C., Govindan, R., & Estrin, D. (2000). Directed diffusion: A scalable and robust communication paradigm for sensor networks. In *Proceedings of the Sixth Annual International Conference on Mobile Computing and Networks (MobiCOM 2000),* Boston, MA (pp. 56-67).

Intanagonwiwat, C., Govindan, R., Estrin, D., Heidemann, J., & Silva, F. (2003). Directed diffusion for wireless sensor networking. *IEEE/ACM Transactions on Networking, 11*(1), 2-16.

Kumar, M., Jeya, K., & Rajesh, R. S. (2008). Performance analysis of energy-aware QoS routing protocol for ad hoc wireless sensor network. *International Journal of Computer Science and Network Security, 8*(4), 94–100.

Lindsey, S., & Raghavendra, C. S. (2002). PEGASIS: Power efficient gathering in sensor information systems. In *Proceedings of the IEEE Aerospace Conference*, Big Sky, Montana (pp. 1125-1130).

Liu, Z., Kwiatkowska, M. Z., & Constantinou, C. (2005). A biologically inspired QoS routing algorithm for mobile ad hoc networks. In *Proceedings of the 19th International Conference on Advanced Information Networking and Applications,* Vancouver, Canada (pp. 426-431).

Medhi, D., & Ramasamy, K. (2007). *Network Routing Algorithms, Protocols, and Architectures*. San Mateo, CA: Morgan Kaufmann Publishers.

Peng, S. (2008). *A bio-inspired QoS and energy-aware routing algorithm for wireless sensor networks*. Master's thesis, University of Guelph.

Peng, S., Yang, S. X., Gregori, S., & Tian, F. (2008). An adaptive QoS and energy-aware routing algorithm for wireless sensor networks. In *Proceedings of the IEEE International Conference on Information and Automation (ICIA'2008),* Zhangjiajie, China (pp. 578-583).

Schoonderwoerd, R., Holland, O., Bruten, J., & Rothkrantz, L. (1996). Ant-based load balancing in telecommunications networks. *Adaptive Behavior, 5*, 169–207. doi:10.1177/105971239700500203

Seo, M., Choi, W., Kim, Y., & Park, J. (2007). Lifetime prediction routing protocol for wireless sensor networks. *IEICE (International Energy Information Community) transactions on communications, 90*(12), 3680-3681.

Sohraby, K., Minoli, D., & Znati, T. (2007). *Wireless Sensor Networks: Technology, Protocols, and Applications.* New York: John Wiley and Sons.

Tavares, J., Velez, F. J., & Ferro, J. M. (2008). Application of wireless sensor networks to automobiles. *Measurement Science Review, 8*(3), 65–70. doi:10.2478/v10048-008-0017-8

Yu, Y., Krishnamachari, B., & Prasanna, V. K. (2004). Energy-latency tradeoffs for data gathering in wireless sensor networks. In *Proceedings of the IEEE INFOCOM Conference on Computer Communications,* Hong Kong, China (pp. 244-255).

KEY TERMS AND DEFINITIONS

Delay: It can be measured in either one-way or round-trip delay. One-way delay calculations require expensive sophisticated test gears; however, measuring round-trip delay is easier and requires less expensive equipment. To get a general measurement of one-way delay, measure round-trip delay and divide the result by two.

Delay Jitter: It is the variation in latency over time from one end point to another end point. For example, if the delay of transmissions varies too widely in a VoIP call, the call quality is greatly degraded. The amount of jitter tolerable on the network is affected by the depth of the jitter buffer on the network node in the path. The more jitter buffer available, the more the network can reduce the effects of jitter.

Load Balancing: In computer networking, it is a technique to spread work between two or more computers, network links, CPUs, hard drives, or other resources, in order to get optimal resource utilization, maximize throughput, and minimize response time.

Metaheuristic: It is an iterative generation process which guides a subordinate heuristic by combining intelligently different concepts for exploring and exploiting the search space, learning strategies are used to structure information in order to find efficiently near-optimal solutions.

Packet Loss: It means losing packets along the data path, which severely degrades the communication application.

Pheromone: It is a chemical that triggers a natural behavioral response in another member of the same species. There are alarm pheromones, food trail pheromones, sex pheromones, and many others that affect behavior or physiology. Their use among insects has been particularly well documented. In addition, some vertebrates and plants communicate by using pheromones.

Chapter 22
Queuing Delay Analysis of Multi-Radio Multi-Channel Wireless Mesh Networks

Chengzhi Li
University of Houston, USA

Wei Zhao
University of Macau, China

ABSTRACT

Wireless mesh networking is becoming an economical means to provide ubiquitous Internet connectivity. In this chapter, we study wireless communications over multi-radio and multi-channel wireless mesh networks with IEEE 802.11e based ingress access points for local clients and point-to-point wireless links over non-overlapping channels for wireless mesh network backbones. We provide a set of algorithms to analyze the performance of such wireless mesh networks with wideband fading channels in various office building and open space environments and commonly-used Regulated and Markov On-Off traffic sources. Our goal is to establish a theoretical framework to predict the probabilistic end-to-end delay bounds for real-time applications over such wireless mesh networks.

I. INTRODUCTION

Wireless mesh networking is a new technology that complements infrastructure-based wired networks to provide ubiquitous Internet connectivity. Generally, wireless mesh networks (WMNs) are composed of wireless mesh routers and clients. The mesh routers with gateway/routing functions form wireless mesh backbones and provide multi-hop connectivity between clients and the Internet or between clients. Unlike nodes in traditional wireless networks such as mobile ad hoc networks (MANETs), mesh routers are static and have no power constraint and their locations can be carefully selected. The unique features and characteristics of WMNs, which distinguish WMNs from wired networks, include

a. Low initial investment
b. Extensive coverage areas
c. Ease of deployment and expansion
d. Fault tolerance

DOI: 10.4018/978-1-61520-680-3.ch022

With extensive research efforts on wireless mesh networking by academic and industrial communities (Akyildiz, Wang, & Wang, 2005; Bruno, Conti, Gregori, Wijting, Kneckt, & Damle, 2005; Faccin, Wijting, Kneckt, & Damle, 2006; Kyasanur, So, Chereddi, & Vaidya, 2006; Lee, Zheng, Ko, & Shrestha, 2006) as well as the commercial deployments of wireless mesh networks (WMNs) around the world, wireless mesh networking technology is becoming a vital component of our daily life.

In wireless communications, wireless channels are error-prone and their capacities are physically limited. There are many factors that affect the performance of wireless channels, e.g., signal power attenuation, inter-channel and co-channel interference, thermal noise, Doppler frequency, shadowing, and multipath channel fading. Therefore, QoS provisioning for many applications, which have diverse performance requirements in terms of minimum data rate, delay/delay jitter bound, and packet loss rate over wireless networks, poses a very difficult challenge. Moreover, it has been revealed (Gupta & Kumar, 2000) that the throughput of per source-destination pair in a multihop wireless network with a single shared channel scales with the number of network nodes n as $O(1/n^{0.5})$. It has also been demonstrated (Ganguly, Navda, Kim, Kashyap, Niculescu, Izmailov, Hong, & Das, 2006; Niculescu, Ganguly, Kim, & Izmailov, 2006) that the performance of VoIP applications over single channel WMNs degrades quickly as the lengths of VoIP traffic routes increase. Thus, the QoS capability of multihop wireless networks with a single shared channel is very limited.

With the rapid evolution of radio technologies, commercial multi-radio products have emerged in the market, e.g., BelAir Networks' BelAir200 mesh router with up to four radios and Motorola's Motomesh node with up to four radios and Strix's OWS mesh router with up to six radios. Moreover, there are 27 non-overlapping channels for IEEE 802.11 based wireless networks, i.e., 3 non-overlapping channels for IEEE 802.11b/g standards in 2.4 GHz frequency band and 24 non-overlapping channels with IEEE 802.11a standard in 5 GHz frequency band. These factors make it natural to consider WMNs with multi-radio and multi-channel mesh routers as a feasible solution to mitigate the inherent capacity limitation of conventional single channel wireless networks for QoS provisioning. It is worth noting that the asymptotic capacity of multi-channel and multi-radio wireless mesh networks has been theoretically characterized (Kyasanur & Vaidya, 2005) and experimentally verified (Kodialam & Nandagopal, 2005).

Generally, one of crucial performance metrics for QoS provisioning over multihop wireless networks is the end-to-end delay experienced by traffic flows. Following the seminal work (Gupta & Kumar, 2000), the wireless communication and networking research community has made extensive efforts to understand the throughput-delay scaling law. The results provided in (Bansal & Liu, 2003; Gamal, Mammen, Prabhakar, & Shah, 2004; Needly & Modiano, 2005; Moraes, Sadjadpour, & Garcia-Luna-Aceves, 2004; Moraes, Sadjadpour, & Garcia-Luna-Aceves, 2004; Lin, Sharma, Mazumdar, & Shroff, 2006; Perevalov & Blum, 2003; Perevalov & Blum, 2006) have established the asymptotic relationship between the average delay and maximum feasible throughput of per source-destination pair in wireless networks with network nodes extending to infinity. While these elegant results shed light on the deployment of wireless networks as well as the design of protocols and algorithms from a high level perspective, the asymptotic features may not match real wireless networks that usually have finite nodes. Therefore, they may not be suitable for practical WMNs. There exist two papers (Chen & Yang, 2006; Bisnik & Abouzeid, 2006) that address the multihop delay analysis for WMNs with finite nodes. In (Bisnik & Abouzeid, 2006), WMNs are modeled as G/G/1 queuing networks. Based on the diffusion approximation approach to characterize the traffic arrivals at each mesh router, the average multihop delay and maximum achievable throughput, which is similar to the well known asymptotic capacity of wireless multihop wireless ad-hoc networks provided in (Gupta & Kumar, 2000), are derived for WMNs. It is worth noting that the performance metrics, such as the average multihop delay and maximum achievable throughput, cannot be used for design and analysis of algorithms and protocols for supporting applications such as VoIP and real-time video streaming with the bounded delay or minimum date rate requirements. In (Chen & Yang, 2006), the authors extend the effective capacity model developed in (Wu & Negi, 2003) to multihop wireless communications and derive a probabilistic bound on delays experienced by traffic flows over multiple point-to-point wireless links. It is worth noting that the results developed in (Chen & Yang, 2006) rely on two assumptions, i.e., the traffic distortion due to multihop traveling can be ignored, and the delays experienced by traffic flows at each hop are independent and identically distributed (i.i.d.) random variables, which may not be true in realistic WMNs. By now, to the best of authors' knowledge, no analytic algorithms and models exist for predicting the probabilistic bounds (not average values) on the end-to-end delays experienced by traffic flows over wireless mesh backbones consisted of static multi-radio mesh routers. Hence, new algorithms and models are needed. This motivates us to write this chapter that is based on our preliminary results described in (Li & Zhao, 2008).

In this chapter, we consider multi-radio and multi-channel WMNs with IEEE 802.11 HCCA MAC based ingress access points for local clients and enough non-overlapping channels for point-to-point wireless links of wireless mesh backbones. In Section II, we briefly describe several components of multi-radio and multi-channel WMNs and some theoretical background about large deviations technique. In Section III, we estimate the probabilistic bound on the delays experienced by a traffic flow at an IEEE 802.11 HCCA MAC based ingress access point. We analyze the multihop delays experienced by various traffic flows over the backbones of multi-radio and multi-channel WMNs. By leveraging the *large deviations* technique, we turn probabilistic problems into deterministic optimization problems and derive a set of algorithms to evaluate the probabilistic bounds on the end-to-end delays experienced by traffic flows. In Section IV, we present the experimental results to demonstrate the feasibility of the proposed theoretical framework described in Section III. Finally, we offer our conclusion and point out potential future research directions in Section V.

II. BACKGROUND

WMN Model

Figure 1 shows an example with 21 wireless mesh routers for the backbones of broadband (high data rate) WMNs considered in this Chapter. Without loss of generality, we assume that each mesh router has multiple radios as the commercial multi-radio mesh router OWS or BelAir200. One of the radios is configured to communicate with local clients and the others are dedicated to forward traffic over the wireless backbone. In Figure 1, the number in a bracket denotes the channel assigned for local clients to access the wireless backbone, while the number above or under a dash line denotes the channel dedicated to the point-to-point wireless link between two adjacent mesh routers. By now, many routing protocols (Draves, Padhye, & Zill, 2004; Tang, Xue, & Zhang, 2005; Kodialam & Nandagopal, 2003) and channel assignment algorithms (Ramachandran, Belding, Almeroth, & Buddhikot 2006; Das, Alazemi, Vijayakumar, & Roy, 2005; Alicherry, Bhatia, & Li, 2005; Vedantham, Kakumanu, Lakshmanan, & Sivakumar, 2006; Kodialam & Nandagopal, 2005; Kyasanur & Vaidya, 2005; Xing, Chen, Ma, & Liang, 2007)

Figure 1. A static mesh network backbone

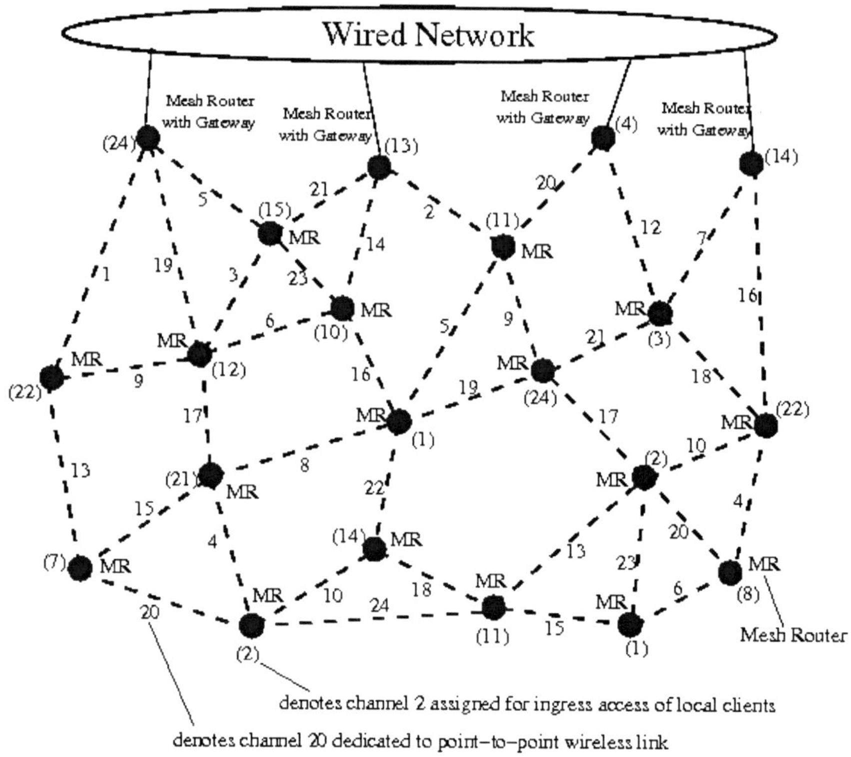

have been proposed to migrate or eliminate contention and interference from wireless communications over backbones of multi-radio and multi-channel WMNs. Therefore, we make the following assumptions:

1. There are enough non-overlapping wireless channels such as 27 non-overlapping IEEE 802.11 channels and a channel assignment algorithm such as that described in (Xing, Chen, Ma, & Liang, 2007);
2. All radios in each mesh router operate in different non-overlapping channels and can transmit or receive simultaneously without interference from each other. It is worth noting that the interference between radios of a mesh router operating at non-adjacent channels can be alleviated or eliminated by the elaborate wireless card shielding and antenna separations (Adya, Bahl, Padhye, Wolman, & Zhou, 2004; Robinson, Papagiannaki, Diot, Guo, & Krishnamurthy, 2005). For example, it has been exhibited (Adya, Bahl, Padhye, Wolman, & Zhou, 2004) that two Netgear WAB501 cards for IEEE 802.11a with a separation 6 inches in the same box and operating at non-adjacent channels (e.g., 56 and 64, 52 and 60) do not interfere each other. Similar empirical results for IEEE 802.11a radios can also be found in (Ramachandran, Sheriff, Belding-Royer, & Almeroth, 2006). Thus, considering the rapid innovation in wireless technologies, this assumption is reasonable for coming commercial multi-radio mesh routers;
3. The wireless mesh backbone is built from multiple point-to-point wireless links and each of these links uses the full spectrum of a channel to carry traffic flows without contention from other wireless links. Contrary to mobile nodes in an ad-hoc wireless network, the locations of static mesh routers can be

carefully selected. Therefore, this assumption is reasonable for well planned WMN backbones with enough of non-overlapping channels;

4. There exists a routing algorithm such as that described in (Draves, Padhye, & Zill, 2004) to provide routes for traffic flows from their sources to the corresponding destinations;
5. All mesh routers of WMNs are identical. It is worth to point out that this assumption is for notational simplification. All results described in this chapter can be extended to WMNs with non-identical mesh routers without any technical difficulty.

MAC at Access Point

Figure 2 exhibits the functions of the IEEE 802.11 medium access control (MAC) protocol that has been specified in the IEEE 802.11 standard (IEEE Std, 2007) and widely used in wireless networks. These functions can be categorized into two classes. One is the distributed contention-based class based on the Carrier Sense Multiple Access with Collision Avoidance (CSMA/CA) mechanism and includes the Distributed Coordination Function (DCF) and the enhanced Distributed Channel Access (EDCA). The other is the centralized contention-free class based on polling scheme and includes the Point Coordination Function (PCF) and the Hybrid-Coordination-Function Controlled Channel Access (HCCA). The DCF and PCF were originally proposed for the legacy IEEE 802.11 wireless networks for data transmission (IEEE Std, 1999), while the EDCA and HCCA are the QoS enhancements of the DCF and PCF respectively. Generally, the DCF is unsuitable for applications with QoS requirements, while the EDCA only provides service differentiation, not delay or throughput guarantees. In addition, the PCF may not be able to provide predictable services, because of the unpredictable beacon delays and unknown transmission durations of the polled stations. It has been exhibited that for G729 codec with 40 *ms* packetization interval, a single IEEE 802.11a access point with the HCCA can support more than 295 VoIP calls (Trad, Munir, & Afifi, 2006), while the same access point with the DCF only supports no more than 126 VoIP calls (Garg & Kappes, 2003). This significant advantage of the HCCA over the contention-based channel access functions such as the DCF or EDCA is due to its contention-free nature that mitigates the packet collisions and random backoff idle time. Thus, in this chapter, we assume that the radio of each mesh router, which is dedicated to the AP for local clients, is supported by the IEEE 802.11 MAC with the HCCA. The HCCA channel access scheme is illustrated in Figure 3. More details about the HCCA can be found from (IEEE Std, 2007) and is omitted in this Chapter.

Figure 2. IEEE 802.11 MAC architecture (IEEE Std, 2007)

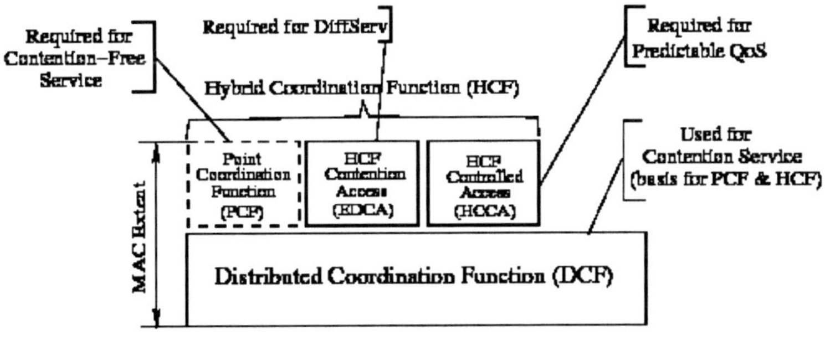

Figure 3. IEEE 802.11 HCCA channel access scheme (IEEE Std, 2007)

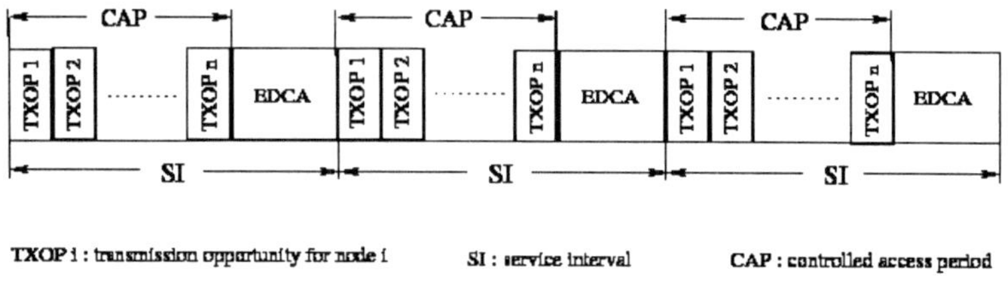

Point-to-Point Wireless Link

IEEE 802.11a/g and IEEE 802.16 are well known standards that have been adopted by WMNs for high data rate wireless communications. These standards use Orthogonal Frequency Division Multiplexing (OFDM), which is a bandwidth-efficient parallel data transmission technique, to combat the frequency selective multipath fading of wideband channels. In OFDM scheme, a wideband channel is divided into multiple narrowband sub-channels, and a high rate data stream is split into multiple lower-rate data streams that are transmitted simultaneously and in parallel over these narrowband sub-channels. For example, IEEE 802.11a standard (IEEE Std, 1999) can provide data rate up to 54 Mbps over a channel of 20 MHz bandwidth in the 5-GHz unlicensed national information infrastructure (UNII) band. The 20 MHz bandwidth channel is divided into 52 sub-channels of about 300 KHz bandwidth. Since the symbol duration substantially increases for narrowband sub-channels, the relative amount of time dispersion caused by multipath delay spread is decreased and the frequency selective multipath fading is mitigated or eliminated. In this chapter, we consider OFDM-based broadband wireless mesh network backbones with stationary and ergodic frequency selective fading channels that vary at a rate much slower than the symbol rate, so each channel condition remains roughly constant over each packet transmitting time.

Wireless Channel Models: Wireless mesh networks will be deployed in various environments. In this chapter, we consider five channel types proposed in (Medbo & Schramm, 1998) for these environments. The first type is Rayleigh fading channels with 50 *ms* root mean square (RMS) delay spread for the typical office environment. The second type is Rayleigh fading channels with 100 *ms* RMS delay spread for the large open space and office environment. The third type is Rayleigh fading channels with 150 *ms* RMS delay spread for the large open space environment. The forth type is Rician fading channels with 140 *ms* RMS delay spread for the large open space environment. The fifth type is Rayleigh fading channels with 250 *ms* RMS delay spread for the large open space environment. Since the impulse response of a wireless channel contains all information necessary to analyze or simulate any type of radio transmission through the wireless channel (Rappaport, 2002), we use the following exponentially decaying finite impulse response (FIR) filter, proposed in (Chayat, 1997), to characterize slowly fading frequency selective Rayleigh channels.

$$h(t) = \sum_{k=0}^{k_{\max}} h_k \delta(t - kT_s), \tag{1}$$

where k_{max} is the minimum integer that is not smaller than $10/\psi$; $\psi = T_s/\tau_{rms}$ is the number of possible multipath components during each sampling time interval; T_s is the length of one sampling time interval; τ_{rms} is the RMS delay spread of the frequency selective fading channel; $h_k \sim CN(0, \sigma_k^2/2)$, $k=0,1,\cdots,k_{max}$, are independent circular symmetric complex Gaussian random variables with zero mean and variance $\sigma_k^2 / 2 = \frac{1 - e^{-\psi}}{2(1 - e^{-k_{\max}\psi})} e^{-k\psi}$;

and δ is the Dirac delta function. Similarly, we use the following exponentially decaying FIR filter to characterize slowly fading frequency selective Rician channels.

$$\hat{h}(t) = \frac{\sqrt{K}}{\sqrt{1+K}} e^{\theta\sqrt{-1}} + \frac{1}{\sqrt{1+K}} h(t), \qquad (2)$$

where K is the Rician factor and $\theta \in [0, 2\pi]$ $\frac{\sqrt{K}}{\sqrt{1+K}} e^{\theta\sqrt{-1}}$ characterizes the direct line-of-sight (LOS) path with phase θ between the transmitter and the receiver; and $h(t)$ is defined by Equation (1). When K=0, Rician channels become Rayleigh channels and Equation (2) degenerates to Equation (1). Furthermore, the background noise is modeled as the additive white Gaussian noise (AWGN) and the transmission power is assumed to be constant. Thus, the signal to noise ratio (SNR) at time t can be determined by

$$SNR(t) = \overline{SNR} \times |\, h(t)\,|^2, \qquad (3)$$

where $\overline{SNR}$ is the average SNR, and $h(t)$ is determined by Equation (1) or Equation (2).

Wireless Channel Capacity Process: To characterize the capacity process of a point-to-point link over a wireless channel with a stop-and-go ARQ protocol, let $\{C[t], t>0\}$ denote the total amount of traffic that can be successfully transmitted over the wireless channel during the time interval $[0,t]$. Based on the mechanism of a stop-and-go ARQ protocol, a successfully transmitted data packet means that the data packet and the corresponding ACK packet are successfully received. Moreover,

$$C[t] = F(t, mc, L_{data}, SNR_{data}(x)\ x \in [0,t],\ mc^*, L_{ack}, SNR_{ack}(x)\ x \in [0,t])'$$

where L_{data} and L_{ack} are the lengths for data packet and ACK packet, respectively; *mc* and *mc** are the modulation and coding schemes used for data packet and ACK packet, respectively.

Wireless Link Effective Capacity: To evaluate probabilistic bounds on the delays experienced by traffic arrivals over a point-to-point wireless link, we use the *effective channel capacity* model defined in (Li, Che, &Li, 2007) to probabilistically lower bound the available channel capacity of the wireless point-to-point link. An effective channel capacity function for a point-to-point link over a fading channel is a non-negative real function S^ε such that for any time interval with length x

$$S^\varepsilon(x) = \sup \left\{ \begin{matrix} X \in [0,\infty) \mid \Pr\{C[t+x] \\ -C[t] < X\} \le \varepsilon, \forall t \ge 0 \end{matrix} \right\}. \qquad (4)$$

For IEEE 802.11a channels, an OFDM PHY layer simulator has been provided in (Heiskala & Terry, 2001) to simulate packet transmissions under various channel environments. By an extension of this OFDM simulator, the corresponding channel capacity process $C[t]$ can be simulated under various channel environments. From the obtained channel capacity process $C[t]$, $S^\varepsilon(x)$ can be estimated; see Section IV for more details.

Moment Bound

In this chapter, we adopt the following moment bound approach to evaluate the tail probability of a random variable X such as delay or backlog length experienced by traffic arrivals.

$$\Pr\{X \ge a\} \le \frac{E[X^k]}{a^k}, \forall a \ge 0 \text{ and } k = 1, 2, \cdots \qquad (5)$$

Several well known properties for the moment of independent random variables have been described in (Bucklew, 1990).

Proposition 1: Let X be a random variable, if 0 is an interior point in $\{s : E[e^{sX}] < \infty\}$,

$$E[X^k] = \left.\frac{d^k E[e^{sX}]}{ds^k}\right|_{s=0}, \ for\ k = 1, 2, \cdots \quad . \tag{6}$$

Let $X_1, \cdots, X_n$ be independent random variables and $X = \sum^n X_i$. If $E[X_i^{k_i}]$ exists for all $0 \le k_i \le k$, $i = 1, \cdots, n$, then

$$E[X^k] = \sum_{k_1+\cdots+k_n=k} k! \prod_{i=1}^{n} \frac{E[X_i^{k_i}]}{k_i!} . \tag{7}$$

Traffic Models

Now, we study the commonly-used regulated and on-off Markov traffic models. Let $\{A_{i,0}[t], t > 0\}$ denote flow-i source traffic process, i.e., $A_{i,0}[t]$ is the total traffic generated by the traffic source during $[0,t]$. Without loss of generality, we make two basic assumptions about traffic sources.

I. Stationary: For any flow i and any non-negative numbers t_1, t_2, τ, and x, the traffic source process $A_{i,0}$ satisfies

$$\Pr\left\{A_{i,0}[t_1 + \tau] - A_{i,0}[t_1] > x\right\} = \Pr\left\{A_{i,0}[t_2 + \tau] - A_{i,0}[t_2] > x\right\}$$

II. Independence: Traffic source processes $A_{i,0}$ and $A_{j,0}$ are stochastically independent for $i \neq j$.

Regulated Traffic: A traffic flow $A_{i,0}$ is *regulated* by a nondecreasing, nonnegative, sub-additive function A_i^* if

$$\forall\, t, \tau \ge 0 : A_{i,0}[t + \tau] - A_{i,0}[t] \le A_i^*(\tau) . \tag{8}$$

One example for such traffic is the leaky bucket controlled traffic with $A_i^*(\tau) = \beta + \rho\tau$, where β is the bucket size and ρ is the token generation rate. Let $X_i(\tau) = A_{i,0}[t+\tau] - A_{i,0}[t]$, according to (Kelly 1996), the moment-generating function of $X_i(\tau)$ is given as:

$$E[e^{sX_i(\tau)}] = 1 + \frac{\rho_i}{A_i^*(\tau)}(e^{sA_i^*(\tau)} - 1), \tag{9}$$

Where

$$\rho_i = \lim_{\tau\to\infty} \frac{A_i^*(\tau)}{\tau} .$$

For n identical and independent traffic sources with a regulated function A^* and an average traffic arriving rate ρ, let $\hat{X}(\tau)$ denote the total traffic arrivals from these n traffic sources during a time interval with length τ. Based on Formula (9),

$$E\left[e^{s\hat{X}(\tau)}\right] = \left[1 + \frac{\rho\tau}{A^*(\tau)}\left(e^{sA^*} - 1\right)\right]^n$$

By Proposition 1 and simple calculus manipulations, we have a formula for the kth moment of $\hat{X}(\tau)$ as follows:

$$E\left[\left(\hat{X}(\tau)\right)^k\right] = \sum_{i=0}^{n-1} \frac{n!\left(A^*(\tau)\right)^k}{(n-i)!\,i!}\left(1 - \frac{\rho\tau}{A^*(\tau)}\right)^i \left(\frac{\rho\tau}{A^*(\tau)}\right)^{n-i} (n-i)^k. \tag{10}$$

Markov On-Off Traffic: A stationary On-Off traffic source is generally characterized as a two-state Markov chain which is described in terms of the probability transition matrix $P = \begin{pmatrix} p_1 & 1-p_1 \\ 1-p_2 & p_2 \end{pmatrix}$ and the stationary distribution vector $\pi = (\pi_1, \pi_2)$. Here, 1- p_1 and 1- p_2 are the transition probabilities from state 'On' to state 'Off' and from state 'Off' to state 'On', respectively. p_1 and p_2 are the probabilities for staying in state 'On' and in state 'Off', respectively. π satisfies $\pi = \pi P$.

In state 'On', the traffic arrival is generated at the peak rate R, while in state 'Off', no traffic arrival occurs. Let $\{A[t], t > 0\}$ denote an on-off traffic source process, i.e., $A[t]$ is the total traffic generated by the on-off source in $[0, t]$. Let $Y(\tau) = A[t+\tau] - A[t]$, according to (Chang, 2000), the moment-generating function of Y(τ) is given as:

$$E[e^{sY(\tau)}] = \pi[\Phi(s)\mathrm{P}]^{\tau-1}\Phi(s)\mathrm{I}, \quad (11)$$

where $\Phi(s) = \begin{pmatrix} e^{Rs} & 0 \\ 0 & 1 \end{pmatrix}$ and $\mathrm{I} = \begin{pmatrix} 1 \\ 1 \end{pmatrix}$. By simple calculus manipulations, we have

$$E\left[\left(Y(\tau)\right)^k\right] = R^k\left(\pi_1\sum_{h=0}^{\tau-1} a_{\tau-1,h}(h+1)^k + \pi_2\sum_{h=1}^{\tau-1} b_{\tau-1,h}h^k\right), \quad (12)$$

where coefficients $a_{\tau\text{-}1,h}$ and $b_{\tau\text{-}1,h}$, $h=0, \cdots, \tau\text{-}1$, are determined by the following iterative equations:

$$a_{0,0} = 1; \qquad b_{0,0} = \; ;0$$
$$a_{1,0} = 1 - p_{1,} a_{1,1} = p_1; \qquad b_{1,0} = p_2, b_{1,1} = 1 - p_2;$$
$$\vdots$$
$$a_{j,0} = (1-p_1)b_{j-1,0}, \cdots, a_{j,h}$$
$$= p_1 a_{j-1,h-1} + (1-p_1)b_{j-1,h}, \cdots, a_{j,j} = p_1 a_{j-1,j-1};$$
$$b_{j,0} = p_2 b_{j-1,0} \cdots, b_{j,h} = (1-p_2)a_{j-1,h-1} + p_2 b_{j-1,h}, \cdots, b_{j,j} = (1-p_2)a_{j-1,j-1};$$
$$\vdots$$
$$a_{\tau-1,0} = (1-p_1)b_{\tau-2,0}, \cdots, a_{\tau-1,h} = p_1 a_{\tau-2,h-1} + (1-p_1)b_{\tau-2,h}, \cdots, a_{\tau-1,\tau-1} = p_1 a_{\tau-2,\tau-2};$$
$$b_{\tau-1,0} = p_2 b_{\tau-2,0} \cdots, b_{\tau-1,h} = (1-p_2)a_{\tau-2,h-1} + p_2 b_{\tau-2,h}, \cdots, b_{\tau-1,\tau-1} = (1-p_2)a_{\tau-2,\tau-2}.$$

For n identical and independent Markov On-Off traffic sources, let $\hat{Y}(\tau)$, denote the total traffic arrivals from these n traffic sources during a time interval with lengthτ. Based on Formula (11), $E[e^{s\hat{Y}(\tau)}] = \left[\pi\left[\Phi(s)\,\mathrm{P}\right]^{\tau-1}\Phi(s)\,\mathrm{I}\right]^n$.

By Proposition 1 and the generalized Leibniz's law, we have a formula for the kth moment of $\hat{Y}(\tau)$ as the following:

$$E[(\hat{Y}(\tau))^k] = \sum_{q_1+\cdots+q_n=k} k!\prod\frac{E[(Y(\tau))^{q_i}]}{q_i!}, \quad (13)$$

where $E[(Y(\tau))^n]$ is defined by Formula (12).

A Challenge Problem

Consider traffic flow i with the path [*mesh router* i_1, *mesh router* i_2, ⋯⋯, *mesh router* i_{p+1}] over a WMN and let $A_{i,j}[t]$ denote the total flow-i traffic arriving at its jth hop, i.e., mesh router i_j, during $[0,t]$. To simplify notations, we also use $[l(i,1), l(i,2), \cdots, l(i,p)]$ to denote flow-i path, where $l(i,j)$ is the wireless link from mesh router i_j to mesh router i_{j+1}, $j=1,2,\cdots,p$. A challenge for supporting delay-sensitive applications such as VoIP, video streaming, and interactive gaming over WMNs, is to predict the end-to-end delay experienced by traffic flow i, i.e., for ε in (0,1), to find $D^{\varepsilon,i}$ such that for all $t>0$,

$$\Pr\begin{pmatrix}\text{end to end delay suffered by} \\ \text{any tagged flow-}i\text{ arrival} > \mathfrak{D}^{\varepsilon,i}\end{pmatrix} \leq \varepsilon\,.$$

III. A THEORETIC FRAMEWORK FOR PREDICTING END-TO-END DELAY

Now, we provide a set of algorithms for predicting the end-to-end delays experienced by traffic flows over multi-radio and multi-channel WMNs.

Ingress Access Point Delay

To analyze the delays experienced by traffic at ingress access points, we need to study the packet drop probability experienced by traffic flows at an ingress access point and the capacity of an ingress access point.

Packet Drop Probability: Generally, the packet transmission error probability depends on the adopted modulation and coding scheme (*mc*), the packet size (*L*), and the SNR. Without loss of generality, we denote it as *PER*(SNR, *mc*, *L*). It is worth noting that two experiment-based packet error probability formulas for OFDM system are recently derived in (Awoniyi, 2005). In this chapter, according to the time and space diversity, we assume that at each ingress access point, packet transmission error events during one SI time interval are independent. Let

$$p_{i,1} = PER(SNR_i, mc_i^1, L_{CF\_Poll})$$

be the transmission error probability of a CF_Poll frame with size L_{CF_Poll} for flow i. Let

$$p_{i,2} = PER(SNR_i, mc_i^2, L_{i,data})$$

be the transmission error probability of a data frame with size $L_{i,data}$ for flow i. Let

$$p_{i,3} = PER(SNR_i, mc_i^3, L_{ack})$$

be the transmission error probability of an ACK packet with size L_{ACK} for flow i. The probability for a flow-i data packet needed to be retransmitted is bounded by

$$p_i = 1 - (1 - p_{i,1})(1 - p_{i,2})(1 - p_{i,3}) \, .$$

If the number of retransmissions of a packet at each ingress access point is limited by n_{drop}, the probability of a flow-i packet being dropped is given as:

$$p_{i,drop} = p_i^{n_{drop}+1} \, . \tag{14}$$

In this chapter, we use $n_{drop} = 4$. The obtained results can be extended to $n_{drop} > 4$ or $n_{drop} < 4$ without any technical difficulty.

Capacity of Ingress Access Point: Based on the reference admission control algorithm provided in (IEEE std, 2007) for IEEE 802.11 MAC with the HCCA, n traffic flows can be supported by an ingress access point if the following inequality holds:

$$\sum_{i=1}^{n} \frac{TXOP_i}{SI} + \frac{TXOP_{retrans.}}{SI} \le 1 - \frac{T_{cp}}{T}, \tag{15}$$

where $TXOP_i$ is the time duration reserved for flow i and includes the transmission time for the data, ACK and CF_Poll frames, as well as the required inter frame spaces; SI denotes the length of a service interval; T_{cp} denotes the available contention-based channel access time during each beacon interval; T is the beacon interval length; and $TXOP_{retrans}$. is the time reserved for the retransmissions of corrupted packets during each SI time interval.

Due to the space and time diversity, it is a rare event that all transmissions during one SI time interval fail.

Thus, we can exploit the multi-user diversity and stochastic multiplexing gains to compute $TXOP_{retrans}$..

Without loss of generality, we assume that the packet size of a traffic flow is a constant, but the different traffic flows may have different packet sizes. Also, based on the sizes of TXOPs, we categorized these n traffic flows into M classes, $S_1, S_2, \ldots\ldots, S_M$ such that

$$TXOP_{i_1} = TXOP_{i_2} \text{ if } i_1, i_2 \in S_i$$

for some i in {1,2,…,M}. Therefore, for a given $\varepsilon_0 \in (0,1)$ and during one SI time interval, the number of class-i data packets that corrupt in their first transmission is not larger than $n_{i,1}^{\varepsilon_0}$ with probability

$1 - \varepsilon_0$, where $n_{i,1}^{\varepsilon_0}$ is given as follows:

$$n_{i,1}^{\varepsilon_0} = \min\left\{ k \mid \sum_{j=k+1}^{|S_i|} \sum_{\substack{S \subset S_i \\ |S|=j}} \prod_{h \in S} p_h \prod_{h \notin S_i \setminus S} (1 - p_h) \le \varepsilon_0 \right\} \tag{16}$$

Furthermore, let S_i^h denote the set of these $n_{i,h}^{\varepsilon_0}$ class-i packets that are needed to be retransmitted after their h^{th} retransmissions, where $h=1,\ldots\ldots, n_{drop}$. Then, with probability $1 - \varepsilon_0$, the number of class-i data packets that corrupt in their h^{th} transmissions is not larger than $n_{i,h+1}^{\varepsilon_0}$, where $n_{i,h+1}^{\varepsilon_0}$ is given as follows:

$$n_{i,h}^{\varepsilon_0} = \min\left\{ k \mid \sum_{j=k+1}^{|S_i^h|} \sum_{\substack{S \subset S_i^h \\ |S|=j}} \prod_{m \in S} p_m \prod_{l \in S_i^h \setminus S} (1 - p_l) \le \varepsilon_0 \right\} \quad (17)$$

Therefore

$TXOP_{retrans.} = \sum_{i=1}^{M} \sum_{h=1}^{n_{drop}} n_{i,h}^{\varepsilon_0} \times TXOP_{k_i}$, where $k_i \in S_i$.

After the above discussion, we begin to evaluate the probabilistic bound on the delays experienced by traffic flows at ingress access points.

Ingress Access Delay: According to the HCCA channel access scheme, during any time interval $[t, t+\tau]$, $\frac{\ddot{A}}{SI} \times L_{i,data}$ amount of flow-i traffic will be either successfully delivered or dropped. A dropped packet can be treated as experiencing an infinite delay. In the following lemma, we provide a formula to evaluate the probabilistic bound $D_{l(i,0)}^{\varepsilon_{i,0}}$ with violation probability $\varepsilon_{i,0}$ on the delays experienced by flow-i traffic at the ingress access point.

Lemma 1: $D_{l(i,0)}^{\varepsilon_{i,0}}$ is given as follows:

$$D_{l(i,0)}^{\varepsilon_{i,0}} = \inf\left\{ d \mid \sup_{x \in [0,t]} \inf_{n \ge 0} \left(\frac{E\left[\left(A_{i,0}[t] - A_{i,0}[x]\right)^n\right]}{\left(\frac{t-x+d}{SI} L_{i,data}\right)^n} \right) \le \eta, \forall t > 0 \right\},$$

where $L_{i,\mathrm{data}}$ is the packet size of flow i; $\varepsilon_{i,0} = p_{i,drop} + \eta + n_{drop} \times \varepsilon_0$; $p_{i,drop}$ is the probability of a flow-i packet being dropped at the ingress access point after experiencing n_{drop} retransmissions and can be determined by Formula (14); $n_{drop} \times \varepsilon_0$ is the probability of a flow-i packet being dropped before n_{drop} retransmissions.

Proof: The proof is similar to and much simpler than that of Theorem 1 in the following, and omitted.

Point-to-Point Wireless Link Delay

Now, we consider the jth hop of flow i, i.e., the point-to-point wireless link $l(i,j)$ from mesh router i_j to mesh router i_{j+1}. For a tagged traffic arrival of flow i arriving at mesh router i_j at time t, let $\tau(t)$ denote the last time before time t when there is no traffic arrival at mesh router i_j to traverse wireless link $l(i, j)$. To simplify notations, let $\mathbf{A}^{l(i,j)}[\tau(t), t]$ denote all traffic arrivals that arrive at mesh router i_j during $[\tau(t), t]$ to traverse over wireless link $l(i,j)$, i.e.,

$$\mathbf{A}^{l(i,j)}[\tau(t), t] = \sum_{g \in Q_{l(i,j)}} \left(A_{g,g_m}[t] - A_{g,g_m}[\tau(t)] \right) \quad (18)$$

where $Q_{l(i,j)}$ denotes the set of all traffic flows traversing wireless link $l(i,j)$, and for $g \in Q_{l(i,j)}$, g_m is defined by $l(g, g_m) = l(i,j)$, i.e., $l(i,j)$ is the g_m^{th} hop of flow g. Hence, the tagged traffic arrival is successfully received or dropped before time $t+d$ if the total traffic arriving at mesh router i_j during $[\tau(t),t]$ to traverse $l(i,j)$ are successfully received or dropped by mesh router i_{j+1} before time $t+d$. Moreover, the available channel capacity during $[\tau(t), t+d]$ for wireless link $l(i,j)$ is $C[t+d] - C[\tau(t)]$, where $C[t]$ is characterized by Formula (4). Thus, a sufficient condition for the tagged traffic arrival being successfully transmitted before time $t+d$ is the following:

$$\overbrace{C[t+d] - C[\tau(t)]}^{\substack{\text{Available Channel} \\ \text{Capacity During} \\ [\tau(t),t+d]}} - \overbrace{\mathbf{A}^{l(i,j)}[\tau(t), t]}^{\substack{\text{Cumulative Traffic} \\ \text{Arrivals HAve To Be} \\ \text{Transmitted During} \\ [\tau(t),t]}} \ge 0. \quad (19)$$

Let $D_{l(i,j)}^{\varepsilon_{i,j}}$ denote a probabilistic bound with violation probability $\varepsilon_{i,j}$ for delays experienced by traffic flow i when traversing wireless link $l(i,j)$. That is, for all $t > 0$

$$\Pr\left\{ \left(C[t + D_{l(i,j)}^{\varepsilon_{i,j}}] - C[\tau(t)] \right) - A^{l(i,j)}[\tau(t), t + D_{l(i,j)}^{\varepsilon_{i,j}}] \ge 0 \right\} \ge 1 - \varepsilon_{i,j}.$$

By the large deviation technique, we provide the following theorem to estimate the probabilistic bound on the delays experienced by traffic arrivals at the point-point wireless link *l*(*i,j*).

Theorem 1: $D_{l(i,j)}^{\varepsilon_{i,j}}$ is given as follows:

$$D_{l(i,j)}^{\varepsilon_{i,j}} = \inf\left\{ d \mid \sup_{x\in[0,t]} \inf_{n\geq 0} \left(\frac{E\left[\left(\mathbf{A}^{l(i,j)}[x,t]\right)^n\right]}{\left(S^{\eta_1}(t-x+d)\right)^n} \right) \leq \eta_2, \forall t > 0 \right\}, \tag{20}$$

where $\varepsilon_{i,j} = \eta_1 + \eta_2 + p_{l(i,j),drop}$; $p_{l(i,j),drop}$ is the probability of a traffic arrival being dropped at the point-to-point wireless link *l*(*i,j*) and can be determined by a formula that is similar to Formula (14); η_1 is related with $S^{\eta_1}(t-x+d)$ that is defined in Formula (4); and $\mathbf{A}^{l(i,j)}[x,t]$ is defined in Formula (18).

Proof: Consider a tagged traffic arrival of flow *i* arriving to the transmitter of wireless link *l*(*i,j*) at time *t*.

First, if this tagged traffic arrival will not dropped due to multiple transmission errors, according to Inequality (19), the probability of violating a delay bound $D_{l(i,j)}^{\varepsilon_{i,j}}$ for this tagged traffic arrival is upper bounded by the following:

$$\begin{aligned}
&\Pr\left\{\left(C[t+D_{l(i,j)}^{\varepsilon_{i,j}}] - C[\tau(t)]\right) - \mathbf{A}^{l(i,j)}[\tau(t),t] < 0\right\} \\
&\overset{(a)}{\leq} \Pr\left\{\begin{array}{l}\left\{\left(C[t+D_{l(i,j)}^{\varepsilon_{i,j}}] - C[\tau(t)]\right) - \mathbf{A}^{l(i,j)}[\tau(t),t] < 0\right\} \\ \cap \left\{\left(C[t+D_{l(i,j)}^{\varepsilon_{i,j}}] - C[\tau(t)]\right) - S^{\eta_1}(t-\tau(t)+D_{l(i,j)}^{\varepsilon_{i,j}}) \geq 0\right\}\end{array}\right\} \\
&\quad + \Pr\left\{\left(C[t+D_{l(i,j)}^{\varepsilon_{i,j}}] - C[\tau(t)]\right) - S^{\eta_1}(t-\tau(t)+D_{l(i,j)}^{\varepsilon_{i,j}}) < 0\right\} \\
&\overset{(b)}{\leq} \Pr\left\{\mathbf{A}^{l(i,j)}[\tau(t),t] - S^{\eta_1}(t-\tau(t)+D_{l(i,j)}^{\varepsilon_{i,j}}) > 0\right\} \\
&+ \Pr\left\{\left(C[t+D_{l(i,j)}^{\varepsilon_{i,j}}] - C[\tau(t)]\right) - S^{\eta_1}(t-\tau(t)+D_{l(i,j)}^{\varepsilon_{i,j}}) < 0\right\} \\
&\overset{(c)}{\leq} \Pr\left\{\sup_{x\in[0,t]}\left[\mathbf{A}^{l(i,j)}[x,t] - S^{\eta_1}(t-x+D_{l(i,j)}^{\varepsilon_{i,j}})\right] > 0\right\} \\
&+ \Pr\left\{\sup_{x\in[0,t]}\left[\left(C[t+D_{l(i,j)}^{\varepsilon_{i,j}}] - C[\tau(t)]\right) - S^{\eta_1}(t-\tau(t)+D_{l(i,j)}^{\varepsilon_{i,j}})\right] < 0\right\} \\
&\overset{(d)}{\approx} \sup_{x\in[0,t]} \Pr\left\{\left[\mathbf{A}^{l(i,j)}[x,t] - S^{\eta_1}(t-x+D_{l(i,j)}^{\varepsilon_{i,j}})\right] > 0\right\} \\
&+ \sup_{x\in[0,t]} \Pr\left\{\left(C[t+D_{l(i,j)}^{\varepsilon_{i,j}}] - C[x]\right) - S^{\eta_1}(t-x+D_{l(i,j)}^{\varepsilon_{i,j}}) < 0\right\} \\
&\overset{(e)}{\leq} \sup_{x\in[0,t]} \inf_{n\geq 0} \frac{E\left[\left(\mathbf{A}^{l(i,j)}[x,t]\right)^n\right]}{\left(S^{\eta_1}(t-x+D_{l(i,j)}^{\varepsilon_{i,j}})\right)^n} + \eta_1 \\
&\overset{(f)}{\leq} \eta_1 + \eta_2,
\end{aligned} \tag{21}$$

where (a) is supported by that for any two random events X and Y, $\Pr\{X\} \leq \Pr\{X \cap Y\} + \Pr\{Y^c\}$, where Y^c denotes the complement of Y; (b) is based on that

$$\begin{aligned}
&\left\{\left(C[t+D_{l(i,j)}^{\varepsilon_{i,j}}] - C[\tau(t)]\right) < \mathbf{A}^{l(i,j)}[\tau(t),t]\right\} \\
&\cap \left\{\left(C[t+D_{l(i,j)}^{\varepsilon_{i,j}}] - C[\tau(t)]\right) \geq S^{\eta_1}(t-\tau(t)+D_{l(i,j)}^{\varepsilon_{i,j}})\right\} \\
&\subseteq \left\{\mathbf{A}^{l(i,j)}[\tau(t),t] > S^{\eta_1}(t-\tau(t)+D_{l(i,j)}^{\varepsilon_{i,j}})\right\};
\end{aligned}$$

(c) comes from $\tau(t) \in [0,t]$; (d) follows the approximation described in Equation (11) in (Knight & Shroff, 1999); (e) is due to Formula (4) and Inequality (5); (f) is obtained from Formula (20).

Second, considering that this tagged traffic arrival is dropped after n_{drop} transmission errors. Since a dropped traffic arrival can be treated as experiencing an infinite delay, the delay bound violation probability for the tagged traffic arrival in this case is the dropped probability $p_{l(i,j),drop}$.

Therefore, combining aforementioned cases, the probability of violating delay bound $D_{l(i,j)}^{\varepsilon_{i,j}}$ for the tagged traffic arrival experiencing at the point-to-point wireless link *l*(*i,j*) is

$$\varepsilon_{i,j} = \eta_1 + \eta_2 + p_{l(i,j),drop}.$$

Remark: $D_{l(i,j)}^{\varepsilon_{i,j}} = D_{l(g,g_m)}^{\varepsilon_{i,j}}$ for $l(i,j) = l(g,g_m)$, i.e., the traffic flows that traverse the same point-to-point wireless link with the first come first serve scheduling discipline share the same probabilistic delay bound.

Characterization of Down Stream Traffic

To apply Theorem 1 to estimate probabilistic bounds on delays experienced by traffic arrivals at down stream mesh routers, the traffic characteristics such as moments at down stream mesh routers are needed. Usually, traffic flows will be distorted after traversing one hop. Without reshaping the distorted traffic at down stream mesh routers, the characteristics of down stream traffic arrivals are not the same as that of upstream traffic arrivals. According to the IEEE 802.11 MAC with the HCCA medium access mechanism at ingress access points and the probabilistic bounds for delays experienced by traffic flows at upstream mesh routers, i.e., $D_{l(i,h)}^{\varepsilon_{i,h}}$, $h = 1,\cdots, j-1$,

$$\Pr\left\{\begin{matrix}\text{delay suffered by any traffic arrival} \\ \text{of flow } i \text{ over } l(i,h) \ > D_{l(i,h)}^{\varepsilon_{i,h}}\end{matrix}\right\} \le \varepsilon_{i,h},$$

we provide the following algorithms to characterize traffic flows at downstream mesh routers.

Based on Lemma 1, we have the following algorithm to characterize traffic flows at their first hop (ingress access points).

Lemma 2: For any flow i, $\forall t > x \ge 0$,

$$\Pr\left\{\left(A_{i,1}[t] - A_{i,1}[x]\right) > \left(A_{i,0}[t] - A_{i,0}[x - D_{l(i,0)}^{\varepsilon_{i,0}}]\right)\right\} \le \varepsilon_{i,0}, \quad (22)$$

where $D_{l(i,0)}^{\varepsilon_{i,0}}$ and $\varepsilon_{i,0}$ are determined in Lemma 1.

Proof: The proof is similar to that of Lemma 3.

Remark: For a Markov On-Off traffic source, if SI equals the minimum packet inter-arrival time, e.g., *SI* = packetization time of voice signal, we have a simpler characterization of $A_{i,1}[t]$ - $A_{i,1}[x]$ as follows:

$$A_{i,1}[t] - A_{i,1}[x] \le A_{i,0}[t] - A_{i,0}[x - SI].$$

Considering the point-to-point wireless link $l(i,j)$ that is on the jth hop of flow-i path. Since there may be other traffic flows that share this link with flow i, according to traffic flows` paths, we partition all flows, which traverse wireless link $l(i, j)$ and are denoted by $Q_{l(i,j)}$, into $\mathbf{K}$ sets, i.e., $\mathbf{W}_m$ $m = 1,2,\cdots,\mathbf{K}$, with the following properties.

1. All flows in $\mathbf{W}_m$ share the same path [mesh router q_1, mesh router q_2,……, mesh router q_{km}], i.e., mesh router q_h is the hth hop for all flows in $\mathbf{W}_m$ for h=1,2,……,k_m;
2. $\mathbf{W}_{m_1} \cap \mathbf{W}_{m_2}$ is empty if $m_1 \ne m_2$ and for any integer $m \le \mathbf{K}$;
3. Mesh router q_{k_m} = mesh router i_j and $l(q, k_m) = l(i, j)$ for $q \in \mathbf{W}_m$;
4. $Q_{l(i,j)} = \bigcup_{m=1}^{\mathbf{K}} \mathbf{W}_m$.

Lemma 3: For any given $\mathbf{W}_m$, $\forall t > x \ge 0$,

$$\Pr\{\sum_{q\in\mathbf{W}_m}\left(A_{q,k_m}[t] - A_{q,k_m}[x]\right) > \sum_{q\in\mathbf{W}_m}(A_{q,1}[t] - A_{q,1}[x - \sum_{h=1}^{k_m-1} D_{l(q,h)}^{\varepsilon_{q,h}}])\} \le \sum_{h=1}^{k_m-1}\varepsilon_{q,h}, \quad (23)$$

where $D_{l(q,h)}^{\varepsilon_{q,h}}$ and $\varepsilon_{q,h}$ are determined by Theorem 1.

Proof: Consider the first traffic arrival (tagged traffic arrival) at mesh router q_{k_m} from flows in $\mathbf{W}_m$ during [x, t]. Let $\kappa(x)$ and s(x) denote its arrival times at mesh router q_{k_m} and mesh router q_1, respectively. So $\kappa(x) \ge x$ and no traffic arrival at mesh router q_{k_m} from flows in $\mathbf{W}_m$ during $[x, \kappa(x))$. It is worth noting that at each mesh router, the traffic arrivals from flows in $\mathbf{W}_m$ are served in the first-in first-out order. Hence, at mesh router q_{k_m}, for the traffic arrivals of flows in $\mathbf{W}_m$ with arriving times in [x, t], their corresponding arriving times at mesh router q_1 are not earlier than $s(x)$. So

$$\begin{aligned}&\sum_{q\in\mathbf{W}_m}\left(A_{q,k_m}[t] - A_{a,k_m}[x]\right) \\ &= \sum_{q\in\mathbf{W}_m}\left(A_{q,k_m}[t] - A_{q,k_m}[\kappa(x)]\right) \\ &\le \sum_{q\in\mathbf{W}_m}\left(A_{q,1}[t] - A_{q,1}[s(x)]\right).\end{aligned}$$

According to the aforementioned definition of $D_{l(q,h)}^{\varepsilon_{q,h}}$ and the fact that $\kappa(x) - s(x)$ is the total delay experienced by the tagged traffic arrival before arriving at mesh router q_{k_m}, we have

$$\Pr\left\{s(x) < \kappa(x) - \sum_{h=1}^{k_m-1} D_{l(q,h)}^{\varepsilon_{q,h}}\right\} \le \sum_{h=1}^{k_m-1} \Pr\left\{\begin{matrix}\text{delay suffered by the tagged} \\ \text{arrival at its h}^{th} \text{ hop } > D_{l(q,h)}^{\varepsilon_{q,h}}\end{matrix}\right\} \le \sum_{h=1}^{k_m-1} \varepsilon_{q,h}, \quad (24)$$

Thus, we have

$$\begin{aligned}
&\Pr\left\{\sum_{q\in\mathbf{W}_m}\left(A_{q,k_m}[t] - A_{q,k_m}[x]\right) > \sum_{q\in\mathbf{W}_m}(A_{q,1}[t] - A_{q,1}[x - \sum_{h=1}^{k_m-1} D_{l(q,h)}^{\varepsilon_{q,h}}])\right\} \\
&\overset{(a)}{=}\Pr\left\{\sum_{q\in\mathbf{W}_m}\left(A_{q,k_m}[t] - A_{q,k_m}[\kappa(x)]\right) > \sum_{q\in\mathbf{W}_m}(A_{q,1}[t] - A_{q,1}[x - \sum_{h=1}^{k_m-1} D_{l(q,h)}^{\varepsilon_{q,h}}])\right\} \\
&\overset{(b)}{\le}\Pr\left\{\sum_{q\in\mathbf{W}_m}\left(A_{q,k_m}[t] - A_{q,k_m}[\kappa(x)]\right) > \sum_{q\in\mathbf{W}_m}(A_{q,1}[t] - A_{q,1}[\kappa(x) - \sum_{h=1}^{k_m-1} D_{l(q,h)}^{\varepsilon_{q,h}}])\right\} \\
&\overset{(c)}{\le}\Pr\left\{\sum_{q\in\mathbf{W}_m}\left(A_{q,k_m}[t] - A_{q,1}[s(x)]\right) > \sum_{q\in\mathbf{W}_m}(A_{q,1}[t] - A_{q,1}[\kappa(x) - \sum_{h=1}^{k_m-1} D_{l(q,h)}^{\varepsilon_{q,h}}])\right\} \\
&\overset{(d)}{\le}\Pr\left\{s(x) < \kappa(x) - \sum_{h=1}^{k_m-1} D_{l(q,h)}^{\varepsilon_{q,h}}\right\} \\
&\overset{(e)}{\le}\sum_{h=1}^{k_m-1}\varepsilon_{q,h},
\end{aligned} \quad (25)$$

where (a) is due to no traffic arriving at mesh router q_{k_m} from flows in $\mathbf{W}_m$ during $[x, \kappa(x))$ (b) comes from $\kappa(x) \ge x$;(c) is supported by the definition of $s(x)$ and the fact that all traffic arriving at mesh router q_1 from flows in $\mathbf{W}_m$ during [0, $s(x)$) have arrived at mesh router q_{k_m} at time $\kappa(x)$, i.e., $\sum_{q\in\mathbf{W}_m} A_{q,k_m}[\kappa(x)] \ge \sum_{q\in\mathbf{W}_m} A_{q,1}[s(x)]$; (d) is based on that $A_{q,k_m}[t] \le A_{q,1}[t]$ and $A_{q,1}[s(x)] \ge A_{q,1}[\kappa(x) - \sum_{h=1}^{k_m-1} D_{l(q,h)}^{\varepsilon_{q,h}}]$, if $s(x) \ge \kappa(x) - \sum_{h=1}^{k_m-1} D_{l(q,h)}^{\varepsilon_{q,h}}$; (e) follows Inequality (24).

According to Lemma 3, the aggregated traffic arrivals at the point-to-point wireless link $l(i, j)$, which is denoted by $\mathbf{A}^{l(i,j)}[x,t]$ and defined in Equation (18) and used in Theorem 1, can be probabilistically characterized by the corresponding traffic flows in their ingress mesh routers as follows:

Corollary 1: $\Pr\left\{\mathbf{A}^{l(i,j)}[x,t] > \hat{\mathbf{A}}^{l(i,j)}[x,t]\right\} \le \sum_{m=1}^{\mathbf{K}}\sum_{h=1}^{k_m}\varepsilon_{q,h}$ where $q \in \mathbf{M}_m$; k_m is determined by $l(q,k_m)=l(i,j)$ and

$$\hat{\mathbf{A}}^{l(i,j)}[x,t] = \sum_{q\in Q_{l(i,j)}}\left(A_{q,1}[t] - A_{q,1}[x - \sum_{h=1}^{k_m-1} D_{l(q,h)}^{\varepsilon_{q,h}}]\right) \quad (26)$$

Proof: The proof is based on Formulas (18) and (26), and Lemma 3, and is simple and omitted.

Now, according to the above results, the probabilistic bounds on the delays experienced by traffic flows at the downstream wireless links can be evaluated only based on the characteristics of their traffic sources as follows.

Corollary 2: For wireless link $l(i,j)$ with traffic arrivals from upstream mesh routers, $D_{l(i,j)}^{\varepsilon_{i,j}}$ is given as follows:

$$D_{l(i,j)}^{\varepsilon_{i,j}} = \inf\left\{d \mid \sup_{x\in[0,t]}\inf_{n\ge 0}\left(\frac{E\left[\left(\hat{\mathbf{A}}^{l(i,j)}[x,t]\right)^n\right]}{\left(S^{\eta_1}(t-x+d)\right)^n}\right) \le \eta_2, \forall t > 0\right\}$$

where $\varepsilon_{i,j} = \eta_1 + \eta_2 + d_{l(i,j),drop} + \sum_{m=1}^{\mathbf{K}}\sum_{h=1}^{k_m}\varepsilon_{q,h}$; $q \in \mathbf{W}_m$ and k_m is determined by $l(q,k_m)=l(i,j)$; η_1 and $d_{l(i,j),drop}$ are the same as that in Theorem 1.

Proof: The proof is similar to that of Theorem 1 and Corollary 1 and omitted.

Probabilistic Bound on End-to-End Delay

After obtaining probabilistic bounds on local delays experienced by traffic arrivals at ingress access points and point-to-point wireless links, a probabilistic bound on the end-to-end delays experienced by flow-*i* traffic arrivals traversing through the path [*mesh router* i_1,*mesh router* i_2,… …,*mesh router* i_{p+1}] is given in the following.

Theorem 2: A bound $\mathfrak{D}^{\varepsilon,i}$ with violation probability ε for the end-to-end delays experienced by flow-*i* traffic is given as follows:

$$\mathfrak{D}^{\varepsilon,i} = D_{l(i,0)}^{\varepsilon_{i,0}} + \sum_{h=1}^{i_p} D_{l(i,h)}^{\varepsilon_{i,h}}, \quad \varepsilon = \varepsilon_{i,0} + \sum_{h=1}^{i_p} \varepsilon_{i,h},$$

where $D_{l(i,0)}^{\varepsilon_{i,0}}$ and $\varepsilon_{i,0}$ are determined by Lemma 1, while $D_{l(i,h)}^{\varepsilon_{i,h}}$ and $\varepsilon_{i,h}$ are determined by Theorem 1 and Corollary 2.

Proof: The proof is simple and omitted.

IV. APPLICATIONS

In this section, we present several examples based on the theoretical framework described in the previous sections. Our goal is to show how to apply the obtained results to predict the probabilistic bounds on the end-to-end delays experienced by traffic flows over WMNs.

System setting: We consider an IEEE 802.11a based WMN backbone with the parameters of PHY and MAC layers provided in (IEEE Std, 1999). We assume the existence of a routing and channel assignment algorithm, such as those proposed in (Draves, Padhye, & Zill, 2004; Xing, Chen, Ma, & Liang, 2007), as well as enough of non-overlapping wideband channels and radios per mesh router for interference free multihop wireless communications over this WMN backbone.

Overhead of MAC and PHY layers: Continuous data frame transmissions between two adjacent mesh routers over an IEEE 802.11a channel are illustrated in Figure 4. As described in the specifications of PHY and MAC layers of IEEE 802.11a standard, the overhead of MAC and PHY layers for a data packet is 33 bytes and 6 bits, while overhead of PHY layer for an ACK packet is 46 bits.

Traffic parameters: We consider the Regulated and Markov On-Off traffic. For Markov On-Off traffic, we consider the output traffic from a G729 encoder with 40 ms packetization interval which digitizes every 40 ms voice signal sample into a data packet with 40 bytes. Thus, the corresponding peak rate is 8 Kbps. Without loss of generality, we assume that a voice activity detector that stops sending data during silence durations is used and the average talkspurt and silence durations of a speaker are 1.34 *s* and 1.67 *s*, respectively. Thus, based on the relationships, i.e., average talkspurt duration =(packetization interval)/(1-p_1) and average silence duration = (packetization interval)/(1-p_2), the probability transition matrix ***P*** and the stationary distribution vector π can be determined for the discrete-time On-Off Markov traffic model described in Section II-E. For the Regulated traffic, we assume that the peak rate is 6 Mbps, the average rate is 0.15 Mbps, and the burst size is 8.192 kb. Note that the overhead of RTP/UTP/IP protocols for a data packet is 40 bytes.

Effective Channel Capacity

For each of the non-overlapping IEEE 802.11a wideband channels between adjacent mesh routers, based on its channel capacity process $C[t]$ obtained by a system-level simulator built on the top of the OFDM simulator for IEEE 802.1a PHY provided in (Heiskala & Terry, 2001), we estimate the effective capacity defined by Equation (4).

Figure 4. Continuous data frame transmissions over a point-to-point wireless link (IEEE Std, 2007)

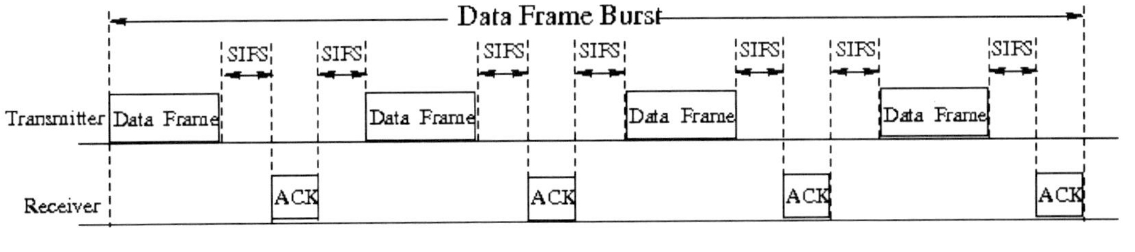

Figure 5. Effective capacities with SNR = 20 dB, packet size=512 Bytes and ε =10^-3 (Rician factor K = 5 dB for Channel D)

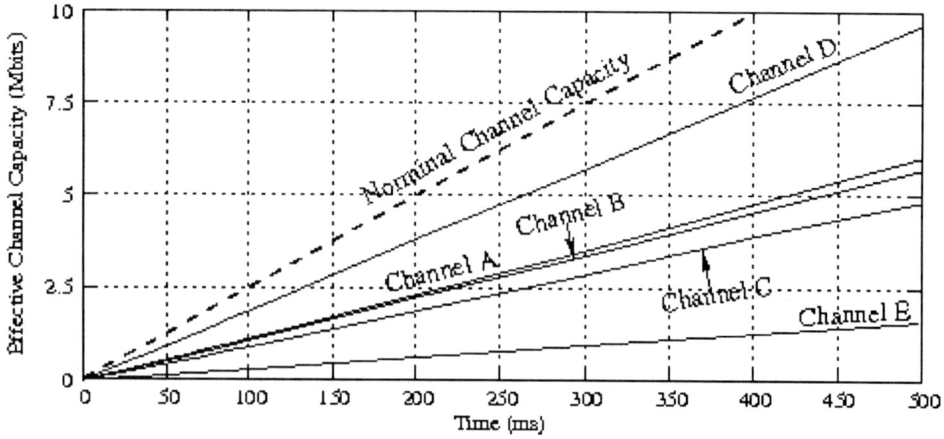

Figures 5-8 depict the obtained effective channel capacity of an IEEE 802.11a wideband channel with 64-QAM modulation and 3/4 coding rate versus the time interval length for five fading channel types, three average SNR values, various packet sizes, and three violation probabilities $\varepsilon = 10^{-1}$, 10^{-2}, 10^{-3}. For each effective channel capacity curve in these figures, 10^3 random realizations of the IEEE 802.11a fading channel with 10^3 ms burst transmission period per realization are performed. For a given time interval length x and $\varepsilon = 10^{-1}, 10^{-2}, 10^{-3}$, about 10^4 sample points for $C[t+x]-C[t]$ are collected from different time interval $[t, t+x]$ and different channel realization. Let J denote the set of all sample points and sort J in the increasing order. Thus, $S^{\varepsilon}(x)$ can be estimated as the kth sample point of the sorted set J, where k is the largest integer bounded by $\varepsilon|J|$. As a benchmark, we also include the corresponding channel nominal capacity determined by *time* x *nominal data rate* in Figures 5-8 and the average channel capacity in Figure 6, where the nominal data rate is the maximum channel data rate after taking into account of the PHY and MAC layer overhead.

We make the following observations. First, according to Figure 5, the impact of the rms delay spread of an IEEE 802.11a fading channel

Figure 6. Effective capacities for Channel C with SNR = 20 dB and packet size = 512 Bytes

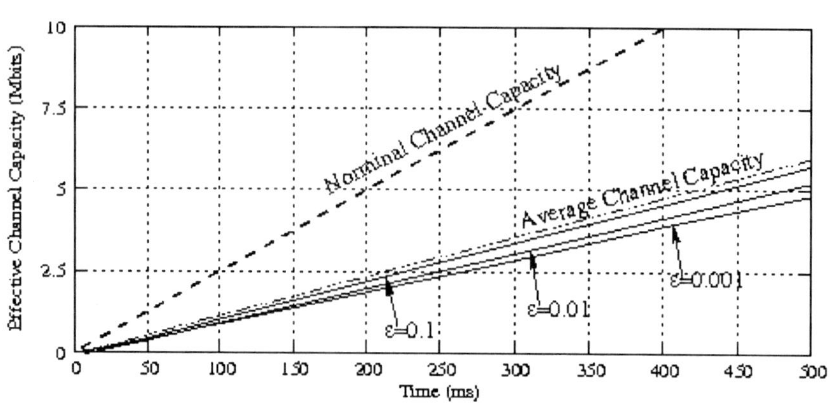

Figure 7. Effective capacities for Channel C with $\varepsilon = 10^{-3}$ *and packet size =512Bytes*

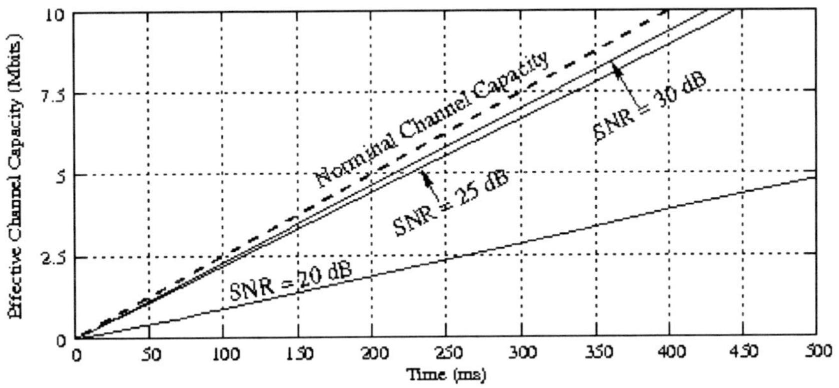

Figure 8. Effective capacities for Channel C with SNR = 25 dB and $\varepsilon = 10^{-}$

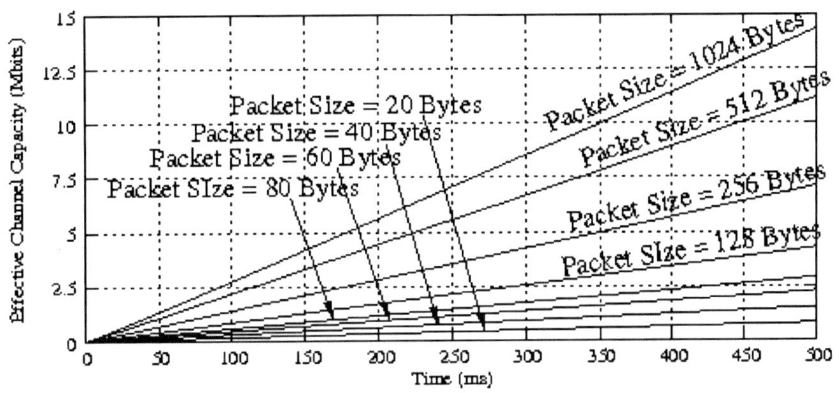

on the effective capacity is significant. For any given time interval length, the effective channel capacity decreases with the increase of the rms delay spread. The larger the rms delay spread, the more severe the inter symbol interference becomes and, consequently, the more that decoding errors will occur. In addition, it is worth noting that Channel D outperforms other channels in Figure 5. This is due to the line-of-sight path of Rician channels. Second, from Figure 6, the impact of the required violation probability on the effective channel capacity is visible. The effective channel capacity decreases as the required violation probability decreases. The essential reason for this scenario is apparent. The smaller the violation probability, the more conservative the available channel capacity lower bound, and thus the smaller the effective channel capacity. Third, according to Figure 7, the impact of the average SNR on the effective channel capacity is substantial. The effective channel capacity increases with the increase of the average SNR. The larger the SNR, the smaller the probability of packet transmission error and consequently, the higher the channel capacity. Finally, according to Figure 8, the impact of packet payload size on the effective channel capacity is significant due to the overheads of network, MAC, and PHY layers.

Overall, the effective channel capacity increases with the increase of the average SNR value and the packet payload size, but decreases as the required violation probability decreases. Figures 5-8 show *Effective Channel Capacities for IEEE 802.11a with 64-QAM and 3/4 Coding Rate.*

Figure 9. Traffic flows in a mesh network backbone

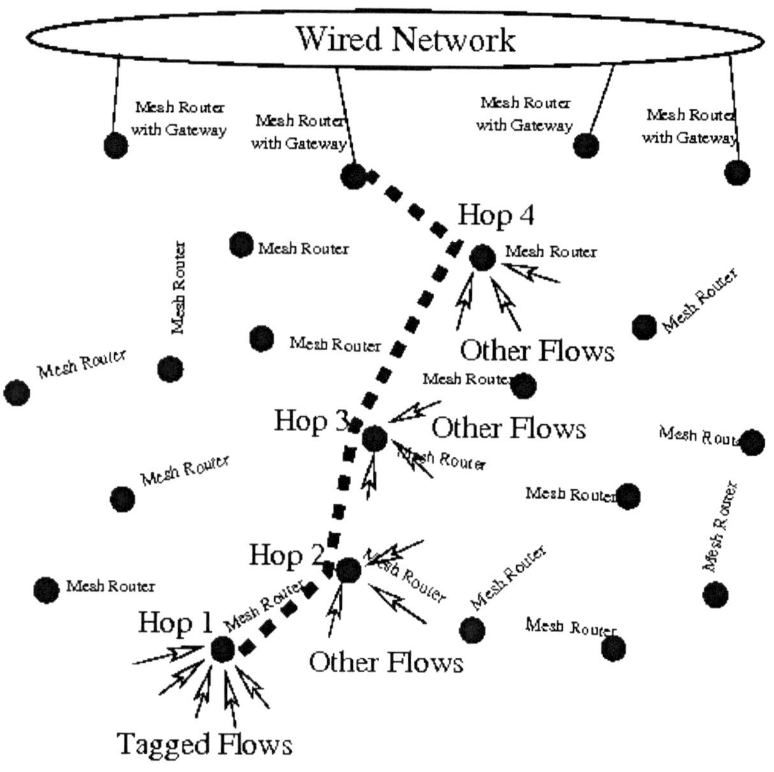

Probabilistic Delay Bounds

In this subsection, we apply Theorem 2 obtained in Section III to predict the end-to-end delays experienced by traffic flows over a WMN. We consider a simple WMN depicted in Figure 9. We focus on the end-to-end delays experienced by the tagged flows that traverse four hops. To simplify experimental study, we assume that if the number of total flows finally entering the wired network is N, there are $N/4$ flows entering the wireless mesh backbone at hop k, k = 1,2,3,4. In addition, for Markov On-Off traffic, we set the beacon interval length T = 200 ms and the service interval length SI = 40 *ms*.

For Regulated traffic, we set T = 200 ms and SI = 20 *ms*. To make our examples succinct, we assume that all wireless channels are the IEEE 802.11a channels with 64-QAM modulation and 3/4 coding rate in environment type C and the average SNR = 25 dB, which are the same as those discussed in the previous subsection. We also assume that there only exists one traffic class, i.e., M=1 and S_1=n. Moreover, by the system-level simulator, when the average SNR = 25 dB, $p_{i,1}$ = $p_{i,3}$ = 0.05, $p_{i,2}$= 0.06 for a data frame with size = 40 bytes, and $p_{i,1}$ = $p_{i,3}$ = 0.05, $p_{i,2}$= 0.09 for a data frame with size = 512 bytes. Thus, based on Formula (14), we obtained $p_{i,\text{drop}}$ = 0.000082 for a data frame with size = 40 bytes and $p_{i,\text{drop}}$ = 0.000182 for a data frame with size = 512 bytes. According to IEEE 802.11a standard, $TXOP_i$ = 0.056407 ms for Markov On-Off traffic flows and $TXOP_i$ = 0.12633 ms for Regulated traffic flows. According to Formulas (16) and (17), the bounds with violation probability ε_0=10^{-4} on the numbers of the data frames that corrupt in their h^{th} transmission during a service interval are obtained and listed in Table 1.

Figures 10-11 describe the relationship between the probabilistic delay bounds which are evaluated by using Theorem 2 and the number of

Table 1. Retransmissions

Regulated Traffic Source				
n	$n_{1,1}^{0.0001}$	$n_{1,2}^{0.0001}$	$n_{1,3}^{0.0001}$	$n_{1,4}^{0.0001}$
5	3	1	0	0
10	5	3	1	0
15	6	3	1	0
20	7	4	2	0
Markov On-Off Traffic Source				
n	$n_{1,1}^{0.0001}$	$n_{1,2}^{0.0001}$	$n_{1,3}^{0.0001}$	$n_{1,4}^{0.0001}$
25	6	3	1	0
50	10	4	2	0
75	13	4	2	0
100	15	5	2	0

total Markov On-Off or Regulated traffic flows that finally enter the wired network. From this figure, we can see that the probabilistic delay bound increases as the required violation probability decreases. The smaller the violation probability, the more conservative the delay bound.

V. CONCLUSION AND FUTURE RESEARCH DIRECTION

In this chapter, we focus on the wireless communications over multi-radio and multi-channel WMNs with IEEE 802.11 HCCA MAC based ingress access points and enough non-overlapping channels for point-to-point wireless links without

Figure 10. Probabilistic Delay Bounds (Channel Type C with SNR = 25 dB)-Markov On-Off Traffic (Packet Payload Size = 40 Bytes)

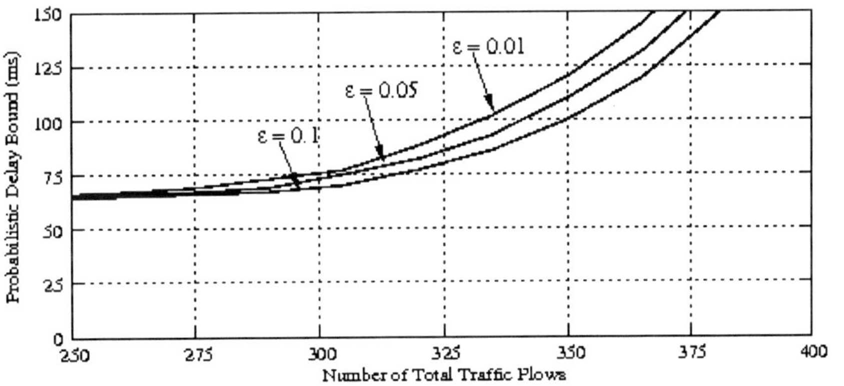

Figure 11. Probabilistic Delay Bounds (Channel Type C with SNR = 25 dB)-Regulated Traffic (Packet Payload Size = 512 Bytes)

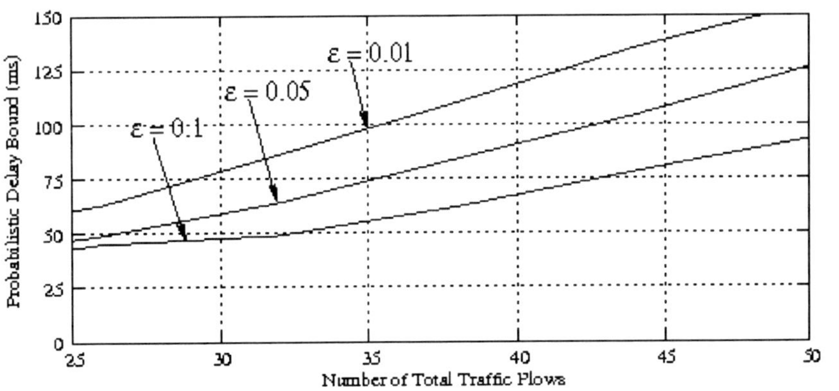

co-channel interference. We develop a theoretic model for the end-to-end performance analysis of such WMNs. By leveraging the large deviations technique, we turn probabilistic problems into deterministic optimization problems and derive a set of algorithms to analyze the end-to-end delays experienced by traffic flows. Consequently, we provide a theoretic framework to predict the probabilistic delay bounds for real-time applications over multi-radio and multi-channel WMNs with enough non-overlapping channels.

The success of the wireless communication technology such as the IEEE 802.11 wireless networks has generated an explosive demand for the wireless spectrum and creates a shortage of spectrum resource. Currently, by governmental agencies such as the Federal Communication Commission (FCC) in USA and the European Telecommunications Standards Institute (ETSI) in Europe, the spectrum is divided into distinct frequency bands for various telecommunication services/users. Generally, all frequency bands can be categorized into licensed bands for licensees with exclusive use and unlicensed bands for public use. For example, there are 27 unlicensed non-overlapping channels for all IEEE 802.11 based wireless networks. However, the measurements obtained by the FCC Spectrum Policy Task Force reveal that at any location, many of the licensed frequency bands between 300 MHz and 3GHz such as TV bands are used sporadically. Recently, a new kind of WMNs, named cognitive WMNs, has been proposed to exploit the idle licensed frequency bands for unlicensed users and enhance the performance of WMNs. Roughly, a cognitive wireless mesh network consists of mesh routers equipped with multiple cognitive radios. Each cognitive radio is able to sense the external environment, learn from the history, and intelligently select available frequency channels for operating. Since the opportunities, i.e., unused licensed bands, vary with time and location and the active licensees of licensed bands need to be protected from harmful interference, the behavior of cognitive WMNs, which depends on the idle patterns of licensed bands as well as the related network topology and traffic load, are much complicated. Therefore, to model cognitive WMNs and predict their performance including queuing delays experienced by traffic flows is an open and very important research topic.

REFERENCES

Adya, A., Bahl, P., Padhye, J., Wolman, A., & Zhou, L. (2004). A Multi-Radio Unification Protocol for IEEE 802.11 Wireless Networks. In *Proceedings of IEEE BROADNETS 2004*.

Akyildiz, I. F., Wang, X., & Wang, W. (2005). Wireless mesh networks: A survey. *Computer Networks and ISDN Systems*, *47*(4), 445–487.

Alicherry, M., Bhatia, R., & Li, L. (2005). Joint channel assignment and routing for throughput optimization in multi-radio wireless mesh networks. In *Proceedings of ACM MOBICOM 2005*.

Awoniyi, O. (2005). *Packet Error Rate in OFDM-based Wireless Networks and Its Impact on VoIP and TCP-Based Data Applications*. Doctoral dissertation, Stanford University, California.

Bansal, N., & Liu, Z. (2003). Capacity, Delay and Mobility in Wireless Ad-Hoc Networks. In *Proceedings of IEEE INFOCOM 2003*.

Bisnik, N., & Abouzeid, A. (2006). Delay and Throughput in Random Access Wireless Mesh Networks. *Proceedings of the IEEE, ICC*, 2006.

Bruno, R., Conti, M., Gregori, E., Wijting, C., Kneckt, J., & Damle, A. (2005). Mesh Networks: Commodity Multihop Ad-Hoc Networks. *IEEE Communications Magazine*, *43*(3), 123–131. doi:10.1109/MCOM.2005.1404606

Bucklew, J. A. (1990). *Large Deviation Techniques in Decision, Simulation and Estimation*. New York: John Wiley.

Chang, C. S. (2000). *Performance guarantees in communication networks*. New York: Springer.

Chayat, N. (1997). *Tentative Criteria for Comparison of Modulation Methods*. IEEE P802.11-97/96.

Chen, J., & Yang, Y. (2006). Multi-Hop Delay Performance in Wireless Mesh Networks. In *Proceedings of IEEE GLOBECOM 2006*.

Das, A. K., Alazemi, H. M. K., Vijayakumar, R., & Roy, S. (2005). Optimization models for fixed channel assignment in wireless mesh networks with multiple radio. In *Proceedings of IEEE SECON 2005*.

de Moraes, R. M., Sadjadpour, H. R., & Garcia-Luna-Aceves, J. J. (2004). Throughput Delay Analysis of Mobile Ad-Hoc Networks with a Multi-Copy Relaying Strategy. In *Proceedings of IEEE SECON 2004*.

de Moraes, R. M., Sadjadpour, H. R., & Garcia-Luna-Aceves, J. J. (2006). Mobility-Capacity-Delay Trade-Off in Wireless Ad-Hoc Networks. *Journal of Ad-Hoc Networks*, *4*(5), 607–620. doi:10.1016/j.adhoc.2005.06.005

Draves, R., Padhye, J., & Zill, B. (2004). Routing in multi-radio, multi-hop wireless mesh networks. In *Proceedings of ACM MOBICOM 2004*.

Faccin, S. M., Wijting, C., Kneckt, J., & Damle, A. (2006). Wireless mesh networks: Concept and system design. *IEEE Wireless Communications*, *13*(2), 10–17. doi:10.1109/MWC.2006.1632476

Gamal, A., Mammen, J., Prabhakar, B., & Shah, D. (2004). Throughput-Delay Trade-Off in Wireless Networks. In *Proceedings of IEEE INFOCOM 2004*.

Ganguly, S., Navda, V., Kim, K., Kashyap, A., Niculescu, D., & Izmailov, R. (2006). Performance optimizations for deploying voip services in mesh networks. *IEEE Journal on Selected Areas in Communications*, *24*(11), 2147–2158. doi:10.1109/JSAC.2006.881594

Garg, S., & Kappes, M. (2003). Can I add a VoIP Call? *Proceedings of the IEEE, ICC*, 2003.

Gupta, P., & Kumar, P. R. (2000). The capacity of wireless networks. *IEEE Transactions on Information Theory*, *46*(2), 388–404. doi:10.1109/18.825799

Heiskala, J., & Terry, J. (2001). *OFDM Wireless LANs: A Theoretical and Practical Guide*. Indianapolis, IN: Sams.

Kelly, F. (1996). Notes on effective bandwidths. In F. P. Kelly, S. Zachary, & I. B. Ziedins (Eds.), *Stochastic Networks: Theory and Applications* (pp. 141-168). Oxford, UK: Oxford University Press.

Knightly, E., & Shroff, N. (1999). Admission control for statistical QoS: Theory and practice. *IEEE Network*, *13*(2), 20–29. doi:10.1109/65.768485

Kodialam, M., & Nandagopal, T. (2003). Characterizing achievable rates in multi-hop wireless networks: The joint routing and scheduling problem. In *Proceedings of ACM MOBICOM 2003*.

Kodialam, M., & Nandagopal, T. (2005). Characterizing the capacity region in multi-radio multi-channel wireless mesh networks. In *Proceedings of ACM MOBICOM 2005*.

Kodialam, M., & Nandagopal, T. (2005). On the capacity region of multi-radio multi-channel wireless mesh networks. In *Proceedings of IEEE WIMESH 2005*.

Kyasanur, P., So, J., Chereddi, C., & Vaidya, N. H. (2006). Wireless mesh networks: A challenge and protocols. *IEEE Wireless Communications*, *13*(2), 30–36. doi:10.1109/MWC.2006.1632478

Kyasanur, P., & Vaidya, N. H. (2005). Capacity of multi-channel wireless networks: Impact of number of channels and interference. In *Proceedings of ACM MOBICOM* 2005.

Lee, M. J., Zheng, J., Ko, Y. B., & Shrestha, D. M. (2006). Emerging standards for wireless mesh technology. *IEEE Wireless Communications*, *43*(4), 56–63. doi:10.1109/MWC.2006.1632481

Li, C., Che, H., & Li, S. (2007). A Wireless Channel Capacity Model for Quality of Service. *IEEE Transactions on Wireless Communications*, *6*(1), 356–366. doi:10.1109/TWC.2007.05282

Li, C., & Zhao, W. (2008). Stochastic Performance Analysis of Multi-Radio Multi-Channel Wireless Mesh Networks. In *Proceedings of the Fourth Internal Conference on Wireless and Mobile Communications 2008*.

Lin, X., Sharma, G., Mazumdar, R. R., & Shroff, N. B. (2006). Degenerate Delay-Capacity TradeOffs in Ad-Hoc Networks with Brownian Mobility. *IEEE Transactions on Information Theory*, *52*(6), 2777–2784. doi:10.1109/TIT.2006.874544

Medbo, J., & Schramm, P. (1998). *Channel Models for HIPERLAN/2*. ETSI/BRAN doc. No. 3ERIO85B.

Neely, M. J., & Modiano, E. (2005). Capacity and delay tradeoffs for ad-hoc mobile networks. *IEEE Transactions on Information Theory*, *51*(6), 1917–1936. doi:10.1109/TIT.2005.847717

Niculescu, D., Ganguly, S., Kim, K., & Izmailov, R. (2006). Performance of VoIP in a 802.11 Wireless Mesh Network. In *Proceedings of IEEE INFOCOM 2006.*

Perevalov, E., & Blum, R. (2003). Delay Limited Capacity of Ad-Hoc Networks: Asymptotically Optimal Transmission and Relaying Strategy. In *Proceedings of IEEE INFOCOM 2003.*

Perevalov, E., & Blum, R. (2006). Capacity Region, Minimum Energy and Delay for a Mobile Ad-Hoc Network. In *Proceedings of IEEE Internal Symposiumon Modeling and Optimization in Mobile Ad-Hoc Wireless Networks 2006.*

Ramachandran, K., Sheriff, I., Belding-Royer, E., & Almeroth, K. (2008). A Multi-Radio 802.11 Mesh Network Architecture. *Journal of Mobile Networks & Applications*, *13*(1), 132–146. doi:10.1007/s11036-008-0026-8

Ramachandran, K. N., Belding, E. M., Almeroth, K. C., & Buddhikot, M. M. (2006). Interference-aware channel assignment in multi-radio wireless mesh networks. In *Proc. of IEEE INFOCOM 2006.*

Rappaport, T. S. (2002). *Wireless Communications Principles and Practice (2nd ed.)*. Englewood Cliffs, NJ: Prentice-Hall.

Robinson, J., Papagiannaki, K., Diot, C., Guo, X., & Krishnamurthy, L. (2005). Experimenting with a Multi-Radio Mesh Networking Testbed. In *Proceedings of IEEE WINMEE 2005.*

Sharma, G., Mazumdar, R., & Shroff, N. (2007). Delay and capacity trade-offs in mobile ad hoc networks: A global perspective. *IEEE/ACM Transactions On Networking,* 15(2), 981-992.

IEEE Std. (1999). *Part 11: Wireless LAN Medium Access Control (MAC) and Physical Layer (PHY) Specifications*. ANSI/IEEE Std 802.11.

IEEE Std. (2000). *Part 11: Wireless LAN MAC and PHY Specifications: Higher-Speed Physical Layer in the 5 GHz Band.* ANSI/IEEE Std 802.11a.

IEEE Std. (2007). *Part 11: Wireless LAN Medium Access Control (MAC) and Physical Layer (PHY) Specifications.* ANSI/IEEE Std 802.11.

Tang, J., Xue, G., & Zhang, W. (2005). Interference-aware topology control and QoS routing in multi-channel wireless mesh networks. In *Proceedings of ACM MOBIHOC 2005*.

Trad, A., Munir, F., & Afifi, H. (2006). Capacity Evaluation of VoIP in IEEE 802.11e WLAN Environment. In *Proceedings of IEEE CCNC 2006.*

Vedantham, R., Kakumanu, S., Lakshmanan, S., & Sivakumar, R. (2006). Component based channel assignment in single radio, multi-channel ad-hoc networks. In *Proc. of ACM MOBICOM 2006.*

Wu, D., & Negi, R. (2003). Effective capacity: A wireless link model for support of quality of service. *IEEE Transactions on Wireless Communications*, *2*(4), 630–643.

Xing, K., Chen, X., Ma, L., & Liang, Q. (2007). Superimposed code based channel assignment in multi-radio multi-channel wireless mesh networks. In *Proceedings of ACM MOBICOM 2007.*

ADDITIONAL READING

Anastasopoulos, M. P., Panagopoulos, A. D., & Cottis, P. G. (2008). A distributed routing protocol for providing QoS in Wireless Mesh Networks operating above 10 GHz. *Wireless Communications and Mobile Computing, 8*(10), 1233–1245. doi:10.1002/wcm.562

Andreopoulos, Y., Mastronarde, N., & Van Der Schaar, M. (2006). Cross-layer optimized video streaming over wireless multihop mesh networks. *IEEE Journal on Selected Areas in Communications, 24*(11), 2104–2115. doi:10.1109/JSAC.2006.881614

Baumann, R., Heimlicher, S., & Plattner, B. (2008). Routing in large-scale wireless mesh networks using temperature fields. *IEEE Network, 22*(1), 25–31. doi:10.1109/MNET.2008.4435899

Hu, H., Zhang, Y., & Chen, H. H. (2008). An effective QoS differentiation scheme for wireless mesh networks. *IEEE Network, 22*(1), 66–73. doi:10.1109/MNET.2008.4435905

Huang, J. H., Wang, L. C., & Chang, C. J. (2008). QoS provisioning in a scalable wireless mesh network for intelligent transportation systems. *IEEE Transactions on Vehicular Technology, 57*(5), 3121–3135. doi:10.1109/TVT.2008.918701

Koksal, C. E., & Balakrishnan, H. (2006). Quality-Aware Routing Metrics for Time-Varying Wireless Mesh Networks. *IEEE Journal on Selected Areas in Communications, 24*(11), 1984–1994. doi:10.1109/JSAC.2006.881637

Lam, R. K., Chiu, D. M., & Lui, J. C. S. (2007). On the access pricing and network scaling issues of wireless mesh networks. *IEEE Transactions on Computers, 56*(11), 1456–1469. doi:10.1109/TC.2007.70753

Liu, T., & Liao, W. (2008). Location-Dependent Throughput and Delay in Wireless Mesh Networks. *IEEE Transactions on Vehicular Technology, 57*(2), 1188–1198. doi:10.1109/TVT.2007.905389

Park, B. N., Lee, W., Ahn, S., & Ahn, S. (2006). QoS-driven wireless broadband home networking based on multihop wireless mesh networks. *IEEE Transactions on Consumer Electronics, 52*(4), 1220–1228. doi:10.1109/TCE.2006.273137

KEY TERMS AND DEFINITIONS

Effective Channel Capacity: The effective capacity with violation probability ε of a point-to-point wireless link over a given channel during a time interval is the amount of information that can successfully be transmitted through the wireless link during the time interval with probability larger than 1- ε.

Network Traffic: In telecommunication, network traffic is data in a network. Usually, the data is encapsulated in packets.

Probabilistic Delay Bound: A probabilistic delay bound consists of two components, i.e., a delay bound D and a violation probability ε, such that the experienced delay is no larger than D with probability 1-ε.

Queuing Delay: In wireless communications, the queuing delay is the time a packet waits in a transmitter until it has successfully been received by a receiver. It is a key component of wireless network delay.

Real-Time Communications: Real-time communications is any mode of telecommunications in which all users can exchange information instantly or with tolerable latency.

Wireless Mesh Network: A wireless mesh network is a communications network made up of wireless mesh routers organized in a mesh topology. Moreover, a wireless mesh network can be seen as a type of wireless ad hoc network, where all radio nodes are static and don't experience direct mobility.

Wireless Mesh Router: It is a combination base station (access point) and router in one device. Also wireless mesh routers are called *mesh nodes* and typically installed on street light poles, from which they obtain their power.

Chapter 23
Scalable Wireless Mesh Network Architectures with QoS Provisioning

Jane-Hwa Huang
National Chiao-Tung University, Taiwan

Li-Chun Wang
National Chiao-Tung University, Taiwan

Chung-Ju Chang
National Chiao-Tung University, Taiwan

ABSTRACT

The wireless mesh network (WMN) is an economical solution to enable ubiquitous broadband services due to the advantages of robustness, low infrastructure costs, and enhancing coverage by low power. The wireless mesh network also has a great potential for realizing green communications since it can save energy and resources during network operation and deployment. With short-range communications, the transmission power in the wireless mesh networks is lower than that in the single-hop networks. Nevertheless, wireless mesh network should face scalability issue since throughput enhancement, coverage extension, and QoS guarantee are usually contradictory goals. Specifically, the multi-hop communications can indeed extend the coverage area to lower the infrastructure cost. However, with too many hops to extend coverage, the repeatedly relayed traffic will exhaust the radio resource and degrade the quality of service (QoS). Furthermore, as the number of users increases, throughput and QoS (delay) degrade sharply due to the increasing contention collisions. In this chapter, from a network architecture perspective we investigate how to overcome the scalability issue in WMNs, so that the tradeoff between coverage and throughput can be improved and the goal of QoS provisioning can be achieved. We discuss main QoS-related research directions in WMNs. Then, we introduce two available scalable mesh network architectures that can relieve the scalability issue and support QoS in WMNs for the wide-coverage and dense-urban coverage. We also investigate the optimal tradeoff among throughput, coverage, and delay for the proposed WMNs by an optimization approach to design the optimal system parameters.

DOI: 10.4018/978-1-61520-680-3.ch023

INTRODUCTION

The wireless mesh network (WMN) plays an important role in the next-generation wireless systems for enabling ubiquitous Internet access, thanks to the advantages of robustness, low infrastructure costs, and enhancing the coverage with low transmission power (Pabst et al., 2004; Jun & Sichitiu, 2003; Lee, Zheng, Ko, & Shrestha, 2006; Lee et al., 2006; Zhang & Wolff, 2004; Fowler, 2001; Lewis, 2003; Qiu et al., 2004; Akyildiz, Wang, & Wang, 2005). In the near future, large-scale broadband network deployment for wireless Internet access will continue at a rapid pace. Traditionally, large-scale network deployment is a very challenging task due to the costly and time-consuming cabling engineering works. Attractively, as shown in Fig. 1, the mesh nodes of the WMNs (including the access points/relay stations/client stations) interconnected via wireless links can forward other node's traffic toward/from the central gateway. The cable connection is required only from the central gateway to the Internet. Clearly, the WMN can be rapidly deployed on a large scale with less cabling engineering works. In addition, WMN will be an economical solution to provide wireless broadband services.

The WMN also has a great potential for realizing green communications since it can save energy and resources during network operation and deployment. Generally, enhancing data rate and coverage will increase the energy consumption of communication and network equipments, which in turn increases the associated CO_2 emission. By contrast, with short range communications, the WMNs have lower transmission power than the single-hop networks. Furthermore, fewer cable connections and less cabling engineering for WMNs further reduce the resource and energy consumption for network deployment.

The advantages of wireless mesh networking technology can be summarized from the following aspects:

Figure 1. Generic multi-hop wireless mesh network architecture with extended network coverage

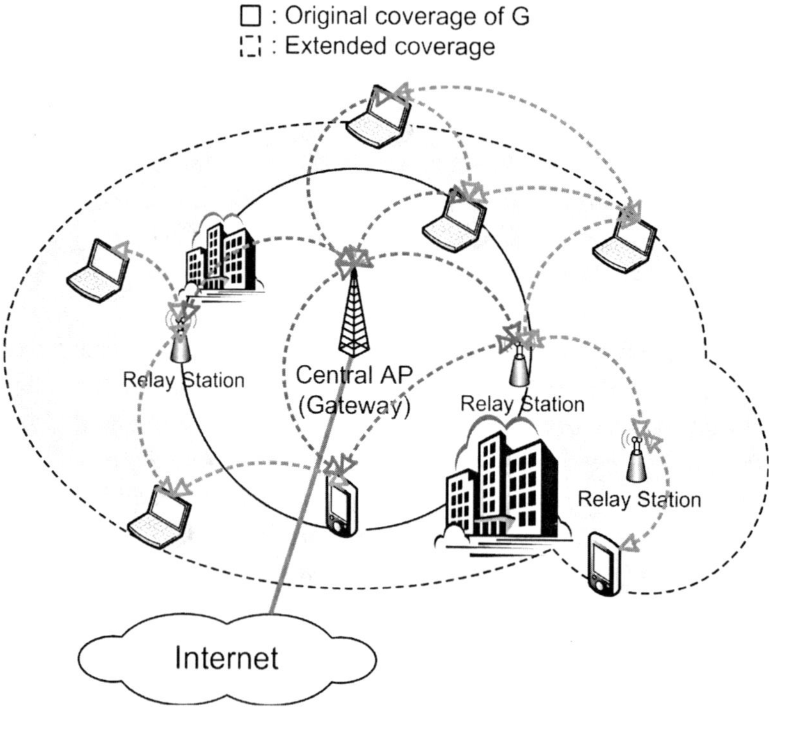

- **First, Rapid deployment:** Since every device is able to act as a wireless relay/router, WMN can be rapidly deployed in a large-scale area with a minimal cabling engineering work so as to lower the infrastructure and deployment costs (Pabst et al., 2004; Jun & Sichitiu, 2003; Fowler, 2001).
- **Second, Reliable communication:** It is well known that mesh networking technology can combat shadowing and severe path loss to extend service coverage.
- **Third, Low transmission power:** By means of short range communications to improve the transmission rate and then the energy efficiency, WMN can realize the goal of low-power communication system. In addition, with a lower interference power, the same frequency channel can be spatially reused by other links at a shorter distance.
- **Fourth, Robustness:** Due to multiple paths between the source node to the destination node, an appealing feature of WMNs is its robustness (Qiu et al., 2004; Akyildiz et al., 2005). If some nodes fail, the mesh network can continue operating with slightly degraded performance by forwarding data traffic via the alternative nodes. For example, as shown in Fig 2, if the original route ($S-B-G1$) is broken, the traffic can be forwarded by the alternative route ($S-R1-G1$). Even more, if the gateway ($G1$) malfunctions, the node S can still connect to the Internet by another gateway $G2$ via the route ($S-R2-G2$), as shown in the figure.
- **Fifth, Heterogeneous network architecture:** Wireless mesh networks can concurrently support a variety of wireless radio access technologies. Therefore, WMN provides the flexibility to achieve a heterogeneous wireless network with many different radio access technologies (MeshNetworks, 2009; MeshDynamics, 2009; Lewis, 2003). Figure 3 shows an example of integrated heterogeneous wireless mesh network, where the 802.16 (WiMAX) and 802.11 (WiFi) radio access technologies are used for the wireless metropolitan area network (WMAN) and the wireless local area network (WLAN), respectively.

Figure 2. An appealing feature in WMN: robustness. If the original route ($S-B-G1$) is broken, the traffic can be forwarded by the alternative route ($S-R1-G1$) or ($S-R2-G2$)

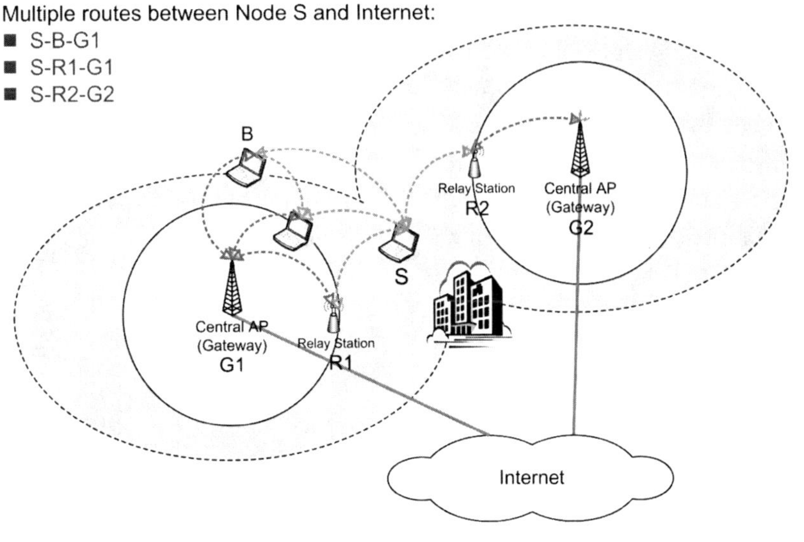

Figure 3. An integrated 802.11/16/Cellular (WLAN/WMAN) WMN. The gateway has a cable connection to the Internet. Multi-mode mesh nodes are interconnected via wireless links to forward other node's traffic to/from the gateway.

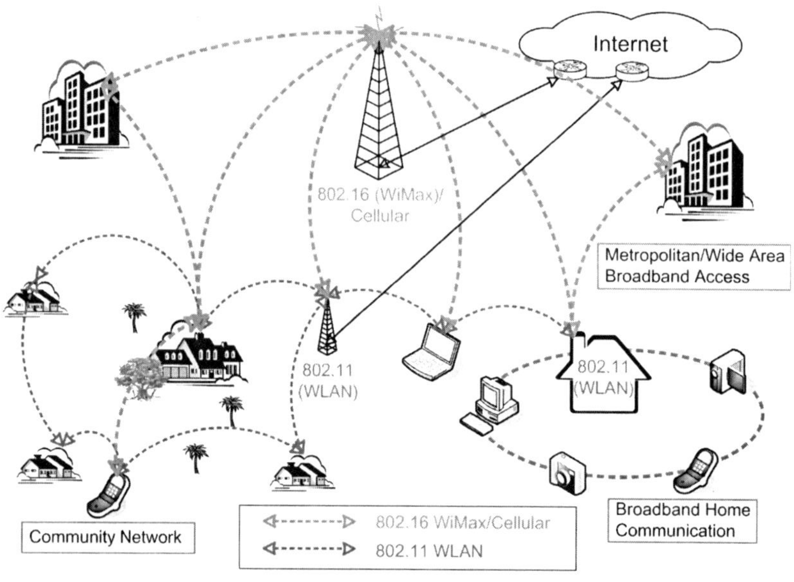

Due to these advantages, the wireless mesh network is a key enabling technology for the next-generation wireless systems.

However, multi-hop mesh networking suffers from the scalability issue. That is, to serve more users in the WMN, extending the system coverage with more hops may degrade system throughput and increase delay (Holland & Vaidya, 2002; Gupta & Kumar, 2000; Jun & Sichitiu, 2003; Akyildiz et al., 2005). Specifically, the multi-hop communications can indeed extend the coverage by more hops and longer hop distance. However, as the number of hops increases, the repeatedly relayed traffic will rapidly exhaust the radio resource, thereby resulting in inefficient usage of frequency spectrum and degrading the quality of service (QoS), e.g., longer delay and higher jitter. In the meanwhile, longer hop distance will also lead to lower data rate and higher delay in the wireless relay link between nodes. As the number of users increases, more contention collisions occur, which further degrades the throughput performance. Therefore, it is important to design a scalable WMN that can extend the coverage without sacrificing the system overall throughput and the quality of service (QoS).

The goal of this chapter is to address the scalability issue of the wireless mesh networks from a network architecture perspective. This chapter introduces two scalable-WMN deployment strategies with QoS support in the typical WMN application scenarios, i.e., *dense-urban coverage* and *wide-area coverage* as shown in Figs. 4 and 5 (Huang, Wang, & Chang, 2008b, 2006a). The proposed WMNs are scalable thanks to the following two factors. First, the suggested frequency planning can reduce collisions as coverage and users increase. Second, the proposed network structure can facilitate the management of QoS, throughput, and coverage in WMNs. Recently, deploying public wireless local area networks (WLANs) in the dense-urban area (i.e., the Manhattan environment) is a hot topic (Pabst et al., 2004). Figure 4 shows a scalable cluster-based WMN. In a cluster, access points (APs) operating at different channels will communicate with neighboring APs via wireless links, and only the central gateway connects to the Internet through cables. By doing so, network deployment in the urban area becomes easier due to less cabling engineering work. Rapidly providing wide-area coverage is another important application scenario of WMNs. Figure 5 shows a scalable ring-based WMN. The mesh cell is divided into several rings allocated with different channels. Nodes in the inner rings will relay data for nodes in the outer rings toward the central gateway. This mesh cell structure can extend the coverage of central gateway to lower cost.

This chapter also investigates the optimal tradeoff among capacity, coverage, and QoS for the proposed scalable WMNs. Most traditional WMNs are not scalable to users and coverage, since the throughput and QoS (delay) are not guaranteed with increasing collisions. By contrast, the proposed WMNs are scalable in terms of users and coverage, since the delay and throughput can be ensured by the multi-channel frequency planning with properly designing the deployment parameters. We apply the mixed-integer nonlinear programming (MINLP) optimization approach to determine the optimal deployment parameters, aiming to maximize the capacity and coverage of the scalable WMNs subject to the QoS requirement.

The rest of this chapter is organized as follows. First, we discuss QoS-related research directions for the WMNs. We also survey the typical multi-hop operation schemes, with the objective to highlight the most suitable one to realize a scalable WMN with QoS support. Then, we introduce the scalable-WMN deployment strategies for the dense-urban and wide-area coverage. Furthermore, we perform optimization designs to determine the optimal deployment parameters for the proposed WMNs. Finally, we summarize this chapter and discuss the future works.

QOS-RELATED RESEARCH DIRECTIONS FOR WMNS

Here, we discuss the scalability and QoS-related performance issues in the WMNs. Due to great popularity and implementation simplicity of carrier sense multiple access (CSMA) protocol, most of the following discussion mainly focuses on the WMNs using the contention-based medium access

Figure 4. Scalable cluster-based WMN for dense-urban coverage, where several APs allocated with different channels form a cluster

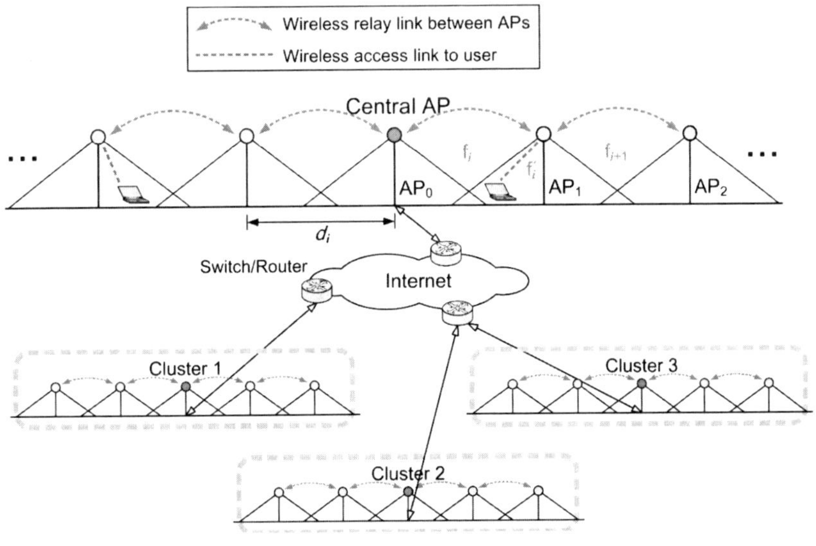

Figure 5. Scalable ring-based WMN for wide-area coverage, where the mesh cell is divided into several rings allocated with different channels. The users in the ring A_i connect to the central gateway via the i-hop communications

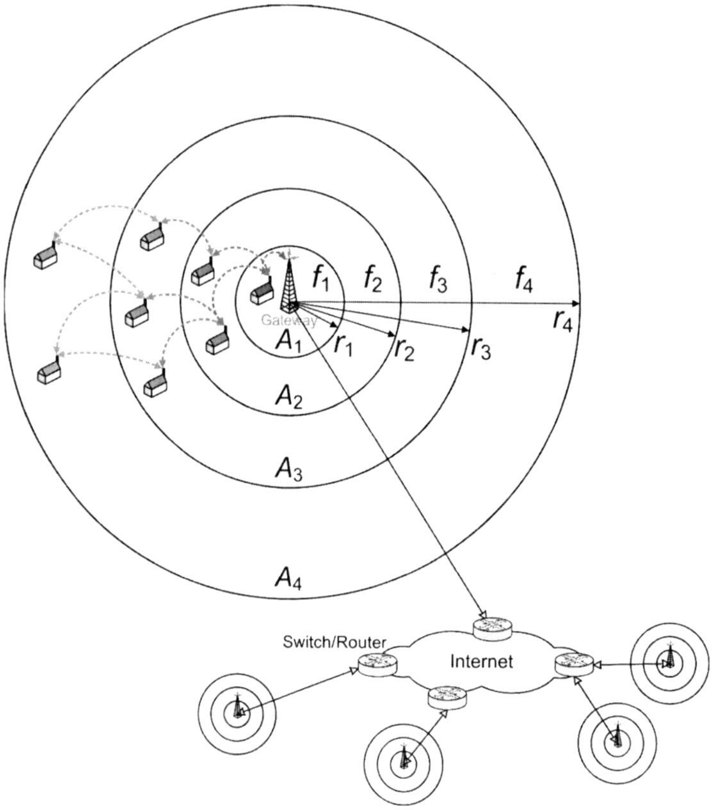

control (MAC), such as CSMA MAC protocol with (request-to-send/clear-to-send) RTS/CTS.

Scalability

The WMNs may face the scalability issue (Huang et al., 2006a, 2008b; Huang, Wang, & Chang, 2008c). As the coverage increases, each user may experience significant MAC throughput and delay degradation due to increasing collisions from more contending users. Consequently, one may fail to find a reliable routing path and then often lose its end-to-end connection.

Because the scalability issue, the WMN using the legacy distributed CSMA MAC protocol can not achieve a reasonable throughput as the network size increases. The results in (Holland & Vaidya, 2002) showed that even with only one user in a multi-hop network, the throughput drops sharply as the number of hops increases from one. Then, it stabilizes at a very low throughput as the number of hops becomes larger (e.g., larger than seven in (Holland & Vaidya, 2002)). This phenomenon is due to the fact that the adjacent hop nodes have to contend for the channel to relay traffic. Moreover, the authors in (Gupta & Kumar, 2000) pointed out that with *k* users in an ad hoc network, the user throughput is scaled like $O\left(1/\sqrt{k\log k}\right)$. The study in (Jun & Sichitiu, 2003) also showed that the throughput per user in a WMN decreases sharply as $O\left(1/k\right)$ due to the throughput bottleneck at the central gateway.

Scalability is a quite desirable feature for WMNs. A scalable WMN architecture can extend the coverage of gateway without sacrificing the system overall throughput and QoS. The low scalability of WMNs mainly lies in the throughput degradation due to a lot of users contending for the same channel (Holland & Vaidya, 2002; Jun & Sichitiu, 2003). How to design scalable mesh network architecture and develop an enhanced MAC protocol to improve throughput and QoS performances as well as overcome the scalability issue are important research topics.

Optimal Tradeoff Among Throughput, Coverage, and QoS

All the performance of throughput, coverage, and QoS are major concerns in the design of WMNs. However, it is challenging to manage the interactions among throughput, coverage, and QoS in a distributed WMN. For example, more hops and longer hop distance can easily extend system coverage. However, the repeatedly relayed traffic with more hops will exhaust most of radio resources and thus degrade QoS. The longer hop distance also lowers the data rate of the relay link between nodes. Therefore, from the standpoint of throughput per user, a smaller coverage with fewer hops and shorter hop distance is preferred due to fewer contending users and higher data rate in the relay link. Another concern in WMNs is delay consisting of contention delay and queuing delay in each hop. From a queueing delay perspective, a longer hop distance may be better due to fewer hops. From a contention delay viewpoint, however, a shorter hop distance is preferred due to fewer contending users.

Obviously, there exist interactions among the throughput, coverage, and QoS. Understanding these interactions and then developing suitable mesh network architecture to facilitate the management of QoS, throughput, and coverage of a WMN is an interesting but challenging issue. Furthermore, designing the system parameters (e.g., the hop distance, the maximum number of hops in a WMN, and the transmission power) to obtain the optimal tradeoff among throughput, coverage, and QoS is also a key issue in WMNs.

Differentiated Services and QoS

One interesting issue in WMNs is to support various services with differentiated QoS requirements. In the literature, some delay guarantee mechanisms are proposed for WLANs. One method is to use the point coordination function (PCF) in IEEE 802.11 WLAN. Then, the delay-sensitive traffic can be sent in the contention free periods. Another well-known mechanism is the EDCA in IEEE 801.11e, which groups services into four access categories (ACs) with different priorities. In EDCA, arbitration inter-frame space (AIFS) is employed instead of distributed coordination function inter-frame space (DIFS). The higher-priority service class can use shorter AIFS and smaller contention window size to lower delay. The EDCA also defines transmission opportunity (TXOP) as a time interval during which a particular station can transmit multiple frames consecutively without contending. After successful contention, the higher-priority station can obtain a larger TXOP to transmit more frames. However, these methods mainly focus on the single-hop wireless networks rather than the multi-hop WMNs.

In WMNs, mesh nodes connect to the central gateway in a multi-hop fashion. At each hop, the packet may be delivered at different transmission rates with different packet loss rates, delay and jitter. Load unbalancing also results in different queuing delay at each node. Since there is no a central control coordinator in the WMNs, it is challenging to provide end-to-end QoS guarantees for different service types as network size and the number of users increase.

Power Unfairness Problem and Power Control

In addition to the bottleneck issue near the central gateway, it is necessary to further resolve the power unfairness problem (Huang, Wang, & Chang, 2008a). Specifically, the inner users near the gateway have to consume more power to relay traffic for others, which induces the power unfairness problem for the inner users. When the users close to the gateway deplete their battery energy, the whole mesh network will not function normally. As the number of users increases, such a power unfairness problem will become even more serious for the inner users. Therefore, while extending the coverage area to serve more users, how to achieve power fairness among users to prolong the lifetime of WMN is an important task. In our previous work (Huang, Wang, & Chang, 2008a), we demonstrated that the proposed scalable ring-based WMN can also achieve power fairness. In the ring-based WMN, we can adjust the ring width of each ring to control the contention level and the hop distance. By reducing the inner ring width, the users in the inner ring can transmit with higher data rate and power efficiency. In result, the power unfairness in the WMN can be resolved. However, the work in (Huang, Wang, & Chang, 2008a) considers a stationary user case. In a mobile environment, how to achieve the power fairness among users is still a challenging task.

The impact of power control also needs to be investigated. In WMNs, higher transmit power can increase the transmission range and the data rate in the relay link. However, it also increases contention collisions and lowers the efficiency of spatial frequency reuse. Hence, determining the proper transmit power to achieve the best tradeoff among power efficiency, QoS, throughput, and coverage is important for WMNs.

Cognitive Radio

Recently, cognitive radio (CR) technique attracts numbers of researchers' attention, since cognitive radio can significantly increase spectrum utilization and system capacity (Akyildiz et al., 2006). In the traditional wireless networks, the operational spectrum is usually assigned by a fixed spectrum allocation policy. According to the statistics of Federal Communications Commission (FCC), by the fixed spectrum allocation policy the spectrum will be underutilized, and the spectrum utilization varies from 15% to 85% depending on the geographical environment (FCC, 2003). On the contrary, with the cognitive and reconfigurable capabilities, the cognitive radio can identify and exploit the unused spectrum, namely, white space spectrum or spectrum hole (Haykin, 2005). In addition, the cognitive radio can self-configure the transmitter parameters according the surrounding environment.

CR technique can improve throughput and QoS of the multichannel WMNs. For example, with CR technique, the mesh user can autonomously exploit the unused channel to mitigate contention level. By the cognitive capability, the mesh user can also sense the contention level of the neighboring nodes. Then, the mesh user can intelligently determine the routing path by selecting the user with less contention level as its next-hop to reduce contention delay and improve throughput. How to develop the CR-based channel selection and CR-based routing path selection mechanisms to improve QoS and throughput of WMN are very interesting issues in the CR-based WMNs.

Cooperative Communications

Different from the conventional WMNs, the node in the cooperative communication system collaborates with the relaying nodes to deliver data traffic, through distributed transmission and processing (Kramer, Gastpar, & Gupta, 2005; Nosratinia, Hunter, & Hedayat, 2004).

Figure 6 shows a typical example of three-node cooperative relaying. As shown in the figure, in the first phase, the source node (*S*) delivers the data frame. In the second phase, the cooperative relay (*R*) forwards the data frame to the destination node (*D*). With such a two-phase cooperative transmission, the data frames can be delivered by not only the source node but also the cooperative relay. Then, the destination node can combine these signals transmitted from different nodes. By doing so, the source node along with several single-antenna cooperative relays form a virtual antenna array system to combat server shadowing and fading, thereby improving link reliability and capacity. Figure 7 illustrates other types of cooperative relaying, including the single-stage and multi-stage cooperation strategies. Clearly, the multi-stage cooperation strategy can have lower transmission distance to further increase the transmission data and reduce the transmission power. However, in a distributed wireless multihop network, achieving the multi-stage cooperation is much more complicated than achieving the single-stage cooperation.

Figure 6. Two-phase cooperative communication. In phase I, the source (S) sends the data, and in phase II, the relay (R) forwards the frame to the destination (D). The source and the relay, each with a single antenna, form a virtual antenna array for the frame transmissions to the destination.

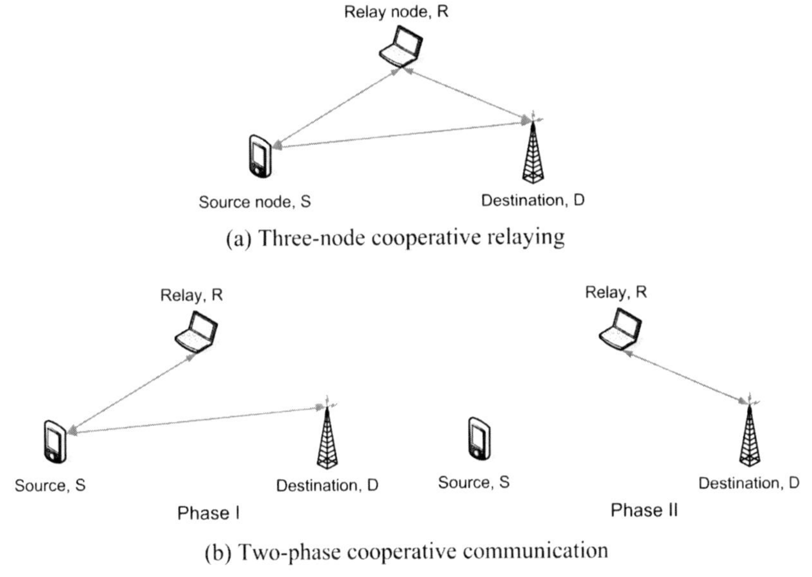

Figure 7. Different cooperation strategies: Single-stage and multi-stage cooperative relaying

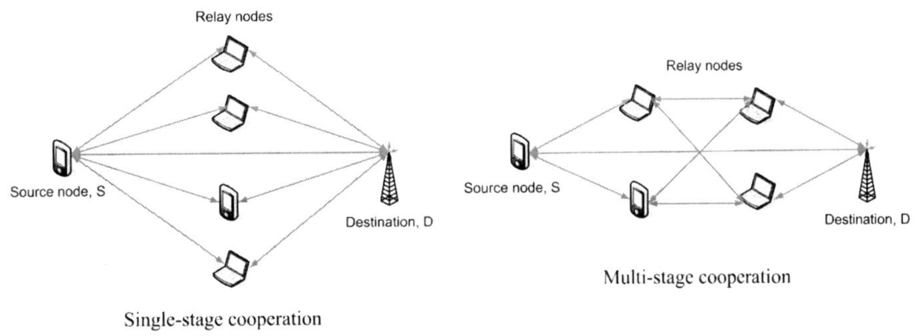

Many essential issues are still open in designing a practical scalable cooperative communication system, such as the scalable network architecture design, cooperation strategy design, performance analysis and optimization, resource management and scheduling, MAC and routing protocol design.

Cross-Layer Design

Cross-layer design can improve the network performance and scalability. For example, one can exploit the physical layer network architecture with the multi-channel frequency planning to improve MAC throughput and reduce delay. Indeed, in addition to the impacts of network architecture and frequency planning, there are many interactions among the transport, routing, MAC, and physical layer protocols in WMNs. The transmission power and rate in the physical layer will influence MAC throughput and routing decisions. The link selection in the routing layer will affect the contention situation at the MAC layer. Besides, according to the end-to-end delay information provided by the transport layer, the MAC protocol can adjust the backoff window size to reduce delay. In a distributed WMN, understanding the cross-layer interactions and then designing a scalable mesh network architecture to provide QoS is a very interesting issue.

Noteworthily, cross-layer design will face the issues of incompatibility with the existing protocols and loss of design abstraction (Kawadia & Kumar, 2005; Maharshi, Tong, & Swami, 2003; Toumpis & Goldsmith, 2003; Li & Baoyu, 2003). Any protocol modification may result in the unexpected impacts on the whole system performance and the difficulty in network management. To avoid these potential problems, some design principles are suggested in (Kawadia & Kumar, 2005).

Figure 8. Examples of single-channel and multi-channel multihop operations in WMN. (a) Single-channel WMN. Since all the nodes contend for the same channel, the collisions may often occur. (b) Multi-channel Single-interface WMN. The nodes in this network can operate on different channels to reduce collision. (c) Multi-channel Multi-interface WMN. Each node can concurrently communicate with different nodes to enhance throughput.

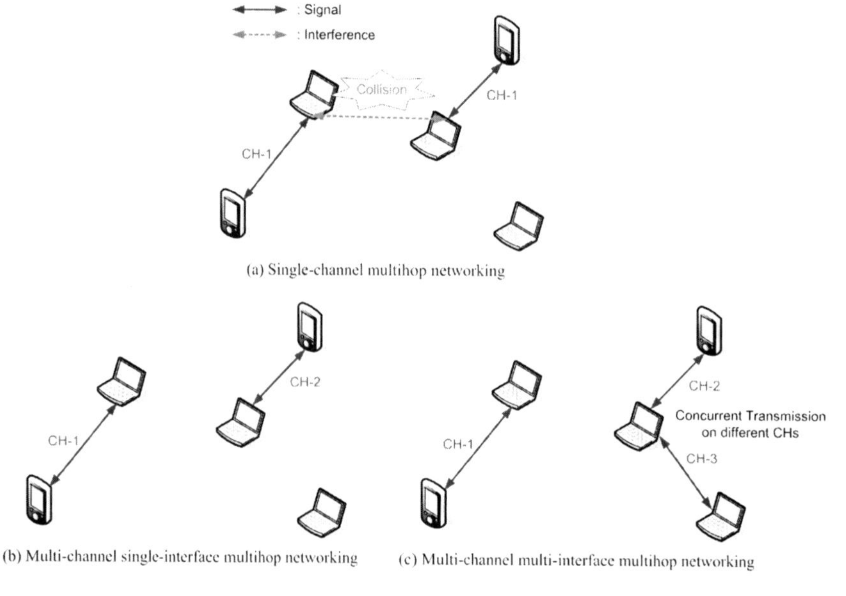

SCALABLE MESH NETWORK ARCHITECTURE: MULTI-CHANNEL OR SINGLE-CHANNEL

The mesh network architecture can be classified into two categories: *single-channel* and *multi-channel* multi-hop schemes, as shown in Fig. 8.

Single-Channel Mesh Network

In the *single-channel multi-hop scheme* (see Fig. 8 (a)), all the nodes contend for the same channel. Since only one node in the contention region can successfully transmit at a time, the single-channel multi-hop networks suffer from a severe scalability issue. That is, if the network coverage becomes larger with more users contending for the same channel, the increasing collisions will significantly degrade the throughput.

To improve the throughput and scalability of the single-channel system, one possible solution is to develop an enhanced RTS/CTS mechanism to reduce the number of exposed nodes as in (Ju, Rubin, & Kuan, 2003). The authors in (Choudhury, Yang, Ramanathan, & Vaidya, 2002) developed a MAC protocol using directional antenna to mitigate the exposed node problem, thereby increasing the transmission opportunity and the throughput of each node. However, directional transmissions may lead to more hidden terminals than the cases using omni-directional antenna. Therefore, resolving the hidden node problem is the major issue in the MAC protocols using directional antenna. Besides, one can exploit power control to reduce the interference, by which users will adjust their transmission power according to the hop distance and the transmission rate (Zhong & Kravets, 2003). However, with power control the hidden node problem may become worse since the users using higher transmission power may fail to sense the communications with lower power level, but will interfere with them.

Multi-Channel Mesh Network

In the *multi-channel multi-hop system*, the nodes can dynamically switch to distinct channels, and thus different nodes can simultaneously deliver their frames at different channels. Accordingly, the WMN becomes more scalable since the number of contending users is reduced and thus system throughput can be improved.

The multi-channel WMNs will operate at either *single-interface* or *multi-interface* fashion:

Single-Interface Multi-Channel WMN

As shown in Fig. 8 (b), this scheme has the advantage of lower hardware cost. However, with only one interface operating at one channel at a time, the node cannot overhear the RTS/CTS exchanges at different channels. Hence, the traditional RTS/CTS mechanism cannot resolve the multi-channel hidden node problem. In this scheme, the main issues are how to coordinate transmissions among nodes and avoid the multi-channel hidden node problems.

The multi-channel MAC (MMAC) in (So & Vaidya, 2004) and slotted-seeded channel hopping (SSCH) scheme in (Bahl, Chandra, & Dunagan, 2004) were proposed to coordinate the nodes each with one interface to dynamically switch between multiple channels. In MMAC, all nodes tune to the default channel at the beginning of each slot. At the default channel, the source can transmit a channel negotiation message to the destination. By doing so, each node is aware of the channel usages within its transmission range to avoid the multi-channel hidden node problem. In SSCH, all nodes will periodically switch their channels at every slot boundary saccording to their pseudo-random hopping sequences. If there are packets to be sent, the source node can follow the destination node's channel-hopping sequence to deliver data.

Network-wide clock synchronization and channel switching overhead are two major concerns in the time-slotted MMAC and SSCH systems. In a distributed WMN with numbers of nodes and hops, synchronization among nodes is not a trivial task. The synchronization overhead also significantly degrades the throughput, especially as network size increases. Moreover, because nodes have to switch channels at every slot boundary, the channel switching overhead (higher than 224 μs(So & Vaidya, 2004)) may further lower the channel utilization.

Multi-Interface Multi-Channel WMN

Referring to Fig. 8 (c), this scheme has the following three advantages. First, with multiple interfaces, each node can concurrently deliver and receive data at different channels to improve throughput per user. Second, without the slotted time structure, the nodes do not need to synchronize with each other. Third, this scheme can work well even if employing the legacy IEEE 802.11 MAC protocol.

How to fully utilize available channels and multiple interfaces of each node is a key issue in the multichannel WMNs. A typical method is the dynamic channel assignment (DCA) protocol proposed in (Wu, Lin, Tseng, & Sheu, 2000), which can coordinate the on-demand transmissions among nodes each with multiple interfaces. In DCA, each node has one *control interface* and several *data interfaces*. Each node uses the control interface fixed at the common control channel to exchange RTS/CTS-like channel negotiation with the destination. After successful negotiation, the data interface switches to the agreed channel to deliver/receive data and acknowledge (ACK) frames. By dedicating one interface to the control channel, each node can be aware of the statuses of all available channels.

However, dedicating one channel to exchange control messages may lower overall channel utilization. Even worse, the control channel may become the bottleneck due to severe collisions as the number of contending users increases, thereby wasting the frequency spectrum of data channel and degrading overall throughput. Hence, improving overall channel utilization and resolving the bottleneck issue of control channel due to contention collisions are essential issues in the multi-channel WMNs with a dedicated control channel.

The multi-interface channel assignment protocol in (Kyasanur & Vaidya, 2005) provides a simple rule to efficiently exploit multiple channels and multiple interfaces. In this scheme, all the interfaces of each node are divided into two groups: fixed and switchable interfaces. The *fixed interfaces* are assigned to some *fixed channels* to receive data. Different nodes can use a different set of fixed channels to fully utilize all available channels. The *switchable interfaces* can be dynamically switched to different channels. The sender will switch the switchable interface to the receiver's fixed channel to deliver data. Without a dedicated common channel, this method avoids the channel utilization degradation and the bottleneck issue for the common channel.

However, if one node has asymmetric traffic (e.g., heavy incoming traffic for one node), the fixed interfaces may be overloaded while the switchable interfaces are always idle. Therefore, according to the traffic of each node, how to adaptively change the number of fixed and switchable interfaces to improve throughput, and dynamically choose the fixed channels for each node to achieve load balancing are important tasks.

To conclude, the multi-interface multi-channel multihop networking is a rather promising solution to achieve a scalable WMN with QoS provisioning. At different channels, one node can send and receive data in parallel to improve throughput and reduce delay. In addition, the operation independence of multiple interfaces at a node can facilitate the design of enhanced MAC for a scalable WMN.

In general, spectrum and hardware costs will be the major concerns in the multi-channel with multi-interface wireless mesh networks. However, there are multiple channels available for the wireless networks. For example, there are twelve non-overlapping channels for the IEEE 802.11a WLAN, three channels for the IEEE 802.11b/g WLAN, and 75MHz of spectrum reserved for the dedicated short range communication (DSRC) in intelligent transport systems (ITS). The price of interfaces also goes down very rapidly since the WLAN has become an off-the-shelf product. In addition, many WLAN equipment vendors have also developed IEEE 802.11 a/b/g multi-mode WLAN devices with multiple interfaces.

SCALABLE MULTI-CHANNEL WMN ARCHITECTURE WITH QOS PROVISIONING

Now we introduce the scalable multi-channel mesh network architectures for the dense-urban coverage and wide-area coverage – the ring-based WMN and the cluster-based WMN as shown in Fig. 4 and Fig. 5. These multi-channel WMNs are more scalable since the frequency planning reduces collisions as the number of users increases. Moreover, with the capability of designing the system parameters to control the contention situation, the proposed mesh network architectures can also facilitate the management of coverage, throughput, and QoS.

Scalable Cluster-Based WMN for Dense-Urban Coverage

Cluster-Based Network Architecture

Figure 4 shows the proposed cluster-based WMN for the dense-urban coverage (Huang et al., 2008b). In each cluster, only the central gateway AP_0 connects to the Internet through cables, while other APs can be lightweight APs and also act as wireless relays for forwarding neighboring AP's traffic to the gateway. Hence, the cabling engineering work for deploying this WMN is reduced.

This cluster-based WMN operates in a multi-channel fashion. Assume that each AP has multiple radio interfaces. Therefore, one AP (like AP_i) can concurrently provide data access for users at channel f_i', receive the forwarded traffic from AP_{i+1} at channel f_{i+1}, and delivery to AP_{i-1} at channel f_i. To avoid co-channel interference and improve throughput, frequency planning is also applied to ensure a sufficient reuse distance between the two co-channel APs.

Scalability and QoS of Cluster-Based WMN

The cluster-based WMN is scalable to the users and coverage of a cluster since frequency planning with multiple channels can reduce collisions. The delay and throughput can be ensured by properly designing the deployment parameters including the number of APs in a cluster and cell radius of each AP. In the following, we discuss how to determine the optimal deployment parameters so as to optimize the tradeoffs among delay, throughput, and coverage.

Optimal Access Point Placement

1. Problem Formulation

All the performance issues of throughput, coverage, and QoS will impact the design of WMNs. From the cost viewpoint, a larger cell is preferred due to fewer APs. From the throughput standpoint, however, a smaller cell is better since fewer users contend for the channel. The small-sized cell also leads to higher relay link capacity between APs. The frame delay consists of contention delay and queuing delay in each relay node. From the queueing delay perspective, a longer separation distance between APs may be better due to fewer hops. From the contention delay viewpoint, a smaller cell coverage is preferred due to fewer contending users. In the following, we formulate an optimization problem to determine the best number of APs in a cluster and the optimal cell radius of each AP subject to the constraints on delay, throughput, and coverage.

Referring to Fig. 9, we discuss the constraints in the optimal AP placement problem:

- The access link capacity $R(r_i)$ for one user communicating with AP_i should be greater than its demanded traffic R_D. That is,

$$f_i' \tag{1}$$

where r_i is the cell radius of AP_i. This constraint guarantees the minimum throughput for each user.

- The relay link capacity $H(d_i)$ between AP_i and AP_{i-1} should be larger enough to accommodate the carried traffic load $H_{r,i}$ of AP_i. Hence,

$$H(d_i) \geq H_{r,i} \tag{2}$$

where d_i is the separation distance between AP_i and AP_{i-1}.

- The frame delay $D_f(i)$ for the user in the cell of AP_i should meet the delay requirement D_{req}. Accordingly,

$$D_f(i) \leq D_{req}. \tag{3}$$

- The cell radius r_i of an access point should be designed from two folds. First, r_i should be less than r_{MAX} to maintain an acceptable data rate in the access link. Second, it should be larger than r_{MIN} to lower the handoff probability. Hence,

$$r_{MIN} \leq r_i \leq r_{MAX}. \tag{4}$$

- The separation distance $d_i = r_i + r_{i-1}$ between APs should be less than the maximal reception range d_{MAX} of the employed wireless system. Therefore,

$$d_i \leq d_{MAX}. \tag{5}$$

2. MINLP Optimization Approach

From the above considerations, the optimal AP placement issue can be formulated as a mixed-integer nonlinear programming (MINLP) problem with the decisions variables n (the number of APs in the single side of one cluster) and $r_0, r_1, \ldots, r_n$ (cell radii of APs). The objective function is to maximize the capacity of a cluster of APs. In this scalable WMN, the optimal coverage and

Figure 9. A cluster of APs in the dense-urban environment. d_i is the separation distance between AP_i and AP_{i-1}

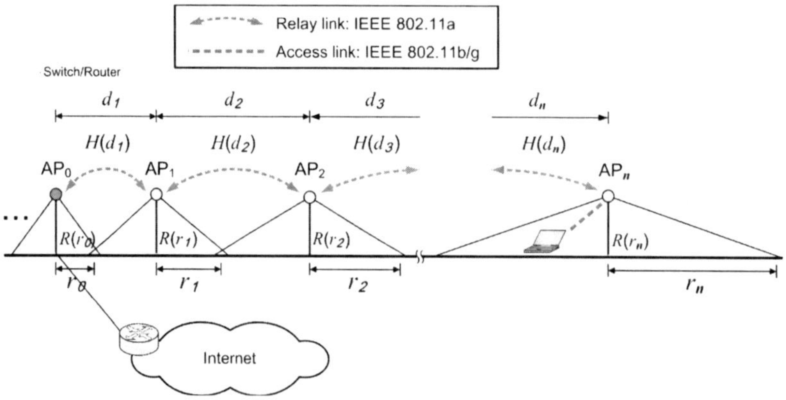

capacity will be achieved simultaneously since frequency planning resolves the collision issue and so improve the capacity. The optimal deployment parameters can be determined by solving the following optimization problem:

$10 \sim 12$

$$= \underset{n,r_0,r_1\&,r_n}{\mathrm{MAX}}\ 2\left|r_0 + 2\sum_{i=1}^{n} r_i\right| \rho R_D \qquad (6)$$

$R(r_i) \geq R_D$

$H(d_i) \geq H_{r,i}$

$D_f(i) \leq D_{req}$

$r_{MIN} \leq r_i \leq r_{MAX}$

$d_i \leq d_{MAX}$

where there are $(2n+1)$ APs in a cluster, the total coverage of a cluster is $2\left|r_0 + 2\sum_{i=1}^{n} r_i\right|$, ρ (users/m) is the user density, and R_D is the traffic demand of a user. The physical/MAC cross-layer analytical model to evaluate $R(r_i)$, $H(d_i)$, and $D_f(i)$ is provided in (Huang et al., 2008b; Huang, Wang, & Chang, 2006b).

Scalable Ring-Based WMN for Wide-Area Coverage

Ring-Based Network Architecture

Figure 5 shows a scalable ring-based WMN for the wide-area coverage (Huang et al., 2006a), where stationary mesh users with the relay capability form a multihop WMN to extend the cell coverage. The mesh cell is divided into several rings $A_i, i = 1, 2, ..., n$, determined by n concentric circles centered at the gateway with radii $r_1 < r_2 < \cdots < r_n$. The user in ring A_i connects to the gateway via an *i*-hop communication and only the gateway connects to the Internet directly. Clearly, this WMN can be rapidly deployed in a large-scale area.

The ring-based WMN operates in a multichannel with multi-interface fashion. In a mesh cell, the rings are allocated with different channels to avoid inter-ring co-channel interference and reduce the contending users. We also assume that each user is equipped with two radio interfaces. Therefore, the user in ring A_i can concurrently communicates with the users in rings A_{i-1} and A_{i+1} at different channels f_i and f_{i+1}, respectively. The suggested ring-based network architecture with frequency planning can resolve the contention collision issue as the coverage and the number of users increase. The frequency planning is simple because it only needs to design each ring width to ensure a sufficient co-channel reuse distance without interference. In addition, the WMN can work well even if employing the legacy CSMA MAC protocol, which in turn avoids the complexity and compatibility issues.

Ring-Based Frequency Planning

Now we explain the ring-based frequency assignment by an example of three-cell WMN as shown in Fig. 10. In this example, channels A_2 and $\sim A_2$ are assigned to the sectors in the innermost rings $1 \sim 3$ and $4 \sim 6$ of each cell. Channels A_1 are repeatedly allocated to the middle rings A_2 and $7 \sim 9$ of the cells with four buffer rings. Channels A_3 are allocated to rings A_4 of the cells, respectively. With four buffer rings, the channels $10 \sim 12$ are reused in the outer ring A_5. This example shows that twelve available channels can ensure four buffer rings between two co-channel rings, and with a sufficient reuse distance the channels allocated to the inner rings can be spatially reused in the outer rings. By such a three-cell pattern, we can deploy multiple cells to cover any area as shown in Fig. 10.

In a WMN, since the inner users near the gateway will relay heavy traffic for others, we also suggest sectorizing the congested inner rings and allocating different channel to each sector as shown in Fig. 10. Partitioning the inner rings can reduce the contending users and significantly improve the throughput. Apparently, if more non-overlapping channels are available, more inner rings can be sectorized to enhance cell capacity and coverage.

Scalability, QoS, and Robustness of Ring-Based WMN

Most traditional WMNs are not scalable to cell coverage because throughput and QoS (delay) are not guaranteed with increasing collisions. By contrast, the suggest ring-based WMN is more scalable in terms of coverage since the ring-based frequency planning can reduce the number of contending users to resolve the contention issue. Then, delay and throughput can be ensured by properly designing the ring widths in a mesh cell. The remaining challenge lies in how to determine the optimal number of rings and the associated ring widths to achieve the optimal tradeoff among delay, throughput, and coverage in the ring-based WMN.

As mentioned earlier, due to multiple paths for mesh node, an appealing feature of WMN is its robustness. Different form the conventional WMNs, the ring-based WMN can easily provide capacity margin for each mesh node by decreasing the ring width (and then the hop distance) to increase the relay link capacity. By doing so, even if some nodes near the central gateway fail, throughput and delay can still be ensured.

Capacity and Coverage Maximization with QoS Support for Ring-Based WMN

1. Problem Formulation

In the following, we formulate an optimization problem to determine the best number of rings in a cell and the optimal width of each ring so as to achieve the optimal tradeoff among capacity, coverage, and QoS in the ring-based WMN. We discuss the constraints in the capacity and coverage maximization problem for the ring-based WMN as shown in Fig. 5:

- The relay link capacity $H_i(d)$ for a user in ring A_i should be greater than its carried traffic load R_i. Consequently,

Figure 10. Example of a three-cell WMN with twelve available channels. Four buffer rings between two co-channel rings are ensured, and the congested inner rings (e.g., $1 \sim 3$ A_6) are sectorized. By the cellular concept, we can deploy many cells in an arbitrary area.

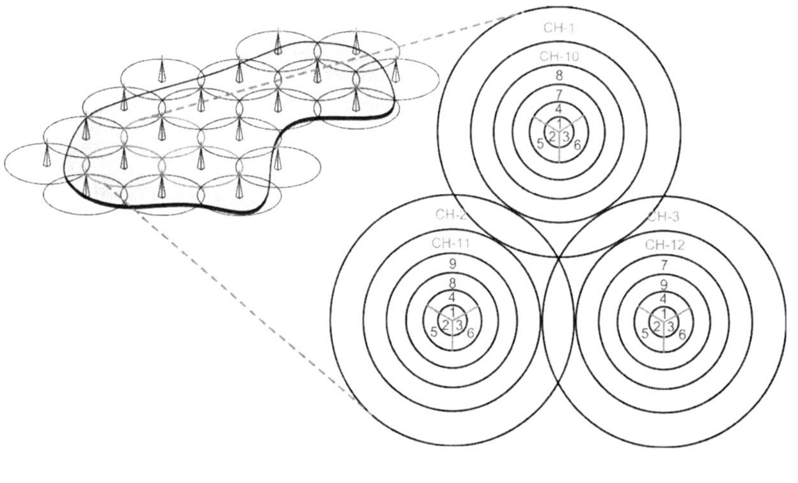

$$H_i(d) \geq R_i \tag{7}$$

where d is the average separation distance between the node and the next-hop node. This constraint guarantees the minimum throughput for each user.

- The frame delay A_2 for the user in ring A_i should meet the delay requirement D_{req}. That is,

$$D_f(i) \leq D_{req}. \tag{8}$$

- The ring width $(r_i - r_{i-1})$ should be less than the maximum reception range d_{MAX}. In addition, to ensure a sufficient co-channel reuse distance, the ring width should be greater than the a distance threshold d_{MIN} Accordingly,

$$d_{MIN} \leq (r_i - r_{i-1}) \leq d_{MAX} \tag{9}$$

where d_{MIN} is a system parameter, which depends on the number of buffer rings and the propagation environment.

2. MINLP Optimization Approach

According to the above considerations, the optimal capacity and coverage issue with the delay requirement in the ring-based WMN can be formulated as an MINLP problem with the decision variables n (the number of rings in a mesh cell) and $r_1, r_2, ..., r_n$. The objective function is to maximize the capacity of a mesh cell as follows.

$$\mathop{\mathrm{MAX}}_{n, r_1, r_2 \&, r_n} \rho\pi r_n^2 R_D \quad \text{(Overall throughput of a mesh cell)} \tag{10}$$

subject to

$$H_i(d) \geq R_i$$

$$D_f(i) \leq D_{req}$$

$$d_{MIN} \leq (r_i - r_{i-1}) \leq d_{MAX}.$$

In (10), the cell radius r_n is defined as the cell coverage, ρ is the user density, $\rho\pi r_n^2$ is the total number of users in a mesh cell, and R_D is the demanded traffic of each user. A cross-layer analytical model to evaluate $H_i(d)$ and $D_f(i)$ is detailed in (Huang et al., 2006a).

Performance Evaluations for Cluster-Based WMN and Ring-Based WMN

Figure 11 investigates the interactions among delay, capacity, and coverage in the proposed cluster-based and ring-based WMNs. The numerical results are analytically derived by means of the MINLP optimization approach and the cross-layer analytical models developed in (Huang et al., 2008b, 2006b, 2006a).

Figure 11 (a) illustrates the frame delay against the capacity and coverage of a cluster under various delay requirements. In the cluster-based WMN, the IEEE 802.11a WLAN is used for data forwarding between APs, while the IEEE 802.11b/g WLAN for data access between APs and users. This figure shows that the frame delay can be dramatically improved from 8×10^{10} to 0.1 (s), while the optimal capacity of a cluster merely decreases from 36.6 to 36.3 Mbps if the number of APs in a cluster is $n = 5$. In the meanwhile, the optimal coverage reduces from 1830 to 1816 (m). This phenomenon of extreme delay can be explained by the fact that the relay link between APs is fully utilized if no delay constraint is imposed. Hence, the sojourn time of data frame at an AP will grow toward a very large value (Gross & Harris, 1998). However, by shortening the hop distance between two APs, the link capacity and delay can be improved at the cost of a smaller coverage of a cluster as shown in the figure. Figure 11 (a) also shows that the delay requirement $D_{req} = 0.01$ (s) can be fulfilled at the expense that the optimal cell capacity decreases to 28.1 Mbps with the coverage of 1404 (m) at $n = 4$.

Figure 11 (b) shows the frame delay versus the cell capacity and coverage under different delay

Figure 11. Optimal tradeoff among capacity, coverage, and delay: (a) In the cluster-based WMN, the user density is $\mathrm{m}^{-2}\ \mathrm{m}^{-1}$ *and the demanded traffic of each user is* $R_D = 0.4$ *(Mbps); (b) In the ring-based WMN, both rings* A_1 *and* A_2 *are sectorized,* $\rho = \left(0.01\right)^2\ \mathrm{m}^{-2}$ *and* $R_D = 0.5$ *(Mbps).*

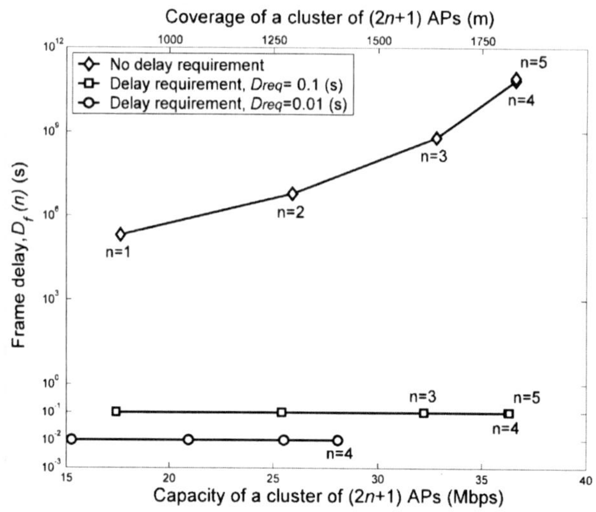

(a) Frame delay versus capacity and coverage of a cluster of $(2n+1)$ APs under various delay requirements.

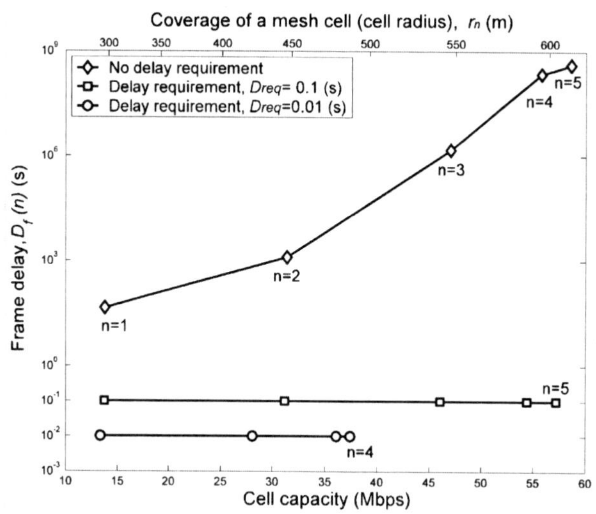

(b) Frame delay versus capacity and coverage of a ring-based mesh cell under various delay requirements.

requirements in the ring-based WMN, where the most-congested rings A_1 and A_2 of each cell are sectorized as shown in Fig. 10. In this example, the IEEE 802.11a WLAN is used for forwarding data between nodes. To meet the delay requirement $D_{req} = 0.1$ (s), the optimal cell capacity slightly decreases from 58.6 to 57.2 (Mbps) and the cell coverage reduces from 610 to 603 (m), when the number of rings in a cell is $n = 5$. However, for the more stringent requirement $D_{req} = 0.1$ (s), the optimal cell capacity will diminish to 37.4 Mbps at $n = 4$ with the coverage of 488 (m).

From the above figures, we investigate the interactions among delay, capacity, and coverage in WMNs. It is demonstrated that by properly designing the system parameters, the optimal capacity and coverage for the considered scalable WMNs can be achieved concurrently. In addition, QoS (delay) can be supported at the cost of lower capacity and coverage. Detailed performance evaluations for the cluster-based WMN and the ring-based WMN are provided in (Huang et al., 2008b, 2006b, 2006a).

SUMMARY

The wireless mesh network (WMN) is a promising technology in the next-generation communication system to support the ubiquitous broadband services with low transmission power. The objective of this chapter is to investigate the scalability issue of WMNs from a network architecture perspective. We concluded that the multi-channel multi-interface mesh network architecture is a rather viable solution to achieve a scalable WMN with QoS support, because of the advantages of better throughput and delay, and easier MAC protocol design.

This chapter has also introduced two scalable multi-channel WMN architectures for the dense-urban and wide-area coverage with QoS support. The considered WMN architecture can relieve the scalability issue for WMN since the multi-channel frequency planning can reduce collisions and improve throughput by reducing the number of contending users at a radio channel. Moreover, the proposed network architecture can facilitate the management of interactions among coverage, throughput, and QoS. Subject to the QoS requirement, the optimization approach has been proposed to maximize the capacity and coverage for the proposed WMNs. Performance evaluation demonstrated that by the proposed scalable WMN architecture with appropriate system parameter design, the goals of cell capacity enhancement and QoS provisioning can be fulfilled at a slight cost of coverage performance.

Many important research problems related to scalability and QoS in WMNs are still open and need to be further investigated, such as the power fairness problem, differentiated services and QoS provisions, etc., as discussed in this chapter. Furthermore, when the advanced techniques such as multi-input multi-output (MIMO), cooperative communication, cognitive radio (CR) and network coding are incorporated into WMNs, developing new scalable network architecture and novel MAC protocols are also very interesting topics and worthwhile for further investigation.

ACKNOWLEDGMENT

This work was supported in part by the MoE ATU Plan, the Program for Promoting Academic Excellence of Universities (Phase II), and the National Science Council under Grant 97W803C, Grant NSC 96-2752-E-009-014-PAE, Grant NSC96-2221-E-009-061, and Grant NSC96-2221-E-009-193.

REFERENCES

Akyildiz, I. F., Lee, W.-Y., Vuran, M. C., & Mohanty, S. (2006). NeXt generation/dynamic spectrum access/cognitive radio wireless networks: a survey. *Computer Networks*, *50*, 2127–2159. doi:10.1016/j.comnet.2006.05.001

Akyildiz, I. F., Wang, X., & Wang, W. (2005, March). Wireless mesh networks: a survey. *Computer Networks*, *47*, 445–487. doi:10.1016/j.comnet.2005.01.002

Bahl, P., Chandra, R., & Dunagan, J. (2004). SSCH: slotted seeded channel hopping for capacity improvement in IEEE 802.11 ad hoc wireless networks. In *Proc. ACM Annual International Conference on Mobile Computing and Networking (MobiCom)* (pp. 216-230).

Choudhury, R., Yang, X., Ramanathan, R., & Vaidya, N. (2002). Using directional antennas for medium access control in ad hoc networks. In *Proc. ACM MobiCom'02* (pp. 59-70).

FCC. (2003, December). *ET Docket No 03-222 Notice of proposed rule making and order.*

Fowler, T. (2001, January). Mesh networks for broadband access. *IEEE Review*, *47*(1), 17–22. doi:10.1049/ir:20010108

Gross, D., & Harris, C. M. (1998). *Fundamentals of queuing theory* (3rd ed.). New York: John Wiley & Sons, Inc.

Gupta, P., & Kumar, P. R. (2000, March). The capacity of wireless networks. *IEEE Transactions on Information Theory*, *46*, 388–404. doi:10.1109/18.825799

Haykin, S. (2005). Cognitive radio: brain-empowered wireless communications. *IEEE Journal on Selected Areas in Communications*, *23*(2), 201–220. doi:10.1109/JSAC.2004.839380

Holland, G., & Vaidya, N. H. (2002, March-May). Analysis of TCP performance over mobile ad hoc networks. *Wireless Networks*, *8*(2-3), 275–288. doi:10.1023/A:1013798127590

Huang, J.-H., Wang, L.-C., & Chang, C.-J. (2006a, November). Capacity and QoS for a scalable ring-based wireless mesh network. *IEEE Journal on Selected Areas in Communications*, *24*(11), 2070–2080. doi:10.1109/JSAC.2006.881622

Huang, J.-H., Wang, L.-C., & Chang, C.-J. (2006b, October). Wireless mesh networks for intelligent transportation systems. In *Proc. IEEE International Conference on Systems, Man, and Cybernetics.*

Huang, J.-H., Wang, L.-C., & Chang, C.-J. (2008a, November). Power fairness in a scalable ring-based wireless mesh network with variable ring-width design. In *Proc. IEEE Globecom'08.*

Huang, J.-H., Wang, L.-C., & Chang, C.-J. (2008b, September). QoS provisioning in a scalable wireless mesh network for intelligent transportation systems. *IEEE Transactions on Vehicular Technology*, *57*(5), 3121–3135. doi:10.1109/TVT.2008.918701

Huang, J.-H., Wang, L.-C., & Chang, C.-J. (2008c, March). Throughput-coverage tradeoff in a scalable wireless mesh network. *Journal of Parallel and Distributed Computing*, *68*(3), 278–290. doi:10.1016/j.jpdc.2007.10.003

Ju, H., Rubin, I., & Kuan, Y. (2003). An adaptive RTS/CTS control mechanism for IEEE 802.11 MAC protocol. In *Proc. IEEE VTC'03.*

Jun, J., & Sichitiu, M. (2003, October). The nominal capacity of wireless mesh networks. *IEEE Wireless Commun. Mag.*, *10*(5), 8–14. doi:10.1109/MWC.2003.1241089

Kawadia, V., & Kumar, P. R. (2005, February). A cautionary perspective on cross layer design. *IEEE Wireless Commun. Mag.*, *12*(1), 3–11. doi:10.1109/MWC.2005.1404568

Kramer, G., Gastpar, M., & Gupta, P. (2005, Sept.). Cooperative strategies and capacity theorems for relay networks. *IEEE Transactions on Information Theory*, *51*(9), 3037–3063. doi:10.1109/TIT.2005.853304

Kyasanur, P., & Vaidya, N. H. (2005). Routing and interface assignment in multichannel multiinterface wireless networks. In *Proc. IEEE WCNC'05.*

Lee, M. J., Zheng, J., Ko, Y.-B., & Shrestha, D. M. (2006, Apr.). Emerging standards for wireless mesh technology. *IEEE Wireless Commun. Mag.*, *13*(2), 56–63. doi:10.1109/MWC.2006.1632481

Lewis, B. (2003, July). *Mesh networks in fixed broadbandwireless access.* Retrieved from http://grouper.ieee.org/groups/802/16/docs/03/C80216-03 10r1.pdf

Li, X., & Bao-yu, Z. (2003, August). Study on cross-layer design and power conservation in ad hoe network. In *Proc. Fourth International Parallel and Distributed Computing, Applications and Technologies* (pp. 324-328).

Maharshi, A., Tong, L., & Swami, A. (2003, August). Cross-layer design of multichannel reservation MAC under rayleigh fading. *IEEE Transactions on Signal Processing, 51*(8), 2054–2067. doi:10.1109/TSP.2003.814465

MeshDynamics. (2009). Retrieved from http://www.meshdynamics.com

MeshNetworks. (2009). Retrieved from http://www.motorola.com/mesh

Nosratinia, A., Hunter, T. E., & Hedayat, A. (2004, October). Cooperative communication in wireless networks. *IEEE Communications Magazine*, 74–80. doi:10.1109/MCOM.2004.1341264

Pabst, R. (2004, Sept.). Relay-based deployment concepts for wireless and mobile broadband radio. *IEEE Communications Magazine, 42*(9), 80–89. doi:10.1109/MCOM.2004.1336724

Qiu, L., et al. (2004, November). *Troubleshooting multihop wireless networks.* Microsoft Research Tech. Report, MSR-TR-2004-11. Retrieved from ftp://ftp.research.microsoft.com/pub/tr/TR-2004-11.pdf

So, J., & Vaidya, N. (2004, May). Multi-channel MAC for ad hoc networks: handling multi-channel hidden terminals using a single transceiver. In *Proc. ACM International Symposium on Mobile Ad Hoc Networking and Computing (MobiHoc)* (pp. 222-233).

Toumpis, S., & Goldsmith, A. J. (2003, May). Performance, optimization, and cross-layer design of media access protocols for wireless ad hoc networks. In *Proc. IEEE ICC'03* (pp. 2234-2240).

Wu, S.-L., Lin, C.-Y., Tseng, Y.-C., & Sheu, J.-P. (2000). A new multi-channel MAC protocol with ondemand channel assignment for multi-hop mobile ad hoc networks. In *Proc. International Symposium on Parallel Architectures, Algorithms and Networks (ISPAN).*

Zhang, M., & Wolff, R. (2004, February). Crossing the digital divide: cost-effective broadband wireless access for rural and remote areas. *IEEE Communications Magazine, 42*(2), 99–105. doi:10.1109/MCOM.2003.1267107

Zhong, R., & Kravets, R. (2003). On demand power management for ad hoc networks. In *Proc. IEEE INFOCOM'03* (pp. 481-491).

Chapter 24
Towards Designing High-Throughput Routing Metrics for Wireless Mesh Networks

T. Nyandeni
Council for Scientific and Industrial Research (CSIR), Defence, Peace, Safety and Security (DPSS), South Africa

C. Kyara
Council for Scientific and Industrial Research (CSIR), MERAKA, South Africa

P. Mudali
University of Zululand, South Africa

S. Nxumalo
University of Zululand, South Africa

N. Ntlatlapa
Council for Scientific and Industrial Research (CSIR), MERAKA, South Africa

M. Adigun
University of Zululand, South Africa

ABSTRACT

Routing is an essential mechanism for proper functioning of large networks, and routing protocols make use of routing metrics to determine optimal paths. The design of routing metrics is critical for achieving high throughput and we begin this chapter by proposing the design principles for routing metrics. These design principles are for ensuring the proper functioning of the network and achieving high throughput. We continue by giving a detail analysis of the existing routing metrics. We also look at the pitfalls of the existing routing metrics. We conclude the chapter by outlining the future research directions.

DOI: 10.4018/978-1-61520-680-3.ch024

INTRODUCTION

Routing in ad-hoc wireless networks has been an active area of research for decades. Most of the research work in this area was highly motivated by the need to consider energy constraints enforced by battery powered nodes and their mobility. The main objective was to provide routes that are flexible against dynamic topology. WMNs have a bit different characteristics from an ordinary ad-hoc network. Most of the nodes in WMNs are stationary and therefore changes that are caused by a dynamic topology are of less concern. Therefore there is a need for the focus to shift from maintaining network connectivity to finding high-throughput routes between nodes, so as to provide users with maximal end-to-end throughput. Supporting Quality of Service (QoS) to enable a rich range of applications is foreseen to be very important for the success of wireless mesh networks (WMN) (Akyildiz, I. Wang X. et al, 2005). Routing is about finding the best path (route) between source and destination(s). Finding this path between source and destination(s) involves two steps:

i. Assigning cost metrics to links and paths
ii. Propagating routing information

The second step, route information propagation, is the responsibility of the routing protocol. Routing protocols have received much attention over the past decade (Koksal, C. 2008). There are two widely accepted types of routing protocols: proactive and reactive. Proactive routing protocols establish paths before they are required. Proactive routing protocols calculate routing tables and maintain them before they are even required. Examples of proactive routing protocols include, Destination-Sequenced Distance Vector Routing (DSDV, (Perkins, C. & Bhagwat, P. 1994)), Fisheye State Routing (FSR, (Gerla, M. Hong, X. et al 2002)), and Optimized Link State Routing (OLSR, (Jacquet, P. Muhlethaler, P. et al 2002)). On the other hand, Reactive routing protocols, do not establish paths before they are required. Route discovery follows the communication request. Examples of reactive protocols include Ad Hoc On Demand Distance Vector (AODV, (Perkins, C. & Royer, E. 1999)) and Dynamic Source Routing (DSR, (Johnson, D. Maltz, D. et al 2002)). The Hybrid approach combines properties of both the reactive and proactive routing protocols and is not as well established as the other two types.

In this chapter we address the issue of assigning the cost metrics to links and paths. A routing protocol needs a method for differentiating different paths according to their quality. This differentiation is the responsibility of routing metric (cost metric, path selection metric). Basically the routing metric is the cost of forwarding a packet along the link. The problem of defining a cost metric is significantly harder in wireless networks than in traditional wired networks, because the notion of a "link" between nodes is not well-defined. This chapter focuses on studying how high throughput can be achieved in WMNs through the use of routing metrics.

Background on Hybrid Routing

The IEEE 802.11s working group proposes an Extensible Path Selection Framework. This framework enables flexible implementation of path selection protocols (routing protocols) and metrics within this standard. This framework specifies a default mandatory protocol and metric for all implementations. This framework also allows vendors to implement any protocol or metric to meet special application needs. A mesh point (MP) may include multiple implementations (e.g., including the default protocol, optional protocols, future standard protocols, etc) (IEEE 2006). Unfortunate only one protocol can be active on a particular link at a time. The default path selection protocol for IEEE 802.11s standard is hybrid wireless mesh protocol (HWMP). Every 802.11s device must implement HWMP to ensure interoperability.

HWMP is an example of a hybrid routing protocol. HWMP integrates the flexibility of on-demand routing capabilities with extensions to enable efficient proactive routing to mesh portals (gateways to the Internet). HWMP is based on the modified version of AODV called Radio Metric AODV (RM_AODV). The main difference between the original AODV and RM_AODV is the routing metric. The original AODV uses *hop count* while RM_AODV uses the newly proposed *airtime link* metric. The combination of on-demand and proactive routing capabilities allow MPs to perform the discovery and maintenance of optimal routes (according to airtime link metric) themselves or to additionally leverage the formation of a tree structure based on a root node (preferable mesh portal point (MPP)) to quickly establish paths to root nodes. HWMP uses a single set of protocol primitives and processing rules taken from AODV (Perkins, C. & Royer, E. 1999) for all routing related functions.

Design Criteria for High-Throughput Routing Metrics

Designing routing metrics is critical for network performance. Usually the design of routing metric is specific to the unique characteristics of application being considered. Nevertheless the design has to meet certain minimum requirements to ensure proper functioning. In this section we focus on the requirements that a routing metric has to meet. We call these requirements design criteria. We have divided the requirements into two groups, the first group is for ensuring the proper functioning of the routing metric and the network and the second group is for ensuring that high throughput is achieved. The first group is formulated based on the work done by Yang, Y. et al, (2006).

Group A: Requirements ensuring proper functioning of the routing metric and network

i. **Route Stability:** A network can be badly affected by unstable path weights. Frequent changes to path weights can cause high volumes of route update messages. This will then degrade an overall network throughput. Route stability can be achieved by the type of path characteristics used as the part of routing metric. A routing metric used can either be load-sensitive or topology-dependent (Yin, S. Xiong, Y. et al 2006). Load-sensitive routing metric assign a weight to a route based on the traffic load on that particular route. Topology-dependent routing metrics assign a weight to a link based on the topology properties of the path, such as the hop count, link capacity of the link. It is still not very clear which type (load-sensitive or topology-dependent) of routing metric yields high network throughput. The only advice that we can give is to select the routing metric that will be more stable.
ii. **Loop-free routing:** There are two main factors that affect the routing efficiency. These factors are routing loops and the maintaining of routing information. The latter is solved by on-demand routing. Routing loops increases packet delivery delay and decrease packet delivery ratio. This may then lead to the decrease on network throughput. It is very pivotal to avoid use of routing metrics that increases chance of routing loops. This can avoided if it is considered at the design stage of routing metric.

Group B: Requirements for achieving high throughput

i. **Physical and MAC attributes:** Generally, each layer has its own state parameters that can be provided to other layers. The methodology of layered protocol design does not necessarily lead to an optimum solution for wireless networks. Cross-layer design can drastically improve network performance (Akyildiz, I. Wang X., et al 2005). Incorporating PHY/MAC attributes into routing metric may render better, high-throughput routes and further improve the overall network throughput (Hou, et al, 2007).

ii. **Mesh client attributes:** Usually mesh clients are battery powered and may move arbitrarily. Most of the existing studies have focused on MAC and routing on mesh routers, without considering the characteristics of mesh clients. There is need for considering the end-to-end performance requirement and constraints of mesh clients into designing routing metric.

iii. **Application Layer QoS demands:** Network users, usually access different types of applications such as VoIP and file sharing. These applications have different QoS demands; therefore there is a need for a routing metric to dynamically change performance metrics when selecting an optimum route for specific application.

iv. **Good performance for minimum weight paths:** The major goal of any routing protocol is to route packets through some optimum path, based on certain routing metric. The optimum path can either be the path with the minimum or maximum weight. This depends on the nature of the routing metric being used. For example if delay is used as the weight of links of the path, then the optimal path should have a minimum weight. In a case of the packet delivery ratio, the optimum path will be the one with the maximum weight. To ensure the maximum utilization of network resources of the mesh network, the path weight selected must have good performance in terms of high throughput. In wireless networks, the bandwidth is shared among neighboring nodes. Inter-flow interference can result in bandwidth starvation for some nodes since those nodes can always experience busy channel. Intra-flow interference increases the bandwidth consumption of the flow at each of the nodes along the path and causes the throughput of the flow to degrade sharply and the delay at each hop to increase dramatically as the hop count of the flow increases.

Review of Routing Metrics for WMN

In this section we study sixteen routing metrics and determine if these metrics meet our design criteria. We also look at the advantages and shortcomings of these routing metrics. The routing metrics we discus in this section have been adopted by different routing protocols like, (Biswas, S. and Morris, R. 2005; koksal, C. and Balakrishnan, H. 2006; Perkins, C. and Bhagwat, P. 1994; Jacquet, P. Muhlethaler, P. et al, 2002; Gerla, M. Hong, X. and Pei, G. 2002; Perkins, C. and Royer, E. 1999) for ad-hoc wireless networks.

The HOP Count (HOP) (Dijkstra, 1959): Hop count, is the most basic metric and is taken from the generalized Dijkstra's Algorithm (Dijkstra, 1959). This is a greedy algorithm that computes the shortest paths from a given source node to every other node in the network. The minimum hop-count metric chooses arbitrarily among the different paths of the same minimum length, regardless of the often large differences in throughput among those paths, and ignoring the possibility that a longer path might offer higher throughput. Link quality for this metric is a binary concept; either the link exists or it does not exist at all. The advantage of this metric is that it is simple to use. Once the topology is known, it is easy to compute and minimize the hop count between a source and a destination. Moreover, computing the hop count requires no additional measurements. The primary disadvantage of this metric is that it does not take packet loss or bandwidth into account. It has been shown in (D. De Couto, et al., 2003) that a route that minimizes the hop count does not necessarily help in maximizing the throughput of a flow. For example, a four-hop path over reliable or fast links can exhibit better performance than a two-hop path over a slow link.

Per-Hop Packet Pair Delay (PktPair) (Keshav, 1991): PktPair measures delay between a pair of back-to-back probes to a neighboring node. To calculate this metric, a node sends two probe packets back-to-back to each neighbor every 2 seconds. The first probe packet is smaller than the second packet. The neighbor calculates the delay between the receipt of the first and the second packets. It then reports this delay back to the sending node. The sender maintains an exponentially weighted moving average of these delays for each of its neighbors. The objective of the routing algorithm is to minimize the sum of these delays. Like the Per-hop Round Trip Time (RTT) (Adya, et al., 2004) metric, this metric also measures several facets of link quality. The main advantage of this metric over RTT is that it is not affected by queuing delays at the sending node, since both packets in a pair will be delayed equally. In addition, using a larger packet for the second probe makes the metric more sensitive to the link bandwidth than the RTT metric. This metric has several disadvantages. First, it is subject to overheads even greater than those of the RTT metric, since two packets are sent to each neighbor, andthe second packet is larger. Second, we discovered that the metric is not completely immune to the phenomenon of self interference.

Path Predicted Transmission Time (PPTT) (Yin et al., 2006): PPTT estimates end-to-end delay of real-time traffics. PPTT is traffic-aware by taking explicit consideration of both self-traffic and neighboring traffics interfering with the real-time flow. The optimal route must be the one with minimal PPTT. This improves the QoS level for the coming real-time flow. By selecting route with minimum end-to-end delay, it improves the overall network throughput.

Expected Transmission Count (ETX) (De Couto, et al., 2003): ETX estimates the number of retransmissions needed to send unicast packets by measuring the loss rate of broadcast packets between pairs of neighboring nodes. The derivation of ETX starts by measuring the underlying packet loss probability for both forward and reverse directions. To compute ETX, each node sends a probe every second which contains the number of probes received by each neighboring node in the previous 10 seconds. Based on the probes, the node can calculate loss rate of probes on the links to and from its neighbors. Since the 802.11 MAC does not retransmit broadcast packets, these counts allow the sender to estimate the number of times the 802.11 ARQ mechanisms will retransmit a unicast packet. The formula that is used to calculate ETX is:

$$ETX = \sum_{k-1}^{\infty} k * s(k) = \frac{1}{1-p} \qquad (1)$$

Parameter k is the number of attempts to send a packet and s (k) is the probability that the packet will be sent successfully from x to y. P denotes the probability that the packet from x to y was not successfully sent and is calculated as $p = 1 - (1 - p_f) * (1 - p_r)$. S(k) is calculated as: $s(k) = p^{k-1} * (1 - p)$.The term p_f represents packet loss probability in the forward direction while p_r represents packet loss probability in the reverse direction. The path metric is the sum of the ETX values for each link in the path. The routing protocol selects the path with minimum path metric. ETX improves the throughput, while its drawback is that it fails under variability link conditions.

Modified ETX (mETX) (Koksal and Balakrishnan, 2006): mETX was proposed as an improvement to ETX, which does not cope well with short-term channel variations because it uses the mean loss ratios in making routing decisions. mETX gives better evaluation of a multi-channel path. Modified ETX is calculated as:

$$mETX = \exp\left(\mu\sum + \frac{1}{2}\sigma_{\sum}^{2}\right) \qquad (2)$$

where $\mu\sum$ represents the impact of slowly varying and static components in the channel, while the term $\sigma_{\sum}^{2}$ symbolizes the impact of relatively fast channel variations. To estimate these two parameters, bit level information is necessary. Counting only the packet losses is not enough; as the result, probe packets are being used for estimation. This routing metric may achieve high throughput but it is complex. This also applies to effective number of retransmission (ENT) (Koksal and Balakrishnan, 2006). ENT was proposed to find routes that satisfy certain higher layer protocol requirement. The main challenge of ENT is to find a path with high throughput while ensuring that the end-to-end packet loss rate visible to higher layers does not exceed a specific threshold. The main drawback of these routing metric is complex channel state estimation method it employs.

Expected Transmission Time (ETT) (Draves et al., 2004): ETT was proposed as improvement of ETX. ETT considers the differences in link transmission rates. The ETT of a link is defined as the expected MAC layer duration for a successful transmission of a packet at link. The weight of a path p is simply the summation of the ETTs of the links on the path. ETT is calculated as:

$$ETT_l = ETX_l * \frac{s}{b_l} \qquad (3)$$

Where b_l is the transmission rate of link l and s is the packet size. Essentially, by introducing b_l into the weight of a path, the ETT metric captures the impact of link capacity on the performance of the path. Similar to ETX, ETT is also isotonic. However, the remaining drawback of ETT is that it still does not fully capture the intra-flow and inter-flow interference in the network. Weighted Cumulative ETT (WCETT) was proposed as an improvement on ETT. WCETT tries to route packets on path that has least number of nodes transmitting on the same channel. This helps to reduce intra-flow interference. For a path p, WCETT is defined as:

$$WCETT = (1-\beta)\sum_{linkl \in p} ETT_l + \beta \max_{1 \le j \le k} x_j \qquad (4)$$

Where β is a tunable parameter subject to $0 \le \beta \le 1$. x_j is the number of times channel j is used along path p and captures the intra-flow interference. The $\max\limits_{1 \le j \le k} x_j$ component in the above equation counts the maximum number of times that the same channel appears along a path.

It captures the intra-flow interference of a path since it essentially gives low weights to paths that have more diversified channel assignments on their links and hence lower intra-flow interference. WCETT has two main drawbacks. The first drawback is that it does not explicitly consider the effects of inter-flow interference, although it does capture intra-flow interference. Therefore, WCETT may route flows to dense areas where congestion is more likely and may even result in starvation of some nodes due to congestion. Different researchers have tried to improve on this routing metric. Ma and Denko (2007) propose the variant of this routing metric, called WCETT-LB (Weighted Cumulative Expected Transmission Time with Load Balancing). WCETT-LB introduces load balancing features at the mesh points and supports global load-aware routing. Integration of a load-balancing scheme can improve the performance of the entire network. The load-balancing component consists of two parts: congestion level and traffic concentration level at each node in a particular path. WCETT-LB is computed as follows:

$$WCETT - LB(p) = WCETT(p) + L(p) \quad (5)$$

where

$$L(p) = \sum_{nodel=p} \frac{QL_i}{b_i} + \min(ETT) N_i \quad (6)$$

QL_i is the average queue length at a node in a particular path, b_i is the transmission rate at a node and N_i is used for considering the traffic concentration of each node. There are other variants of ETT that have been proposed on the literature. These include Exclusive Expected Transmission Time (EETT) (Jiang et al., 2007). EETT of a link *l* represents the busy degree of the channel used by link *l*. Other WCETT variants are Metric of Interference and Channel Switching (MIC) (Yang et al, 2005), Multi-Channel Routing Protocol (MCR) (Kyasanur and Vaidya, 2005), Interference-aware routing metric (i-AWARE) (Subramanian et. al, 2006), Adjusted Expected Transmission Delay (AETD) (Zhou et al., 2006) and Exclusive Expected Transmission Time (EETT) (Jiang et al., 2007). These variants add components of the interference or channel switching to the original ETT. Another ETT variant is the Improved Expected Transmission Time (iETT) Biaz, S, & Qi, B. 2008). The iETT routing metric is designed to take into account (a) the discrepancy of link loss rates within one path and (b) the MAC layer overheads when computing an expected packet transmission time (instead of simply using packet/bandwidth). By being able to capture the two characteristics, iETT chooses a route with better performance.

Adjusted Expected Transmission Delay (AETD) (Zhou et al., 2006): The key idea of AETD is to consider both delay and jitter of candidate paths when making the routing decision. It is designed to select a route on which hops operating on the same frequency channel are separated as far as possible. In this way, interference and channel contention may be minimized along the preferred route and network throughput may be improved. When a sequence of packets is transmitted from a source node to a destination, the achieved throughput is determined by the following features of the selected route:

- **ETD:** The expected end-to-end transfer delay of a single packet
- **EDJ:** The lower bound of the expected delay jitter between consecutive packet transmissions

Both ETD and EDJ of a perfect route have to be small. ETD is affected by the following: (1) the hop count of the route; and (2) the bandwidth and link quality of each hop along the route that determine the per-hop transmission rate and transmission time. A shorter path does not necessarily have smaller end-to-end transfer delay (De Couto, et al., 2003; Tang, et al., 2005; Kodialam, M. and Nandagopal, T., 2003). EDJ is affected by: (1) the channel diversity of the route; and (2) the bandwidth and link quality of each hop along the path that determine the per-hop transmission rate and transmission time. A more channel-diverse route experiences less interference as packet transmissions on different channels do not interfere with each other. In the extreme case when the route is perfectly channel-diverse, i.e., when packet transmissions on any two hops along the route do not interfere with each other — either because they are far apart from each other or because they operate on different frequency channels, packet transmissions on each hop may proceed successfully at the same time without encountering any channel contention and the consequent contention resolution procedure. Hence, a very short delay jitter between consecutive packet transmissions is expected under such scenario, which equals the maximum single-hop transmission time along the route.

$$ETD_r = \sum_{hi \in H_r} ETT_{hi} \quad (7)$$

Where $H_r = \{h_1, h_2, \cdots\cdots, h_k\}$ indicate the corresponding hop sequence along the route, and h_i represent the hop between nodes (i-1) and i. r denotes the expected packet transmission time over hop h_i.

$$EDJ_{r(i)} = \begin{cases} ETT_{hk} \quad if \\ ETT_{hi+1} + EDJ_{r(i+1)} \\ if \ \exists i+1 < j \leq \min\{i+m+1, k\} \\ such \ that \ C_{hi+1} = C_{hj}, \\ \max\{ETT_{hi+1}, EDJ_{r(i+1)}\} \quad else \end{cases} \tag{8}$$

where m is the interference distance (measured in hops) in the hop-distance-based interference model.

$$AETD = (1-\alpha) \times ETD + \alpha \times EDJ \tag{9}$$

Airtime Link Metric (IEEE 802.11, 2006): This path selection metric was proposed to take into consideration amount of channel resources that are consumed by transmitting a frame over particular link. The airtime cost of each link is determined by:

$$c_a = \left[O_{ca} + O_p + \frac{B_t}{r}\right]\frac{1}{1-e_{pt}} \tag{10}$$

Where O_{ca} is a channel access overhead, O_p is MAC protocol overhead and B_t is the number of bits of a test frame, listed in table 2, r, and e_{pt} are the bit rate in Mbps and the frame error rate for the test frame size B_t respectively. The values of these parameters depend on the used IEEE 802.11 transmission technology such as IEEE 802.11b or IEEE 802.11g (Bahr, M, 2006), (see table 1). The r represents the rate at which a MP would transmit the test frame under current conditions. The principal advantage of airtime link metric is that it takes into account the quality of different links (Shen, Q and Fang, X, 2006). The airtime link metric is not network load and interference aware, this has lead on the new path selection metric (Multi-Metric) being proposed by Shen, Q. & Fangu, X. (2006).

Multi-Metric (Shen, Q and Fang, X, 2006): Multi-metric takes the residual bandwidth and frame delivery ratio (FDR) of the link into consideration when selecting an optimal path. The use of FDR is simply because it is sensitive to interference. The use of FDR helps multi-metric on addressing issues of interference that are neglected by airtime link metric. The path's cost is given by linear combination (Eq. 11) of minimum residual bandwidth, maximum load and FDR of the link.

$$C_a = \alpha.Min\_Bw - \beta.Max\_Load + \gamma.FDR \tag{11}$$

Where, Min_Bw is the minimal residual bandwidth, and α is its weighted factor. Mac_Load is the maximum load of node in route, and β is its weighted factor. γ is the weighted factor FDR. α, β and γ must satisfy the following constrain of $|\alpha| + |\beta| + |\gamma| = 1$ according to (Shen, Q and Fang, X, 2006). This path selection metric neglects the issues of load balancing. It is pivotal for a path selection metric to be able to identify potential bottleneck nodes

Table 1. Airtime cost contents (IEEE 802.11, 2006)

Parameter	Value (802.11a)	Value (802.11b)
O_{ca}	75μs	335μs
O_p	110μs	364μs
B_t	8224	8224

Table 2. Summary of Routing Metrics: Part 1

Metric	Definition	Benefit	Based on	Drawback	PHY/MAC Attribute	Performance Metric	Mesh Client Attributes	Higher Layer QoS Demands
Hop Count	Number of hops	Simplicity	Number of hops	Chooses poor links. Does not consider QoS demands of the flow.	Null	Null	Null	Null
RTT	Per-hop Round Trip Time	Incorporate multiple factors	Measured RTT	Self interference. Does not consider QoS demands of the flow.	Channel contention	Delay	Null	Null
PktPair	Per-hop Packet Pair Delay	Reduces self interference	Measured RTT	High overhead. Does not consider QoS demands of the flow.	Channel contention	Delay	Null	Null
PPTT	Path Transmission Time	Support real-time and reduces self interference	Estimated Transmission count	Complex. Does not consider QoS demands of the flow.	Channel contention	End-to-end delay	Null	Null
ETX	Expected Number Count	Improves throughput	Estimated Transmission count	Fails under variability link conditions. Does not consider QoS demands of the flow.	Null	End-to-end delay	Null	Null

and avoid paths that are made of such node.

Both Table 2 and 3 (Figure 1 gives the summary in percentages) give the summary of all routing metrics that we have discussed. The summary is based on our design criteria. So far we have discussed sixteen routing metrics that have been proposed for wireless mesh networks. Our analysis shows that these routing metrics can be grouped into three groups as follows: 1) routing metrics that are based on transmission count and time, 2) routing metrics that are based on round trip time and 3) routing metrics that are based on packet loss rate. Most of these routing metrics are based on estimated transmission time. Most of routing metrics that we have illustrated in this chapter do not take advantage of service differentiation, and QoS demands of the flows.

Jiang, H., Zhuang, W., et al (2006), suggests that routing metric should find a path that satisfies multiple metrics, so as to meet QoS demands of various flows. The main pitfall of these routing metrics is that they find path that satisfy only performance metric. These routing metrics do not consider mesh client attributes (Fig. 1).

FUTURE RESEARCH DIRECTION

Despite the fact that, vast literature exist on the area of wireless routing, there are still open research issues on routing that should be addressed in order to achieve high-throughput and stable WMNs. Many of these routing metrics have not been implemented in real-life networks. There are number of directions along which the routing metrics can be designed. In this section we present what we believe should be research direction to be taken while studying routing metrics for WMNs.

- **Cross-layer routing:** Theoretical results has demonstrated the advantage of cross-layer design and optimization in WMNs. Exploring the cross-layer based routing metrics will promote the integration of routing and MAC design, which will then yield high throughput routes. This point needs to be considered while exploring cross-layer based routing metrics. Considering physical and MAC (PHY/MAC) attribute on designing a routing metric may produce better and higher throughput routes, which would further improve the network performance. However, cross-layer design makes hardware to be expensive and very challenging to design.
- **Composite metrics:** Many of the routing metrics that we discussed in chapter find paths that satisfy only one performance metric, i.e. delay or transmission time. There is

Figure 1. Relationship between routing metric and design principles

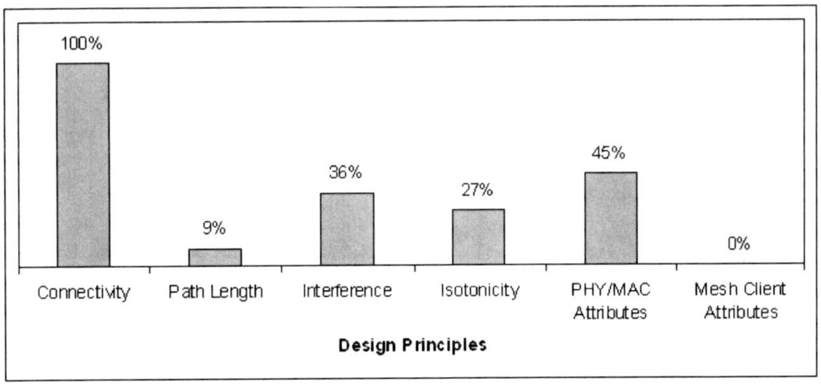

Table 3. Summary of routing metrics: Part 2

Metric	Definition	Benefit	Based on	Drawback	PHY/MAC Attribute	Performance Metric	Mesh Client Attributes	Higher Layer QoS Demands
mETX	Modified ETX	Works with variability links	Estimated Transmission count	Complex error estimation method. Does not consider QoS demands of the flow.	Link-level channel conditions (Johnson, D. et al, 2002)	End-to-end delay	Null	Null
ENT	Effective Number of Transmission	Provides controlled QoS	Estimated packet loss rate	Not compasable	Link-level channel conditions (25)	Packet Loss rate	Null	Yes
ETT and WC-ETT	Expected Transmission Time	Reduces Interference	Estimated Transmission Time	High overhead. Does not consider QoS demands of the flow.	Null	End-to-end delay	Null	Null
WCETT-LB	Weighted cumulative ETT with load balancing	Avoid congestion	Estimated Transmission count	High overhead. Does not consider QoS demands of the flow.	Null	End-to-end delay	Null	Null
Quantized Loss Rate	Packet loss rate	Eliminates lossy links	Packet loss probability	Selects low bandwidth paths. Does not consider QoS demands of the flow.	Null	Packet loss ratio	Null	Null

a need for routing metric that can find path that satisfy multiple performance metrics, i.e. delay AND PDR. Therefore, composite routing metric should be explored. The greatest challenge about composite routing metrics is combining together performance metrics into one composite link metric. Composite metrics are also known as Multi-dimensional metrics. The second challenge will be to find the shortest path in a network with composite metric.

- **Higher layer QoS demand:** Since different types of applications demand different network resources and quantities. It is pivotal to have routing metric that can be able to dynamically adjust its parameters as it is searching the optimal path for different types of applications. Certain type of application might be sensitive to delay, while another type is highly sensitive to packet delivery ratio (PDR), so the routing metric must be able to adjust itself accordingly.

CONCLUSION

WMNs have emerged as a network paradigm for wide range of applications (Akyildiz, I. Wang X. et al, 2005). End-to-end optimization of certain QoS measures such as throughput and delay plays an important role in designing algorithms, protocols and architectures for next generation wireless networks. The primary problem of WMNs is the poor performance of QoS mechanisms. This results in low network throughput. This problem can be solved by the routing metric. In this chapter we have provided the design criteria to be considered when designing routing metrics in order to achieve high throughput. We have also discussed the details of the existing WMN routing metrics and the relationship between them. Finally, we outlined several research opportunities in which future research can follow.

REFERENCES

Adya, A., Bahl, P., Padhye, J., Wolman, A., & Zhou, L. (2003, July). A multi-radio unification protocol for IEEE 802.11 wireless networks, Tech. Rep. MSR-TR-2003-44, Microsoft Research, Akyildiz, I., Wang, X., & Wang, W. (2005). Wireless mesh networks: a survey. *Computer Networks*, *47*, 445–487.

Bahr, M. (2006). Proposed Routing for IEEE 802.11s WLAN Mesh Networks. In *Proceedings of the 2nd Annual International Workshop on Wireless Internet* (Vol. 220).

Biaz, S., & Qi, B. (2008). *iETT: A Quality Routing Metric for Multi-rate Multi-hop Networks* (Tech. Rep.) Department of Computer Science and Software Engineering, Auburn University, USA.

Biswas, S., & Morris, R. (2005). ExOR: Opportunistic multi-hop routing for wireless networks. In *Proceedings of ACM SIGCOMM*, August 2005.

Casetti, C., Chiasserini, C., & Fiore, M. (2007). Relay quality awareness in mesh networks routing. In *Proceedings of Tyrrhenian International Workshop on Digital Communications (TWD '07)*, Ischia Island, Italy.

DeCouto, D., Aguayo, D., Bicket, J., & Morris, R. A high throughput path metric for multi-hop wireless routing. In *Proceedings of ACM MobiCom, 2003.*

Dijkstra, E. (1959). A note on two problems in connection with graphs. *Numer. Math.*, *1*, 269–271. doi:10.1007/BF01386390

Draves, R., Padhye, J., & Patra, R. (2004). Comparison of routing metrics for static multi-hop wireless networks. In *Proceedings of ACM SIGCOMM.*

Draves, R., Padhye, J., & Zill, B. (2004). Routing in Multi-Radio, Multi-Hop Wireless Mesh Networks. In *Proceedings of MobiCom'04,* Philadelphia, Pennsylvania, USA.

Gerla, M., Hong, X., & Pei, G. (2002). *Fisher state routing protocol for ad-hoc networks*. Internet Draft, draft-ietf-manet-fsr-03.txt

Hou, J., Park, K., Kim, T., & Kung, L. (2008) Medium Access Control and Routing Protocols for Wireless Mesh Networks. In E. Hossain & K. Leung (Eds.), *Wireless Mesh Networks: Architectures and Protocols* (pp. 76 - 111). New York: Springer.

Iannone, L., Khalili, R., Salamatian, K., & Fdida, S. (2004). Cross-layer routing in wireless mesh networks. In *1st International Symposium on Wireless Communication Systems* (Vol. 2004, pp. 319-323).

IEEE 802.11. (2006, March). *Working Group of the LAN/MAN Committee*. IEEE P802.11s/DO.O1.

Jacquet, P., Muhlethaler, P., Qayyum, A., Laouiti, A., Viennot, L., & Clausen, T. (2002). *Optimized link state routing protocol*. Internet Draft, draft-ietf-manet-olsrr-07.txt

Jiang, H., Zhuang, W., Shen, X., Abdrabou, A., & Wang, P. (2006). Differentiated Services for Wireless Mesh Backbone. *IEEE Communications Magazine*, *44*(7), 113–119. doi:10.1109/MCOM.2006.1668391

Jiang, W. Liu, S. Zhu, Y. Zhang, Z. (2007). *Optimized Routing Metrics for Large-Scale Multi-Radio Mesh Networks*. Tsinghua University, Beijing 100084, China.

Johnson, D., Maltz, D., Hu, Y., & Jetcheva, J. (2002). *The dynamic source routing protocol (DSR)*. Internet Draft, draft-ietf-manet-dsr-07.txt

Keshav, S. (1991, September). A Control-Theoretic Approach to Flow Control. *ACM SINGCOMM Computer Communication Review*, *21*(4), 3–15. doi:10.1145/115994.115995

Kim, B., & Lee, H. (2006). Performance Comparison of Route Metrics for Wireless Mesh Networks. *IEICE Transactions on Communications (Institute of Electronics, Information and Communication Engineers) . E (Norwalk, Conn.)*, *89-B*(11), 3124–3127.

Kodialam, M., & Nandagopal, T. (2003). Characterizing achievable rates in multi-hop wireless mesh networks: the joint routing and scheduling problem. In *Proceedings of 9th ACM Mobile Computing and networking*, San Diego, CA, USA

Koksal, C. (2008). Quality of Service Routing Metrics in Wireless Mesh Networks. In E. Hossain & K. Leung (Eds.), *Wireless Mesh Networks: Architectures and Protocols* (pp. 227 - 243). New York: Springer.

Koksal, C., & Balakrishnan, H. (2006). Quality aware routing metric for time-varying wireless mesh networks. *IEEE Journal on Selected Areas in Communications*, *24*, 1984–1994. doi:10.1109/JSAC.2006.881637

Lee, S., Yinzhe, Y., & Nelakudithi, S. (2004, March). Proactive Vs Reactive Approaches to Failure Resilient Routing. In Twenty-third Annual Joint Conference of the IEEE Computer and Communications Societies.

Ma, L., & Denko, M. (2007). A Routing Metric for Load-Balancing in Wireless Mesh Networks. In *Proceedings of the 21st International Conference on Advanced Information Networking and Applications Workshops (AINAW'07)*.

Mogre, P., Hollick, M., & Steinmetz, R. (2007). QoS in Wireless Mesh Networks: Challenges, Pitfalls, and Roadmap to its Realization. In *Proceedings of Network and Operating System Support for Digital Audio and Video (NOSSDAV 2007)*.

Park, V., & Corson, M. (1997). A highly adaptive distributed routing algorithm for mobile networks. In *Proceedings of IEEE INFOCOM*.

Perkins, C., & Bhagwat, P. (1994). Destination sequence distances vector routing. In *Proceedings of ACM SIGCOMM.*

Perkins, C., & Royer, E. (1999). Ad hoc on demand distance vector routing. In *Proceedings of IEEE WMCSA.*

Shen, Q., & Fang, X. (2006). A Multi-metric AODV Routing in IEEE 802.11s. In . *Proceedings of ICCT, 06*, 1–4.

Simó, J., Osuna, P., Seoane J., & Martínez, A. (2006). *An IEEE 802.11 static mesh network design for isolated rural areas in developing regions.* IEEE INFOCOMM 2006

Singh, S., Kim, J., Perkins, C., Pedro, M., & Ruiz, T. (2005). *Ad hoc network auto-configuration: definition and problem statement.* Internet-Draft draft-singh-autoconf-adp-00.txt

Subramanian, A. P., Buddhikot, M. M., & Miller, S. (2006). *Interference Aware Routing in Multi-Radio Wireless Mesh Networks* [Tech. Rep.]. State University of New York at Stony Brook, 2006.

Sun, Y., Sheriff, I., Belding-Royer, E., & Almeroth, K. (2005). An Experimental Study of Multimedia Traffic Performance in Mesh Networks. In *Proceedings of International Conference On Mobile Systems, Applications And Services* (pp. 25-30).

Tang, J., Xue, G., & Zhang, W. (2005). Interference-aware topology control and QoS routing in multi-channel wireless mesh networks. In *Proceedings of the 6th ACM Mobile as hoc networking and computing* (pp. 68 -77).

Yang, Y., Wang, J., & Kravets, R. (2006). *Design Routing Metrics for Mesh Networks* [Tech. Rep.] University of Illinois at Urbana-Champaign, 2006.

Yarvis, M., Conner, W., Krishnamurthy, L., Chhabra, J., Elliott, B., & Mainwaring, A. (2002). Real world experiences with an interactive ad-hoc sensor network. In *Proceedings of the International Workshop on Ad Hoc Networking.*

Yin, S., Xiong, Y., Zhang, Q., & Lin, X. (2006). Prediction-based routing for real time communications in wireless multi-hop networks. In *Proceedings of the 3rd international conference on Quality of service in heterogeneous wired/wireless networks.*

Zhang, G., Wu, Y., & Liu, Y. (2007). Stability and sensitivity for congestion control in wireless mesh networks with time varying link capacities. *Ad Hoc Networks, 5*(6), 769–785. doi:10.1016/j.adhoc.2006.12.002

Zhou, W., Zhang, D., & Qiao, D. (2006). Comparative Study of Routing Metrics for Multi-Radio Multi-Channel Wireless Networks. In *Proceedings Wireless Communications and Networking Conference (WCNC 2006)* (pp. 270-275).

Section 6
Future

Chapter 25
Quality of Service (QoS) Provisioning in Cognitive Wireless Ad Hoc Networks:
Architecture, Open Issues and Design Approaches

Kok-Lim Alvin Yau
Victoria University of Wellington, New Zealand

Peter Komisarczuk
Victoria University of Wellington, New Zealand

Paul D. Teal
Victoria University of Wellington, New Zealand

ABSTRACT

Cognitive Radio (CR) is a next-generation wireless communication technology that improves the utilization of the overall radio spectrum through dynamic adaptation to local spectrum availability. In CR networks, unlicensed or Secondary Users (SUs) may operate in underutilized spectrum owned by licensed or Primary Users (PUs) conditional upon the PU encountering acceptably low interference levels. A Cognitive Wireless Ad Hoc Network (CWAN) is a multihop self-organized and dynamic network that applies CR technology for ad-hoc mode wireless networks that allow devices within range of each other to discover and communicate in a peer-to-peer fashion without necessarily involving infrastructure such as base stations or access points. Research into Quality of Service (QoS) in CWAN is still in its infancy. To date, there is only a perfunctory attempt to improve the data-link and network layers of the Open Systems Interconnection (OSI) reference model for CR hosts, and so this is the focus of this chapter. We present a discussion on the architecture, open issues and design approaches related to QoS provisioning in CWAN. Our discussion aims to establish a foundation for further research in several unexplored, yet promising areas in CWAN.

DOI: 10.4018/978-1-61520-680-3.ch025

INTRODUCTION

Traditional static spectrum allocation policies have been to grant each wireless service, such as radio and TV stations, exclusive usage of certain frequency bands, whilst leaving several spectrum bands unlicensed for a wide range of purposes. Examples of unlicensed bands include the Industrial, Scientific and Medical (ISM) and Unlicensed National Information Infrastructure (UNII). In practice, the precious and limited unlicensed radio spectrums are shared by many wireless applications including Bluetooth, WiFi, WiMAX, and Zigbee. Other devices such as microwave ovens and cordless phones also operate in those bands. The unlicensed wireless devices are prohibited from using the licensed spectrum bands. However with the tremendous growth in ubiquitous low-cost wireless applications that utilize the unlicensed spectrum bands, network-wide performance of wireless communication networks will inevitably degrade in the future because of the increasing competition for spectrum especially in populated urban areas.

The Federal Communications Commission (FCC) Spectrum Policy Task Force (2002) pointed out that the current static spectrum allocation has led to overall low spectrum utilization where up to 70% of the allocated licensed spectrum remains unused (these are called white space), at any one time, even in a crowded area. Hence, the main reason of spectrum scarcity among the unlicensed users is, in fact, because of the spectrum allocation policy that is inefficient. White space is defined by time, frequency and maximum transmission power at a particular location. Consequently, Dynamic Spectrum Access (DSA) has been proposed so that unlicensed spectrum users or Secondary Users (SU)s are allowed to use the white space of licensed users' or Primary Users (PU)s' spectrum conditional on the interference to the PU being below an acceptable level. This function is realized using Cognitive Radio (CR) technology that enables an SU to change its transmission and reception parameters including operating frequencies. This enables the SUs to search for and use white space in the licensed spectrum. According to Cabric, Mishra & Brodersen (2004), the SUs are expected to operate over a wide range of non-contiguous frequency bands: 400-800MHz (UHF TV bands) and 3-10GHz. The time scale of the spectrum occupancy varies from milliseconds to hours depending on the activity levels of the PUs. An example of emerging standards based CR network is the IEEE 802.22 Wireless Regional Area Network (WRAN). The IEEE 802.22 working group has been working towards developing CR-based Medium Access Control-Physical (MAC-PHY) air interface for SUs to operate in TV bands, in this approach the SU access to spectrum is controlled by a centralized base station. As an alternative to this infrastructure oriented solution we can consider a cooperative peer to peer models such as traditional ad hoc networks. The ad hoc networks provide a dynamic mechanism to interconnect nodes through the provision of network relay functions and such networks can be mobile or fixed in nature.

The WRAN is a single-hop infrastructure-based static network which means that an SU can only have direct communication with the base station and without a base station, the SU would not function. This type of solution is not suitable for Cognitive Wireless Ad Hoc Network (CWAN), which is the focus of this chapter. In contrast the CWAN is a multihop self-organized and dynamic network that applies CR technology. The SUs are potentially mobile, capable of communicating among themselves, and nodes can act as relays to create multiple hop networks. Quality of Service (QoS) provisioning in CWAN is a daunting challenge as the capacity of the wireless channel on which the SUs are operating is apt to change dependent on the spectrum utilization of PUs, as well as any nodal mobility or adaptation actions to combat poor wireless conditions. Nodal mobility and network adaptation are currently being addressed in traditional wireless ad hoc network solutions. To date, a number of projects have considered the design of QoS architectures for wireless ad hoc networks; but unfortunately none of them can be directly applied to CWAN. A QoS architecture details a framework for the provision of QoS guarantees on an end-to-end basis for various traffic types with different priority levels such as video, voice, and data. Typical QoS parameters that need to be considered include bandwidth, end-to-end delay, packet loss rate and jitter.

In CR networks, research has been focusing on single-hop and static networks much like the WRAN model outlined above. There has been only a perfunctory attempt to provide QoS guarantee's based on an end-to-end basis in CWAN. In addition, the main CR research focus has been largely limited to the physical layer. In this chapter, we discuss a cross-layer QoS architecture for CWAN called C^2net that covers particularly the network and data link layers, its open issues, and design approaches. The C^2net is a cross-layer QoS architecture based on the Next-Steps in Signaling (NSIS) framework from the IETF that provides end-to-end QoS guarantee in CWAN. Our discussion aims to motivate new research interests in this field.

The chapter is organized as follows. The Background section provides an overview of the traditional spectrum allocation policy, CR networks, the cognition cycle, QoS architecture, and NSIS framework. The next section introduces C^2net, a QoS architecture based on the application of the NSIS framework, as well as the cross-layer paradigm in C^2net. A cross-layer design and its open issues are discussed in this section. The design approaches section discusses context-awareness and intelligence as one of the key solutions to the open issues in CR networks. Also, an application of context-awareness and intelligence in addressing the open issues in CR networks is presented using an example in this section. Finally, future research areas and conclusion are presented.

BACKGROUND

In this section the fundamental concepts of CR networks are discussed including the spectrum allocation policy, cognition cycle, QoS architecture, and NSIS framework.

Spectrum Allocation Policy

Traditionally, radio spectrum has been partitioned into ranges of *licensed* and *unlicensed* frequency bands. The licensed frequency bands are normally sold through auctions that could bring considerable revenue to the government. Some small areas of the spectrum are allocated to unlicensed users who contend for access to this free resource. Unlicensed users are forbidden to access any of the licensed bands that have been purchased. Many popular wireless communication systems, such as the IEEE 802.11, have been operating in unlicensed bands without incurring any cost. As an analogy, the spectrum allocation policy is like a swimming competition where the limited pool (radio spectrum) is divided into many lanes (frequency bands). Each contestant (spectrum user) is assigned a lane that is used throughout its communication session. The contestant is forbidden to cross over into other lanes or interfere with the other contestants and the contestant does not generally occupy the whole of the lane. The lane that represents the unlicensed spectrum is typically crowded with many competitors that jostle for space. As the number of unlicensed users increases, it is inevitable that the unlicensed lane becomes more congested. As a consequence, the QoS of the unlicensed users is adversely affected.

Cognitive Radio Networks

CR aims to improve the utilization of radio spectrum, which is one of the scarcest resources in wireless communications. Without sufficient spectrum, QoS provisioning to support many sophisticated wireless applications would not be

Figure 1. An illustration of DSA. An SU exploits white spaces across various channels. Each location has different spectrum utilization by the PU

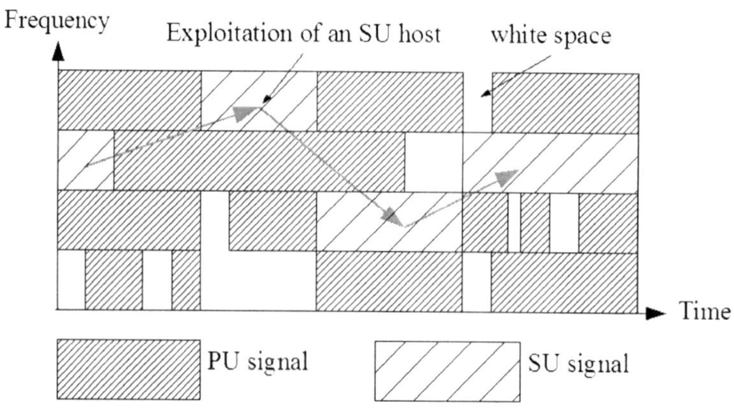

achievable. CR enables the SUs to search for and use the white space in licensed spectrum bands when the unlicensed spectrum is highly utilized. Since the white space is a limited resource, it is postulated that there will be an intense competition for spectrum usage among the SUs for the available white space. Hence, not only do the SUs have to search for white space, they also need to use the white space efficiently. Now, let's take an analogy. Suppose you are driving to school or work during the peak hours. While driving straight ahead, you find that the lane becomes congested. To arrive on time, you carefully switch to a nearby lane that is less congested, while ensuring that you don't collide with the other road users. The same principle is applicable to CR. If its current licensed or unlicensed bands are fully utilized, an SU switches its operating frequency to another band without interfering with the PU activity. This occurs when the licensed channel is underutilized or contains white space. Through accessing the white space in licensed spectrum dynamically, the overall spectrum utilization improves. In CR networks, one of the most important tasks is therefore to create a "friendly" environment for the coexistence between the PUs and the SUs using CR technology to enable Dynamic Spectrum Access (DSA) as shown in Figure 1.

In Figure 1, the spectrum utilization from the PU is represented by the time and frequency axes. An SU host switches its channel across various frequency bands from time to time in order to utilize the white space in the licensed channels it is sensing. Since the white space is location dependent, for a successful communication, the white space must be available at both the SU transmitter and receiver. In mobile networks, this is particularly important if the SU nodes are moving at high speed as from moment to moment each location may have different spectrum utilization by the PU. However, since the transmission range of the PU is often large, such as transmission for the TV bands, the spectrum utilization of the PU at various locations may not differ by much, and thus collaboration in channel sensing for white space among the SUs is an effective means to avoid collision with the PU's transmissions.

Dynamic spectrum access (DSA) can be realized in three different ways as defined by Doyle & Forde (2007), these are the *current regime*, the *common regime* and the *market-based regime*. In the current regime, SUs are capable of sensing and utilizing temporary white spaces at licensed spectrum without incurring any cost providing that there is no harmful interference to the PUs. Hence, whenever a PU makes use of their allocated spectrum, which has been classified as white space by SUs, the SUs must vacate the spectrum as soon as possible. The common regime supports equal right for all entities to spectrum access much like the current unlicensed spectrum bands; hence there is no concept of PU and SU. In the market-based regime, spectrum is sold as blocks of white spaces by the PU that provides exclusive access to SU purchasers. Hence, the market-based regime provides better guarantee of white space availability and is more reliable. Doyle & Forde (2007), report that the market-based approach is backed by several prominent regulators such as the FCC, the UK Office for Communication (Ofcom), and the EU Commission Radio Spectrum Policy Group. In comparison to the current regime, the market-based regime provides some guarantee to white space access though it comes at a price.

Our primary design focus for CWANs are around deployment in a complex wireless communication and a broadband access scenario comprised of various heterogeneous mobile and stationary units in a densely populated urban area. Consumers may access the CWAN using consumer devices, laptops, mobile phones, PDAs, vehicular intelligent transportation systems and so on, in a single or a multihop manner, for example to allow extension of hot spot coverage. Certain unlicensed frequency bands such as the ISM and UNII bands are highly utilized in metropolitan areas; however, with CR technology, an SU could search for and utilize unused licensed bands. The focus of our work in C^{2}net is to provide stable QoS assurance to high priority flows such as video and audio traffic. This scenario, as shown in Figure 2, is may be useful for telecom operators to extend wireless access among subscribers that are outside base station coverage for example.

Cognition Cycle

Generally speaking, what an SU node does effects its operating environment. The SU's action could affect the environment for better or for worse, or maintain the status quo (which is to have no effect); and this in turn affects the SU's next course of action. For instance, if an SU node fails to transmit well in a channel, it switches to another channel with more white space or better transmission properties. Its transmission over the white space affects the operating environment by reducing the amount of white space in that channel. Given a particular operating environment, the state-action-effect association can be learnt so that the SU node knows what to do for the best performance when the environment reoccurs. This idea is portrayed in a cognition cycle. The adage 'practice makes perfect' is the concept that the cognition cycle was founded upon. While making a perfect system is a far more difficult endeavor, a cognition cycle aims to achieve a system with better performance as time goes by. Although the cognition cycle has not been extensively applied in network protocol design, it has great potential for system enhancement. In CR network, the cognition cycle was first introduced by Mitalo III & Maguire (1999). A simplified version is shown in Figure 3.

The cognition cycle is comprised of five main states: *observe*, *learn*, *plan*, *decide*, and *act*. In the *observe* state, the SU node *i* receives information about the dynamic operating environment at time instance t-1, t, t+1, ... The observation can also be an internal event such as instantaneous queue size. The *learn* state provides knowledge on the operating environment through observing the consequences of its prior actions, states or both. The knowledge or learning outcome can be shared among SUs by explicit message exchange. The *plan* state draws up a long term course of actions; while the *decide* state determines the next action that can improve the performance. In the *act* state, there may be various actions including message exchange, a backoff mechanism, a sensing operation and even "cease to act".

The cognition cycle is a key concept to enable DSA. To identify *high* quality white space among the available channels, an SU observes its state or

Figure 2. CWAN deployment scenario. PUs and SUs (including the CRs) are operating in overlapping frequencies

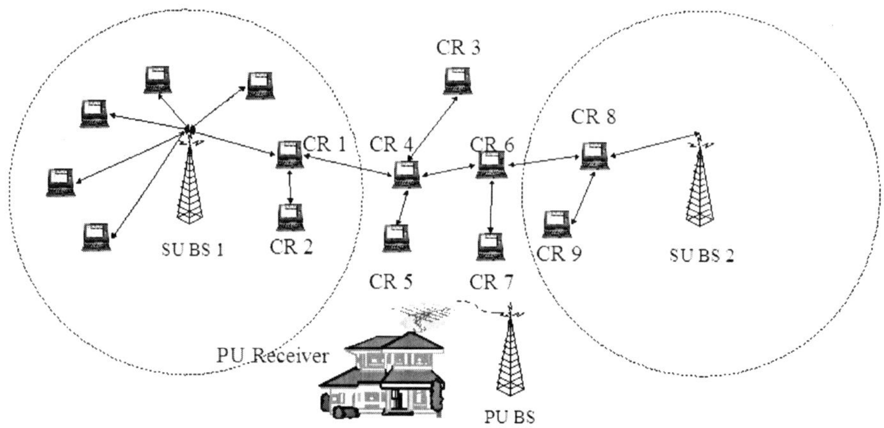

channel conditions such as the PU traffic pattern, its utilization level and time of use; as well as the channel quality. The high quality white space improves the probability of successful packet transmission. The SU's action is to choose a channel for data transmission, after which it waits for an acknowledgement, which is the effect, from its receiver. This state-action-effect association or knowledge, which is used to plans and decides for its next channel for data transmission, is learnt as time goes by so that the best possible matching can be achieved. In this chapter, we implement the cognition cycle through the application of context-awareness and intelligence mechanisms. Context-awareness enables a CR host to be aware of its operating environment; while intelligence helps the host to learn the optimal action for each possible condition.

Doyle & Forde (2007) suggest two levels of cognition cycle: node-level and network-level. At node-level, each node runs a cognition cycle and makes its own unilateral decision in a non-cooperative manner. The node-level cognition cycle can be used in distributed networks such as Mobile Ad Hoc Networks (MANETs) so that each SU can determine its own channel for data

Figure 3. The cognition cycle

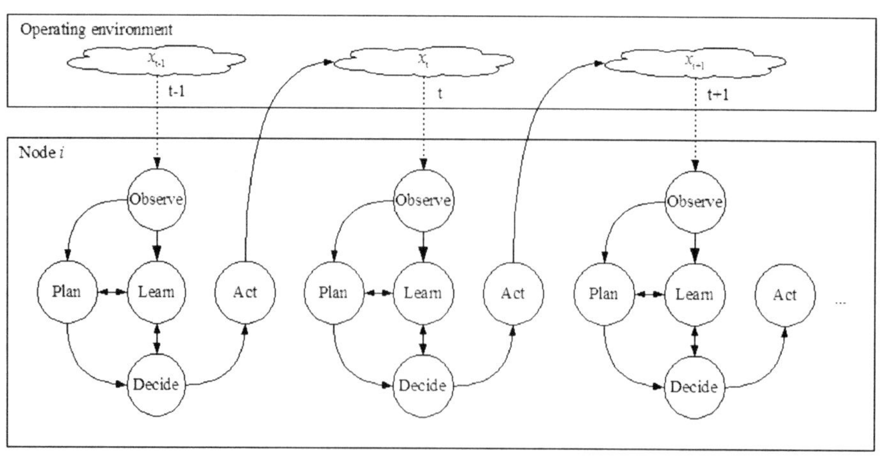

transmission. Conversely, at network-level, actions are made in a multilateral and cooperative fashion; hence, it is more suitable to be applied at the base station. An example of the application of a network-level cognition cycle is at the base station of the WRAN. In WRAN, each unlicensed Customer-Premises Equipment (CPE) or the SU is associated with one of the SU base stations. The base station coordinates and instructs its CPEs to operate in certain frequency bands. A network-level cognition cycle at the base station helps it to choose channels with high quality white space for data transmission. This improves the SUs' throughput and delay performance.

Quality of Service Architecture

Two very early QoS architectures have been proposed for static wired networks, namely Integrated Services (IntServ) by Braden, Clark & Shenker (1994), and Differentiated Services (DiffServ) by Blake (1998). This chapter reviews some of the key concepts in IntServ and DiffServ architectures. The readers should refer to the aforementioned references for more detailed descriptions.

The IntServ architecture provides a per-flow granularity in QoS guarantee. This requires every intermediate node of a flow to perform resource reservation and admission control mechanisms. A signaling protocol called Resource Reservation Protocol (RSVP) is used to reserve and maintain resources (or states), such as bandwidth, for each flow at intermediate nodes. The realization of IntServ in wireless networks is questionable because of four issues: 1) scalability concerns as a result of storing state information for each flow at all intermediate nodes; 2) the high amount of overhead in RSVP signaling; 3) resource reservation that is difficult to adapt to dynamic topology in MANETs; and 4) complex implementation of QoS functions at each node such as admission control and state information maintenance.

In contrast, DiffServ provides per-class granularity in QoS guarantee. DiffServ limits complicated QoS functions such as admission control, packet classification and conditioning to the source node. A source node classifies a packet from its various flows according to their QoS requirements based on their traffic priority class, marks the DiffServ Codepoint (DSCP) field in the packet IP header, and conditions the packet based on a traffic policy. Intermediate nodes that receive the packet match the DSCP with Per-Hop Behaviour (PHB) and forward the packet accordingly. The PHB identifies how a packet should be forwarded according to its priority class. Thus, DiffServ ameliorates the aforementioned scalability and complexity issues of IntServ. However, two disadvantages of DiffServ are: 1) per-class granularity only provides long-term QoS guarantee for each flow; and 2) there is no QoS signaling to ensure QoS is supported on an end-to-end basis.

Based on IntServ and DiffServ frameworks, various QoS architectures for wireless ad hoc networks have been proposed. Lee, Ahn, & Zhang (2000) propose INSIGNIA that adopts the IntServ framework and hence inherited its disadvantages; while Ahn, Campbell, Veres, & Sun (2002) propose SWAN that applies the DiffServ model. As DiffServ does not provide end-to-end signaling, a source node in SWAN sends a probing message to its destination to estimate available resources along its route, such as bottleneck bandwidth and end-to-end delay. The resource information is required to perform admission control. Xiao, Seah, Lo, & Chua (2000) propose FQMM and He, & Wahab (2006) propose HQMM which are hybrid models that embrace both IntServ and DiffServ concepts. The hybrid model provides per-flow granularity to a small amount of high priority flows, while the rest of the flows are treated as per-class granularity. None of these QoS architectures can be adopted in CWAN because of the additional requirement to cope with the PUs. In the next few sections, C^2net, which is a QoS architecture that adopts the hybrid model for CR networks, will be discussed.

Next-Steps in Signaling (NSIS) Framework

Recently, NSIS framework has been proposed by Hancock, Karagiannis, Loughney, & Bosch (2005) as the end-to-end QoS signaling protocol to supplement the DiffServ model. Using NSIS, resource reservation along a route comprised of different QoS models can be made. Hence, the NSIS is particularly suitable for C^{2}net, which is a hybrid QoS model of IntServ and DiffServ. Architecturally, NSIS is comprised of two components (see Figure 4), namely the NSIS Transport Layer Protocol (NTLP) and the NSIS Signaling Layer Protocols (NSLPs) as described by Fu et al (2005). The NTLP has a messaging component called General Internet Signaling Transport (GIST), which is a successor to RSVP, that uses standard transport protocols such as User Datagram Protocol (UDP), Transmission Control Protocol (TCP), Stream Control Transmission Protocol (SCTP), and Datagram Congestion Control Protocol (DCCP) for sending signaling messages. NSLPs provide application-specific functions such as QoS provisioning and security. In this chapter, we focus on the QoS NSLP. Four types of signaling messages are defined in QoS NSLP, namely RESERVE, QUERY, RESPONSE, and NOTIFY. The RESERVE creates, refreshes, modifies and deletes a flow's resource reservation state information at a node; QUERY probes available resources along a route, such as bandwidth; RESPONSE serves as acknowledgment or confirmation of received QoS NSLP signaling message; and NOTIFY conveys error conditions.

An example of NSIS signaling scenario for QUERY message is shown in Figure 5. Suppose, node 1 is the source node and node 4 is the destination node. Node 2 and 3 are intermediate nodes in a route that help to relay packets to destination node 4. Using its QoS NSLP, node 1 creates QUERY message that contains its flow requested bandwidth and probes bandwidth availability along its route. The GIST encapsulates the QoS NSLP message and transports the packet using one of the transport protocols until the destination node 4 is reached. Upon receiving the QUERY packet, the QoS NSLPs of the intermediate node 2 and 3 update their available bandwidth in the packet respectively. Hence, the key design component in the NSIS framework in a QoS architecture is the QoS NSLPs. In the next section, we discuss about this component extensively.

C^{2}NET: A CROSS-LAYER QUALITY OF SERVICE ARCHITECTURE FOR COGNITIVE WIRELESS AD HOC NETWORKS

Research into CWAN is still in its infancy. Thus far, research has been focusing on single hop

Figure 4. Components in NSIS framework

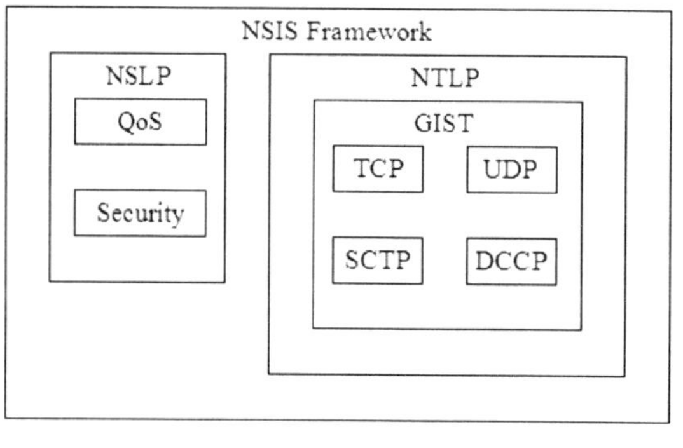

Figure 5. NSIS signaling scenario for QUERY message

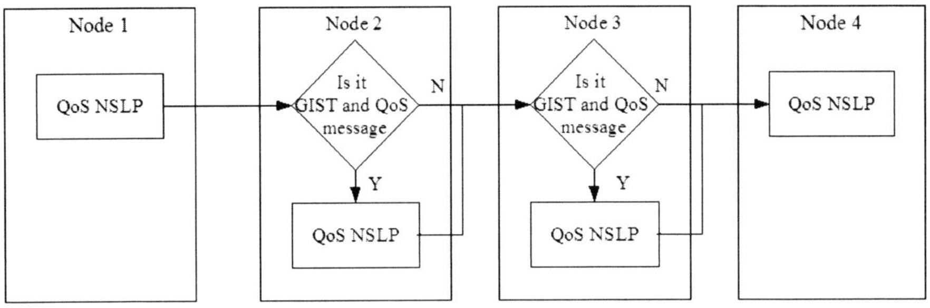

and centralized networks. In view of this, there is a substantial need to design a QoS architecture for CWAN in order to provide end-to-end QoS guarantee. In this section, we present the C²net architecture based on Next Steps in Signaling (NSIS) framework and cross-layer approach. The main objective is to provide stable QoS assurance to high priority flows. In addition, we present a cross-layer feature of C²net, namely joint Dynamic Channel Selection (DCS) and topology management.

A QoS Architecture Based on NSIS Framework

C²net is a hybrid model of IntServ and DiffServ. In this architecture, a small number of high priority flows, such as voice and video, adopts the IntServ model; while the other flows adopt the DiffServ model. From an economic point of view, consumers prefer to send best-effort flows at the lowest possible price; while high priority flows may incur some charges with occasional packet loss being acceptable as long as the perceived quality is not significantly degraded. Thus, the DiffServ model applies the current regime, while IntServ uses the market-based regime. In the market-based regime, nodes have exclusive access to white spaces in a deterministic manner; hence, the small number of high priority flows achieve better QoS guarantee.

The QoS NSLP is the key component in NSIS framework for QoS provisioning. The flowchart for QoS NSLP in C²net at each intermediate node is shown in Figure 6, in conjunction with other QoS elements. For brevity, RESPONSE and NOTIFY are ignored. In general, there are two types of channels in CWAN, namely, the common control channel and data channels. Both the common control and data channels are located in the licensed or unlicensed spectrum. Each node is equipped with two transceivers: control transceiver is tuned to a particular common control channel; while the data transceiver is tuned to one of the data channels for data transmission. During normal operation, all nodes are constantly listening to the common control channel. The common control channel is meant for control message exchanges, such as Request-to-Send/Clear-to-Send (RTS)/(CTS) messages, data channel negotiation messages, and notification to vacate a data channel upon detection of PU activity. During the data channel negotiation, the sender and receiver nodes choose a data channel among all the available channels for data transmission, after which the data transceiver is tuned to the negotiated data channel. The SUs constantly explore the data channels in search of high quality white space. In Figure 6, procedures at the control channel are related to QoS NSLP; while procedures at the data channel are for all data packets.

Upon receiving control messages on the common control channel, the GIST messages that carry QoS information are processed in the QoS NSLP. The QUERY message processing checks for available bandwidth at the node. Two types

Figure 6. Flowchart for QoS elements at each node in C^2net architecture. Solid line indicates control flow; while dotted line indicates data flow

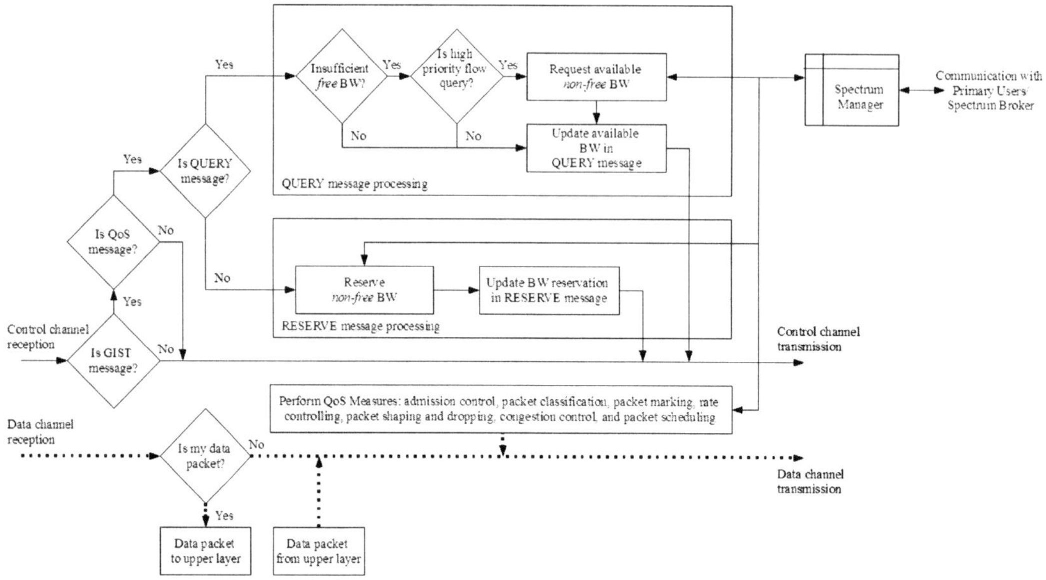

of data channels are the free unlicensed and non-free licensed channels. If the channel availability of the free unlicensed channels is insufficient and the flow has a high priority level, the node requests bandwidth from non-free licensed channels through its spectrum manager using a market-based regime. The spectrum manager at the SU determines the amount of white space to be purchased during the resource reservation process later; and communicates with the PUs or a spectrum broker to know about the available white space or bandwidth that could be purchased through spectrum trading. Available bandwidth is updated in the QUERY message, and the QUERY message is sent to the next hop, which implements a similar procedure, using the common control channel. The QUERY message is also used for state refreshment, modification and deletion. In Figure 6, the RESERVE message processing is implemented for high priority flows only. In this process, the spectrum manager at each node is requested to purchase the required white spaces for high priority flows. A description of spectrum trading is proposed by Buddhikot et al (2005).

Whether the reservation is successful is indicated in the RESERVE message which is transmitted from the destination to its sender. The state is reserved in a soft manner such that if the QUERY message or data packets from a flow are not received after a certain time interval, the state is withdrawn. In this case, the spectrum manager stops the purchase of white spaces for the flow. For a source node of a flow, rather than receiving the GIST messages, it creates a QUERY message, as shown in Figure 5, based on the profile of its traffic flow such as bandwidth requirement.

On the data channel, DiffServ QoS measures such as admission control, packet classification, packet marking, rate control, packet shaping and dropping are performed to ensure that the rate and burst profile for each flow is compliant with the Traffic Conditioning Agreement (TCA) as stipulated in the Service Level Agreement (SLA). The purpose is to ensure that the QoS of the high priority flows are not jeopardized. A detailed description of the implementation of the QoS measures is given by Blake (1998). Additionally, if the spectrum manager has reserved white spaces for a high priority flow, its packets will be forwarded using the reserved resources.

The NSIS framework enables end-to-end signaling mechanism for C^2net comprised of IntServ and DiffServ models. The key design is the QoS NSLP that manages resources at each intermediate node including spectrum purchasing for higher priority flows from the PUs. As the market-based regime provides a better guarantee of white space availability, higher priority flows enjoy a better QoS guarantee. However, there are other various factors that affect the end-to-end QoS provisioning at the data link and network layers. A Cross-layer approach is adopted to address the issues.

The Cross-Layer Paradigm

The cross-layer paradigm has overcome the traditional layered approach through joint design of multiple components at various layers of the Open Systems Interconnection (OSI) reference model. Zhang & Zhang (2008) provide a good discussion on cross-layer design in multihop wireless networks. Various cross-layer designs are possible; however, due to space limitation, we focus on an open issue in CWAN: joint Dynamic Channel Selection (DCS) and topology management. So why is the cross-layer paradigm potentially important in CR networks? In CR networks, an SU node has to be aware of its operating environment. The DCS scheme, which resides in the MAC or data link layer in the OSI model, must sense for white space across various channels and choose a channel dynamically for data transmission. To enable the functions at the upper layer to be aware of their operating environment, functions such as routing and topology management in the network layer must cooperate with the DCS in the lower layer. Joint DCS and topology management performs channel selection dynamically in the presence of dynamic topology comprised of mobile hosts, as well as dynamic PU activity. At the time this chapter was written, little or no effort has been made to investigate this joint design in CR networks. The next subsection discusses the joint DCS and topology management as well as its open issues.

Joint Dynamic Channel Selection and Topology Management

In CWAN, the SUs may operate in under-utilized channels owned by the PU conditional upon acceptable interference with the PU nodes. A problem arises as to what is the best strategy to select an available channel among the licensed channels for data transmission from an SU node. The objective is to reduce the packet loss of high priority flows for stable end-to-end QoS provisioning, as well as maximizing overall throughput, in the presence of nodal mobility.

For stable, reliable and robust transmissions, some nodes in a neighbourhood that are relatively stable, in terms of mobility characteristics, are selected to form a Dominating Set (DS) such as the SU nodes of CR1, CR4, CR6, and CR8 in Figure 2. Nodal stability is determined using Link Expiration Time (LET), associativity in Hello messages, or both. For instance, a node is relatively stable if it is capable of serving as a DS node for the longest time interval compared to its one-hop neighbour nodes. Nodes within a DS connect among themselves to form a backbone topology, which is connected to the SU base station, throughout the network, while non-DS nodes establish links with DS nodes. Various clustering algorithms in wireless networks utilize the DS concept in order to improve network scalability through reduction of routing overhead as stated by Bao, & Aceves (2003). As an added advantage for CR networks, the DS provides a means of coordination for cooperative sensing. Cooperative sensing has been proposed in CR networks to mitigate the effects of unreliable spectrum sensing outcome without imposing higher sensitivity requirements at each SU node. A DS node performs decision fusion on sensing outcomes from its neighbour nodes to improve sensing accuracies. The decision fusion is a decision making process where local sensing outcomes at neighbour nodes are combined to reach a more accurate result.

The channels in a licensed spectrum have differing amounts of white spaces (termed White Space Capacity, WSC henceforth). In C^2net, nodes do not choose and switch between available channels for data transmission in a random manner. This would introduce instability in channel availability among the nodes. Specifically, it is not possible for an SU node to determine the available bandwidth in its selected channel for data transmission if its neighbour nodes constantly switch their channels. Instead, channel selection is performed based on nodal stability, backbone connectivity, channel quality, and WSC in each available channel. The DS, which forms a connected backbone, is relatively stable and has higher authority in channel selection, so that channels with better quality and higher amount of WSC are chosen. Non-DS nodes choose the remaining available channels. In view of the dynamic nature of the network topology, nodal stability, connectivity, channel quality and WSC within a channel, this information must be maintained continuously.

Traditionally, the backbone topology throughout the network is formed using the Minimum Dominating Set (MDS) as proposed by Bao & Aceves (2003). In general, the amount of routing overhead increases with the number of DS nodes. To reduce the amount of routing overhead, the least possible number of DS nodes forms the MDS backbone. In C^2net, the main purpose is to provide stable data transmission. Therefore, it forms a Connected Dominating Set (CDS) instead of MDS. The CDS ensures the connectivity of the DS nodes at the backbone topology. It should be noted that the type of information carried, which is routing overheads in MDS and data packets in CDS, differentiates the backbone functionalities in C^2net from that of traditional schemes. Other possible considerations in DS node selection are energy levels at the mobile host, Signal-to-Noise Ratio (SNR) in various channels and so on. Ensuring connectivity in the backbone helps to alleviate congestion and packet loss since the DS nodes have higher authority to select channels with better quality and high amount of WSC.

Again, consider the snapshot of a mobile topology in Figure 2. Suppose, based on nodal stability, CR1, CR4, CR7, and CR8 are relatively stable and become DS nodes. Since nodes are either DS nodes, or direct neighbour to a DS node, it is a valid CDS. However, there is no connectivity between the DS nodes. As CR6 does not have the authority to select channels with better quality and high amount of WSC for data transmission, it becomes a bottleneck node and congestion occurs. Thus, CR6 is chosen as a DS node although it does not have higher stability than CR7. In this case, the DS nodes are connected, and hence form a valid CDS. The connectivity of the backbone topology (CR1-CR4-CR6-CR8) is thus maintained. In short, the most stable node within a subset to fulfill the connectivity requirement is chosen to become the DS node in backbone maintenance.

The open issues in this joint design are: 1) backbone construction and maintenance; and 2) DCS, such that DS nodes select a channel better quality and higher amount of WSC, as well as switching to a better channel when a PU increases its activity in its channel.

DESIGN APPROACHES

The key elements in ensuring a successful CR deployment are *context-awareness* and *intelligence*, which can be achieved through solving the cognition cycle. Various design approaches are possible in solving the cognition cycle. This section provides an insight into achieving context-awareness and intelligence as mechanisms to solve problems and open issues in CR networks.

CONTEXT-AWARENESS AND INTELLIGENCE AS THE KEY SOLUTIONS TO OPEN ISSUES IN COGNITIVE RADIO NETWORKS

CR technology has brought about a paradigm shift in the way a host defines its operating policy, which is a set of decision rules that determine how a node should behave in various scenarios. Traditionally, the policy is hard-coded into the host. A common policy is the *if-then-else* conditional statement as shown in Figure 7. When a host encounters a particular condition or state, it performs its corresponding action. For instance, using a fixed lookup table, a host chooses its modulation technique, such as Quadrature Amplitude Modulation (QAM) and Binary Phase Shift Keying (BPSK), according to different levels of Signal-to-Noise Ratio (SNR). There are two major drawbacks in the strict and static self-defined policy. Firstly, the policy, which might not be optimal in all conditions, cannot be changed on the fly. Secondly, the condition in the *if* statements might not cover all kinds of circumstances. In CR networks, a host must be aware of its operating environment. It senses the channels, detects and uses the white space. It is expected that a CR host takes the optimal actions in a wide range of conditions. In fact, the CR host might not have encountered some of them before. It is therefore more appropriate if the *if-then-else* policy is adjusted dynamically on the fly through the capability of context-awareness and intelligence.

Figure 7. The if-then-else strict and static self-defined policy

```
if     (condition C1, event E1) then (action A1);
elseif (condition C2, event E2) then (action A2);
elseif (condition C3, event E3) then (action A3);
...
else   (condition C4, event En) then (action An);
end if;
```

A key question is: "*What is the cognitive radio context?*". The context is captured by the condition in the *if* statement. All the elements in the operating environment that a CR host resides may not necessarily be important unless network-wide performance can be improved by tackling them. Therefore, the context, which is the information that characterizes the important factor(s), is very much dependent on the schemes or designs a researcher is focusing on. An example is DCS where the important factors are the packet error rate and channel utilization by the PUs at each available channel.

Hence, the main focus in CR network research is to design a practical and yet simple technique to achieve context-awareness and intelligence that succeed the cognition cycle concept. The well-researched context-awareness and intelligence methodology can be applied in various schemes and designs, such as topology management, DCS, scheduling, and routing.

Application of Context-Awareness and Intelligence in Addressing Open Issues in Cognitive Radio Networks

In this section, a context-awareness and intelligence methodology, specifically Reinforcement Learning (RL) is applied to address the DCS scheme as an example. To date, research has been focused on how an SU exploits and uses the white spaces (Hoyhtya, Pollin & Mammela, 2008; Xin, Ma, & Shen, 2007; Bian & Park, 2007). However, using RL, we are able to achieve the next level of enhancement, that is how an SU exploits and uses the *high* quality white spaces. In practice, the SUs are expected to operate over a wide range of non-contiguous frequency bands: 400-800 MHz (UHF TV bands) and 3-10 GHz as according to Cabric, Mishra & Brodersen (2004), where the time scale of the spectrum occupancy varies from milliseconds to hours. Due to channel heterogeneity, the properties of the white space at different channels vary with carrier frequency and time-varying channel condition, hence resulting in various packet error rate in different channel. The traditional hard-coded policy is insufficient to address the channel heterogeneity issue because the packet error rate and the amount of WSC in a channel affect the network performance in a complex manner. For instance, how would an SU choose a data channel given that one channel has high amount of WSC and packet error rate, while the other has just the opposite, low amount of WSC and packet error rate. The static policy is less likely to be applicable in all conditions, which are the combinations of various levels of WSC and packet error rates. Hence, context-awareness and intelligence must be achieved. Not only is an SU able to sense the white spaces, but also to infer their quality so that packet transmission successful rate is high.

REINFORCEMENT LEARNING AS A DECISION MODEL TO ACHIEVE CONTEXT-AWARENESS AND INTELLIGENCE

Q-learning (Sutton & Barto, 1998; Watkins, 1989) is an on-line algorithm in Reinforcement Learning (RL) that determines an optimal policy using only simple modeling. Each node in the network is a learning agent or host as shown in Figure 3. Q-learning is used to learn the channel conditions such as the PU traffic pattern, its utilization level and time of use; as well as the channel quality. As time progresses, the host learns to carry out proper actions given a particular condition or state.

In Q-learning, the learnt action value or Q-value, $Q(state, action)$ is updated using delayed reward and maintained in a two-dimensional lookup Q-table with size $|state| \times |action|$. For every state-action pair, the Q-value represents the expected amount of reward that a host receives. For each state, an appropriate action would be rewarded and its Q-value is increased. In contrast, inappropriate action would be punished and its Q-value is decreased. In other words, the Q-value indicates the appropriateness of the selection of an action in a state. The Q-value tackles both the current and future rewards, which is discounted over the future, that the host receives for each state-action pair.

Denote state by s, action by a, reward by r, learning rate by α and discount factor by γ. At time t+1, the Q-value of a chosen action in a state at time t is updated as follows:

$$Q_{t+1}(s_t, a_t) = (1-\alpha)Q_t(s_t, a_t) + \alpha\left[r_{t+1}(s_{t+1}) + \gamma \max_{a \in A} Q_t(s_{t+1}, a)\right]$$

where $0 \le \gamma \le 1$ and $0 \le \alpha \le 1$. If $\alpha = 1$, the agent will forget all its previous learnt Q-value, giving a single-shot learning. The higher the value of γ, the greater the agent relies on the future reward $Q_t(s_{t+1}, a_{t+1})$ compared to the immediate reward $r_{t+1}(s_{t+1})$. The expected future reward is obtained by choosing an action that maximize the future Q-value given the next state. The Q-value is not dependent on the expected future reward if it is excluded from the equation. Changes in the Q-value will lead to changes in an agent action. The RL searches for an optimal policy that maximizes its accumulated reward through choosing the action with maximum Q-value for any time instance.

An important aspect of RL is the tradeoff between exploration and exploitation (Sutton & Barto, 1998; Watkins, 1989). The update of the Q-value in (1) does not cater for the actions that are never chosen. Choosing the best overall action, or the greedy action at all times is termed exploitation. To improve the estimates of all the Q-values, non-optimal actions are chosen once in a while so that better actions may be discovered, which is a procedure called exploration. The balance between exploitation and exploration depends on the accuracy of the Q-value estimation and level of dynamic behaviour in the environment. Examples of tradeoff methodologies are ε-greedy and softmax approach. In the ε-greedy approach, an agent chooses the greedy action as its next action with probability $1-\varepsilon$, and random action with a small probability ε.

To achieve context-awareness and intelligence, the DCS scheme is modeled using RL. The state s represents set of node i's neighbor nodes; while the action a represents an available channels that a node has chosen for data transmission. The reward $r(s,a)$ is the constant value to be rewarded (or incurred) for successful (or unsuccessful) data packets transmission. At each decision epoch t, based on its receiver neighbour node, node i chooses a channel for data transmission. The Q-value is updated as follows:

$$Q_{t+1}(s_t,a_t) = (1-\alpha)Q_t(s_t,a_t) + \alpha r_{t+1}(s_{t+1})$$

with the $\max_{a \in A} Q_t(s_{t+1},a)$ in (1) omitted to indicate no dependency on future discounted rewards. The greedy action is to choose the action with the best Q-value. At the beginning of every attempt to transmit a data packet, a node chooses to either continue or change an action or channel. In order to reduce the number of channel switching, a node switches channel only if the Q-value of the current action is lower than the other option or during exploration.

The throughput achieved by the RL and Random scheme is investigated for various levels of channel utilization by the PU. In this simulation, we showed that the RL method helps a CR host to choose a channel with low level of PU activity for data transmission with constant packet error rate of 0.1 across all the available channels. This means that, for every packet that the sender transmits, the receiver detects error in the packet with a probability of 0.1. We assume that there are only two nodes in a static network, a sender and a receiver. We further assume that there are three available channels. All the three channels can reach the receiver from the transmitter. The RL-based DCS is compared against a Random DCS scheme, where the channel for data transmission is chosen randomly without learning. Figure 8 shows that the RL scheme outperforms the Random for all levels of channel utilization by the PU. Hence, the RL scheme learns well and helps the SU node to choose a channel with low PU utilization such that packet successful transmission rate is high.

In short, the RL enables a CR host to achieve context-awareness and intelligence. The RL-based DCS scheme is capable of choosing a channel with high amount of WSC. Since a channel with high quality white space increases packet successful transmission probability, the RL-based DCS can also literally help a CR host to avoid channels that incur high level of packet error rate.

FUTURE RESEARCH

CR improves utilization of the overall radio spectrum through dynamic adaptation to local spectrum availability. The IEEE 802.22 Working Group was formed in November 2004 to define the first worldwide wireless standard based on CR. The IEEE 802.22 is a centralized and single-hop network that exploits the TV spectrums. Having long range coverage, which is contributed by better propagation characteristics in TV bands, the IEEE 802.22 is targeted at rural areas. In future, it is anticipated that the IEEE 802.22 will expand its functionality progressively to cover multi-hop and mobility support, as well as to define a more market-based model. To increase its market share, IEEE 802.22 will be enhanced to target both the urban and rural area. The current draft is not suitable for application in urban area due to its long range transmission that results in high level of interference and low spatial reuse. Hence, power control may be designed or new licensed spectrum may be opened for this purpose. A market-based regime is most suitable to utilize the licensed spectrum for Private Land and Com-

Figure 8. The mean throughput of an SU sender against mean of channel utilization by PU for RL and random schemes

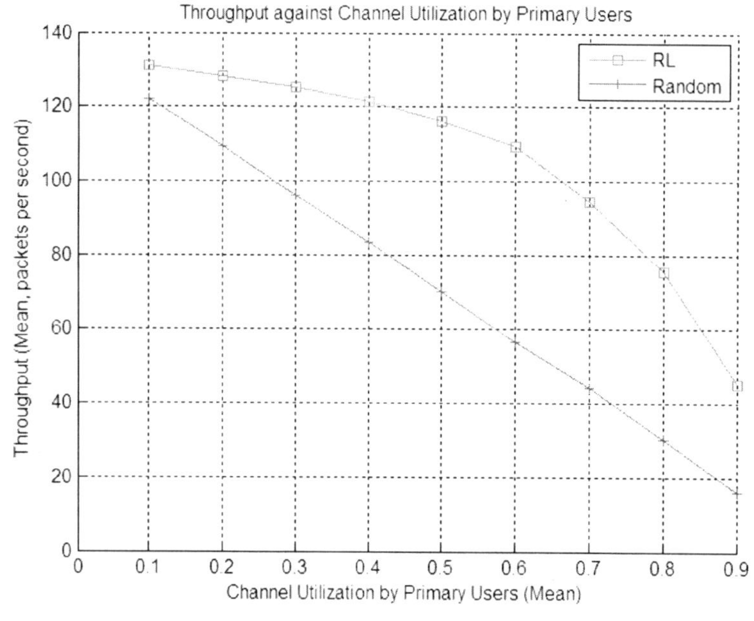

mercial Mobile Radio Service (PLMRS/CMRS). Only with these enhancements, will it be able to address the spectrum scarcity issues in urban area. On supporting mobility, it has been opposed thus far as it may interfere protected contour and area of licensed users, and difficult to trace the hosts that interfere with the protected contour and area. Unless these issues are solved, IEEE 802.22 would have to inter-operate with other standards such as IEEE 802.16 and IEEE 802.11 to support mobility.

Nevertheless, research into CWAN with mobility support is of paramount importance. Current research focuses on centralized, static and single-hop networks much like the IEEE 802.22 without end-to-end QoS provisioning. In view of this, we proposed C^2net as a unified solution to provide QoS based on an end-to-end semantic. C^2net is a cross-layer QoS architecture based on an NSIS framework. Proper design of a component in NSIS called QoS NSLP enables a hybrid model of IntServ and DiffServ, as well as multiple regimes of spectrum access including the current and market-based regime. We also proposed a joint DCS and topology management design that is imperative to end-to-end QoS provisioning.

The key elements that will determine the success or failure of various schemes in CR networks are context-awareness and intelligence. We define context-awareness and intelligence as the capability of a CR host to sense, learn, and response accordingly in an efficient manner with respect to its operating environment without adhering to a strict and static self-defined policy. Context-awareness and intelligence can be achieved using a wide range of techniques such as reinforcement learning. Well-researched context-awareness and intelligence method can be applied to design various schemes in CR networks including the cross-layer designs. Therefore, future research could be pursued for various context-awareness and intelligence techniques and their applicability to addressing a wide range of problems in CR networks.

CONCLUSION

A cross-layer Quality of Service (QoS) architecture called C^2net is proposed for Cognitive Wireless Ad Hoc Networks (CWAN). The main objective of C^2net is to provide and maintain a stable QoS to high priority flows throughout its connection. C^2net is a hybrid model of Integrated Service (IntServ) and Differentiated Service (DiffServ) that applies Next Steps in Signaling (NSIS) framework as the QoS signaling protocol. The IntServ model fulfills the stringent QoS requirements of a flow at reasonable cost by purchasing white spaces from PU if there is spectrum scarcity among the unlicensed spectrums. The DiffServ model provides services for lower priority packets. A cross-layer design, namely topology management and dynamic channel selection, is presented. The key elements in the schemes are context-awareness and intelligence, which can be achieved by solving the cognition cycle. As an example, the context-awareness and intelligence is achieved using Reinforcement Learning (RL) to design a Dynamic Channel Selection (DCS) scheme. In this chapter, we have introduced the concept of context-awareness and intelligence, as well as new research interests in QoS provisioning in CWAN.

REFERENCES

Ahn, G.-S., Campbell, A. T., Veres, A., & Sun, L.-H. (2002). Supporting service differentiation for real-time and best-effort traffic in stateless wireless ad hoc networks (SWAN). *IEEE Transactions on Mobile Computing*, *1*(3), 192–207. doi:10.1109/TMC.2002.1081755

Bao, L., & Aceves, J. J. G.-L. (2003). Topology management in ad hoc networks. In *Proceedings of 4th ACM International Symposium on Mobile Ad Hoc Networking & Computing* (pp. 129-140). New York: ACM.

Bellman, R. E. (Ed.). (2003). *Dynamic Programming*. Princeton, NJ: Princeton University Press.

Bian, K., & Park, J.-M. (2007). Segment-based channel assignment in cognitive radio ad hoc networks. In *Proc. of Cognitive Radio Oriented Wireless Networks and Communications (CrownCom'07)* (pp. 327-335). Washington, DC: IEEE.

Blake, S. (1998). *An architecture for differentiated services*. Internet Engineering Task Force (IETF) RFC2475.

Braden, B., Clark, D. & Shenker, S. (1994). *Integrated services in the internet architecture: An overview*. Internet Engineering Task Force (IETF) RFC1663.

Buddhikot, M. M., Kolodzy, P., Miller, S., Ryan, K., & Evans, J. (2005). DIMSUMNET: a new directions in wireless networking using coordinated dynamic spectrum access. In *Proceedings of 6th International Symposium on World of Wireless Mobile and Multimedia Networks (WoWMoM'05)* (pp. 78-85). Washington, DC: IEEE.

Cabric, D., Mishra, S. M., & Brodersen, R. W. (2004). Implementation issues in spectrum sensing for cognitive radios. In *Proceedings of Asilomar Conference on Signals, Systems and Computers* (pp. 772-776). Washington, DC: IEEE.

Chen, T., Zhang, H., Maggio, G. M., & Chlamtac, I. (2007). Topology management in CogMesh: A cluster-based cognitive radio mesh network. In *Proc. of Intl. Conference on Communications (ICC'07)* (pp. 6516-6521). Washington, DC: IEEE.

Denardo, E. V. (Ed.). (1982). *Dynamic programming: models and applications*. Upper Saddle River, NJ: Prentice Hall.

Doyle, L., & Forde, T. (2007). The wisdom of crowds: cognitive ad hoc networks. In Q. H. Mahmoud (Ed.), *Cognitive networks: Towards self-aware networks* (pp. 203-220). Hoboken, NJ: John Wiley and Sons Ltd.

FCC Spectrum Policy Task Force. (2002). *Report of the spectrum efficiency working group* (Tech. Rep. No. 02-155). Federal Communications Comm.

Fu, X., Schulzrinne, H., Bader, A., Hogrefe, D., Kappler, C., & Karagiannis, G. (2005). NSIS: A new extensible IP signaling protocol suite. *IEEE Communications Magazine*, *43*(10), 133–141. doi:10.1109/MCOM.2005.1522137

Hancock, R., Karagiannis, G., Loughney, J., & Bosch, S. V. D. (2005). *Next Steps in Signaling (NSIS): Framework*. Internet Engineering Task Force (IETF) RFC4080.

He, Y., & Wahab, H. A. (2006). HQMM: A hybrid QoS model for mobile ad-hoc networks. In *Proceedings of 11th Symposium on Computers and Communications (ISCC'06)* (pp. 194-200). Washington, DC: IEEE.

Hoyhtya, M., Pollin, S., & Mammela, A. (2008). Performance improvement with predictive channel selection for cognitive radios. In *Proceedings of 1st International Workshop on Cognitive Radio and Advanced Spectrum Management (CogART'08)* (pp. 1-5). Washington, DC: IEEE.

Lee, S.-B., Zhang, A. X., & Campbell, A. T. (2000). INSIGNIA: An IP-based quality of service framework for mobile ad hoc networks. *Journal of Parallel and Distributed Computing . Special Issue on Wireless and Mobile Computing and Communications*, *60*(4), 374–406.

Mitalo, J. III, & Maguire, G. Q. (1999). Cognitive radio: Making software radios more personal. *IEEE Personal Communications*, *6*, 13–18. doi:10.1109/98.788210

Puterman, M. L. (Ed.). (1994). *Markov decision processes: discrete stochastic dynamic programming*. New York: Wiley-Interscience.

Sutton, R. S., & Barto, A. G. (1998). *Reinforcement Learning: An Introduction.* Cambridge, MA: MIT Press.

Watkins, C. (1989). *Learning from Delayed Rewards.* PhD Thesis, University of Cambridge, UK.

Xiao, H., Seah, W. K.-G., Lo, A., & Chua, K.-C. (2000). A flexible quality of service model for mobile ad-hoc networks. In *Proceedings of Vehicular Technology Conference (VTC'00-Spring)* (pp. 445-449). Washington, DC: IEEE.

Xin, C., Ma, L., & Shen, C.-C. (2007). Path-centric channel assignment in cognitive radio wireless networks. In *Proceedings of Cognitive Radio Oriented Wireless Networks and Communications (CrownCom'07)* (pp. 313-320). Washington, DC: IEEE.

Zhang, Q., & Zhang, Y.-Q. (2008). Cross-layer design for QoS support in multihop wireless networks. *Proceedings of the IEEE*, *96*(1), 64–76. doi:10.1109/JPROC.2007.909930

ADDITIONAL READING

Akyildiz, I. F., Lee, W.-Y., Vuran, M. C., & Mohanty, S. (2006). Next generation/dynamic spectrum access/cognitive radio wireless networks: A survey. *International Journal of Computer and Telecommunications Network*, *50*(13), 2127–2159.

Akyildiz, I. F., Lee, W.-Y., Vuran, M. C., & Mohanty, S. (2008). A survey on spectrum management in cognitive radio networks. *IEEE Communications Magazine*, *46*(4), 40–48. doi:10.1109/MCOM.2008.4481339

Brodersen, R. W., Wolisz, A., Cabric, D., Mishra, S. M., & Willkomm, D. (2004). *CORVUS: A cognitive radio approach for usage of virtual unlicensed spectrum.*

Brown, T. X., & Sicker, D. C. (2007). Can cognitive radio support broadband wireless access? In *Proceedings of 2nd IEEE International Symposium on New Frontiers in Dynamic Spectrum Access Networks (DySPAN'07)* (pp. 123-132). Washington, DC: IEEE.

Challapali, K., Cordeiro, C., & Birru, D. (2006). Evolution of Spectrum-Agile Cognitive Radios: First Wireless Internet Standard and Beyond. In *Proceedings of 2nd Annual International Wireless Internet Conference (WICON'06)*. New York: ACM.

Chowdhury, K. R., & Akyildiz, I. F. (2008). Cognitive Wireless Mesh Networks with Dynamic Spectrum Access. *IEEE Journal on Selected Areas in Communications*, *26*(1), 168–181. doi:10.1109/JSAC.2008.080115

Clancy, C., Hecker, J., Stuntebeck, E., & Shea, T. O. (2007). Applications of machine learning to cognitive radio networks. *IEEE Wireless Communications*, *14*(4), 47–52. doi:10.1109/MWC.2007.4300983

Cordeiro, C., Challapali, K., & Birru, D. (2006). IEEE 802.22: An introduction to the first wireless standard based on cognitive radios. *The Journal of Communication*, *1*(1), 38–47.

Cordeiro, C., Challapali, K., & Ghosh, M. (2006). Cognitive PHY and MAC layers for dynamic spectrum access and sharing of TV bands. In *Proceedings of 1st International Workshop on Technology and Policy for Accessing Spectrum (TAPAS'06)*. New York: ACM.

Deng, S., Chen, J., He, H., & Tang, W. (2007). Collaborative strategy for route and spectrum selection in cognitive radio networks. In *Proceedings of Future Generation Communication and Networking (FGCN'07)* (pp. 168-172). Washington, DC: IEEE.

Ghasemi, A., & Sousa, E. S. (2008). Spectrum sensing in cognitive radio networks: requirements, challenges and design trade-offs. *IEEE Communications Magazine*, *46*(4), 32–39. doi:10.1109/MCOM.2008.4481338

Haykin, S. (2005). Cognitive radio: brain-empowered wireless communications. *IEEE Journal on Selected Areas in Communications*, *23*(2), 201–220. doi:10.1109/JSAC.2004.839380

Hincapie, R., Tang, J., Xue, G., & Bustamante, R. (2008). QoS routing in wireless mesh networks with cognitive radios. In *Proceedings of IEEE Global Communications Conference (GLOBECOM'08)*. Washington, DC: IEEE.

Hou, Y. T., Shi, Y., & Sherali, H. D. (2008). Spectrum sharing for multi-hop networking with cognitive radios. *IEEE Journal on Selected Areas in Communications*, *26*(1), 146–155. doi:10.1109/JSAC.2008.080113

Ji, Z., & Liu, K. J. R. (2007). Dynamic spectrum sharing: A game theoretical overview. *IEEE Communications Magazine*, *45*(5), 88–94. doi:10.1109/MCOM.2007.358854

Krenik, W., & Batra, A. (2005). Cognitive radio techniques for wide area networks. In *Proceedings of 42nd Design Automation Conference* (pp. 409-412). Washington, DC: IEEE.

Krenik, W., Wyglinski, A. M., & Doyle, L. E. (2007). Cognitive radios for dynamic spectrum access. *IEEE Communications Magazine*, *45*(5), 64–65. doi:10.1109/MCOM.2007.358850

Krishnamurthy, S., Thoppian, M., Venkatesan, S., & Prakash, R. (2005). Control channel based MAC-layer configuration, routing and situation awareness for cognitive radio networks. In *Proceedings of Military Communications Conference (MILCOM'05)* (pp. 455-460). Washington, DC: IEEE.

Lee, C.-H., & Wolf, W. (2008). Energy efficient techniques for cooperative spectrum sensing in cognitive radios. In *Proceedings of Consumer Communications and Networking Conference (CCNC '08)* (pp. 968-972). Washington, DC: IEEE.

Lei, G., Wang, W., Peng, T., & Wang, W. (2008). Routing metrics in cognitive radio networks. In *Proceedings of 4th International Conference on Circuits and Systems for Communications (ICCSC '08)* (pp. 265-269). Washington, DC: IEEE.

Sengupta, S., Ahmad, M. Z., & Chatterjee (2007). Initializing mesh architecture for cognitive radio based IEEE 802.22. In *Proceedings of Cognitive Radio Oriented Wireless Networks and Communications (CrownCom '07)* (pp. 519-523). Washington, DC: IEEE.

Sengupta, S., Brahma, S., Chatterjee, M., & Shankar, N. S. (2007). Enhancements to cognitive radio based IEEE 802.22 air-interface. In *Proceedings of IEEE International Conference on Communications (ICC '07)*. Washington, DC: IEEE.

Sengupta, S., Chatterjee, M., & Chandramouli, R. (2008). A coordinated distributed scheme for cognitive radio based IEEE 802.22 wireless mesh networks. In *Proceedings of IEEE International Conference on Communications Workshops (ICC '08)* (pp. 461-465). Washington, DC: IEEE.

Su, H.,\& Zhang, X. (2008). Cross-layer based opportunistic MAC protocols for QoS provisioning over cognitive radio wireless networks. *IEEE Journal on Selected Areas in Communications, 26*(1), 118–129. doi:10.1109/JSAC.2008.080111

Wang, L.-C., Chen, A., & Wei, D. S. L. (2007). A Cognitive MAC Protocol for QoS Provisioning in Overlaying Ad Hoc Networks. In *Proceedings of Consumer Communications and Networking Conference (CCNC '07)* (pp. 1139-1143). Washington, DC: IEEE.

Wright, D. (2007). Dynamic Spectrum Access in the IEEE 802.22 Wireless Regional Area Network. In *Proceedings of 2nd International Conference on Access Networks and Workshops (AccessNets '07)* (pp.1-6). Washington, DC: IEEE.

Yang, Z., Cheng, G., Liu, W., Yuan, W., & Cheng, W. (2008). Local coordination based routing and spectrum assignment in multi-hop cognitive radio networks. *Journal in Mobile Networks and Applications, 13*(1-2), 67–81. doi:10.1007/s11036-008-0025-9

Chapter 26
Evolution of QoS Control in Next Generation Mobile Networks

Alberto Díez Albaladejo
Fraunhofer FOKUS, Germany

Fabricio Gouveia
Fraunhofer FOKUS, Germany

Marius Corici
Fraunhofer FOKUS, Germany

Thomas Magedanz
Technische Universität Berlin, Germany

ABSTRACT

Next Generation Mobile Networks (NGMNs) constitute the evolution of mobile network architectures towards a common IP based network. One of the main research topics in wireless networks architectures is QoS control and provisioning. Different approaches to this issue have been described. The introduction of the NGMNs is a major trend in telecommunications, but the heterogeneity of wireless accesses increases the challenges and complicates the design of QoS control and provisioning. This chapter provides an overview of the standard architectures for QoS control in Wireless networks (e.g. UMTS, WiFi, WiMAX, CDMA2000), as well as, the issues on this all-IP environment. It provides the state-of-the-art and the latest trends for converging networks to a common architecture. It also describes the challenges that appear in the design and deployment of QoS architectures for heterogeneous accesses and the available solutions. The Evolved Core from 3GPP is analyzed and described as a suitable and promising solution addressing these challenges.

DOI: 10.4018/978-1-61520-680-3.ch026

INTRODUCTION

In the last few years both the Internet and telecommunication world are passing through an evolutionary phase: they are merging. Each is a successful paradigm by itself. The Internet is based on the Internet Protocol (IP) and provides many of the most of today's used services like World Wide Web, email, instant messaging, file sharing, etc., with Best Effort (BE) transport and no Quality of Services (QoS). There are no guaranties that the resources, like bandwidth, will be delivered for a particular session. Mobile networks offer voice services with great mobility (cellular networks). Making calls and offering other telecommunication services using the Internet or using Internet services in cellular networks are a trend today. This global trend increases the demand for integrated services, which at their turn increase the complexity of the networks, challenging the current network architectures and QoS control systems.

The main motivation of this book chapter is to describe first how QoS control mechanisms function on some of the most used wireless technologies including cellular technologies, and then describe the challenges that arise while converging distinct networks (very heterogeneous by technology), as well as on how end-to-end QoS should be approached. Finally it is presented a well suited architecture for coping with these issues and offering a platform for managing and controlling heterogeneous networks and services.

QOS CONTROL IN WIRELESS NETWORKS 3G (THIRD GENERATION)

ITU-T QoS Specifications

In order to support end-to-end QoS solutions in the converging world of Internet and Telecommunications, Next Generation Networks (NGN) have to offer a common set of IP packet transfer performance parameters and QoS objectives (Song, 2007). With this objective in mind the Telecommunication Standardization Sector (ITU-T) that coordinates standards for telecommunications on behalf of the International Telecommunication Union (ITU), started in 2002 to prepare international standards (recommendations) to help with 3G definition.

ITU-T has produced recommendations for defining standard performance parameters for packet transfer in IP-based networks, network-interface-to-network-interface (NI–NI) objectives, different QoS classes and many other standards for performance objectives and QoS parameters.

ITU-T specified the Resource and Admission Control Function (RACF) in order to provide the required NGN independence between service and transport stratum. The RACF is the element that determines resources availability in the transport layer and appropriately controls the network element. It defines different QoS control scenarios for the User Equipment (UE) with different QoS signaling capabilities, which are:

- The UE cannot signal QoS (No signaling capability)
- The UE has QoS SIP signaling capability
- The UE can reserve resources directly in the transport layer (e.g., RSVP)

QoS control in the RACF is done in pull or in push mode, which are described in the Policy and Charging Control (PCC) architecture section. Finally, the RACF is also responsible for defining Network Address and Port Translation (NAPT) control function.

QoS in UMTS

The Universal Mobile Telecommunications System (UMTS) started to be specified in Release 99 of The Third Generation Partnership Project (3GPP) standards and defines mechanisms for QoS support considering an end-to-end QoS architecture. There are mainly four UMTS QoS or traffic classes that are specified. These traffic classes are handled according to the operator's requirements on each of them. The 3G traffic classes are:

- Conversational class, which is used for voice and real time multimedia messaging
- Streaming class, for streaming type of applications, Video on Demand (VoD) etc.
- Interactive class, for interactive type of applications, eCommerce, Web browsing, etc.
- Background class, for background type applications, email, File Transfer protocol (FTP), etc.

These classes assign different treatment for the packets with respect to various QoS aspects such as flow priority, guaranteed or maximum bit rate and packet drop precedence. In addition, policy can be used to specify the forwarding of packets based on various classification criteria. The policy controls the set of configuration parameters and forwarding for each class or admission conditions for reservations of flows (depending on the QoS scheme utilized - e.g. RSVP, DiffServ).

UMTS makes use of the Packet Data Protocol (PDP) context for controlling users' sessions. The PDP context carries session information from the subscribers during an activated session. In order to use GPRS services the session information is exchanged between the Serving General Packet Radio Service (GPRS) Support Node (SGSN) and the Gateway GPRS Support Node (GGSN). This procedure allocates a PDP context data structure in the SGSN the user is currently connected and the GGSN serving the users access point. The data stored contains the user's IP address, International Mobile Subscriber Identity (IMSI) and tunnel endpoint IDs at both the GGSN and the SGSN, as well as the Quality of Service settings.

QoS in CDMA 2000

The CDMA2000 access defined by the Third Generation Partnership Project 2 (3GPP2) (3GPP2 C.S0024-B, 2006) is a mobile digital technology which competes with UMTS. Several enhancements of CDMA2000 technology exist which provide optimizations and higher transmission rates. In the current standards 1xEV-DO Rev A and Rev B advanced support for link efficiency and queue management is included. CDMA2000 supports definition of QoS parameters in two levels: per flow to state air interface resources required for applications and definition of filters to define the traffic flow classification and QoS treatment through the establishment of Traffic Flow Templates (TFT).

The network architecture for CDMA2000 differs slightly from one revision to the other but in the 1xEV-DO standard for the packet switching leg it includes the base station (BTS), the Radio Network Controller (RNC) and the Packet Data Serving Node (PDSN) as depicted in Figure 1. These network entities allow providing QoS based on subscribers profile and per application. The PDSN implements the TFT and provides packet classification (e.g. Differentiated Service Code Points (DSCP) packet marking) and traffic shaping and policing based on user profile that is acquired from the AAA server. The QoS per application is implemented in the different entities of the network. The mobile device and base station include several PHY and MAC mechanisms that permit Multi-Flow Packet Application both in the uplink and the downlink, including a QoS aware scheduler. The PDSN applies the TFTs for packet marking and classification in both directions.

The TFT is implemented using the Resource Reservation Protocol (RSVP) between the mobile equipment and the PDSN while the Flow Specification uses CDMA2000 signaling between the mobile equipment and the access network.

Since the first versions of CDMA2000, interworking of the architecture with the one defined by

Figure 1. CDMA2000 Network architecture for the PS leg

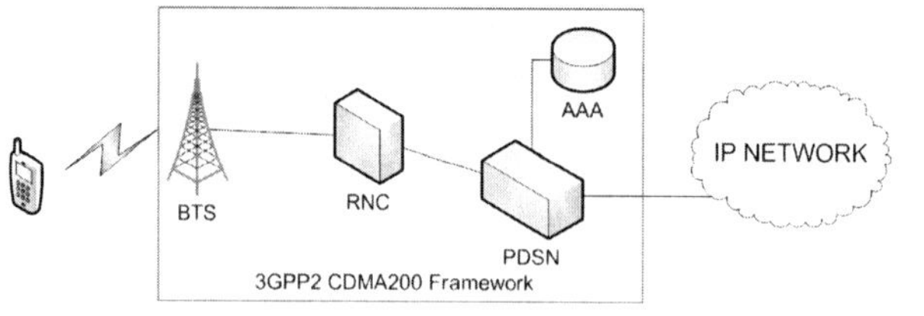

3GPP for GPRS and UMTS has been considered by 3GPP2 and therefore a convergence path for both networks is left open.

QoS in Long Term Evolution

3GPP has defined a new access technology named Long Term Evolution (LTE) which is packet only and provides higher data rates, reduced latency, improved capacity and coverage with reduced operational costs and spectrum allocation compatibility. The new access network for LTE is called E-UTRAN (Evolved UTRAN).

The E-UTRAN contains one logical node the eNodeB, which at its turn includes the base station and implements the physical, MAC and Radio Link Control (RLC) layers of the LTE. The eNodeB connects to other eNodeBs, the Mobility Management Entity (MME) and the Serving Gateway. These two last entities are part of the 3GPP access network architecture and the Evolved Packet Core (EPC), as described in the following sections. The eNodeB communicates with the MME for mobility control and has a user plane interface with the Serving Gateway which is used for the IP data flows. The communication with these entities is done using the GPRS Tunneling Protocol (GTP) control and user planes.

The eNodeB performs radio bearer management including the initial admission control and bearer allocation, operations executed under the control of the MME. The eNodeB controls the uplink and downlink radio resource management and the data packet scheduling executed in the PHY, MAC and Radio Link Control (RLC) layers. The eNodeB negotiates with the MME the GTP tunnel establishment and the Evolved Packet System (EPS) bearer establishment including the QoS parameters associated.

The EPS bearer is the E-UTRAN equivalent of the PDP Context of UMTS and is the lowest level of granularity for QoS. Within these QoS parameters the QoS Class Identifier (QCI) is included, this is mapped to a Layer 2 Packet Delay Budget (L2PDB) and Layer 2 Packet Loss Rate (L2PLR) which are then used in the eNodeB to derive the appropriate waiting queues and MAC Hybrid Automatic Repeat-Request (HARQ) parameters.

Wireless QoS provisioning in LTE is done in the MAC layer were data scheduling, priority handling and HARQ mechanisms are implemented and where the shared channel used to transport user control or traffic data (UL-SCH) is managed.

With this, LTE implements a very simplified architecture in which only one node, the eNodeB, performs the QoS related functions with the assistance of the MME. The QoS provisioning is based in scheduling and queue prioritization, the parameters to apply per-user and per-flow (EPS bearer) are received from the MME on session establishment, modification and release. The MME itself, being part of a more general 3GPP access network as it also performs functions for interconnecting GERAN (GPRS Radio Access Network), UTRAN with e-UTRAN has interfaces to subscriber repositories and other network functions (Serving Gateway, SGSN etc.) from which it receives per PDP context or EPS bearer QoS parameters.

Figure 2. QoS control in WiFi

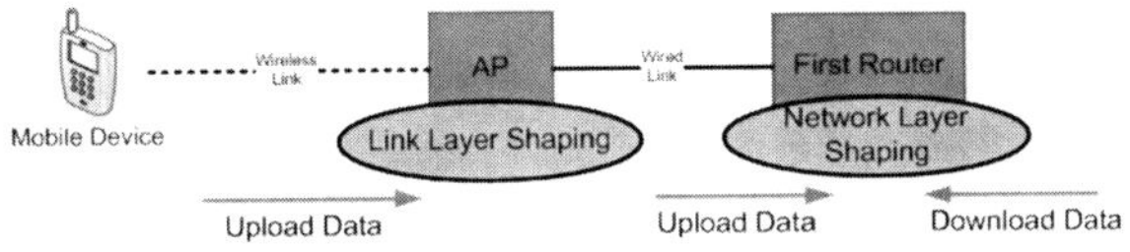

The Serving Gateway being part of the EPC may include static pre-provisioned QoS policies or support dynamic QoS provisioning based in the Policy and Charging Control architecture (PCC) explained in a following section.

QOS CONTROL FOR WIRELESS (NON-CELLULAR) NETWORKS WI-FI

For the 802.11 (IEEE 802.11 group, 2007) WiFi deployments the QoS control is done typically at the network layer on the first router from the mobile device to the network. Usually the router is different of the Access Point (AP) and the AP has only the role of bridging the information from the wireless link to a wired link. While this mechanism enables the download traffic to be shaped according to the specific rules of the operator, for the upload traffic this mechanism does not prove enough security. The upload traffic can be shaped using this mechanism based on the IP address and the momentary data traffic parameters for each mobile station. But it does not offer protection against Denial of Service attacks based on link layer messages. For example a malicious mobile station may transmit multiple requests for attaching to the WiFi access which by default have to be responded. This will cause a congestion of the wireless link which can not be observed by the network layer QoS control. Therefore the network layer filter does not protect completely the communication over the wireless link and cannot guarantee the amount of resources required. A link-layer shaping has to be deployed on the access point for the upload data, as depicted in Figure 1.

The WiFi wireless technology defines its own mechanisms for link layer QoS control. These mechanisms enable the mobile devices to part the communication channel according to some specific rules for all the upload packets including those which are not received by the network layer shaping mechanism of the first router.

For WiFi two types of QoS control services are defined: the Distributed Coordination Function (DCF) which supports delay-insensitive data and the Point Coordination Function (PCF) which supports delay sensitive transmissions.

The first mechanism is based on the CSMA/CA where the stations compete for the transmission environment. For collisions, it considers a random idle time between some specified boundaries. The second function is a centralized polling-based approach which avoids collisions by polling the mobile nodes individually for transmission. This function offers a better primitive for service differentiation between stations.

However, the mechanism leads to a longer time usage of the channel for mobile nodes with a lower rate transmission. Also stations, being the only ones that know the type of traffic transmitted, do not take part in the decision of polling. The decision is taken by the access point, making it static to the type of quality of service required.

In order to solve these problems, two new functions were introduced in a Quality of Service standard for WiFi. First one, the Enhanced Distributed Channel Access (EDCA), is similar to the DCF, but it considers a dynamic value for the idle time depending on the type of traffic the station has. For this, four types of priority are defined. The classification of packets is done by the mobile node before queuing them for transmission.

This QoS mechanism relies on the good behavior of the mobile stations as they take the decision on the classification of the upload packets. A malicious mobile device may transmit all its packets having the highest priority which enables it to pre-empt packets transmitted by the other stations which may create a Denial of Service attack on the wireless link although the packets may be re-shaped by the network layer QoS filter from the first router.

The second one, the Hybrid Coordination Controlled Access (HCCA) extends the rules of the EDCA by introducing a polling mechanism for the stations, depending on their resource reservation and the time slot prior reserved. The HCCA has as requisite that the stations pre-reserve a time slot for transmission as part of a window of transmission.

The HCCA is similar to the QoS solution for UMTS (3GPP TS 23.107, 2007) as it also considers a set of priority classes and a channel time reservation for each specific station. The reservation is done considering two main parameters: the maximum bandwidth and the guaranteed bandwidth for each data flow for each signaled service session. This enables to discriminate between different flows and their priorities, not considering the whole traffic of the mobile device as to have a specific priority.

The HCCA offers enough protection for the data traffic of each mobile station. In order to implement a HCCA mechanism for QoS control, the stations have to be allowed and must signal their QoS requirements to the WiFi access point which has to be done using another signaled service. Therefore the HCCA is based on an external signaling service which as seen in practice is not desired by the device manufacturers. In fact, to this date, the HCCA is not implemented in any WiFi access points and mobile devices.

In conclusion the QoS control over the wireless link of the WiFi technology is based more on the network layer traffic shaping which do not offer enough guarantees of the required resources. Because of this the WiFi accesses may not be regarded as completely controlled by the network which makes it vulnerable to easy Denial of Service attacks. The WiFi network may be considered only in a limited amount as able to guarantee the QoS required by the mobile devices.

WiMAX

WiMAX or IEEE 802.16 (IEEE 802.16e, 2005) has been standardized by IEEE evolving from wired Ethernet standard as a Wireless Metropolitan Area Network (WMAN) access technology. It can be used as a radio data link for fixed wireless access or for mobile stations as well, with the ability of covering rural areas providing a good data transfer rate. Since 2007 Mobile WiMAX has been adopted by ITU-T as one of the IMT-2000 technologies turning into one of the major global cellular standards as well.

The WiMAX physical and MAC layers support QoS, offering a robust reliable link. Several options are possible when deploying a WiMAX network. Most extended is the usage of a Time Division Duplex (TDD) which maximizes the usage of available bandwidth and a Point-to-Multipoint (PMP) configuration in which a Base Station (BS) coordinates the traffic to several Subscriber Stations (SS) in the coverage area. Time Division Multiple Access (TDMA) is used to share the uplink channel between the SS.

A time frame is divided in its downlink and uplink subframes and preceded by the downlink and uplink maps that include the information about the boundaries of the intervals assigned to the SSs.

Therefore, the BS schedules at the beginning of each time frame the uplink and downlink in order to meet the QoS requirements needed. For the downlink, the BS selects the parameters for the packet queues and controls them. The downlink scheduler takes the data for the queues and it's able to distribute the downlink subframes to meet the QoS requirements of each SS. For the uplink also the BS manages and schedules the time interval given to each SS. In order to request bandwidth the SS have several mechanisms available: including implicit requests at connection setup, explicit bandwidth requests messages, piggybacked requests and poll-requests bits.

There are four different classes for traffic prioritization implemented in the WiMAX MAC: Unsolicited Grant Services which is used for Constant Bit Rates (CBR), Real-Time Polling Services (rtPS) for variable bit rates, non-real-time Polling Services (nrtPS), which provides a better than best effort service, and Best Effort (BE).

The BS is able to queue and schedule packets according to the QoS requested and ensuring that the QoS requirements are satisfied and the interface is shared fairly between the SSs. This design allows prioritizing packet transmission and reducing latency and jittering.

The priorities are associated with the service flows to which the packets belong to. The services flows are managed over the air interface.

The WiMAX Forum (WiMAX Forum, 2007) also provides a WiMAX network reference model. It is divided in an Access Service Network (ASN) and a Connectivity Service Network (CSN). The model is depicted in Figure 3.

The ASN represents the complete set of network functions needed to provide radio access to the subscriber station. The CSN represents the network functions which provide IP connectivity services to the WiMAX subscribers. This separation considers that two differentiate business entities are present the Network Access Provider and the Network Service Provider and that these can belong to different organizations.

To cope with the QoS provisioning and the associated functions of Admission Control (AC) and resources assignment, this architecture includes a Policy Framework that allows per-subscriber QoS profiles, local policies and admission control policies.

The Policy Framework as defined by the WiMAX specification (version 1.2) includes a Service Flow Agent (SFA), a Service Flow Manager (SFM) and a Policy Function (PF).

The SFM and SFA are part of the ASN network. The SFM is included in the Base Station (BS) and the SFA in the Gateway (ASN-Gw). The SFM is responsible for the creation, admission, activation, modification and deletion of service flows. It contains the information about the local resource usage and includes an AC function. The SFA downloads the subscribers QoS profile from an AAA server in the CSN and evaluates service requests against this profile.

The PF is part of the CSN and includes a database with general policy rules and application dependant policy rules. The PF evaluates service requests against these policies.

In the next version of the specifications (1.5) the WiMAX Policy Framework converges (Taaghol, 2008) to the 3GPP Policy and Charging Control (PCC) integrating in the Evolved Packet Core (EPC) covered in the following sections of this book chapter.

Figure 3. WiMAX network architecture

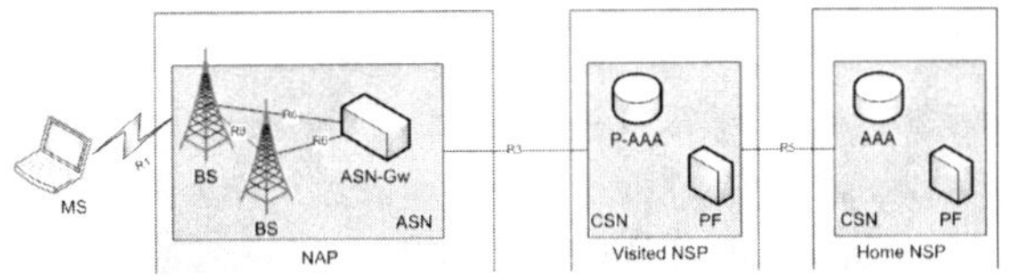

CONVERGENCE OF QOS CONTROL

Integration of Heterogeneous Networks

Integration of diverse access networks into one convergent architecture presents several challenges associated with the diversity of approaches for QoS control that these access networks have.

In general a QoS control architecture is motivated by a set of principles (Aurrecoechea, 2008):

- **Transparency to the applications:** The QoS control architecture has to hide the details the underlying QoS parameters and procedures to the application level.
- **Integration:** QoS has to be configurable, predictable and maintainable all over the network. This can only be achieved if each resource provides QoS configurability, guarantees and maintenance.
- **Separation:** The actual media transfer has to be separated from the control and the management of the QoS control architecture.

Following this principles it is clear that the QoS control architecture has to communicate both with the application level to inform about the session parameters and the required service level and with the transport and access networks in order to be able to guarantee a service level along the path.

The functionality of a QoS architecture has to include QoS provisioning, QoS control and QoS management. A convergent QoS architecture has to target all of these functionalities and cope with the following requirements for heterogeneous integration. QoS provisioning involves QoS mapping, admission control and resource reservation. When connecting to different access networks the problem parameter mapping arises. Each access networks defines its own service levels and QoS classes. An abstract classification has to be done to the heterogeneous QoS control which can be mapped to the concrete access specific QoS classes in order to provide a certain service level. This function is referred as QoS mapping.

To relation the QoS required by services with the available resources the QoS architecture has also to provide an admission control framework that can deny establishment of sessions beyond the actual resource capacity of each network.

Every specific access network has its own over-the-air reservation procedures. The QoS architecture has to be able to trigger and control these procedures when necessary.

Not all access networks allow the same level of granularity for QoS specifications. In IP based networks it is normal that the level of granularity is the IP flow, but many protocols and frameworks allow the concept of IP flow aggregate. The convergent QoS control architecture has to provide both the IP flow level and the session level or IP flow aggregate.

For each IP flow the QoS control mechanisms include the flow scheduling, flow shaping, flow policing and flow control. Depending on the level of trust between entities this functions have to be done in all the entities in the path of the flow or just when entering and leaving a domain. When untrusted access networks connect to a convergent architecture QoS control has to be performed at the edge of the core network with the untrusted access network.

QoS management also establishes requirements for monitoring, availability, degradation maintenance and scalability but this are not specific of the heterogeneous environment.

Following these criteria the heterogeneity associated with the different access networks the convergent network connects to, can be successfully overcome. The difference in the architectures these access networks integrate, and the different QoS methods they implement in the lower layers can be abstracted over the IP level and with the use of a convergent QoS control framework QoS can be guaranteed along the path in the NGN.

End-to-End Assurance

The current global Internet service is based on best-effort service. This service does not guarantee anything, even delivering the IP packets within the network. Considering a packet sent to the Internet for delivery to a destination host, the network does not guarantee any specified delivery time, delivery speed, the available bandwidth, or even if the packet will be dropped if it faces congestion. Delay is not a problem if we consider delivering of an email message, where seconds or minutes will have a small impact on the end user. But if the transmission delay in a voice-over-IP (VoIP) call is large, or delays vary too much, or too many packets are lost, the quality will become unacceptable.

QoS is specific to the service being executed. Each service may be expressed by a set of parameters that are specific to it. Jitter is a parameter that is applied to packet switched networks, Cell Loss Ratio (CLR) to Asynchronous Transfer Mode (ATM) and these parameters would be meaningless in a Circuit Switched (CS) analog network, which makes the provision of QoS very heterogeneous and dependent of the architectures and services. Other characteristic is that QoS is an end-to-end issue. This means that all entities in the path between the parties are concerned to make the service possible and all the segments are involved in the process of QoS guarantee.

On the route that packets follow, from now on referred as the data path of a service, each intermediary node forwards the packets to the next one, considering the specific local routing rules. This enables the packets to get closer to their intended destination. In order to reserve resources on this path a mechanism for signaling and enforcement on each node has to be introduced.

Integrated Service (IntServ) (Braden, 1994) framework was defined to support the end-to-end QoS for multiple applications and to guarantee the resource reservation for the specific flows of them. IntServ considers that for each data flow a signaling on the data path has to be introduced. For example a multimedia session containing a video flow with a bandwidth of 346kbps with a delay of 100ms and an audio flow of 80kbps and a delay of 120ms will have 4 flows, 2 video ones and 2 audio ones for the bi-directional data traffic. Because of the different data path and parameters, each flow has to have its resources reserved separately in order to receive continuous data at the other parties of the session. Thus for resource reservation, each flow has to be separately signaled.

Although IntServ provides the necessary resource allocation on the end-to-end path the number of messages exchanged during the provisioning is directly proportional with the number of flows, which introduces a scalability problem on the network devices. Also the logic for processing RSVP is complex and has to be introduced in all the nodes of the network in order to make the service available.

By keeping a soft state reservation, the resources cannot be re-used immediately after a session is terminated which reduces the capacities of the network and increases the probability of a Denial of Service attack.

The Differentiated Services (DiffServ) (Blake, 1998) consider a complete aggregation of the flows on each border of a DiffServ Domain. A small number of classes is used and the classification is done only at the border of the domains. The core network routers have only to aggregate the flows and to route them using a priority system. No state or reservation is maintained in the intermediary routers, reducing the scalability problem compared to the Integrated Services.

Compared to the IntServ, the DiffServ reduces the computation load on the mobile devices. They do not have to signal the resource reservation. Each data packet receives its classification at the First hop Router and it is forwarded then to the DiffServ domain.

Figure 4. End-to-end QoS assurance

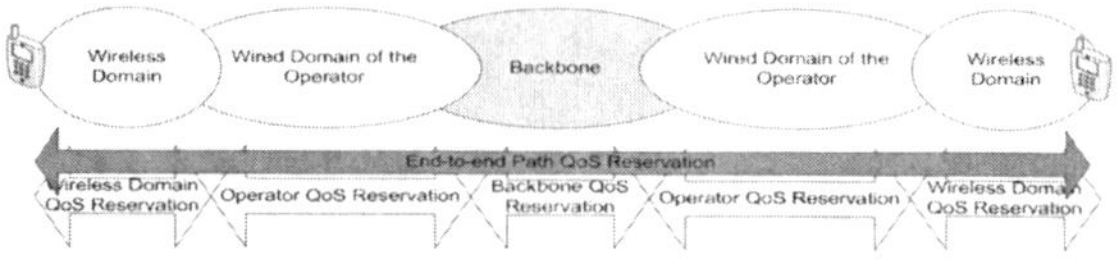

In order to ensure the QoS on the wireless link (e.g. between the host and the First hop Router) another QoS mechanism has to be deployed together with the Diff-Serv. Therefore the QoS reservation path is separated into at least two reservations: one between the mobile device and the First-hop router and one for the DiffServ domain. Multiple other solutions for QoS reservation for specific links from the end-to-end path are also deployed like the RSVP-Traffic Engineering which addresses the bulk traffic that is exchanged between different backbone nodes.

Therefore the end-to-end QoS reservation path is split between different domains depending on the traffic type. This allows a scalable management of the network as it considers that between two nodes of the end-to-end path the QoS reservation protocol is appropriately chosen.

As depicted in Figure 4, the end-to-end resource reservation path is split between different QoS reservations depending on the domain through which the packets are transported. For each of these domains an end-to-end QoS has to be assured. As previously described there are several mechanisms already deployed for the wired domain of the operators and the backbone network. The main problem of the end-to-end QoS assurance at this moment is the resource reservation over the end domains, especially over the wireless domains, as the same mobile user may want to exchange data in time over different wireless links because of its mobility.

Also the wireless domains are open to multiple mobile devices, whose number may vary largely over a period of time while the operator domain and the backbone are considered as fixed networks and may scale the links according to different analysis of the data traffic. Thus, the end-to-end resource reservation can be reduced to the resource reservation on the last link for all the entities involved in a service as being the link on which the concurrent access can not be completely predicted.

To reserve resources in all the last domains a mechanism of signaling the QoS requirements from one domain to another is to be deployed. The IETF introduced the Policy Decision Framework (Boyle, 2000) which enables a specific QoS to be reserved on a node or on a set of nodes. The architecture contains two entities: the Policy Decision Point (PDP) and the Policy Enforcement Point (PEP).

The Policy Decision Point receives a trigger from the exterior, typically from the path on which

Figure 5. End-to-end policy based QoS assurance

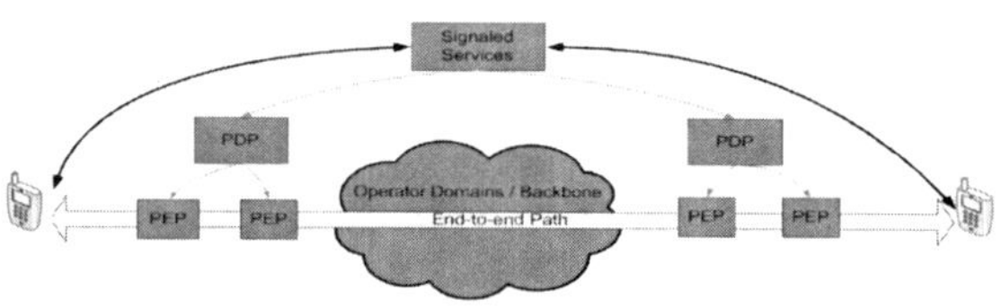

the service was signaled. Then by applying the policies of the operator, it enforces a specific QoS on the Policy Enforcement Point which is placed on the end-to-end data path, as depicted in Figure 5. Using this mechanism the signaling of a specific service is transmitted to a policy engine located in both last link domains which at their turn trigger the resource reservation.

As the end devices are only signaling the services of the user, the full control of the QoS reservation is passed to the network of the operator which enables it to decide which type of QoS can be enforced. In this way, the parted resources, which are the wireless links at the borders of the end-to-end path, can be managed by the operator according to its own internal policies which allows the optimization of the full QoS reservation.

Vertical Handover and QoS

The most important challenge in the wireless IP networks is the efficient allocation of the resources while the users are mobile. For example, in an IntServ network whenever the user changes the point of attachment, a new resource reservation is required on the new path. The resource reservation has to be initiated at least on the local path, which increases the load on the wireless interface and induces a high delay, unacceptable for time constraint traffic.

There is no guarantee that the same level of resources can be reserved in the new network, making the handover decision a static one as related to the network load. This implies that even if the process of handover is possible the new network may not be able to provide the same resources, thus a new negotiation at the service level and a new resource provisioning is necessary.

From the perspective of the network providers a scalable resource management solution is needed. The access networks can be scaled as the bandwidth capacities of the wireless environment are known. But the mobility of the users can not be determined.

In environments where cells vary in size and the mobile nodes have different velocities, the typical Poisson handoff distributions cannot be applied. The possibility that the terminals connect simultaneously to more than one access network increases the complexity of the network decision and resource reservation. As to be protected from the starvation of some terminals a load balancing between the networks of the same operator is necessary in dense networks environment as users and as access networks.

Being the only area where the network provider does not have control and having users that can connect freely to the base stations of the network, leads to the conclusion that the most problematic area of the end-to-end resource reservation is the wireless network.

The EPC-PCC solution offers the possibility of renegotiation of the session profiles during the session. The architecture allows, by introduction of enforcement points for policy based autonomic decisions. However the network was designed for static users and for users which desire to hand over the sessions from one terminal to another, a session mobility type without having the time constraints of the user mobility.

The latency of the session provisioning and the intensive signaling ensure that the user preferences are taken into account. But in a wireless environment where the handovers require a low latency, special signaling is required in order to offer the desired service continuity.

Especially the multimedia traffic is very affected by the bandwidth and latency fluctuations. In order to be able to provide a continuous stream of the packets pertaining to a real-time flow, the resources have to be fully reserved on the path, thus this necessary new resource reservation has to be optimized as to handle the required user satisfaction.

Without a mobility management the network is not aware of the next point of attachment of the mobile node and cannot respond adequately to the service adaptation requirement. In the current context the only entity aware of the mobility is the terminal which has to re-signal the session on the new network and by this increasing the end-to-end delay of the session provisioning in the new network.

Without a triggering mechanism able to announce the network of imminent changes of point of attachment of the mobile node, the network cannot handle the resource reservation but after the new IP context is determined. Therefore the delay of the provisioning in the new network adds the complete delay of session provisioning, both service profile negotiation and resource reservation.

CONVERGENT QOS NGMN ARCHITECTURE

This section addresses the 3GPP solution to cope with the heterogeneity of future networks. It explains the Policy and Charging Control (PCC) architecture, which is part of the Evolved Packet System (EPS) (Lescuyer & Lucidarme, 2008). EPS represents the latest evolution of the UMTS standard, providing an evolution step with a new radio interface and an evolved architecture for both the access and the core network parts. It envisages improving performance metrics like reduced latency and improved spectrum efficiency, as well as, providing a unified and simplified architecture based on Packet Switched technology (PS).

Policy and Charging Control Architecture

The PCC architecture (3GPP TS 23.203, 2008) has been defined by 3GPP since Release 7 to provide QoS control, access control and charging to their packet data based networks. Already in Release 6 3GPP had defined an architecture for the exchange of policy and QoS information between the Radio Access Network (RAN) and their IP based service layer, i.a. IP Multimedia Subsystem (IMS). This architecture was made of two components: the Policy Decision Function (PDF) and the Policy Enforcement Point (PEP). The PDF could be deployed as an internal subfunction of the P-CSCF entity of IMS or as a separate entity connected to the P-CSCF through a Diameter interface named Gq. The PEP was a subfunction of the GGSN in the 3GPP RAN. The policies to enforce in the RAN are sent from the PDF to the PEP through a COPS interface named Go. With this policy and QoS framework an interconnection of the service layer and the radio access was done in order to assure QoS levels for IMS based services.

In Release 7 the framework evolved to a more complex architecture with the addition of the charging functions and the separation from the entities from the IMS domain. Therefore the P-CSCF is no longer mentioned in the specifications but instead a generic Application Function AF that can be any kind of operators services function, which communicates through a Diameter interface very similar to Gq named Rx with the Policy and Charging Rules Function (PCRF). This entity performs the policy decisions and communicates with the Policy and Charging Enforcement Function (PCEF) which is situated in the RAN gateway through a Diameter interface named Gx. The RAN Gateway is a generalization of the GGSN in order to also provide QoS control and policy enforcement for other access networks that connect to the 3GPP Release 7 architecture. A new functional entity is mentioned in the specifications the Subscribers Profile Repository (SPR) from which the PCRF is able to download the subscribers profile in order to apply subscriber specific policies. The PCRF connects to the SPR through the Sp which was never defined in the specifications.

The evolution from Release 6 to Release 7 gave the policy framework of 3GPP more importance as it is separated from the IMS specifications and its made more abstract and generic in order to cope with the requirements of different access networks.

Figure 6. The PCC architecture non-roaming and roaming configurations

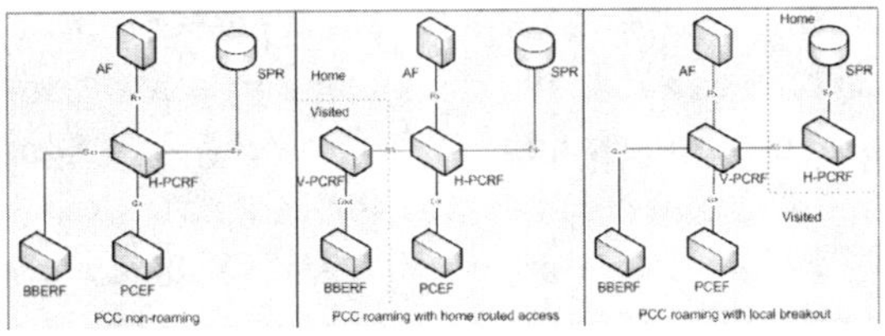

In Release 8 which is the current stable version of the 3GPP architecture the PCC has gained even more importance. The PCRF remains being the main entity of the architecture, executing the policy control decisions based on the information provided by the PCEF, the SPR and the AF; but it is split in two functionalities the Home-PCRF (H-PCRF) and the Visited-PCRF (V-PCRF) in order to better support roaming scenarios.

The main difference between the H-PCRF and the V-PCRF is that the H-PCRF can access the subscriber profile and the V-PCRF cannot. Thus the V-PCRF communicates with the H-PCRF in order to receive subscriber specific policies through an interface S9 which has not yet been completely specified.

The PCEF which performs service data flow detection and policy enforcement has also a counterpart the Bearer Binding and Event Reporting Function (BBERF) which only performs bearer binding and event reporting to the PCRF. These two functionalities are deployed in the access networks gateways where the packet marking for prioritization and the QoS parameters for the RAN is set.

The need for these separated functionalities V-PCRF, H-PCRF and PCEF, BBERF is brought because of the different roaming configurations which 3GPP considers. The current architecture with the non-roaming and roaming options are depicted in Figure 6.

The most relevant functionality of the PCC is policy control. Policy control includes gating control, binding, event reporting and QoS control. Gating control allows performing access control in this architecture based on service data flows. The decision about access control: to open or to close the gate; is taken by the V-PCRF and enforced by the PCEF.

Binding refers to bearer binding and session binding. It's a functionality performed by the BBERF and the PCEF in order to associate the service data flow to the appropriate bearers that transport that data flow.

Event reporting is performed in both the BBERF and the PCEF. The PCRF can subscribe to events of the bearer level and the BBERF and PCEF report when such events occur. In the same way the AF can subscribe to events of the PCRF. This scheme permits the PCRF or the AF to react upon these events and trigger procedures in the user plane. An example of these events is the BBERF reporting to the PCRF a loss of bearer or a QoS or Radio Access Type change for a subscriber or a certain data flow. The PCRF can decide then what to do with this specific data flow or report to the AF when it has subscribed to this specific event.

The QoS control comprises the authorization and enforcement of the QoS parameters procedures taken by the PCRF based on the information provided by the SPR, i.a. subscriber profile, the AF, i.a. service parameters or the BBERF or PCEF i.a. bearer level parameters.

Not considering the roaming scenario, two modes of operation are described for the PCC: the PUSH and PULL modes.

In the PUSH mode the service signaling is received by the AF first and then it communicates the PCRF, through the Rx interface, the service parameters that the user requests. The PCRF may then request the subscribers profile from the SPR and perform the policy decision based on the operator's policies, the session parameters and the subscriber profile. This decision outputs one or more PCC Rules and QoS Rules. These are the basic set of information about the TFTs and parameters for the bearer binding to be enforced in the gateways. The PCC Rules are sent to the PCEF through the Gx interface and the QoS Rules to the BBERF through the Gxx interface. These rules include the events subscription of the PCRF for these service flows and the gate status.

The PCEF and BBERF then enforce the TFTs and open or close the gate associated with the flows. In case a subscribed event occurs they inform the PCRF about it.

In the PULL mode a bearer request is received from the RAN and triggers a QoS Rules or PCC Rules requests from the BBERF or the PCEF (maybe not both are deployed in a concrete network). These requests include the identification parameters of the subscribers (e.g. IP address) and the IP bearer details. The PCRF then may request the subscribers profile from the SPR and perform the policy decision sending the QoS Rules or PCC rules back to the BBERF and PCEF if the session is authorized.

The PCC architecture as described currently in the 3GPP specifications includes the necessary abstraction to cover different access networks in one QoS control framework that can apply the procedures needed for end-to-end QoS support. The PCC is part of a broader architecture for interconnection of packet switching heterogeneous networks into one unique homogeneous Evolved Packet Core (EPC) which provides the flexibility and abstraction needed to be a generic All-IP, NGN network with QoS support.

To achieve this and as part of the PCC standardization 3GPP has established a set of QoS Class Identifiers (QCI) which give standard QoS characteristics in terms of traffic type, priority, packet delay budget and packet error loss rate. It uses 8 different QCIs that range from Guaranteed Bit Rate services with a delay of 100ms and a packet error rate of 10^{-2} for conversational voice to a low priority non guaranteed bit rate with delay of 300ms and packet error rate of 10^{-6} for TCP based services like email or web access. These QCIs are converted into access specific QoS classes and priorities at the gateways of the RAN which know the standardized requirements of associated with each QCI and the mapping into their specific access methods.

Evolved Packet Core

The Evolved Packet Core (EPC) (3GPP TS 23.401, 2008; 3GPP TS 23.402, 2009) is the core network architecture designed by 3GPP to cope with the challenges of their new access system: LTE. The EPC standardization includes the connection of all the 3GPP packet radio access technologies GPRS, UMTS and LTE to a common core network that manages the QoS control, security and mobility issues.

To this same core network also the non-3GPP access can connect to in order to provide convergent network architecture that aligns with the Next Generation Mobile Networks paradigm. This core network is an IP only nucleus with support for network access control, packet routing and transfer functions, mobility management, security, radio resource management and network management.

Logically the EPC is a layer between the radio access systems and the services provided over IP. To a certain extent the EPC comes from the necessity to provide mobility, security and QoS below the services layer which was represented by the IMS. The EPC provides a level of abstraction from the access technology which guarantees the seamless delivery of services to the users.

To provide QoS and policy control the EPC includes the PCC functionality and the PCRF is an important entity of the whole architecture as it provides the signaling connection between the services domain and the EPC.

Figure 7. The EPC architecture (non-roaming)

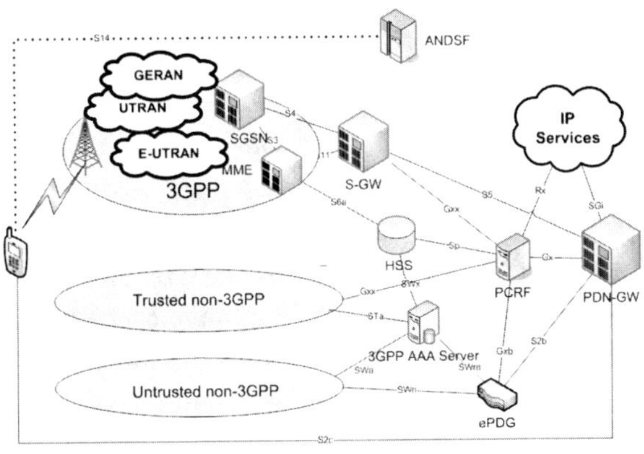

The main functional entities of the EPC for the non-roaming case are depicted in Figure 7.

The Packet Data Network (PDN) Gateway (PDN-Gw) is the main gateway of the EPC. Through it, all the data of the subscriber is routed towards the services network. It also performs mobility anchoring functionality when the user roams from a 3GPP network to a non-3GPP network and implements the PCC functionality of the PCEF providing QoS enforcement.

Another gateway is needed as the specific gateway for each radio access network. For the 3GPP networks this is the Serving Gw, for the non-3GPP networks it can be either the evolved Packet Data Gateway (ePDG) or the non-3GPP trusted Gateway (e.g. the ASN-Gw in WiMAX or the HRPDGw in HRPD). These entities include the PCC functionality of the BBERF for QoS rules enforcement and event reporting.

The Serving Gw provides mobility anchoring when the user roams within different 3GPP accesses e.g. UMTS and LTE.

The ePDG and the trusted and untrusted non-3GPP network connect to the 3GPP AAA Server for authorization, authentication and accounting. The 3GPP access networks connect directly to the Home Subscriber Server (HSS) for these purposes.

For the access network selection and discovery between heterogeneous access technologies the Access Network Discovery and Selection Function (ANDSF) entity has been defined although it is not completely described in the current release of the specifications (Release 8).

The 3GPP AAA Server and the PCRF split their functionalities in two modes 3GPP AAA Proxy and V-PCRF for the visiting network and 3GPP AAA Server and H-PCRF for the home network when the user is roaming in the EPC.

The full deployment of the PCC architecture, even though its optional, is the key that permits the EPC to provide QoS control including gating, policing, traffic control, flow control, packet marking etc. The resource reservation and scheduling is done by the access network specific entities and architectures as covered in the previous sections.

The definition of standard QCI classes for the EPC as discussed previously allows for a coherent QoS deployment in a core network to which very different mobile technologies connect to.

Different access technologies are supported in the EPC. Three different groups of access networks permit a consistent EPC definition. The 3GPP access networks comprise GERAN for GPRS, UTRAN for UMTS acces and E-UTRAN for the new LTE access. These are the best described in the specifications where full QoS control is applied until the air interface. GERAN, UTRAN and E-UTRAN include radio resource reservation with specific bandwidth and QoS requirements per session and per flow.

The EPC considers for the 3GPP accesses two QoS control schemes: the bearer level QoS control and the service level QoS control.

The bearer level QoS scheme is associated with an EPS bearer (a PDP context for GERAN or UTRAN). The EPS bearer is associated with a Service Data Flow aggregate that is transported together in a GTP tunnel. An EPS bearer receives always the same QoS treatment (e.g. scheduling policy, queue management policy, rate shaping policy etc.). The BBERF and the PCEF (if GTP is supported until the PDN-Gw) can perform bearer binding and associate a TFT for the EPS bearers.

The QoS parameters associated with the bearer level QoS control are the QCI, the Allocation and Retention Priority (ARP) which is a priority to reject the EPS bearer in case of resource limitations, the guaranteed bit rate (GBR) and the maximum bit rate (MBR). Also two parameters can be available in the subscriber's profile that limits the amount of data a user can generate per PDN and globally (Aggregate Maximum Bit Rate, AMBR).

The Service level QoS control applies when the Rx interface between the PCRF and the operator's services exist and conveys the same parameters of the bearer level QoS control in the form of PCC and QoS rules and the information needed to perform the bearer binding that permits the BBERF and PCEF to associate the service flows with the specific bearers that convey those services.

For 3GPP access networks the BBERF is part of the Serving Gw. Depending on the configuration of the architecture which allows for different options the bearer binding and TFT assignment has to be done in the Serving-Gw or can be done in the PDN-Gw. In the most general case it will be done in the Serving-Gw that includes the connection to the access network specific entities (MME and SGSN) which will apply the resource allocation procedures (with the associated queue management and scheduling) over the air interface and not only packet marking, filtering and queuing.

FUTURE RESEARCH DIRECTIONS

The first deployments of the EPC and LTE are envisaged to begin end of 2009, but there is still work in progress and many issues that need to be addressed to bring the described advantages of this architecture into life. These topics are not necessarily required, but in our view are necessary to release the full potential of the convergent NGN architecture in this new context. They can be summarized as:

- The current work on the EPC architecture includes the study of the necessary interaction between the different entities of the EPC (and PCC) to support the vertical handover with QoS parameters continuity;
- There is ongoing work to consider QoS parameters adaption to different access networks;
- Study of QoS management for heterogeneous mobile networks based on the EPC;
- Study the roaming implications in QoS interactions in the EPC i.a. the PCC;
- Analysis of the challenges associated with QoS control upon the integration of fixed networks (DSL and cable) in the EPC, which are still work in progress in the 3GPP standards;
- Analysis of possible simplifications to the EPC and PCC architectures reducing the required procedures associated with QoS control.
- Analysis of the impacts of terminal multiplicity and session mobility bring to the provisioning and QoS control in the EPC.
- Impacts of the usage of dual mode terminals for the connection of a subscriber to multiple PDN at the same time and the mobility of IP flows associated to a session between PDNs in the QoS control architecture of the EPC.

CONCLUSION

This study presented QoS control schemes of some of the today's most used wireless technologies. The different wireless technologies provide data connectivity with different QoS schemes and procedures. Since the All-IP network paradigm is driving NGN, the provision of QoS over IP has become of great interest in the research community. As the complexity of these networks increase with the addition of new access technologies and multimedia services requirements, the need for convergent solutions becomes more important. In this context, convergent networks that allow heterogeneous accesses to be connected to a same IP core that is able to cope with the heterogeneity of QoS schemes and parameters are needed.

The QoS control in the access networks is based on the use of different classes, which control the specific QoS parameters for each access network. The E-UTRAN is based on a QoS Class Identifier (QCI), which is a scalar used to reference node specific parameters for controlling packet forwarding treatment (e.g. scheduling, admission thresholds) and that have been pre-configured by the operator owning the node (e.g. eNodeB). CDMA 2000 uses the same classes as UMTS, since they share the 3GPP model. WiFi uses link layer mechanisms for offering QoS to the services. They are based on the Distributed Coordination Function (DCF) which supports delay-insensitive data and the Point Coordination Function (PCF) which supports delay sensitive transmissions.

In this book chapter we described the features and advantages of the EPC as a solution to the different issues described before. The EPC addresses the challenges of heterogeneity while providing an integrated solution for QoS control: the PCC. The PCC presents a common QoS control mechanism which enforces its decisions on a unified base in the core network and specific to each access technology on the wireless link. This mechanism enables an easy integration of new access technologies into the overall system and also enables the service providers to reserve and to release resources using a single mechanism independent on the access technology to which the mobile devices are connected to.

QoS control in a converged scenario is taught and there are still many topics open. The PCC has evolved in order to cope with most of the aspects of QoS control convergence. Still there are some topics that need further work in order to accomplish the desired connected anytime, anywhere with any device paradigm.

REFERENCES

Blake, S., Black, D., Carlson, M., Davies, E., Wang, W., & Weiss, W. (1998). *An architecture for differentiated service*. IETF, Network Working Group.

Boyle, J., Cohen, R., Herzog, S., Rajan, R., & Sastry, A. (2000). *The COPS (Common Open Policy Service) protocol*. IETF, Network Working Group.

Braden, R., Clark, D., & Shenker, S. (1994). *Integrated services in the internet architecture: an overview*. IETF, Network Working Group.

3GPP2 C.S0024-B. (2006). *CDMA2000 High Rate Packet Data Air Interface Specification, V1.0.* Aurrecoechea, C., Campbel, A.T., & Hauw, L. (2008). A survey of QoS architectures. *Springer Journal of Multimedia Systems, 6*(3).

3GPP TS 23.107. (2007). *Quality of Service (QoS) Concept and Architecture, V8.0.0.*

3GPP TS 23.203. (2008). *Policy and charging control architecture, V8.4.0.*

3*GPP TS 23.401. (2008).* General Packet Radio Service (GPRS) enhancements for Evolved Universal Terrestrial Radio Access Network (E-UTRAN) access, V8.4.1.

3GPP TS 23.402. (2009). *Architecture enhancements for non-3GPP accesses, V8.4.1.*

IEEE. 802.16e. (2005). *Air Interface for Fixed and Mobile Broadband Wireless Access Systems.*

IEEE. 802.11. (2007). *Part 11: Wireless LAN Medium Access Control (MAC) and Physical Layer (PHY) Specifications.* IEEE Standards for Information Technology– Telecommunications and Information Exchange between Systems – Local and Metropolitan Area Network – Specific Requirements, IEEE.

Lescuyer, P., & Lucidarme, T. (2008). *Evolved Packet System. The LTE and SAE Evolution of 3G UMTS.* Hoboken, NJ: John Wiley and Sons, Ltd.

Song, J., Chang, M. Y., & Lee, S. S. (2007). Overview of ITU-T NGN QoS Control. *IEEE Communications Magazine*, 116–123. doi:10.1109/MCOM.2007.4342866

Taaghol, P., Salkintzis, A. K., & Iyer, J. (2008). Seamless Integration of Mobile WiMAX in 3GPP Networks. *IEEE Communications Magazine*, *46*(10). doi:10.1109/MCOM.2008.4644122

WiMAX Forum. (2007). *Network Architecture Stage 2 and 3.* Release 1.0 (Revision 1.2).

Compilation of References

3GAmericas. (2006). *Mobile Broadband: The Evolution of UMTS/HSPA, 3GPP Release 7 and Beyond.* Retrieved June 30, 2008, from http://www.3gamericas.org

3GPP TS 125.302. (2002, March). *Services provided by the physical layer, V5.0.0.*

3GPP TS 22 105 Version 7.1.0 (2006, December). *UMTS; Services and Services Capabilities.*

3GPP TS 23.107 Version 7.1.0. (2007, June). *UMTS; Multimedia Messaging Service (MMS).*

3GPP TS 23.107. (2007). *Quality of Service (QoS) Concept and Architecture, V8.0.0.*

3GPP TS 23.110: UTRAN access stratum: services and functions, V6.4.0 (2004-12).

3GPP TS 23.203. (2008). *Policy and charging control architecture, V8.4.0.*

3GPP TS 23.402. (2009). *Architecture enhancements for non-3GPP accesses, V8.4.1.*

3GPP TS 25.211. (2007, October). *Physical channels and mapping of transport channels onto physical channels (FDD), V7.3.0.*

3GPP TS 26.346. (2005). *Multimedia Broadcast/Multicast Service (MBMS); Protocols and Codecs – Version 6.1.0.* Third Generation Partnership Project.

3GPP TS 36.104 version 8.3.0 (Release 8). (n.d.). *Base Station (BS) radio transmission and reception.*

3GPP, R6, TS 125.301 (2005). *Radio interface protocol architecture.*

3GPP. (2001). *Digital cellular telecommunications system (Phase 2+): Multiplexing and multiple access on the radio path (3GPP TS 05.02 version 8.9.0 Release 1999).* Retrieved from http://www.3gpp.org

3GPP. (2008, December). *TS 23.203 Policy and Charging Control Architecture v8.4.0* (3GPP Technical Specification).

3GPP. (2008, December). *TS 23.401 General Packet Radio Service enhancements for Evolved Universal Terrestrial Radio Access Network V8.4.1* (3GPP Technical Specification).

3GPP. (2008, February). *TR 23.822 Framework for Gq'/Rx Harmonization V0.2.0* (3GPP Technical Recommendation).

3GPP. TS 23.246 v8.2.0. (2008). Multimedia Broadcast/Multicast Service, Architecture and functional description (Release 8).

3GPP. TS 25.301, v8.2.0. (2008). *Radio interface protocol architecture.* A. Demers, S. Keshav, S. Shenkar. (1990). *Analysis and Simulation of a Fair Queueing Algorithm, Internet. Res., & . Exper., 1*(1), 3–26.

3GPP2 C.S0014-0 version 1.0. (1999). Enhanced Variable Rate Codec (EVRC).

3GPP2 C.S0024-B. (2006). *CDMA2000 High Rate Packet Data Air Interface Specification, V1.0.* Aurrecoechea, C., Campbel, A.T., & Hauw, L. (2008). A survey of QoS architectures. *Springer Journal of Multimedia Systems, 6*(3).

3GPPTS 23.401. (2008). General Packet Radio Service (GPRS) enhancements for Evolved Universal Terrestrial Radio Access Network (E-UTRAN) access, V8.4.1.

3rd Generation Partnership Project. (1999). *Technical Specification Group GSM/EDGE Radio Access Network; Radio subsystem link control (Release 1999).* Al-Agha, K. (2008, November). *Internet new generation and Post-IP architecture.* Tutorial presented at the Second IEEE / IFIP International Conference on New Technologies, Mobility, and Security IEEE/IFIP NTMS 2008, Tangier, Morocco.

Abedi, S. (2005). Efficient Radio Resource Management for Wireless Multimedia Communications: A Multidimensional QoS-Based Packet Scheduler. *IEEE Transactions on Wireless Communications, 4*(6), 2811–2822. doi:10.1109/TWC.2005.858032

Acampora, A. S., & Naghshineh, M. (1994). An Architecture and Methodology for Mobile-Executed Handoff in Cellular ATM Networks. *IEEE Journal on Selected Areas in Communications, 12,* 1365–1374. doi:10.1109/49.329339

Acampora, A. S., & Naghshineh, M. (1994). Control and Quality-of-Service Provisioning in High Speed Microcellular Networks. *IEEE Personal Communications,* 36-43.

Adya, A., Bahl, P., Padhye, J., Wolman, A., & Zhou, L. (2004). A Multi-Radio Unification Protocol for IEEE 802.11 Wireless Networks. In *Proceedings of IEEE BROADNETS 2004.*

Aghdasi, H. S., Abbaspour, M., Moghadam, M. E., & Samei, Y. (2008). An energy efficient and high-quality video transmission architecture in wireless video-based sensor networks. *Sensors, 8,* 4529–4559.

Agrawal, D. P., Li, W., & Wang, H. (2005). *Dynamic admission control and QoS for 802.16 Wireless MAN.* Paper presented at the Wireless Telecommunications Symposium 2005.

Agrawal, P., Anvekar, D. K., & Naredran, B. (1996). Channel management policies for handovers in cellular networks. *Bell Labs Technical Journals, 1*(2), 97–110. doi:10.1002/bltj.2018

Ahn, G.-S., Campbell, A. T., Veres, A., & Sun, L.-H. (2002). Supporting service differentiation for real-time and best-effort traffic in stateless wireless ad hoc networks (SWAN). *IEEE Transactions on Mobile Computing, 1*(3), 192–207. doi:10.1109/TMC.2002.1081755

Ahn, K., & Kim, S. (2003). Optimal bandwidth allocation for bandwidth adaptation in wireless multimedia networks. *Computers & Operations Research, 30,* 1917–1929. doi:10.1016/S0305-0548(02)00115-6

Air interface for fixed broadband wireless access systems - amendment for physical and medium access control layers for combined fixed and mobile operation in licensed bands. (2005). *IEEE Standard 802.16e.*

Akkaya, K., & Younis, M. (2005). A survey of routing protocols in wireless sensor networks. *Elsevier Ad Hoc Network Journal, 3*(3), 325–349.

Akyildiz, I. F., Ho, J. S. M., & Lin, Y.-B. (1996). Movement-based location update and selective paging for PCS networks. *IEEE Transactions on Networking, 4*(4), 629–638. doi:10.1109/90.532871

Akyildiz, I. F., Lee, W.-Y., Vuran, M. C., & Mohanty, S. (2006). NeXt generation/dynamic spectrum access/cognitive radio wireless networks: a survey. *Computer Networks, 50,* 2127–2159. doi:10.1016/j.comnet.2006.05.001

Akyildiz, I. F., Wang, X., & Wang, W. (2005, March). Wireless mesh networks: a survey. *Computer Networks, 47,* 445–487. doi:10.1016/j.comnet.2005.01.002

Albaladejo, A., Gouveia, F., Corcici, M., & Magedanz, T. (2008, November). The PCC Rule in the 3GPP IMS Policy and Charging Control Architecture. In *Proceedings of 2008 IEEE Global Communications Conference* (pp. 1-5).

Alfano, F., McCann, P., & Towle, T. (2006). IMS Service-Based Bearer Control. *Bell Labs Technical Journal, 10*(4), 151–166. doi:10.1002/bltj.20131

Alicherry, M., Bhatia, R., & Li, L. (2005). Joint channel assignment and routing for throughput optimization in multi-radio wireless mesh networks. In *Proceedings of ACM MOBICOM 2005.*

Allen, A. O. (1990). *Probability, Statistics, and Queueing Theory with Computer Science Applications* (2nd ed.). New York: Academic Press.

Almes, G., Kalidindi, S., & Zekauskas, M. (1999). *A One-way Packet Loss Metric for IPPM*. IETF RFC 2680.

Anue Systems. (n.d.). *Network Emulation*. Retrieved from http://www.anuesystems.com

AlQahtani, S. A., & Mahmoud, A. S. (2008). Performance analysis of two throughput-based call admission control schemes for 3G WCDMA wireless networks supporting multi-services. *Computer Communications*, *31*, 49–57. doi:10.1016/j.comcom.2007.10.029

Alwan, A., Bagrodia, R., Bambos, N., Gerla, M., Kleinrock, L., Short, J., &Villasenor, J.,(1996, April). Adaptive Mobile Multimedia Networks. *IEEE Personal Communications, 34-5.1*

Andrews, J. G., Chen, R., Ghosh, A., & Wolter, D. R. (2005). Broadband wireless access with WiMax/802.16: current performance benchmarks and future potential. *IEEE Communications Magazine*, *43*(2), 129136.

Apostolopoulos, G., Kama, S., Williams, D., Guerin, R., Orda, A., & Przygienda, T. (1999, August). *RFC 2676: QoS Routing Mechanisms and OSPF Extensions* (experimental).

Arena, G. S. M. (2009). Retrieved June 30, 2008 from http://www.gsmarena.com

Arlitt, M. F., & Williamson, C. L. (1996). *Web Server Workload Characterization: The Search for Invariants*. In *Proc. SIGMETRICS*, Philadelphia, PA, USA.

Armitage, G. (2000). *Quality of Service in IP Networks*. Indianapolis, IN: MacMillan Technical Publishing

Arunachalam, A. V. (1999). *Quality of Service (QoS) classes for BWA*. A contribution to the IEEE 802.16 Broadband Wireless Access Working Group.

Asokan, R., & Natarajan, A. M. (2007, July). An approach for reducing the end-to-end delay and increasing network lifetime in mobile ad hoc networks. *International Journal of Information Technology*, *4*(2), 121–127.

Asokan, R., & Natarajan, A. M. (2008, March). Performance evaluation of energy and delay aware quality of service (QoS) routing protocols in mobile ad hoc networks. *Int. Journal of Business Data Communications and Networking*, *4*(1), 52–63.

Asokan, R., & Natarajan, A. M. (2008, October). Performance Analysis of Quality of Service Enabled Temporally Ordered Routing Algorithm using Ant Colony Optimization in Mobile Ad hoc Networks. *Journal of Advances in Engineering Science*, *2*(2), 11–18.

Asokan, R., Natarajan, A. M., & Venkatesh, C. (2008, July). Ant Based Dynamic Source Routing Protocol to Support Multiple Quality of Service (QoS) Metrics in Mobile Ad Hoc Networks. *International Journal of Computer Science and Security*, *2*(3), 48–56.

Association, G. S. M. (2007, February). *HSPA mobile broadband today*. Retrieved June 30, 2008, from http://www.gsmworld.com/hspa

ATIS. (2006). *ATIS-0800004, A Framework for Qos Metrics and Measurements Supporting IPTV Services*. Washington, DC: Alliance for Telecommunications Industry Solutions.

Awoniyi, O. (2005). *Packet Error Rate in OFDM-based Wireless Networks and Its Impact on VoIP and TCP-Based Data Applications*. Doctoral dissertation, Stanford University, California.

Ayyagari, D., & Empremides, A. (1999). Admission control with priorities: approaches for multi-rate wireless systems. *ACM Mobile Networks and Applications*, *4*, 209–218. doi:10.1023/A:1019103016426

Azarian, K., El Gamal, H., & Schniter, P. (2008). On the optimality of the ARQ-DDF protocol. *IEEE Transactions on Information Theory*, *54*(4), 1718–1724. doi:10.1109/TIT.2008.917722

Babich, F., Comisso, M., D'Orlando, M., & Vatta, F. (2006). Distortion Estimation Algorithms (DEAs) for wireless video streaming. In *Proceedings of the IEEE Global Telecommunications Conference GLOBECOM*, San Francisco, USA (pp. 1- 5).

Babich, F., D'Orlando, M., & Vatta, F. (2008). Video Quality Estimation in Wireless IP Networks: Algorithms and Applications. *ACM Transactions on Multimedia Computing, Communications, and Applications (TOMCCAP)*, 4(1).

Babich, F., D'Orlando, M., & Vatta, F. (2008b). Applications of Video Distortion Estimation Algorithms for Efficient Video Streaming. In *Proceedings of the IEEE Global Telecommunications Conference, GLOBECOM*, New Orleans, USA.

Badis, H., & Agha, K. A. (2005). QOLSR, QoS Routing for Ad Hoc Wireless Networks using OLSR. *Wiley European Trans. Telecommunications, 15*(4), 427–442. doi:10.1002/ett.1067

Bae, T. M., Thang, C. T., Kim, D. Y., Ro, Y. M., Kang, J. W., & Kim, J. G. (2006). Multiple Region-of-Interest Support in Scalable Video Coding. *Electronics and Telecommunications Research Institute Journal, 28*(2), 239–242.

Bahl, P., Chandra, R., & Dunagan, J. (2004). SSCH: slotted seeded channel hopping for capacity improvement in IEEE 802.11 ad hoc wireless networks. In *Proc. ACM Annual International Conference on Mobile Computing and Networking (MobiCom)* (pp. 216-230).

Bahr, M. (2006). Proposed Routing for IEEE 802.11s WLAN Mesh Networks. In *Proceedings of the 2nd Annual International Workshop on Wireless Internet* (Vol. 220).

Bai, X., Shami, A., & Ye, Y. (2008). Robust QoS control for single carrier PMP mode IEEE 802.16 systems. *IEEE Transactions on Mobile Computing, 7*(4), 416–429. doi:10.1109/TMC.2007.70738

Balakrishnan, H., Padmanabhan, V., Seshan, S., & Katz, R. (1996). A comparison of mechanisms for improving TCP performance over wireless links. In *Proc. of ACM SIGCOMM'96,* Stanford, CA.

Bansal, N., & Liu, Z. (2003). Capacity, Delay and Mobility in Wireless Ad-Hoc Networks. In *Proceedings of IEEE INFOCOM 2003.*

Bao, L., & Aceves, J. J. G.-L. (2003). Topology management in ad hoc networks. In *Proceedings of 4th ACM International Symposium on Mobile Ad Hoc Networking & Computing* (pp. 129-140). New York: ACM.

Baoxian, Z., & Hussein, T. M. (2005, October). QoS routing for wireless ad hoc networks: problems, algorithms, and protocols. *IEEE Communications Magazine, 42*(10), 110–117. doi:10.1109/MCOM.2005.1522133

Baras, J. S., & Mehta, H. (2003, March). PERA: A probabilistic emergent routing algorithm for mobile Adhoc Networks. In *Proceedings of WiOpt 03: Modeling and optimization in mobile adhoc and wireless networks.*

Barceló, F., & Jordán, J. (2000). Channel holding time distribution in public telephony systems (PARM and PCS). *IEEE Transactions on Vehicular Technology, 49*(5), 1615–1625. doi:10.1109/25.892545

Barford, P., Bestavros, A., Bradley, A., & Crovella, M. (1998). *Changes in web client access patterns: Characteristics and caching implications* (Tech. Rep. No. 1998-023). Boston, MA: Boston University.

Barile, M., Camarda, P., Dell'Aquila, R., & Vitti, N. (2006, September). Parametric Models for Speech Quality Estimation in GSM Networks. In *Proceedings of software in telecommunications and computer networks, SoftCOM 2006* (pp. 204 - 208).

Bar-Noy, A., Kessler, I., & Sidi, M. (1994). Mobile users: To update or not to update. In *IEEE Proc. INFOCOM*, Toronto, Ontario, Canada, June 14-16 (Vol. 1, pp. 570-576).

Barolli, L., Koyama, A., Suganuma, T., & Shiratori, N. (2003, June). GAMAN: A GA based QoS routing method for mobile ad hoc networks. *Journal of Interconnection Networks, 4*(3), 251–270. doi:10.1142/S0219265903000866

Barriac, V. (2003, October). Non-intrusive evaluation of voice quality in IP environments: On-going work in standardization. In *ITU-T Workshop*, Oct. 2003. Retrieved from http://www.itu.int/ITU-T/worksem/qos/index.html

Barry, M., Campbell, A.-T., & Veres, A. (2001). Distributed Control Algorithms for Service Differentiation in Wireless Packet Net-works. In *Proceeding of IEEE INFOCOM 2001*, Anchorage, Alaska, (vol. 1, pp. 582–90).

Barzilai, T., Kandlur, D., Mehra, A., & Saha, D. (1998). Design and implementation of an RSVP-based quality of service architecture for an integrated services Internet. *IEEE Journal on Selected Areas in Communications*, *16*(3), 397–413. doi:10.1109/49.669047

Beigy, H., & Meybodi, M. R. (2002a). An adaptive limited fractional guard channel policy using continuous action learning automata. In *Proceedings of the 10th International Conference on Software, Telecommunications and Computer Networks (SoftCom 2002)*, Croatia, Italy (pp. 536-540).

Beigy, H., & Meybodi, M. R. (2003a). Adaptive uniform fractional guard channel algorithm: the steady state analysis. In *Proceedings of the 9th International Symposium on Wireless Systems and Networks (ISWSN'03)*, Dhahran, Saudi Arabia (pp. 177-181).

Beigy, H., & Meybodi, M. R. (2003b). An optimal prioritized channel assignment scheme for using in mobile transaction environments. In *Proceedings of the 8th Annual International Computer Society of Iran Computer Conference (*CISCC-2003), Iran (pp. 68-74).

Beigy, H., & Meybodi, M. R. (2003c). An optimal channel assignment scheme. In *Proceedings of the 11th Iranian Conference on Electrical Engineering (ICEE-2003)*, Shiraz, Iran (Vol. 2, pp. 634-641).

Beigy, H., & Meybodi, M. R. (2003c). User based call admission control policies for cellular mobile systems: A survey. *The CSI Journal on Computer Science and Engineering*, *1*(2b), 27–40.

Beigy, H., & Meybodi, M. R. (2004a). A new fractional guard channel policy. *Journal of High Speed Networks*, *13*(1), 25–36.

Beigy, H., & Meybodi, M. R. (2005a). A general call admission policy for next generation wireless networks. *Computer Communications*, *28*, 1798–1813. doi:10.1016/j.comcom.2004.12.009

Beigy, H., & Meybodi, M. R. (2005b). An adaptive call admission algorithm for cellular networks. *Computers & Electrical Engineering*, *31*, 132–151. doi:10.1016/j.compeleceng.2004.12.002

Beigy, H., & Meybodi, M. R. (in press). Learning automata based dynamic guard channel algorithms. *Journal of High Speed Networks*.

Beigy, H., Meybodi, M. R. (2004b). Adaptive uniform fractional guard channel algorithms. *Iranian journal of Electrical and Computer Engineering*, *3*(1), 47-53.

Bellman, R. E. (Ed.). (2003). *Dynamic Programming*. Princeton, NJ: Princeton University Press.

Bengsch, A., Kopp, H., Petry, A., Daher, R., & Tavangarian, D. (2004). Efficiency of Wireless Local Area Networks in High Speed Moving Clients. In *Proc. Innovative Internet Community Systems (I2CS)*, Guadalajara, Mexico.

Beres, E., & Adve, R. (2008). Selection cooperation in multi-source cooperative networks. *IEEE Transactions on Wireless Communications*, *7*(1), 118–127. doi:10.1109/TWC.2008.060184

Berger, I. (2007). Standards for Car Talk. *IEEE – The Institute*, Vol. 31, No. (1).

Berlemann, L., Hiertz, G. R., Hoymann, C., & Mangold, S. (2006, May). *Coexistence and interworking of IEEE 802.16 and IEEE 802.11(e)*. Paper presented at the Vehicular Technology Conference 2006 (VTC 2006), Melbourne, Australia.

Bernet, Y., Ford, P., Yavatkar, R., Baker, F., Zhang, L., Speer, M., et al. (2000, November). *RFC 2998: A Framework for Integrated Services Operation over Diffserv Networks* .

Bertrand, G. (2007 May). *The IP multimedia subsystem in next generation networks*. Retrieved from http://www.irisa.fr/armor/lesmembres/Alia/IMS_principes_et_architecture/

Bertsekas, D. P. (1999). *Nonlinear Programming (2nd Ed.)*. Belmont, MA: Athena Scientific.

Beuran, R. (2004a). *Mesure de la Qualite dans les Reseaux Informatiques*. Ph. D. Thesis, Politehnica University of Bucharest, România & Jean Monet University of St. Etienne, France.

Beuran, R., & Ivanovici, M. (2004c). *User-Perceived Quality Assessment for VoIP Applications*. Technical report (delivered to U4EA Technologies), *CERN-OPEN-2004-007*, CERN, Geneva, Switzerland.

Beuran, R., Ivanovici, M., & Buzuloiu, V. (2004b). File Transfer Performance Evaluation. *Scientific Bulletin of Politehnica University Bucharest, C Series (. Electrical Engineering, 66*(2-4), 3–14.

Beuran, R., Ivanovici, M., Davies, N., & Dobinson, B. (2004). Evaluation of the Delivery QoS Characteristics of Gigabit Ethernet Switches. In *CERNOPEN-2005-002*, CERN, Geneva, Switzerland.

Beuran, R., Ivanovici, M., Dobinson, B., Davies, N., & Thompson, P. (2003). Network Quality of Service Measurement System for Application Requirements Evaluation. In *International Symposium on Performance Evaluation of Computer and Telecommunication Systems*, July 20-24, 2003, Montreal, Canada (pp. 380-387).

Beuran, R., Nakata, J., Okada, T., Miyachi, T., Chinen, K., Tan, Y., & Shinoda, Y. (2007). Performance Assessment of Ubiquitous Networked Systems. In *5th International Conference on Smart Homes and Health Telematics (ICOST2007),* Nara, Japan (pp. 19-26S).

Beuran, R., Nguyen, L. T., Latt, K. T., Nakata, J., & Shinoda, Y. (2007). QOMET: A Versatile WLAN Emulator. In *21st IEEE International Conference on Advanced Information Networking and Applications (AINA-07)*, Niagara Falls, Ontario, Canada (pp. 348-353).

Bharghavan, V. (1999). Fair Queuing in Wireless Networks: Issues and Approaches. *IEEE Personal Comm., 6*(1), 44–53. doi:10.1109/98.752787

Bharghavan, V., Demers, A., Shenker, S., & Zhang, L. (1994). MACAW: A Media Access Protocol for Wireless LANs. In *Proc. ACM SIGCOMM'94* (pp. 212–225).

Bian, K., & Park, J.-M. (2007). Segment-based channel assignment in cognitive radio ad hoc networks. In *Proc. of Cognitive Radio Oriented Wireless Networks and Communications (CrownCom'07)* (pp. 327-335). Washington, DC: IEEE.

Bianchi, G. (2000). Performance analysis of the IEEE 802.11 distributed coordination function, *in IEEE Journal of Selected Area on Communication*, vol. 18, pp. 535-547, March 2000.

Bianchi, G., Fratta, L., & Oliveri, M. (1996). Performance evaluation and enhancement of the CSMA/CA MAC protocol for 802.11 wireless LANs, in proceeding of *IEEE PIMRC 1996*, Taipei, Taiwan, Oct. 1996, (pp. 392-396).

Biaz, S., & Qi, B. (2008). *iETT: A Quality Routing Metric for Multi-rate Multi-hop Networks* (Tech. Rep.) Department of Computer Science and Software Engineering, Auburn University, USA.

Birattari, M., Stützle, T., Paquete, L., & Varrentrapp, K. (2002). A racing algorithm for configuring metaheuristics. In *Proceedings of the Genetic and Evolutionary Computation Conference (GECCO 2002),* New York, USA (pp. 11-18).

Bisnik, N., & Abouzeid, A. (2006). Delay and Throughput in Random Access Wireless Mesh Networks. *Proceedings of the IEEE, ICC*, 2006.

Biswas, S., & Morris, R. (2005). ExOR: Opportunistic multi-hop routing for wireless networks. In *Proceedings of ACM SIGCOMM*, August 2005.

Blake, S. (1998). *An architecture for differentiated services*. Internet Engineering Task Force (IETF) RFC2475.

Blake, S., Black, D., Carlson, M., Davies, E., Wang, Z. & Weiss, W. (1998, December). *An architecture for differentiated services*. IETF RFC 2475.

Bletsas, A., Khisti, A., Reed, D. P., & Lippman, A. (2006). A simple cooperative diversity method based on network path selection. *IEEE Journal on Selected Areas in Communications, 24*(3), 659–672. doi:10.1109/JSAC.2005.862417

Bolot, J. C. (1993, September). End-to-end packet delay and loss behaviour in the Internet. In *Proceedings of SIGCOMM 1993*, New York, USA.

Borin, J. F., & Fonseca, N. L. (2008a). Scheduler for IEEE 802.16 Networks. *IEEE Communications Letters, 12*(4), 274–276. doi:10.1109/LCOMM.2008.072110

Borin, J. F., & Fonseca, N. L. (2008b). A standard-compliant scheduler for WiMAX networks. In *4th IEEE Broadband Wireless Workshop* (pp. 1-5).

Borin, J. F., & Fonseca, N. L. (2008c). Simulator for WiMAX networks. *Simulation Modelling Practice and Theory, 16*(7), 817–833. doi:10.1016/j.simpat.2008.05.002

Bossom, R., Brignolo, R., Ernst, T., Evensen, K., Frötscher, A., Höfs, W., et al. (2008). Communication for eSafety (COMeSafety). Overall Framework - Proof of Concept Implementation, Version 2.0. European ITS Communication Architecture.

Boumerdassi, S., & Beylot, A. (1999). Adaptive channel allocation for wireless PCN. *ACM Mobile Networks and Applications, 4*, 111–146. doi:10.1023/A:1019186412792

Boyd, S., & Vandenberghe, L. (2004). *Convex Optimization (1st ed.).* Cambridge, UK: Cambridge University Press.

Boyer, J., Falconer, D. D., & Yanikomeroglu, H. (2004). Multihop diversity in wireless relaying channels. *IEEE Transactions on Communications, 52*(10), 1820–1830. doi:10.1109/TCOMM.2004.836447

Boyle, J., Cohen, R., Herzog, S., Rajan, R., & Sastry, A. (2000). *The COPS (Common Open Policy Service) protocol.* IETF, Network Working Group.

Bozinovski, M., Popovski, P., & Gavrilovska, L. (2000). Novel strategy for call admission control in mobile cellular network. In *Proceedings of IEEE Conference on Vehicular Technology (VTCC 2000)* (Vol. 4, pp. 1597-1602).

Bozinovski, M., Popovski, P., & Gavrilovska, L. (2000). QoS-Based policy for call admission control in mobile cellular networks. In *Proceedings of IEEE Conference on Wireless Communications and Networking (WCNC 2000)* (Vol. 2, pp. 502-506).

Braden, B., Clark, D. & Shenker, S. (1994). *Integrated services in the internet architecture: An overview.* Internet Engineering Task Force (IETF) RFC1663.

Braden, R., Zhang, L., Berson, S., Herzog, S., & Jamin, S. (1997). *Resource ReSerVation Protocol (RSVP) - Version 1 Functional Specification.* IETF RFC 2205.

Brady, P. (1969). A model for generating On-Off speech patterns in Two-Way conversations. *The Bell System Technical Journal, 48*, 2445–2472.

Breslau, L., Estrin, D., Fall, K., Floyd, S., Heidemann, J., & Helmy, A. (2000). Advances in Network Simulation. *IEEE Computer, 33*(5), 59–67.

Broadband Forum TR-059. (2003). *DSL Evolution - Architecture Requirements for the Support of QoS-Enabled IP Services.* Retrieved from http://www.broadband-forum.org/technical/download/TR-059.pdf

Broadband Forum TR-069. (2003). *CPE WAN Management Protocol.* Retrieved from http://www.broadband-forum.org/technical/download/TR-069.pdf

Broadband Forum TR-126. (2006). *Triple-Play Services Quality of Experience (QoE) Requirements.* Retrieved from http://www.broadband-forum.org/technical/download/TR-126.pdf

Broadband Forum TR-135. (2003). *Data Model for TR-069 Enabled STB.* Retrieved from http://www.broadband-forum.org/technical/download/TR-135.pdf

Broom, S. R. (2006, November). VoIP Quality Assessment: Taking Account of the Edge-Device. *IEEE Transactions on Audio, Speech, And Language Processing, 14*(6), 1977–1983. doi:10.1109/TASL.2006.883233

Brun, O., Bockstal, C., & Garcia, J. M. (2006). A simple formula for end-to-end jitter estimation in packet-switching networks. In *Proceedings of the International Conference on Networking, International Conference on Systems and International Conference on Mobile Communications and Learning Technologies (ICNI-CONSMCL'06),* Morne, Mauritius (pp. 14).

Brunner, M. (2004, April). *Requirements for Signaling Protocols* (IETF Request For Comments 3726).

Bruno, R., Conti, M., Gregori, E., Wijting, C., Kneckt, J., & Damle, A. (2005). Mesh Networks: Commodity Multihop Ad-Hoc Networks. *IEEE Communications Magazine, 43*(3), 123–131. doi:10.1109/MCOM.2005.1404606

Bu, T., Chan, M. C., & Rainjee, R. (2005, March). *Designing wireless radio access networks for third generation cellular networks*. Paper presented at the 24th Annual Joint Conference of the IEEE Computer and Communications Societies.

Bucciol, P., Masala, E., Filipi, E., & De Martin, J.C. (2007). Cross Layer Perceptual ARQ for Video Communication over 802.11e Wireless Networks. *Advances in Multimedia, 2007*(1), 1-12.

Bucklew, J. A. (1990). *Large Deviation Techniques in Decision, Simulation and Estimation*. New York: John Wiley.

Buddhikot, M. M., Kolodzy, P., Miller, S., Ryan, K., & Evans, J. (2005). DIMSUMNET: a new directions in wireless networking using coordinated dynamic spectrum access. In *Proceedings of 6th International Symposium on World of Wireless Mobile and Multimedia Networks (WoWMoM'05)* (pp. 78-85). Washington, DC: IEEE.

Cabric, D., Mishra, S. M., & Brodersen, R. W. (2004). Implementation issues in spectrum sensing for cognitive radios. In *Proceedings of Asilomar Conference on Signals, Systems and Computers* (pp. 772-776). Washington, DC: IEEE.

Cali, F., Conti, M., & Gregori, E. (2000, December). Dynamic Tuning of the IEEE 802.11 protocol to achieve a theoretical throughput limit. *IEEE/ACM Transaction on Networking*, 8(6), 785-790.

Callejo-Rodrigues, M., & Enriquez-Gabeiras, J. (2008, October). Bridging the Standardization Gap to Provide QoS in Current NGN Architectures. *IEEE Communications Magazine*, *46*(10), 132–137. doi:10.1109/MCOM.2008.4644130

CalTech. (2002). *FAST Kernel for Large Data Transfers*. California Institute of Technology. Retrieved from http://netlab.caltech.edu/FAST/

Câmara, D., & Loureiro, A. A. F. (2001). GPS/ant-like routing in ad hoc networks. *Telecommunication Systems*, *18*(1-3), 85–100. doi:10.1023/A:1016739402641

Campbell, A. T., & Gomez-Castellanos, J. (2000). IP micro mobility protocols. *SIGMOBILE Mob. Computer. Communication. Rev.*, *4*(4), 45–53.

Campbell, A., & Coulson, G. (1997). QoS adaptive transports: Delivering scalable media to the desktop. *IEEE Network*, *11*(2), 18–27. doi:10.1109/65.580910

Campbell, A., Gomez, J., Kim, S., Valko, A., Wan, C.-Y., & Turanyi, Z. (2000). Design, Implementation, and Evaluation of Cellular IP. *IEEE Personal Communications*, *7*(4), 42–49. doi:10.1109/98.863995

Campbell, A., Gomez, J., Kim, S., Wan, C.-Y., Turanyi, Z., & Valko, A. (2002). Comparison of IP micromobility protocols. *IEEE Wireless Communications*, *9*(1), 72–82. doi:10.1109/MWC.2002.986462

Canton, A. F., & Chahed, T. "End-to-End Reliability in UMTS: TCP over ARQ", IEEE, 2001.

Cao, M., Ma, W., Wang, X., & Zhang, Q. (2007). Analysis of IEEE 802.16 mesh mode scheduler performance. *IEEE Transactions on Wireless Communications*, *6*(4), 1455–1464. doi:10.1109/TWC.2007.348342

Cao, M., Ma, W., Wang, X., Zhang, Q., & Zhu, W. (2005, May). *Modelling and performance analysis of the distributed scheduler in IEEE 802.16 mesh mode*. Paper presented at the MobiHoc '05, Urbana-Champaign, Illinois, USA.

Casetti, C., Chiasserini, C., & Fiore, M. (2007). Relay quality awareness in mesh networks routing. In *Proceedings of Tyrrhenian International Workshop on Digital Communications (TWD'07),* Ischia Island, Italy.

Cavalli, V., & Verharen, E. (2002). *TF-STREAM Real Time Multimedia Applications*. TERENA Technical Report.

Cerutti, I., Fumagalli, A., & Gupta, P. (2008). Delay models of single-source single-relay cooperative ARQ protocols in slotted radio networks with Poisson frame arrivals. *IEEE/ACM Trans. Networking*, *16*(2), 371–382. doi:10.1109/TNET.2007.900360

Chakareski, J., & Frossard, P. (2005). Distributed packet scheduling of multiple video streams over shared communication resources. In *Proceedings of Multimedia Signal Processing* (pp. 1- 4).

Chakareski, J., Apostolopoulos, J. G., Wee, S., Tan, W., & Girod, B. (2005). Rate distortion hint tracks for adaptive video streaming. *IEEE Transactions on Circuits and Systems for Video Technology*, *15*(10), 1257–1269. doi:10.1109/TCSVT.2005.854227

Chakchai, S.-In., Jain, R., & Tamimi, A. K. (2009). Scheduling in IEEE 802.16e mobile WiMAX networks: key issues and a survey. *IEEE Journal on Selected Areas in Communications*, *27*(2), 156–171. doi:10.1109/JSAC.2009.090207

Chan, A., & Liew, S. C. (2007). VOIP Capacity over Multiple IEEE 802.11 WLANs. In *IEEE International Conf. on Communications (ICC)* (pp. 3251-3258). Washington, DC: IEEE.

Chandra, S., & Sahoo, A. (2007). An efficient call admission control for IEEE 802.16 networks. In *15th IEEE Workshop on Local & Metropolitan Area Networks* (pp. 188-193).

Chandran, R., Raghunathan, K. S., Venkatesan, S., & Prakash, R. (2001, February). A feedback-based scheme for improving TCP performance in ad hoc wireless networks. *IEEE Personal Communications Magazine*, *8*(1), 34–39. doi:10.1109/98.904897

Chang, B. J., & Liang, Y. H. (2004, October). *Analysis of OVSF code tree for code assignment in WCDMA cellular communications.* Paper presented at the IEEE/ACM MASCOTS 2004.

Chang, B. J., Hsiao, W. C., & Hwang, R. H. (2007, February). *Multiple classes of QoS guarantee in distributed multicast routing.* Paper presented at the 6th International Conference on Advanced Communication Technology.

Chang, B.-J., Chen, Y.-L., & Chou, C.-M. (2007). Adaptive hierarchical polling and cost-based call admission control in IEEE 802.16 WiMAX networks. In *Wireless Communications and Networking Conference* (pp. 1954-1958).

Chang, C. (2005). A Mobile-IP Based Mobility System for Wireless Metropolitan Area Networks. In *Int. Conf. on Parallel Processing Workshops*, (pp. 429-435).

Chang, C. S. (2000). *Performance guarantees in communication networks.* New York: Springer.

Chang, C., Chang, C. J., & Lo, K. R. (1999). Analysis of hierarchical cellular systems with reneging and dropping for waiting new calls and handoff calls. *IEEE Transactions on Vehicular Technology*, *48*(4), 1080–1091. doi:10.1109/25.775357

Chang, K., & Kim, D. (2001). Optimal prioritized channel allocation in cellular mobile systems. *Computers & Operations Research*, *28*, 345–356. doi:10.1016/S0305-0548(99)00121-5

Chao, H. J., & Guo, X. (2002). *Quality of Service in High-Speed Networks.* Hoboken, NJ: John Wiley & Sons, Inc.

Chauveau, D. (2005, June). *NGN, strategies for shaping future regulation: the value added of standardisation.* NGN Workshop.

Chayat, N. (1997). *Tentative Criteria for Comparison of Modulation Methods.* IEEE P802.11-97/96.

Chellappan, B., Moh, T.-S., & Moh, M. (2009). On Supporting Multiple Quality-of-Services Classes in Mobile WiMAX Handoff. Proceedings of *IEEE International Conference on Computing in Engineering, Science, and Information*, Fullerton, CA, Apr 2009.

Chellappan, B., Moh, T.-S., & Moh, M. (2009). On Supporting Multiple Quality-of-Services Classes in Mobile WiMAX Handoff. Proceedings of *IEEE International Conference on Computing in Engineering, Science, and Information*, Fullerton, CA, Apr 2009.

Chen, B. B., & Chan, M. C. (2006). Resource Management in Heterogeneous Wireless Networks with Overlapping Coverage. In *IEEE Int. Conf. on Communications Systems Software and Middleware (Comsware)* (pp. 1-10). Washington, DC: IEEE.

Chen, C.-L., Lee, J.-W., Wu, C.-Y., & Kuo, Y.-H. (2009). *Fairness and QoS Guarantees of WiMAX OFDMA*

Scheduling with Fuzzy Controls. EURASIP Journal on Wireless Communications and Networking, Article ID 512507.

Chen, D., & Varshney, P. K. (2004). QoS Support in Wireless Sensor Networks: A Survey. In *Proceedings of the International Conference on Wireless Networks (ICWN '04),* Las Vegas, USA (pp. 227-233).

Chen, G. C., & Lee, S. Y. (2001). Modeling of static and dynamic guard channel schemes for mobile transactions. *IEICE Transactions on Information and Systems . E (Norwalk, Conn.), 84-D,* 87–99.

Chen, I., Guo, Q., & Jiao, W. (2005, December). *An integrated QoS control architecture for IEEE 802.16 broadband wireless access systems.* Paper presented at the Global Telecommunications Conference 2005 (GLOBECOM '05).

Chen, J., & Yang, Y. (2006). Multi-Hop Delay Performance in Wireless Mesh Networks. In *Proceedings of IEEE GLOBECOM 2006.*

Chen, J., Jiao, W., & Wang, H. (2005). A service flow management strategy for IEEE 802.16 broadband wireless access systems in TDD mode. In *IEEE International Conference on Communications* (pp. 3422-3426).

Chen, K., Shah, S. H., & Nahrstedt, K. (2002). Cross Layer Design for Data Accessibility in Mobile Ad Hoc Networks. *Journal of Wireless Communications, 21,* 49–75. doi:10.1023/A:1015509521662

Chen, L. W., Tseng, Y. C., Wang, D. W., Wang, Y. C., & Wu, J. J. (2009). Exploiting Spectral Reuse in Routing, Resource Allocation, and Scheduling for IEEE 802.16 Mesh Networks. *IEEE Transactions on Vehicular Technology, 58*(1), 301–313. doi:10.1109/TVT.2008.923685

Chen, L., & Heinzelman, W. B. (2005, March). QoS-Aware Routing Based on Bandwidth Estimation for Mobile Ad Hoc Networks. *IEEE Journal on Selected Areas in Communications, 23*(3), 561–572. doi:10.1109/JSAC.2004.842560

Chen, M., Kwon, T., Mao, S., Yuan, Y., & Leung, V. C. M. (2008). Reliable and energy-efficient routing protocol in dense wireless sensor networks. *International Journal of Sensor Networks, 4*(1-2), 104–117. doi:10.1504/IJSNET.2008.019256

Chen, S., & Nahrstedt, K. (1998). An overview of Quality-of-Service routing for the Next Generation high-speed Networks: Problems and Solutions. *IEEE Network Magazine.*

Chen, S., & Nahrstedt, K. (1999, August). Distributed Quality-of-Service Routing in Ad Hoc Networks. *IEEE Journal on Selected Areas in Communications, 17*(8), 1488–1505. doi:10.1109/49.780354

Chen, T., Zhang, H., Maggio, G. M., & Chlamtac, I. (2007). Topology management in CogMesh: A cluster-based cognitive radio mesh network. In *Proc. of Intl. Conference on Communications (ICC'07)* (pp. 6516-6521). Washington, DC: IEEE.

Chen, W.P., Hsiao, Y.C., Hou, J.C., Ge, Y., & Fitz, M.P. (2001, April). Syndrome: A light-weight approach to improving TCP performance in mobile wireless networks. *The journal of Wireless communications and Mobile computing.*

Chen, X., Li, B., & Fang, Y. (2005, March). A dynamic multiple-threshold bandwidth reservation (DMTBR) scheme for QoS provisioning in multimedia wireless networks. *IEEE Transactions on Wireless Communications, 4*(2). doi:10.1109/TWC.2004.842981

Chen, Y., & Hsieh, F. (2007). A Cross Layer Design for Handoff in 802.16e Network with IPv6 Mobility. *Wireless Communications and Networking Conference,* 2007, 3844 – 3849.

Che-Sheng Chiu. Chen-chiu Lin, "Comparative Downlink Shared Channel Performance Evaluation of WCDMA Release 99 and HSDPA", 2004 IEEE International Conference on Networking, Sensing & Control, Taipei, Taiwan, March 21-23, 2004.

Cheung, G., & Tan, W. (2006). Packet scheduling of streaming video with flexible reference frame using dynamic programming and integer rounding. In *Proceedings of the IEEE International Conference of Multimedia Expo* (pp. 1969-1972).

Chiang, M., Low, S. H., Calderbank, A. R., & Doyle, J. C. (2007). Layering as optimization decomposition: A mathematical theory of network architectures. *Proceedings of the IEEE*, *95*(1), 255–312. doi:10.1109/JPROC.2006.887322

Cho, C., Ko, Y., & Kwang, H. L. (1997). Fuzzy adaptive guard channel assignment strategy for handoff in PCS system. *Proceedings of 6th IEEE International Conference on Fuzzy Systems, (pp.* 1511-1518).

Cho, S., Kwun, J., Park, C., Cheon, J., Lee, O., & Kim, K. (2006). Hard Handoff Scheme Exploiting Uplink and Downlink Signals in IEEE 802.16e Systems. *IEEE Vehicular Tec. Conference 2006, 3*, 1236 – 1240.

Choi, J., Kwon, T., Choi, Y., & Naghshineh, M. (2000). Call admission control for multimedia services in mobile cellular networks: A Markov decision approach. In *Proceedings of 5th IEEE Symposium on Computers and Communications (ISCC 2000)* (pp. 594-599).

Choi, L., Ivrlac, M., Steinbach, E., & Nossek, J. (2005). Analysis of distortion due to packet loss in streaming video transmission over wireless communication links. In *IEEE International Conference on Image Processing (ICIP)* (Vol. 1, pp. 189-192).

Choi, S., & Shin, K. G. (1998). Predictive and adaptive bandwidth reservation for handoffs in QoS-sensitive cellular networks. *ACM SIGCOMM Computer Communication Review*, *28*(4), 155–166. doi:10.1145/285243.285278

Choi, S., Hwang, G., Kwon, T., Lim, A., & Cho, D. (2005). Fast Handover Scheme for Real-Time Downlink Services in IEEE 802.16e BWA System. *IEEE 61st Vehicular Tech. Conf.*, *3*, 2028–2032.

Choi, S., Hwang, G., Kwon, T., Lim, A., & Cho, D. (2005). Fast Handover Scheme for Real-Time Downlink Services in IEEE 802.16e BWA System. *IEEE 61st Vehicular Tech. Conf.*, *3*, 2028–2032.

Choi, Y.-J., & Bahk, S. (2006). Delay-Sensitive Packet Scheduling for a Wireless Access Link. *IEEE Transactions on Mobile Computing*, *5*(10), 1374–1383. doi:10.1109/TMC.2006.149

Chou, C. L., Lin, J. C., & Liu, C. H. (2008, March). *Performance evaluation for scheduling algorithms in WiMAX network*. Paper presented at the 22nd International Conference on Advanced Information Networking and Applications.

Chou, P. A., & Miao, Z. (2006). Rate-distortion optimized streaming of packetized media. *IEEE Transactions on Multimedia*, *8*(2), 390–404. doi:10.1109/TMM.2005.864313

Choudhury, R., Yang, X., Ramanathan, R., & Vaidya, N. (2002). Using directional antennas for medium access control in ad hoc networks. In *Proc. ACM MobiCom'02* (pp. 59-70).

Christensen, T. K., Nielsen, B. F., & Iversen, V. B. (2004). Phase-type models of channel-holding times in cellular communication systems. *IEEE Transactions on Vehicular Technology*, *53*(3), 725–733. doi:10.1109/TVT.2004.825803

Chu, G., Wang, D., & Mei, S. (2002). A QoS architecture for the MAC protocol of IEEE 802.16 BWA system. In *IEEE International Conference on Communications, Circuits and Systems and West Sino Expositions* (pp. 435-439).

Chuberre, N., et al. (2004). Satellite Digital Multimedia Broadcasting for 3G and beyond 3G systems. In *13th IST Mobile & Wireless Communication Summit 2004*, Lyon, France.

Chuberre, N., et al. (2005). Relative Positioning of the European Satellite Digital Multimedia Broadcast (SDMB) Among Candidate Mobile Broadcast Solutions. In *IST Mobile & Wireless Communications Summit 2005*, Dresden, Germany.

Chung, S. P., & Chiu, C. L. (2002). Joint call admission control / congestion for wireless integrated voice/data Networks. *Computer Communications*, *25*, 1653–1664. doi:10.1016/S0140-3664(02)00050-6

Cicconetti, C., Lenzini, L., Mingozzi & Eklund, C. (2006, March). *Quality of service support in IEEE 802.16 networks*. IEEE Network Magazine, pp. 50-55.

Ciobotaru, M., Ivanovici, M., Beuran, R., & Stancu, S. (2005). Versatile FPGAbased Hardware Platform for Gigabit Ethernet Applications. In *6th Annual Postgraduate Symposium*, Liverpool, UK.

Ciobotaru, M., Stancu, S., LeVine, M., & Martin, B. (2005) GETB - A Gigabit Ethernet Application Platform: its Use in the ATLAS TDAQ Network. In *Real Time 2005*, Stockholm, Sweden.

Clark, A. D. (2001). Modeling the Effects of Burst Packet Loss and Recency on Subjective Voice Quality. In *Proceedings of IP Telephony Workshop*, Columbia, USA.

Cole, R. G., & Rosenbluth, J. H. (2001, April). Voice over IP Performance Monitoring. *Computer Communication Review, ACM Sigcomm Computer . Communication Review, 31*(2).

Coltun, R., Ferguson, D., Moy, J., & Lindem, A. (2008, July). *RFC 5340: OSPF for IPv6.*

Control, T. (2009). *Linux Advanced Routing & Traffic Control.* Retrieved January 2009, from http://lartc.org/

Courville, N., & Bischl, H. (2005). Critical Issues of Onboard Switching in DVB-S/RCS Broadband Satellite Networks. *IEEE Wirelesss Communications, 12*(5), 28–36. doi:10.1109/MWC.2005.1522101

Cover, T. M., & El Gamal, A. A. (1979). Capacity theorems for the relay channel. *IEEE Transactions on Information Theory, 25*(5), 572–584. doi:10.1109/TIT.1979.1056084

Cox, D. R. (1962). *Renewal Theory.* London: Methuen & Co.

Cranley, N., & Davis, M. (2007). An experimental investigation of IEEE 802.11e TXOP facility for real-time video streaming. In *Proceedings of Global Communications Conference* (pp. 2075-2080).

Crawley, E., Nair, R., Rajagopalan, B., & Sandick, H. (1998, August). *RFC 2386: A Framework for QoS-based Routing in the Internet.*

Cruz, R. L. (1991). A Calculus for network delay, Part 1: Network elements in isolation. *IEEE Transactions on Information Theory, 37*(1), 114–121. doi:10.1109/18.61109

Cruz-Perez, A., & Ortigoza-Guerrero, L. (2007). Fractional resource reservation in mobile cellular systems. In Y. Zhang, H. Hu & M. Fujise (Ed.), *Resource, Mobility, and Security Management in Wireless Networks and Mobile Communications* (pp. 122-138). New York: Auerbach Publishing.

Cui, Y., & Nahrstedt, K. (2003). Layered peer-to-peer streaming. In *Proceedings of ACM NOSSDAV* (pp. 162–171).

Culler, D. (2003). 10 emerging technologies that will change the world. *Technology Review, 106*(1), 33–49.

Curet, D., Gouleau, E., Relier, S., Roux, C., Clement, P., & Cherry, G. (2002). *RTP Payload Format for MPEG-4 FlexMultiplexed Streams* [IETF draft].

Dadkhah Chimeh, J., Hakkak, M., Bakhshi, H., & Azmi, P. "Throughput Evaluation in UMTS", 2008 Second International Conference on Future Generation and Networking, 2008.

Daher, R., & Tavangarian, D. (2006a). Resource Reservation and Admission Control in IEEE 802.11 WLANs. In *Proc. Third International Conference on Quality of Service in Heterogeneous Wired/Wireless Networks (QShine 2006)*, Waterloo, Canada.

Daher, R., & Tavangarian, D. (2006b). QoS-oriented Load Balancing for WLANs. In *Proc. First International Workshop on Operator-assisted (Wireless Mesh) Community Networks 2006 (OpComm'06)*, Berlin, Germany.

Daher, R., Krohn, M., Gladisch, A., Arndt, M., & Tavangarian, D. (2008a). *Vehicular Communication Networks: Novel Concept for Wireless Backbone Infrastructure of Roadside Networks.* Internal Report, University of Rostock, Rostock, Germany.

Daher, R., Krohn, M., Gladisch, A., Arndt, M., & Tavangarian, D. (2008b). *Cluster-Oriented Routing Protocol (CORP) for Hierarchical Roadside-Backbone Infrastructures of Vehicular Communication Networks.* Internal Report, University of Rostock, Rostock, Germany.

Dai, L., & Letaief, K. B. (2008). Throughput maximization of ad-hoc wireless networks using adaptive cooperative diversity and truncated ARQ. *IEEE Transactions on Communications, 56*(11), 1907–1918. doi:10.1109/TCOMM.2008.041164

Das, A. K., Alazemi, H. M. K., Vijayakumar, R., & Roy, S. (2005). Optimization models for fixed channel assignment in wireless mesh networks with multiple radio. In *Proceedings of IEEE SECON 2005.*

Das, S. K., Sen, S. K., & Jayaram, R. (1998). A novel of load balancing scheme for the tele-traffic hot spot problem in cellular networks. *ACM/Baltzer Journal on Wireless Networks*, 4(4), 325-340.

Das, S., Klein, T., Rajkumar, A., Rangarajan, S., Turner, M., & Viswanathan, H. (2006). System Aspects and Handover Management for IEEE 802.16e. *Bell Labs Technical Journal, 11*(1), 123–142. doi:10.1002/bltj.20148

Dastis, V., Hollick, M., Mogre, P. S., Schwingenschlogl, C., & Steinmetz, R. (2006, May). *Performance analysis of the real-time capabilities of coordinated centralized scheduling in 802.16 mesh mode.* Paper presented at the Vehicular Technology Conference 2006 (VTC 2006-Spring).

Davie, B., Charny, A., Bennet, J., Benson, K., Boudec, J. L., Courtney, W., et al. (2002, March). *RFC 3246: An Expedited Forwarding PHB (Per-Hop Behavior)*.

DD-WRT. (2009). *DD-WRT.* Retrieved January 2009, from http://www.dd-wrt.com/dd-wrtv3/

De Martin, J. C., & Quaglia, D. (2001). Distortion-based packet marking for MPEG video transmission over DiffServ networks. In *Proceedings of the IEEE International Conference on Multimedia and Expo* (pp. 399- 402).

de Moraes, R. M., Sadjadpour, H. R., & Garcia-Luna-Aceves, J. J. (2004). Throughput Delay Analysis of Mobile Ad-Hoc Networks with a Multi-Copy Relaying Strategy. In *Proceedings of IEEE SECON 2004.*

de Moraes, R. M., Sadjadpour, H. R., & Garcia-Luna-Aceves, J. J. (2006). Mobility-Capacity-Delay Trade-Off in Wireless Ad-Hoc Networks. *Journal of Ad-Hoc Networks, 4*(5), 607–620. doi:10.1016/j.adhoc.2005.06.005

De Rango, F., Malfitano, A., & Marano, S. (2006, November). *PER evaluation for IEEE 802.16 - SC and 802.16e protocol in HAP architecture with user mobility under different modulation schemes.* Paper presented at the Global Telecommunications Conference 2006 (GLOBECOM '06).

De Rango, F., Malfitano, A., & Marano, S. (2007, December). *Parametric Markov chain model in HAP architecture with IEEE 802.16 protocol.* Paper presented at the Int. Symposium on Wireless Personal Multimedia Communications, Jaipur, India.

De Rango, F., Malfitano, A., & Marano, S. (2008, November). *Performance comparison of Markov chain based models in a IEEE 802.16e scenario.* Paper presented at the MILCOM 2008, San Diego.

De Rango, F., Malfitano, A., & Marano, S. (2009). *Instant weighed probability model to guarantee QoS in IEEE 802.16e scenario.* Paper presented at the WCNC 2009, Budapest, Hungary.

DeCouto, D., Aguayo, D., Bicket, J., & Morris, R. A high throughput path metric for multi-hop wireless routing. In *Proceedings of ACM MobiCom, 2003.*

Demichelis, C., & Chimento, P. (2002). *Instantaneous Packet Delay Variation Metric for IPPM.* IETF RFC 3393.

Denardo, E. V. (Ed.). (1982). *Dynamic programming: models and applications.* Upper Saddle River, NJ: Prentice Hall.

Dharmaraju, D., Chowdhury, A. R., Hovareshti, P., & Baras, J. S. (2002, August). INORA-a unified signaling and routing mechanism for QoS support in mobile ad hoc networks. In *Proceedings of 31st International Conference on Parallel Processing Workshops* (pp. 86–93).

Di Caro, G., & Dorigo, M. (1998). AntNet: Distributed stigmergetic control for communications networks. [JAIR]. *Journal of Artificial Intelligence Research, 9*, 317–365.

Di Caro, G., & Dorigo, M. (1998, December). AntNet: Distributed stigmergetic control for communications networks. *Journal of Artificial Intelligence Research, 9*, 317–365.

Di Caro, G., Ducatelle, F., & Gambardella, L. (2005). AntHocNet: An adaptive nature-inspired algorithm for routing in mobile ad hoc networks. *European Transactions on Telecommunications . Special Issue on Self-Organization in Mobile Networking, 16*(5), 443–455.

Diamond, J., Hossain, E., & Niyato, D. (2007). IEEE 802.16/WiMAX-based broadband wireless access and its application for telemedicine/e-health services. *IEEE Wireless Communications, 14*(1), 72–83. doi:10.1109/MWC.2007.314553

Dijkstra, E. (1959). A note on two problems in connection with graphs. *Numer. Math., 1*, 269–271. doi:10.1007/BF01386390

Donnelly, R. A., Jr. (2004). *The Complete Idiot's Guide to Statistics*. New York: Alpha Books.

Dorigo, M., & Gambardella, L. M. (1997). Ant colony system: A cooperative learning approach to the traveling salesman problem. *IEEE Transactions on Evolutionary Computation, 1*(1), 53–66. doi:10.1109/4235.585892

Dorigo, M., & Stützle, T. (2004). *Ant Colony Optimization*. Cambridge, MA: MIT Press.

Dorigo, M., Maniezzo, V., & Colorni, A. (1996). The ant system: Optimization by a colony of cooperating agents. *IEEE Transactions on Systems, Man, and Cybernetics-Part B, 26*(1), 29–41. doi:10.1109/3477.484436

Dovrolis, C. (2002). Proportional Differentiated Services: Delay Differentiation and Packet Scheduling. *IEEE Transactions on Networking, 10*(1), 12–26. doi:10.1109/90.986503

Doyle, L., & Forde, T. (2007). The wisdom of crowds: cognitive ad hoc networks. In Q. H. Mahmoud (Ed.), *Cognitive networks: Towards self-aware networks* (pp. 203-220). Hoboken, NJ: John Wiley and Sons Ltd.

Draves, R., Padhye, J., & Patra, R. (2004). Comparison of routing metrics for static multi-hop wireless networks. In *Proceedings of ACM SIGCOMM*.

Draves, R., Padhye, J., & Zill, B. (2004). Routing in Multi-Radio, Multi-Hop Wireless Mesh Networks. In *Proceedings of MobiCom'04,* Philadelphia, Pennsylvania, USA.

Du, H. (2007). *Efficient radio resource management for satellite digital multimedia broadcasting*. Unpub. doctoral dissertation, University of Surrey.

Du, H. (2008). Adaptive QoS-aware Resource Allocation Scheme for Satellite Multimedia Broadcasting. In *Proc. IEEE WoWMoM'08*, Newport Beach, CA, USA, June 23-27.

Du, H., Evans, B. G., & Chlamtac, I. (in press). Hierarchical Packet Scheduling for Satellite Multimedia Broadcasting: An Adaptive QoS-aware Design . *IEEE Wireless Commun. Mag.*

Du, H., Fan, L., & Evans, B. G. (2006). Two-level Channel Multiplexing: A Novel Radio Resource Allocation Strategy for SDMB system. In *Proc. IEEE ICC'06*, Istanbul, Turkey (pp. 4445-4450).

Du, H., Fan, L., & Evans, B. G. (2007a). Adaptive Multi-dimensional QoS-based Packet Scheduling scheme for Multimedia Broadcasting over Geostationary Satellite Networks. In *Proc. IEEE GLOBECOM'07* (pp. 4613-4618), Washington DC, USA, November 26-30.

Du, H., Fan, L., & Evans, B. G. (2007b). Combined Delay and Rate Differentiation Packet Scheduling for Multimedia Content Delivery in Satellite Broadcast/Multicast Systems. In *Proc. IEEE ICC'07*, Glasgow, Scotland, UK (pp. 4444-4449).

Du, H., Fan, L., & Evans, B. G. (2007c). Cross-Layer Joint Priority Packet Scheduling Scheme for Multimedia QoS Provisioning in Satellite Broadcast/Multicast Systems. In *Proc. AIAA ICSSC'07*, Seoul . *Korea & World Affairs*, (April): 10–13.

Du, H., Fan, L., Mudugamuwa, U., & Evans, B. G. (2007). A Cross-Layer Packet Scheduling Scheme for Multimedia Broadcasting via Satellite Digital Multimedia Broadcasting System. *IEEE Communications Magazine, 45*(8), 94–101. doi:10.1109/MCOM.2007.4290320

DVB Project DVB-HN A109. (2007). A109 *(TM3690r2), DVB-HN (Home Network) Reference Model Phase 1*. Retrieved from http://www.dvb.org/technology/standards/a109.tm3690r2.DVB-HN_ref_model.pdf

DVB Project DVB-HN CR. (2008).*CM639R2 DVB-HN Commercial Requirements Phase 1*. Retrieved from http://www.dvb.org/technology/standards/a132.Technical%20requirements%20for%20DVB-QoS.pdf

DVB Project HLTR. (2008, September). *High-level Technical Requirements for QoS for DVB Services in the Home Network DVB TM IPI*. Retrieved from http://www.dvb.org/technology/standards/a132.Technical%20requirements%20for%20DVB-QoS.pdf

Eardley, P., Mihailovic, A., & Suihko, T. (2000). A Framework for the Evaluation of IP Mobility Protocols. In *The 11th IEEE International Symposium on Personal, indoor and mobile radio communications, 2000*. (Vol. 1, pp. 451–457).

Einhaus, M., & Klein, O. (2006). Performance Evaluation of a Basic OFDMA Scheduling Algorithm for Packet Data Transmissions. In *11th IEEE Symposium on Computers and Communications* (pp. 695-702).

Elahi, A. (2000). *Network Communication Technology*. London: Thomson Learning.

Electronic Communications Committee. (2008a). *ECC/DEC/(08)01. ECC Decision of 14 March 2008 on the harmonised use of the 5875-5925 MHz frequency band for Intelligent Transport Systems (ITS).*

Electronic Communications Committee. (2008b). *ECC/REC (08)01. ECC Recommendation (08)01 Use of the Band 5855-5875 MHz for Intelligent Transport Systems (ITS).*

Elias, P. (1955). Coding for two noisy channels. In *Proceedings of the 3rd London Symposium on Information Theory*, London, UK (pp. 61–76).

Ericsson. (2008, December). *Measuring Speech Quality* [White paper]. Retrieved from http://www.ericsson.com/tems

ETSI EN 302 304. (2004). Digital Video Broadcasting: Transmission System for Handheld Terminals (DVB-H).

ETSI report. (2002, April). *2nd ETSI TIPHON VoIP Speech Quality Test Event*. Sophia Antipolis, France.

ETSI Technical report. (July, 2000). *Telecommunications and Internet Protocol Harmonization Over Network (TIPHON); End to end Quality of Service in TIPHON Systems*. Part 1 of general aspects of Quality of Service (QoS), France, TR 101 329-1, V 3.1.1.

ETSI TISPAN. (2007, December). *TR 182 022 Architectures for QoS handling* (Technical Report).

ETSI TISPAN. (2008a, June). *ES 28 018 H.248 Profile for controlling Border Gateway Functions (BGF) in the Resource and Admission Control Subsystem (RACS)* (ETSI Standard).

ETSI TISPAN. (2008b, July). *ES 282 003 Resource and Admission Control Subsystem (RACS) Functional Architecture* (ETSI Standard).

ETSI TS 102 034. (2007). *Transport of MPEG 2 Transport Stream (TS) Based DVB Services over IP Based Networks*. European Telecommunications Standardization Institute.

ETSI TS 102 585 V1.1.1. (2007). *System Specifications for Satellite services to Handheld devices (SH) below 3 GHz*.

ETSI Working Group. (n.d.). *Telecommunications and Internet converged Services and Protocols for Advanced Networking (TISPAN)*. Retrieved from http://www.etsi.org/tispan/

ETSI. (1992). *Satellite Earth Stations (SES), Transmit/receive VSATs used for data communications operating in the FSS 11/12/14-GHzbands*. (DE/SES-2002), ETS 300159.

EU FP5 IST MoDiS. (n.d.). Retrieved from http://www.ist-modis.org

EU FP6 IST MEASTRO. (n.d.). Retrieved from http://ist-maestro.dyndns.org

Faccin, S. M., Wijting, C., Kneckt, J., & Damle, A. (2006). Wireless mesh networks: Concept and system design. *IEEE Wireless Communications*, *13*(2), 10–17. doi:10.1109/MWC.2006.1632476

Fall, K., & Floyd, S. (1996). Simulation-based Comparisons of Tahoe, Reno, and SACK TCP. *Computer Communication Review*, *26*(3), 5–21. doi:10.1145/235160.235162

Fan, L., Du, H., Mudugamuwa, U., & Evans, B. G. (2006). Novel Radio Resource Management Strategy for Multimedia Content Delivery in SDMB system. In *Proc. AIAA ICSSC'06*, 2006-5476, San Diego, California, USA.

Fang, Y. (2003). Movement-based mobility management and trade-off analysis for wireless mobile networks. *IEEE Transactions on Computers*, *52*(6), 791–803. doi:10.1109/TC.2003.1204834

Fang, Y. (2003). Thinning schemes for call admission control in wireless networks. *IEEE Transactions on Computers*, *52*(5), 686–688. doi:10.1109/TC.2003.1197136

Fang, Y., & Zhang, Y. (2002). Call admission control schemes and performance analysis in wireless mobile networks. *IEEE Transactions on Vehicular Technology*, *51*, 371–382. doi:10.1109/25.994812

Fang, Y., Chlamtac, I., & Lin, Y.-B. (1997). Modeling PCS networks under general call holding time and cell residence time distributions. *IEEE/ACM Transactions on Networking*, 5(6), 893-905.

Farserotu, J., & Prasad, R. (2000). A Survey of Future Broadband Multimedia Satellite System, Issues and Trends. *IEEE Communications*, *38*(6), 128–133. doi:10.1109/35.846084

Favalli, L., Lanati, M., Monighini, S. L., & Savazzi, P. (2007). Resource assignment in multiservice 802.16e. In *The 18th Annual IEEE International Symposium on Personal, Indoor and Mobile Radio Communications* (pp. 1-5).

FCC Spectrum Policy Task Force. (2002). *Report of the spectrum efficiency working group* (Tech. Rep. No. 02-155). Federal Communications Comm.

FCC. (2003, December). *ET Docket No 03-222 Notice of proposed rule making and order.*

Federal Communications Commission. (2004). *FCC 03-324 A1. Amend Rules Regarding Dedicated Short Range Communications Services and rules for Mobile Service for Dedicated Short Range Communications of Intelligent Transportation Services.* FCC 03-324 A1.

Filin, S. A., Moiseev, S. N., & Kondakov, M. S. (2008). Fast and Efficient QoS-Guaranteed Adaptive Transmission Algorithm in the Mobile WiMAX System. *IEEE Transactions on Vehicular Technology*, *57*(6), 3477–3487. doi:10.1109/TVT.2008.919930

Floyd, S. (2002). *High-Speed TCP for Large Congestion Windows* [Internet draft].

Fluckiger, F. (1995). *Understanding Network Multimedia - Applications and Technologies.* Upper Saddle River, NJ: Prentice Hall.

Fowler, T. (2001, January). Mesh networks for broadband access. *IEEE Review*, *47*(1), 17–22. doi:10.1049/ir:20010108

Franz, W., Hartenstein, H., & Mausve, M. (2005). *Inter-Vehicle-Communications Based on Ad-hoc Networking Principles.* The FleetNet Project, University of Karlsruhe.

Friderikos, V., Mihailovic, A., & Aghvami, H. (2004). Analysis of Cross Issues between QoS Routing and Micromobility protocols. *IEEE Proceedings Communications*, *151*(3), 258–262. doi:10.1049/ip-com:20040574

Frossard, P., & Verscheure, O. (2001). Joint Source/FEC Rate Selection for Quality-Optimal MPEG-2 Video Delivery. *IEEE Transactions on Image Processing*, *10*(12), 1815–1825. doi:10.1109/83.974566

Fu, X., Karl, H., & Kappler, C. (2002). QoS-Conditionalized Handoff for Mobile IPv6. In *Networking2002* (Vol. 2345, pp. 721–730).

Fu, X., Schulzrinne, H., Bader, A., Hogrefe, D., Kappler, C., & Karagiannis, G. (2005). NSIS: A new extensible IP signaling protocol suite. *IEEE Communications Magazine*, *43*(10), 133–141. doi:10.1109/MCOM.2005.1522137

Fujimoto, K., Ata, S., & Murata, M. (2002, November). Adaptive playout buffer algorithm for enhancing perceived quality of streaming applications. In *Proceedings of IEEE Globecom2002.*

Furcht, B., et al. (2003). *The Handbook of Video Databases: Design and Applications.* Boca Raton, FL: CRC Press.

Gakhar, K., Gravey, A., & Leroy, A. (2005, October). *IROISE: A new QoS architecture for IEEE 802.16 and IEEE 802.11e interworking.* Paper presented at the 2nd International Conference on Broadband Networks (BroadNets 2005), Boston, MA.

Gamal, A., Mammen, J., Prabhakar, B., & Shah, D. (2004). Throughput-Delay Trade-Off in Wireless Networks. In *Proceedings of IEEE INFOCOM 2004.*

Gambardella, L. M., & Dorigo, M. (1996). Solving symmetric and asymmetric TSPs by ant colonies. In *Proceedings of IEEE Conference on Evolutionary Computation (ICEC'96),* Nagoya, Japan (pp. 622-627).

Gambardella, L. M., Taillard, E., & Agazzi, G. (1999). MACS-VRPTW: A multiple ant colony system for vehicle routing problems with time windows. In D. Corne, M. Dorigo, & F. Glover (Eds.), *New Ideas in Optimization* (pp. 63-76). London: McGraw-Hill.

Ganguly, S., Navda, V., Kim, K., Kashyap, A., Niculescu, D., & Izmailov, R. (2006). Performance optimizations for deploying voip services in mesh networks. *IEEE Journal on Selected Areas in Communications, 24*(11), 2147–2158. doi:10.1109/JSAC.2006.881594

Ganz, A., & Wongthavarawat, K. (2003, October). *IEEE 802.16 based last mile broadband wireless military networks with quality of service support.* Paper presented at the Military Communications Conference, MILCOM 2003.

Garcia-Molina, H. (1982). Elections in a distributed computing system. *IEEE Transactions on Computers, 31*(1). doi:10.1109/TC.1982.1675885

Garg, S., & Kappes, M. (2003). Can I add a VoIP Call? *Proceedings of the IEEE, ICC,* 2003.

GB 20600—2006. (2006). Framing Structure, Channel Coding and Modulation For Digital Television Terrestrial Broadcasting System, Chinese National Standard.

Geetha, S., & Jayaparvathy, R. (2007, November). *Resource allocation and game theoretic scheduling in IEEE 802.16 fixed broadband wireless access systems.* Paper presented at the 4th International Conference Innovations in Information Technology, Dubai.

Gerasimov, I., & Simon, R. (2002, April). A bandwidth reservation mechanism for on-demand ad hoc path finding. In *Proceedings of the IEEE/SCS 35th Annual Simulation Symposium,* CA (pp. 27-34).

Gerla, M., & Tsai, J. T. (1995). Multicluster, Mobile, Multimedia Radio Network. *ACM Baltzer Journal of Wireless Networks, 1*(3), 255–265. doi:10.1007/BF01200845

Gerla, M., Hong, X., & Pei, G. (2002). *Fisher state routing protocol for ad-hoc networks.* Internet Draft, draft-ietf-manet-fsr-03.txt

Ghaderi, M., & Boutaba, R. (2006). Call admission control in mobile cellular networks: A comprehensive survey. *Wireless Communications and Mobile Computing, 6*(1), 69–93. doi:10.1002/wcm.246

Ghanbari, M. (1996). Postprocessing of late cells for packet video. *IEEE Transactions on Circuits and Systems for Video Technology, 6*(6), 669–678. doi:10.1109/76.544738

Gheorghisor, I., & Leung, K. H. (2008, May). *Broadband wireless networks for airport surface communications.* Paper presented at the Integrated Communications, Navigation and Surveillance Conference (ICNS 2008), Bethesda, MD.

Giambene, G., et al. (2007). *Adaptive Resource Management and Optimization in Satellite Networks: Optimization and Cross-Layer Design.* New York: Springer.

Giles, T., Markendahl, J., Zander, J., & Zetterberg, P. (2004). Cost Drivers and Deployment Scenarios for Future Broadband Wireless Networks – Key research problems and directions for research. In *IEEE 59th Vehicular Technology Conf.* (Vol. 4, pp. 2042-2046). Washington, DC: IEEE.

Girod, B., Kalman, M., Liang, Y., & Zhang, R. (2002). Advances in Channel-Adaptive Video Streaming [Invited paper]. *Wireless Communications and Mobile Computing, 2*(6), 549–552. doi:10.1002/wcm.87

Glisic, S. G. (2006). *Advanced Wireless Networks, 4G Technologies*. Hoboken, NJ: John Wiley & Sons, Ltd.

Goyal, P., Vin, H. M., & Chen, H. (1997). Start-time Fair Queuing: A scheduling Algorithm for Integrated Service Packet Switching Networks. *IEEE/ACM Trans. Networking, 5*(5), 690–704. doi:10.1109/90.649569

Gozalvez, J., Martin-Sacristan, D., Lucas-Estan, M., Monserrat, J. F., & Gonzalez-Delicado, J. J. Gozalvez, & D., Marhunda, M. (2007). SPHERE – A Simulation Platform for Heterogeneous Wireless Systems. In *IEEE TridentCom* (pp.1-10). Washington, DC: IEEE.

Gross, D., & Harris, C. M. (1998). *Fundamentals of queuing theory* (3rd ed.). New York: John Wiley & Sons, Inc.

Gu, D., & Zhang, J. (2003, June). QoS enhancement in IEEE802.11 wireless local area networks. *IEEE Communications Magazine, 41*(6), 120–124. doi:10.1109/MCOM.2003.1204758

Guern, R. (1988). Queuing-blocking system with two arrival streams and guard channels. *IEEE Transactions on Communications, 36*, 153–163. doi:10.1109/26.2745

Gui, C., & Mohapatra, P. (2003, June). SHORT: self-healing and optimizing routing techniques for mobile ad hoc networks. In *Proceedings of the 4th ACM International Symposium on Mobile Ad Hoc Networking & Computing* (pp. 279–290).

Gundavelli, S., Leung, K., Devarapalli, V., Chowdhury, K., & Patil, B. (2008, August). *RFC 5213: Proxy Mobile IPv6.*

Gunduz, D., & Erkip, E. (2007). Opportunistic cooperation by dynamic resource allocation. *IEEE Transactions on Wireless Communications, 6*(4), 1446–1454. doi:10.1109/TWC.2007.348341

Günes, M., Sorges, U., & Bouazizi, I. (2002). ARA – the ant-colony based routing algorithm for MANETs. In *Proceedings of the ICPP International Workshop on Ad Hoc Networks (IWAHN)*, Vancouver, Canada (pp. 79–85).

Gunes, M., Sorges, U., & Bouazizi, I. (2002, August). ARA - The Ant-colony based Routing Algorithm for MANETs. In *Proceedings of International Conference on Parallel Processing Workshops* (pp. 79-85).

Guo, X., Ma, W., Guo, Z., & Hou, Z. (2007). Dynamic bandwidth reservation admission control scheme for the IEEE 802.16e broadband wireless access systems. In *Wireless Communications and Networking Conference* (pp. 3420-3425).

Gupta, P., & Kumar, P. R. (2000). The capacity of wireless networks. *IEEE Transactions on Information Theory, 46*(2), 388–404. doi:10.1109/18.825799

Gupta, P., & Kumar, P. R. (2000, March). The capacity of wireless networks. *IEEE Transactions on Information Theory, 46*, 388–404. doi:10.1109/18.825799

Han, K., & Choi, S. (2006). Performance Analysis of Sleep Mode Operation in IEEE 802.16e Mobile Broadband Wireless Access Systems. *Vehicular Technology Conference, 3*, 1141-1145.

Han, M., & Nilsson, A. A. (2000). Population-based call admission control in wireless cellular networks. In *Proceedings of the 2000 IEEE International Conference on Communications* (pp. 1519-1523).

Hancock, R., Karagiannis, G., Loughney, J., & Bosch, S. V. D. (2005). *Next Steps in Signaling (NSIS): Framework.* Internet Engineering Task Force (IETF) RFC4080.

Haring, G., Marie, R., Puigjaner, R., & Trivedi, K. (2001). Loss formulas and their application to optimization for cellular networks. *IEEE Transactions on Vehicular Technology, 50*, 664–673. doi:10.1109/25.933303

Hartenstein, H., & Laberteaux, K. P. (2008). A Tutorial Survey on Vehicular Ad Hoc Networks. *IEEE Communications Magazine, vol. 41*, no. (6).

Hawa, M., & Petr, D. D. (2002). Quality of Service scheduling in cable and broadband wireless access systems. In *10th International Workshop on Quality of Service* (pp. 247-255).

Haykin, S. (2005). Cognitive radio: brain-empowered wireless communications. *IEEE Journal on Selected*

Areas in Communications, 23(2), 201–220. doi:10.1109/JSAC.2004.839380

He, J., Guild, K., Yang, K., & Chen, H.-H. (2007). Modeling Contention Based Bandwidth Request Scheme for IEEE 802.16 Networks. *IEEE Communications Letters, 11*(7), 698–700.

He, Y., & Wahab, H. A. (2006). HQMM: A hybrid QoS model for mobile ad-hoc networks. In *Proceedings of 11th Symposium on Computers and Communications (ISCC'06)* (pp. 194-200). Washington, DC: IEEE.

Heidemann, J., Silva, F., Intanagonwiwat, C., Govindan, R., Estrin, D., & Ganesan, D. (2001). Building efficient wireless sensor networks with low-level naming. In *Proceedings of the eighteenth ACM Symposium on Operating Systems Principles*, Banff, Alberta, Canada (pp. 146-159).

Heinanen, J., Baker, F., Weiss, W., & Wroclawski, J. (1999, June). *RFC 2597: Assured Forwarding PHB Group* .

Heinzelman, W., Chandrakasan, A., & Balakrishnan, H. (2000). Energy-efficient communication protocol for wireless microsensor networks. In *Proceedings of the Hawaii International Conference System Sciences (HICSS'00),* Hawaii, USA (pp. 4-7).

Heiskala, J., & Terry, J. (2001). *OFDM Wireless LANs: A Theoretical and Practical Guide.* Indianapolis, IN: Sams.

Hempel, M., Sharif, H., Wang, W., Mahasukhon, P., & Zhou, T. (2008, May). *Throughput vs. distance tradeoffs and deployment considerations for a multi-hop IEEE 802.16e railroad test bed.* Paper presented at the Vehicular Technology Conference (VTC Spring 2008), Singapore.

Hilt, V., & Camarillo, G. (2008, July). *A Session Initiation Protocol (SIP) Event Package for Session-Specific Session Policies - draft-ietf-sipping-policy-package-05* (IETF Internet).

Hilt, V., Camarillo, G., & Rosenberg, J. (2008, November). *A Framework for Session Initiation Protocol (SIP) Session Policies - draft-ietf-sip-session-policy-framework-05* (IETF Internet Draft).

Ho, C.-J., & Lea, C.-T. (1999). Improving call admission policies in wireless networks. *Wireless Networks, 5*, 257–265. doi:10.1023/A:1019107329380

Hoene, C. (2005, December). *Internet telephony over wireless links.* PhD thesis, Technical University of Berlin, Germany.

Holland, G., & Vaidya, N. H. (2002, March-May). Analysis of TCP performance over mobile ad hoc networks. *Wireless Networks, 8*(2-3), 275–288. doi:10.1023/A:1013798127590

Holma, H., & Toskala, A. (2002). *WCDMA for UMTS: Radio Access For Third Generation mobile Communications* (2nd Ed.). Hoboken, NJ: John Wiley & Sons Ltd.

Holma, H., Toskala, A., KRanta-aho, K., & Pirskanen, K. (2007). High Speed Packet Access Evolution in 3GPP Release 7. *IEEE Communications, 45*(12), 29-35.

Homer, S., & Selman, A. (2000). *Computability and Complexity Theory.* New York: Springer.

Hong, D., & Rappaport, S. (1986). Traffic modeling and performance analysis for cellular mobile radio telephone systems with prioritized and non-prioritized handoffs procedure. *IEEE Transactions on Vehicular Technology, 35*, 77–92. doi:10.1109/T-VT.1986.24076

Hong, X., Xu, K., & Gerla, M. (2002). Scalable routing protocols for mobile ad hoc networks. *IEEE Network, 16*(4), 11–21. doi:10.1109/MNET.2002.1020231

Hossain, E., & Niyato, D. (2006). Queue-aware uplink bandwidth allocation and rate control for polling service in IEEE 802.16 broadband wireless networks. *IEEE Transactions on Mobile Computing, 5*(6), 668–679. doi:10.1109/TMC.2006.85

Hossain, E., & Niyato, D. (2007). QoS-aware bandwidth allocation and admission control in IEEE 802.16 broadband wireless access network: A non-cooperative game theoretic approach. *Computer Networks: The International Journal of Computer and Telecommunications Networking, 51*(11), 3305–3321.

Hossain, E., & Niyato, D. (2007). Radio resource management games in wireless networks: an approach to bandwidth allocation and admission control for polling service in IEEE 802.16. *IEEE Wireless Communications*, *14*(1), 27–35. doi:10.1109/MWC.2007.314544

Hoßfeld, T., & Binzenhöfer, A. (2008). Analysis of Skype VoIP traffic in UMTS: End-to-end QoS and QoE measurements. *Computer Networks*, *52*, 650–666. doi:10.1016/j.comnet.2007.10.008

Hoßfeld, T., Tran-Gia, P., & Fiedler, M. (2007). Quantification of Quality of Experience for Edge-Based Applications. In *LNCS 4516* (pp. 361-373).

Hou, F., She, J., Ho, P.-H., & Shen, X. (2008). Performance Analysis of Weighted Proportional Fairness Scheduling in IEEE 802.16 Networks. In *IEEE International Conference on Communications* (pp. 3452-3456).

Hou, J., & Fang, Y. (2001). Mobility-based call admission control schemes for wireless mobile networks. *Wireless Communications and Mobile Computing*, *1*, 269–282. doi:10.1002/wcm.18

Hou, J., Park, K., Kim, T., & Kung, L. (2008) Medium Access Control and Routing Protocols for Wireless Mesh Networks. In E. Hossain & K. Leung (Eds.), *Wireless Mesh Networks: Architectures and Protocols* (pp. 76 - 111). New York: Springer.

Hoyhtya, M., Pollin, S., & Mammela, A. (2008). Performance improvement with predictive channel selection for cognitive radios. In *Proceedings of 1st International Workshop on Cognitive Radio and Advanced Spectrum Management (CogART'08)* (pp. 1-5). Washington, DC: IEEE.

Hu, F., & Sharma, N. K. (2003). Multimedia call admission control in mobile networks: A dynamical reservation-pool approach. *Computer Networks*, *43*(3), 263–288. doi:10.1016/S1389-1286(03)00271-8

Hu, R. Q., Paranchych, D., Fong, M.-H., & Wu, G. (2007). On the evolution of handoff management and network architecture in WiMAX. In *Proc. of IEEE Mobile WiMAX Symposium*, 2007.

Hu, R. Q., Paranchych, D., Fong, M.-H., & Wu, G. (2007). On the evolution of handoff management and network architecture in WiMAX. In *Proc. of IEEE Mobile WiMAX Symposium*, 2007.

Hua, Y. (2007). Advances in Recursive Per-Pixel End-to-End Distortion Estimation for Robust Video Coding in H.264/AVC. *IEEE Transactions on Circuits and Systems for Video Technology*, *17*(7), 845–856. doi:10.1109/TCSVT.2007.897116

Huang, C., Chuang, Y., & Yang, D. (2008). Implementation of call admission control scheme in next generation mobile communication networks using particle swarm optimization and fuzzy logic systems. *Expert Systems with Applications*, *35*, 1246–1251. doi:10.1016/j.eswa.2007.08.023

Huang, J., Subramanian, V. G., Berry, R., & Agrawal, R. (2007). Joint Scheduling and Resource Allocation in Uplink OFDM Systems. In *41st Asilomar Conference on Signals, Systems and Computers* (pp. 265-269).

Huang, J.-H., Wang, L.-C., & Chang, C.-J. (2006a, November). Capacity and QoS for a scalable ring-based wireless mesh network. *IEEE Journal on Selected Areas in Communications*, *24*(11), 2070–2080. doi:10.1109/JSAC.2006.881622

Huang, J.-H., Wang, L.-C., & Chang, C.-J. (2006b, October). Wireless mesh networks for intelligent transportation systems. In *Proc. IEEE International Conference on Systems, Man, and Cybernetics*.

Huang, J.-H., Wang, L.-C., & Chang, C.-J. (2008a, November). Power fairness in a scalable ring-based wireless mesh network with variable ring-width design. In *Proc. IEEE Globecom'08*.

Huang, J.-H., Wang, L.-C., & Chang, C.-J. (2008b, September). QoS provisioning in a scalable wireless mesh network for intelligent transportation systems. *IEEE Transactions on Vehicular Technology*, *57*(5), 3121–3135. doi:10.1109/TVT.2008.918701

Huang, J.-H., Wang, L.-C., & Chang, C.-J. (2008c, March). Throughput-coverage tradeoff in a scalable wireless mesh network. *Journal of Parallel and Distributed Computing*, *68*(3), 278–290. doi:10.1016/j.jpdc.2007.10.003

Huang, V., & Zhuang, W. (2004). QoS-Oriented Packet Scheduling for Wireless CDMA network. *IEEE Transactions on Mobile Computing, 3*(1), 73–85. doi:10.1109/TMC.2004.1261818

Huang, W.-J., Hong, Y.-W. P., & Kuo, C.-C. J. (2008). Lifetime maximization for amplify-and-forward cooperative networks. *IEEE Transactions on Wireless Communications, 7*(5), 1800–1805. doi:10.1109/TWC.2008.061075

Hunter, T. E., & Nosratinia, A. (2006). Diversity through coded cooperation. *IEEE Transactions on Wireless Communications, 5*(2), 283–289. doi:10.1109/TWC.2006.1611050

Huynh, P.-Q., Jangyodsuk, P., & Moh, M. (2010). Supporting Video Streaming and VoIP over Mobile WiMAX Networks by enhanced FMIPv6-based Handover. Accepted to be presented at *the Fourth International Conference on Information Systems, Technology and Management (ICISTM-10),* to be held in Bangkok, Thailand, March 2010.

Hwang, Y., Lee, H., & Varshney, P. (2003). An Adaptive QoS Routing Protocol with Dispersity for Ad hoc Networks. In *Proceedings of the 36th Hawaii International Conference on System Sciences.*

Iannone, L., Khalili, R., Salamatian, K., & Fdida, S. (2004). Cross-layer routing in wireless mesh networks. In *1st International Symposium on Wireless Communication Systems* (Vol. 2004, pp. 319-323).

IEEE 802.11. (2006, March). *Working Group of the LAN/MAN Committee.* IEEE P802.11s/DO.O1.

IEEE 802.16 Amendment 4: mobility enhancements. (2007). *Draft amendment to IEEE standard for local and metropolitan area network.* IEEE 802.16e-03/07.

IEEE 802.16 Working Group on Broadband Wireless Access. (2004). *Part 16: Air Interface for Fixed Broadband Wireless Access Systems.*

IEEE 802.16 Working Group on Broadband Wireless Access. (2005). *Part 16: Air Interface for Fixed and Mobile Broadband Wireless Access Systems Amendment 2: Physical and Medium Access Control Layers for Combined Fixed and Mobile Operation in Licensed Bands and Corrigendum 1.*

IEEE 802.16m-08/003r7. (2009). *The IEEE 802.16m System Description Document.*

IEEE P1609.0 D0.2 (2007). *IEEE Trial Use Standard for Wireless Access in Vehicular Environments (WAVE) – Architecture.*

IEEE P802.16Rev2/D4. (2008, May). *Draft Standard for Local and Metropolitan Area Networks Part 16: Air Interface for Fixed and Mobile Broadband Wireless Access Systems.*

IEEE Standard for Local and Metropolitan Area Networks - Part 16: Air Interface for Fixed Broadband Wireless Access Systems. (2004). *IEEE Standard 802.16* .

IEEE Std 802.11-2007 (2007). *IEEE Standard for Information technology - Telecommunications and information exchange between systems - Local and metropolitan area networks - Specific requirements Part 11: Wireless LAN Medium Access Control (MAC) and Physical Layer (PHY) Specifications.* (Revision of IEEE Std 802.11-1999)

IEEE Std 802.11e. (2005). *IEEE Standard for Information Technology -Telecommunications and information exchange between systems –Local and metropolitan area networks – Specific requirements part 11: Wireless LAN Medium Access Control (MAC) and Physical Layer (PHY) specifications Amendment 8: medium access control(MAC) quality of service enhancements.*

IEEE Std. (1999). *Part 11: Wireless LAN Medium Access Control (MAC) and Physical Layer (PHY) Specifications.* ANSI/IEEE Std 802.11.

IEEE Std. (2000). *Part 11: Wireless LAN MAC and PHY Specifications: Higher-Speed Physical Layer in the 5 GHz Band.* ANSI/IEEE Std 802.11a.

IEEE Std. (2007). *Part 11: Wireless LAN Medium Access Control (MAC) and Physical Layer (PHY) Specifications.* ANSI/IEEE Std 802.11.

IEEE. 802.11. (2007). *Part 11: Wireless LAN Medium Access Control (MAC) and Physical Layer (PHY) Specifications.* IEEE Standards for Information Technology–

Telecommunications and Information Exchange between Systems – Local and Metropolitan Area Network – Specific Requirements, IEEE.

IEEE. 802.16e. (2005). *Air Interface for Fixed and Mobile Broadband Wireless Access Systems.*

IEEE802.11 Wireless LAN Medium Access Control (MAC) and Physical Layer (PHY) Specifications. (1999). IEEE Standard 802.11, June 1999.

IEEE802.11e Wireless LAN Medium Access Control (MAC) Enhancements for Quality of service (QoS). (2005). Final 802.11e standard, July 2005.

IEEE802.11e/D5.0. (2003). Draft Supplement to Part 11: Wireless Medium Access Control (MAC) and physical layer (PHY) specifications: Medium Access Control (MAC) Enhancements for Quality of Service (QoS).

IEEE802.16-2004. (2004). IEEE Standard for Local and metropolitan area networks Part 16: Air Interface for Fixed Broadband Wireless Access Systems.

IEEE802.16e-2005. (2005). IEEE Standard for Local and metropolitan area networks Part 16: Air Interface for Fixed and Mobile Broadband Wireless Access Systems Amendment for Physical and Medium Access Control Layers for Combined Fixed and Mobile Operation in Licensed Bands.

IEEE802.16e-2005. (2006, February 28). IEEE Standard for Local and Metropolitan Area Networks Part 16: Air Interface for Fixed and Mobile Broadband Wireless Access Systems - Amendment 2: Physical and Medium Access Control Layers for Combined Fixed and Mobile Operation in Licensed Bands and Corrigendum 1.

Imai, N., Morikawa, H., & Aoyama, T. (2001). Prefetching Architecture for Hot-Spotted Networks. ICC 2001. In *IEEE International Conference on Communications (ICC 2001).*, Helsinki, Finland.

Intanagonwiwat, C., Govindan, R., & Estrin, D. (2000). Directed diffusion: A scalable and robust communication paradigm for sensor networks. In *Proceedings of the Sixth Annual International Conference on Mobile Computing and Networks (MobiCOM 2000),* Boston, MA (pp. 56-67).

Intanagonwiwat, C., Govindan, R., Estrin, D., Heidemann, J., & Silva, F. (2003). Directed diffusion for wireless sensor networking. *IEEE/ACM Transactions on Networking, 11*(1), 2-16.

Intel. (2006). *Wireless Broadband EDUCAUSE 2006.*

Iptables. (2009). *Iptables network firewall.* Retrieved from January 2009, From http://www.netfilter.org/

Ishikawa, Y., & Umeda, N. (1997). Capacity design and performance of call admission control in CDMA systems. *IEEE Journal on Selected Areas in Communications, 15,* 1627–1635. doi:10.1109/49.634800

ITE. (2004). *Institute of Transportation Engineers, Traffic Management Data Dictionary (TMDD) and Message Sets for External Traffic Management Center Communications (MS/ETMCC).* [Online]. Available:Retrieved from http://www.ite.org/tmdd

ITU-R. (1998). *Recommendation BT.500, Subjective quality assessment methods of television pictures.* International Telecommunication Union.

ITU-R. (1998). *Recommendation J.140, Subjective assessment of picture quality in digital cable television systems.* International Telecommunication Union.

ITU-R. (2000). *Methodology for the subjective assessment of the quality of television pictures.* ITU-R Rec. BT.500–10.

ITU-R. (2002). *Methodology for the Subjective Assessment of the Quality of Television Pictures* (Rec BT.500-11. Question ITU-R 211/11). Geneva, Switzerland.

ITU-R. (2004). Draft New Recommendation for Subjective Assessment of Streaming Multimedia Images by Nonexpert Viewers. (Document 6Q/ 57-E.) Source: EBU, Geneva, Switzerland.

ITU-T Recommendation G.1010. (2001, November). *Quality of service and performance: End-user multimedia QoS categories.*

ITU-T Recommendation G.107. (2008). *The E-Model, a computational model for use in transmission planning.* International Telecommunication Union– Telecommunication.

ITU-T Recommendation G.1070. (2007). *Opinion model for video-telephony applications.* International Telecommunication Union– Telecommunication.

ITU-T Recommendation G.114 (2003). *One-way transmission time.*

ITU-T Recommendation H.264 and ISO/IEC 14496-10. (2007). *Advanced Video Coding for Generic Audiovisual Services (MPEG-4 AVC).* International Telecommunication Union– Telecommunication and International Organization for Standardization/International Electrotechnical Commission, Joint Technical Committee 1. Version 7: April 2007.

ITU-T Recommendation J.144. (2004). *Objective perceptual video quality measurement techniques for digital cable television in the presence of a full reference.* International Telecommunication Union– Telecommunication.

ITU-T Recommendation J.148 (2003). *Requirements for an objective perceptual multimedia quality model.* International Telecommunication Union– Telecommunication.

ITU-T Recommendation P.564. (2007). *Conformance testing for voice over IP transmission quality assessment models.* International Telecommunication Union– Telecommunication.

ITU-T Recommendation P.800. (1996, August). *Methods for subjective determination of transmission quality.*

ITU-T Recommendation P.862. (2001). *Perceptual evaluation of speech quality (PESQ): An objective method for end-to-end speech quality assessment of narrowband telephone networks and speech codecs.* International Telecommunication Union– Telecommunication.

ITU-T Recommendation Y.1541 Amendment 3. (2008). *Network performance objectives for IP-based services, Amendment 3: Revised Appendix VIII - Effects of IP network performance on digital television transmission QoS.* International Telecommunication Union– Telecommunication.

ITU-T Recommendation Y.1541. (2006). *Network performance objectives for IP-based services.* International Telecommunication Union– Telecommunication.

ITU-T. (1990). *Recommendation G.726, 40, 32, 24, 16 kbit/s Adaptive Differential Pulse Code Modulation (ADPCM).* International Telecommunication Union.

ITU-T. (1993). *Recommendation G.711. Pulse Code Modulation (PCM) of voice frequencies.* International Telecommunication Union.

ITU-T. (1996). *Methods for subjective determination of transmission quality.* ITU-T Rec. P.800.

ITU-T. (1996). *Recommendation G.729, Coding of speech at 8 kbit/s using Conjugate-Structure Algebraic-Code-Excited Linear-Prediction (CSACELP).* International Telecommunication Union.

ITU-T. (1996). *Recommendation P.800, Methods for subjective determination of transmission quality*, International Telecommunication Union.

ITU-T. (1996). *Recommendation P.910, Subjective Video Quality Assessment Methods for Multimedia Applications.* International Telecommunication Union.

ITU-T. (1998). *Recommendation P.861, Objective quality measurement of telephone-band (300-3400 Hz) speech codecs.* International Telecommunication Union.

ITU-T. (1999). [*Internet Protocol][IP][Data Communication Service - IP Packet Transfer and Availability Performance Parameters.* International Telecommunication Union.]. *Recommendation, I*, 380.

ITU-T. (1999). *Subjective video quality assessment methods for multimedia applications.* ITU-T Rec. P.910.

ITU-T. (2000). *Recommendation J.143, User requirements for objective perceptual video quality measurements in digital cable television.* International Telecommunication Union.

ITU-T. (2001). *Rec. Y. 2111 Resource and Admission Control Functions in Next Generation Networks.*

ITU-T. (2001). *Recommendation P.862, Perceptual evaluation of speech quality (PESQ), an objective method for end to end speech quality assessment of narrow-band telephone networks and codecs.* International Telecommunication Union.

ITU-T. (2001). *Recommendation Y.1541. Network Performance Objectives for IP-Based Services.* International Telecommunication Union draft.

ITU-T. (2004). *Objective Perceptual Video Quality Measurement Techniques for Digital Cable Television in the Presence of a Full Reference.* (Recommendation ITU-T J.144). Geneva, Switzerland.

ITU-T. (2005). *Advanced Video Coding for Generic Audiovisual Services*, v3, (Rec. H.264| ISO/ IEC IS 14496-10 AVC). Geneva, Switzerland.

Ivanovici, M. (2005). Objective Performance Evaluation for MPEG-4 Video Streaming Applications. In *Scientific Bulletin of Politehnica University Bucharest, C Series (. Electrical Engineering, 67*(3), 55–64.

Ivanovici, M. (2006). *Network Quality Degradation Emulation—an FPGA-based approach to application performance assessment.* Ph. D. thesis, Politehnica University of Bucharest, România.

Ivanovici, M., & Beuran, R. (2006). *User-perceived Quality Assessment for Multimedia Applications.* In *The 10th International Conference on Optimization of Electrical and Electronic Equipment (OPTIM'06),* Brasov, Romania, May 18-19 (vol. 4, pp. 55-60).

Ivanovici, M., Beuran, R., & Davies, N. (2005). Assessing Application Performance in Degraded Network Environments - an FPGA-based approach. In *Communicating Process Architectures*, Eindhoven, Netherlands.

Iwata, A., Chiang, C. C., Pei, G., Gerla, M., Chen, T. W. (1999). Scalable routing strategies for ad hoc wireless networks. *IEEE Journal on Selected Areas in Communications, vol. 17*, no. 8(8), pp. 1369-1379.

J. J. Blum, J. J., Eskandarian, A., & Hoffman, L. J. (2004). Challenges of Intervehicle Ad Hoc Networks. *IEEE Transactions. on Intelligent Transport.t. Systems,* vol. 5, no. (4), pp. 347–51.

Jacobson, V. (1988). Congestion Avoidance and Control. In *ACM Computer Communication Review, SIGCOMM '88 Symposium,* Stanford, CA, USA.

Jacobson, V., Nichols, K., & Poduri, K. (1999, June). *RFC 2598: An Expedited Forwarding PHB.*

Jacquet, P., Muhlethaler, P., Clausen, T., Laouiti, A., Qayyum, A., & Viennot, L. (2001, December). Optimized Link State Routing Protocol for Ad hoc Networks. In *Proceedings of the IEEE International Multi Topic Conference* (pp. 62-68).

Jain, R. (2006, October). Internet 3.0: Ten problems with current Internet architecture and solutions for the Next Generation. In *Proceedings of IEEE Military Communications Conference (Milcom 2006)*, Washington, DC, USA.

Jalalim, A., Padovani, R., & Pankai, R. (2000). Data throughput for CDMA HDR a high efficiency-high data rate personal communication wireless system. *IEEE International Vehicular Technology Conference* (pp. 1854-1858).

Jang, H., Jee, J., Han, Y., Park, S. D., & Cha, J. (2007). *Mobile IPv6 Fast Handovers over IEEE 802.16e Networks.* IETF Internet Draft, March 2008 (replaces a 2007 version).

Jedrzycki, C., & Leung, V. C. M. (1996). Probability distribution of channel holding time in cellular telephony systems. In *Proc. Vehicular Technology Conf. (VTC'96)*, August 1996, (pp. 247-251).

Jelassi, S., & Youssef, H. (2007, October). EVOM: A software based platform for voice transmission and quality assessment over wireless ad-hoc networks. In *the proceeding of 10th IEEE/ACM Annual International Symposium on Modeling, Analysis and Simulation of Wireless and Mobile Systems (MSWiM 2007)*, Chania, Crete Island, Greece.

Jeruchim, M. C., Balaban, P., & Chanmugan, K. S. Simulation of Communication Systems: modeling, methodology and techniques, Kluwer Academic/Plenum Publishers, 2000.

Jiang, H., Zhuang, W., Shen, X., Abdrabou, A., & Wang, P. (2006). Differentiated Services for Wireless Mesh Backbone. *IEEE Communications Magazine, 44*(7), 113–119. doi:10.1109/MCOM.2006.1668391

Jiang, W. Liu, S. Zhu, Y. Zhang, Z. (2007). *Optimized Routing Metrics for Large-Scale Multi-Radio Mesh Networks.* Tsinghua University, Beijing 100084, China.

Jiang, W., & Schulzrinne, H. (1999). *QoS Measurement of Internet Real-Time Multimedia Services.* Technical Report, Department of Computer Science, Columbia University.

Jing, Y., & Hassibi, B. (2006). Distributed space-time coding in wireless relay networks. *IEEE Transactions on Wireless Communications, 5*(12), 3524–3536. doi:10.1109/TWC.2006.256975

Jing, Y., & Jafarkhani, H. (2008). Single and multiple relay selection schemes and their diversity orders, In *Proc. IEEE Inter. Conf. on Commun.*, Beijjing, China.

Johnson, D. B., & Maltz, D. A. (1996). Dynamic source routing in ad hoc wireless networks. *Mobile Computing, 353*(5), 153–181. doi:10.1007/978-0-585-29603-6_5

Johnson, D., Maltz, D., Hu, Y., & Jetcheva, J. (2002). *The dynamic source routing protocol (DSR).* Internet Draft, draft-ietf-manet-dsr-07.txt

Johnson, D., Perkins, C., & Arkko, J. (2004, June). *RFC 3775: Mobility Support in IPv6.*

Ju, H., Rubin, I., & Kuan, Y. (2003). An adaptive RTS/CTS control mechanism for IEEE 802.11 MAC protocol. In *Proc. IEEE VTC'03.*

Jun, J., & Sichitiu, M. (2003, October). The nominal capacity of wireless mesh networks. *IEEE Wireless Commun. Mag., 10*(5), 8–14. doi:10.1109/MWC.2003.1241089

Kaloxylos, A., Passas, N., & Triantafyllopoulou, D. K. (2007, April). *A cross-layer optimization mechanism for multimedia traffic over IEEE 802.16 networks.* Paper presented at the European Wireless 2007.

Kaloxylos, A., Passas, N., Salkintzis, K., & Triantafyllopoulou, D. K. (2007). A heuristic cross-layer mechanism for real-time traffic over IEEE 802.16 networks. *International Journal of Network Management, 17*(5), 1–14.

Karaliopoulos, M., Henrio, P., Narenthiran, K., Angelou, E., & Evans, B. G. (2004). Packet scheduling for the delivery of multicast and broadcast services over S-UMTS . *International Journal of Satellite Communication and Networking, 22*, 503–532. doi:10.1002/sat.797

Karlsson, A., Heikkilä, G., Minde, T. B., Nordlund, M., & Timus, B. (1999, June). Radio link parameter based speech quality index – SQI. In *Proceedings of IEEE Workshop on Speech Coding.*

Karn, P. (1990, April). MACA -a New Channel Access Method for Packet Radio. In *Proceedings of ARRL/CRRL Amateur Radio 9th Computer Networking Conference* (pp. 134–140).

Katevenis, M., Sidiropoulos, S., & Courcoubetis, C. (1991). Weighted Round-Dobin Cell Multiplexing in a General Purpose ATM Switch Chip. *IEEE Journal on Selected Areas in Communications, SAC-9*(8), 1265–1279. doi:10.1109/49.105173

Katzela, I., & Naghshineh, M. (1996). Channel assignment schemes for cellular mobile telecommunication systems: A comprehensive survey. *IEEE Personal Communications, 3*, 10–31. doi:10.1109/98.511762

Kawadia, V., & Kumar, P. R. (2005, February). A cautionary perspective on cross layer design. *IEEE Wireless Commun. Mag., 12*(1), 3–11. doi:10.1109/MWC.2005.1404568

Kelly, F. (1996). Notes on effective bandwidths. In F. P. Kelly, S. Zachary, & I. B. Ziedins (Eds.), *Stochastic Networks: Theory and Applications* (pp. 141-168). Oxford, UK: Oxford University Press.

Kelly, F. P., Maulloo, A. K., & Tan, D. K. H. (1998). Rate control in communication networks: shadow prices, proportional fairness and stability. *The Journal of the Operational Research Society, 49*(3), 237–252.

Keshav, S. (1991, September). A Control-Theoretic Approach to Flow Control. *ACM SINGCOMM Computer Communication Review, 21*(4), 3–15. doi:10.1145/115994.115995

Khandani, A. E., Abounadi, J., Modiano, E., & Zheng, L. (2007). Cooperative routing in static wireless networks. *IEEE Transactions on Communications, 55*(11), 2185–2192. doi:10.1109/TCOMM.2007.908538

Kihl, M., Sichitiu, M., Ekeroth, T., & Rozenberg, M. (2007). Reliable Geographical Multicast Routing in

Vehicular Ad-Hoc Networks. In *Proceedings of the 5th international Conference on Wired/Wireless internet Communications. (Lecture Notes in Computer Science, vol. 4517)*. Berlin, Germany: Springer-Verlag, Berlin.

Kim, B., & Lee, H. (2006). Performance Comparison of Route Metrics for Wireless Mesh Networks. *IEICE Transactions on Communications (Institute of Electronics, Information and Communication Engineers) . E (Norwalk, Conn.)*, *89-B*(11), 3124–3127.

Kim, K., & Koo, I. (2005). *CDMA Systems Capacity Engineering*. Norwood, MA: Artech House.

Kim, K., Kim, C., & Kim, T. (2005). A Seamless Handover Mechanism for IEEE 802.16e Broadband Wireless Access. In *Proceedings of 5th International Conference on Computational Science - ICCS 2005* (LNCS 3515(II), pp. 527-534).

Kim, P. S., & Kim, Y. J. (2006). New authentication mechanism for vertical handovers between IEEE 802.16e and 3G wireless networks. *International Journal of Computer Science and Network Security*, *6*(9B), 138–143.

Kim, S., & Yeom, I. (2007). TCP-Aware Uplink Scheduling for IEEE 802.16. *IEEE Communications Letters*, *11*(2), 146–148. doi:10.1109/LCOMM.2007.061509

Kleinrock, L. (1975). *Queueing systems Vol. I: theory*. Hoboken, NJ: John Wiley & Sons.

Knightly, E., & Shroff, N. (1999). Admission control for statistical QoS: Theory and practice. *IEEE Network*, *13*(2), 20–29. doi:10.1109/65.768485

Ko, Y., Vorobyov, S. A., & Ardakani, M. (2009a). How much multiuser diversity gain is required over large-scale fading? In *Proc. IEEE Inter. Conf. on Commun.*, Dresden, Germany.

Ko, Y., Vorobyov, S. A., & Ardakani, M. (2009b). How much multiuser diversity is required for homogeneous multiuser systems? *IEEE Trans. Wireless Commun.*, revised.

Kodialam, M., & Nandagopal, T. (2003). Characterizing achievable rates in multi-hop wireless mesh networks: the joint routing and scheduling problem. In *Proceedings of 9th ACM Mobile Computing and networking*, San Diego, CA, USA

Kodialam, M., & Nandagopal, T. (2005). Characterizing the capacity region in multi-radio multi-channel wireless mesh networks. In *Proceedings of ACM MOBICOM 2005*.

Kodialam, M., & Nandagopal, T. (2005). On the capacity region of multi-radio multi-channel wireless mesh networks. In *Proceedings of IEEE WIMESH 2005*.

Koetter, R., & Medard, M. (2003). An algebraic approach to network coding. *IEEE/ACM Trans. Netw.*, *11*(5), 782–795.

Kohler, E. (2000). *The Click Modular Router*. Ph.D. dissertation delivered at MIT.

Koksal, C. (2008). Quality of Service Routing Metrics in Wireless Mesh Networks. In E. Hossain & K. Leung (Eds.), *Wireless Mesh Networks: Architectures and Protocols* (pp. 227 - 243). New York: Springer.

Koksal, C., & Balakrishnan, H. (2006). Quality aware routing metric for time-varying wireless mesh networks. *IEEE Journal on Selected Areas in Communications*, *24*, 1984–1994. doi:10.1109/JSAC.2006.881637

Koodli, R. (2005). Fast Handovers for Mobile IPv6. *IETF RFC 2068*.

Korcyl, K., Sladowski, G., Beuran, R., Dobinson, R. W., Meirosu, C., Ivanovici, M., & Maia, M. L. (2004). Network performance measurements as part of feasibility studies on moving part of the ATLAS Event Filter to off-site Institutes. In *Grid Computing: First European Across Grids Conference*, Santiago de Compostela, Spain (pp. 206 - 213).

Kozamernik F., Sunna P., Wyckens E. and & Pettersen, D. I. (2005). Subjective quality of internet video codecs — phase II evaluations using SAMVIQ. *European Broadcasting Union (EBU) Technical Review*, 301.

Kozatchok, I., & Pierre, S. (2002). User tracking and mobility management algorithms for wireless networks. *The Computer Journal*, *45*(5), 525–539. doi:10.1093/comjnl/45.5.525

Kramer, G., Gastpar, M., & Gupta, P. (2005, Sept.). Cooperative strategies and capacity theorems for relay networks. *IEEE Transactions on Information Theory, 51*(9), 3037–3063. doi:10.1109/TIT.2005.853304

Krashinsky, R., & Balakrishnan, H. (2005). Minimizing Energy for Wireless Web Access with Bounded Slowdown. *Wireless Networks, 11*(1-2), 135–148. doi:10.1007/s11276-004-4751-z

Krohn, M., Daher, R., Gladisch, A., Arndt, M., & Tavangarian, D. (2008a). *Host Abstraction Routing Platform (HARP) for Hierarchical Roadside-Backbone Infrastructures of Vehicular Communication Networks.* Internal Report, University of Rostock, Rostock, Germany.

Krohn, M., Unger, H., & Tavangarian, D. (2007). Advanced Wireless Network Architectures for Highways. In *Proceedings of Wireless Congress*, Munich, Germany.

Krohn, M., Unger, H., & Tavangarian, D. (2008b). A Distributed Proxy System for High Speed Clients. In *Proceedings of 13th Joint International and National CSI Computer Conference*, Kish Island, Iran.

Ksentini, A., Nafaa, A., Guéroui, A., & Naimi, M. (2007, June). ETXOP: A Novel IEEE 802.11 MAC protocol with Admission Control for Sensitive Multimedia Applications. *Elsevier's Performance Evaluation Journal, 64*(5), 419–443.

Kulavaratharash, M. D., & Aghvami, A. H. (1999). Teletraffic performance evaluation of microcellular personal communication networks (PCN's) with prioritized handoff procedures. *IEEE Transactions on Vehicular Technology, 48*, 137–152. doi:10.1109/25.740076

Kumar, M., Jeya, K., & Rajesh, R. S. (2008). Performance analysis of energy-aware QoS routing protocol for ad hoc wireless sensor network. *International Journal of Computer Science and Network Security, 8*(4), 94–100.

Kuo, G. S., & Yao, H. J. (2006 October). *An integrated QoS-aware mobility architecture for seamless handover in IEEE 802.16e Mobile BWA Networks.* Paper presented at the Military Communications Conference 2006 (MILCOM 2006).

Kuo, G.-S., & Ko, P.-C. (2000). Dynamic RSVP for mobile IPv6 in wireless networks. In *IEEE 51st Vehicular technology conference, 2000* (Vol. 1, pp. 455–459).

Kurihara, T. M. 2009. IEEE 1609 Working Group – Project Status Report (2009-05-11), doc: IEEE 802.11-09-0093-02-000p.

Kutzner, K., Tchouto, J.-J., Bechler, M., Wolf, L., Bochow, B., & Luckenbach, T. (2003). Connecting Vehicle Scatternets by Internet-Connected Gateways. In *Proceedings of Multiradio Multimedia Communications (MMC) - Communication Technology for Vehicles.*, Dortmund, Germany.

Kwon, D. H., Park, J., & Suh, Y. J. (2006, September). *An integrated handover scheme for fast mobile IPv6 over IEEE 802.16e systems.* Paper presented at the Vehicular Technology Conference 2006 (VTC-2006 Fall), Montreal, Canada.

Kwon, T., Choi, Y., & Naghshineh, M. (1998). Optimal distributed call admission control for multimedia services in mobile cellular networks. In *Proceedings of 5th International Workshop on Mobile Multi-Media Communications(MoMuc98).*

Kyasanur, P., & Vaidya, N. H. (2005). Capacity of multi-channel wireless networks: Impact of number of channels and interference. In *Proceedings of ACM MOBICOM* 2005.

Kyasanur, P., & Vaidya, N. H. (2005). Routing and interface assignment in multichannel multiinterface wireless networks. In *Proc. IEEE WCNC'05.*

Kyasanur, P., So, J., Chereddi, C., & Vaidya, N. H. (2006). Wireless mesh networks: A challenge and protocols. *IEEE Wireless Communications, 13*(2), 30–36. doi:10.1109/MWC.2006.1632478

L7-filter. (2009). *Application Layer Packet Classifier for Linux.* Retrieved January 2009, from http://l7-filter.sourceforge.net/

Lamm, R., & Schneider, S. J. (2001). *Investigation into the Development of a Wireless Ethernet Backbone, 10-9233 – Internal Report.* Southwest Research Institute.

Retrieved from http://www.swri.org/3pubs/ IRD2001/10-9233.htm

Lampropoulos, G., Kaloxylos, A., Passas, N., & Merakos, L. (2006.) A Seamless Service Continuity Scheme for Enhanced Network Performance in UMTS/WLAN Networks. In *Proc. 17th IEEE Int. Symp. On Personal, Indoor and Mobile Radio Comm.* (pp.1-5). Washington, DC: IEEE.

Laneman, J. N., & Wornell, G. W. (2003). Distributed space–time-coded protocols for exploiting cooperative diversity in wireless networks. *IEEE Transactions on Information Theory, 49*(10), 2415–2425. doi:10.1109/TIT.2003.817829

Laneman, J. N., Tse, D. N. C., & Wornell, G. W. (2004). Cooperative diversity in wireless networks: Efficient protocols and outage behavior. *IEEE Transactions on Information Theory, 50*(1), 3062–3080. doi:10.1109/TIT.2004.838089

Le, L. B., & Hossain, E. (2007). Multihop cellular networks: Potential gains, research challenges, and a resource allocation framework. *IEEE Communications Magazine, 45*(9), 66–73. doi:10.1109/MCOM.2007.4342859

Le, L. B., & Hossain, E. (2008a). An analytical model for ARQ cooperative diversity in multihop wireless networks. *IEEE Transactions on Wireless Communications, 7*(5), 1786–1791. doi:10.1109/TWC.2008.060798

Le, L. B., & Hossain, E. (2008b). Cross-layer optimization frameworks for multihop wireless networks using cooperative diversity. *IEEE Transactions on Wireless Communications, 7*(7), 2592–2602. doi:10.1109/TWC.2008.060962

Le, L. B., Hossain, E., & Zorzi, M. (2007). Queueing analysis for GBN and SR ARQ protocols under dynamic radio link adaptation with non-zero feedback delay. *IEEE Transactions on Wireless Communications, 6*(9), 3418–3428. doi:10.1109/TWC.2007.06020038

Lee, D. H., Kyamakya, K., & Umondi, J.P. (2006). Fast Handover Algorithm for IEEE 802.16e Broadband Wireless Access System. *1st Int. Symp. on Wireless Pervasive Computing*, 2006

Lee, H., Kwon, T., & Cho, D.-H. (2005). An enhanced uplink scheduling algorithm based on voice activity for VoIP services in IEEE 802.16d/e system. *IEEE Communications Letters, 9*(8), 691–693. doi:10.1109/LCOMM.2005.1496584

Lee, K., Trong, S. T., Lee, B., & Kim, Y. (2008). QoS-Guaranteed IPTV Service Provisioning in IEEE 802.11e WLAN-based Home Network. In *Proceeding of Network Operations and Management Symposium Workshops*, Brazil (pp. 71-76).

Lee, M. J., Zheng, J., Ko, Y. B., & Shrestha, D. M. (2006). Emerging standards for wireless mesh technology. *IEEE Wireless Communications, 43*(4), 56–63. doi:10.1109/MWC.2006.1632481

Lee, S., & Park, S. (1998). Handoff with dynamic guard channels in ATM-based mobile networks. In *Proc. of 12th International Conference on Information Networking (ICOIN-12)* (pp. 439-442).

Lee, S., Narlikar, G., Pal, M., Wilfong, G., & Zhang, L. (2006, April). *Admission control for multihop wireless backhaul networks with QoS support.* Paper presented at the IEEE WCNC 2006, Las Vegas NV.

Lee, S., Yinzhe, Y., & Nelakudithi, S. (2004, March). Proactive Vs Reactive Approaches to Failure Resilient Routing. In Twenty-third Annual Joint Conference of the IEEE Computer and Communications Societies.

Lee, S.-B., Zhang, A. X., & Campbell, A. T. (2000). INSIGNIA: An IP-based quality of service framework for mobile ad hoc networks. *Journal of Parallel and Distributed Computing . Special Issue on Wireless and Mobile Computing and Communications, 60*(4), 374–406.

Lee, Y. (2006). Measured TCP Performance in CDMA 1x EV-DO Network. In *PAM conference*.

Leong, C. W., & Zhuang, W. (2002). Call admission control for voice and data traffic in wireless communications. *Computer Communications, 25*, 972–979. doi:10.1016/S0140-3664(01)00434-0

Leontaris, A., & Cosman, P. C. (2003). Video compression with intra/inter mode switching and a dual frame buffer. In *Proceedings of the IEEE Data Compression Conference* (pp. 63–72).

Lescuyer, P., & Lucidarme, T. (2008). *Evolved Packet System. The LTE and SAE Evolution of 3G UMTS.* Hoboken, NJ: John Wiley and Sons, Ltd.

Leung, K. K., Mukherjee, S., & Rittenhouse, G. E. (2005). Mobility Support for IEEE 802.16d Wireless Networks. *2005 IEEE Wireless Comm., & . Networking Conf., 3*, 1446–1452.

Levine, D. A., Akylidiz, I. F., & Naghsineh, M. (1997). A resource estimation and call admission algorithm for wireless multi-media networks using the shadow cluster concept. *IEEE/ACM Transactions on Networking*, 5, 1-12.

Lewis, B. (2003, July). *Mesh networks in fixed broadband wireless access.* Retrieved from http://grouper.ieee.org/groups/802/16/docs/03/C80216-03 10r1.pdf

Li, B., Qin, B. Y., Low, C. P., & Gwee, C. L. (2007, December). A Survey on Mobile WiMAX (Wireless Broadband Access). *IEEE Communications Magazine, 45*(12), 70–75. doi:10.1109/MCOM.2007.4395368

Li, B., Xu, D., Nahrstedt, K., & Liu, J. W. S. (1998). End-to-end QoS support for adaptive applications over the Internet. In *Proceedings of SPIE (Symp. on Voice, Video Data Communications)*, Boston, MA (pp. 166–176).

Li, C., & Zhao, W. (2008). Stochastic Performance Analysis of Multi-Radio Multi-Channel Wireless Mesh Networks. In *Proceedings of the Fourth Internal Conference on Wireless and Mobile Communications 2008.*

Li, C., Che, H., & Li, S. (2007). A Wireless Channel Capacity Model for Quality of Service. *IEEE Transactions on Wireless Communications, 6*(1), 356–366. doi:10.1109/TWC.2007.05282

Li, J., Kameda, H., & Li, K. (2000). Optimal dynamic mobility management for PCS networks. *IEEE/ACM Transactions on Networking, 8*(3), 319-327.

Li, J., Montuno, D. Y., Wang, J., & Zhao, Y. Q. "Performance Evaluation of the Radio Link Control Protocol in 3G UMTS", *Proc. Of the 4th IASTED International Multi-Conference, Wireless and Optical Communication*, Banf Canada, pp. 524-529, July, 2003.

Li, J., Pan, Y., & Jia, X. (2002). Analysis of dynamic location management for PCS networks. *IEEE Transactions on Vehicular Technology, 51*(5), 1109–1119. doi:10.1109/TVT.2002.800632

Li, Q., & van der Schaar, M. (2004). Providing Adaptive QoS to Layered Video over Wireless Local Area Networks through Real-Time Retry Limit Adaptation. *IEEE Transactions on Multimedia, 6*(2), 278–290. doi:10.1109/TMM.2003.822792

Li, Q., Lin, Z., Liu, F., Tao, J., & Zeng, Z. (2005, July) *Achieving QoS for IEEE 802.16 in Mesh Mode.* Paper presented at the 8th Internal Conference on Computer Science and Informatics, Salt Lake City, Utah.

Li, S.-Y. R., Yeung, R. W., & Cai, N. (2003). Linear network coding. *IEEE Transactions on Information Theory, 49*(2), 371–381. doi:10.1109/TIT.2002.807285

Li, X., & Bao-yu, Z. (2003, August). Study on cross-layer design and power conservation in ad hoe network. In *Proc. Fourth International Parallel and Distributed Computing, Applications and Technologies* (pp. 324-328).

Li, Y., Vucetic, B., Zhou, Z., & Dohler, M. (2007). Distributed adaptive power allocation for wireless relay networks. *IEEE Transactions on Wireless Communications, 6*(3), 948–958. doi:10.1109/TWC.2007.05256

Liang, Y. J., Apostolopoulos, J. G., & Girod, B. (2008). Analysis of Packet Loss for Compressed Video: Effect of Burst Losses and Correlation Between Error Frames. *IEEE Transactions on Circuits and Systems for Video Technology, 18*(7), 861–874. doi:10.1109/TCSVT.2008.923139

Liang, Y., Apostolopoulos, J. G., & Girod, B. (2003). Analysis of Packet Loss for Compressed Video: Does Burst Loss Matter? In *Proceedings of the IEEE International Conference on Acoustics, Speech, and Signal Processing (ICASSP)*, Hong Kong, China.

Liang, Y., Veeravalli, V. V., & Poor, H. V. (2007). Resource allocation for wireless fading relay channels: Max-min solution. *IEEE Transactions on Information Theory*, *53*(10), 3432–3453. doi:10.1109/TIT.2007.904996

Liao, W. H., & Tseng, Y. C. Wang, S.L., & Sheu, J.P. (2001). A multi-path QoS routing protocol in a wireless mobile ad hoc network. In *Proceedings of IEEE Intl. Conf. on Networking* (pp. 158–167).

Liao, W.-T., Chen, J.-C., & Chen, M.-S. (2001). Adaptive recovery techniques for real-time audio streams. In *INFOCOM 2001*.

Lin, C. R. (2001, April).On-Demand QoS Routing in Multihop Mobile Networks. *Proceedings of IEEE International Conference on Computer Communications*,3, 1735–1744.

Lin, C. R., & Liu, J. S. (1999, August). QoS routing in ad hoc wireless networks. *IEEE Journal on Selected Areas in Communications*, *17*(8), 1426–1438. doi:10.1109/49.779924

Lin, R., & Gerla, M.,(1997).Asynchronous Multimedia Multihop Wireless Networks. *IEEE INFOCOM 1997*.

Lin, X., Sharma, G., Mazumdar, R. R., & Shroff, N. B. (2006). Degenerate Delay-Capacity TradeOffs in Ad-Hoc Networks with Brownian Mobility. *IEEE Transactions on Information Theory*, *52*(6), 2777–2784. doi:10.1109/TIT.2006.874544

Lin, Y. H., Mai, Y. T., & Yang, C. C. (2007, February). *Cross-layer QoSframework in the IEEE 802.16 network*. Paper presented at the 9th International Conference on Advanced Communication Technology.

Lin, Y.-B., Mohan, S., & Noerpel, A. (1994). Queuing priority channel assignment strategies for handoff and initial access for a PCS network. *IEEE Transactions on Vehicular Technology*, *43*, 704–712. doi:10.1109/25.312778

Lin, Z., Erkip, E., & Stefanov, A. (2006). Cooperative regions and partner choice in coded cooperative systems. *IEEE Transactions on Communications*, *54*(7), 1323–1334. doi:10.1109/TCOMM.2006.877963

Lincke, S. J. (2005). Wireless Load Sharing with Heterogeneous Services and Adaptive Placement. In *Proc. IEEE Wireless Telecommunications Symp.* (pp. 79-84). Washington, DC: IEEE.

Lincke, S. J. (2007). An Investigation of Seamless Mobility from the Operational Perspective. In *Proc. World Wireless Congress* (pp. 42-47).

Linden, J. (n.d.). *VoIP: Better than PSTN?* Retrieved from http://www.analogzone.com/nett0307.pdf

Lindsey, S., & Raghavendra, C. S. (2002). PEGASIS: Power efficient gathering in sensor information systems. In *Proceedings of the IEEE Aerospace Conference*, Big Sky, Montana (pp. 1125-1130).

Liu, J., & Singh, S. (2001, July). ATCP: TCP for mobile ad hoc networks. *IEEE JSAC*, *19*(7), 1300–1315.

Liu, Q. (2005a). Cross-Layer Scheduling with Prescribed QoS Guarantees in Adaptive Wireless Networks . *IEEE Journal on Selected Areas in Communications*, *23*(5), 1056–1066. doi:10.1109/JSAC.2005.845430

Liu, Q. (2005b). Queuing with Adaptive Modulation and Coding over Wireless Links: Cross-Layer Analysis and Design. *IEEE Transactions on Wireless Communications*, *4*(3), 1142–1153. doi:10.1109/TWC.2005.847005

Liu, Q. (2006). A Cross-Layer Scheduling Algorithm with QoS Support in Wireless Networks. *IEEE Transactions on Vehicular Technology*, *55*(3), 839–847. doi:10.1109/TVT.2006.873832

Liu, Z., Kwiatkowska, M. Z., & Constantinou, C. (2004). A Swarm Intelligence routing algorithm for MANETs. *Proc. of the IASTED Intl. Conference*, 484–489.

Liu, Z., Kwiatkowska, M. Z., & Constantinou, C. (2005). A biologically inspired QoS routing algorithm for mobile ad hoc networks. In *Proceedings of the 19th International Conference on Advanced Information Networking and Applications*, Vancouver, Canada (pp. 426-431).

Lochert, C., Mauve, M., Fusler, H., & Hartenstein, H. (2005). Geographic Routing in City Scenarios. In *ACM SIGMOBILE Mobile Computing and Communications Review*, (pp. pages 69–72).

Lu, S., Bharghavan, V., & Srikant, R. (1998). Fair Scheduling in Wireless Packet Networks. *ACM SIGCOMM*.

Luby, M., Stockhammer, T., & Watson, M. (2008). Application Layer FEC In IPTV Services. *Institute of*

Electrical and Electronics Engineers Communications Magazine, *46*(5), 94–101.

Ludwig, R., Ekstrom, H., Willars, P., & Lundin, N. (2006, September). An Evolved 3GPP QoS Concept. In *Proceedings of 2006 IEEE Vehicular Technology Conference* (pp. 388–392).

Luis, M. G. (2008). *TcpDump Libpcap*. Retrieved October 13, 2008, http://www.tcpdump.org/tcpdumpman.html.

Luo, J., Blum, R. S., Cimini, L. J., Greenstein, L. J., & Haimovich, A. M. (2007). Decode-and-Forward cooperative diversity with power allocation in wireless networks. *IEEE Transactions on Wireless Communications*, *6*(3), 793–799. doi:10.1109/TWC.2007.05272

Ma, L., & Denko, M. (2007). A Routing Metric for Load-Balancing in Wireless Mesh Networks. In *Proceedings of the 21st International Conference on Advanced Information Networking and Applications Workshops (AINAW'07).*

Ma, W., & Fang, Y. (2004). Dynamic hierarchical mobility management strategy for mobile IP networks. *IEEE Journal on Selected Areas in Communications*, *22*(4), 664–676. doi:10.1109/JSAC.2004.825968

Madan, R., Mehta, N. B., Molisch, A. F., & Zhang, J. (2009). Energy-efficient decentralized cooperative routing in wireless networks. *IEEE Transactions on Automatic Control*, *54*(3), 512–527. doi:10.1109/TAC.2009.2012979

Madsen, A. H., & Zhang, J. (2005). Capacity bounds and power allocation for wireless relay channels. *IEEE Transactions on Information Theory*, *51*(6), 2020–2040. doi:10.1109/TIT.2005.847703

Maharshi, A., Tong, L., & Swami, A. (2003, August). Cross-layer design of multichannel reservation MAC under rayleigh fading. *IEEE Transactions on Signal Processing*, *51*(8), 2054–2067. doi:10.1109/TSP.2003.814465

Majumdar, A., Sachs, D. G., Kozintsev, I. V., Ramchandran, K., & Yeung, M. M. (2002). Multicast and unicast real-time video streaming over wireless LANs. *IEEE Journal on Selected Areas in Communications*, *12*, 524–534.

Malden Electronics Ltd. (n.d.). *Speech Performance Analyser*. Retrieved from http://www.malden.co.uk

Malden Electronics. (2008, December). *MultiDSLA Application – Wireless handover* [white paper]. Retrieved from http://www.malden.co.uk/documents/MultiDSLAWirelessHandover.pdf

Malkowski, M., & Claßen, D. (2007). Performance of Video Telephony Services in UMTS using Live Measurements and Network Emulation. *Wireless Personal Communications Journal.*

Mangold, S., Chois, S., May, P., Klein, O., Hertz, G., & Sibor, L. (2002). IEEE 802.11 wireless LAN for quality of service. In *Proceedings of European Wireless conference*, Florence, Italy, February 2002.

Manner, J., Karagiannis, G., & McDonal, A. (2008, February). *NSLP for Quality-of-Service Signaling - draft-ietf-nsis-qos-nslp-16* (IETF Internet Draft).

Maric, I., & Yates, R. D. (2004). Cooperative multihop broadcast for wireless networks. *IEEE Journal on Selected Areas in Communications*, *22*(6), 1080–1088. doi:10.1109/JSAC.2004.830912

Marsh, I., Grönvall, B., & Hammer, F. (2006, May). The design and implementation of a quality-based handover trigger. In *5th International IFIP-TC6 Networking Conference (NETWORKING 2006)*, Coimbra, Portugal.

Martinez-Bauset, J., Gimenenz-Guzman, J. M., & Pla, V. (2008). Optimal admission in multimedia mobile networks with handover prediction. *IEEE Wireless Communications*, 38-44.

Masip-Bruin, X., Yannuzzi, M., Serral-Gracia, R., Domingo-Pascual, J., Enriquez-Gabreiras, J., & Callejo, M. (2007, February). The EuQoS System: A Solution for QoS Routing in Heterogeneous Networks. *IEEE Communications Magazine*, *45*(2), 96–103. doi:10.1109/MCOM.2007.313402

Mathis, M., Semske, M., Mahdavi, J., & Ott, T. (1997). The macroscopic behavior of the TCP congestion avoidance algorithm. *Computer Communication Review, 27*(3).

Matolak, D. W., Sen, I., & Wang, B. (2007, April). *Performance of IEEE 802.16 OFDMA Standard Systems in*

Airport Surface Area Channels. Paper presented at the Integrated Communications, Navigation and Surveillance Conference.

Matskani, E., Sidiropoulos, N. D., Luo, Z.-Q., & Tassiulas, L. (2007). Joint multiuser downlink beamforming and admission control: A semidefinite relaxation approach. In *Proc. IEEE Inter. Conf. Accous. Speech and Sig. Proc.*, Honolulu, Hawaii, USA.

Matskani, E., Sidiropoulos, N. D., Luo, Z.-Q., & Tassiulas, L. (2008). Convex approximation techniques for joint multiuser downlink beamforming and admission control. *IEEE Transactions on Wireless Communications*, *7*(7), 2682–2693. doi:10.1109/TWC.2008.070104

Medbo, J., & Schramm, P. (1998). *Channel Models for HIPERLAN/2*. ETSI/BRAN doc. No. 3ERIO85B.

Meddahi, A., & Afifi, H. (2006, November). Packet-E-Model: E-Model for VoIP quality evaluation. *Elsevier Computer Networks Journal*, *50*, 2659–2675. doi:10.1016/j.comnet.2005.10.008

Meddahi, A., Afifi, H., & Vanwormhoudt, G. (2006, June). "MOSQoS": Subjective VoIP Quality Feedback Control and Dynamic QoS Adaptation. In *Proceedings of IEEE ICC*, Istanbul, TURKEY (pp. 2034-2039).

Medhi, D., & Ramasamy, K. (2007). *Network Routing Algorithms, Protocols, and Architectures*. San Mateo, CA: Morgan Kaufmann Publishers.

Meer, J. van der, Mackie, D., Swaminathan, V., Singer, D., & Gentric P. (2002). *Transport of MPEG-4 Elementary Streams* [IETF draft].

Melvin, H. (2004, October). *The use of synchronized time in voice over Internet Protocol (VoIP) applications*. Ph.D. Thesis, University College Dublin, Ireland.

MeshDynamics. (2009). Retrieved from http://www.meshdynamics.com

Miao, Z., & Ortega, A. (2002). Expected run-time distortion based scheduling for delivery of scalable media. In *Proceedings of International Packet Video Workshop*, Pittsburgh, PA.

Miras, D. (2002). *A Survey on Network QoS Needs of Advanced Internet Applications* [working document]. Internet2 QoS Working Group.

Mirhakkak, M., Schult, N., & Thomson, D. (2000). Dynamic Quality-of-Service for mobile ad-hoc networks. In *Proceedings of the 1st ACM International Symposium on Mobile Ad Hoc Networking and Computing*, Boston, Massachusetts (pp. 137-148).

Misra, A., & Banerjee, S. (2002).MRPC: Maximizing Network Lifetime for Reliable Routing in Wireless Environments. *Proc. IEEE Wireless Communication and Networking Conf.*, Orlando.

Mitalo, J. III, & Maguire, G. Q. (1999). Cognitive radio: Making software radios more personal. *IEEE Personal Communications*, *6*, 13–18. doi:10.1109/98.788210

Mogre, P., Hollick, M., & Steinmetz, R. (2007). QoS in Wireless Mesh Networks: Challenges, Pitfalls, and Roadmap to its Realization. In *Proceedings of Network and Operating System Support for Digital Audio and Video (NOSSDAV 2007)*.

Moh, M., Chellappan, B., Moh, T.-S., & Vanesgopal, S. (2010). Handoff Mechanisms for IEEE 802.16 Networks Supporting Intelligent Transportation Systems. to appear in M.-T. Zhou, Y. Zhang, and L. Yang (Eds.), *Wireless Technologies for Intelligent Transportation Systems*. Hauppauge, NY: Nova Science Publisher Inc.

Mohammadi, A., Akl, S. G., & Behnamfar, F. (2008). Optimal linear-time QoS-based scheduling for WiMAX. In *21st Canadian Conference on Electrical and Computer Engineering* (pp. 1811-1814).

Monteiro, J. M., & Nunes, M. S. (2007, July). A Subjective Quality Estimation Tool for the Evaluation of Video Communication Systems. In *proceedings of IEEE Symposium on Computers and Communications (ISCC07), Mediawin Worshop*, Aveiro, Portugal (pp. 75-80).

Monteiro, J. M., Calafate, C. T., & Nunes, M. S. (2008, September). Evaluation of the H.264 Scalable Video Coding in Error Prone IP Networks. *Institute of Electrical and Electronics Engineers Transactions on Broadcasting*, *54*(3), 652–659.

Morgenshtern, V. I., & Bolcskei, H. (2007). Crystallization in large wireless networks. *IEEE Transactions on Information Theory, 53*(10), 3319–3349. doi:10.1109/TIT.2007.904789

Morley, C. D., & Grover, W. D. (2000). Strategies to Maximize Carried Traffic in Dual-Mode Cellular Systems. *IEEE Transactions on Vehicular Technology, 49*(2), 357–366. doi:10.1109/25.832966

MPEG. (n.d.). *Motion Picture Experts Group*. Retrieved from http://www.mpeg.org/

MPEG4IP. (n.d.). *MPEG4IP Project*. Retrieved from http://mpeg4ip.sourceforge.net

Murphy, L., Noonan, J., Perry, P., & Murphy, J. (September, 2007). An application-quality-based mobility management scheme. In *Proceedings of 9th IFIP / IEEE International Conference on Mobile and Wireless Communications Networks (MWCN 2007)*, Cork, Ireland.

Murthy, S. R. C., & Manoj, B. S. (2004). *Ad hoc Wireless Networks*. Upper Saddle River, NJ: PrinticePrentice Hall Inc.

Nabar, R. U., Bolcskei, H., & Kneubuhler, F. W. (2004). Fading relay channels: Performance limits and space–time signal design. *IEEE Journal on Selected Areas in Communications, 22*(6), 1099–1109. doi:10.1109/JSAC.2004.830922

Nafaa, A., & Ksentini, A. (2008). On Sustained QoS Guarantees in Operated IEEE 802.11 Wireless LANs. [TPDS]. *IEEE Transactions on Parallel and Distributed Systems*, (August): 2008.

Nafaa, A., Ksentini, A., Mehoua, A., Ishibashi, A., Iraqi, Y., & Boutaba, R. (2005, July/August). Sliding Contention Window (SCW): Towards Backoff Range-based Service Differentiation over IEEE 802.11 Wireless LAN Networks. *IEEE Network Magazine, Special Issue on Wireless Local Area Networking: QoS provision & Resource Management, 19*(4).

Naghshineh, M., & Schwartz, M. (1996). Distributed Call Admission Control in Mobile/Wireless Networks. *IEEE Journal on Selected Areas in Communications, 12*, 711–717. doi:10.1109/49.490422

Nakhjiri, M. (2007, September). *Use of EAP-AKA, IETF HOKEY and AAA mechanisms to provide access and handover security and 3G-802.16M interworking.* Paper presented at the 18th Annual IEEE International Symposium on Personal, Indoor and Mobile Radio Communications (PIMRC'07).

Nanda, S. (1993). Teletraffic models for urban and suburban microcells: Cell sizes and handoff rates. *IEEE Transactions on Vehicular Technology, 42*, 673–682. doi:10.1109/25.260744

Narasimhan, R. (2008). Delay-limited throughput of cooperative multiple access channels with hybrid-ARQ. In *Proc. IEEE Sym. Info. Theory*, Toronto, Canada.

Navlakha G., & Ferguson J. (2008). *IPerf Tool.*

Nechiporenko, T., Phan, K. T., Tellambura, C., & Nguyen, H. H. (2009). (to appear). Capacity of Rayleigh fading cooperative systems under adaptive transmission. *IEEE Transactions on Wireless Communications.*

Neely, M. J., & Modiano, E. (2005). Capacity and delay tradeoffs for ad-hoc mobile networks. *IEEE Transactions on Information Theory, 51*(6), 1917–1936. doi:10.1109/TIT.2005.847717

Networks, R. (n.d.). *Helix Server*. Retrieved from http://www.realnetworks.com/ & https://helixcommunity.org/

Neves, P., Fontes, F., Monteiro, J., Sargento, S., & Bohnert, T. M. (2008). Quality of service differentiation support in WiMAX networks. *International Conference on Telecommunications (ICT) 2008.*

Ng, T. C.-Y., & Yu, W. (2007). Joint optimization of relay strategies and resource allocations in a cooperative cellular network. *IEEE Journal on Selected Areas in Communications, 25*(2), 328–339. doi:10.1109/JSAC.2007.070209

Ng, T. S. E., et al. (1998). Packet Fair Queueing Algorithms for Wireless Networks with Location-Dependent Errors. In *IEEE INFOCOM* (pp. 1103-1111).

Nguyen, L. T., Beuran, R., & Shinoda, Y. (2007). Performance Analysis of IEEE 802.11 in Multi-hop Wireless

Networks. In *3rd International Conference on Mobile Ad-hoc and Sensor Networks (MSN 2007)*, Beijing, China (LNCS 4864, pp. 326-337). Berlin, Germany: Springer-Verlag.

Niculescu, D., Ganguly, S., Kim, K., & Izmailov, R. (2006). Performance of VoIP in a 802.11 Wireless Mesh Network. In *Proceedings of IEEE INFOCOM 2006.*

Nie, J., Wen, J., & Zeng, L. (2007, November). *A bandwidth based Adaptive fuzzy logic handoff in IEEE 802.16 and IEEE 802.11 hybrid networks.* Paper presented at the International Conference on Convergence Information Technology.

NIST. (n.d.). *NIST Net network emulator.* National Institute of Standards and Technology. Retrieved from http://www-x.antd.nist.gov/nistnet/

Niu, Z., Yao, W., Ni, Q., & Song, Y. (2007). DeReQ: a QoS routing algorithm for multimedia communications in vehicular ad hoc networks. In *Proceedings of the 2007 international Conference on Wireless Communications and Mobile Computing (IWCMC '07),* Honolulu, Hawaii, USA.

Niyato, D., & Hossain, E. (2007). QoS-aware bandwidth allocation and admission control in IEEE 802.16 broadband wireless access networks: A non-cooperative game theoretic approach. *Computer Networks, 55*(11), 3305–3321. doi:10.1016/j.comnet.2007.01.031

Nosratinia, A., Hunter, T. E., & Hedayat, A. (2004). Cooperative communication in wireless networks. *IEEE Communications Magazine, 42*(10), 68–73. doi:10.1109/MCOM.2004.1341264

Oh, S., & Tcha, D. (1992). Prioritized channel assignment in a cellular radio network. *IEEE Transactions on Communications, 40,* 1259–1269. doi:10.1109/26.153371

Oh, S.-M., & Kim, J.-H. (2005). The Analysis of the Optimal Contention Period for Broadband Wireless Access Network. In *Third IEEE International Conference on Pervasive Computing and Communications Workshops* (pp. 215-219).

Okamoto, J., et al. (2004). *An Objective Quality Assessment Method for Arbitrary Video Streams.* Picture Coding Symposium 2004, San Francisco, USA.

Okanishi, S., Kon, M., Chiku, S., Sugiyama, M., Sakurai H. (2008). Traffic Network System. *NEC Technical Journal (special iss.: (ITS), vol.3* no.(1).

Oliveria, C., Kim, J. B., & Suda, T. (1998). An adaptive bandwidth reservation scheme for high-speed multimedia wireless network. *IEEE Journal on Selected Areas in Communications, 16*(6), 858–874. doi:10.1109/49.709449

Olson, P. (2006). Perception-Response Time to Unexpected Roadway Hazards. *Hum. Factors, vol. 28,* no. (1), pp. 91–96.

Olsson, R. A., & Keen, A. W. (2004). *The JR Programming Language Concurrent Programming in an Extended Java.* New York: Kluwer Academic Publishers.

Ong, E., Yang, X., Lin, W., Lu, Z., & Yao, S. (2004). Video quality metric for low bitrate compressed videos. In *Proceedings of the IEEE International Conference on Image Processing,* Genova, Italy (Vol. 5, pp. 3531–3534).

OPTICOM GmbH. Germany. (n.d.). *PEVQ – Advanced Perceptual Evaluation of Video Quality* [white paper]. Retrieved from http://www.opticom.de

Orlik, P. V., & Rappaport, S. S. (1998). Traffic performance and mobility modeling of cellular communications with mixed platforms and highly variable mobilities. *Proceedings of the IEEE, 86,* 1464–1479. doi:10.1109/5.681374

Ott, J., Wenger, S., Sato, N., Burmeister, C., & Rey, J. (2004). *Extended RTP Profile for Real-time Transport Control Protocol (RTCP)-Based Feedback* (RFC 4585). Internet Engineering Task Force.

Pabst, R. (2004, Sept.). Relay-based deployment concepts for wireless and mobile broadband radio. *IEEE Communications Magazine, 42*(9), 80–89. doi:10.1109/MCOM.2004.1336724

Pack, S., & Cho, Y. (2003, 7-10 September). Performance analysis of hierarchical mobile ipv6 in IP-based cellular networks. In *14th IEEE proceedings of personal, indoor and mobile radio communications (PIMRC)* (Vol. 3, p. 2818-2822).

Pack, S., Nam, M., & Choi, Y. (2004). A study on optimal hierarchy in multi-level hierarchical mobile IPv6 networks. In *Proceedings of GLOBECOM.*

Pack, S., Shen, X. S., Mark, J. W., & Pan, J. (2007). Adaptive route optimization in hierarchical mobile IPv6 networks. *IEEE Transactions on Mobile Computing, 6*(8), 903–914. doi:10.1109/TMC.2007.1010

Padhye, J, Firoiu, V., Towsley, D.F., & Kurose, J. (2000). Modeling TCP Reno Performance: A Simple Model and Its Empirical Validation. *IEEE/ACM Transactions on Networking, 8*(2), 133-45.

Padhye, J., & Floyd, S. (2001). Identifying the TCP Behavior of Web Servers. In *SIGCOMM 2001.*

Padhye, J., Firoiu, V., Towsley, D., & Kurose, J. (1998). Modeling TCP throughput: a simple model and its empirical validation. *Computer Communication Review, 28*(4), 303–314. doi:10.1145/285243.285291

Pahalawatta, P. V., & Katsaggelos, A. K. (2007). Review of content-aware resource allocation schemes for video streaming over wireless networks. *Wireless Communications and Mobile Computing, 7,* 131–142. doi:10.1002/wcm.469

Pandey, A., et al. (2002). Application of MIMO and proportional fair scheduling to CDMA downlink packet data channels. In *IEEE International Vehicular Technology Conference* (pp. 1046-1050).

Papoulis, A., & Pillai, S. U. (2002). *Probability, Random Variables and Stochastic Processes*, fourth edition, McGrawHill.

Parekh, A. K., & Gallager, R. G. (1993). A generalized processor sharing approach to flow control in integrated services networks: the single-node case. *IEEE/ACM Transactions on Networking, 1*(3), 344-357.

Park, K., & Willinger, W. (Eds.). (2000). *Self-similar network traffic and performance evaluation.* Hoboken, NJ: John Wiley & Sons, Inc.

Park, V. D., & Corson, M. S. (1997, April). A highly adaptive distributed routing algorithm for mobile wireless networks. *Proceedings of IEEE INFOCOM '97,*3, 405– 1413.

Paskalis, S., Kaloxylos, A., & Zervas, E. (2001). An Efficient QoS Scheme for Mobile Hosts. In *LCN'01: Proceedings of the 26th annual IEEE conference on local computer networks* (p. 630). Washington, DC: IEEE Computer Society.

Paskalis, S., Kaloxylos, A., Zervas, E., & Merakos, L. (2002). Evaluating the RSVP mobility proxy concept. In *the 13th IEEE international symposium on Personal, indoor and mobile radio communications, 2002.* (Vol. 1, pp. 270–274).

Paskalis, S., Kaloxylos, A., Zervas, E., & Merakos, L. (2003). An Efficient RSVP-Mobile IP Interworking Scheme. *Mobile Networks and Applications, 8*(3), 197–207. doi:10.1023/A:1023385429587

Passas, N., Salkintzis, A. K., & Xergias, S. (2008). Centralized resource allocation for multimedia traffic in IEEE 802.16 mesh networks. *Proceedings of the IEEE, 96*(1), 54–63. doi:10.1109/JPROC.2007.909929

Paul, J., & Gaorge, K. (2005). Pheromone Aided Robust Multipath and Multi-priority Routing in Wireless MANETs. In *Proceedings of the 2nd ACM international workshop on Performance Evaluation of Wireless Ad hoc, Sensor and Ubiquitous Networks*, 106-113.

Paul, P., & Raghavan, S. V. (2002). Survey of QoS Routing. In *ICCC '02: Proceedings of the 15th international conference on computer communication* (pp. 50–75). Washington, DC: International Council for Computer Communication.

Paxson, V., Almes, G., Mahdavi, J., & Mathis, M. (1998). *Framework for IP Performance Metrics.* IETF RFC 2330.

Peha, J. M., & Sutivong, A. (2001). Admission Control Algorithms for Cellular Systems. *Wireless Networks, 7,* 117–125. doi:10.1023/A:1016629421079

Pelton, J. N. (1989). International VSAT Applications and ISDN. *IEEE Communications Magazine*, *27*(5), 60–61.

Peng, F., & Deyun, Z. (2006, February). Hybrid Optimize Strategy based QoS Route Algorithm for Mobile Ad hoc Networks. *Journal of Computer Science*, *2*(2), 160–165. doi:10.3844/jcssp.2006.160.165

Peng, S. (2008). *A bio-inspired QoS and energy-aware routing algorithm for wireless sensor networks*. Master's thesis, University of Guelph.

Peng, S., Yang, S. X., Gregori, S., & Tian, F. (2008). An adaptive QoS and energy-aware routing algorithm for wireless sensor networks. In *Proceedings of the IEEE International Conference on Information and Automation (ICIA'2008)*, Zhangjiajie, China (pp. 578-583).

Perevalov, E., & Blum, R. (2003). Delay Limited Capacity of Ad-Hoc Networks: Asymptotically Optimal Transmission and Relaying Strategy. In *Proceedings of IEEE INFOCOM 2003*.

Perevalov, E., & Blum, R. (2006). Capacity Region, Minimum Energy and Delay for a Mobile Ad-Hoc Network. In *Proceedings of IEEE Internal Symposiumon Modeling and Optimization in Mobile Ad-Hoc Wireless Networks 2006*.

Perez-Romero, J., Sallent, O., Agusti, R., Garcia, N., et al. (2006). Network Controlled Cell-Breathing for Capacity Improvement in Heterogeneous CDMA/TDMA Scenarios. In *IEEE Wireless Communications and Networking Conf.* (Vol. 1, pp. 36-41). Washington, DC: IEEE.

Perkins, C. (2002, January). *RFC 3220: IP Mobility Support for IPv4*.

Perkins, C. E., & Royer, E. M. (1999, February). Ad-hoc On Demand Distance Vector Routing. In *Proceedings of the 2nd IEEE Workshop on Mobile Computing Systems and Applications* (Vol. 3, pp. 90-100).

Perkins, C., & Bhagwat, P. (1994). Destination sequence distances vector routing. In *Proceedings of ACM SIGCOMM*.

Perkins, C., & Royer, E. (1999). Ad hoc on demand distance vector routing. In *Proceedings of IEEE WMCSA*.

Perkins, C.E., Royer, E.M., & Das, S.R. (2000, July). Quality of Service for Ad hoc On Demand Distance Vector Routing. *IETF internet draft, draft-ietf-manet-aodvqos-00.txt*.

Peterson, L., Shenker, S., & Turner, J. (April, 2005). Overcoming the Internet impasse through virtualization. *IEEE Computer Magazine, 38*(4).

Phan, T. K., Le, L. B., Vorobyov, S. A., & Le-Ngoc, T. (2009a). Centralized and distributed power allocation in multi-user wireless relay networks. In *Proc. IEEE Inter. Conf. Commun.*, Dresden, Germany.

Phan, T. K., Le, L. B., Vorobyov, S. A., & Le-Ngoc, T. (2009c). (to appear). Power allocation and admission control in multi-user relay networks via convex programming: Centralized and distributed schemes. *EURASIP J. Wireless Commun. and Networking*.

Phan, T. K., Le-Ngoc, T., Vorobyov, S. A., & Tellambura, C. (2008). Power allocation in wireless relay networks: A geometric programming based approach. In *Proc. IEEE Global Telecommun. Conf.*, New Orleans, LA, USA.

Phan, T. K., Le-Ngoc, T., Vorobyov, S. A., & Tellambura, C. (2009b). (to appear). Power allocation in wireless multiuser relay networks. *IEEE Transactions on Wireless Communications*.

Podolsky, M., McCanne, S., & Vetterli, M. (1998). *Soft ARQ for layered streaming media, Technical Report UCB/CSD-98-1024*. University of California, Computer Science Department, Berkeley, CA.

Pong, D., & Moors, T. (2003). Call Admission Control for IEEE 802.11 Contention Access Mechanism. In *Proceeding of IEEE Globecom 2003*, San Francisco, California, USA (pp. 174-8).

Pragad, A. D., Friderikos, V., Pangalos, P., & Aghvami, A. H. (2007). The Impact of Mobility Agent based Micro-Mobility on the Capacity of Wireless Access Networks. *In Proceedings of Globecom 2007*, Washington DC.

Prasant, M., Jian, L., & Chao, G. (2003, March). QoS in ad hoc wireless network. *IEEE Wireless Communications Magazine*, 44-52.

Prez-Romero, J. Sallent, O., Agusti, R., & Diaz-Guerra, M. A. (2005). *Radio Resource Management Strategies in UMTS*. New York: Wiley.

Puterman, M. L. (Ed.). (1994). *Markov decision processes: discrete stochastic dynamic programming*. New York: Wiley-Interscience.

Python. (n.d.). *Python Programming Language*. Retrieved from http://www.python.org

Qiu, L., et al. (2004, November). *Troubleshooting multi-hop wireless networks*. Microsoft Research Tech. Report, MSR-TR-2004-11. Retrieved from ftp://ftp.research.microsoft.com/pub/tr/TR-2004-11.pdf

Qiu, S., Rui, H., & Zhang, L. (2006). No-reference Perceptual Quality Assessment for Streaming Video Based on Simple End-to-end Network Measures. In *proceedings of the International Conference on Networking and Services*, California, USA (pp. 53–53).

Rahnema, M. (1993). Overview of the GSM system and protocol architecture. *IEEE Communications Magazine*.

Rajagopalan, S., & Shen, C. C. (2005). ANSI: A Unicast Routing Protocol for Mobile Networks using Swarm Intelligence. In *Proceedings of the International Conference on Artificial Intelligence* (pp. 24-27).

Raju, G. V. S., Hernandes, G., & Zou, Q. (2000). Quality of Service Routing in Ad Hoc Networks. In . *Proceedings of the IEEE Wireless Communications and Networking Conference*, *1*, 263–265.

Ramachandran, K. N., Belding, E. M., Almeroth, K. C., & Buddhikot, M. M. (2006). Interference-aware channel assignment in multi-radio wireless mesh networks. In *Proc. of IEEE INFOCOM 2006*.

Ramachandran, K., Sheriff, I., Belding-Royer, E., & Almeroth, K. (2008). A Multi-Radio 802.11 Mesh Network Architecture. *Journal of Mobile Networks & Applications*, *13*(1), 132–146. doi:10.1007/s11036-008-0026-8

Ramanathan, P., & Agrawal, P. (1998). Adapting Packet Fair Queuing Algorithms to Wireless Networks. *ACM MOBICOM*(pp. 1-9).

Ramanathan, P., Sivalingam, K. M., Agrawal, P., & Kishore, S. (1999). Dynamic resource allocation schemes during handoff for mobile multimedia wireless networks. *IEEE Journal on Selected Areas in Communications*, *17*, 1270–1283. doi:10.1109/49.778185

Ramirez, C. L., & Fernandez, V. M. (2007). QoS in Vehicular and Intelligent Transport Networks Using Multipath Routing. ISIE 2007 In. *IEEE International Symposium on Industrial Electronics (ISIE 2007).* (pp. 2556-2561).

Ramjee, R., Towsley, D., & Nagarajan, R. (1997). On optimal call admission control in cellular networks. *Wireless Networks*, *3*, 29–41. doi:10.1023/A:1019172226345

Ramjee, R., Varadhan, K., Salgarelli, L., Thuel, S., Wang, S.-Y., & La Porta, T. (2002). HAWAII: a Domain-based Approach for supporting Mobility in Wide-Area Wireless Networks. *IEEE/ACM Transactions on Networking*, *10*(3), 396–410.

Rappaport, S., & Purzynski, C. (1996). Prioritized resource assignment for mobile cellular communication systems with mixed services and platform types. *IEEE Transactions on Vehicular Technology*, *45*, 443–458. doi:10.1109/25.533760

Rappaport, T. S. (2002). *Wireless Communications Principles and Practice (2nd ed.)*. Englewood Cliffs, NJ: Prentice-Hall.

Rath, H. K., Bhorkar, A., & Sharma, V. (2006). An Opportunistic Uplink Scheduling Scheme to Achieve Bandwidth Fairness and Delay for Multiclass Traffic in Wi-Max (IEEE 802.16) Broadband Wireless Networks. In *IEEE Global Telecommunications Conference* (pp. 1-5).

Reijs, V. (2002). *Perceived Quantitative Quality of Applications*. HEAnet. Retrieved from http://www.heanet.ie

Reinbold, P., & Bonaventure, O. (2003). IP Micro-mobility Protocols. *IEEE Communications Surveys and Tutorials*, *5*(1).

Rejaie, R., Handley, M., & Estrin, D. (2000). Layered quality adaptation for Internet video streaming. In *IEEE*

Journal on Selected Areas in Communications - Special Issue Internet QoS, 18(12), 2530–2543.

Rey, J., Leon, D., Miyazaki, A., Varsa, V., & Hakenberg, R. (2005). *RTP Retransmission Payload Format* (RFC 4588). Internet Engineering Task Force.

RFC 3095. (n.d.). *ROHC Framework and four profiles: RTP, UDP, ESP, and uncompressed.*

Ribeiro, A., Cai, X., & Giannakis, G. B. (2005). Symbol error probabilities for general cooperative links. *IEEE Transactions on Wireless Communications, 4*(3), 1264–1273. doi:10.1109/TWC.2005.846989

Richards, A., Antoniades, M., Witana, V., & Rogers, G. (1998). Mapping User Level QoS from a Single Parameter. In *Proc. Second IFIP/IEEE International Conference on Management of Multimedia Networks and Services (MMNS'98)*, November 1998, Versailles, France.

Rix, A., Beerends, J., Kim, D., Kroon, P., & Ghitza, O. (2006, November). Objective assessment of speech and audio quality: Technology and Applications. *IEEE Transactions on Audio, Speech, and Language Processing, 14*(6), 1890–1901. doi:10.1109/TASL.2006.883260

Rizzom, L. (n.d.). *Dummynet: a simple approach to the evaluation of network protocols.*

Robinson, J., Papagiannaki, K., Diot, C., Guo, X., & Krishnamurthy, L. (2005). Experimenting with a Multi-Radio Mesh Networking Testbed. In *Proceedings of IEEE WINMEE 2005.*

Rodríguez-Dagnino, R. M. (1998). Handoff analysis in wireless multimedia networks. In *SPIE, ITCom Conference on Performance and Control of Network Systems II* (Vol. 3530, pp. 76-84).

Rodríguez-Dagnino, R. M., & Leyva-Valenzuela, C. A. (1999). Performance analysis in broadband wireless networks. In *SPIE, ITCom Conference on Performance and Control of Network Systems III* (Vol. 3841, pp. 220-228).

Rodríguez-Dagnino, R. M., & Takagi, H. (2001). Handover Distributions for wireless cellular networks with Pareto call holding times. In *SPIE, ITCom Conference on Internet Performance and Control of Network Systems II* (Vol. 4523, pp. 202-210).

Rodríguez-Dagnino, R. M., & Takagi, H. (2002). Counting handovers in a random number of talkspurts and silence periods. In *SPIE, ITCom Conference on Internet Performance and Control of Network Systems III* (Vol. 4865, pp. 202-212).

Rodríguez-Dagnino, R. M., & Takagi, H. (2003). Counting handovers in a cellular mobile communication network: Equilibrium renewal process approach. *Performance Evaluation, 52*(2-3), 153–174. doi:10.1016/S0166-5316(02)00178-5

Rodríguez-Dagnino, R. M., & Takagi, H. (2005). Distribution of the number of handovers in a cellular mobile communication network delayed renewal process approach. *Journal of the Operations Research Society of Japan, 48*(3), 207–225.

Rodríguez-Dagnino, R. M., & Takagi, H. (2007). Modeling of handover counting and location management for wireless mobile networks. In *SPIE, ITCom Conference on Next-Generation Communications and Sensor Networks* (Vol. 6773, pp. 1-13).

Rodríguez-Dagnino, R. M., & Takagi, H. (2007). Movement-based location management for general cell residence times. *IEEE Transactions on Vehicular Technology, 56*(5), 2713–2722. doi:10.1109/TVT.2007.900377

Rodríguez-Dagnino, R. M., Hernández-Lozano, G., & Takagi, H. (2000). Wireless handover distributions in mixed platforms with multimedia services. In *SPIE, ITCom Conference on Internet Quality and Performance and Control of Network Systems* (Vol. 4211, pp. 59-69).

Rodríguez-Dagnino, R. M., Ruiz-Cedillo, J. J., & Takagi, H. (2002). Dynamic mobility management for cellular networks: A delayed renewal process approach. *IEICE Transactions on Communications . E (Norwalk, Conn.), 85-B*(6), 1069–1074.

Rodríguez-Dagnino, R. M., Ruiz-Cedillo, J. J., & Takagi, H. (2005). Mobility management for general distributed location areas. In *SPIE, ITCom Conference on Performance, Quality of Service, and Control of Next-Generation Communication and Sensor Networks III* (Vol. 6011, pp. 1-13).

Romdhani, L. Ni, Qi., & Turletti, T. (2003). Adaptive EDCF: Enhanced Service Differentiation for IEEE 802.11 802.11 Wireless Ad-Hoc Networks. In *Proceedings of IEEE WCNC 2003*, New Orleans, USA, March 2003.

Rosenberg, J., & Schulzrinne, H. (1999). *An RTP Payload Format for Generic Forward Error Correction* (RFC 2733). Internet Engineering Task Force.

Rosenbrock, K., Sanmugam, R., Bradner, S. & Klensin, J. (2001, June). *3GPP-IETF Standardization Collaboration* (IETF Request For Comments 3113).

Roth, M., & Wicker, S. (2003, June). Termite: Ad hoc Networking with Stigmergy. In . *Proceedings of the IEEE Global Communications Conference, 5*, 2937–2941.

Rothenberg, C., & Roos, A. (2008, March). A Review of Policy-Based Resource and Admission Control Functions in Evolving Access and Next Generation Networks. *Journal of Network and Systems Management, 16*(1), 14–45. doi:10.1007/s10922-007-9096-3

Rouil, R., & Golmie, N. (2006). Adaptive Channel Scanning for IEEE 802.16e. *IEEE Military Comm. Conf, 2006*, 1–6.

Rouil, R., & Golmie, N. (2007). Seamless Mobility in WiMAX. *WiMAX Forum Conference.*

Roychoudhuri, L., Al-Shaer, E., & Settimi, R. (2006, March). Statistical measurement approach for on-line audio quality assessment. In *Proceedings of Passive and Active Measurement (PAM'06).*

Roychoudhuri, L., Al-Shaer, E., Hamed, H., & Brewster, G. (2003, May). Audio transmissions over the Internet: experiments and observations. In *IEEE Proceedings of International Conference on Communications (ICC 2003)*, Anchorage, Alaska, USA.

Royer, E. M., & Toh, C. H. (1999, April). A review of current routing protocols for ad hoc mobile wireless networks. *IEEE Personal Communications, 6*(2), 46–55. doi:10.1109/98.760423

Salamah, M., & Lababidi, H. (2005). Dynamically adaptive channel reservation scheme for cellular networks. *Computer Networks, 49*, 787–796. doi:10.1016/j.comnet.2005.02.005

Sallent, O., Pérez-Romero, J., Agusti, R., & Casadevall, F. (2003). Provisioning multimedia wireless networks for better QoS: RRM strategies for 3G W-CDMA. *IEEE Communications Magazine, 41*(2), 100–106. doi:10.1109/MCOM.2003.1179558

Sanchez, J. J., Morales-Jimenez, D., Gomez, G., & Enbrambasaguas, J. T. (2007). Physical Layer Performance of Long Term Evolution Cellular Technology. In *16th IST Mobile & Wireless Communications Summit* (pp. 1-5). Washington, DC: IEEE.

Sandberg, U. (1995). Building direct-to-home television and entertainment networks forEurope using the new digital video broadcast standards. In *International Broadcast Convention IBC* (pp. 175-177).

Saquib, M., & Yates, R. (1995). Optimal call admission to a mobile cellular network. *Proceedings of IEEE Vehicular Technology Conference* (pp. 190-194).

Sat, B., & Wah, B. W. (2006). Analysis and evaluation of the Skype and Google-Talk VoIP system. In *Proceedings of IEEE international conference on Multimedia and Exposition.*

Sayenko, A., Alanen, O., & Hämäläinen, T. (2008). Scheduling solution for the IEEE 802.16 base station. *Computer Networks, 52*(1), 96–115. doi:10.1016/j.comnet.2007.09.021

Scaglione, A., & Hong, Y. W. (2003). Opportunistic large arrays: Cooperative transmission in wireless multihop ad hoc networks to reach far distances. *IEEE Transactions on Signal Processing, 51*(8), 2082–2092. doi:10.1109/TSP.2003.814519

Scaglione, A., Goeckel, D., & Laneman, J. N. (2006). Cooperative communications in mobile ad-hoc networks: Rethinking the link abstraction. *IEEE Signal Processing Magazine, 23*(5), 18–29. doi:10.1109/MSP.2006.1708409

Schierl, T., Stockhammer, T., & Wiegand, T. (2007, September). Mobile Video Transmission Using Scalable Video Coding. *Institute of Electrical and Electronics Engineers Transactions on Circuits and Systems for Video Technology, 17*(9), 1204–1217.

Schmidt, J. C., & Rose, K. (2007). First-order distortion estimation for efficient video streaming at moderate to high packet loss rates. In *Proceedings of Packet Video Workshop* (pp. 318-325).

Schmidt, J. C., & Rose, K. (2008). Macroblock-based retransmission for error-resilient video streaming. In *Proceedings of the IEEE International Conference on Image Processing* (pp. 2308-2311).

Schoch, E., Kargl, F., Weber, M. & Leinmüller, T. (2008). Communication Patterns in VANETs. *IEEE Communications Magazine, vol. 46*, no. (11).

Schoonderwoerd, R., Holland, O. E., Bruten, J. L., & Rothkrantz, L. J. (1996). Ant-based load balancing in telecommunications networks. *Adaptive Behavior, 5*, 169–207. doi:10.1177/105971239700500203

Schulzrinne, H., Casner, S., Frederick, R., & Jacobson, V. (1996). *A Transport Protocol for Real-Time Applications.* Request for Comments (standard) RFC 1889, Internet Engineering Task Force (IETF).

Schulzrinne, H., Casner, S., Frederick, R., & Jacobson, V. (2003). *RTP: A transport protocol for real-time applications* (STD 0064, RFC 3550). Internet Engineering Task Force.

Seeling, P., Reisslein, M., & Kulapala, B. (2004). Network Performance Evaluation Using Frame Size and Quality Traces of Single-Layer and Two-Layer Video: A Tutorial. *IEEE Communications Surveys and Tutorials, 6*(2), 58–78. doi:10.1109/COMST.2004.5342293

Sendonaris, A., Erkip, E., & Aazhang, B. (2003a). User cooperation diversity-Part I: System description. *IEEE Transactions on Communications, 51*(11), 1927–1938. doi:10.1109/TCOMM.2003.818096

Sendonaris, A., Erkip, E., & Aazhang, B. (2003b). User cooperation diversity-Part II: Implementation aspects and performance analysis. *IEEE Transactions on Communications, 51*(11), 1939–1948. doi:10.1109/TCOMM.2003.819238

Seo, M., Choi, W., Kim, Y., & Park, J. (2007). Lifetime prediction routing protocol for wireless sensor networks. *IEICE (International Energy Information Community) transactions on communications, 90*(12), 3680-3681.

Serrador, A., Galvano, G., Ferreira, L., & Correia, L. M. (2006). Parameters for the Definition of Scenarios for CRRM Performance Evaluation. In *IEEE MELECON.* (pp. 575-578). Washington, DC: IEEE.

Servis, V. (2001). *Measuring speech quality over VoIP networks.* The TOLLY Group.

Setton, E., & Girod, B. (2008). *Peer-to-Peer Video Streaming.* Berlin, Germany: Springer.

Setton, E., Noh, J., & Girod, B. (2006). Congestion-distortion optimized peer-to-peer video streaming. In *Proceedings of the IEEE International Conference on Image Processing (ICIP)*, Atlanta, GA.

Sharma, G., Mazumdar, R., & Shroff, N. (2007). Delay and capacity trade-offs in mobile ad hoc networks: A global perspective. *IEEE/ACM Transactions On Networking,* 15(2), 981-992.

Shen, Q., & Fang, X. (2006). A Multi-metric AODV Routing in IEEE 802.11s. In . *Proceedings of ICCT, 06*, 1–4.

Sheng, M., Li, J., & Shi, Y. (2003, January). Routing protocol with QoS guarantees for ad-hoc network. *Electronics Letters, 39*, 143–145. doi:10.1049/el:20030017

Shenker, S., Partridge, C., & Guerin, R. (1997, September). *RFC 2212: Specification of Guaranteed Quality of Service .*

Sheu, S.-T., & Sheu, T.-F. (2001, October). A bandwidth allocation/sharing/extension protocol for multimedia over IEEE 802.11 ad hoc wireless LANs. *IEEE Journal on Selected Areas in Communications, 19*, 2065–2080. doi:10.1109/49.957320

Shimura, A., Aizono, T., & Sakaibara, T. (2002).A proposal of Quality of Service (QoS) Control Method to

Realize Highly-Reliable Communications for Roadside System. In *IEEE Intelligent Vehicle Symposium*, IEEE. (vol.2, pp. 370- 377).

Shin, I.-H., & Lee, C.-W. (2004). A QoS Guaranteed Fast Handoff Algorithm for Wireless LAN. In *IEEE international conference on communications* (Vol. 7, pp. 3827–3832 Vol.7).

Shokrollahi, A. (2006). Raptor codes. *Institute of Electrical and Electronics Engineers Transactions on Information Theory, 52*(6), 2551–2567.

Shunra. (2004). *The Virtual Enterprise: Eliminating the risk of delivering distributed IT services* [white paper]. Retrieved from http://www.shunra.com

Simena. (n.d.). *Network Emulators*. Retrieved from http://www.simena.net

Simó, J., Osuna, P., Seoane J., & Martínez, A. (2006). *An IEEE 802.11 static mesh network design for isolated rural areas in developing regions.* IEEE INFOCOMM 2006

Simulator Homepage, N. S. (n.d.). Retrieved August 23, 2008, from http://www.isi.edu/nsnam/ns

Singer, D., Belknap, W., & Franceschini, G. (2001). *ISO Media File Format Specification: MP4 Technology Under Consideration for ISO/IEC.*

Singh, S., & Raghavendra, C.S. (1998, July). PAMAS - power aware multi-access protocol with signaling for ad hoc networks. *ACM Computer Communication Review.*

Singh, S., Kim, J., Perkins, C., Pedro, M., & Ruiz, T. (2005). *Ad hoc network auto-configuration: definition and problem statement.* Internet-Draft draft-singh-autoconf-adp-00.txt

Singh, V., & Sharma, V. (2006). Efficient and fair scheduling of uplink and downlink in IEEE 802.16 OFDMA networks. In *Wireless Communications and Networking Conference* (pp. 984-990).

Sirkeci-Mergen, B., & Scaglione, A. (2007a). On the power efficiency of cooperative broadcast in dense wireless networks. *IEEE Journal on Selected Areas in Communications, 25*(2), 497–507. doi:10.1109/JSAC.2007.070223

Sirkeci-Mergen, B., & Scaglione, A. (2007b). Randomized space-time coding for distributed cooperative communication. *IEEE Transactions on Signal Processing, 55*(10), 5003–5017. doi:10.1109/TSP.2007.896061

Sirkeci-Mergen, B., Scaglione, A., & Mergen, G. (2006). Asymptotic analysis of multistage cooperative broadcast in wireless networks. *IEEE Transactions on Information Theory, 52*(11), 5166–5166. doi:10.1109/TIT.2006.883627

Sivakumar, R., Sinha, P., & Bharghavan, V. (1999, August). CEDAR: a core extraction distributed ad hoc routing algorithm. *IEEE Journal on Selected Areas in Communications, 17*(8), 1454–1465. doi:10.1109/49.779926

Skorin-Kapov, L., Mosmondor, M., Dobrijevic, O., & Matijasevic, M. (2007, July). Application-Level QoS Negotiation and Signaling for Advanced Multimedia Services in the IMS. *IEEE Communications Magazine, 45*(7), 108–116. doi:10.1109/MCOM.2007.382669

So, J., & Vaidya, N. (2004, May). Multi-channel MAC for ad hoc networks: handling multi-channel hidden terminals using a single transceiver. In *Proc. ACM International Symposium on Mobile Ad Hoc Networking and Computing (MobiHoc)* (pp. 222-233).

Sohraby, K., Minoli, D., & Znati, T. (2007). *Wireless Sensor Networks: Technology, Protocols, and Applications.* New York: John Wiley and Sons.

Soliman, H., Castelluccia, C., & El Malk, K. L. B. (2005, August). *RFC 4140: Hierarchical Mobile IPv6 Mobility Management (HMIPv6).*

Song, J., Chang, M. Y., & Lee, S. S. (2007). Overview of ITU-T NGN QoS Control. *IEEE Communications Magazine*, 116–123. doi:10.1109/MCOM.2007.4342866

Song, J., Yang, Z., Yang, L., Gong, K., Pan, C., Wang, J., & Wu, Y. (2007). Technical Review on Chinese Digital Terrestrial Television Broadcasting Standard and Measurement on Some Working Modes. *IEEE Transactions on Broadcasting, 53*(1), 1–7. doi:10.1109/TBC.2007.891835

Song, S., & Chen, C. W. (2007, September). Scalable H.264/AVC Video Transmission Over MIMO Wireless

Systems with Adaptive Channel Selection Based on Partial Channel Information. *Institute of Electrical and Electronics Engineers Transactions on Circuits and Systems for Video Technology, 17*(9), 1218–1226.

Song, W., Zhuang, W., & Cheng, Y. (2007, January-February). Load Balancing for Cellular/WLAN Integrated Networks. *IEEE Network*, 27–33. doi:10.1109/MNET.2007.314535

Srivastava, V., & Motani, M. (2005). Cross-Layer Design: A Survey and the Road Ahead. *IEEE Communications Magazine, 43*(12), 112–119. doi:10.1109/MCOM.2005.1561928

Std, S. A. E. J1746. (2001). *ISP-Vehicle Location Referencing Standard.*

Stewart, R. (Ed.). (2007). *Stream Control Transmission Protocol.* IETF RFC 4960.

Stibor, L., Zang, Y., & Reumerman, H.-J. (2007). Neighborhood Evaluation of Vehicular Ad-Hoc Network Using IEEE 802.11p. In *13th European Wireless Conference*, Paris, France.

Stockhammer, T., Wiegand, T., & Wegner, S. (2002). Optimized transmission of H.26L/JVT coded video over packet-switched networks. In *Proceedings of the IEEE International Conference on Image Processing* (pp. 173–176).

Streltsov, S., & Vakiki, P. (1996, March). Variance reduction algorithms for parallel replicated simulation of uniformized Markov chains. *Discrete Event Dynamic Systems: Theory and Applications, 6*, 159–180. doi:10.1007/BF01797237

Su, H., & Zhang, X. (2007). Clustering-Based Multichannel MAC Protocols for QoS Provisioning over Vehicular Ad Hoc Networks. *IEEE Transactions on Vehicular Technology, vol. 56*, no. (6), part 1.

Subramanian, A. P., Buddhikot, M. M., & Miller, S. (2006). *Interference Aware Routing in Multi-Radio Wireless Mesh Networks* [Tech. Rep.]. State University of New York at Stony Brook, 2006.

Sun, L., & Ifeachor, E. (2004, June). New models for perceived voice quality prediction and their applications in play-out buffer optimization for VOIP networks. In *Proceedings of IEEE International Conference on Communication (ICC'04).*

Sun, L., & Ifeachor, E. C. (2006, August). Voice Quality Prediction Models and Their Application in VoIP Networks. *IEEE Transactions on Multimedia, 8*(4).

Sun, W., Yamaguchi, H., Yukimasa, K., and & Kusumoto, S. (2006). GVGrid: A QoS Routing Protocol for Vehicular Ad Hoc Networks. In *Fourteenth IEEE International Workshop on Quality of Service (IWQoS 2006).* New Haven, CT: Yale University, New Haven, CT, USA.

Sun, Y., Sheriff, I., Belding-Royer, E., & Almeroth, K. (2005). An Experimental Study of Multimedia Traffic Performance in Mesh Networks. In *Proceedings of International Conference On Mobile Systems, Applications And Services* (pp. 25-30).

Sutton, R. S., & Barto, A. G. (1998). *Reinforcement Learning: An Introduction.* Cambridge, MA: MIT Press.

Suzuki, T., Kutsuna, T., & Tasaka, T. (2008) QoE Estimation from MAC-Level QoS in Audio-Video Transmission with IEEE 8O2.lle EDCA. In *Proceedings of the Personal, Indoor and Mobile Radio Communications* (pp. 1-6).

Taaghol, P., Salkintzis, A. K., & Iyer, J. (2008). Seamless Integration of Mobile WiMAX in 3GPP Networks. *IEEE Communications Magazine, 46*(10). doi:10.1109/MCOM.2008.4644122

Takagi, H., & Rodríguez-Dagnino, R. M. (2007). Counting the number of renewals during a random interval in a discrete-time delayed renewal process. *Operations Research Letters, 35*(1), 119–124. doi:10.1016/j.orl.2005.12.002

Takahashi, A., Hands, D., & Barriac, V. (2008). Standardization Activities in the ITU for a QoE Assessment of IPTV. *Institute of Electrical and Electronics Engineers Communications Magazine, 46*(2), 78–84.

Takahashi, A., Kurashima, A., & Yoshino, H. (2006, November). Objective assessment methodology for estimating conversational quality in VoIP. *IEEE Transactions on Audio, Speech, and Language Processing, 14*(6).

Takahashi, A., Yoshino, H., & Kitawaki, N. (2004, July). Perceptual QoS assessment technologies for VoIP. *IEEE Communications Magazine*, *42*(7), 28–34. doi:10.1109/MCOM.2004.1316526

Talukdar, A. K., Badrinath, B. R., & Acharya, A. (2001). MRSVP: a resource reservation protocol for an integrated services network with mobile hosts. *Wireless Networks*, 7(1), 5–19. doi:10.1023/A:1009035929952

Talwalkar, R. A., & Ilyas, M. (2008). Analysis of Quality of Service (QoS) in WiMAX networks. In *16th IEEE International Conference on Networks (ICON)*.

Tan, W. L., Lam, F., & Lau, W. C. (2007). An Empirical Study on 3G Network Capacity and Performance. In *INFOCOM 2007. 26th IEEE International Conference on Computer Communications* (pp. 1514-1522). Washington, DC: IEEE.

Tang, J., Xue, G., & Zhang, W. (2005). Interference-aware topology control and QoS routing in multi-channel wireless mesh networks. In *Proceedings of the 6th ACM Mobile as hoc networking and computing* (pp. 68 -77).

Tannious, R., & Nosratinia, A. (2008). Spectrally-efficient relay selection with limited feedback. *IEEE Journal on Selected Areas in Communications*, *26*(8), 1419–1428. doi:10.1109/JSAC.2008.081008

Tarchi, D., Fantacci, R., & Bardazzi, M. (2006). Quality of Service management in IEEE 802.16 wireless metropolitan area networks. In *IEEE International Conference on Communications* (pp. 1789-1794).

Tarokh, V., Jafarkhani, H., & Calderbank, A. R. (1999). Space-time block coding from orthogonal designs. *IEEE Transactions on Information Theory*, *45*(5), 1456–1467. doi:10.1109/18.771146

Tarokh, V., Seshadri, N., & Calderbank, A. R. (1998). Space-time codes for high data rate wireless communication: Performance criterion and code construction. *IEEE Transactions on Information Theory*, *44*(2), 744–765. doi:10.1109/18.661517

Tavares, J., Velez, F. J., & Ferro, J. M. (2008). Application of wireless sensor networks to automobiles. *Measurement Science Review*, *8*(3), 65–70. doi:10.2478/v10048-008-0017-8

Taylor-Fuller, D., & Lincke, S. J. (in press). The Development of a Graphic Multidimensional Markov Chain Modeler to Diagram Large State Spaces. In *Proc. 41st Annual Simulation Symposium*, Association of Computing Machinery (ACM).

Technologies, T. (n.d.). *Next Generation Network (NGN) Services* [white paper]. Retrieved from http://www.mobilein.com/NGN_Svcs_WP.pdf

Telatar, I. E. (1998). Capacity of multi-antenna Gaussian Channels. *European Trans. Telecommunications*, *10*(6), 585–595. doi:10.1002/ett.4460100604

The Internet Traffic Archive. (n.d.). Retrieved from http://ita.ee.lbl.gov/index.html

TIA-1099. (2006). Forward Link Only Air Interface Specification for Terrestrial Mobile Multimedia Multicast.

Tian, C., & Yuan, D. (2007, September). *A novel cross-layer scheduling algorithm for IEEE 802.16 WMAN*. Paper presented at the I International Workshop on Cross Layer Design (WCLD '07).

Tian, X., & Ji, C. (2001). Bounding the performance of dynamic channel allocation with QoS provisioning for distributed admission control in wireless networks. *IEEE Transactions on Vehicular Technology*, *50*, 388–397. doi:10.1109/25.923051

Tian, Y., Xu, K., & Ansari, N. (2005, March). TCP in wireless environments: problems and solutions. *IEEE Communications Magazine*, *43*(3), 27–32. doi:10.1109/MCOM.2005.1404595

Tierney, B. (2005). *TCP Tuning and Network Troubleshooting*. Retrieved from http://www.ONLamp.com

Tipper, D., Ramaswamy, S., & Dahlberg, T. (1999). PCS Network Survivability. In *1999 IEEE Wireless Communication & Networking Conf.* (Vol. 2, pp. 1028). Washington, DC: IEEE

Toh, C. K. (2001, June). Maximum battery life routing to support ubiquitous mobile computing in wireless ad hoc networks. *IEEE Communications Magazine, 39*(6), 138–147. doi:10.1109/35.925682

Toril, M., & Wille, V. (2008, June). Optimization of the Assignment of Base Station to Base Station Controllers in GERAN. *IEEE Communications Letters, 12*(6), 477–479. doi:10.1109/LCOMM.2008.072044

Toumpis, S., & Goldsmith, A. J. (2003, May). Performance, optimization, and cross-layer design of media access protocols for wireless ad hoc networks. In *Proc. IEEE ICC'03* (pp. 2234-2240).

Trad, A., Munir, F., & Afifi, H. (2006). Capacity Evaluation of VoIP in IEEE 802.11e WLAN Environment. In *Proceedings of IEEE CCNC 2006.*

Tripathi, N. D. (2001). Simulated Base analysis of the radio interface performance of an IS-2000 system for various data services. *IEEE.*

TS 25 401. (2007, October). *UTRAN overall description.*

Tseng, C.-C., Lee, G.-C., Liu, R.-S., & Wang, T.-P. (2003). HMRSVP: a hierarchical mobile RSVP protocol. *Wireless Networks, 9*(2), 95–102. doi:10.1023/A:1021833430898

Tzeng, S. S., & Lu, H. Y. (2008). Dual-threshold admission control for non-real-time traffic in wireless data networks. *Computer Communications, 31*, 2428–2434. doi:10.1016/j.comcom.2008.03.010

Vacirca, F., Vendictis, A. D., & Baiocchi, A. (2003). Investigating Interactions between ARQ Mechanisms and TCP over Wireless Links. *IEEE GLOBECOM.*

Vacirca, F., Vendictis, A. D., Todini, A., & Baiocchi, A. "On the Effects of ARQ Mechanisms on TCP Performance in Wireless Environments", *Globecom, IEEE*, pp. 671-671, 2003.

van den Branden Lambrecht, C. J. (1997). *Survey of Image and Video Quality Metrics based on Vision Models.* Presentation.

Van der Schaar, M., & Andreopoulos, Y. (2005). Rate-distortion-complexity modeling for network and receiver aware. *IEEE Transactions on Multimedia, 7*(3), 471–479. doi:10.1109/TMM.2005.846790

Vazquez-Avila, L. J., Cruz-Perez, F. A., & Ortigoza-Guerrero, L. (2006). Performance Analysis of Fractional Guard Channel Policies in Mobile Cellular Networks. *IEEE Transactions on Wireless Communications, 5*(2), 301–305. doi:10.1109/TWC.2006.1611053

Vedantham, R., Kakumanu, S., Lakshmanan, S., & Sivakumar, R. (2006). Component based channel assignment in single radio, multi-channel ad-hoc networks. In *Proc. of ACM MOBICOM 2006.*

Veres, A., Campbell, A-T, Barry, M., & Sun, L (2001). Supporting service differentiation in wireless packet networks using distributed control,(in *IEEE Journal of Selected Area on Communication*, vol. 19, no. 10, October 2001, (pp. 2081-2093).

Video Quality Experts Group (VQEG). (n.d.). *Final report from the Video Quality Experts Group on the validation of objective models of video quality assessment.* Retrieved from http://www.vqeg.org

VQEG. (2008). *The Video Quality Experts Group.* Retrieved on October 1, 2008, from http://www.vqeg.org

VSC. (2005). Vehicle Safety Communications Project: Final Rep. Submitted to NHTSA and FHWA in response to cooperative agreement DTFH61-01-X-001.

Wan, S., Tang, J., & Wolff, R. S. (2008). Reliable Routing for Roadside to Vehicle Communications in Rural Areas. ICC '08. In *IEEE International Conference on Communications (ICC '08).*

Wang, H. S., & Moayeri, N. (1995). Finite-state Markov channel – A useful model for radio communication channels. *IEEE Transactions on Vehicular Technology, 44*(1), 163–171. doi:10.1109/25.350282

Wang, H., He, B., & Agrawal, D. P. (2007). Above packet level admission control and bandwidth allocation for IEEE 802.16 wireless MAN. *Simulation Modelling Practice and Theory, 15*(14), 366–382. doi:10.1016/j.simpat.2006.11.010

Wang, L., Liu, F., Ju, Y., & Ruangchaijatupon, N. (2007). Admission Control for Non-preprovisioned Service Flow in Wireless Metropolitan Area Networks. In *Fourth Eu-*

ropean Conference on Universal Multiservice Networks (pp. 243-249).

Wang, L.-C., & Lin, W.-J. (2004). Throughput and fairness enhancement for OFDMA broadband wireless access systems using the maximum C/I scheduling. In *IEEE 60th Vehicular Technology Conference* (pp. 4696-4700).

Wang, X., Fan, P., & Pan, Y. (2008). A more realistic thinning scheme for call admission control in multimedia wireless networks. *IEEE Transactions on Computers, 57*(8), 1143–1148. doi:10.1109/TC.2008.29

Wang, X., Fan, P., Li, J., & Pan, Y. (2009). Modeling and Cost Analysis of Movement-Based Location Management for PCS Networks with HLR/VLR Architecture and General Location Area and Cell Residence Time Distributions. *IEEE Transactions on Vehicular Technology, 57*, 3815–3831. doi:10.1109/TVT.2008.919613

Wang, Z. (2001). *Internet QoS: Architectures and Mechanisms for Quality of Service (The Morgan Kaufmann Series in Networking).* San Francisco: Morgan Kaufmann

Wanstedt, S., Pettersson, J., Xianchun, T., & Heikkila, G. (2002). Development of an objective speech quality measurement model for the AMR CODEC. In *Proceedings of MESAQIN 02*, Prague, Czech Republic.

Watkins, C. (1989). *Learning from Delayed Rewards.* PhD Thesis, University of Cambridge, UK.

Wei, S., Goeckel, D. L., & Valenti, M. C. (2006). Asynchronous cooperative diversity. *IEEE Transactions on Wireless Communications, 5*(6), 1547–1557. doi:10.1109/TWC.2006.1638675

Wennstrom, A., Alferedsson, S., & Brunstorm, A. (2004). TCP over Wireless networks. Karlstad, Sweden: Karlstad University Press.

Werner, M., Kamps, K., Tuisel, U., Beerends, J. C., & Vary, P. (2003, September). *Parameter-based speech quality measures for GSM.* In Proceedings of 14-th IEEE PIMRC 2003.

Wiegand, T. (2003). Rate-Constrained coder control and comparison of video coding standards. *IEEE Transactions on Circuits and Systems for Video Technology, 13*(7).

Wiles, B. C., & Walker, J. (n.d.). *Speak Freely.* Retrieved from http://www.speakfreely.org

WiMAX Forum. (2006, August). *Mobile WiMAX - Part I: A Technical Overview and Performance Evaluation* [White Paper]. Retrieved from http://www.wimaxforum.org/news/downloads/Mobile_WiMAX_Part1_Overview_and_Performance.pdf

WiMAX Forum. (2007). *Network Architecture Stage 2 and 3.* Release 1.0 (Revision 1.2).

WiMAX Forum. (n.d.a). *Network Working Group Document, Release 1.0.*

WiMAX Forum. (n.d.b). *Universal services interface (USI): An Architecture for Internet+ Service Model.* Network Working Group Document, Release 1.5 (draft).

Winkler, S. (1999). Issues in Vision Modelling for Perceptual Video Quality Assessment. *Signal Processing, 78*(2), 231–252. doi:10.1016/S0165-1684(99)00062-6

Winkler, S. (2001). Visual Fidelity and Perceived Quality: Towards Comprehensive Metrics. In *Proc. SPIE Human Vision and Electronic Imaging, San Jose, California* (vol. 4299, pp. 114-125).

Winkler, S. (2005). *Digital Video Quality: Vision Models and Metrics.* Hoboken, NJ: J. Wiley & Sons.

Winkler, S., & Mohandas, P. (2008). The Evolution of Video Quality Measurement: From PSNR to Hybrid Metrics. *Institute of Electrical and Electronics Engineers Transactions on Broadcasting, 54*(3), 1–9.

Winkler, S., Sharma, A., & McNally, D. (2001). Perceptual Video Quality and Blockiness Metrics for Multimedia Streaming Applications. In *Proc. 4th International Symposium on Wireless Personal Multimedia Communications*, Aalbord, Denmark (pp. 553-556).

Wireshark. (2008). *Wireshark: Network Protocol Analyzer.* Retrieved October 2008, from http://www.wireshark.org/

Wisely, D. (2009). *IP for 4G.* New York: Wiley.

Wong, W. K., et al. (2003). Scheduling Algorithm in a Point-to-Multipoint Broadband Wireless Access

Network. In *IEEE International Vehicular Technology Conference (VTC 2003)*(pp. 1593-1597).

Wongthavarawat, K., & Ganz, A. (2003). IEEE 802.16 based last mile broadband wireless military networks with Quality of Service support. In *IEEE MILCOM* (pp. 779-784).

Woo, K., Yu, C., Youn, H. Y., & Lee, B. (2001). Non-Blocking, Localized Routing Algorithm for Balanced Energy Consumption in Mobile Ad Hoc Networks. In *Proceedings of Int'l Symp. on Modeling, Analysis and Simulation of Computer and Telecommunication Systems* (pp. 117-124).

Wroclawski, J. (1997a, September). *RFC 2210: The Use of RSVP with IETF Integrated Services.*

Wroclawski, J. (1997b, September). *RFC 2211: Specification of the Controlled-Load Network Element Service.*

Wu, D., & Negi, R. (2003). Effective capacity: A wireless link model for support of quality of service. *IEEE Transactions on Wireless Communications, 2*(4), 630–643.

Wu, D., Hou, Y. T., Li, B., Zhu, W., Zhang, Y.-Q., & Chao, H. J. (2000). An end-to-end approach for optimal mode selection in Internet video communication. *IEEE Journal on Selected Areas in Communications, 18*(6), 977–995. doi:10.1109/49.848251

Wu, S., Wong, K. Y. M., & Li, B. (2002). A dynamic call admission policy with precision QoS guarantee using stochastic control for mobile wireless networks. *IEEE Transactions on Networking, 10*, 257–271. doi:10.1109/90.993306

Wu, S.-L., Lin, C.-Y., Tseng, Y.-C., & Sheu, J.-P. (2000). A new multi-channel MAC protocol with ondemand channel assignment for multi-hop mobile ad hoc networks. In *Proc. International Symposium on Parallel Architectures, Algorithms and Networks (ISPAN).*

Xiao, F. (2007). *DCT-based Video Quality Evaluation.* Final project for EE392J, Digital Video Processing, Stanford University.

Xiao, H., Seah, W. K.-G., Lo, A., & Chua, K.-C. (2000). A flexible quality of service model for mobile ad-hoc networks. In *Proceedings of Vehicular Technology Conference (VTC'00-Spring)* (pp. 445-449). Washington, DC: IEEE.

Xiao, L., Fuja, T. E., Kliewer, J., & Costello, D. J. Jr. (2007). A network coding approach to cooperative diversity. *IEEE Transactions on Information Theory, 53*(10), 3714–3722. doi:10.1109/TIT.2007.904990

Xiao, L., Johansson, M., & Boyd, S. (2004). Simultaneous routing and resource allocation via dual decomposition. *IEEE Transactions on Communications, 52*(7), 1136–1144. doi:10.1109/TCOMM.2004.831346

Xiao, Y., & Li, H. (2004). Evaluation of Distributed Admission Con-trol for the IEEE 802.11e EDCA . *IEEE Communications Magazine, 42*(9), S20–S24. doi:10.1109/MCOM.2004.1336720

Xiao, Y., Li, H., & Choi, S. (2004). Protection and Guarantee for Voice and Video Traffic in IEEE 802.11e Wireless LANs, proceeding of *IEEE INFOCOM* 2004.

Xin, C., Ma, L., & Shen, C.-C. (2007). Path-centric channel assignment in cognitive radio wireless networks. In *Proceedings of Cognitive Radio Oriented Wireless Networks and Communications (CrownCom'07)* (pp. 313-320). Washington, DC: IEEE.

Xing, K., Chen, X., Ma, L., & Liang, Q. (2007). Superimposed code based channel assignment in multi-radio multi-channel wireless mesh networks. In *Proceedings of ACM MOBICOM 2007.*

Xu, K., Tang, K., Bagrodia, R., Gerla, M., & Bereschinsky, M. (2003, October). Adaptive bandwidth management and QoS provisioning in large scale ad-hoc networks. In *Proceedings of IEEE MILCOM 2003*, Boston, Massachusetts.

Xu, Q., Mak, T., Ko, J., & Sengupta, R. (2007). Medium Access Control Protocol Design for Vehicle–Vehicle Safety Messages. *IEEE Transactions on Vehicular Technology, vol. 56*, no. (22).

Xue, Q., & Ganz, A. (2003, February). Ad hoc QoS on-demand routing (AQOR) in mobile ad hoc networks. *Journal of Parallel and Distributed Computing, 63*(2), 154–165. doi:10.1016/S0743-7315(02)00061-8

Yamagishi, K., & Hayashi, T. (2008). Parametric Packet-Layer Model for Monitoring Video Quality of IPTV Services. In *IEEE International Conference on Communications 2008*, Beijing, China (pp. 110–114).

Yang, K., Ou, S., Chen, H., & He, J. (2007). A multi-hop peer-communication protocol with fairness guarantee for IEEE 802.16-based vehicular networks. In *IEEE Trans. On Vehicular Technology* (pp. 3358-3370).

Yang, O., & Lu, J. (2006). New scheduling and CAC scheme for real-time video application in fixed wireless networks. In *3rd IEEE Consumer Communications and Networking Conference* (pp. 303-307).

Yang, Y., Wang, J., & Kravets, R. (2006). *Design Routing Metrics for Mesh Networks* [Tech. Rep.] University of Illinois at Urbana-Champaign, 2006.

Yao, Y., Cai, X., & Giannakis, G. B. (2005). On energy efficiency and optimum resource allocation of relay transmissions in the low-power regime. *IEEE Transactions on Wireless Communications*, *4*(6), 2917–2927. doi:10.1109/TWC.2005.858294

Yarvis, M., Conner, W., Krishnamurthy, L., Chhabra, J., Elliott, B., & Mainwaring, A. (2002). Real world experiences with an interactive ad-hoc sensor network. In *Proceedings of the International Workshop on Ad Hoc Networking.*

Yavuz, E. A., & Leung, V. C. M. (2006). Computationally efficient method to evaluate the performance of guard channel based call admission control in cellular networks. *IEEE Transactions on Vehicular Technology*, *55*(4), 1412–1424. doi:10.1109/TVT.2006.877470

Yeom, I., & Narasimha Reddy, A. L. (1998). ENDE: An End-to-end Network Delay Emulator. *Journal on Multimedia Tools and Applications.*

Yeung, K. L., & Nanda, S. (1996). Channel management in microcell/macrocell cellular radio systems. *IEEE Transactions on Vehicular Technology*, *45*, 601–612. doi:10.1109/25.543716

Yih-Guang, J., Yang-Han, L., Ming-Hsueh C., Hsien-Wei, T., Jheng-Yao, L., & Chih-Wei, S. (2007, July). *Mean Traffic Bit Rate with ON-SID Modeling of VoIP Traffic*, IEEE 802.16 Broadband Wireless Access Working Group.

Yin, L., Li, B., Zhang, Z., & Lin, Y. (2000). Performance analysis of a dual-threshold reservation (DTR) scheme for voice/data integrated mobile wireless networks. In *Proceedings of the IEEE Wireless Communications and Networking Conference (WCNC-2000)* (pp. 258-262).

Yin, S., Xiong, Y., Zhang, Q., & Lin, X. (2006). Prediction-based routing for real time communications in wireless multi-hop networks. In *Proceedings of the 3rd international conference on Quality of service in heterogeneous wired/wireless networks.*

Yong, J., Guangwei, B., Peng, Z., Hang, S., & Junyuan, W. (2008). Performance Evaluation of a Hybrid FEC/ARQ for Wireless Media Streaming. In *Proceedings of the IEEE International Conference on Circuits and Systems for Communications (ICCSC)*, (pp. 90–94).

Yoon, C. H., & Kwan, C. (1993). Performance of personal portable radio telephone systems with and without guard channels. *IEEE Journal on Selected Areas in Communications*, *11*, 911–917. doi:10.1109/49.232300

Yousefi, S., Fathy, M., & Benslimane, A. (2007). Performance of beacon safety message Dissemination in Vehicular Ad hoc NETworks (VANETs). *Journal of Zhejiang University – Science A., Vol. 8*, no. (12). Zhejiang, China: Zhejiang University Press.

Yu, O. T. W., & Leung, V. C. M. (1997). Adaptive Resource Allocation for Prioritized Call Admission over an ATM-Based Wireless PCN. *IEEE Journal on Selected Areas in Communications*, *15*, 1208–1225. doi:10.1109/49.622906

Yu, O., & Leung, V. (1996). Self-tuning prioritized call handling mechanism with dynamic guard channels for mobile cellular systems. In *Proceedings of IEEE Vehicular Technology Conference*, (pp. 1520-1524).

Yu, Y., Krishnamachari, B., & Prasanna, V. K. (2004). Energy-latency tradeoffs for data gathering in wireless sensor networks. In *Proceedings of the IEEE INFOCOM Conference on Computer Communications*, Hong Kong, China (pp. 244-255).

Yuan, W., & Nahrstedt, K. (2006). Energy-Efficient CPU Scheduling for Multimedia Applications. *ACM Transactions on Computer Systems*, *24*(3), 1–40. doi:10.1145/1151690.1151693

Zhai, F., Luna, C. E., Eisenberg, Y., Pappas, T. N., Berry, R., & Katsaggelos, A. K. (2003). A Novel Cost–Distortion Optimization Framework for Video Streaming over Differentiated Services Networks. In *Proceedings of the IEEE International Conference on Image Processing*, Barcelona, Spain (pp. 14-17).

Zhai, H., Chen, X., & Fang, Y. (2004). How Well Can the IEEE 802.11 Wireless LAN Support Quality of Service? *in IEEE Transaction on Wireless Communications*, 2004.

Zhai, H., Chen, X., & Fang, Y. (2006). A Call Admission and Rate Control Scheme for Multimedia Support over IEEE 802.11 Wireless LANs, *in ACM Wireless Networks (Winet)*, July 2006.

Zhang, G., Wu, Y., & Liu, Y. (2007). Stability and sensitivity for congestion control in wireless mesh networks with time varying link capacities. *Ad Hoc Networks*, *5*(6), 769–785. doi:10.1016/j.adhoc.2006.12.002

Zhang, L. (1991). VirtualClock: A New Traffic Control Algorithm for Packet Switching Networks. *IEEE . ACM Transactions on Computer Systems*, *9*(2), 101–124. doi:10.1145/103720.103721

Zhang, M., & Wolff, R. (2004, February). Crossing the digital divide: cost-effective broadband wireless access for rural and remote areas. *IEEE Communications Magazine*, *42*(2), 99–105. doi:10.1109/MCOM.2003.1267107

Zhang, Q., & Kassam, S. A. (1999). Finite-state Markov model for Reyleigh fading channels. *IEEE Transactions on Communications*, *47*(11), 1688–1692. doi:10.1109/26.803503

Zhang, Q., & Zhang, Y.-Q. (2008). Cross-layer design for QoS support in multihop wireless networks. *Proceedings of the IEEE*, *96*(1), 64–76. doi:10.1109/JPROC.2007.909930

Zhang, R., Regunathan, S. L., & Rose, K. (2000). Video coding with optimal inter/intra-mode switching for packet loss resilience. *IEEE Journal on Selected Areas in Communications*, *18*(6), 966–976. doi:10.1109/49.848250

Zhang, X., Lee, S. B., Gahng-Seop, A., & Campbell, A. T. (2000, April). INSIGNIA: an IP based quality of service framework for mobile ad hoc networks. *Journal of Parallel and Distributed Computing*, *60*(4), 374–406. doi:10.1006/jpdc.1999.1611

Zhang, X., Su, H., & Chen, H. (2006). Cluster-Based Multi-Channel Communications Protocols in Vehicle Ad Hoc Networks. *IEEE Wireless Communications, vol. 13*, no. (5), pp. 44-51.

Zhang, Y., Yang, L. T., & Ma, J. (2008). Unlicensed Mobile Access Technology: Protocols, Architectures, Security, Standards and Applications. Published byBoca Raton, FL: CRC Press.

Zhao, B., & Valenti, M. C. (2005). Practical relay networks: A generalization of hybrid-ARQ. *IEEE Journal on Selected Areas in Communications*, *23*(1), 7–18. doi:10.1109/JSAC.2004.837352

Zhao, Y., Adve, R., & Lim, T. J. (2007). Improving amplify-and-forward relay networks: Optimal power allocation versus selection. *IEEE Transactions on Wireless Communications*, *6*(8), 3114–3123.

Zheng, L., & Tse, D. (2003). Diversity and multiplexing: A fundamental tradeoff in multiple antenna channels. *IEEE Transactions on Information Theory*, *49*(5), 1073–1096. doi:10.1109/TIT.2003.810646

Zheng, Z. H., & Lam, W. H. (2002). Performance analysis of dynamic channel assignment with queuing and guard channel scheme combined scheme for handoff prioritization. *Electronics Letters*, *38*, 1728–1729. doi:10.1049/el:20021177

Zhifeng, P. L., Lin, Z., Erkip, E., & Panwar, S. (2006). Cooperative wireless communications: A cross-layer design. *IEEE Communications Magazine*, *13*(4), 84–92. doi:10.1109/MWC.2006.1678169

Zhong, L., Liu, F., Wang, X., & Ji, Y. (2007). Fast Handover Scheme for Supporting Network Mobility in IEEE 802.16e BWA System. In *Int. Conf. on Wireless*

Communications, Networking, and Mobile Computing, 21-25 Sept., (pp. 1757-1760).

Zhong, R., & Kravets, R. (2003). On demand power management for ad hoc networks. In *Proc. IEEE INFOCOM'03* (pp. 481-491).

Zhou, W., Zhang, D., & Qiao, D. (2006). Comparative Study of Routing Metrics for Multi-Radio Multi-Channel Wireless Networks. In *Proceedings Wireless Communications and Networking Conference (WCNC 2006)* (pp. 270-275).

Zhu, H., & Cao, G. (2005). rDCF: A relay-enabled medium access control protocol for wireless ad hoc networks. In *Proc. IEEE INFOCOM,* Miami, FL, USA.

Ziouva, E., & Antonakopoulos, T. (2002). CSMA/CA Performance under High Traffic Conditions: Throughput and Delay Analysis . *Elsevier's Computer Communications Journal, 25*(3), 313–321.

Zonoozi, M. M., & Dassanayake, P. (1997). User mobility modeling and characterization of mobility patterns. *IEEE Journal on Selected Areas in Communications, 15,* 1239–1252. doi:10.1109/49.622908

About the Contributors

Sasan Adibi is currently a Member of Technical Staff, Advanced Technology at Research In Motion (RIM). He is also expected to graduate from University of Waterloo in 2010 with a Ph.D. degree from Electrical and Computer Engineering Department. He has an extensive research background mostly in the areas of Quality of Service (QoS) and Security. He is the first author of +25 journal/conference/book chapter/white paper publications. He also +9 years of high-tech industry-based experience, having worked in numerous high-tech companies, including Nortel Networks and Siemens Canada.

Raj Jain is a Fellow of IEEE, a Fellow of ACM, a winner of ACM SIGCOMM Test of Time award, CDAC-ACCS Foundation Award 2009, and ranks among the top 50 in Citeseer's list of Most Cited Authors in Computer Science. Dr. Jain is currently a Professor of Computer Science and Engineering at Washington University in St. Louis. Previously, he was one of the Co-founders of Nayna Networks, Inc - a next generation telecommunications systems company in San Jose, CA. He was a Senior Consulting Engineer at Digital Equipment Corporation in Littleton, Mass and then a professor of Computer and Information Sciences at Ohio State University in Columbus, Ohio. He is the author of ``Art of Computer Systems Performance Analysis,'' which won the 1991 ``Best-Advanced How-to Book, Systems'' award from Computer Press Association. His fourth book entitled " High-Performance TCP/IP: Concepts, Issues, and Solutions," was published by Prentice Hall in November 2003.

Shyam Parekh is a Distinguished Member of Technical Staff in the Network Performance & Reliability department of Bell Labs at Alcatel-Lucent. He holds a PhD in Electrical Engineering (1986) and an MA in Statistics (1984) from UC Berkeley, and a BE in Electrical & Electronics Engineering (1980) from BITS, Pilani, India. He has worked and published extensively on performance optimization and architecture of broadband wired and wireless networks. He has also contributed broadly in the areas of novel analytical and simulation techniques. He has an ongoing affiliation as a visiting faculty with the EECS department of UC Berkeley. He has been the Co-Chair of the Application Working Group of the WiMAX Forum and a Principal Investigator for an NSF funded Future Internet Design (FIND) project. He is a Senior Member of IEEE and a member of the Alcatel-Lucent Technical Academy.

Mostafa Tofighbakhsh (Tom Tofigh) is a Principal Member of Technical Staff in the Radio Technology Architecture group at AT&T Labs. He holds a JD (1995) and completed his PHD requirements for electrical engineering & computer science at GWU(1990). He taught graduate courses from 1996-2001 at various universities as an adjunct including GWU and South Eastern University. He chaired the WiMax Forum application working group forum 2004 – 2009. He has contributed broadly in major technical conferences and he is currently involved in application performance studies and cross layer optimization and radio layer APIs. He is a Senior Member of ACM, IEEE and a member of many industry standard forums.

* * *

M. O. Adigun is currently Professor and Head of Department of Computer Science University of Zululand, South Africa. His research interest centers on the Software Engineering of the Wireless Internet and Mobile Computing using the Wireless Mesh Network. He brings to his research a unique combination of expertise in Agent-based Systems, Mobile Commerce Security and Utility computing. He is currently advancing a novel OGSA-based framework known as Grid-based Utility Infrastructure for SMME-enabling Technologies. As a Research Niche Area leader of the theme: Software Infrastructure for E-Commerce and E-Business, Dr. Adigun nurtured a one-man niche area into a 21 research personnel strong activity area between 2001 to 2008. Dr Adigun's research has been funded by the South African National Research Foundation and the Department of Trade and Industry for up to eight and six years respectively. Dr Adigun received the THRIP Excellence Award in 2004 for his contribution to the growth of research development in a historically disadvantaged institution. He has brought his research interest to bear on the development of the community by his commitment to the SMME-enabling technology research focus.

A. Hamid Aghvami received the M.Sc. and Ph.D. degrees from the University of London, London, U.K. in 1978 and 1981, respectively. He joined the academic staff of King's College London, London, U.K., in 1984. In 1989, he was promoted to the position of Reader, and in 1993, he was promoted to Professor of telecommunications engineering. He is currently the Director of the Centre for Telecommunications Research, King's College London. He carries out consulting work on digital radio communications systems for both British and international companies. He is the author of more than 400 technical papers and has given invited talks all over the world on various aspects of personal and mobile radio communications and giving courses on the subject worldwide. He was a Visiting Professor with NTT Radio Communication Systems Laboratories in 1990 and a Senior Research Fellow with BT Laboratories from 1998 to 1999. He was an Executive Advisor with Wireless Facilities Inc., San Diego, CA, from 1996 to 2002. He is the Managing Director of Wireless Multimedia Communications Ltd. (his own consultancy company). He leads an active research team working on numerous mobile and personal communications projects for third- and fourth-generation systems; these projects are supported by both the government and industry. Prof. Aghvami is a Fellow of the Institute of Electrical and Electronics Engineers, the Royal Academy of Engineering and the Institution of Engineering and Technology. He was a member of the Board of Governors of the IEEE Communications Society from 2001 to 2003. He is a Distinguished Lecturer of the IEEE Communications Society and has been member, Chairman, and Vice Chairman of the technical program and organizing committees of a large number of international conferences. He is also the Founder of the International Conference on Personal, Indoor, and Mobile Radio Communications (PIMRC).

Alberto Diez Albaladejo received his M.S. degree in telecommunications engineering from the University of Malaga, Spain. He joined the Next Generation Network Infrastructures competence center from Fraunhofer FOKUS on 2007. His research interests include network architecture and design, seamless integration of different technologies and interworking across domains.

R. Asokan received the B.E. degree in Electronics and Communication from Bharathiyar University and M.S. degree in Electronics and Control from Birla Institute of Technology. He received M.Tech. Degree in Electronics and Communication from Pondicherry University with distinction. He completed Ph.D degree in wireless networks from Anna University Chennai. He has 21 years of teaching experience. He has published more than 50 papers in International and National conference proceedings and journals. His areas of interest include mobile networks and network security. At present he is working as Professor in Department of ECE at Kongu Engineering College, Perundurai, Erode , Tamilnadu, India

Fulvio Babich received the doctoral degree in electrical engineering, from the University of Trieste, in 1984. After graduation he worked in the Research and Development Department of Telettra, working on optical communications. He subsequently joined Zeltron, where he held the position of Company Head associated with Home System European Projects. Since 1992 he has been with the Department of Electrical Engineering (DEEI) at the University of Trieste, where he is Professor of Digital Communication and Telecommunication Networks. He has been engaged in numerous research activities, including channel coding, joint source and channel coding, adaptive transmission techniques and channel modeling, publishing more than 100 papers on international journals and conference proceedings, and being the main guest editor of a recent special issue on Wireless Video. He has served as TPC member in numerous conferences, and as co-chair of the Communication Theory Symposium at ICC 2005, Seoul. His current research interests are in the field of wireless networks and multimedia wireless communications. Fulvio Babich is a Senior Member of IEEE.

Hamid Beigy received the B.S. and M.S. degrees in Computer Engineering from the Shiraz University in Iran, in 1992 and 1995, respectively. He also received the Ph.D. degree in Computer Engineering from the Amirkabir University of Technology in Iran, in 2004. Currently, he is an Assistant Professor in Computer Engineering Department at the Sharif University of Technology, Tehran, Iran. His research interests include, channel management in cellular networks, learning systems, parallel algorithms, and soft computing.

Razvan Beuran received the B.Sc. degree in Computer Science in 1999, and the M.Sc. degree in Electrical Engineering in 2000 from "Politehnica" University, Bucharest, Romania. He received the joint Ph.D. degree in Electrical Engineering and Computer Science from "Politehnica" University, Bucharest, Romania and "Jean Monnet" University, Saint Etienne, France in 2004. From 1999 to 2005 he was with "Politehnica" University, Bucharest, Romania as a Research Assistant, and then as a Teaching Assistant. From 2001 to 2005 he was also with CERN, Geneva, Switzerland as a Researcher, and then as a Project Associate. In 2006 he was a post-doc Research Fellow with the Japan Institute of Science and Technology, Ishikawa, Japan, where he is currently a Project Researcher. Since 2006 he is Researcher with the National Institute of Information and Communications Technology, Hokuriku Research Center, Ishikawa, Japan. His research topics include: quality testing and measurement in wired and wireless networks, network emulation, and network reliability and dependability in connection with disaster situations. He is an IEEE member.

Juliana Freitag Borin is a final year PhD candidate at the Institute of Computing of the University of Campinas – Brazil. She holds an MSc in Computer Science (2004) from the University of Campinas and a BA in Informatics (2002) from the State University of Western Paraná – Brazil. Her research interests include quality of service provisioning, design and performance evaluation of medium access protocols, traffic modeling and control, and multimedia services. Results of her work have been published in reputed international conferences and journals. Currently, Juliana is a mentor for the 2009 Google Summer of Code ns-3 Network Simulator project.

Jennifer Brandner earned her BS in Computer Science and Mathematics from the University of Wisconsin – Parkside in 2008. She is an information technology analyst at SC Johnson's global headquarters. Her interests include network traffic overflow algorithms, network communication protocols, systems simulation, and computational models.

Fabricio Carvalho de Gouveia (M.Sc) received his graduation in Electrical Engineering from the University Regional of Blumenau (Brazil) in 2000 and his M.Sc. in Telecommunications from the Federal University from Parana (Brazil) in 2003. He is employed as a Research Associate at the research center for Next Generation Network Infrastructure at FOKUS Fraunhofer Institute, where he is working towards his Ph.D in the Field of Next Generation Networks (NGN).

Chung-Ju Chang was born in Taiwan, ROC, in August 1950. He received the B.E. and M.E. degrees in electronics engineering from National Chiao-Tung University (NCTU), Hsinchu, Taiwan, in 1972 and 1976, respectively, and the Ph.D degree in electrical engineering from National Taiwan University (NTU), Taiwan, in 1985. From 1976 to 1988, he was with Telecommunication Laboratories, Directorate General of Telecommunications, Ministry of Communications, Taiwan, as a Design Engineer, Supervisor, Project Manager, and then Division Director. In the meantime, he also acted as a Science and Technical Advisor for the Minister of the Ministry of Communications from 1987 to 1989. In 1988, he joined the Faculty of the Department of Communication Engineering, College of Electrical Engineering and Computer Science, National Chiao-Tung University, as an Associate Professor. He has been a Professor since 1993. He was Director of the Institute of Communication Engineering from August 1993 to July 1995, Chairman of Department of Communication Engineering from August 1999 to July 2001, and the Dean of the Research and Development Office from August 2002 to July 2004. Also, he was an Advisor for the Ministry of Education to promote the education of communication science and technologies for colleges and universities in Taiwan during 1995 - 1999; he is acting as a Committee Member of the Telecommunication Deliberate Body, Taiwan. He serves as Editor for IEEE Communications Magazine and Associate Editor for IEEE Transactions on Vehicular Technology. His research interests include performance evaluation, wireless communication networks, and broadband networks. Dr. Chang is a member of the Chinese Institute of Engineers (CIE) and an IEEE Fellow.

Bhuvaneswari Chellappan is a Software Engineer in the Silicon Valley, California. She graduated from San Jose State University with MS in Computer Science. During her graduate program, her research focus was in WIMAX. Her interests include server-side applications and databases. She likes to develop high quality software applications with emphasis on good design principles.

Marius-Iulian Corici received the Diploma-Engineer in the Science of Systems and Computers - Computers Engineering of University Politehnica" of Bucharest in 2005 with the diploma paper "VDSat: Nomadic Satellite-Based VoIP Infrastructure". In the last four years, he is a researcher in the competence center for Next Generation Network Infrastructures (NGNI) at Fraunhofer Fokus Institut

Berlin, Germany. His interests include the network infrastructures supporting the mobility of the mobile devices and service continuity through vertical handovers.

Floriano De Rango received the degree in computer science engineering in October 2000, and a Ph.D. in electronics and communications engineering in January 2005, both at the University of Calabria, Italy. From January 2000 to October 2000 he worked in the Telecom Research LAB C.S.E.L.T. in Turin as visiting scholar student. From March 2004 to November 2004 he was visiting researcher at the University of California at Los Angeles (UCLA). From November 2004 until September 2007 he has been a Research Fellow in the D.E.I.S. Department, University of Calabria where he is now Assistant Professor. He was recipient of Young Researcher Award in 2007. He served as reviewer and TPC member for many International Conferences such as IEEE VTC, ICC, WCNC, Globecom, Med Hoc Net, SPECTS, WirelessCOM, WinSys and reviewer for many journals such as IEEE Communication Letters, JSAC, IEEE Trans.on Vehicular Technology, Computer Communication, Eurasip JWCN, WINET etc. His interests include Satellite networks, IP QoS architectures, Adaptive Wireless Networks, Ad Hoc Networks and Pervasive Computing. He has co-authored more than 130 papers among International Journal and Conferences Proceedings.

Marco D'Orlando received his Master degree (summa cum Laude) in Telecommunication Engineering from the University of Trieste in December 2003. Then he joined the Telecommunication Group at the Department of Electrical Engineering (DEEI), University of Trieste, and in March 2008 he earned a Ph.D. degree in Information Engineering discussing the thesis: "Multimedia over Wireless IP Networks: Distortion Estimation and Applications". His research activities focus on video distortion estimation, error-robust video communication and on wireless access protocols with QoS support. He is author of more than 10 technical papers published on international journals and conference proceedings. Marco D'Orlando now works in a private company involved in network security and continues his research activities in the Protocol Laboratory of the DEEI. He has been involved in the review activities for the following conferences: IEEE Global Telecommunication Conference (Globecom) 2006, IEEE International Communications Conference (ICC) 2005 and ICC2006, International Conference on Image Processing (ICIP) 2006. Marco D'Orlando is a student Member of the IEEE.

Hongfei Du (S'05-M'07) received the B.Eng degree in electronic information engineering from the Department of Electronic Engineering, Beijing University of Aeronautics & Astronautics, Beijing, China, in 2003. He received the M.Sc, M.Phil, and Ph.D degrees in Wireless Communications from University of Surrey, United Kingdom, in 2004, 2005 and 2007, respectively. From 2007-2008, he was with CREATE-NET international research institute, Italy, as a member of research staff then project leader, coordinating and conducting EU research projects on middleware/software implementation, system architecture and protocol design for the convergence between heterogeneous broadcast and mobile networks. From 2008, he is with School of Computing Science & School of Engineering Science, Simon Fraser University, as a postdoctoral researcher and Ebco-Epic Fellow, working on adaptive video transmission over mobile WiMAX networks. Hongfei has been involved in extensive research projects in the area of mobile broadcasting convergence, mobile communications and satellite communications systems and has also served as a TPC and reviewer for many leading journals and international conferences/workshops including IEEE Wireless Communication, IEEE Transaction on Vehicular Technology, ICC, Globecom, etc. His research interests lie in the area of mobile and satellite multimedia broadcasting, focusing on radio resource management, packet scheduling, quality-of-service support, scalable video coding and cross-layer design.

Vasilis Friderikos graduated from the Aristotle University of Thessaloniki, Greece, Department of Electrical and Computer Engineering - with major in Telecommunications - in 1998. He completed the M.Sc by Research in Telecommunications (with Distinction) at the Centre for Telecommunications Research (London) in 1999. During his Ph.D he was working as a Research Associate in a Mobile-VCE research programme on algorithmic aspects of QoS enabled pure IP based mobile/wireless networks. He is currently a Lecturer at King's College London and his research interests revolve around cross layer optimization algorithms with emphasis on scheduling and routing for single or multi hop wireless networks.

Richard Good received his BSc (Hons) degree from the University of Cape Town in 2005. He has submitted his PhD dissertation at the same institution. He is an active open source software contributor and has developed various open source IMS tools. He has acted as TPC for a number of technical conferences. His research interests include next generation resources management, service provisioning and QoS in heterogeneous networks.

Stefano Gregori received the Laurea degree and the Doctorate degree in Electronic Engineering from the University of Pavia, Italy. After graduating, he was assistant professor at the School of Engineering and Computer Science, University of Texas at Dallas, USA. Currently, he is associate professor at the School of Engineering of the University of Guelph, Canada. He served as Chair of the Circuits, Devices, and Systems Symposium for the 2008 and 2009 Canadian Conference on Electrical and Computer Engineering. His research interests are in the design, analysis, and characterization of integrated circuits with analog and digital applications.

Jane-Hwa Huang received the B.S., M.S., and Ph.D degrees in electrical engineering from the National Cheng-Kung University, Taiwan, R.O.C., in 1994, 1996, and 2003, respectively. He joined the Department of Communication Engineering, National Chiao-Tung University, Taiwan, as a Postdoctoral Researcher from 2004 to January 2006, and a Research Assistant Professor since January 2006. His current research interests are in the areas of wireless networks, wireless multi-hop communications, vehicular communication networks, and radio resource management.

Mihai Ivanovici received the B.Sc. degree from the "Transilvania" University, Braşov, România, then his M.Sc. and Ph.D. degrees from the "Politehnica" University, Bucureşti, România, in 2001, 2002 and 2006 respectively. The title of his Ph.D. thesis was "Network Quality Degradation Emulation – An FPGA-based Approach to Application Performance Assessment" and the research was carried out at CERN (European Organization for Nuclear Research), Geneva, Switzerland, between 2002 and 2005, where he was a project associate. In 2007, 2008 and 2009 he was a postdoc/invited researcher at the SIC (Signals, Images and Communications) Laboratory, University of Poitiers, France. He is a member of the Image Processing and Analysis Laboratory from "Politehnica" University, Bucureti, România and represents this institution in the ATLAS experiment at CERN, Geneva, Switzerland, where he is an associated member. He is a member of the following societies: IEEE Communications, IEEE Signal Processing and IEEE Engineering in Biology and Medicine. Currently, he is a lecturer at the Faculty of

Electrical Engineering and Computer Science, within the "Transilvania" University, Braşov, România, where he leads the research group of the MIV Imaging Venture laboratory, within the Department of Electronics and Computers. His research interests include network emulation, the assessment of QoS and QoE for multimedia applications, digital signal and image processing and analysis and their applications in medicine. He is the author/co-author of more than 20 publications, including two books.

Sofiene Jelassi was born in Bizerte, Tunisia, in 1979. He received his Bachelor of Science and Master of Science degrees in Computer Science from Faculté des Sciences de Monastir (FSM), University of Monastir, Tunisia in 2003 and 2005, respectively. He is currently a doctoral student at the Ecole Nationale des Sciences de l'Informatique (ENSI), University of Manouba, Tunisia. His research interests include ad-hoc wireless networks, heterogeneous networks, multimedia content delivering, conversational application integration, seamless mobility provision, and perceptual quality assessment.

Peter Komisarczuk researches, lectures and consults in networking and distributed systems. He has published in the areas of telecommunications, broadband networks, Next Generation Networks, wireless networks and Grid computing and he is a member of the Victoria University Distributed Systems Research Group. Peter has worked extensively in industry at Ericsson Ltd, the Fujitsu Telecommunication Research Centre (UK) Ltd. and Nortel Networks (UK) Ltd, in the areas of next generation "intelligent networks", broadband access, optical networks and Internet technology. He has a PhD from the University of Surrey (1998) and an MSc in Modern Electronics from Nottingham University (1984). Peter is a Chartered Engineer (CEng), and an active member of the IET, IEEE, and NZCS.

Adlen Ksentini is an Associate Professor at the University of Rennes 1, France. He is a member of the CNRS IRISA laboratory of Rennes. He received an M.S in telecommunication and multimedia networking from the University of Versailles. He obtained his Ph.D thesis in computer science from the University of Cergy-Pontoise in 2005, on QoS provisioning in IEEE 802.11-based networks. His others interests include: Mobility and QoS support in IEEE 802.16, QoS support in the newly IEEE 802.11s Mesh networks, multimedia transmission. Dr. Ksentini is involved in several industrial projects and the FP6 IST-ANEMONE, which aim at to realize a large scale testbed supporting mobile user on heterogeneous wireless technologies. Dr. Ksentini is a co-author of over 20 technical journal or international conference.

C. Kyara is currently working towards a MEng in Computer Engineering at the University of Pretoria. His research is focused on Heterogeneous Wireless Mesh Networks.

Long Bao Le received the B.Eng. degree with highest distinction from Ho Chi Minh City University of Technology, Vietnam, in 1999, the M.Eng. degree from Asian Institute of Technology (AIT), Thailand, in 2002 and the Ph.D. degree from University of Manitoba, Canada, in 2007. He is currently a Postdoctoral Research Associate at Massachusetts Institute of Technology, USA. His current research interests include cognitive radio, link and transport layer protocol issues, cooperative diversity and relay networks, stochastic control and cross-layer design for communication networks.

Tho Le-Ngoc obtained his B.Eng. (with Distinction) in Electrical Engineering in 1976, his M.Eng. in Microprocessor Applications in 1978 from McGill University, Montreal, and his Ph.D. in Digital Communications 1983 from the University of Ottawa, Canada. During 1977-1982, he was with Spar Aerospace Limited as a Design Engineer and then a Senior Design Engineer, involved in the development and design of the microprocessor-based controller of Canadarm (of the Space Shuttle), and SCPC/FM, SCPC/PSK, TDMA satellite communications systems. During 1982-1985, he was an Engineering Manager of the Radio Group in the Department of Development Engineering of SRTelecom Inc., developed the new point-to-multipoint DA-TDMA/TDM Subscriber Radio System SR500. He was the System Architect of this first digital point-to-multipoint wireless TDMA system. During 1985-2000, he was a Professor in the Department of Electrical and Computer Engineering of Concordia University. Since 2000, he has been a Professor in the Department of Electrical and Computer Engineering of McGill University. His research interest is in the area of broadband digital communications. He is the recipient of the 2004 Canadian Award in Telecommunications Research, and recipient of the IEEE Canada Fessenden Award 2005. He holds a Canada Research Chair (Tier I) on Broadband Access Communications, and a Bell Canada/NSERC Industrial Research Chair on Performance & Resource Management In Broadband xDSL Access Networks.

Chengzhi Li is currently a visiting scholar at the University of Houston. He received his B.S. degree in Applied Mathematics and M.S. degree in Operations Research from Fuzhou University and Xiamen University, China respectively. He received his Ph.D. in Computer Engineering from Texas A&M University in 1999. From 1999 to 2001, he was a postdoctoral fellow at Rice University. From 2001 to 2003, he was a research scientist at the University of Virginia. From 2003 to 2005, he was a visiting assistant professor at the University of Texas at Arlington. From 2006 to 2008, he was with Texas A&M University. His research areas encompass wireline and wireless networking, control theory, numerical analysis, and applied functional analysis. His research was partially supported by the National Science Foundation under Grant No. 0324988, 0329181, and 0081761.

Jie Liang (S'99-M'04) received the B.E. and M.E. degrees from Xi'an Jiaotong University, China, in 1992 and 1995, the M.E. degree from National University of Singapore (NUS), in 1998, and the Ph.D. degree from the Johns Hopkins University, Baltimore, MD, in 2003, respectively. Since May 2004, he has been an Assistant Professor at the School of Engineering Science, Simon Fraser University, Burnaby, BC, Canada. From 2003 to 2004, he was with the Video Codec Group of Microsoft Digital Media Division, Redmond, WA. Dr. Liang's research interests include image and video coding, multirate signal processing, and joint source channel-coding.

Susan Lincke earned her PhD in Computer Science from Illinois Institute of Technology and she is an Assoc. Prof. at University of Wisconsin-Parkside. She has 17 years of telecommunications industry experience, including at MCI, Motorola, and GE. Her areas of research include wireless telecommunications, analytic models, and network and information systems security.

iangchuan Liu (S'01-M'03-SM'08) received the BEng degree (cum laude) from Tsinghua University, Beijing, China, in 1999, and the PhD degree from The Hong Kong University of Science and Technology in 2003, both in computer science. He was a recipient of Microsoft Research Fellowship (2000), a recipient of Hong Kong Young Scientist Award (2003), and a co-inventor of one European patent and two US patents. He co-authored the Best Student Paper of IWQoS'08 and the Best Paper (2009) of IEEE Multimedia Communications Technical Committee (MMTC). He is currently an Assistant Professor in the School of Computing Science, Simon Fraser University, British Columbia, Canada, and was an Assistant Professor in the Department of Computer Science and Engineering at The Chinese University of Hong Kong from 2003 to 2004. His research interests include multimedia systems and networks, wireless ad hoc and sensor networks, and peer-to-peer and overlay networks. He is an Associate Editor of IEEE Transactions on Multimedia, and an editor of IEEE Communications Surveys and Tutorials. He is a Senior Member of IEEE and a member of Sigma Xi.

Andrea Malfitano was born in Neuchatel (Switzerland), on December 25, 1977and accomplish degree in Computer Science Engineering with final grade: 110\110 on July 17, 2006 presenting the final thesis with the subsequent title: "Channel state analysis and medium access protocol for wireless mesh networks with base station on the HAP". He actually is a Ph.D student at University of Calabria and collaborates with a telecommunication factory. He has published several papers and has been reviewer of important international conferences eg: PIRMC 2007, WTS 2007, WTS 2009 and Globecom 2009. He has also some experiences in teaching in Telecommunication courses. His main research interests are about topics related to physical and MAC (Medium Access Control) level of IEEE 802.16 protocol stack. He in particular researches about signal impairment effects due to environment, transmission technique provided by IEEE 802.16 protocol, transmission channel modeling, scheduling issues, mechanisms to support QoS and call admission control topics.

Thomas Magedanz (PhD) is professor in the electrical engineering and computer sciences faculty at the Technische Universität Berlin, Germany. In addition, he is director of the Next Generation Network Infrastructures (NGNI) division of the Fraunhofer Institute FOKUS, which provides various testbeds and tools in the context of converging networks and open Service Delivery Platforms. Since more than 20 years Prof. Magedanz is working in the convergence field of fixed and mobile telecommunications, the Internet and information technologies. Under his leadership many service development platforms, toolkits and testbeds have been developed, such as the Grasshopper Mobile Agent platform, the OSA/ Parlay Playground, the Open Source IMS Core System, and most recently the Open SOA Telco Playground. In the course of his research activities he published more than 200 technical papers/articles. In addition, Prof Magedanz is senior member of the IEEE, and editorial board member of several journals. In 2007, Prof. Magedanz joined the European FIRE (Future Internet Research and Experimentation) Expert Group.

Salvatore Marano, graduated in electronics engineering at the University of Rome in 1973. In 1974 he joined the Fondazione Ugo Bordoni. Between 1976 and 1977 he worked at the ITT Laboratory in Leeds, United Kingdom. Since 1979 he has been an Associate Professor at the University of Calabria, Italy. He was reviewer for many journals such as IEEE Communication Letters, JSAC, IEEE Trans. on Vehicular Technology, IEEE Transaction on Wireless Comm., European Transaction on Telecommunication Journal. His research interests include performance evaluation in mobile communication systems, satellite systems and 3g/4G networks. He published more than 160 papers among international conferences and journals.

André Marquet works as product manager for Wit-Software, a leading innovative company that creates smart applications and services for telecommunication and media companies, prior he worked as a system architect for Nokia Siemens Networks, where he was responsible for designing the web interface technologies for IPTV and to follow up QoE monitoring solutions. In 2006 Mr. Marquet served as IT manager for ICEP (now AICEP), an agency of the Portuguese government, and before that, from 2004 to 2006 he served as a pre-sales for EFACEC SA, in international offices. He began his career with ADETTI, working as a researcher while completing his Master's degree in computer and telecommunications from ISCTE of Lisbon, after having earned an engineering degree from the same University in 2003. Mr. Marquet has been granted with international and national patents related to automatic video quality estimation.

Nuno Martins is enrolled as System Architect at Nokia Siemens Networks (previously on Siemens) since 2004 in the IPTV business unit, responsible for technical system concepts in several IPTV related areas, namely on the quality of experience monitoring/assessment. Mr. Martins holds a Master degree issued by ISCTE, Lisbon, in 2006 in Computer Science and Telecommunications Engineering, Telecommunications field, specialization in vide o quality estimation models and transmission over QoE aware 3.5G networks. He was previously involved on international investigation projects with INESC-INOV and ADETTI in optimization techniques for video distribution over IP networks and video estimation models, holding several publications in international conferences and patents, including in video quality estimation methods. He is a non-permanent member of the IEEE ICC reviewers' board.

Mohammad Reza Meybodi received the B.S. and M.S. degrees in Economics from Shahid Beheshti University in Iran, in 1973 and 1977, respectively. He also received the M. S. and Ph.D. degree from Oklahoma University, USA, in 1980 and 1983, respectively in computer science. Currently he is a Full Professor in computer engineering department, Amirkabir University of Technology, Tehran, Iran. Prior to current position, he worked from 1983-1985 as an assistant professor at Western Michigan University, and from 1985-1991 as an associate professor at Ohio University, USA. His research interests include, channel management in cellular networks, learning systems, parallel algorithms, soft computing and software development.

Melody Moh obtained her BSEE from National Taiwan University, MS and Ph.D., both in computer science, from Univ. of California - Davis. She joined San Jose State University in 1993, and has been a Professor since Aug 2003. Her research interests include mobile, wireless networking and network security. She has published over 90 refereed technical papers in international journals and conferences, and has consulted for various companies.

Teng-Sheng Moh received Ph. D. in Computer Science from University of California, Davis. He is current a faculty member at the Computer Science Dept, San Jose State University.

Jânio M. Monteiro received the Electrical Engineer and Computers Engineer degree in 1995 from the FEUP, University of Porto, Portugal, and the M. Sc. degree in Electronics Engineer and Computers in 2003, from the IST, Technical University of Lisbon, Portugal. He is currently doing his Ph.D. course at the same institution, working in the area of Video Transmission over IP Networks. He is a Assistant at the University of Algarve, where he teaches communication networks and telecommunications in graduate courses since 1997. He is a researcher at INESC-ID, a research institute in Lisbon, since 2002.

P. Mudali is currently working towards a PhD in Computer Science at the University of Zululand. His research is focused on Topology Control for Wireless Ad Hoc Networks.

Abdelhamid Nafaa is a Marie Curie Research Fellow under the EU-FP6 EIF Marie Curie action that seeks broader synergy in the European research space. He has been granted the Marie Curie award to undertake independent research work at UCD in the area of multimedia services distribution over carrier-grade networks. Before joining UCD, Dr. Nafaa was a professor assistant at University of Versailles-SQY and acted as Technology Consultant for an U.S. and a European –based companies in the area of reliable multimedia communication over WiFi technology and IMS-based multicasting in DVB-S2 satellite networks, respectively. He obtained his Masters and PhD degrees in 2001 and 2005 from the University of Versailles-SQY where he was involved in several national and European projects: NMS, IST-ENTHRONE1, IST-ATHENA, and IST-IMOSAN. Dr. Nafaa is now involved in a successful FP7 proposal CARMEN that aim to develop a mixed WiFi/WiMax wireless mesh networks to support carrier-grade services. Dr. Nafaa is a co-author of over 25 technical journal or international conference papers on multimedia communications.

A. M. Natarajan received the B.E. degree in Electrical Engineering, and M.Sc.(Engg.) in Applied Electronics and Servo Mechanism from Madras University and Ph.D degree in System Engineering from Madras University. He has 39 years of teaching experience. He received the "Best Engineering college principal award in India for the Year 2000" from Indian Society for Technical Education, New Delhi. He has published more than 100 papers in International and National journals and conference proceedings. He has published 10 books. His areas of research include systems engineering and mobile networks. At present he is working as Chief Executive and Professor of Electronics and Communication Engineering Bannari Amman Institute of Technology Sathyamangalam, Tamilnadu, India.

Mário S. Nunes graduated with the Electronics Engineer degree in 1975, Ph.D. degree in Electronics Engineer and Computers in 1987, and the Aggregation degree in the same area in 2006, all from the Instituto Superior Técnico, Technical University of Lisbon, Portugal. He is now Associated Professor at Instituto Superior Técnico, where he teaches in telecommunications and networking areas in graduate and postgraduate courses. He has been responsible for the INESC participation in several european projects, namely RACE, ACTS and IST programs in the areas of fixed and wireless networks. Since 2001 he is Director of INESC Inovação, where he is coordinator of the Telecom Area. He is author of two books and submitted 10 patents. He is a Senior Member of IEEE.

S. Nxumalo is currently working towards a MSc in Computer Science at the University of Zululand. His research is focused on Quality of Service and Routing Metrics for Wireless Ad Hoc Networks.

T. Nyandeni is currently working towards a PhD in Computer Science at the University of Zululand. His research is focused on Quality of Service and Routing Metrics for Wireless Ad Hoc Networks.

Shanghong Peng received the B.Sc. degree in Testing & Measuring Technique and Instrumentations in 1994 from Chongqing University, China and the M.Sc. degree in systems and computer engineering in 2008 from the University of Guelph, Canada. When served as a senior network test engineer and supervisor at the China Telecom. Guangzhou Research Institute for more than ten years, she was responsible for planning, development, testing, maintenance, optimization, and standardization in the information and communication technology field such as PSTNs, wired and wireless access networks. Currently she is a research associate in the School of Engineering at the University of Guelph, Canada.

Her research interests include wireless communications, artificial intelligence, network optimization, and testing methodology.

Guy Pujolle is currently a Professor at the Pierre et Marie Curie University (Paris 6) and a member of the Scientific Advisory Board of Orange/France Telecom Group. He was appointed by the Education Ministry to found the Department of Computer Science at the University of Versailles, where he spent the period 1994-2000 as Professor and Head. He was Head of the MASI Laboratory (University Pierre et Marie Curie - Paris 6), 1981-1993, Professor at ENST (Ecole Nationale Supérieure des Télécommunications), 1979-1981, and member of the scientific staff of INRIA (Institut National de la Recherche en Informatique et Automatique), 1974-1979.

Khoa T. Phan received the B.Sc. degree with First Class Honors from the University of New South Wales (UNSW), Sydney, NSW, Australia, in 2005 and the M.Sc. degree from the University of Alberta, Edmonton, AB, Canada, in 2008. He is currently at the Department of Electrical Engineering, California Institute of Technology (Caltech), Pasadena, CA, USA. His current research interests are mathematical foundations, control, and optimization of communications networks. He is also interested in network economics, applications of game theory, and mechanism design in communications networks. He has been awarded several prestigious fellowships including the Australian Development Scholarship, the Alberta Ingenuity Fund Student Fellowship, the iCORE Graduate Student Award, and most recently the Atwood Fellowship to name a few.

A. Dev Pragad graduated from King's College London, UK, majoring in Computer Systems and Electronics in 2005. He obtained first class honours in his BEng and graduated with over seven awards for outstanding academic performance. Following the BEng, Dev joined the Centre for Telecommunications Research to pressure his PhD in the autumn of 2005. As part his PhD, he was involved in the Mobile VCE Core 4 research programme. His research work focused on Mobility and QoS issues in future IP based mobile networks. At the completion of the Core 4 research programme of Mobile VCE in 2009, Dev was awarded the Outstanding Researcher Award for his exceptional research contributions to Mobile VCE. His researched interest includes optimisation of IP based mobile networks, mobility management in future mobile IP networks, mechanisms for optimal performance of Mobility and QoS mechanisms.

Guy Pujolle is the French representative at the Technical Committee on Networking at IFIP. He is an editor for International Journal of Network Management, WINET, Telecommunication Systems and Editor in Chief of the indexed Journal "Annals of Telecommunications". He was an editor for Computer Networks (until 2000), Operations Research (until 2000), Editor-In-Chief of Networking and Information Systems Journal (until 2000), Ad Hoc Journal and several other journals. Guy Pujolle is a pioneer in high-speed networking having led the development of the first Gbit/s network to be tested in 1980.

Ramón M. Rodríguez-Dagnino is a full Professor at the Tecnologico de Monterrey (ITESM), Monterrey, México, and Director of the Telecommunications Management Master Program (2000--). He received his Ph. D. from the University of Toronto, 1993, and his M. Sc. from the Research and Advanced Studies Center (CInvEstAv) in Mexico City, 1984. He worked at the R&D Center of TelMex (Mexican Telephone Co.) from 1984 to 1989. He was the Chair of the Electronics and Telecommunications Center at ITESM (2000--2001), and member of the Academic University Council during the 2000--2001 academic years. He has won the ITESM Best Teaching and Research Award twice, in 1998 and 2001. He is the Chairman of the IEEE-MTTS-17 Chapter in Mexico. His research interests include teletraffic modeling, multimedia network design, and electromagnetics. He has served as a technical reviewer of IEEE journals and conferences, and in the Program Committee of SPIE conferences. He is a member of IEEE, SPIE, AMS, the Mexican Academy of Sciences (AMC), and the Mexican National System of Research (SNI).

Nelson Luis Saldanha da Fonseca received his Electrical Engineer (1984) and MSc in Computer Science (1987) degrees from The Pontificial Catholic University of Rio de Janeiro, Brazil, and the MSc (1993) and Ph.D (1994) degrees in Computer Engineering from The University of Southern California. He received the title of "Livre Docente" in Computer Networks from the University of Campinas in 1999. He is a Full Professor at Institute of Computing of the University of Campinas, Campinas - Brazil and has been affiliated to it since 1995. Currently, he is Head of the Computer Systems Department and Associate Chair for Graduate Studies. He lectured at Department of Informatics and Telecommunications, University of Trento, Italy (2004 and 2007) and at the University of Pisa (2007). He held Lecturer positions at Pontificial Catholic University (1985 - 1987) and worked in the Computer Communications group at IBM Rio Scientific Center (1989). He received the Medal of the Chancelor of the University of Pisa (2007). He is the recipient of the 2003 State University of Campinas Zeferino Vaz award for academic productivity in Computer Science, the Elsevier Computer Network Journal Editor of Year 2001 award, the 1994 University of Southern California International Book award and the Brazilian Computer Society First Thesis and Dissertations award. He is listed in Marqui's Who is Who in the World, Who's Who in Science and Engineering. Nelson Fonseca has published 200+ refereed papers and book chapter. He has supervised 40+ graduate students. He is the EiC of IEEE Communications Surveys and Tutorials. He served as EiC of the IEEE Communications Society Electronic Newsletter (2004-2007), Associate EiC of IEEE Communications Surveys and Tutorials (2006) and Editor of the Global Communications Newsletter (1999-2002). He is on the editorial board of Computer Networks, IEEE Communications Surveys and Tutorials, the IEEE Communications Magazine. He served as Associate editor for IEEE Transaction on Multimedia (1999-2004), for the Brazilian Journal of Telecommunications (2001-2004) and for the Journal of the Brazilian Computer Society (2002-2007). He co-edited Teletraffic Engineering in the Internet Era, Elsevier, 2001, and organized several special issues to Computer Networks Journal, IEEE Journal of Selected Areas in Communications, IEEE Communications Magazine Journal of the Brazilian Telecommunications Society and Journal of the Brazilian Computer Society. Dr. Fonseca co-chaired over 15 conferences, most of them IEEE sponsored conferences. Currently, he is the IEEE ComSoc Director for Latin America. He served as IEEE ComSoc Director for On-Line Services (2002-2003).

Hideaki Takagi is Professor in the School of Systems and Information Engineering at the University of Tsukuba, Japan. He received his B.S. and M.S. degrees in Physics from the University of Tokyo in 1972 and 1974, respectively. In 1974 he joined IBM Japan as a Systems Engineer. From 1979 to 1983, he studied at the University of California, Los Angeles, and received his Ph.D. degree in Computer Science. From 1983 to 1993, he was with IBM Research, Tokyo Research Laboratory. He moved to the University of Tsukuba in October 1993 as Professor at the Institute of Policy and Planning Sciences. Prior to the current position, he was Chair of the Doctoral Program in Policy and Planning Sciences in the 1997--1998, Chair of the Institute in 2000--2001, and the Vice President of the University of Tsukuba in 2002--2003. His research interests include enumerative combinatorics, probability theory, queueing theory and stochastic processes as applied to the performance evaluation of computer communication networks. He is the author of research monographs Analysis of Polling Systems (The MIT Press, 1986), and Queueing Analysis : A Foundation of Performance Evaluation, Volumes 1--3 (Elsevier, 1991--1993). He is IEEE Fellow (1996) and IFIP Silver Core Holder (2001). He served as editors for IEEE Transactions on Communications (1986--1993) and IEEE/ACM Transactions on Networking (1992--1994). He currently serves as editors for Performance Evaluation (from 1984 onwards) and Queueing Systems (from 1988 onwards) journals.

Djamshid Tavangarian is a member of the Faculty of CS & EE at the University of Rostock/Germany, where he represents the teaching and research area of Computer Architecture. He studied EE & IT at the Technical University of Berlin, finished his Ph.D. at the University of Dortmund and his professorship work at the University of Frankfurt-Germany in CS. After an industrial work at the Hewlett-Packard Company he worked at the University of Hagen/Germany (The Distance University of Germany) and was responsible for the field of computer architecture and design of integrated circuits. Connected with research contracts he worked at the Universities of Berkeley (UCB) and Santa Barbara (UCSB) in the USA, too. The current main topics of his research activities concentrate on computer architectures for local-area and wide-area computing systems, pervasive computing, adaptive and embedded systems, wireless communication systems and especially, eLearning and multimedia architectures for mobile distance learning. He is the program chair and organiser of the international series of IEEE Pervasive Learning (PerEL), organiser of the Workshop series "Pervasive University, PERU", and was the program chair of the two in the German speaking area most important eLearning Conferences DeLFI 2005 and GMW 2005 in Rostock. He is consulter of the Federal and State Ministries of Educations and governmental institutions in Germany in the fields of multimedia-based and mobile eLearning. Dr. Tavangarian was the coordinator and leader of a number of single and joint eLearning projects; he is holder of different scientific awards as well as author, co-author, and editor of more than 300 scientific publications. He is member of several scientific organizations and is currently the Dean of the Faculty of Computer Science and Electrical Engineering of the University of Rostock.

Robil Daher is a scientific assistant at the Chair of Computer Architecture at the University of Rostock (Germany). He received his B.Sc. degree in Electronic Engineering from Tishreen University (Syria) in 1996, and his Ph.D. from Rostock University in 2007 in the field of load balancing and QoS for wireless networks. In 1997 he is awarded certificate and prize by Ministry of Higher Education (Syria) for excellent achievements and also for being the best student among the graduates. His research interests include vehicular communication networks, wireless ad hoc networks, heterogeneous wireless networks, resource and mobility management, QoS and load balancing, and routing protocols. He is also interested in inter-planetary communication networks and bionic-inspired solutions for performance enhancement of wireless networks. He is organiser of several workshops and author/co-author of several scientific publications. He is member of several scientific organizations and has recently established the community "Routing Lexicon" for studying and classification of routing mechanisms and protocols of different technologies. He is the head of the workgroup wireless networks at the Chair of Computer Architecture and currently works as a team manager in the project Wi-Roads (Wireless Infrastructure Networks for high-speed Roads). Additionally, He currently works on his next book "theory of load distribution".

Paul Teal received the B.E. degree in electrical engineering from the University of Sydney, NSW, Australia, in 1989, and the Ph.D. degree from the Australian National University, Canberra in 2002. He joined Telecom Australia (now Telstra) in 1988, working on telecommunications network design and network management systems design. In 1991-2, in New Zealand, he worked on design of industrial control and telemetry systems. In 1993-6, he worked on voice processing systems and call centres in the roles of both Designer and Consultant. In 1997-2006 he was a research scientist at Industrial Research Limited in New Zealand. From 2006 has been Senior Lecturer in Statistical Signal Processing at Victoria University of Wellington, New Zealand. His research interests include applications of signal processing to communications, to acoustics, and to biomedical devices. Of particular interest are blind source separation and statistical learning.

Rath Vannithamby received his B.S., M.S. and Ph.D. degrees in Electrical and Computer Engineering from the University of Toronto, Ontario, Canada, in 1994, 1996 and 2001 respectively. He is currently a Senior Research Scientist, Manager in Corporate Technology Group at Intel Corporation, Hillsboro, Oregon, USA and manages a lab with eight scientists and leads the MAC and signaling layer standardization of next generation WiMAX system. Prior to joining Intel, he was with Ericsson Inc., San Diego, California, USA and was involved in CDMA standardization and high level system design. Dr. Vannithamby is a member of IEEE and IEEE/TCPC. He has published over 30 papers, and has over 60 patents pending. He has served on technical program committee for major wireless communication conferences including ICC, Globecom, VTC, and WCNC. He has also served as a guest editor for EURASIP Journal of Wireless Communications and Networking special issue on Radio Resource Management for 3G+ Systems. He has previously given tutorials on 3G systems in major IEEE conferences and has an online CDMA2000 tutorial in IEEE/ComSoc. He has written a book chapter on VoIP support over WiMAX that is currently under publication process by Wiley publishers. His current research interests are in the area of Radio Resource Management techniques, QoS provisioning, Cross-layer design and MAC/Signaling Layer Protocols for high-speed wireless access networks using OFDMA technologies including 4G and IEEE 802.16.

Francesca Vatta received a Laurea in Ingegneria Elettronica in 1992 from University of Trieste, Italy. From 1993 to 1994 she has been with Iachello S.p.A., Olivetti group, Milano, Italy. Since 1995 she has been with the Department of Electrical Engineering (DEEI) of the University of Trieste where she received her Ph.D. degree in telecommunications, in 1998. In November 1999 she became assistant professor at University of Trieste. Starting in 2002, she spent several months as visiting scholar at the University of Notre Dame, Notre Dame, IN, U.S.A., cooperating with the Coding Theory Research Group under the guidance of Prof. D. J. Costello, Jr. She is author of more than 60 papers published on international journals and conference proceedings. Her current research interests are in the area of channel coding concerning, in particular, the analysis and design of concatenated coding schemes for wireless applications.

Muthaiah Venkatachalam is the lead system architect in the Wireless Standards and Advanced Technology group at Intel Corporation. He currently leads the MAC layer definition, design and specification for the next generation mobile WiMAX. He has played a significant role in the evolution of broadband wireless technology by actively participating and contributing to standards development at IEEE and WiMAX Forum. He is currently chairing the Femto Cell and Self Optimization work in WiMAX Forum.

In the past he was the chair of the "Idle mode and Paging" work and the "Location based services" work in WiMAX Forum; as well as the MAC Rapporteur Group Chair in IEEE 802.16m. He has served as an editor for Elsevier Journal of "Computer Networks" and as an editorial board member for "Special Issue on Media and Stream Processing" in the International Journal of Embedded Systems. He has also served as an organizing committee member for 5th, 6th, 7th and 8th Workshops on Media and Streaming Processors. He has several publications with 3 issued patents and 60+ patents pending. Previously at Intel, he has led the efforts on developing network processor based IP and ATM traffic management solutions; processing architectures for Intel's IXP23xx Network processor family; and system architectures for broadband access, wireless access platforms and metropolitan optical networking systems.

Neco Ventura is the Head of the Centre for Broadband Networks and the Director of the Communications Research Group in the Department of Electrical Engineering at the University of Cape Town. His current research interests are centered on Next Generation architectures, infrastructures, specifically in QoS and mobility support across heterogeneous networks.

Sergiy A. Vorobyov received the M.S. and Ph.D. degrees in systems and information processing from National University of Radioelectronics, Kharkiv, Ukraine, in 1994 and 1997, respectively. Since 2006, he has been with the Department of Electrical and Computer Engineering, University of Alberta, Edmonton, Alberta, Canada, as an Assistant Professor. Since his graduation, he also occupied various research and faculty positions in National University of Radioelectronics, Kharkiv, Ukraine; Institute of Physical and Chemical Research (RIKEN), Japan; McMaster University, Ontario, Canada; Duisburg-Essen University and Darmstadt University of Technology, Germany; and Joint Research Institute, Heriot-Watt and Edinburgh Universities, UK. His research interests include statistical and array signal processing, applications of optimization and linear algebra methods in signal processing and communications, estimation and detection theory, sampling theory and applications, and cooperative and cognitive systems. He is a recipient of the 2004 IEEE Signal Processing Society Best Paper Award, 2007 Alberta Ingenuity New Faculty Award, and other research awards. He serves as an Associate Editor for the IEEE Transactions on Signal Processing and IEEE Signal Processing Letters. He is a member of Sensor Array and Multi-Channel Signal Processing Technical Committee of IEEE Signal Processing Society.

David Waiting holds a Masters and PhD degree in Electrical Engineering both obtained from the University of Cape Town, South Africa. He currently holds a position in the Telkom Group, one of the largest telecommunications service providers in Africa, where his responsibilities include integrating core network technologies in their fixed and mobile networks. David has been published many times on his work in various fields relating to the IP Multimedia Subsystem, and has developed several open source tools that are used extensively worldwide for IMS research.

Li-Chun Wang received the B.S. degree in electrical engineering from the National Chiao-Tung University, Hsinchu, Taiwan, R.O.C., in 1986, the M.S. degree in electrical engineering from the National Taiwan University, Taipei, Taiwan, in 1988, and the M.Sc. and Ph.D. degrees in electrical engineering from Georgia Institute of Technology, Atlanta, in 1995 and 1996, respectively. From 1990 to 1992, he was with Chunghwa Telecom . In 1995, he was affiliated with Northern Telecom in Richardson, Texas. From 1996 to 2000, he was with AT&T Laboratories, where he was a Senior Technical Staff Member in the Wireless Communications Research Department. Since August 2000, he has joined the Department of Communication Engineering of National Chiao-Tung University in Taiwan as an Associate Professor and has been promoted to a full professor since August 2005. Dr. Wang was a corecipient of the Jack Neubauer Best Paper Award from the IEEE Vehicular Technology Society in 1997. His current research interests are in the areas of cellular architectures, radio network resource management, cross-layer optimization for cooperative and cognitive wireless networks. He is the holder of four U.S. patents with three more pending.

Simon X. Yang received the B.Sc. degree in engineering physics from Beijing University, China in 1987, the first of two M.Sc. degrees in biophysics from Chinese Academy of Sciences, Beijing, China in 1990, the second M.Sc. degree in electrical engineering from the University of Houston, Houston, USA in 1996, and the Ph.D. degree in electrical and computer engineering from the University of Alberta, Edmonton, Canada in 1999. Currently he is a Professor and the Head of the Advanced Robotics & Intelligent Systems (ARIS) Laboratory at the University of Guelph in Canada. His research interests include intelligent systems, robotics, sensors and multi-sensor fusion, wireless sensor networks, control systems, soft computing, and computational neuroscience. Prof. Yang serves as an Associate Editor of IEEE Transactions of Neural Networks, IEEE Transactions on Systems, Man, and Cybernetics, Part B, and several other journals. He has involved in the organization of many conferences and other professional activities.

Kok-Lim Alvin Yau received the B. Eng. degree in Electrical and Electronics Engineering (first-class honors) from the Universiti Teknologi Petronas, Malaysia in 2005. He received the MSc (Electrical Engineering) from the National University of Singapore in 2007. He was awarded the 2007 Professional Engineer Board of Singapore Gold Metal for being the best graduate of the MSc degree in 2006/07. He is currently pursuing his PhD degree at the School of Engineering and Computer Science, Victoria University of Wellington, New Zealand under the supervision of Dr. Peter Komisarczuk and Dr. Paul Teal. His research interests include wireless networks, quality of service, context-awareness, and cognitive radio networks. He worked as Design Engineer (Intern) for eight months beginning from end of 2003, and as Design Engineer at the end of 2005 with Intel Malaysia. He worked as a Postgraduate Research Intern for Institute for Infocomm Research, Singapore during the summer of 2006.

Habib Youssef received a Diplôme d'Ingénieur en Informatique from the Faculté des Sciences de Tunis, University of El-Manar, Tunisia in June 1982 and a Ph.D. in computer science from the University of Minnesota, USA, in January 1990. From September 1990 to January 2001 he was a Faculty member of the computer engineering department of King Fahd University of Petroleum & Minerals (KFUPM), Saudi Arabia (Assistant Professor from 1990 to 1995 and Associate Professor from September 1995 to January 2001). From February 2001 to June 2002, he was a Maître de Conférences en informatique at the Faculté des Sciences de Monastir (FSM), University of Monastir, Tunisia. From July 2002 to August 2005, he served as the Director of the Institut Supérieur d'Informatique et Mathématiques of the University of Monastir. He is currently serving as a Professor of computer science and Director of the Institut Supérieur d'Informatique et des Technologies de Communication, Hammam Sousse, University of Sousse, Tunisia. Habib Youssef has over 130 publications to his credit in the form of books, book chapters, and journal and conference papers. He is the author with S. Sait of two books, (1) "VLSI Physical Design Automation: Theory and Practice", McGraw-Hill 1995, (also co-published by IEEE Press 1995), and reprinted with corrections by World Scientific in 1999, and (2) "Iterative Computer Algorithms with Applications in Engineering", IEEE CS Press 1999, and since 2003 published by John Wiley & Sons, which has also been translated into Japanese. His main research interests are computer networks, performance evaluation of computer systems, and algorithms for combinatorial optimization.

Wei Zhao is currently the Rector of the University of Macau. Before joining the University of Macau, he served as the Dean of the School of Science at Rensselaer Polytechnic Institute. Between 2005 and 2006, he served as the director for the Division of Computer and Network Systems in US National Science Foundation when he was on leave from Texas A&M University, where he served as Senior Associate Vice President for Research and Professor of Computer Science. He was the founding director of the Texas A&M Center for Information security and Assurance, which has been recognized as a Center of Academic Education by the National Security Agency. Dr. Zhao completed his undergraduate program in physics at ShaanXi Normal University, Xian, China, in 1977. He received the MS and PhD degrees in Computer and Information Sciences at the University of Massachusetts at Amherst in 1983 and 1986, respectively. Since then, he has served as a faculty member at Amherst College, the University of Adelaide, and Texas A&M University. As an elected IEEE Follow, Wei Zhao has made significant contributions in distributed computing, real-time systems, computer networks, and cyber space security. His research was partially supported by the National Science Foundation under Grant No. 0324988, 0329181, and 0081761.

Index

Symbols

A

D

E

F

G

H

I

J

L

M

N

O

P

Q

R

S